Lecture Notes in Computer Science 2719

Edited by G. Goos, J. Hartmanis, and J. van Leeuwen

Springer-Verlag Berlin Heidelberg GmbH

Jos C.M. Baeten   Jan Karel Lenstra
Joachim Parrow   Gerhard J. Woeginger (Eds.)

Automata, Languages and Programming

30th International Colloquium, ICALP 2003
Eindhoven, The Netherlands, June 30 – July 4, 2003
Proceedings

Springer

Volume Editors

Jos C.M. Baeten
Technische Universiteit Eindhoven, Dept. of Mathematics and Computer Science
P.O. Box 513, 5600 MB Eindhoven, The Netherlands
E-mail: j.c.m.baeten@tue.nl

Jan Karel Lenstra
Georgia Institute of Technology, School of Industrial and Systems Engineering
765 Ferst Drive, Atlanta, GA 30332-0205, USA
E-mail: jkl@isye.gatech.edu

Joachim Parrow
Uppsala University, Department of Information Technology
P.O. Box 337, 75105 Uppsala, Sweden
E-mail: joachim@docs.uu.se

Gerhard J. Woeginger
University of Twente
Faculty of Electrical Engineering, Mathematics and Computer Science
P.O. Box 217, 7500 AE Enschede, The Netherlands
E-mail: g.j.woeginger@math.utwente.nl

Cataloging-in-Publication Data applied for
A catalog record for this book is available from the Library of Congress

Bibliographic information published by Die Deutsche Bibliothek
Die Deutsche Bibliothek lists this publication in the Deutsche Nationalbibliographie;
detailed bibliographic data is available in the Internet at <http://dnb.ddb.de>.

CR Subject Classification (1998): F, D, C.2-3, G.1-2, I.3, E.1-2

ISSN 0302-9743

ISBN 978-3-540-40493-4   ISBN 978-3-540-45061-0 (eBook)

DOI 10.1007/978-3-540-45061-0

http://www.springer.de

© Springer-Verlag Berlin Heidelberg 2003

Originally published by Springer-Verlag Berlin Heidelberg New York in 2003

Typesetting: Camera-ready by author, data conversion by PTP-Berlin GmbH
Printed on acid-free paper   SPIN: 10928936   06/3142   5 4 3 2 1 0

Preface

The 30th International Colloquium on Automata, Languages and Programming (ICALP 2003) was held from June 30 to July 4 on the campus of the Technische Universiteit Eindhoven (TU/e) in Eindhoven, The Netherlands. This volume contains all contributed papers presented at ICALP 2003, together with the invited lectures by Jan Bergstra (Amsterdam), Anne Condon (Vancouver), Amos Fiat (Tel Aviv), Petra Mutzel (Vienna), Doron Peled (Coventry) and Moshe Vardi (Houston).

Since 1972, ICALP has been the main annual event of the European Association for Theoretical Computer Science (EATCS). The ICALP program can be divided into two tracks, viz. track A (algorithms, automata, complexity, and games) and track B (logics, semantics, and theory of programming).

In response to the Call for Papers, the program committee received 212 submissions: 131 for track A and 81 for track B. The committee met on March 14 and 15 in Haarlem, The Netherlands and selected 84 papers for inclusion in the scientific program. The selection was based on originality, quality and relevance to theoretical computer science. We wish to thank all authors who submitted extended abstracts for consideration, and all referees and subreferees who helped in the extensive evaluation process.

The EATCS Best Paper Award for Track A was given to the paper "The Cell Probe Complexity of Succinct Data Structures" by Anna Gál and Peter Bro Miltersen and the award for Track B was given to the paper "A Testing Scenario for Probabilistic Automata" by Mariëlle Stoelinga and Frits Vaandrager.

ICALP 2003 was a special ICALP. Two other computer science conferences co-located with ICALP this time: the 24th International Conference on Application and Theory of Petri Nets (ATPN 2003) and the Conference on Business Process Management (BPM 2003). During ICALP 2003 the following special events took place: the EATCS Distinguished Service Award was given to Grzegorz Rozenberg (Leiden), and the Lifetime Achievement Award of the NVTI (Dutch Association for Theoretical Computer Science) was given to N.G. de Bruijn (Eindhoven). Several high-level workshops were held as satellite events of ICALP 2003, coordinated by Erik de Vink. These included the following workshops: Algorithms for Massive Data Sets, Foundations of Global Computing (FGC), Logic and Communication in Multi-Agent Systems (LCMAS), Quantum Computing, Security Issues in Coordination Models, Languages and Systems (SecCo), Stochastic Petri Nets, Evolutionary Algorithms, the 1st International Workshop on the Future of Neural Networks (FUNN), and Mathematics, Logic and Computation (workshop in honor of N.G. de Bruijn's 85th birthday). In addition, there was a discussion forum on Education Matters — the Challenge of Teaching Theoretical Computer Science organized by Hans-Joerg Kreowski.

The scientific program of ICALP 2003 and satellite workshops showed that theoretical computer science is a vibrant field, deepening our insights into the foundations and future of computing and system design in many application areas.

The sponsors of ICALP 2003 included the municipality of Eindhoven, Sodexho, Océ, the research school IPA, the European Educational Forum, Springer-Verlag, Elsevier, Philips Research, Atos Origin, Pallas Athena, Pearson Education Benelux, and ABE Foundation. We are very grateful to the Technische Universiteit Eindhoven for supporting and hosting ICALP 2003. The organizing committee consisted of Jos Baeten, Tijn Borghuis, Erik Luit, Emmy van Otterdijk, Anne-Meta Oversteegen, Thieu Rietjens, Karin Touw and Erik de Vink, all of the TU/e. Thanks is owed to them, and to everybody else who helped, for their outstanding effort in making ICALP 2003 a success.

June 2003

Jos Baeten
Jan Karel Lenstra
Joachim Parrow
Gerhard Woeginger

Program Committee

Track A

Harry Buhrman, CWI Amsterdam
Jens Clausen, DTK Lyngby
Martin Dyer, Leeds
Lars Engebretsen, KTH Stockholm
Uri Feige, Weizmann
Philippe Flajolet, INRIA Rocquencourt
Kazuo Iwama, Kyoto
Elias Koutsoupias, UCLA
Jan Karel Lenstra, Georgia Tech, Co-chair
Stefano Leonardi, Roma
Rasmus Pagh, Copenhagen
Jean-Eric Pin, CNRS and Paris 7
Uwe Schoening, Ulm
Jiri Sgall, CAS Praha
Micha Sharir, Tel Aviv
Vijay Vazirani, Georgia Tech
Ingo Wegener, Dortmund
Peter Widmayer, ETH Zürich
Gerhard Woeginger, Twente, Co-chair

Track B

Samson Abramsky, Oxford
Eike Best, Oldenburg
Manfred Broy, TU München
Philippe Darondeau, INRIA Rennes
Rocco De Nicola, Firenze
Rob van Glabbeek, Stanford
Ursula Goltz, Braunschweig
Roberto Gorrieri, Bologna
Robert Harper, Carnegie Mellon
Holger Hermanns, Twente
Kim Larsen, Aalborg
Jean-Jacques Levy, INRIA Rocquencourt
Flemming Nielson, DTU Lyngby
Prakash Panangaden, McGill
Joachim Parrow, Uppsala, chair
Amir Pnueli, Weizmann
Davide Sangiorgi, INRIA Sophia
Bernhard Steffen, Dortmund
Björn Victor, Uppsala

Referees

Karen Aardal
Parosh Abdulla
Luca Aceto
Jiri Adamek
Pankaj Agarwal
Susanne Albers
Alessandro Aldini
Jean-Paul Allouche
Noga Alon
André Arnold
Lars Arvestad
Vincenzo Auletta
Giorgio Ausiello
Holger Austinat
Yossi Azar
Marie-Pierre Béal
Christel Baier
Amotz Bar-Noy
Peter Baumgartner
Danièle Beauquier
Luca Becchetti
Marek Bednarczyk
Gerd Behrmann
Michael Bender
Thorsten Bernholt
Vincent Berry
Jean Berstel
Philip Bille
Lars Birkedal
Markus Blaeser
Bruno Blanchet
Luc Boasson
Chiara Bodei
Hans Bodlaender
Beate Bollig
Viviana Bono
Michele Boreale
Ahmed Bouajjani
Peter Braun
Franck van Breugel
Mikael Buchholtz
Daniel Bünzli
Marzia Buscemi
Nadia Busi

Julien Cassaigne
Didier Caucal
Amit Chakrabarti
Christian Choffrut
Marek Chrobak
Mark Cieliebak
Mario Coppo
Robert Cori
Flavio Corradini
Cas Cremers
Vincent Cremet
Maxime Crochemore
Mary Cryan
Artur Czumaj
Peter Damaschke
Ivan Damgaard
Zhe Dang
Olivier Danvy
Pedro D'Argenio
Giorgio Delzanno
Jörg Derungs
Josee Desharnais
Alessandra Di Pierro
Volker Diekert
Martin Dietzfelbinger
Dino Distefano
Stefan Droste
Abbas Edalat
Stefan Edelkamp
Stephan Eidenbenz
Isaac Elias
Leah Epstein
Thomas Erlebach
Eric Fabre
Rolf Fagerberg
Francois Fages
Stefan Felsner
Paolo Ferragina
Jiří Fiala
Amos Fiat
Andrzej Filinski
Bernd Finkbeiner
Alain Finkel
Thomas Firley

Paul Fischer
Hans Fleischhack
Emmanuel Fleury
Wan Fokkink
Cédric Fournet
Gudmund Frandsen
Martin Fränzle
Thomas Franke
Séverine Fratani
Ari Freund
Alan Frieze
Toshihiro Fujito
Naveen Garg
Olivier Gascuel
Michael Gatto
Stéphane Gaubert
Cyril Gavoille
Blaise Genest
Dan Ghica
Jeremy Gibbons
Oliver Giel
Inge Li Gørtz
Leslie Goldberg
Mikael Goldmann
Roberta Gori
Mart de Graaf
Serge Grigorieff
Martin Grohe
Jan Friso Groote
Roberto Grossi
Claudia Gsottberger
Joshua Guttman
Johan Håstad
Stefan Haar
Lisa Hales
Mikael Hammar
Chris Hankin
Rene Rydhof Hansen
Sariel Har-Peled
Jerry den Hartog
Gustav Hast
Anne Haxthausen
Fabian Hennecke
Thomas Hildebrandt

Yoram Hirshfeld
Thomas Hofmeister
Jonas Holmerin
Juraj Hromkovic
Michaela Huhn
Hardi Hungar
Thore Husfeldt
Michael Huth
Oscar H. Ibarra
Keiko Imai
Purush Iyer
Jan Jürjens
Radha Jagadeesan
Jens Jägersküpper
Petr Jančar
Klaus Jansen
Thomas Jansen
Mark Jerrum
Tao Jiang
Magnus Johansson
Georgi Jojgov
Jørn Justesen
Erich Kaltofen
Viggo Kann
Haim Kaplan
Juhani Karhumaki
Anna Karlin
Joost-Pieter Katoen
Claire Kenyon
Rohit Khandekar
Joe Kilian
Josva Kleist
Bartek Klin
Jens Knoop
Stavros Kolliopoulos
Petr Kolman
Jochen Konemann
Guy Kortsarz
Juergen Koslowski
Michal Koucký
Daniel Král'
Jan Krajíček
Dieter Kratsch
Matthias Krause
Michael Krivelevich

Werner Kuich
Dietrich Kuske
Salvatore La Torre
Anna Labella
Ralf Laemmel
Jim Laird
Cosimo Laneve
Martin Lange
Ruggero Lanotte
Francois Laroussinie
Thierry Lecroq
Troy Lee
James Leifer
Arjen Lenstra
Reinhold Letz
Francesca Levi
Huimin Lin
Andrzej Lingas
Luigi Liquori
Markus Lohrey
Sylvain Lombardy
Michele Loreti
Roberto Lucchi
Gerald Luettgen
Eva-Marta Lundell
Parthasarathy
  Madhusudan
Jean Mairesse
Kazuhisa Makino
Oded Maler
Luc Maranget
Alberto
  Marchetti-Spaccamela
Martin Mareš
Frank Marschall
Fabio Martinelli
Andrea Masini
Sjouke Mauw
Richard Mayr
Colin McDiarmid
Pierre McKenzie
Michael Mendler
Christian Michaux
Kees Middelburg
Stefan Milius

Peter Bro Miltersen
Joe Mitchell
Eiji Miyano
Faron Moller
Franco Montagna
Christian Mortensen
Peter Mosses
Tilo Muecke
Markus Mueller-Olm
Madhavan Mukund
Haiko Muller
Ian Munro
Andrzej Murawski
Anca Muscholl
Hiroshi Nagamochi
Seffi Naor
Margherita Napoli
Uwe Nestmann
Rolf Niedermeier
Mogens Nielsen
Stefan Nilsson
Takao Nishizeki
Damian Niwinski
John Noga
Thomas Noll
Christian N.S. Pedersen
Gethin Norman
Manuel Núñez
Marc Nunkesser
Anna Östlin
David von Oheimb
Yoshio Okamoto
Paulo Oliva
Nicolas Ollinger
Hirotaka Ono
Vincent van Oostrom
Janos Pach
Catuscia Palamidessi
Anna Palbom
Mike Palis
Alessandro Panconesi
Christos Papadimitriou
Andrzej Pelc
David Peleg
Holger Petersen

Seth Pettie
Iain Phillips
Giovanni Pighizzini
Henrik Pilegaard
Sophie Pinchinat
G. Michele Pinna
Conrad Pomm
Ely Porat
Giuseppe Prencipe
Corrado Priami
Guido Proietti
Pavel Pudlák
Rosario Pugliese
Uri Rabinovich
Theis Rauhe
Andreas Rausch
António Ravara
Klaus Reinhardt
Michel A. Reniers
Arend Rensink
Christian Retoré
James Riley
Martin Roetteler
Maurice Rojas
Marie-Francoise Roy
Oliver Ruething
Bernhard Rumpe
Wojciech Rytter
Géraud Sénizergues
Nicoletta Sabatini
Andrei Sabelfeld
Kunihiko Sadakane
Marie-France Sagot
Louis Salvail
Bruno Salvy
Christian Salzmann
Peter Sanders
Miklos Santha
Martin Sauerhoff
Daniel Sawitzki
Andreas Schaefer

Norbert Schirmer
Konrad Schlude
Philippe Schnoebelen
Philip Scott
Roberto Segala
Helmut Seidl
Peter Selinger
Nicolas Sendrier
Maria Serna
Alexander Shen
Natalia Sidorova
Detlef Sieling
Marc Sihling
Hans Simon
Alex Simpson
Michael Sipser
Martin Skutella
Michiel Smid
Pawel Sobocinski
Eljas Soisalon-Soininen
Ana Sokolova
Frits Spieksma
Renzo Sprugnoli
Jiří Srba
Rob van Stee
Angelika Steger
Christian Stehno
Ralf Steinbrueggen
Colin Stirling
Leen Stougie
Martin Strecker
Werner Struckmann
Hongyan Sun
Ichiro Suzuki
Tetsuya Takine
Hisao Tamaki
Amnon Ta-Shma
David Taylor
Pascal Tesson
Simone Tini
Takeshi Tokuyama

Mauro Torelli
Stavros Tripakis
john Tromp
Emilio Tuosto
Irek Ulidowski
Yaroslav Usenko
Frits Vaandrager
Frank Valencia
Vincent Vanackère
Moshe Vardi
Helmut Veith
Laurent Viennot
Alexander Vilbig
Jørgen Villadsen
Erik de Vink
Paul Vitanyi
Berthold Voecking
Walter Vogler
Marc Voorhoeve
Tjark Vredeveld
Stephan Waack
Igor Walukiewicz
Dietmar Wätjen
Birgitta Weber
Heike Wehrheim
Elke Wilkeit
Tim Willemse
Harro Wimmel
Peter Winkler
Carsten Witt
Philipp Woelfel
Ronald de Wolf
Derick Wood
Jürg Wullschleger
Shigeru Yamashita
Wang Yi
Heisung Yoo
Hans Zantema
Gianluigi Zavattaro
Pascal Zimmer
Uri Zwick

Table of Contents

Invited Lectures

Algorithms

Process Algebra

Approximation Algorithms

Languages and Programming

Complexity

Optimization and Games

Graphs and Bisimulation

Online Problems

Verification

Around the Internet

Temporal Logic and Model Checking

Graph Problems

Logic and Lambda-Calculus

Data Structures and Algorithms

Types and Categories

Probabilistic Systems

Sampling and Randomness

Scheduling

Geometric Problems

Author Index 1197

On Equivalent Representations of Infinite Structures

Arnaud Carayol and Thomas Colcombet

IRISA, Campus universitaire de Beaulieu,
35042 Rennes Cedex, France {Arnaud.Carayol, Thomas.Colcombet}@irisa.fr

Abstract. According to Barthelman and Blumensath, the following families of infinite graphs are isomorphic: (1) prefix-recognisable graphs, (2) graph solutions of VR equational systems and (3) MS interpretations of regular trees. In this paper, we consider the extension of prefix-recognisable graphs to prefix-recognisable structures and of graphs solutions of VR equational systems to structures solutions of positive quantifier free definable (PQFD) equational systems. We extend Barthelman and Blumensath's result to structures parameterised by infinite graphs by proving that the following families of structures are equivalent: (1) prefix-recognisable structures restricted by a language accepted by an infinite deterministic automaton, (2) solutions of infinite PQFD equational systems and (3) MS interpretations of the unfoldings of infinite deterministic graphs. Furthermore, we show that the addition of a *fuse* operator, that merges several vertices together, to PQFD equational systems does not increase their expressive power.

1 Introduction

The automatic verification of properties on infinite structures is an important technique for proving behavioural properties on programs. A natural encoding of a program behaviour is an infinite directed graph where vertices are states of the machine, and edges mimic the transition steps of the program. Properties on the program can then be expressed as logical formulas referring to this graph (or its unfolding when considering *e.g* temporal logics). The problem of model-checking is then to decide the satisfaction of a formula over the graph. This problem is usually undecidable. However, on certain families of infinite graphs and for some given logics the model-checking problem is decidable.

In this work, we are dealing with monadic second-order (MS) logic: an extension of first-order logic which allows quantification over sets of vertices. The first decidability result for this logic over an infinite graph was provided by Büchi for the infinite semi-line. Rabin extended this result to the infinite binary tree.

With the work of Muller and Schupp on pushdown graphs [MS85], the focus of study shifted from infinite graphs to families of infinite graphs.

Since then, many families of graphs have been presented with various decidability and structural properties. Those families can be classified according to their representation into three categories.

J.C.M. Baeten et al. (Eds.): ICALP 2003, LNCS 2719, pp. 599–610, 2003.
© Springer-Verlag Berlin Heidelberg 2003

The equational representation describes an infinite structure as the solution of an equational system. The family of structures (or graphs) obtained in this way depends on the choice of the operators. The most famous examples are hyperedge replacement equational structures (HR) [Cou89] and the vertex replacement equational graphs (VR) [Cou90]. The VR operators also have been extended into vertex replacement with product operators [Col02].

The transformational representation consists in applying some finite sequence of transformations over an already-known structure. Transformations can be the unfolding of graphs [CW98], Shelah-Muchnik-Walukiewicz tree-like construction [Wal96], or logically defined transformations (FO interpretations, inverse finite or rational mappings [Cau96,Urv02], MS interpretations or general MS-definable transductions [Cou94]).

The internal representation amounts to give an exact description of both the universe and the relations of the structure. The most used universe is the set of words over a given finite alphabet. Relations over words can then be defined by means of many techniques:

Rewriting: Prefix (or suffix) rewriting of words describes the family of pushdown graphs [MS85,Cau92]. When the set of rules is recognisable, it leads to prefix-recognisable graphs [Cau96].

Transductions: Relations recognised by synchronised transductions describe the class of automatic graphs [S n92] and structures [Blu99]. When the transduction is rational, it defines the rational graphs [Mor00].

Structures defined over the universe of closed terms have also been presented [DT90,Blu99,L d02,Col02].

The above mentioned techniques are not independent from each other. Many connections have been stated in the literature. In our case we are specially involved with the following: the graphs solution of VR equational systems are isomorphic to prefix-recognisable graphs [Bar97] and to MS interpretations of infinite regular trees [Blu01].

To some extent, these classes of graphs are defined upon *finite* objects. In particular, a VR-equational graph is the solution of a *finite* system of equations and prefix-recognisable graph is a rewriting system restricted to the language accepted by a *finite* automaton. These two kinds of graphs are equivalent and can be obtained by MS-definable transduction of the unfolding of a *finite* graph.

We generalise this triple equivalence to structures defined by *infinite* objects. We show that interpretation of infinite systems of PQFD equations (which is a natural extension of VR operators introduced in [Bar97]), *PR*-structures restricted by infinite deterministic automaton and MS-definable transductions of the unfoldings of deterministic infinite graphs are equivalent. Furthermore, this equivalence is effective in the sense that MS-definable transductions link the system of equations, the automaton and the graph.

In [CM02], the authors prove that for describing sets of finite structures the addition of a *fuse* operator — which merges vertices together — to PQFD-like operators does not increase the expressivity of the considered systems. The authors also emphasizes on how this extension unifies the description of HR-equational et VR-equational graphs. We naturally investigated the infinite counterpart of this result and proved under reasonable technical restrictions that the addition of a *fuse* operator to PQFD operators does not increase their expressive power. The two results are however technically significantly different.

The rest of the paper is divided as follows. Section 2 introduces the basic definitions. Section 3 presents structures defined by equational systems and Section 4 defines unfolding and states the first inclusion. Section 5 introduces PR-systems and states the last two inclusions.

2   Definitions

Relational Structures

We define the *global signature* Ξ to be equal to $\bigcup_{n>0} \Xi_n$ where Ξ_n is an infinite set of symbols of *arity* n. For any R in Ξ, $|R|$ designates the arity of R.

A *relational structure* S is a pair $(\mathcal{U}, Val)$ where $\mathcal{U}$ is an 'at most countable' set called the universe and Val associates to a symbol of arity n a subset of $\mathcal{U}^n$. We will write R^S instead of $Val(R)$. Moreover, we suppose that Val has a finite support (i.e. the set of R such that $Val(R) \neq \emptyset$ is finite). A *signature* Σ of S is a *finite* set which contains the support of Val.

The *restriction* of a structure $S = (\mathcal{U}, Val)$ to a universe $\mathcal{U}' \subseteq \mathcal{U}$ is denoted $S|_{\mathcal{U}'}$ and designates the structure $(\mathcal{U}', Val')$ where $Val'(R) = Val(R) \cap (\mathcal{U}')^{|R|}$.

Two structures S and S' of respective universe $\mathcal{U}$ and $\mathcal{U}'$ are *isomorphic* (written $S \approx S'$) if there exists a one to one mapping ρ from $\mathcal{U}$ onto $\mathcal{U}'$ such that for any symbol $R \in \Xi$, $R^S(x_1, \dots, x_n) \Leftrightarrow R^{S'}(\rho(x_1), \dots, \rho(x_n))$.

Graphs

A *directed graph* G (or simply a *graph*) labelled by a finite set E is a relational structure admitting a signature with binary symbols only (identified with E). The universe is denoted by V and its elements are called vertices. A directed graph is *rooted* if its signatures contain an unary relation *root* which is interpreted as a singleton. By slight abuse, we will use the constant *root* in our formulas. The graph is said to be *deterministic* if for any $x, y, z \in V$ and for any relation $e \in E$, if $e(x, y)$ and $e(x, z)$ then $y = z$.

A *path* π in a graph G labelled by E is a finite sequence $v_1 e_1 \dots e_{n-1} v_n$ in $(VE)^*V$ such that for all $i \in [1, n-1]$, $e_i(v_i, v_{i+1})$. For any $w \in E^*$, we write $x \overset{w}{\Longrightarrow} y$ if there exists a path $v_1 e_1 \dots e_{n-1} v_n$ between x and y such that $w = e_1 \dots e_{n-1}$. For $W \subseteq E^*$ a language, $x \overset{W}{\Longrightarrow} y$ holds iff for some $w \in W$, $x \overset{w}{\Longrightarrow} y$.

Given a graph G labelled by E of universe V and a finite set of *fresh* binary symbols $\mathcal{K} = \{k_1, \dots, k_n\}$ (i.e $\mathcal{K} \cap E = \emptyset$), the $\mathcal{K}$-copying of G is the graph G' of universe $V \times [0, n]$ and such that for any relation $R \in E$, $R^{G'} = \{((x_1, 0), \dots, (x_{|R|}, 0)) \mid (x_1, \dots, x_{|R|}) \in R^G\}$ and for $k_i \in \mathcal{K}$, $k_i^{G'} = \{((x, 0), (x, i)) \mid x \in V\}$.

Example 1. Throughout this paper we illustrate all the techniques for describing structures with one example: the step-ladder graph depicted in Figure 1.

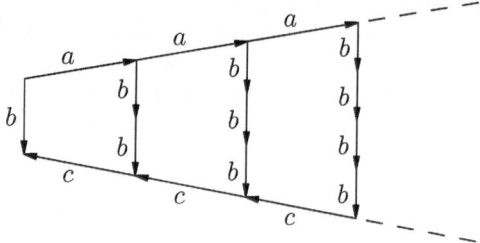

Fig. 1. The step-ladder graph.

Monadic Second Order Logic

In the following, we assume that first order variables are ranged over by $x, y, z \dots$ whereas monadic second order variables are ranged over by $X, Y, Z \dots$ First order variables are interpreted as elements of the universe whereas monadic second order variables are interpreted as subsets of the universe. The atomic predicates of monadic formulas are $x \in X$, $x = y$ and $R(x_1, \dots, x_{|R|})$. Monadic formulas are then inductively defined as $\exists X. \Phi$, $\exists x. \Phi$, $\neg \Phi$ and $\Phi \vee \Psi$ for Φ and Ψ formulas. MS formulas have the usual semantic [Tho97]. If $\Phi(x_1, \dots, x_n)$ is an MS formula and if $(u_1, \dots, u_n)$ is a tuple of elements of $\mathcal{U}$, $\mathcal{S} \models \Phi(u_1, \dots, u_n)$ means that $\mathcal{S}$ models Φ when x_i is interpreted by u_i for all $i \in [1, n]$.

A *MS interpretation* $\mathcal{I}$ is given by a MS formula $\delta(x)$ together with a finite set of formulas $(\Phi_R)_{R \in \Sigma}$ where Φ_R has free variables in $\{x_1, \dots, x_{|R|}\}$. $\mathcal{I}$ associates to each structure $\mathcal{S}$ of universe $\mathcal{U}$ the structure $\mathcal{I}(\mathcal{S})$ of universe $\mathcal{U}_{\mathcal{I}(\mathcal{S})} = \{x \in \mathcal{U} \mid \mathcal{S} \models \delta(x)\}$ and such that if $R \in \Sigma$, $R^{\mathcal{I}(\mathcal{S})} = \{\bar{x} \in (\mathcal{U}')^{|R|} \mid \mathcal{S} \models \Phi_R(\bar{x})\}$ (if $R \notin \Sigma$, $R^{\mathcal{I}(\mathcal{S})} = \emptyset$).

An *MS-definable transduction* $\mathcal{T}$ [Cou94] is the composition of a $\mathcal{K}$-copying operation and an MS interpretation. This transformation preserves the decidability of the MS theory.

3   Equational Systems

In this section we present how to describe infinite structures as solutions of equational systems over a given set of operators. Classical examples of this approach are hyperedge replacement systems [Cou89] or vertex replacement (VR) [Cou90] graphs.

For the rest of this section, we fix a signature Σ. For $\mathcal{V}$ a given set of variable names, we write $B^+(\mathcal{V})$ the set of positive boolean formulas over variables $\mathcal{V}$. Those formulas are built upon predicates of the signature applied to variables in $\mathcal{V}$, of the boolean connectives $\wedge$ and $\vee$, and the constants $\mathbf{t}$ (true) and $\mathbf{f}$ (false). Quantifiers as well as negation are not allowed.

We use the set of symbols $PQFD = PQFD_0 \cup PQFD_1 \cup PQFD_2$ with:

$$PQFD_0 = \{\mathbf{one}\} \qquad PQFD_2 = \{\oplus\}$$
$$PQFD_1 = \{\mathbf{pqfd}[(\phi_R)_{R \in \Sigma}] \mid \forall R \in \Sigma, \ \phi_R \in B^+(x_1, \dots, x_{|R|})\}$$

Symbols in $PQFD_i$ have arity i. The mapping $\_$ gives their semantic:

Singleton structure one: $\mathcal{U}_{\mathbf{one}} = \{0\}$ and $R^{\mathbf{one}} = \emptyset$ for any symbol R ,

Positive quantifier-free definable interpretation pqfd$[(\phi_R)]$:

given a relational structure $\mathcal{S}$, $\mathcal{U}_{\mathbf{pqfd}[(\phi_R)](\mathcal{S})} = \mathcal{U}_\mathcal{S}$,

and $R^{\underline{\mathbf{pqfd}[(\phi_R)](\mathcal{S})}}(u_1, \ldots, u_{|R|})$ iff $\mathcal{S} \models \phi_R(u_1, \ldots, u_{|R|})$,

Disjoint union $\oplus$: given two structures $\mathcal{S}_1$ and $\mathcal{S}_2$,

$\mathcal{U}_{\mathcal{S}_1 \oplus \mathcal{S}_2} = \{1\} \times \mathcal{U}_{\mathcal{S}_1} \cup \{2\} \times \mathcal{U}_{\mathcal{S}_2}$ and,

for any symbol R, $R^{\mathcal{S}_1 \oplus \mathcal{S}_2} = \{((1, x_1), \ldots, (1, x_{|R|})) \mid R^{\mathcal{S}_1}(x_1, \ldots, x_{|R|})\}$
$\cup \{((2, x_1), \ldots, (2, x_{|R|})) \mid R^{\mathcal{S}_2}(x_1, \ldots, x_{|R|})\}$.

A similar set of operators has been introduced by Barthelman [Bar97].

Let us emphasize that this set of operators provides a strict and natural extension to relational structures of vertex replacement (VR) graph operators. Let us illustrate how to obtain VR systems with PQFD systems on graphs. The usual definition of VR operators works over coloured directed graphs: directed graphs labelled by a finite set E and extended with a mapping which associates to each vertex a color belonging to some given finite set C of colors. In our case, we can encode such a graph into a structure over the signature $\Sigma = C \cup E$ where symbols in C and E have respective arity 1 and 2 and encode respectively the fact that a vertex has a given color, and the presence of an edge between two nodes. We can now introduce the four VR operators, and their equivalent PQFD expression.

Single vertex constant of color c — simply written $\mathbf{c}$ — represents the graph with one vertex of colour c and no edge.

It can be expressed as $\mathbf{pqfd}[\phi_c](\mathbf{one})$ with $\phi_{c'} = \mathbf{f}$ for any $c' \neq c$, $\phi_c = \mathbf{t}$ and $\phi_a = \mathbf{f}$ for any $a \in E$.

Disjoint union — written $\oplus$ as for structures — performs the disjoint union of two coloured graphs.

It can naturally be encoded by the disjoint union of structures $\oplus$.

Renaming color c_1 into color c_2 of a coloured graph G — written $\mathbf{ren}_{c_1,c_2}(G)$ — changes the color mapping in such a way that every vertex of color other than c_1 keeps its original color, and vertices of original color c_1 have new color c_2.

Let us suppose for simplicity that $c_1 \neq c_2$. The renaming operator can be encoded by $\mathbf{pqfd}[(\phi_R)_{R \in \Sigma}](G)$ with $\phi_{c_1} = \mathbf{f}$, $\phi_{c_2} = c_1(x_1) \vee c_2(x_1)$, $\phi_c = c(x_1)$ for $c \in C - \{c_1, c_2\}$, and $\phi_a = a(x_1, x_2)$ for $a \in E$.

Adding edges labelled by a between color c_1 and color c_2 to a graph G — written $\mathbf{add}_{c_1,c_2,a}(G)$ — adds to the coloured graph G all possible edges labelled by a with as origin a vertex of color c_1 and as destination a vertex of color c_2.

The edge-adding operator can be encoded by $\mathbf{pqfd}[(\phi_R)_{R \in \Sigma}](G)$ with $\phi_a = a(x_1, x_2) \vee (c_1(x_1) \wedge c_2(x_2))$, $\phi_b = b(x_1, x_2)$ for $b \in E - \{a\}$ and $\phi_c = c(x_1)$ for any $c \in C$.

PQFD operators can be used in equational systems: One can equip structures with the partial order of inclusion defined by $\mathcal{S} \subseteq \mathcal{S}'$ iff $\mathcal{U}_\mathcal{S} \subseteq \mathcal{U}_{\mathcal{S}'}$ and $R^\mathcal{S} \subseteq R^{\mathcal{S}'}$

for any symbol R. This ordering is a complete partial order (cpo) admitting the only structure of empty universe $\perp$ as smallest element. The semantic of operators is continuous with respect to this cpo. It means that a (even infinite) system of equations using PQFD operators admits a unique smallest solution.

Example 2. Let us illustrate infinite VR systems of equations for producing the graph of Figure 1. Let us first introduce the intermediate coloured graphs X_n presented in Figure 2.

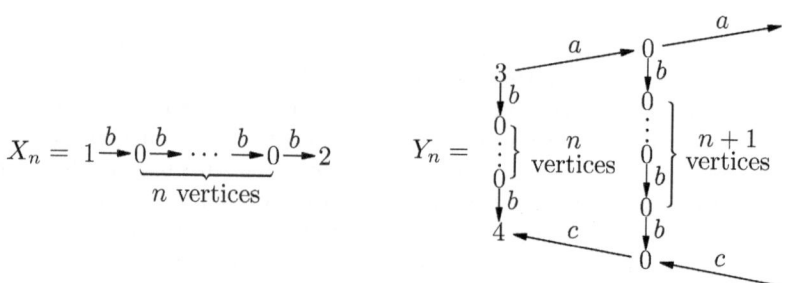

Fig. 2. The graphs X_n and Y_n.

The X_n graphs can be defined by the following recursive equations:

$$X_0 = \mathbf{add}_{1,2,b}(1 \oplus 2) \quad \text{and} \quad X_{n+1} = \mathbf{ren}_{3,2}(\mathbf{ren}_{2,0}(\mathbf{add}_{2,3,b}(X_n \oplus 3))) \quad (1)$$

We can now define the Y_n coloured graphs (notice Y_0 is isomorphic to the graph of Figure 1). They satisfy the following equation:

$$Y_n = \mathbf{ren}_{2,4}(\mathbf{ren}_{1,3}(\mathbf{ren}_{4,0}(\mathbf{ren}_{3,0}(\mathbf{add}_{1,3,a}(\mathbf{add}_{4,2,c}(X_n \oplus Y_{n+1})))))) \quad (2)$$

In fact the coloured graphs X_n and Y_n are the smallest possible graphs satisfying the equations (1) and (2): the step-ladder graph is the smallest solution of this equational system. Let us notice that, though infinite, this equational system can be represented by an infinite graph as depicted in Figure 3.

This process of encoding the equational system into a rooted graph is general. Formally, a rooted graph $\mathcal{E}$ is a PQFD-equational system if its edges:

- are labelled by $\{\oplus_1, \oplus_2\} \cup PQFD_0 \cup PQFD_1$ and
- for all element x of $\mathcal{U}_S$,
  - if there is an edge labelled by **one** of target x then no edges originate from x.
  - else, either two edges originate from x, and are labelled by respectively $\oplus_1$ and $\oplus_2$,
    or only one edge has origin x, and this edge is labelled by **one** or $\mathbf{pqfd}[(\phi_R)]$ for some ϕ_R.

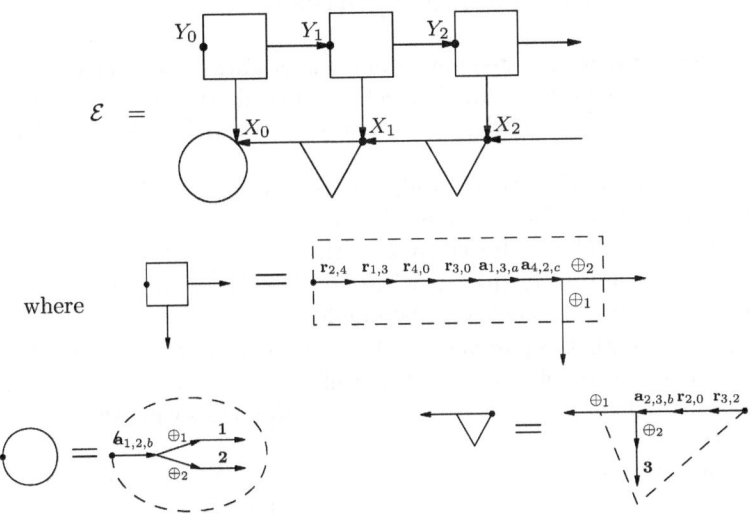

Fig. 3. An infinite VR equational system $\mathcal{E}$ describing the graph of Figure 1

The solution of such a system is defined as follows: let $\sigma^{\mathcal{E}}$ be the smallest function from vertices of $\mathcal{E}$ to structures satisfying:

- If $\mathbf{one}^{\mathcal{E}}(x, y)$ then $\sigma^{\mathcal{E}}(x) = \underline{\mathbf{one}}$,
- if $\mathbf{pqfd}[(\phi_R)]^{\mathcal{E}}(x, y)$ then $\sigma^{\mathcal{E}}(x) = \underline{\mathbf{pqfd}}[(\phi_R)](\sigma^{\mathcal{E}}(y))$,
- and if $\oplus_1{}^{\mathcal{E}}(x, y)$ and $\oplus_2{}^{\mathcal{E}}(x, z)$ then $\sigma^{\mathcal{E}}(x) = \sigma^{\mathcal{E}}(y)\underline{\oplus}\sigma^{\mathcal{E}}(z)$.

then the semantic of the equational system $\mathcal{E}$, written $[\![\mathcal{E}]\!]$ is the graph $\sigma^{\mathcal{E}}(root)$.

We will also make use of another operator: for $p \in \Sigma_1$ a unary symbol, the operator $\mathbf{fuse}_p$ applied to a structure $\mathcal{S}$ keeps the structure unchanged but collapses all the elements x satisfying $p^{\mathcal{S}}(x)$ into a single one. Formally, we define the equivalence relation $\equiv_p$ over $\mathcal{U}_{\mathcal{S}}$ by $x \equiv_p y$ iff $x = y$ or $p^{\mathcal{S}}(x)$ and $p^{\mathcal{S}}(y)$. The classes of equivalence for $\equiv_p$ of an element x is written $[x]_p$. The semantic of $\mathbf{fuse}_p$ is then defined by $\mathbf{fuse}_p(\mathcal{S}) = \mathcal{S}'$ with $\mathcal{U}_{\mathcal{S}'} = \{[x]_p \mid x \in \mathcal{U}_{\mathcal{S}}\}$ and for any n-ary symbol R, $R^{\mathcal{S}'} = \{([v_1]_p, \ldots, [v_n]_p) \mid R^{\mathcal{S}}(v_1, \ldots, v_n)\}$. The set of operators $PQFD$ increased with $\mathbf{fuse}$ operators is written $PQFD + F$.

In fact, the cpo used has to be slightly changed for the $\mathbf{fuse}$ operators to be continuous. Furthermore, the $\mathbf{fuse}$ operators make it necessary to put some extra restrictions to the system: a $PQFD + F$ equational system is said normalised if there is no predicate $R(y_1, \ldots, y_{|R|})$ such that $y_i = y_j$ for $i \neq j$ in any formula appearing in a $\mathbf{pqfd}$ operator.

4   The Transformational Approach

Successively applying a finite number of transformations to a relational structure is a second technique for obtaining new relational structures. In this work, we are basically using two such transformations: MS-definable transduction and unfolding.

MS-definable transduction has already been presented. We define here a version of unfolding suitable for deterministic rooted graphs only. Given a deterministic rooted graph G labelled by E with a vertices set V, ρ_G is the function from E^* to V such that $\rho_G(u) = x$ with $root \overset{u}{\Rightarrow} x$ (since the graph is deterministic, there is at most one such x). The unfolding of G is the deterministic rooted graph $Unf(G)$ with a set of vertices $V' = \{u \mid \exists x \in V, \ root \overset{u}{\Rightarrow} x\}$ and such that for all edge symbol a, $a^{Unf(G)}(u, v)$ iff $a^G(\rho_G(u), \rho_G(v))$. The function ρ_G is a morphism of graph and is called the reduction (following the terminology of bisimulation).

We are interested here in transforming a deterministic graph by applying successively an unfolding and a MS-definable transduction.

Example 3. Let G be the graph presented in Figure 4.a with its *root* marked by an unlabelled edge and let $\mathcal{I}$ be the MS interpretation $(\delta, \{\Phi_a, \Phi_b, \Phi_c\})$ with $\delta(x) = \textbf{true}$, $\Phi_a(x_1, x_2) = a(x_1, x_2)$, $\Phi_b(x_1, x_2) = b(x_1, x_2)$ and $\Phi_c(x_1, x_2) = (\exists x_1'.\exists x_2'. x_1' \overset{b^*}{\Longrightarrow} x_1 \ \wedge \ x_2' \overset{b^*}{\Longrightarrow} x_2 \ \wedge \ a(x_2', x_1')) \ \wedge \ \neg(\exists z. a(x_1, z) \ \vee \ a(x_2, z))$ where $x \overset{b^*}{\Longrightarrow} y$ is a MS formula stating that there is a path between x and y using only edges labelled by b. $\mathcal{I}(Unf(G))$ is the step-ladder of Figure 1 (Figure 4.b presents the unfolding of G).

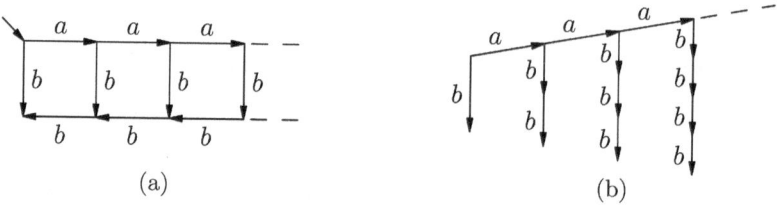

(a)   (b)

Fig. 4. The graph G (a) and its unfolding (b).

Those two transformations are sufficient for expressing $PQFD + F$ equational systems:

Lemma 1. *Given a normalised $PQFD + F$ equational system $\mathcal{E}$, there exists an MS interpretation $\mathcal{I}$ such that $\mathcal{I}(Unf(\mathcal{E}))$ is isomorphic to $[\![\mathcal{E}]\!]$.*

Proof (sketch). The first remark used in the proof is that unfolding preserves the solution of equational systems: $[\![\mathcal{E}]\!] = [\![Unf(\mathcal{E})]\!]$.

For simplicity, let us suppose first that no **fuse** operators are used. Under this hypothesis, each element of $\mathcal{U}_{[\![\mathcal{E}]\!]}$ can be uniquely identified with the **one** operator appearing in $Unf(\mathcal{E})$ which has introduced it (if this operator is removed from the tree, then the element disappears from the structure). Let us call ρ this injective mapping from $\mathcal{U}_{[\mathcal{E}]}$ to $V_{Unf(\mathcal{E})}$. Then, there exists formulas Φ_R for all symbol R of arity n in the signature such that $Unf(\mathcal{E}) \vdash \Phi_R(\rho(x_1), \ldots, \rho(x_n))$ iff $R^{[\mathcal{E}]}(x_1, \ldots, x_n)$ holds. Let $\delta(x)$ be $(\exists y, \mathbf{one}(x, y))$, then the interpretation $\mathcal{I} = (\delta, (\Phi_R))$ is such that $\mathcal{I}(Unf(\mathcal{E}))$ is isomorphic to $[\mathcal{E}]$ (and ρ is the isomorphism).

If **fuse** operators are used, a similar ρ mapping can be provided: the difference is that it maps elements of $\mathcal{U}_{[\mathcal{E}]}$ to either **one** operators or **fuse** operators. The intention is that an element of $\mathcal{U}_{[\mathcal{E}]}$ is uniquely represented by a **one** operator iff no **fuse** operator 'touched' it in the equational system, in the other case, the element is uniquely represented by the closest to the root **fuse** operator in which it was involved. Apart from this distinction, the same technique is applied for providing the interpretation $\mathcal{I}$.

5 Prefix-Recognisable Structures

In this section, we focus on the internal representation of structures. Prefix-recognisable graphs have been introduced by Caucal [Cau96]. A possible description of these graphs is by systems of word rewriting. Blumensath [Blu01] extended this definition to relations of arbitrary arity. Those structures, when restricted to binary relations coincide with prefix-recognisable graphs. We give here a similar (and equivalent) definition of prefix-recognisable structures.

For simplicity, we fix a common *infinite* alphabet A. Let R, R' be two relations over A^* of respective arities k and l, we designate by $R \times R'$ the $(k+l)$-ary relation defined by $(R \times R')(u_1, \ldots, u_k, v_1, \ldots, v_l)$ if and only if $R(u_1, \ldots, u_k)$ and $R'(v_1, \ldots, v_l)$. Let R be a k-ary relation over A^* and U a language of A^*, we designate by $U \cdot R$ the k-ary relation defined by $(U \cdot R)(uv_1, \ldots, uv_k)$ iff $u \in U$ and $R(v_1, \ldots, v_k)$. Let R be a k-ary relation and π a permutation of $[1, k]$, $R_\pi(x_1, \ldots, x_k)$ iff $R(x_{\pi(1)}, \ldots, x_{\pi(k)})$.

Definition 1. *The set of prefix-recognisable (PR) relations over A^* is the smallest set of relations satisfying:*

- *for U a rational subset of A^*, the unary relation U is in PR,*
- *if $R, R' \in PR$ then $R \times R' \in PR$,*
- *for $R, R' \in PR$ of same arity, $R \cup R' \in PR$,*
- *for $R \in PR$ and U a rational subset of A^*, $U \cdot R \in PR$,*
- *for $R \in PR$ and π a permutation of $[1, |R|]$, $R_\pi \in PR$.*

Remark that the definition of each rational language only involves a finite number of letters in A. Thus each relation in PR refers to a *finite* number of letters.

A *PR-structure* is a relational structure of universe A^* with all interpretations in PR.

Prefix-recognisable graphs [Cau96] can be defined as graphs with edges defined by a finite union of relations of the form $U(V \times W)$ (with U, V and W rational languages) and vertices defined by a rational language L. This naturally corresponds to the class of binary PR-systems restricted by a finite automaton. We extend this notion of restriction to infinite deterministic automaton.

In this article, we will use the term *automaton* to designate a rooted deterministic graph labelled by a finite subset of A. Moreover, we will assume that this graph comes with a set of vertices *Final*. As for finite automaton, we associate to every automaton $\mathcal{A}$ a language $\mathcal{L_A} \subseteq A^*$ consisting of all words corresponding to the labelling of a path from *root* to an element in *Final*.

A *PR-system* $\mathcal{R}$ is a pair $(\mathcal{S}, \mathcal{A})$ where $\mathcal{S}$ is a PR-structure and $\mathcal{A}$ is an automaton. In the following, $\mathcal{R}$ will also designate the structure obtained by restricting $\mathcal{S}$ to $\mathcal{L_A}$.

Example 4. Our example graph of Figure 1 can be described by a PR-system $\mathcal{R} = (\mathcal{S}, \mathcal{A})$. The PR-structure $\mathcal{S}$ has three non-empty binary relations a,b and c such that $a^\mathcal{S} = x^*y^* \cdot (\{\varepsilon\} \times (y+z))$, $b^\mathcal{S} = x^* \cdot (\{\varepsilon\} \times x)$ and $c^\mathcal{S} = x^* \cdot (xy^*z \times y^*z)$. The automaton $\mathcal{A}$ is presented in Figure 5.a. Its root is pointed by an unlabelled edge and all its states are in *Final*. The language recognised by $\mathcal{A}$ is the set of prefixes of $\{x^n y^n z \mid n \geq 0\}$. The graph obtained by restricting $\mathcal{S}$ to the language recognised by $\mathcal{A}$ (Figure 5.b) is isomorphic to the step-ladder (Figure 1).

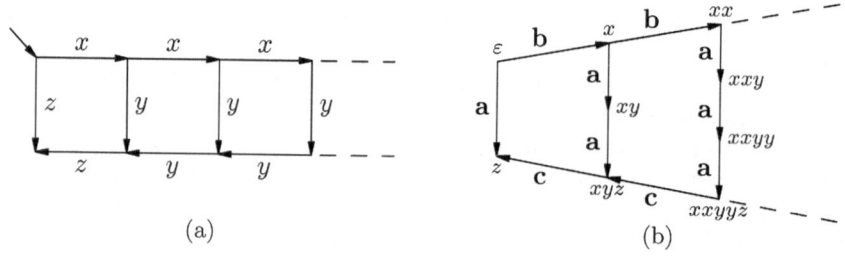

(a)   (b)

Fig. 5. The automaton $\mathcal{A}$ (a) and the PR-system $\mathcal{R}$ (b).

Lemma 2. *For any MS-definable transduction $\mathcal{T}$ and any deterministic graph G, there exists a PR-system $\mathcal{R} = (\mathcal{S}, \mathcal{A})$ such that $\mathcal{T}(Unf(G))$ is isomorphic to $\mathcal{R}$ and $\mathcal{A}$ is obtained from G by an MS-definable transduction.*

Proof (sketch). Let us consider here the simpler case where $\mathcal{T}$ is a *non-erasing* MS interpretation $(\textbf{true}, (\Phi_R)_{R \in \Sigma})$. For T a tree and x one of its nodes, $T_{/x}$ denotes the subtree of T rooted at x.

For every formula Φ_R of arity n, there exists an associated parity automaton $\mathcal{A}_R$. This automaton works on deterministic trees with n distinguished vertices $m_1, \ldots, m_n$ called *marks*. The autaton accepts a tree T iff $T \models \Phi_R(m_1, \ldots, m_n)$. We can always suppose that the states Q of $\mathcal{A}_R$ are enriched with informations about the expected marks: there is a mapping ϕ from Q to $2^{[1,n]}$ such that if a node x of T is assigned a state q in a successful run of $\mathcal{A}_R$ then the marks appearing in $T_{/x}$ are exactly the one of indices in $\phi(q)$.

We want to attach to every node of $Unf(G)$ the set of transitions of $\mathcal{A}_R$ starting a successful run on $Unf(G)_{/x}$. Let $\mathcal{M}_R$ be this application. By definition of the runs of the automaton, the same transitions lead to the same winning runs for any two bisimilar starting nodes (x is bisimilar to y iff $Unf(G)_{/x} \approx Unf(G)_{/y}$). It follows that there is a mapping $\mathcal{B}_R$ attaching transitions to the vertices of graphs, such that $\mathcal{M}_R(Unf(G)) = Unf(\mathcal{B}_R(G))$. Furthermore, we show that this application $\mathcal{B}_R$ is an MS-interpretation (see also [Wal96] for a similar construction).

Finally, we define a n-ary PR-relation R which simulates the run of the parity automaton when $\phi(q) \neq \emptyset$, and prunes the run according to the information provided by $\mathcal{M}_R$ whenever $\phi(q) = \emptyset$.

Lemma 3. *For any PR-system $\mathcal{R} = (\mathcal{S}, \mathcal{A})$, there exists a PQFD-system $\mathcal{E}$ such that $\mathcal{E}$ is obtained by an MS definable transduction from $\mathcal{A}$ and $\mathcal{R}$ is isomorphic to $[\![\mathcal{E}]\!]$.*

Proof (sketch). The proof is syntactical. Let $(P_R)_{R \in \Sigma}$ be PR-relations and let $A_1, \ldots, A_k$ be the finite automata accepting the rational languages involved in the PR-expressions describing the P's relations and let $\mathcal{A}$ be the automaton restricting the PR-system.

We produce a new equational system working over signature Σ enriched by a new symbol for each state of an automaton A_i. The arity of the symbol is the arity of the relation in which $\mathcal{L}_{A_i}$ is used. The equational system is obtained from $\mathcal{A}$ by replacing each edge labelled by a with a **pqfd** operator which simulates simultaneously all a transitions of the A_i's. Disjoint union operators are used to follow the branching structure of $\mathcal{A}$. **one** operators are used for each $Final$ state of $\mathcal{A}$.

6   Conclusion

By combining Lemmas 1,2 and 3, we obtain the following theorem:

Theorem 1. *Let $\mathcal{F}$ be a family of deterministic graphs closed by MS-definable transductions, the following classes of structures are isomorphic:*

- *the solutions of systems of equations over the PQFD operators represented by a graph in $\mathcal{F}$,*
- *the solutions of normalised systems of equations over the PQFD + F operateors represented by a graph in $\mathcal{F}$,*
- *the structures obtained by applying an MS-definable transduction to the unfolding of a deterministic graph in $\mathcal{F}$,*
- *the prefix-recognisable structures restricted to the language accepted by a deterministic automaton in $\mathcal{F}$.*

Le us notice that, according to the third representation, if $\mathcal{F}$ has a decidable MS theory, then it is also the case of the resulting structures.

Removing the normalisation of $PQFD + F$ equational systems is an open question.

Acknowledgements. We would like to thank Didier Caucal for his advices.

References

[Bar97]   K. Barthelmann. On equational simple graphs. Technical Report 9, Universität Mainz, Institut für Informatik, 1997.

[Blu99]   A. Blumensath. Automatic structures. Diploma thesis, RWTH-Aachen, 1999.

[Blu01]   A. Blumensath. Prefix-recognisable graphs and monadic second-order logic. Technical Report AIB-06-2001, RWTH Aachen, May 2001.

[Cau92]   D. Caucal. On the regular structure of prefix rewriting. *TCS*, 106:61–86, 1992.

[Cau96]   D. Caucal. On infinite transition graphs having a decidable monadic theory. In *ICALP'96*, volume 1099 of *LNCS*, pages 194–205, 1996.

[CM02]   B. Courcelle and J.A Makowsky. Fusion in relational structures and the verification of mso logic. In *MSCS*, volume 12, pages 203–235, 2002.

[Col02]   T. Colcombet. On families of graphs having a decidable first order theory with reachability. In *ICALP'02*, 2002.

[Cou89]   B. Courcelle. The monadic second-order logic of graphs ii: infinite graphs of bounded tree width. *Math. Systems Theory*, 21:187–221, 1989.

[Cou90]   B. Courcelle. *Handbook of Theoretical Computer Science*, chapter Graph rewriting: an algebraic and logic approach. Elsevier, 1990.

[Cou94]   B. Courcelle. Monadic-second order definable graph transductions : a survey. *TCS*, vol. 126:pp. 53–75, 1994.

[CW98]   B. Courcelle and I. Walukiewicz. Monadic second-order logic, graph coverings and unfoldings of transition systems. In *Annals of Pure and Applied Logic*, 1998.

[DT90]   M. Dauchet and S. Tison. The theory of ground rewrite systems is decidable. In *Fifth Annual IEEE Symposium on Logic in Computer Science*, pages 242–248, 1990.

[L d02]   C. L ding. Ground tree rewriting graphs of bounded tree width. In *Stacs 02*. LNCS, 2002.

[Mor00]   C. Morvan. On rational graphs. In J. Tiuryn, editor, *FOSSACS'00*, volume 1784 of *LNCS*, pages 252–266, 2000.

[MS85]   D. Muller and P. Schupp. The theory of ends, pushdown automata, and second-order logic. *Theoretical Computer Science*, 37:51–75, 1985.

[S n92]   G. S nizergues. Definability in weak monadic second-order logic of some infinite graphs. In *Dagstuhl seminar on Automata theory: Infinite computations, Warden, Germany*, volume 28, page 16, 1992.

[Tho97]   W. Thomas. Languages, automata, and logic. *Handbook of Formal Language Theory*, 3:389–455, 1997.

[Urv02]   T. Urvoy. On abstract families of graphs. In *DLT 02*, 2002.

[Wal96]   I. Walukiewicz. Monadic second order logic on tree-like structures. In *STACS'96*, volume 1046 of *LNCS*, pages 401–414, 1996.

Adaptive Raising Strategies Optimizing Relative Efficiency

Arnold Schönhage

Institut für Informatik III, Universität Bonn
Römerstrasse 164, D 53117 Bonn, Germany
schoe@cs.uni-bonn.de

Abstract. Adaptive raising by successive *trials* $t_0 < t_1 < \cdots$ until some unknown goal $g > 1$ has been *found* by $t_n \geq g$, causing total cost $T(g) = t_0 + \cdots + t_n$, is studied for optimizing $T(g)/g$. For corresponding games, where player G setting g and 'finder' F choosing $t_0, t_1, \ldots$ are playing mixed strategies, we prove a "Law of optimal adapting factor e". Section 2 is more general about adaptive raising on several tracks, in Sect. 3 we add proofs for the optimal *competitive factors* under corresponding worst case analysis.— Methods and results are similar to those about searching for a point on a line or on many rays, see [1,3,4,5,6].

1   Adaptive Raising for Goals of Unknown Size

The subject of this study is a scheduling problem of a very basic nature, specified in the following way. Assume we have to settle some task, to reach some goal of unknown size, described by a real number $g > 1$ (e.g. measuring computing time, or any sort of cost), and assume that the only moves at our disposal are successive *trials* of increasing (else freely choosable) costs $t_0 < t_1 < \cdots$ until (by some a posteriori testing after each trial) $t_n \geq g$ has been found, to be taken as indication of successful completion of that task, with total cost $T(g) = t_0 + t_1 + \cdots + t_n$. Then the question is how to choose these t_j in order to achieve good *relative* efficiency, i.e. to keep the quotient $T(g)/g$ as small as possible, to achieve the smallest possible *competitive factor*.

When we restrict this question to strategies in geometric progression, like $t_j = a \cdot z^j$ with a fixed factor $z > 1$ and initial $t_0 = a$ in $1 < a \leq z$, the corresponding *worst case* analysis becomes quite simple: the first successful trial settling goals $y = a \cdot z^k + \varepsilon$ with small $\varepsilon > 0$ is $t_{k+1} = a \cdot z^{k+1}$, so such g's enforce total cost $T(g) = a \cdot (z^{k+2} - 1)/(z - 1)$. With $\varepsilon \to 0$, $k \to \infty$, the resulting quotients $T(g)/g$ are increasing to their limit value $z^2/(z - 1)$, which becomes minimal for $z = 2$. The simple doubling strategy $t_j = 2^{j+1}$, for example, guarantees $T(g) < 4 \cdot g - 2$. — Remarkably, this factor 4 remains the best worst case bound of such kind for any other general strategy as well.

Theorem 1.1. *For any unbounded real sequence* $1 < t_0 < t_1 < t_2 < \cdots$ *and* $g > 1$, *use the minimum* n *with* $t_n \geq g$ *to define* $T(g) = t_0 + \cdots + t_n$. *Then* $\rho = \limsup_{g \to \infty} T(g)/g$ *satisfies* $\rho \geq 4$.

J.C.M. Baeten et al. (Eds.): ICALP 2003, LNCS 2719, pp. 611–623, 2003.
© Springer-Verlag Berlin Heidelberg 2003

We shall prove this (and a more general result, see Theorem 2.1) in Sect. 3 below. The main ideas for such proofs can already be found in [4] and in Sect. 2 of [1] on the similar problem of search for a point on a line and its generalization to any finite number of rays. Besides the translation to the problem of adaptive raising, our proof will also present slightly modified versions of those methods.

From a practical point of view, however, the worst case behavior will often not be of major concern. Usually we will rather be interested in strategies with good (or optimal) performance *on the average*. One way to discuss problems of that type would be to analyze such optimizations for various common or plausible probability distributions of goals, regarding g as a random variable. But how to proceed, if even the distribution of g is unknown, or arbitrary?

Here we prefer to study such issues in a *game theoretic* framework with two players F, G. For any constant $c > 1$, we define the *game of adaptive raising* $\Gamma(c)$ by the following rules. First the "goal setter" G plays some $g > 1$ and has to deposit a payment of $c{\cdot}g$ (with an umpire or trustee), later handed out to F. Next the "finder" F plays a sequence of t_j (as in Theorem 1.1) until $t_n \geq g$, and has then to pay $T(g) = t_0 + \cdots + t_n$ to G.

In this setting optimization on the average is captured by the notion of *mixed* strategies, and so the problem of optimal relative efficiency boils down to this basic question: *For which values of c are the games $\Gamma(c)$ advantageous for F?* — Beyond that we are, of course, also interested to know corresponding finder strategies explicitly, e. g. in the sense of clever *randomized* algorithms for F. Perhaps we may even restrict the finder's strategies by imposing some condition of constructiveness, whereas the devilish adversary G may clearly invoke mixed strategies of any sort, relying on pure mathematical existence!

In any case, there is a very simple answer to that basic question, given in our next theorem. In view of its fundamental nature, it deserves an extra naming; we propose to call it the "Law of optimal adapting factor e".

Theorem 1.2. (i) *Game $\Gamma(c)$ is advantageous for F iff $c \geq e = 2.71828\ldots$.*
(ii) *For any $g > 1$, finder F achieves an expected value $E(T(g)) = e{\cdot}g - 1$ by playing the mixed strategy choosing $t_0 = x$ at random in the interval $(1, e]$ with logarithmic distribution dx/x and then proceeding in geometric progression $t_{j+1} = e{\cdot}t_j$ for all j.*

Let us first prove (ii), covering the *if*-part of (i). It will be instructive to discuss such mixed strategies of geometric progressions $t_{j+1} = z{\cdot}t_j$ for arbitrary $z > 1$, then choosing $t_0 = x$ with distribution $(1/\ln z) \cdot dx/x$ on $(1, z]$. So let $g = b{\cdot}z^k$ with unique $b \in (1, z]$. For x in the lower interval $t_0 = x < b$, the final trial is $t_n = x{\cdot}z^{k+1}$ with $n = k + 1$, which (depending on the choice of x) causes costs $T(g|x) = x{\cdot}(z^{k+2} - 1)/(z - 1)$, similarly $T(g|x) = x{\cdot}(z^{k+1} - 1)/(z - 1)$ in the upper interval $b \leq t_0 = x \leq z$. Hence the expected costs are

$$E(T(g)) = \frac{1}{\ln z} \int_1^z T(g|x)\, dx/x = \frac{b{\cdot}z^{k+1} - 1}{\ln z} = \frac{z}{\ln z} \cdot g - \frac{1}{\ln z}, \qquad (1)$$

which yields (ii), minimizing that factor $z/\ln z$ by choosing $z = e$.

The *only-if*-part of (i), to show that F has no strategy to win for any value $c < e$, is relying on the same idea: we exhibit a suitable mixed strategy for the adversary G enforcing sufficiently large expected costs with any strategy of F. Asymptotically, admitting arbitrarily large values of g, such a good mixed strategy for G is furnished by playing $g = y$ with probability dy/y^2, but we cannot argue with this distribution directly, since that density $1/y^2$ would lead to infinite expectation $E(g)$, and due to $T(g) \geq g$ also to $E(T(g)) = \infty$.

To circumvent this difficulty, we approximate that infinite case considering mixed strategies for G with the same distribution dy/y^2 now restricted to some bounded interval $(1, B]$, and playing the top value $g = B$ with the remaining probability $\int_B^\infty dy/y^2 = 1/B$. As we are going to prove *lower* bounds for the expected costs $E(T(g))$ of the finder F, we may additionally favor F by assuming that F *knows* this interval bound B. Then it will suffice to consider *finite* sequences $1 < t_0 < t_1 < \cdots < t_l = B$ as strategies of F, with the quotients $q_0 = t_0$, $q_1 = t_1/t_0$, $\ldots$, $q_l = t_l/t_{l-1}$ satisfying the side condition

$$q_0 \cdot q_1 \cdots q_l = B. \tag{2}$$

With respect to that distribution dy/y^2 on $(1, B]$ plus the atom $1/B$ in its endpoint, the expected deposit of G is $E(c \cdot g) = c \cdot \ln B + c$, and the expected costs for such a strategy of F are readily calculated as

$$E(T(g)) = t_0 + t_1 \cdot 1/t_0 + \cdots + B \cdot 1/t_{l-1} = q_0 + \cdots + q_l.$$

Since the arithmetic mean of the q's, $E(T(g))/(l+1)$, is never less than their geometric mean known from (2), we thus obtain the lower bound

$$E(T(g)) \geq (l+1)B^{1/(l+1)} \geq e \cdot \ln B$$

for any strategy of F, where setting $z = B^{1/(l+1)}$ with $l + 1 = \ln B/\ln z$ again amounts to that crucial quotient $z/\ln z \geq e$. — Altogether this shows for any value $c < e$, that F will suffer an expected loss of at least $(e - c) \cdot \ln B - c > 0$ whenever G is playing such a mixed strategy with a sufficiently large B. □

In addition we point out that these geometric search strategies with factor $z = e$ optimizing the relative efficiency in the game theoretic sense are still fairly efficient also with respect to their worst case behavior, as the resulting worst case ratio $e^2/(e-1) < 4.3003$ is not much greater than the optimal factor 4. Conversely, (1) shows that playing a mixed doubling strategy with $z = 2$, optimizing the worst case behavior, will at the same time achieve an *expected* competitive factor $2/\ln 2 < 2.8854$ less than $1.062 \cdot e$.

In [2], a similar game theoretic approach has been worked out for the "linear search problem" (see [3]), also using such a distribution $\gamma \cdot dy/y^2$ on large finite intervals. The same issue has then found renewed interest as so-called "cow-path problem" (more generally also on $m > 2$ lanes), see [6] and [5], although these papers are using mixed adversary strategies with a different family of distributions, apparently because they are optimizing the *expected competitive ratio* (cf. Lemma 4.2 of [6]) rather than the ratio of expectations.

2   Adaptive Raising on Several Tracks

Now we consider the more general problem of adaptive raising with some goal hidden in any of m "tracks" $R_0, \ldots, R_{m-1}$, modeled as m copies of $[1, \infty) \subset \mathbb{R}$. So the goal is now some $g > 1$ plus some *selection index* $s \in \{0, \ldots, m-1\}$. Then the pure strategies for finder F ("F-strategies", for short) are infinite sequences of pairs (s_j, t_j) of real numbers $t_j > 1$ plus selection indices $s_j < m$ such that the m classes $J_\mu = \{j \in \mathbb{N} : s_j = \mu\}$ are infinite and the subsequences $(t_j : j \in J_\mu)$ are strictly increasing and unbounded for all $\mu < m$. With such a strategy, F is said to have found (g, s) after having arrived at the minimal *stopping index* n with $t_n \geq g$ and $s_n = s$, and with total cost $T(g, s) = t_0 + t_1 + \cdots + t_n$. —
How should finder F choose these sequences $st = ((s_j, t_j) : j \in \mathbb{N})$ to achieve good relative efficiency, i.e. to keep $T(g, s)/g$ as small as possible?

2.1   Worst Case Analysis for m Tracks

Let us first have a brief look at the worst case scenario, where F chooses such a pure strategy, here assumed to be known to adversary G who then can pick some (g, s) to make $T(g, s)/g$ especially large. Again we begin with the case of geometric strategies $t_j = z^{j+1}$ with a fixed factor $z > 1$, cyclically alternating the tracks by choosing $s_j = (j \bmod m)$. Then the disadvantageous goals are $g = z^{k+1} + \varepsilon$ with $s \equiv k \pmod m$ and small $\varepsilon > 0$, leading to stopping index $n = k + m$ and thus to total cost $T(g, s) = (z^{k+m+2} - z)/(z - 1)$. With $\varepsilon \to 0$, $k \to \infty$, the resulting quotients $T(g, s)/g$ are increasing to their limit $z^{m+1}/(z - 1)$. This becomes minimal for $z = 1 + 1/m$, with minimum value

$$\beta_m = (m+1)((m+1)/m)^m < (m + \tfrac{1}{2}) \cdot e < \beta_m + 1/8m. \tag{3}$$

So that cyclic strategy with $t_j = (1 + 1/m)^{j+1}$ guarantees the upper bound $T(g, s) < \beta_m \cdot g - (m+1)$, analogous to the doubling strategy for $m = 1$.

Closely related constants $2\beta_m + 1$ have been found in [4] (see also [1]) as optimal worst case competitive ratios for the linear search problem generalized to $m+1$ rays. As we shall prove in Sect. 3, analogously (worst case) optimality of these β_m does hold for our problem of adaptive raising on m tracks.

Theorem 2.1. *For any F-strategy $((s_j, t_j) : j \in \mathbb{N})$ for adaptive raising on m tracks and $g > 1$, $s < m$, use the minimum n with $t_n \geq g$ and $s_n = s$ to define $T(g, s) = t_0 + \cdots + t_n$. Then $\rho = \limsup_{g \to \infty} T(g, s)/g$ satisfies $\rho \geq \beta_m$.*

2.2   The Game Theoretic Version

Let us now turn to the issue of mixed strategies in corresponding games $\Gamma_m(c)$ under the more natural condition that first player G chooses some secret goal (g, s) and deposits $c \cdot g$ for F, and only then finder F plays some F-strategy $((s_j, t_j) : j \in \mathbb{N})$ until $t_n \geq g$ with $s_n = s$, and finally pays $T(g, s) = t_0 + \cdots + t_n$ to G. As for the simple case $m = 1$ of Sect. 1, again the crucial question is: *For which values of c are the games $\Gamma_m(c)$ advantageous for F?*

We are able to extend our method of proof for Theorem 1.2 such that we shall obtain a complete answer to this question, quite analogous to Theorem 1.2, although things are more complicated by additional technicalities. For the sake of motivation, we therefore first describe and analyze certain mixed F-strategies that will lead to particular functions $f_m(z)$ analogous to the factor $f_1(z) = z/\ln z$ of Sect. 1. Then we shall state the main theorem and continue in 2.3 with the lower bound proof by analyzing corresponding mixed strategies for player G, while all technical details concerning the minimum points z_m of the f_m and the optimal ratios $\alpha_m = f_m(z_m)$ are postponed to Subsection 2.4, including an analysis of the asymptotical behavior of the z_m and of the α_m for $m \to \infty$.

Definition 2.1. *For any $z > 1$, $S(m, z)$ shall denote the mixture of F-strategies with initial index $s_0 = i$ at random, any $i < m$ with probability $1/m$, and $t_0 = x \in (1, z]$ at random with distribution $1/\ln z \cdot dx/x$, then F proceeds geometrically by $t_{j+1} = z \cdot t_j$ with cyclic track selection $s_j = (j + i \bmod m)$.*

In order to compute the expectation $E(T(g, s))$ with respect to $S(m, z)$ for any given goal (g, s), we write $g = b \cdot z^k$ with unique $b \in (1, z]$ and $k \in \mathbb{N}$. Since $S(m, z)$ is symmetric with respect to the probabilities for $s_0 = i$, we may without restriction assume that $s \equiv k \pmod{m}$. Then the stopping index n with $t_n \geq g$ and $k \equiv s = s_n \equiv n + i \pmod{m}$ is easily determined as

$$n(i, x) = k + m - i \quad \text{for } i > 0, \qquad \begin{array}{ll} n(0, x) = k + m & \text{for } x < b, \\ n(0, x) = k & \text{for } x \geq b. \end{array}$$

So we obtain $T(g, s)$ with analogous dependencies on i, x, namely

$$\begin{array}{ll} T(g, s \,|\, i, x) = x(z^{k+m-i+1} - 1)/(z - 1) & \text{for } 0 < i < m, \\ T(g, s \,|\, 0, x) = x(z^{k+m+1} - 1)/(z - 1) & \text{for } x < b, \\ T(g, s \,|\, 0, x) = x(z^{k+1} - 1)/(z - 1) & \text{for } x \geq b. \end{array}$$

Based on these expressions, an elementary calculation yields

$$E(T(g, s)) = 1/m \sum_{i=0}^{m-1} \frac{1}{\ln z} \int_1^z T(g, s \,|\, i, x)\, dx/x$$

$$= \frac{b \cdot (z^{k+m} + \cdots + z^{k+1})}{m \cdot \ln z} - \frac{1}{\ln z} = f_m(z) \cdot g - \frac{1}{\ln z},$$

with $\quad f_m(z) = 1/m \cdot (z^m + \cdots + z)/\ln z = \dfrac{z^{m+1} - z}{m(z - 1) \cdot \ln z} \quad (z > 1). \quad$ (4)

In 2.4 we shall prove that these functions have unique minimum points z_m, with $w_m = z_m^m$ converging to the unique solution ω of the equation $\ln w = 2 - 2/w$ in $w > 1$, with $\omega = 4.92155\ldots$, and shall analyze the growth of the minima $\alpha_m = f_m(z_m)$ for $m \to \infty$. Prepared in this way, we now state our main result.

Theorem 2.2. (i) *Game $\Gamma_m(c)$ is advantageous for F iff $c \geq \alpha_m = f_m(z_m)$, where $z_m > 1$ is the unique minimum point of the function f_m in (4).*
(ii) *For any goal (g, s), $g > 1$, $s < m$, finder F achieves the expected value $E(T(g, s)) = \alpha_m \cdot g - 1/\ln z_m$ by playing strategy $S(m, z_m)$ as defined above.*

Part (ii) and the *if*-part of (i) are clear from the analysis of the strategies $S(m, z)$ that have led us to the f_m. Our lower bound proof for the *only-if*-part, that F cannot achieve any smaller competitive factor, follows in the next section.

Very similar functions $R(m, z) = 1 + f_m(z)\cdot 2/z$ occur in Theorem 3.1 of [6] as competitive factors for analogous probabilistic algorithms for the "cow-path" problem on m lanes ($m \geq 2$), where the minima $r_m^* = \min_{z>1} R(m, z)$ have turned out to be the *optimal* competitive factors for that problem. For $m = 2$, that optimality proof is in [6] (previously also in [2]); for general $m > 2$ we refer to Sect. 3.2 and to Appendix B of [5].

2.3   Good Mixed Strategies for Player G

In obvious generalization of our proof for Theorem 1.2 we let player G select any of the tracks R_s ($s < m$) with probability $1/m$ and then choose $g = y > 1$ with distribution dy/y^2, but again restricting $g = y$ to some large finite interval $(1, B]$ with that distribution for $y < B$, or playing $g = B$ with probability $1/B$. Then the expectation of G's deposit is easily determined as $E(c \cdot g) = c \cdot \ln B + c$.

In the sequel we consider any F-strategy st and derive a lower bound for the expectation of the corresponding random variable $T(g, s)$. Again we may assume that F knows this B, hence it will suffice to discuss *finite* F-strategies $st = ((s_j, t_j) : 0 \leq j < l)$ of any length $l \geq m$ with $s_j < m$, $t_j > 1$, and nonempty index classes $J_\mu = \{j < l : s_j = \mu\}$ for all $\mu < m$, where the subtupels $(t_j : j \in J_\mu)$ are strictly increasing and ending with the maximal value $t_j = B$. Moreover it is convenient to extend such strategies at their lower end by a "preamble" of m dummy pairs $(i - m, t_{i-m})$ with $t_{i-m} = 1$ for $0 \leq i < m$, and to extend the index classes accordingly, setting $J_i' = \{i - m\} \cup J_i$. This admits to define the *preceding selections* for any $j \geq 0$ as

$$p(i, j) = \max\{k \in J_i' : k < j\}. \tag{5}$$

Then the expectation of $T(g, s) = t_0 + t_1 + \cdots + t_n$ with stopping index n as random variable under that mixed strategy of G is easily calculated from the probabilities $\text{prob}(n \geq j)$, namely

$$E(T(g, s)) = \sum_{j=0}^{l-1} t_j \cdot \text{prob}(n \geq j) = 1/m \sum_{j=0}^{l-1} t_j \cdot {\sum_{i<m}}' 1/t_{p(i,j)}, \tag{6}$$

where the prime at the last summation sign shall indicate to omit all terms $1/B$ resulting from $p(i, j) = \max J_i$. Our subsequent analysis requires to generalize formula (6) and such finite strategies to preamble pairs $(i - m, u_i)$ with any $u = (u_0, \dots, u_{m-1}) \in [1, B]^m$, then of course with $t_j > u_i$ for all $j \in J_i$. Let $FF(u, B)$ denote the set of such finite F-strategies st. Imitating (6) we define a *cost function* C and infima $V(u)$,

$$C(u, st) = \sum_{j=0}^{l-1} t_j \cdot {\sum_{i<m}}' 1/t_{p(i,j)} \qquad \text{for} \ \ st \in FF(u, B) \tag{7}$$

$$V(u) = \inf\{C(u, st) : st \in FF(u, B)\}, \tag{8}$$

which will turn out to be *minima* attained for certain optimal strategies st. Finally a lower bound for $V(1, \dots, 1)$ will yield the desired lower bound for $E(T(q, s))$ in (6), but that requires first to show (in a sequence of several "steps") that the optimal strategies for $u = (1, \dots, 1)$ must necessarily be *cyclic* most of the time, with $s_j = (j \bmod m)$ as long as $t_j < B$, after some initial permutation.

Step 1. Let us call $st \in FF(u, B)$ "close to optimal" iff $C(u, st) < V(u) + \frac{1}{2}$. Such st have *bounded length* $l < mB + 1$, since each term of the outer sum in (7) is > 1, whence $l < C(u, st) < V(u) + \frac{1}{2}$, and the shortest $st_0 \in FF(u, B)$ with just one pair (i, B) for each i with $u_i < B$ shows $V(u) \le C(u, st_0) \le mB$.

Step 2. With $\delta = 1/2mB$, any $st \in FF(u, B)$ close to optimal with trials t_j, index classes J_i, and preceding selections as in (5) satisfies

$$j \in J_i \text{ and } (t_j < B \text{ or } p(i,j) \ge 0) \implies t_j - t_{p(i,j)} \ge \delta. \tag{9}$$

Proof. Let $j \in J_i$, $k = p(i, j)$, and assume $t_j < t_k + \delta$. With $t_j < B$, we could then alter st into a shorter strategy st' by omitting the pair (i, t_j), which in (7) would save $t_j/t_k > 1$ at least, and increase less than $l < mB + 1$ other terms t_h/t_j to t_h/t_k by a factor $t_j/t_k < 1 + \delta$. This would imply

$$C(u, st') < (C(u, st) - 1)(1 + \delta) < V(u) - \tfrac{1}{2} + mB \cdot \delta < V(u),$$

contradicting (8). In the other case $t_j = B$ and $k \ge 0$, we could then replace the k-th pair of st by (i, B) and omit the j-th pair, thereby in (7) saving

$$t_k/t_{p(i,k)} + t_j/t_k - t_j/t_{p(i,k)} > 1 - \delta > \tfrac{1}{2}$$

at least, which is impossible for st being close to optimal. □

The main conclusion from Steps 1, 2 is that we can restrict definition (8) to the *compact* subset of strategies $st \in FF(u, B)$ of length $l < mB + 1$ and satisfying (9), compact when considered as a finite collection of closed bounded subsets of some $\mathbb{R}^l$. Since $C(u, st)$ is continuous (and even analytic) in the t_j on each of these components, $V(u)$ is therefore a *minimum*.

Step 3. For any optimal strategy st with $V(u) = C(u, st)$ and for any of its intermediate pairs (i, t_j) with $u_i < t_j < B$, we must have $\partial C(u, st)/\partial t_j = 0$. Moreover, permuting the track selection, i.e. the index classes of an optimal strategy st, we find that $V(u_0, \dots, u_{m-1})$ is a *symmetric* function.

Step 4. For $i < m$, $1 \le u_i < B$, $V(u_0, \dots, u_{m-1})$ *is strictly decreasing in* u_i. *Proof.* Consider an st with $V(u) = C(u, st)$ and for $u_i < u'_i < \min(B, u_i + \delta)$ the strategy $st' \in FF(u', B)$ obtained from st by replacing its preamble pair $(i-m, u_i)$ with $(i-m, u'_i)$. Then (7) implies $C(u, st) - C(u', st') = \sigma \cdot (1/u_i - 1/u'_i)$ with a nonzero sum σ of certain t_j, hence $V(u') \le C(u', st') < C(u, st) = V(u)$.

Step 5. *For any t_0 with $u_0 < t_0 \le B$, we have*

$$V(u_0, u_1, \dots, u_{m-1}) \le t_0 \cdot {\sum_{i < m}}' 1/u_i + V(t_0, u_1, \dots, u_{m-1}), \tag{10}$$

where equality in (10) implies $u_0 \le u_i$ *for all* $i < m$.

Proof. That inequality (10) follows from the definitions (7) and (8), since an initial pair $(0, t_0)$ continued with an optimal strategy st' for $V(t_0, u_1, \ldots, u_{m-1})$ yields a strategy $st \in FF(u, B)$. If such st happens to be optimal with equality in (10), the additional claim $u_0 \leq u_i$ is obvious for $u_i \geq t_0$, else consider $u_1 < t_0$, for example. Then the symmetry of V in u_0, u_1 combined with (10) yields also

$$V(u_0, u_1, u_2, \ldots) \leq t_0 \cdot \sum'_{i<m} 1/u_i + V(u_0, t_0, u_2, \ldots, u_{m-1}),$$

whence $V(u_1, t_0, u_2, \ldots) = V(t_0, u_1, u_2, \ldots) \leq V(u_0, t_0, u_2, \ldots, u_{m-1})$, and because of Step 4 therefore $u_1 \geq u_0$.

Step 6. *If an optimal strategy $st \in FF(u, B)$ has the initial pairs $(0, t_0)$, $(1, t_1)$ with $t_0 < B$ and $t_1 < B$, then $t_0 < t_1$. By symmetry, the same will hold for other index pairs as well.*

Proof. Because of $t_0 < B$ and $s_0 = 0$, there exists $k = \min\{j \in J_0 : j > 0\}$, and by $t_1 < B$ also $h = \min\{j \in J_1 : j > 1\}$, with $t_h > t_1$. For this optimal st the terms in (7) depending on t_0 are

$$t_0(1/u_0 + w) + (t_1 + \cdots + t_k)/t_0 \quad \text{with} \quad w = \sum'_{0<i<m} 1/u_i,$$

where the partial derivative with respect to t_0 must be zero (cf. Step 3) which implies $t_0^2 = (t_1 + \cdots + t_k)/(1/u_0 + w)$. In the same manner we obtain $t_1^2 = (t_2 + \cdots + t_h)/(1/t_0 + w)$, and with this t_1 the t_1-contribution to (7) becomes $H = (t_2 + \cdots + t_h) \cdot 2/t_1$. If $h > k$, then $t_h > t_1$ and $t_0 > u_0$ imply $t_1^2 > t_0^2$, hence $t_1 > t_0$. In any case this argument excludes that $t_0 = t_1$, because otherwise we could exchange the roles of J_0 and J_1 to obtain $h > k$ with $t_1 > t_0$.

To falsify $t_0 > t_1$, we apply (10) (with equality) to st, and to the strategy st' with alternate initial pairs $(0, t_1)$, $(1, t_0)$, obtaining

$$\begin{aligned}
V(u_0, u_1, u_2, \ldots) &= t_0(1/u_0 + w) + t_1(1/t_0 + w) + V(t_0, t_1, u_2, \ldots) \\
&\leq t_1(1/u_0 + w) + t_0(1/t_1 + w) + V(t_1, t_0, u_2, \ldots).
\end{aligned}$$

By symmetry of V therefore $t_0/u_0 + t_1/t_0 \leq t_1/u_0 + t_0/t_1$, and so dividing $(t_0 - t_1)/u_0 \leq (t_0^2 - t_1^2)/t_0 t_1$ by $t_0 - t_1 > 0$ yields $t_0 t_1 \leq (t_0 + t_1)u_0 < 2t_0 u_0$, $t_1 < 2u_0 \leq 2u_1$ by Step 5. This, however, would decrease the t_1-contribution to (7) to $(t_2 + \cdots + t_h)/u_1 < H$, which would replace that H if we omit the pair $(1, t_1)$ from strategy st, contradicting the optimality of st.

Step 7. *If an optimal strategy $st \in FF(u, B)$ is beginning with a pair (μ, B), then $u_i \geq B/4m$ for all $i < m$.*

Proof. It suffices to show $u_0 \geq B/4m$ for $\mu = 0$, say; then Step 5 implies that lower bound for all i. In case of $B/u_0 > 4m$ we could replace that pair $(0, B)$ which contributes $B(1/u_0 + w)$ to (7) with two pairs $(0, t_0)$, $(0, B)$ contributing $D = t_0(1/u_0 + w) + B/t_0 + Bw$. Since $u_0 \leq u_i$ for all i implies $1/u_0 + w \leq m/u_0$, choice of $t_0 = B/2m$ would then lead to $D \leq B/2u_0 + 2m + Bw < B/u_0 + Bw$, contradicting the optimality of st. $\square$

Now we are prepared to derive the desired lower bound for $V(1, \ldots, 1)$, for large $B \gg 4m$. According to Step 5, the selection indices $s_0, \ldots, s_{m-1}$ of the first m pairs of an optimal strategy for $V(1,\dots,1)$ (all $u_i = 1$) must attain m different values, after suitable permutation of the tracks we can therefore assume $s_i = i$ for $i < m$. Then repeated use of Step 6 shows $t_0 < t_1 < \cdots < t_{m-1}$, and combined with Step 5 continuation $t_{m-1} < l_m < t_{m+1} < \cdots$ with cyclic selection $s_j = (j \bmod m)$ until we arrive at the first index k with $t_k = B$. Then Step 7 guarantees $t_j \geq B/4m$ for all $j \geq k - m$. For the quotients $q_j = t_j/t_{j-1}$ (with $q_j = 1$ for the preamble indices $j < 0$) this implies

$$q_{k-1-i} \cdot q_{k-2-i} \cdots q_{-i} \geq B/4m \quad \text{for } 0 \leq i \leq m-1. \tag{11}$$

In formula (7) for this strategy st and $u = (1,\dots,1)$, the preceding selections $p(i,j)$ for any $j < k$ are just the indices $j-1,\dots,j-m$ in some cyclic permutation. Therefore we have

$$V(1,\dots,1) > \sum_{j=0}^{k-1}\sum_{i=0}^{m-1} t_j/t_{j-m+i} = \sum_{i=0}^{m-1}\sum_{j=0}^{k-1} q_j \cdots q_{j-i},$$

and estimating the inner sums on the right-hand side viewed as k-fold arithmetic means by the corresponding geometric means and their lower bounds resulting from (11), we thus obtain $V(1,\dots,1) > \sum_{\mu=1}^{m} k \cdot (B/4m)^{\mu/k}$ as decisive lower bound for the right-hand side of (6). Finally we set $(B/4m)^{1/k} = z$, with $k = (\ln B - \ln(4m))/\ln z$, which yields

$$E(T(g,s)) > 1/m \cdot ((z + z^2 + \cdots + z^m)/\ln z)(\ln B - \ln(4m))$$
$$\geq f_m(z_m) \cdot (\ln B - \ln(4m)).$$

Comparing this with the expected deposit determined at the beginning of this subsection, we see that for any $c < \alpha_m = f_m(z_m)$ player F will loose

$$E(T(g,s)) - E(c \cdot g) > \alpha_m(\ln B - \ln(4m)) - c \cdot \ln B - c > 0$$

for sufficiently large B. This completes our proof of Theorem 2.2.

2.4 Analysis of the Cost Functions and Optimal Ratios

Let us now have a closer look at the functions f_m defined by (4). Taking the logarithmic derivative and multiplying with z we obtain

$$\frac{z \cdot f_m'(z)}{f_m(z)} = \frac{(m+1)z^m - 1}{z^m - 1} - \frac{z}{z-1} - \frac{1}{\ln z} = \frac{m \cdot z^m}{z^m - 1} - \frac{1}{z-1} - \frac{1}{\ln z}, \tag{12}$$

which has a unique zero $z_m \in (1,\infty)$, where f_m attains its unique minimum, since on this domain $1/\ln z$ is decreasing from ∞ to 0, while the other part $h(z) = m + m/(z^m - 1) - 1/(z-1)$ is monotonically *increasing* to its limit m, hence $1/\ln z_m < m$, $z_m^m > e$. Here $h'(z) = -m^2 z^{m-1}/(z^m-1)^2 + 1/(z-1)^2 > 0$ follows from $h'(z)(z^m - 1)^2(z-1)^2/z^m = -m^2(z - 2 + 1/z) + (z^m - 2 + 1/z^m)$, where we use $4m^2 \cdot \sinh^2(y) < 4 \cdot \sinh^2(my)$ with $y = \ln z$.

To study the behavior of these z_m and of the $\alpha_m = f_m(z_m)$ in greater detail, we use $w_m = z_m^m$, so the right-hand side of (12) set to zero yields the equation

$$\frac{w}{w-1} - \frac{1}{m(w^{1/m}-1)} = \frac{1}{\ln w} \quad \text{for } w = w_m = z_m^m, \tag{13}$$

and $\ln w = v$, $w^{1/m} = \exp(v/m)$ with $v < m(w^{1/m}-1) < v/(1-v/2m)$ show

$$1/(m(w^{1/m}-1)) = 1/\ln w - \theta/2m \quad \text{with some } \theta \in (0,1). \tag{14}$$

In the limit for $m \to \infty$, (13) thus becomes $w/(w-1) = 2/\ln w$, and so the solutions w_m of (13) are converging to the simple zero $\omega > 1$ of the function $d(w) = \ln w - 2 + 2/w$, with $d'(w) = 1/w - 2/w^2 \geq 0$ for $w \geq 2$. Leaving aside the zero at 1 and searching between the minimum value $d(2) = -0.306\ldots$ and $d(5) = \ln 5 - 1.6 > 0$, one finds the unique zero $\omega = 4.9215536\ldots$, lying above the inflection point at 4, with $d(4) = -0.1137\ldots$ and $d'(w) < d'(4) = 1/8$ for all $w > 4$. More precisely, (13) and (14) yield $w_m/(w_m-1) = 2/\ln w_m - \theta/2m$,

$$d(w_m) = \ln w_m - 2 + 2/w_m = -\ln(w_m)(1 - 1/w_m) \cdot \theta/2m < 0,$$

thus $w_m < \omega$ and $d(w_m) > -(\ln \omega)(1-1/\omega)/2m = -(\ln \omega)^2/4m > -0.635/m$. For $m \geq 6$, this implies $w_m > 4$, therefore $w_m > \omega - (0.635/m)/d'(4)$, and the same holds for $m < 6$ (as verified numerically), hence

$$\omega - 5.08/m < w_m < \omega \quad \text{for all } m, \tag{15}$$

and $z_m = \sqrt[m]{w_m} = 1 + c/m + O(1/m^2)$ with $c = \ln \omega = 1.593624\ldots$.

Finally let us analyze the growth of the $\alpha_m = f_m(z_m)$ for $m \to \infty$. Since (13) implies $m(z_m-1)\ln w_m/(w_m-1) = m(z_m-1) + \ln w_m - m(z_m-1)\ln w_m$, we can rewrite the expression $\alpha_m = f_m(z_m) = z_m \cdot (z_m^m - 1)/((z_m - 1)\ln(z_m^m))$ resulting from (4) as

$$\alpha_m = \frac{m \cdot z_m}{m(z_m - 1) + \ln w_m - m(z_m - 1)\ln w_m}.$$

With $m(z_m - 1) = \ln w_m + O(1/m)$ from (14) and $\ln w_m = \ln \omega - O(1/m)$ from (15), this yields the following quantitative supplement to our Theorem 2.2, stating *asymptotically linear growth of the optimal ratios*

$$\alpha_m = \frac{m + O(1)}{2 \cdot \ln \omega - \ln^2 \omega + O(1/m)} = m \cdot \tau + O(1), \tag{16}$$

with $\tau = 1/(2 \cdot \ln \omega - \ln^2 \omega) = \omega/(2 \cdot \ln \omega) = 1.54413865\ldots$.

A similar linear growth result has been obtained in Sect. 6 of [6] for the cow-path problem on m lanes, in that case with the doubled factor $\kappa = 2\tau$.

By pursuing the foregoing analysis with greater care, one can show that the convergence $w_m \to \omega$, see (15), is indeed not faster than of order $1/m$, and the final $O(1)$ in (16) can be sharpened to the form $\sigma - O(1/m)$ with a certain constant $\sigma = 1.230388\ldots$, where its approximate value appears as a byproduct of the numerical data given in Table 2 of the Appendix. Moreover we can also infer from that table that the convergence of the values of $m(z_m - 1)$ to their limit $\ln \omega = 1.59362\ldots$ is *from above*.

3   Worst Case Lower Bound Proofs

Our proof of Theorem 2.1 (including that of Theorem 1.1) will be indirect: assuming $\rho < \beta_m$ will lead to a contradiction. We begin with a technical reformulation of that limsup ρ belonging to any given F-strategy $((s_j, t_j) : j \in \mathbb{N})$. The quotients $T(g, s)/g$ are becoming large for goals $g = t_j + \varepsilon$ with positive ε tending to zero and $s = s_j$, then approaching $T_n(j)/t_j$, where we define

$$n(l) := \min\{\nu > l : s_\nu = s_l\}, \tag{17}$$

and write $T_n = t_0 + \cdots + t_n$ for any n. By $t_j \to \infty$, we therefore also have $\rho = \limsup_{j \to \infty} T_{n(j)}/t_j$. Moreover we shall exploit that we may restrict our indirect proof to F-strategies satisfying $t_j \leq t_{j+1}$ for all j, due to the following

Lemma 3.1. *To every F-strategy $st = ((s_j, t_j) : j \in \mathbb{N})$ with limit ratio $\limsup_{j \to \infty} T_n(j)/t_j = \rho$ there exists some $st^* = ((s_j^*, t_j^*) : j \in \mathbb{N})$ with monotonicity $t_j^* \leq t_{j+1}^*$ for all j and $\limsup_{j \to \infty} T_{n^*(j)}^*/t_j^* = \rho^* \leq \rho$.*

Proof. If strategy st does not satisfy this monotonicity, we consider minimal j with $t_j > t_{j+1}$ with selection indices $s_j = i$ and $s_{j+1} = i' \neq i$ (since the subsequence $(t_j : j \in J_i)$ is strictly increasing) and alter st to a new F-strategy st' defined as follows: we exchange t_j and t_{j+1} setting $t_j' := t_{j+1}$, $t_{j+1}' := t_j$, while keeping $t_l' = t_l$ for all other l, and exchange the roles of i and i' above $j+1$, setting $s_l' := i'$ for $l > j + 1$, $s_l = i$, $s_l' := i$ for $l > j + 1$, $s_l = i'$, and $s_l' := s_l$ for all other l. Via (17), this s' will then induce new "next" indices $n'(l)$, but such that most of the quotients $T'_{n'(l)}/t_l'$ remain the same as before, including $T'_{n'(j)}/t_j' = T_{n(j+1)}/t_{j+1}$ and $T'_{n'(j+1)}/t_{j+1}' = T_{n(j)}/t_j$, with the only possible exception of an index $r < j$ with $n(r) = j$, but in that case

$$T_j'/t_r' = (t_0 + \cdots + t_{j-1} + t_{j+1})/t_r < T_{n(r)}/t_r.$$

It may be necessary to carry out an infinite sequence of such changes (then formally to be specified as a recursive definition), but as the t_j are tending to infinity, this will establish that monotonicity up to any fixed index j within a finite number of such steps. Since none of the quotients $T_{n(j)}/t_j$ gets ever increased, the limsup ρ^* of the quotients $T_{n^*(j)}^*/t_j^*$ of the resulting *limit strategy* st^* cannot be greater than the initial ρ. $\square$

So let us now consider an F-strategy $st = ((s_j, t_j) : j \in \mathbb{N})$ for adaptive raising on m tracks with $t_j \leq t_{j+1}$ for all j and such limsup $\rho < \beta_m$. Then we pick some β with $1 \leq \rho < \beta < \beta_m$, choose some large $k \geq \max_i \min J_i$ with $n(l) > k \implies T_{n(l)}/t_l \leq \beta$, and consider for any fixed $j \geq k$ the m indices $q(i) := \max\{q \in J_i : q < j + m\}$, for $i < m$, satisfying $n(q(i)) \geq j + m$. As the J_i form a partition of $\mathbb{N}$, these $q(i)$ are m different integers $< j + m$, so we can use the minimal $\mu < m$ with $q(\mu) \leq j$ to infer from $T_{j+m} \leq T_{n(q(\mu))}$ and $t_{q(\mu)} \leq t_j$ that $T_{j+m}/t_j \leq T_{n(q(\mu))}/t_{q(\mu)} \leq \beta$. Altogether, this implies

$$t_0 + \cdots + t_{k+n+m} \leq \beta \cdot t_{k+n} \quad \text{for all } n \geq 0. \tag{18}$$

From these inequalities we will now conclude that there exists an $a > 0$ and a positive real sequence $x_0, x_1, x_2, \ldots$ satisfying corresponding *equations*

$$a + x_0 + \cdots + x_{n+m} = \beta \cdot x_n \quad \text{for all } n \geq 0. \tag{19}$$

First we normalize (18) by setting $a = (t_0 + \cdots + t_{k-1})/t_k$ and $x_l = t_{k+l}/t_k$ for $l \geq 0$, whence the set $X \subset \mathbb{R}_{>0}^\infty$ of positive sequences $(x_l : l \in \mathbb{N})$ satisfying the linear inequalities $x_0 \leq 1$ and

$$a + x_0 + \cdots + x_{n+m} \leq \beta \cdot x_n \quad \text{for all } n \geq 0 \tag{20}$$

is not empty. Moreover this X is *compact*, since that constant $a > 0$ and $x_0 \leq 1$ imply the bounds $a/(\beta - 1) \leq x_n$ and $x_{n+1} < \beta \cdot x_n$, thus $x_n \leq \beta^n$ by induction, and (20) describes intersections with closed half spaces. So X also contains a sequence $(x_l : l \in \mathbb{N})$ with x_0 being *minimal*, which then must satisfy the equations in (19), because otherwise any defect "$< \beta \cdot x_n$" in (20) for some n would admit to decrease x_n a bit, thereby causing a defect "$< \beta \cdot x_{n-1}$" in the preceding condition, and so forth down to a decrease of x_0 which, however, is excluded by its minimality.

We finish the indirect proof of Theorem 2.1 by subtracting the equations (19) for $n+1$ and n, which yields the linear recurrence $x_{n+m+1} = \beta \cdot x_{n+1} - \beta \cdot x_n$, and proceed as in Sect. 2 of [1] to arrive at the desired contradiction: For $\beta < \beta_m$, see (3), the characteristic polynomial $y^{m+1} - \beta \cdot y + \beta$ of this recurrence has only *simple* roots $v_0, v_1, \ldots, v_m$, and no positive real root; more precisely these are $\lfloor (m+1)/2 \rfloor$ pairs of conjugate roots of distinct moduli, since $|v_\mu| = |v_\nu|$ implies $|v_\mu - 1| = |v_\nu - 1|$, plus one negative root, if m is even. Expressing the x_n from (19) as linear combination $x_n = \sum_{\mu=0}^{m} \gamma_\mu \cdot v_\mu^n$ of the standard solutions thus shows that $x_n > 0$ for all n is impossible. □

References

[1] R. A. Baeza-Yates, J. C. Culberson, and G. J. E. Rawlins, *Searching in the plane.* Information and Computation **106** (1993), 234–252.

[2] A. Beck and D. J. Newman, *Yet more on the linear search problem.* Israel J. Math. **8** (1970), 419–429.

[3] R. Bellman, *A minimization problem.* Bull. AMS **62** (1956), 270. — *An optimal search problem.* — problem 63–9 in *SIAM* Rev. **5** (1963), 274.

[4] S. Gal, *Minimax solutions for linear search problems.* SIAM J. Appl. Math. **27** (1974), 17–30.

[5] M. Y. Kao, Y. Ma, M. Sipser, and Y. Yin, *Optimal constructions of hybrid algorithms.* J. of Algorithms **29** (1998), 142–164; — also in *Proc. 5th ACM-SIAM Symposium on Discrete Algorithms* (1994), 372–381.

[6] M. Y. Kao, J. H. Reif, and S. R. Tate, *Searching in an unknown environment: An optimal randomized algorithm for the cow-path problem.* Information and Computation **131** (1997), 63–80; — also in *Proc. 4th ACM-SIAM Symposium on Discrete Algorithms* (1993), 441–447.

Appendix

Here we display some numerical data illustrating the analysis of Subsection 2.4. Table 1 compares the game theoretic optimal ratios α_m with the worst case ratios β_m for small m, showing savings of about 32 to 41 percent; (16) and (3) imply that the quotients α_m/β_m are converging to $\tau/c = 0.56805\ldots$ Table 2 shows the minimum points z_m of the f_m, the solutions w_m of equation (13), their logarithms and the "approximate logarithms" $m(z_m-1)$ for selected values of m, especially also for large m in order to demonstrate the limiting behavior of the deviations $\sigma_m = \alpha_m - m \cdot \tau$ in (16), and of those $m(z_m - 1)$.

Table 1. Optimal ratios compared

m	α_m	β_m
1	2.7182818	4.0000000
2	4.2848795	6.7500000
3	5.8385908	9.4814814
4	7.3880617	12.2070312
5	8.9356038	14.9299200
6	10.4821048	17.6513846
7	12.0279794	20.3719975

Table 2. Adaptive raising on m tracks: Optimal factors for the geometric progressions and some related quantities with their asymptotical behavior for large values of m.

m	z_m	$w_m = z_m^m$	$\ln w_m$	$m(z_m - 1)$	σ_m
1	2.71828183	2.71828183	1.00000000	1.71828183	1.17414318
2	1.83503707	3.36736104	1.21412936	1.67007413	1.19660227
3	1.54962339	3.72116127	1.31403579	1.64887018	1.20617489
4	1.40922432	3.94385120	1.37215771	1.63689729	1.21150715
5	1.32583930	4.09689133	1.41022847	1.62919651	1.21491062
6	1.27063753	4.20852654	1.43711260	1.62382518	1.21727291
7	1.23140919	4.29355217	1.45711440	1.61986436	1.21900889
8	1.20210283	4.36046644	1.47257903	1.61682262	1.22033866
9	1.17937924	4.41449781	1.48489408	1.61441315	1.22138997
10	1.16124573	4.45903929	1.49493334	1.61245729	1.22224204
20	1.08016646	4.67534563	1.54230309	1.60332911	1.22620345
50	1.03195159	4.81910205	1.57258761	1.59757925	1.22868616
100	1.01595614	4.86963100	1.58301816	1.59561428	1.22953245
1000	1.00159382	4.91629704	1.59255561	1.59382440	1.23030237
10000	1.00015936	4.92102732	1.59351731	1.59364429	1.23037980
100000	1.00001594	4.92150100	1.59361356	1.59362626	1.23038755
1000000	1.00000159	4.92154837	1.59362319	1.59362446	1.23038832

A Competitive Algorithm for the General 2-Server Problem

René A. Sitters[1], Leen Stougie[1,3], and Willem E. de Paepe[2]

[1] Department of Mathematics and Computer Science
[2] Department of Technology Management
Technische Universiteit Eindhoven
P.O.Box 513, 5600 MB Eindhoven, The Netherlands
{r.a.sitters, l.stougie, w.e.d.paepe}@tue.nl
[3] CWI, P.O.Box 94079, 1090 GB Amsterdam, The Netherlands

Abstract. We consider the general on-line two server problem in which at each step both servers receive a request, which is a point in a metric space. One of the servers has to be moved to its request. The special case where the requests are points on the real line is known as the CNN-problem. It has been a well-known open question if an algorithm with a constant competitive ratio exists for this problem. We answer this question in the affirmative sense by providing the first constant competitive algorithm for the general two-server problem on any metric space.

1 Introduction

In the *general k-server problem* we are given servers $s_1, \ldots, s_k$, each of which moving in a metric space M_i. Requests $r \in M_1 \times M_2 \times \ldots \times M_k$ are presented on-line one by one. Thus, a request is a k-tuple $r = (z_1, z_2, \ldots, z_k)$ and it is served by moving one of the servers s_i to his corresponding point z_i. The decision which server to move is irrevocable and has to be taken without any knowledge about future requests. The cost of moving server s_i to z_i is equal to the distance travelled by s_i from his current location to z_i. The objective is to minimize the total cost to serve all given requests. The performance of an on-line algorithm is measured through competitive analysis. An online algorithm is c-competitive if, for any request sequence σ, the algorithm's cost are at most c times the cost of the optimal solution of the corresponding off-line problem plus some additive constant independent of σ.

The general k-server problem is a natural generalization of the well-known *k-server problem* for which $M_1 = M_2 = \ldots = M_k$ and $z_1 = z_2 = \ldots = z_k$ at each time step. The k-server problem was introduced by Manasse, McGeoch and Sleator [9], who proved a lower bound of k on the competitive ratio of any deterministic algorithm for any metric space with at least $k + 1$ points and posed the well-known k-server conjecture saying that there exists a k-competitive algorithm for any metric space. The conjecture has been proved for $k = 2$ [9] and some special metric spaces [2][3]. For $k \geq 3$ the current best upper bound of $2k - 1$ is given by Koutsoupias and Papadimitriou [7].

J.C.M. Baeten et al. (Eds.): ICALP 2003, LNCS 2719, pp. 624–636, 2003.

The *weighted k-server problem* turns out to be much harder. In this problem a weight is assigned to each server and the total cost is the sum of the weighted distances. Fiat and Ricklin [6] prove that for any metric space there exists a set of weights such that the competitive ratio of any deterministic algorithm is at least $k^{\Omega(k)}$. For a uniform metric space, on which the problem is called the weighted paging problem, Feuerstein et al. [5] give a 6.275-competitive algorithm. For $k = 2$ Chrobak and Sgall [4] provided a 5-competitive algorithm and proved that no better competitive ratio is possible.

A weighted k-server algorithm is called competitive if the competitive ratio is independent of the weights. For a general metric space no competitive algorithm is known yet even for $k = 2$. It is easy to see that the general k-server problem is a generalization of the weighted k-server problem as well.

The general 2-server problem in which both servers move on the real line has become well-known as the CNN-problem[1]. Koutsoupias and Taylor [8] emphasize the importance of the CNN-problem as one of the simplest problems in a rich class of so-called sum of task systems [1]. In the sum-problem each system gets a task (request) and only one system has to fulfill its task. Such problems form a richer class than the k-server problem for modelling purposes (see [8]).

Koutsoupias and Taylor [8] prove a lower bound of $6 + \sqrt{17}$ on the competitive ratio of any deterministic on-line algorithm for the general 2-server problem, through an instance of the weighted 2-server problem on the real line. They also conjecture that the *work function algorithm* has constant competitive ratio for the general 2-server problem. It seems to be a bad tradition of multiple-server problems to keep unsettled conjectures. For the general 2-server problem the situation was even worse than for the k-server problem: the question if any algorithm exists with constant competitive ratio remained unanswered.

In this paper we answer this question affirmatively, by designing an algorithm and prove an upper boud of $100,000$ on its competitive ratio. The constant is huge, but our goal was indeed to settle the question. We believe that our result gives new insight in the problem and will lead to more and much better algorithms for the general k-server problem in the near future.

Optimal off-line solutions of metrical task systems can easily be found by dynamic programming (see [1]), which yields a $O(n^2)$ time algorithm for the general two server problem. As a result our algorithm can be implemented to work in polynomial time.

2   A Competitive Algorithm

A request is given by a pair $r_i = (x_i, y_i)$ with x_i a point in metric space $\mathbb{M}_1 = (\mathbb{X}, d_x)$ and y_i in $\mathbb{M}_2 = (\mathbb{Y}, d_y)$. We suppress the sub-indices on the distance since it will always be clear from the context which of the two measures is meant. We denote the two servers as the x- and y-server. The distance $\delta : (\mathbb{M}_1 \times \mathbb{M}_2)^2 \to \mathbb{R}$ is defined as $\delta((x_1, y_1), (x_2, y_2)) = d(x_1, x_2) + d(y_1, y_2)$.

---

[1] The name CNN-problem was suggested by Gerhard Woeginger.

We say that an online algorithm for the general two server problem is *lazy* if at any request only the server that services the request moves, and halts in the request. Our online algorithm is not lazy but it is easy to make it into a lazy algorithm by treating all moves made by the algorithm as virtual, and move a server only for real when it serves the next request. The triangle inequality ensures that the real movement is no more than the sum of the virtual moves. Moreover, we allow that the virtual moves are made to points outside the metric space. This is useful when we want to make a virtual move to a point between two points a_1 and a_2 of the metric space. This can easily be done by adding a new points a to the metric space and defining for any other point z in the metric space $d(z,a) = \min\{d(z,a_1)+d(a_1,a), d(z,a_2)+d(a_2,a)\}$, where choose $d(a,a_1)$ and $d(a,a_2)$ such that their sum is $d(a_1,a_2)$.

A tour is defined as a directed path in the product space $\mathbb{M}_1 \times \mathbb{M}_2$, and we denote the length of a tour T by $|T|$. We say that a tour T serves the request sequence $r_1,\ldots,r_n$ if there is a sequence of pairs $(\bar{x}_1,\bar{y}_1),\ldots,(\bar{x}_n,\bar{y}_n)$, that lie on T in this order, such that for all $j \in \{1,\ldots,n\}$, $\bar{x}_j = x_j$ or $\bar{y}_j = y_j$.

Given a configuration $(\hat{x}_0,\hat{y}_0)$ and a request sequence $r_1, r_2, \ldots, r_n$ we denote X_{0j} $(0 < j \leq n)$ as the length of the path $\hat{x}_0, x_1, \ldots, x_j$, and X_{ij} $(1 \leq i < j \leq n)$ as the length of the path $x_i, x_{i+1}, \ldots, x_j$. We denote Y_{ij} $(0 \leq i < j \leq n)$ in a similar way.

2.1   Basic Properties and a Sketch

We state two important properties of the general 2-server problem, which have inspired the design of our on-line algorithm. They are stated in Lemmas 1 and 2 and illustrated in Figure 2.1. The figure shows a part of an instance of the CNN-problem consisting of 7 requests. Five possible tours are shown. Tour T_D serves all request with the the x-server, starting from the x-coordinate of the first request, hence $|T_D| = X_{1,7}$. Similarly, $|T_E| = Y_{1,7}$. The other three tours each have length much smaller than $\min\{X_{1,7}, Y_{1,7}\}$. Tour T_A is relative far apart from tours T_B and T_C, whereas the tours T_B and T_C are relative close to one another. Lemma 1 states the impossibility of the existence of more than two

Fig. 1. Part of an instance of the CNN-problem and several feasible tours.

small tours, serving the same request sequence, which are mutually relatively far apart. First we give a notion of closeness of two tours which we employ in this paper. Let T_A and T_B be tours in $\mathbb{M}_1 \times \mathbb{M}_2$. We say that they are *connected* if there are points $x \in \mathbb{X}$ and $y \in \mathbb{Y}$ such that x and y are points on T_A as well as on T_B. (We do not impose that (x, y) is on the tours).

Lemma 1. *Given are three tours T_A, T_B and T_C, each serving a request sequence $r_1, r_2, \ldots, r_n$. If non of the three pairs $\{T_A, T_B\}$, $\{T_B, T_C\}$ and $\{T_A, T_C\}$ is connected, then $|T_A| + |T_B| + |T_C| \geq \min\{X_{1n}, Y_{1n}\}$.*

Proof. Assume without loss of generality that the x-server of T_A shares no request with the x-server of T_B and no request with the x-server of T_C. If the y-server of T_A serves all requests then the lemma obviously holds. So assume request r_i is served by the x-server of T_A. Then T_B and T_C must serve point y_i. But this means that the x-servers of T_B and T_C do not share a request. Hence, the three x-servers share no request. Thus, each request must be served by at least two y-servers and for each two consecutive requests there is a y-server that serves both requests. □

Lemma 2. *If T_A and T_B are connected and (x_{j_1}, y_{j_1}) and (x_{j_2}, y_{j_2}) are points on respectively T_A and T_B, then $\delta((x_{j_1}, y_{j_1}), (x_{j_2}, y_{j_2})) \leq |T_A| + |T_B|$.*

Proof. Let $x \in \mathbb{X}$ and $y \in \mathbb{Y}$ be points that connect both tours. Then,

$$\begin{aligned}
\delta((x_{j_1}, y_{j_1}), (x_{j_2}, y_{j_2})) &= d(x_{j_1}, x_{j_2}) + d(y_{j_1}, y_{j_2}) \\
&\leq d(x_{j_1}, x) + d(x, x_{j_2}) + d(y_{j_1}, y) + d(y, y_{j_2}) \\
&\leq |T_A| + |T_B|.
\end{aligned}$$

□

The idea behind our competitive algorithm is to try to remain close to an optimal tour, unless the optimal tour is relatively large. The algorithm works in phases, that are separate except that the end positions of the servers in one phase are their starting positions in the next phase. The algorithm is defined such that in each phase it is successful with respect to at least one of the following two goals:
Keeping the tour relatively short in comparison to an optimal tour that serves the same request sequence presented in the phase;
A substantial decrease in distance between the position of its own servers and those of an optimal tour at the end of the phase in comparison to the start of the phase.

In the beginning of each phase the algorithm chooses a reasonable strategy, which we call BALANCE and which is presented in the following subsection. While applying this strategy the algorithm keeps track of short tours for serving the requests in the phase. If such a short tour emerges then the algorithm tends to move its servers to the server positions of this short tour. If only one such a tour exists or if all short tours are relatively near to each other then the second goal stated above is reached and the phase stops. We know from Lemma 1 that at most two of them may be relatively far apart. In case indeed two such tours exist then the algorithm needs to move its servers to the positions the servers have on one of the short tours. It chooses the one that it is in a certain way nearest.

This choice may turn out to be unfortunate, which is illustrated in Figure 2.1: all requests are given in turn in points 1 and 2. To stay competitive an algorithm, starting the phase from v, must either move to point a or to point b. On an optimal tour the requests could be served at zero cost if this tour started the phase in a or b. If the algorithm moves its servers to a while the servers on an optimal tour appear to be in b then the distance to the optimal tour has even increased after this move. Similarly, if the algorithm moves its servers to b, the optimal servers may turn out to be in a.

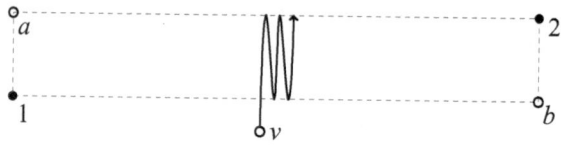

Fig. 2. A difficult choice.

We define the strategy COMPETE, described in Section 2.3, to avert this potential danger. The achievement of COMPETE is that, at the end of the phase, small tours, if any exist, have their servers on positions which are concentrated around a single point in $\mathbb{M}_1 \times \mathbb{M}_2$. Once this is achieved the phase is finished by moving the servers in the direction of the positions on the shortest tour at the end of the phase.

It will be clear from the sketch of the algorithm that in each phase, ONLINE carefully chooses its steps in order to stay competitive. If at some moment the requested points are relatively far away for both servers, then the algorithm is forced to make a large step. If this step is much larger than the sum of all preceding steps in the phase, then the phase is terminated immediately and the request is considered as the first request in the new phase.

The precise description of the algorithm, which is found in Section 2.4, is rather technical. The basic ideas have just been sketched, but their implementation allows freedom for specific choices. The choices we made are motivated by no more and no less than the fact that they allowed us to prove the desired competitiveness. Many other choices are possible, also alternatives for BALANCE and COMPETE and may give even better competitive ratios. However, the main goal of our research was to prove the conjecture that a constant competitive algorithm for the general two-server problem exists. We leave it to future research to find better competitive ratios.

2.2 Algorithm Balance

Algorithm BALANCE is applied at the beginning of each phase of ONLINE and within the subroutine COMPETE. We describe it on a request sequence $r_1, \ldots, r_n$ starting from the positions $(\hat{x}_0, \hat{y}_0)$. Let S_j^x and S_j^y be the total costs made by, respectively, the x- and the y-server after serving request r_j, and $S_j := S_j^x + S_j^y$. Let $(\hat{x}_j, \hat{y}_j)$ be the server positions after serving request r_j.

> BALANCE:
>
> If $S_j^x + d(\hat{x}_j, x_{j+1}) \leq S_j^y + d(\hat{y}_j, y_{j+1})$, then move the x-server to request x_{j+1}. Else move the y-server to request y_{j+1}.

The following lemma gives an upper bound on the cost of BALANCE.

Lemma 3. $S_j \leq 2 \max\{S_j^x, S_j^y\} \leq \min\{X_{0j}, Y_{0j}\} \; \forall j \in \{0, \ldots, n\}$.

Proof. Clearly, $S_j^x \leq X_{0j}$ and $S_j^y \leq Y_{0j}$. Let request r_i, $i \leq j$, be the last request served by the x-server. Then, by definition, $S_j^x = S_i^x \leq S_{i-1}^y + d(\hat{y}_{i-1}, y_i) \leq Y_{0i} \leq Y_{0j}$. Hence, $S_j^x \leq \min\{X_{0j}, Y_{0j}\}$. Similarly it is shown that $S_j^y \leq \min\{X_{0j}, Y_{0j}\}$. □

Lemma 4. $S_{j+1} \leq 3S_j + \min\{d(\hat{x}_0, x_{j+1}), d(\hat{y}_0, y_{j+1})\}, \; \forall j \geq 1$.

Proof. Without loss of generality we assume that $d(\hat{x}_0, x_{j+1}) \leq d(\hat{y}_0, y_{j+1})$. By definition of BALANCE we have $S_{j+1} \leq S_j + \min\{S_j^x + d(\hat{x}_j, x_{j+1}), S_j^y + d(\hat{y}_j, y_{j+1})\} \leq S_j + S_j^x + d(\hat{x}_j, x_{j+1}) \leq S_j + 2S_j^x + d(\hat{x}_0, x_{j+1}) \leq 3S_j + d(\hat{x}_0, x_{j+1})$. □

2.3 Algorithm Compete

We denote the positions of the servers at the beginning of algorithm COMPETE by $(\hat{x}, \hat{y})$. The behavior of the algorithm depends on a parameter $(x^*, y^*) \in \mathbb{M}_1 \times \mathbb{M}_2$, which is regarded as the position of the servers on the alternative short tour. Define $\Delta = \frac{1}{2} \max\{d(\hat{x}, x^*), d(\hat{y}, y^*)\}$. We describe the algorithm in case $\Delta = \frac{1}{2}d(\hat{x}, x^*)$. Interchanging the role of x and y gives the description in case $\Delta = \frac{1}{2}d(\hat{y}, y^*)$. The algorithm works in phases. The only information it takes to the next phase is the current position of its servers. We describe a generic phase on a sequence of requests $r_1, r_2, \ldots$. Occasionally both servers make a move after the release of a request. Let S_j^x and S_j^y be the distance travelled by respectively the x- and the y-server during the current phase, after both servers made their moves upon the release of request r_j and $(\hat{x}_j, \hat{y}_j)$ their positions at the same time.

A phase of COMPETE(x^*, y^*) :

⟨1⟩ Apply BALANCE, until the release of a request r_j with $d(\hat{x}, x_j) \geq \Delta$ in which case go to Step ⟨2⟩.

⟨2⟩ Apply the following three steps.

a. If $d(\hat{y}, y_j) < d(\hat{x}, x_j)$, then serve y_j, else serve x_j.

b. If the x-server has not served any request in the phase, and $j > 1$, then move the x-server over a distance S^y_{j-1} towards x_{j-1}.

c. Start a new phase.

The following lemma shows that if the two alternative short tours remain small, then COMPETE remains competitive with the sum of the two short tours. We exploit that COMPETE starts with its servers in the same position as one of the two short tours.

Lemma 5. *Given any request sequence let T be the tour made by Compete (x^*, y^*) starting from position $(\hat{x}, \hat{y})$. Let T_1 and T_2 be tours, both serving the same request sequence and starting in respectively $(\hat{x}, \hat{y})$ and (x^*, y^*). If $|T_1| + |T_2| < \Delta$, then $|T| \leq 10(|T_1| + |T_2|)$.*

Proof. We assume that $\Delta = \frac{1}{2}d(\hat{x}, x^*)$, the other case being similar. Let $(x^{(1)}, y^{(1)})$ and $(x^{(2)}, y^{(2)})$ be the final positions of respectively T_1 and T_2. We give an auxiliary request $(x^{(2)}, y^{(1)})$ at the end, which is served in T_1 and T_2 at no extra cost. Since we assume $|T_1| + |T_2| < \Delta$ we have $d(\hat{x}, x^{(2)}) \geq 2\Delta - d(x^*, x^{(2)}) > \Delta$ and therefore the last phase will end properly, i.e. with a ⟨2⟩ step.

Consider an arbitrary phase in the algorithm, and suppose this phase contains n requests. We use $(\hat{x}_0, \hat{y}_0)$ for the positions of the servers at the beginning of the phase (in the first phase $(\hat{x}_0, \hat{y}_0) = (\hat{x}, \hat{y})$). Define $S = S^x_n + S^y_n$. Let C^x_1 (C^y_1) and C^x_2 (C^y_2) be the total cost of the x-server (y-server) on, respectively, T_1 and T_2 in the phase and define $C_1 = C^x_1 + C^y_1$, $C_2 = C^x_2 + C^y_2$, and $C = C_1 + C_2$. The positions of the servers in T_1 after serving request r_j, $j \in \{0, \ldots, n\}$ are denoted by (x'_j, y'_j).

We define a potential at the beginning of the phase as $\Psi = 3d(\hat{x}_0, x'_0)$, whence the increase in potential during this phase is $\Delta\Psi = 3d(\hat{x}_n, x'_n) - 3d(\hat{x}_0, x'_0)$. We will prove that $S \leq 10C - \Delta\Psi$. This proves the lemma since taking the sum over all phases yields $|T| \leq 10(|T_1| + |T_2|) - 3d(\hat{x}_N, x'_N) \leq 10(|T_1| + |T_2|)$, where $(\hat{x}_N, x'_N)$ denote the final positions of the x-servers of T and T_1 in the last phase.

Given the condition of the lemma, request r_n is served by the y-servers of T_1, since otherwise $|T_1| \geq d(\hat{x}, x_n) \geq \Delta$. For the same reason, r_n is served by the y-server of T, since otherwise, by definition of Step 2a of COMPETE, $d(\hat{y}, y_n) \geq d(\hat{x}, x_n) \geq \Delta$, implying again $|T_1| \geq \Delta$. Hence, $\hat{y}_0 = y'_0$ and $\hat{y}_n = y'_n = y_n$. To simplify notation we define $y_0 = \hat{y}_0$ $(= y'_0)$. By a similar argument $r_1, \ldots, r_{n-1}$ must be served by the y-server of T_2. We distinguish three cases.

Case 1. The y-server of T_1 serves a request r_k with $k \in \{1, \ldots, n-1\}$. In this case, $C^y_1 \geq d(y_0, y_k) + d(y_k, y_n)$ and $C^y_2 \geq d(y_1, y_k) + d(y_k, y_{n-1})$. By the triangle inequality we have $d(y_0, y_k) + d(y_k, y_n) + C^y_2 \geq d(y_0, y_1) + d(y_{n-1}, y_n)$. Hence, $C^y_1 + 2C^y_2 \geq d(y_0, y_1) + d(y_{n-1}, y_n) + C^y_2 \geq S^y_n$. By the use of BALANCE also $d(y_0, y_1) + C^y_2 \geq S^x_n$. Clearly the increase in potential is bounded by $\Delta\Psi \leq 3(C^x_1 + S^x)$. We obtain

$$S = 4S_n^x + S_n^y - 3S_n^x$$
$$\leq 5C_1^y + 10C_2^y + 3C_1^x - \Delta\Psi$$
$$\leq 10C - \Delta\Psi.$$

Case 2. The y-server of T_1 serves only r_0 and r_n and the x-server of COM-
PETE *serves no request in the phase.* In this case $C_1^y = d(y_0, y_n)$, $x_n' = x_{n-1}$ (for $n \geq 2$), and $S_n^x = S_{n-1}^y$. The increase in potential is $\Delta\Psi \leq 3(C_1^x - S_n^x)$. Therefore,

$$S = S_n^x + S_n^y$$
$$\leq S_n^x + 2S_{n-1}^y + d(y_0, y_n)$$
$$= 3S_n^x + d(y_0, y_n)$$
$$\leq 3C_1^x - \Delta\Psi + d(y_0, y_n)$$
$$\leq 3C - \Delta\Psi.$$

Case 3. The y-server of T_1 serves only r_0 and r_n and the x-server of COM-
PETE *serves a request r_j with $j \in \{1, \ldots, n-1\}$.* Again, $C_1^y = d(y_0, y_n)$ and $x_n' = x_{n-1}$ (for $n \geq 2$). The increase in potential is $\Delta\Psi \leq 3(C_1^x - d(\hat{x}_0, x_1))$. Clearly $S_n^x \leq d(\hat{x}_0, x_1) + C_1^x$, and by definition of BALANCE also $S_{n-1}^y \leq d(\hat{x}_0, x_1) + C_1^x$. We obtain,

$$S = S_n^x + S_n^y$$
$$\leq S_n^x + 2S_{n-1}^y + d(y_0, y_n)$$
$$\leq 6C_1^x - \Delta\Psi + d(y_0, y_n)$$
$$\leq 6C - \Delta\Psi.$$

$\square$

2.4   Algorithm Online

Algorithm ONLINE works in phases. The only information taken from one phase to the next phase is the position of the servers at the end of the phase. Each phase starts by applying the simple BALANCE routine, until it becomes clear that there exists a short tour whose servers positions are not far from the starting position of ONLINE in the phase. At that moment, it makes an extra move to the positions of the servers on a short tour. If there is only one such a tour or every two such tours are relatively near to each other, then the phase ends. Otherwise ONLINE switches to the subroutine COMPETE. We notice that all these short tours do not need to start the phase in the same position as ONLINE. The phase ends with again an extra move. As announced in the sketch of the algorithm the phase can also end prematurely as soon as a request is presented that requires a relatively large move, which is defined in the algorithm by the *exception rule*.

We denote $v_0 = (O, O)$ as the starting position in phase 1, and v_i as the end position in phase i ($i \geq 1$), and hence the starting position in phase $i + 1$. The requests given in phase i are denoted by $r_{i1}, r_{i2}, \ldots$. We describe a generic phase of ONLINE and to facilitate the exposition we suppress the subindex i of the phase. Thus, we denote the requests in the generic phase by $r_1, r_2, \ldots$, the starting position by v_{-1} and the end position by v. We denote S_j ($j \geq 1$) as the cost of serving $r_1, r_2, \ldots r_j$ by BALANCE in Step (1) of the algorithm. Similarly, Z_j we denote the length of the tour by COMPETE in Step (2).

As indicated above the algorithm occasionally makes more moves than necessary to serve a next request r_j. We denote $(\hat{x}_j, \hat{y}_j)$ as the position of ONLINE after all moves are made, and we denote A_j as the cost of ONLINE until this moment. Additionally, we use the notation $(\hat{x}_0, \hat{y}_0)$ for the initial positions v_{-1} of the ONLINE servers in the phase. The constant η appearing in the description has value $\eta = 1/120$. For the tours T_A, T_B, T_C, and T_D in the description, v_A, v_B, v_C, and v_D represent their end points (end position of their servers).

A phase of ONLINE :

Apply the Steps (1),(2) and (3), with the following *exception rule*: If at any moment a request $r_j = (x_j, y_j)$ $(j \geq 2)$ is released for which $\min\{d(\hat{x}_0, x_j), d(\hat{y}_0, y_j)\} \geq 4A_{j-1}$, then return to $v_{-1} = (\hat{x}_0, \hat{y}_0)$ and let (x_j, y_j) be the first request in a new phase.

(1) Apply BALANCE. At the release of a request r_j for which there is a tour T_A, with $|T_A| < \eta S_j$ and $\delta(v_{-1}, v_A) \leq S_j/3$, return to v_{-1} after serving r_j, and continu with Step (2). Let r_{k_1} be this request.

(2) Let T_B and T_C be tours, serving $r_1, \ldots, r_{k_1}$, with $\max\{|T_B|, |T_C|\} < \eta S_{k_1-1}$, and $\delta(v_B, v_C) \geq 16\eta S_{k_1-1}$. If no such tours exist, then define $k_2 = k_1$ and continue with (3). Assume w.l.o.g. that $\delta(v_{-1}, v_B) \leq \delta(v_{-1}, v_C)$. Move the servers to v_B and apply COMPETE(v_C) until $Z_j \geq S_{k_1-1}$. Let r_{k_2} be this request. Move the servers back to v_{-1}.

(3) Let T_D serve $r_1, \ldots, r_{k_2}$ and have minimum length. If $|T_D| < \eta S_{k_1-1}$, then move from v_{-1} towards v_D over a distance $\frac{1}{3} S_{k_1-1}$. Start a new phase from this position.

We emphasize that Z_j, $(k_1 + 1 \leq j \leq k_2)$ is the cost COMPETE(v_C) makes starting from v_B.

3   Competitive Analysis

In the competitive analysis we distinguish two types of phases. The phases of type I are those that terminated prematurely by the exception rule. The other phases are of type III. Notice that the last phase, which we denote by N, is of type III since any phase of type I is followed by at least one more phase.

For the analysis we introduce a potential function Φ that measures the distance of the position of the ONLINE servers to those of the optimal tour. The potential at the beginning of a phase i is $\Phi_{i-1} = 1000 \cdot \delta(v_{i-1}, v_{i-1}^*)$. We define the potential at the end of the last phase N to be zero, i.e., $\Phi_N = 0$.

We will prove in Lemma 8 that in each phase of type III either the ONLINE tour is relatively short with respect to the optimal tour over the requests in the phase or the potential function has decreased substantially over the phase. First, in the next two lemmas we will bound the length of the ONLINE tour in a type III phase in terms of the bound from Lemma 1, which in a sense bounds the length of tours from below.

Consider an arbitrary phase of type III, and suppose it contains n requests $r_1, \ldots, r_n$. As before, we denote X_{ij} $(0 \leq i < j \leq n)$ as the length of the path $x_i, x_{i+1}, \ldots, x_j$ (For $i = 0$ this path is $\hat{x}_0, x_1, \ldots, x_j$). We denote Y_{ij} in a similar way. To simplify notation we write, in the following lemmas, S shortly for S_{k_1-1}, the length of the tour that BALANCE makes in Step (1) of ONLINE. By $|P|$ we denote the length of the ONLINE tour in the phase.

Lemma 6. *For each phase of type III, $\min\{X_{1k_1}, Y_{1k_1}\} \geq (1/12 - \eta/2)S$*

Proof. Assume w.l.o.g. $X_{1k_1} \leq Y_{1k_1}$ and notice that Lemma 3 implies $S_{k_1} \leq 2\min\{X_{0k_1}, Y_{0k_1}\}$. Tour T_A cannot serve merely y-requests, since in that case $\delta(v_{-1}, v_A) \geq d(\hat{y}_0, y_1) - |T_A| = Y_{0k_1} - Y_{1k_1} - |T_A| \geq Y_{0k_1} - 2|T_A| \geq (1/2 - 2\eta)S_{k_1} > S_{k_1}/3$. Now, let x_j be the last x-request served by T_A. In this case we obtain $S_{k_1}/3 \geq \delta(v_{-1}, v_A) \geq d(\hat{x}_0, x_j) - |T_A| \geq X_{0k_1} - 2X_{1k_1} - |T_A| \geq (1/2 - \eta)S_{k_1} - 2X_{1k_1}$. Hence, $X_{1k_1} \geq (1/12 - \eta/2)S_{k_1} \geq (1/12 - \eta/2)S$. $\square$

Lemma 7. *For each phase of type III, $|P| < 191S$.*

Proof. Notice that Lemma 4 and the exception rule imply $S_{k_1} \leq 3S + \min\{d(\hat{x}_0, x_{k_1}), d(\hat{y}_0, y_{k_1})\} \leq 7S$. The cost in Step (1) is at most $2S_{k_1} \leq 14S$ and the cost in Step (3) is at most $S/3$. If COMPETE is not applied in the phase, then only Step (1) and (3) add to the cost, whence $|P| \leq 14\frac{1}{3}S$. So assume COMPETE is applied.

First we bound $\delta(v_{-1}, v_B)$. Applying Lemma 6, $|T_A| + |T_B| + |T_C| \leq 7\eta S + \eta S + \eta S < \min\{X_{1k_1}, Y_{1k_1}\}$. Hence, these three tours do not satisfy the property of Lemma 1. By Lemma 2 the tours T_B and T_C cannot be connected, whence T_A must be connected to T_B or T_C. In the first case we have $\delta(v_{-1}, v_B) \leq \delta(v_{-1}, v_A) + |T_A| + |T_B|$, applying Lemma 3. Similarly if T_A is connected to T_C then $\delta(v_{-1}, v_C) \leq \delta(v_{-1}, v_A) + |T_A| + |T_C|$. Since we assumed $\delta(v_{-1}, v_B) \leq \delta(v_{-1}, v_C)$ we have $\delta(v_{-1}, v_B) \leq \delta(v_{-1}, v_A) + |T_A| + \min\{|T_B|, |T_C|\} \leq S_{k_1}/3 + \eta S_{k_1} + \eta S < 4S$.

Next we bound the cost made by COMPETE. Since by definition of ONLINE $Z_{k_2-1} < S$ the total cost after r_{k_2-1} is served is $A_{k_2-1} < 14S + 4S + S = 19S$. If r_{k_2} is served in Step $\langle 1 \rangle$ of COMPETE then the last step was a BALANCE step in a phase of COMPETE. Assume the phase started in the point (x, y). By Lemma 4,

$$\begin{aligned}
Z_{k_2} &< 3S + \min\{d(x, x_{k_2}), d(y, y_{k_2})\} \\
&\leq 3S + \delta((x, y), (\hat{x}_0, \hat{y}_0)) + \min\{d(\hat{x}_0, x_{k_2}), d(\hat{y}_0, y_{k_2})\} \\
&\leq 3S + 5S + 4 \cdot 19S \\
&= 84S.
\end{aligned}$$

For the third inequality we used the exception rule. Now assume the request is served in Step $\langle 2 \rangle$ of COMPETE. The cost of the Step in $\langle 2 \rangle$a is at most

$Z_{k_2-1}+\min\{d(\hat{x}_{k_1},x_{k_2}),d(\hat{y}_{k_1},y_{k_2})\}$, and the cost in $\langle 2 \rangle$b is no more than Z_{k_2-1}. Therefore,

$$Z_{k_2} \leq 3Z_{k_2-1} + \min\{d(\hat{x}_{k_1},x_{k_2}),d(\hat{y}_{k_1},y_{k_2})\}$$
$$< 3S + \delta((\hat{x}_0,\hat{y}_0),(\hat{x}_{k_1},\hat{y}_{k_1})) + \min\{d(\hat{x}_0,x_{k_2}),d(\hat{y}_0,y_{k_2})\}$$
$$\leq 3S + 4S + 4 \cdot 19S$$
$$= 83S.$$

We conclude that the total cost in the phase is no more than $14S+2(4S+84S)+S/3 = 190\frac{1}{3}S$. $\qquad\square$

In the following crucial lemma we use $|P^*|$ as the length of an optimal tour to serve the request in the phase considered. We use v^*_{-1} and v^* for the starting and finishing positions of the servers on an optimal tour in the phase. In accordance with suppressing the subindex for the phase, we also write Φ_{-1} and Φ for the potential function, respectively at the beginning and at the end of the phase.

Lemma 8. *For each phase of type III,* $2|P| < 10^5|P^*| - \Phi + \Phi_{-1}$.

Proof. First assume the phase considered is not the last phase. We have to show that

$$F \equiv c_1|P^*| + c_2\left(\delta(v_{-1},v^*_{-1}) - \delta(v,v^*)\right) > 2|P|, \qquad (1)$$

with $c_1 = 10^5$ and $c_2 = 10^3$. We distinguish three cases.

Case 1. $|P^*| \geq \eta S$. In this case $\delta(v_{-1},v^*_{-1}) - \delta(v,v^*) \geq -\delta(v,v_{-1}) - \delta(v^*,v^*_{-1}) \geq -S/3 - |P^*|$. Inequality (1) becomes in this case $F \geq c_1|P^*| - c_2(S/3 + |P^*|) = (c_1 - c_2)|P^*|) - c_2S/3 > 382S > 2|P|$.

Case 2. $|P^*| < \eta S$ *and* COMPETE *was not applied.* From the proof of Lemma 7 $|P| \leq 14\frac{1}{3}S$. By definition of ONLINE the endpoint of any tour serving requests $r_1,\ldots,r_{k_1-1}$ and with length smaller than ηS, must be at a distance greater than $S/3$ from point v_{-1}. In particular $\delta(v_{-1},v_D) > S/3$. Since COMPETE was not applied $\delta(v_D,v^*) < 16\eta S$. By the triangle inequality

$$\delta(v_{-1},v^*) \geq \delta(v_{-1},v_D) - \delta(v_D,v^*) > \delta(v_{-1},v_D) - 16\eta S,$$

and

$$\delta(v,v^*) \leq \delta(v_{-1},v_D) - S/3 + \delta(v_D,v^*).$$

Hence,

$$\delta(v_{-1},v^*_{-1}) - \delta(v,v^*) \geq \delta(v_{-1},v^*) - \delta(v,v^*) - \delta(v^*_{-1},v^*)$$
$$> S/3 - 32\eta S - |P^*|$$
$$= 8\eta S - |P^*|.$$

Hence, $F > c_1|P^*| + c_2(8\eta S - |P^*|) = (c_1 - c_2)|P^*| + 8c_2\eta S \geq 8c_2\eta S > 29S > 2|P|$.

Case 3. $|P^*| < \eta S$ *and* COMPETE *was applied.* Let T'_B be an optimal extensions of T_B, i.e. it starts in v_B, serves the requests $r_{k_1+1},\ldots,r_{k_2}$ and has minimum length. Define T'_C similar as T'_B with respect to T_C. Now we apply Lemma 5 with the parameters $(\hat{x}, \hat{y}) = v_B$, $(x^*, y^*) = v_C$, $T_1 = T'_B, T_2 = T'_C$, and $\Delta = \frac{1}{2}\max\{d(\hat{x}, x^*), d(\hat{y}, y^*)\} \geq \delta(v_B, v_C)/4 \geq 4\eta S$. Lemma 5 implies

$$|T'_B| + |T'_C| \geq \min\{\Delta, S/10\} \geq 4\eta S. \tag{2}$$

Now let $\widehat{T}_B$ and $\widehat{T}_C$ be arbitrary tours that serve $r_1, \ldots, r_{k_2}$ and are connected with T_B and T_C respectively. Assume that T_B and $\widehat{T}_B$ both serve the requests x_i and y_j for some $i, j \in \{1, \ldots, k_1\}$. A possible extension of T_B is to move the servers to (x_i, y_j) and serve the requests $r_{k_1+1}, \ldots, r_{k_2}$ similar to $\widehat{T}_B$. This implies $|T'_B| \leq |T_B| + |\widehat{T}_B|$. Similarly $|T'_C| \leq |T_C| + |\widehat{T}_C|$. Together with (2) this yields

$$|\widehat{T}_B| + |\widehat{T}_C| \geq |T'_B| + |T'_C| - |T_B| - |T_C| \geq 2\eta S. \tag{3}$$

Let T be an arbitrary tour that serves $r_1, \ldots, r_{k_2}$ with $|T| < \eta S$. Since $|T_B| + |T_C| + |T| < 3\eta S < \min\{X_{1k_1}, Y_{1k_1}\}$, these three tours do not satisfy the property of Lemma 1. (To apply Lemma 1 strictly we should consider T restricted to the first k_1 requests.) Since T_B and T_C are not connected (using Lemma 2) tour T must be connected with either T_B or T_C. With (3) this implies that either any such tour T is connected with T_B or any such tour is connected with T_C. Hence the optimal tour P^*, and the tour T_D defined by ONLINE, are both connected with T_B or are both connected with T_C. This implies $\delta(v_D, v^*) \leq |T_D| + \max\{|T_B|, |T_C|\} + |P^*| < 2\eta S + |P^*|$. With the triangle inequality we obtain

$$\begin{aligned}
\delta(v_{-1}, v^*_{-1}) &- \delta(v, v^*) \\
&\geq \delta(v_{-1}, v^*) - \delta(v, v^*) - \delta(v^*_{-1}, v^*) \\
&\geq (\delta(v_{-1}, v_D) - \delta(v, v_D)) - 2\delta(v_D, v^*) - |P^*| \\
&> S/3 - 2(2\eta S + |P^*|) - |P^*| \\
&= 2S/5 - 3|P^*|
\end{aligned}$$

Inequality (1) becomes $F > c_1|P^*| + c_2(2S/5 - 3|P^*|) > c_2 \cdot 2S/5 > 382S > 2|P|$.

It is clear that the inequality of the lemma also holds for the last phase of ONLINE, phase N, in case this phase finishes with Step (3). That the inequality als holds if this phase finishes in one of the two other steps is a matter of case checking, which we omit in this extended abstract. If the phase finishes in one of the two other steps, then a much better inequality can be obtained by going through the analysis above. We omit this in this extended abstract.   □

Constant competitiveness of ONLINE is now an easy consequence. We use P_i and P_i^* for the ONLINE and the optimal tour in phase i, respectively.

Theorem 1. ONLINE *is* 100.000-*competitive for the general two server problem.*

Proof. Consider a phase j of type I. Since ONLINE ends the phase in the same position as it started, any increase in potential is caused by the change of positions of the servers on the optimal tour only. Hence, $10^5 \cdot |P_j^*| - \Phi_j + \Phi_{j-1} \geq$

$10^5 \cdot |P_j^*| - 10^3 |P_j^*| \geq 0$. On the other hand, any phase j of type $\mathbb{I}$ is followed by a phase $j+1$ in which the cost of the first step, is at least twice the total cost $|P_j|$ of phase j. The last phase is of type $\mathbb{III}$ implying that the ONLINE cost over all phases of type $\mathbb{I}$ is at most $\frac{1}{2} + \frac{1}{4} + \ldots = 1$ times the cost over all phases of type $\mathbb{III}$. We conclude that

$$\sum_{j \in \mathbb{I} \cup \mathbb{III}} |P_j| \leq 2 \sum_{j \in \mathbb{III}} |P_j| < \sum_{j \in \mathbb{III}} 10^5 \cdot |P_j^*| - \Phi_j + \Phi_{j-1} \leq$$

$$\sum_{j \in \mathbb{I} \cup \mathbb{III}} 10^5 \cdot |P_j^*| - \Phi_j + \Phi_{j-1} = \sum_{j \in \mathbb{I} \cup \mathbb{III}} 10^5 \cdot |P_j^*|.$$

$\square$

References

1. Allan Borodin, Nathan Linial, and Michael Saks, *An optimal online algorithm for metrical task system*, Journal of the ACM **39** (1992), 745–763.
2. Marek Chrobak, Howard Karloff, Tom H. Payne, and Sundar Vishwanathan, *New results on server problems*, SIAM Journal on Discrete Mathematics **4** (1991), 172–181.
3. Marek Chrobak and Lawrence L. Larmore, *An optimal online algorithm for k servers on trees*, SIAM Journal on Computing **20** (1991), 144–148.
4. Marek Chrobak and Jiří Sgall, *The weighted 2-server problem*, The 17th Annual Symposium on Theoretical Aspects of Computer Science (STACS), LNCS 1770, Springer-Verlag, 2000, pp. 593–604.
5. Esteban Feuerstein, Steve Seiden, and Alejandro Strejlevich de Loma, *The related server problem*, Unpublished manuscript, 1999.
6. Amos Fiat and Moty Ricklin, *Competitive algorithms for the weighted server problem*, Theoretical Computer Science **130** (1994), 85–99.
7. Elias Koutsoupias and Christos Papadimitriou, *On the k-server conjecture*, Journal of the ACM **42** (1995), 971–983.
8. Elias Koutsoupias and David Taylor, *The CNN problem and other k-server variants*, The 17th Annual Symposium on Theoretical Aspects of Computer Science 2000 (STACS), LNCS 1770, Springer-Verlag, 2000, pp. 581–592.
9. Mark Manasse, Lyle A. McGeoch, and Daniel Sleator, *Competitive algorithms for server problems*, Journal of Algorithms **11** (1990), 208–230.

On the Competitive Ratio for Online Facility Location*

Dimitris Fotakis

Max-Planck-Institut für Informatik
Stuhlsatzenhausweg 85, 66123 Saarbrücken, Germany
fotakis@mpi-sb.mpg.de

Abstract. We consider the problem of Online Facility Location, where demands arrive online and must be irrevocably assigned to an open facility upon arrival. The objective is to minimize the sum of facility and assignment costs. We prove that the competitive ratio for Online Facility Location is $\Theta(\frac{\log n}{\log \log n})$. On the negative side, we show that no randomized algorithm can achieve a competitive ratio better than $\Omega(\frac{\log n}{\log \log n})$ against an oblivious adversary even if the demands lie on a line segment. On the positive side, we present a deterministic algorithm achieving a competitive ratio of $O(\frac{\log n}{\log \log n})$. The analysis is based on a hierarchical decomposition of the optimal facility locations such that each component either is relatively well-separated or has a relatively large diameter, and a potential function argument which distinguishes between the two kinds of components.

1 Introduction

The (metric uncapacitated) Facility Location problem is, given a metric space along with a facility cost for each point and a (multi)set of demand points, to find a set of facility locations which minimize the sum of facility and assignment costs. The assignment cost of a demand point is the distance to the nearest facility. Facility Location provides a simple and natural model for network design and clustering problems and has been the subject of intensive research over the last decade (e.g., see [17] for a survey and [9] for approximation algorithms and applications).

The definition of *Online Facility Location* [16] is motivated by practical applications where either the demand set is not known in advance or the solution must be constructed incrementally using limited information about future demands. In Online Facility Location, the demands arrive one at a time and must be irrevocably assigned to an open facility without any knowledge about future demands. The objective is to minimize the sum of facility and assignment costs, where each demand's assignment cost is the distance to the facility it is assigned to.

---

* This work was partially supported by the Future and Emerging Technologies programme of the EU under contract number IST-1999-14186 (ALCOM–FT).

J.C.M. Baeten et al. (Eds.): ICALP 2003, LNCS 2719, pp. 637–652, 2003.

We evaluate the performance of online algorithms using *competitive analysis* (e.g., [5]). An online algorithm is *c*-competitive if for all instances, the cost incurred by the algorithm is at most c times the cost incurred by an optimal offline algorithm, which has full knowledge of the demand sequence, on the same instance. We always use n to denote the number of demands.

Previous Work. In the offline case, where the demand set is fully known in advance, there are constant factor approximation algorithms based on Linear Programming rounding (e.g., [18]), local search (e.g., [10]), and the primal-dual schema (e.g., [12]). The best known polynomial-time algorithm achieves an approximation ratio of 1.52 [14], while no polynomial-time algorithm can achieve an approximation ratio less than 1.463 unless $NP = DTIME(n^{O(\log \log n)})$ [10].

Online Facility Location was first defined and studied in [16], where a simple randomized algorithm is shown to achieve a constant performance ratio if the demands, which are adversarially selected, arrive in random order. In the standard framework of competitive analysis, where not only the demand set but also the demand order is selected by an oblivious adversary, the same algorithm achieves a competitive ratio of $O(\frac{\log n}{\log \log n})$[1]. It is also shown a lower bound of $\Omega(\log^* n)$ on the competitive ratio of any online algorithm, where $\log^*$ is the inverse Ackerman function.

Online Facility Location should not be confused with the problem of Online Median [15]. In Online Median, the demand set is fully known in advance and the number of facilities increases online. An $O(1)$-competitive algorithm is known for Online Median [15].

Online Facility Location bears a resemblance to the extensively studied problem of Online File Replication (e.g., [4,2,1,13,8]). In Online File Replication, we are given a metric space, a point initially holding the file, and a replication cost factor. Read requests are generated by points in an online fashion. Each request accesses the nearest file copy at a cost equal to the corresponding distance. In between requests, the file may be replicated to a set of points at a cost equal to the replication cost factor times the total length of the minimum Steiner tree connecting the set of points receiving the file to at least one point already holding the file. Similarly to Facility Location, File Replication asks for a set of file locations which minimize the sum of replication and access costs. The important difference is that the cost of each facility only depends on the location, while the cost of each replication depends on the set of points which hold the file and the set of points which receive the file.

Online File Replication is a generalization of Online Steiner Tree [11]. Hence, there are metric spaces in which no randomized online algorithm can achieve a competitive ratio better than $\Omega(\log n)$ against an oblivious adversary. They are known both a randomized [4] and a deterministic [2] algorithm achieving a competitive ratio of $O(\log n)$ for the more general problem of Online File Allocation. For trees and rings, algorithms of constant competitive ratio are known [1,13,8].

---

[1] Only a logarithmic competitive ratio is claimed in [16]. However, a competitive ratio of $O(\frac{\log n}{\log \log n})$ follows from a simple modification of the same argument.

Contribution. We prove that the competitive ratio for Online Facility Location is $\Theta(\frac{\log n}{\log \log n})$. On the negative side, we show that no randomized algorithm can achieve a competitive ratio better than $\Omega(\frac{\log n}{\log \log n})$ against an oblivious adversary even if the metric space is a line segment. The only previously known lower bound was $\Omega(\log^* n)$ [16]. On the positive side, we present a deterministic algorithm achieving a competitive ratio of $O(\frac{\log n}{\log \log n})$ in every metric space. To the best of our knowledge, this is the first deterministic upper bound on the competitive ratio for Online Facility Location.

As for the analysis, the technique of [2], which is based on a hierarchical decomposition/cover of the optimal file locations such that each component's diameter is not too large, cannot be adapted to yield a sub-logarithmic competitive ratio for Online Facility Location. On the other hand, it is not difficult to show that our algorithm achieves a competitive ratio of $O(\frac{\log n}{\log \log n})$ for instances whose optimal solution consists of a single facility. To establish a tight bound for general instances, we show that any metric space has a hierarchical cover with the additional property that any component either is relatively well-separated or has a relatively large diameter. Then, we prove that the sub-instances corresponding to well-separated components can be treated as essentially independent instances whose optimal solutions consist of a single facility, and we bound the additional cost incurred by the algorithm because of the sub-instances corresponding to large diameter components.

Problem Definition. The problem of Online Facility Location is formally defined as follows. We are given a metric space $\mathcal{M} = (C, d)$, where C denotes the set of points and $d : C \times C \mapsto \mathbb{R}^+$ denotes the distance function which is symmetric and satisfies the triangle inequality. For each point $v \in C$, we are also given the cost f_v of opening a facility at v. The demand sequence consists of (not necessarily distinct) points $w \in C$. When a demand w arrives, the algorithm can open some new facilities. Once opened, a facility cannot be closed. Then, w must be irrevocably assigned to the nearest facility. If w is assigned to a facility at v, w's assignment cost is $d(w, v)$. The objective is to minimize the sum of facility and assignment costs.

Throughout this paper, we only consider unit demands by allowing multiple demands to be located at the same point. We always use n to denote the total number of demands. We distinguish between the case of uniform facility costs, where the cost of opening a facility, denoted by f, is the same for all points, and the general case of non-uniform facility costs, where the cost of opening a facility depends on the point.

Notation. A metric space $\mathcal{M} = (C, d)$ is usually identified by its point set C. For a subspace $C' \subseteq C$, $D(C') = \max_{u,v \in C'} \{d(u, v)\}$ denotes the diameter of C'. For a point $u \in C$ and a subspace $C' \subseteq C$, $d(C', u) = \min_{v \in C'} \{d(v, u)\}$ denotes the distance from u to the nearest point in C'. We use the convention that $d(u, \emptyset) = \infty$. For subspaces $C', C'' \subseteq C$, $d(C', C'') = \min_{u \in C''} \{d(C', u)\}$ denotes the minimum distance between a point in C' and a point in C''. For a point $u \in C$ and a non-negative number r, $B(u, r)$ denotes the ball of center u and radius r, $B(u, r) = \{v \in C : d(u, v) \leq r\}$.

2 A Lower Bound on the Competitive Ratio

In this section, we restrict our attention to uniform facility costs and instances whose optimal solution consists of a single facility. These assumptions can only strengthen the proven lower bound.

Theorem 1. *No randomized algorithm for Online Facility Location can achieve a competitive ratio better than* $\Omega(\frac{\log n}{\log \log n})$ *against an oblivious adversary even if the metric space is a line segment.*

Proof Sketch. We first prove that the lower bound holds if the metric space is a complete binary Hierarchically Well-Separated Tree (HST) [3]. Let T be a complete binary rooted tree of height h such that (i) the distance from the root to each of its children is D, and (ii) on every path from the root to a leaf, the edge length drops by a factor exactly m on every step. The height of a vertex is the number of edges on the path to the root. Every non-leaf vertex has exactly two children and every leaf has height exactly h. The distance from a vertex of height i to each of its children is exactly $\frac{D}{m^i}$. Let f be the cost of opening a new facility, which is the same for every vertex of T.

For a vertex v, let T_v denote the subtree rooted at v. The lower bound is based on the following property of T: The distance from a vertex v of height i to any vertex in T_v is at most $\frac{m}{m-1} \frac{D}{m^i}$, while the distance from v to any vertex not in T_v is at least $\frac{D}{m^{i-1}}$.

By Yao's principle (e.g., [5, Chapter 8]), it suffices to show that there is a probability distribution over demand sequences for which the ratio of the expected cost of any deterministic online algorithm to the expected optimal cost is $\Omega(\frac{\log n}{\log \log n})$.

We define an appropriate probability distribution by considering demand sequences divided into $h + 1$ phases. Phase 0 consists of a single demand at the root v_0. After the end of phase i, if v_i is not a leaf, the adversary proceeds to the next phase by selecting v_{i+1} uniformly at random and independently (u.i.r.) among the two children of v_i. Phase $i + 1$ consists of m^{i+1} consecutive demands at v_{i+1}.

The total number of demands is at most $m^h \frac{m}{m-1}$, which must not exceed n. The optimal solution opens a single facility at v_h and, for each phase i, incurs an assignment cost no greater than $D\frac{m}{m-1}$. Therefore, the optimal cost is at most $f + hD\frac{m}{m-1}$.

Let Alg be any deterministic online algorithm. We fix the adversary's random choices $v_0, \ldots, v_i$ up to phase i, $0 \leq i \leq h - 1$, (equivalently, we fix T_{v_i}), and we consider the expected cost (conditional on T_{v_i}) incurred by Alg for demands and facilities not in $T_{v_{i+1}}$. If Alg has no facilities in T_{v_i} when the first demand at v_{i+1} arrives, the assignment cost of demands at $v_i \in T_{v_i} \setminus T_{v_{i+1}}$ is at least

```
A ← ∅; L ← ∅; /* Initialization */
For each demand w:
    r_w ← d(A,w)/x; B_w ← {u ∈ L ∪ {w} : d(w, u) ≤ r_w}; Pot(B_w) ← Σ_{u∈B_w} d(A, u);
    if Pot(B_w) ≥ f then    /* A new facility is opened */
        if d(A, w) < f then
            Let ν be the smallest integer:
                either there exists exactly one u ∈ B_w such that
                    Pot(B_w ∩ B(u, r_w/2^ν)) > Pot(B_w)/2 ,
                or, for any u ∈ B_w, Pot(B_w ∩ B(u, r_w/2^{ν+1})) ≤ Pot(B_w)/2 .
            Let ŵ be any demand in B_w: Pot(B_w ∩ B(ŵ, r_w/2^ν)) > Pot(B_w)/2 .
        else ŵ ← w;
        A ← A ∪ {ŵ}; L ← L \ B_w;
    else L ← L ∪ {w};    /* w is marked unsatisfied */
    Assign w to the nearest facility in A.
```

Fig. 1. The algorithm Deterministic Facility Location – DFL.

mD. Otherwise, since v_{i+1} is selected u.i.r. among v_i's children, with probability at least $\frac{1}{2}$, there is at least one facility in $T_{v_i} \setminus T_{v_{i+1}}$. Therefore, for any fixed T_{v_i}, the (conditional) expected cost incurred by Alg for demands and facilities not in $T_{v_{i+1}}$ is at least $\min\{mD, \frac{f}{2}\}$ plus the cost for demands and facilities not in T_{v_i}. Since this holds for any fixed choice of $v_0, \ldots, v_i$ (equivalently, for any fixed T_{v_i}), the (unconditional) expected cost incurred by Alg for demands and facilities not in $T_{v_{i+1}}$ is at least $\min\{mD, \frac{f}{2}\}$ plus the (unconditional) expected cost for demands and facilities not in T_{v_i}. Hence, at the beginning of phase i, $0 \le i \le h$, the expected cost incurred by Alg for demands and facilities not in T_{v_i} is at least $i \min\{mD, \frac{f}{2}\}$. For the last phase, Alg incurs a cost no less than $\min\{mD, f\}$ inside T_{v_h}.

For $m = h$ and $D = \frac{f}{h}$, the total expected cost of Alg is at least $\frac{h+2}{2} hD$, while the optimal cost is at most $\frac{2h-1}{h-1} hD$. For the chosen value of h, the quantity $\frac{h^{h+1}}{h-1}$ must not exceed n. Setting $h = \left\lfloor \frac{\log n}{\log \log n} \right\rfloor$ yields the claimed lower bound.

To conclude the proof, we consider the following embedding of T in a line segment. The root is mapped to 0 (i.e., the center of the segment). Let v be a vertex of height i mapped to $\tilde{v}$. Then, v's left child is mapped to $\tilde{v} - \frac{D}{m^i}$ and v's right child is mapped to $\tilde{v} + \frac{D}{m^i}$. It can be shown that, for any $m \ge 4$, this embedding results in a hierarchically well separated metric space.   □

3   A Deterministic Algorithm for Uniform Facility Costs

In this section, we present the algorithm Deterministic Facility Location – DFL (Fig. 1) and prove that its competitive ratio is $O(\frac{\log n}{\log \log n})$.

Outline. The algorithm maintains its *facility configuration* A and the set L of *unsatisfied demands*, which are the demands not having contributed towards opening a new facility so far. A new demand w is marked unsatisfied and added to L only if no new facilities are opened when w arrives. Each unsatisfied demand $u \in L$ can contribute an amount of $d(A, u)$ to the cost of opening a new facility in its neighborhood. We refer to the quantity $d(A, u)$ as the *potential of u*. Only unsatisfied demands and the demand currently being processed have non-zero potential. For a set S consisting of demands of non-zero potential, let $\mathsf{Pot}(S) = \sum_{u \in S} d(A, u)$ be the potential of S.

The high level idea is to keep a balance between the algorithm's assignment and facility costs. For each demand w, the algorithm computes the set B_w consisting of w and the unsatisfied demands at distance no greater than $\frac{d(A, w)}{x}$ from w, where x is a sufficiently large constant. If B_w's potential is less than f, w is assigned to the nearest facility, marked unsatisfied and added to L. Otherwise, the algorithm opens a new facility at an appropriate location $\hat{w} \in B_w$ and assigns w to it. In this case, the demands in B_w are marked satisfied and removed from L. The location $\hat{w}$ is chosen as the center of a smallest radius ball/subset of B_w contributing more than half of B_w's potential.

An Overview of the Analysis. For an arbitrary sequence of n demands, we compare the algorithm's cost with the cost of a fixed offline optimal solution. The optimal solution is determined by k facility locations $c_1^*, c_2^*, \ldots, c_k^*$. The set of optimal facilities is denoted by C^*. Each demand u is assigned to the nearest facility in C^*. Hence, C^* defines a partition of the demand sequence into optimal clusters $C_1, C_2, \ldots, C_k$. Let $d_u^* = d(C^*, u)$ denote the assignment cost of u in the optimal solution, let $\mathsf{S}^* = \sum_u d_u^*$ be the total optimal assignment cost, let $\mathsf{F}^* = kf$ be the total optimal facility cost, and let $\sigma^* = \frac{\mathsf{S}^*}{n}$ be the average optimal assignment cost.

Let ρ, ψ denote a fixed pair of integers such that $\rho^\psi > n$. For any integer j, $0 \le j \le \psi$, let $r(j) = \rho^j \sigma^*$. We also define $r(-1) = 0$ and $r(\psi + 1) = \infty$. We observe that, for any demand u, $d_u^* < r(\psi)$. Let λ be some appropriately large constant, and, for any integer j, $-1 \le j \le \psi + 1$, let $R(j) = \lambda r(j)$. Throughout the analysis of DFL, we use $\lambda = 3x + 2$.

The Case of a Single Optimal Cluster. We first restrict our attention to instances whose optimal solution consists of a single facility c^*. The convergence of A to c^* is divided into $\psi + 2$ phases, where the current phase ℓ, $-1 \le \ell \le \psi$, starts just after the first facility within a distance of $R(\ell + 1)$ from c^* is opened and ends when the first facility within a distance of $R(\ell)$ from c^* is opened. In other words, the current phase ℓ lasts as long as $d(A, c^*) \in (R(\ell), R(\ell + 1)]$.

The demands arriving in the current phase ℓ and the demands remaining in L from the previous phase are partitioned into *inner* demands, whose optimal assignment cost is less than $r(\ell)$, and *outer* demands. The last phase ($\ell = -1$) never ends and only consists of outer demands.

For any outer demand u, $d(A, u)$ is at most $\lambda \sigma^* + (\lambda \rho + 1) d_u^*$ (Ineq. (3)). Hence, the assignment cost of an outer demand arriving in phase ℓ can be charged to its optimal assignment cost. We charge the total assignment cost of inner demands arriving in phase ℓ and the total facility cost incurred by the algorithm in phase ℓ to the optimal facility cost and the optimal assignment cost of the outer demands marked satisfied in phase ℓ.

The set of inner demands is included in a ball of center c^* and radius $r(\ell)$. If $R(\ell)$ is large enough compared to $r(\ell)$ (namely, if λ is chosen sufficiently large), we can think of the inner demands as being essentially located at c^*, because they are much closer to each other than to the current facility configuration A. Hence, we refer to the total potential accumulated by unsatisfied inner demands as the potential accumulated by c^* or simply, the potential of c^*. For any inner demand w, B_w includes the entire set of unsatisfied inner demands. Therefore, the potential accumulated by c^* is always less than f (Lemma 3).

However, a new facility may decrease the potential of c^*, because (i) it may be closer to c^*, and (ii) some unsatisfied inner demands may contribute their potential towards opening the new facility, in which case they are marked satisfied and removed from L. As a result, the upper bound of f on the potential accumulated by c^* cannot be directly translated into an upper bound on the total assignment cost of the inner demands arriving in phase ℓ as in [16].

Each time a new facility is opened, the algorithm incurs a facility cost of f and an assignment cost no greater than $\frac{f}{x}$. The algorithm must also be charged with an additional cost accounting for the decrease in the potential accumulated by c^*, which cannot exceed f. Hence, for each new facility, the algorithm is charged with a cost no greater than $\frac{2x+1}{x}f$.

Using the fact that $R(\ell)$ is much larger than $r(\ell)$, we show that if the inner demands included in B_w contribute more than half of B_w's potential, the new facility at $\hat{w}$ is within a distance of $R(\ell)$ from c^* (Lemma 4). In this case (Lemma 8, Case Isolated.B), the current phase ends and the algorithm's cost is charged to the optimal facility cost. Otherwise (Lemma 8, Case Isolated.A), the algorithm's cost is charged to the potential of the outer demands included in B_w, which is at least $f/2$. The optimal facility cost is charged $O(\psi)$ times and the optimal assignment cost is charged $O(\lambda\rho)$ times. Hence, setting $\psi = \rho = O(\frac{\log n}{\log\log n})$ yields the desired competitive ratio.

The General Case. If the optimal solution consists of $k > 1$ facilities $c_1^*,\ldots,c_k^*$, the demands are partitioned into the optimal clusters $C_1,\ldots,C_k$. The convergence of A to an optimal facility c_i^* is divided into $\psi + 2$ phases, where the current phase ℓ_i, $-1 \le \ell_i \le \psi$, lasts as long as $d(A, c_i^*) \in (R(\ell_i), R(\ell_i + 1)]$. For the current phase ℓ_i, the demands of C_i are again partitioned into inner and outer demands, and the inner demands of C_i can be thought of as being essentially located at c_i^*.

As before, the potential accumulated by an optimal facility c_i^* cannot exceed f. However, a single new facility can decrease the potential accumulated by many optimal facilities. Therefore, if we bound the decrease in the potential of each optimal facility separately and charge the algorithm with the total additional cost, we can only guarantee a logarithmic upper bound on the competitive ratio. To establish a tight bound, we show that the average (per new facility) decrease in the total potential accumulated by optimal facilities is $O(f)$.

We first observe that as long as the distance from the algorithm's facility configuration A to a set of optimal facilities K is large enough compared to the diameter of K, the inner demands assigned to facilities in K are much closer to each other than to A. Consequently, we can think of the inner demands assigned to K as being located at some optimal facility $c_K^* \in K$. Therefore, the total potential accumulated by optimal facilities in K is always less than f (Lemma 3). This observation naturally leads to the definition of an (optimal facility) *coalition* (Definition 2).

Our potential function argument is based on a *hierarchical cover* (Definition 1) of the subspace C^* comprising the optimal facility locations. Given a facility configuration A, the hierarchical cover determines a minimal collection of *active* coalitions which form a partition of C^* (Definition 3).

A coalition is *isolated* if it is well-separated from any other disjoint coalition, and *typical* otherwise. A new facility can decrease the potential accumulated by at most one isolated active coalition. Therefore, for each new facility, the decrease in the total potential accumulated by isolated active coalitions is at most f (Lemma 8, Case Isolated).

On the other hand, a new facility can decrease the potential accumulated by several typical active coalitions. We prove that any metric space has a hierarchical cover such that each component either is relatively well-separated or has a relatively large diameter (i.e., its diameter is within a constant factor from its parent's diameter (Lemma 1). Typical active coalitions correspond to the latter kind of components. Hence, we obtain a bound on the relative length of the interval for which an active coalition remains typical (Lemma 2), which can be translated into a bound of $O(f)$ on the total decrease in the potential accumulated by an active coalition, while the coalition remains typical (potential function component $\Xi_K^{(2)}$ and Lemma 7).

In the remaining paragraphs, we prove the following theorem by turning the aforementioned intuition into a formal potential function argument.

Theorem 2. *For any constant $x \geq 10$, the competitive ratio of Deterministic Facility Location is $O(\frac{\log n}{\log \log n})$.*

Hierarchical Covers and Optimal Facility Coalitions. We start by showing that any metric space has a hierarchical cover with the desired properties.

Definition 1. *A hierarchical cover of a metric space C is a collection $\mathcal{K} = \{K_1, \ldots, K_m\}$ of non-empty subsets of C which can be represented by a rooted tree $T_\mathcal{K}$ in the following sense:*

(A) C belongs to $\mathcal{K}$ and corresponds to the root of $T_\mathcal{K}$.
(B) For any $K \in \mathcal{K}$, $|K| > 1$, $\mathcal{K}$ contains sets $K_1, \ldots, K_\mu$, each of diameter less than $D(K)$, which form a partition of K. The sets $K_1, \ldots, K_\mu$ correspond to the children of K in $T_\mathcal{K}$.

We use $\mathcal{K}$ and its tree representation $T_\mathcal{K}$ interchangeably. By definition, every non-leaf set has at least two children. Therefore, $T_\mathcal{K}$ has at most $2|C| - 1$ nodes. For a set K different from the root, we use P_K to denote the immediate ancestor/parent of K in $T_\mathcal{K}$. Our potential function argument is based on the following property of metric spaces.

Lemma 1. *For any metric space C and any $\gamma \geq 16$, there exists a hierarchical cover T_K of C such that for any set K different from the root, either $D(K) > \frac{D(P_K)}{\gamma^2}$ or $d(K, C \setminus K) > \frac{D(P_K)}{4\gamma}$.*

Proof Sketch. Let C be any metric space, and let $D = D(C)$. We first show that, for any integer $i \geq 0$, C can be partitioned into a collection of *level i groups* $G_1^i, \ldots, G_m^i$ such that (i) for any $j_1 \neq j_2$, $d(G_{j_1}^i, G_{j_2}^i) > \frac{D}{4\gamma^i}$, and (ii) if $D(G_j^i) > \frac{D}{\gamma^i}$, then G_j^i does not contain any subset $G \subseteq G_j^i$ such that both $D(G) \leq \frac{D}{\gamma^{i+1}}$ and $d(G, G_j^i \setminus G) > \frac{D}{4\gamma^i}$. Since the collection of level i groups is a partition of C, for any G_j^i, $d(G_j^i, C \setminus G_j^i) > \frac{D}{4\gamma^i}$.

Level i groups are further partitioned into *level i components* $K_1^i, \ldots, K_{m'}^i$, such that (i) $D(K_j^i) \leq \frac{D}{\gamma^i}$, and (ii) either $D(K_j^i) > \frac{D}{\gamma^{i+1}}$ or $d(K_j^i, C \setminus K_j^i) > \frac{D}{4\gamma^i}$. To ensure a hierarchical structure, we proceed inductively in a bottom-up fashion. We create a single level i component for each level i group G_j^i of diameter no greater than $\frac{D}{\gamma^i}$. We recall that $d(G_j^i, C \setminus G_j^i) > \frac{D}{4\gamma^i}$. If $D(G_j^i) > \frac{D}{\gamma^i}$, G_j^i is partitioned into level i components of diameter in the interval $(\frac{D}{\gamma^{i+1}}, \frac{D}{\gamma^i}]$. For $\gamma \geq 16$, such a partition exists, because G_j^i does not contain any well-separated subsets of small diameter. Finally, we eliminate multiple occurrences of the same component at different levels. $\square$

Definition 2. *A set of optimal facilities $K \subseteq C^*$ with representative $c_K^* \in K$ is a coalition with respect to the facility configuration A if $d(A, c_K^*) \geq \lambda D(K)$. A coalition K is called* isolated *if $d(K, C^* \setminus K) \geq 2\,d(A, c_K^*)$, and* typical *otherwise. A coalition K becomes* broken *as soon as $d(A, c_K^*) < \lambda D(K)$.*

Given a hierarchical cover T_K of the subspace C^* comprising the optimal facility locations, we choose an arbitrary optimal facility as the representative of each set K. The representative of K always remains the same and is denoted by c_K^*. Then, T_K can be regarded as a system of optimal facility coalitions which hierarchically covers C^*. The current facility configuration A defines a minimal collection of active coalitions which form a partition of C^*.

Definition 3. *Given a hierarchical cover T_K of C^*, a coalition $K \in T_K$ is an* active coalition *with respect to A if $d(A, c_K^*) \geq \lambda D(K)$ and for any other coalition K' on the path from K to the root of T_K, $d(A, c_{K'}^*) < \lambda D(K')$.*

Lemma 2. *For any $\gamma \geq 8\lambda$, there is a hierarchical cover T_K of C^* such that if K is a typical active coalition with respect to the facility configuration A, then $\lambda \frac{D(P_K)}{\gamma^2} < d(A, c_K^*) < (\lambda + 1)D(P_K)$.*

Proof. For some $\gamma \geq 8\lambda$, let T_K be the hierarchical cover of C^* implied by Lemma 1. We show that T_K has the claimed property. The root of T_K is an isolated coalition by definition. Hence, we can restrict our attention to coalitions $K \in T_K$ different from the root for which the parent function P_K is well-defined.

Since K is an active coalition, its parent coalition P_K must have become broken. The upper bound on $d(A, c_K^*)$ follows from the triangle inequality and the fact that c_K^* also belongs to P_K.

For the lower bound, we consider two cases. If K has a relatively large diameter $(D(K) > \frac{D(P_K)}{\gamma^2})$, the lower bound on $d(A, c_K^*)$ holds as long as K remains a coalition. If K is relatively well-separated $(d(K, C^* \setminus K) > \frac{D(P_K)}{4\gamma})$ and the lower bound on $d(A, c_K^*)$ does not hold, we conclude that $2\, d(A, c_K^*) < d(K, C^* \setminus K)$ (K is an isolated coalition), which is a contradiction. $\square$

Notation. The set of active coalitions with respect to the current facility configuration A is denoted by $\mathsf{Act}(A)$. For a coalition K, ℓ_K denotes the index of the current phase. Namely, ℓ_K is equal to the integer j, $-1 \le j \le \psi$, such that $d(A, c_K^*) \in (R(j), R(j+1)]$. If $d(A, c_K^*) > R(\psi)$, $\ell_K = \psi$ (the first phase), while if $d(A, c_K^*) \le R(0)$, $\ell_K = -1$ (the last phase). Let $C_K = \bigcup_{c_i^* \in K} C_i$ be the optimal cluster corresponding to K. Since $\mathsf{Act}(A)$ is always a partition of C^*, the collection $\{C_K : K \in \mathsf{Act}(A)\}$ is a partition of the demand sequence. For the current phase ℓ_K, the demands of C_K are partitioned into inner demands $\mathsf{In}(K) = \{u \in C_K : d_u^* < r(\ell_K)\}$ and outer demands $\mathsf{Out}(K) = C_K \setminus \mathsf{In}(K)$. Let also $\Lambda_K = L \cap \mathsf{In}(K)$ be the set of unsatisfied inner demands assigned to K.

We should emphasize that ℓ_K, $\mathsf{In}(K)$, $\mathsf{Out}(K)$, and Λ_K depend on the current facility configuration A. In addition, Λ_K depends on the current set of unsatisfied demands L. For simplicity of notation, we omit the explicit dependence on A and L by assuming that while a demand w is being processed, ℓ_K, $\mathsf{In}(K)$, $\mathsf{Out}(K)$, and Λ_K keep the values they had when w arrived.

Properties. Let K be a coalition with respect to the current facility configuration A. Then, $d(A, c_K^*) \ge \lambda \max\{D(K), r(\ell_K)\}$. The diameter of the subspace comprising the inner demands of K is $D(\mathsf{In}(K)) < 3\max\{D(K), r(\ell_K)\}$. We repeatedly use the following inequalities. Let u be any demand in C_K and let $c_u^* \in K$ be the optimal facility to which u is assigned. Then,

$$d(A, u) \le d(A, c_K^*) + d(c_K^*, c_u^*) + d(c_u^*, u) \le d(A, c_K^*) + D(K) + d_u^* \le \tfrac{\lambda+1}{\lambda} d(A, c_K^*) + d_u^* \tag{1}$$

If u is an inner demand of K ($u \in \mathsf{In}(K)$),

$$d(u, c_K^*) \le d(u, c_u^*) + d(c_u^*, c_K^*) < r(\ell_K) + D(K) \le 2\max\{D(K), r(\ell_K)\} \le \tfrac{2}{\lambda} d(A, c_K^*) \tag{2}$$

If u is an outer demand of K ($u \in \mathsf{Out}(K)$),

$$d(A, u) \le (\lambda + 1)\sigma^* + ((\lambda+1)\rho + 1)d_u^* \tag{3}$$

Proof of Ineq. (3). Since u is an outer demand, it must be the case that $d_u^* \ge r(\ell_K)$. In addition, by Ineq. (1), $d(A, u) \le \tfrac{\lambda+1}{\lambda} d(A, c_K^*) + d_u^*$. If the current phase is the last one ($\ell_K = -1$), then $d(A, c_K^*) \le \lambda\sigma^*$, and the inequality follows. Otherwise, the current phase cannot be the first one (i.e., it must be $\ell_K < \psi$), because $d_u^* < r(\psi)$ and u could not be an outer demand. Therefore, $d(A, u) \le R(\ell_K + 1) = \lambda \rho\, r(\ell_K) \le \lambda \rho\, d_u^*$, and the inequality follows. $\square$

Lemma 3 and Lemma 4 establish the main properties of DFL.

Lemma 3. *For any coalition K, $\mathsf{Pot}(\Lambda_K) = \sum_{u \in \Lambda_K} d(A, u) < f$.*

Proof. In the last phase ($\ell_K = -1$), $\mathsf{Pot}(\Lambda_K) = 0$, because there are no inner demands ($\mathsf{In}(K) = \emptyset$). If $\ell_K \geq 0$, for any inner demand u of K ($u \subset \mathsf{In}(K)$),

$$d(A, u) \geq d(A, c_K^*) - d(c_K^*, u) > 3x \max\{D(K), r(\ell_K)\} \ ,$$

where the last inequality follows from (i) $d(A, c_K^*) \geq \lambda \max\{D(K), r(\ell_K)\}$, because K is a coalition, (ii) $d(u, c_K^*) < 2 \max\{D(K), r(\ell_K)\}$, because of Ineq. (2), and (iii) $\lambda = 3x + 2$.

Let w be the demand in Λ_K which has arrived last, and let A_w be the facility configuration when w arrived. The last time $\mathsf{Pot}(\Lambda_K)$ increased was when w was added to L (and hence, to Λ_K). Since $D(\mathsf{In}(K)) < 3 \max\{D(K), r(\ell_K)\} < \frac{d(A,w)}{x} \leq \frac{d(A_w, w)}{x}$, B_w must have contained the entire set Λ_K (including w). $\mathsf{Pot}(B_w)$ must have been less than f, because w was added to L. Therefore, $\mathsf{Pot}(\Lambda_K) \leq \mathsf{Pot}(B_w) < f$. □

Lemma 4. *Let w be any demand such that $\mathsf{Pot}(B_w) \geq f$, and, for a coalition K, let $\Lambda_K^w = B_w \cap \mathsf{In}(K)$. If there exists an active coalition K such that $\mathsf{Pot}(\Lambda_K^w) > \frac{\mathsf{Pot}(B_w)}{2}$, then $d(\hat{w}, c_K^*) < 8 \max\{D(K), r(\ell_K)\}$.*

Proof. We first consider the case that $d(A, w) \geq f$ and $\hat{w}$ coincides with w. If there exists an active coalition K such that $w \in \mathsf{In}(K)$, the conclusion of the lemma follows from Ineq. (2). For any active coalition K' such that $w \notin \mathsf{In}(K')$, Lemma 3 implies that $\mathsf{Pot}(\Lambda_{K'}^w) < \frac{\mathsf{Pot}(B_w)}{2}$, because $\mathsf{Pot}(B_w \setminus \Lambda_{K'}^w) \geq d(A, w) \geq f$.

We have also to consider the case that $d(A, w) < f$. We observe that any subset of B_w including a potential greater than $\frac{\mathsf{Pot}(B_w)}{2}$ must have a non-empty intersection with Λ_K^w. If $\frac{r_w}{2^\nu} < 6 \max\{D(K), r(\ell_K)\}$, let u be any demand in $\Lambda_K^w \cap B(\hat{w}, \frac{r_w}{2^\nu})$. Since u is an inner demand of K, using Ineq. (2), we show that

$$d(\hat{w}, c_K^*) \leq d(\hat{w}, u) + d(u, c_K^*) < 6 \max\{D(K), r(\ell_K)\} + 2 \max\{D(K), r(\ell_K)\} \ .$$

Otherwise, it must be $\frac{r_w}{2^{\nu+1}} \geq 3 \max\{D(K), r(\ell_K)\} > D(\mathsf{In}(K))$. Therefore, for any $u \in \Lambda_K^w$, $B_w \cap B(u, \frac{r_w}{2^{\nu+1}})$ includes the entire set Λ_K^w and hence, a potential greater than $\frac{\mathsf{Pot}(B_w)}{2}$. Consequently, there must be a single demand $u \in B_w$ such that $\mathsf{Pot}(B_w \cap B(u, \frac{r_w}{2^\nu})) > \frac{\mathsf{Pot}(B_w)}{2}$. Since the previous inequality is satisfied by any demand $u \in \Lambda_K^w$, there must be only one demand in Λ_K^w, and $\hat{w}$ must coincide with it. The lemma follows from Ineq. (2), because $\hat{w}$ is an inner demand of K. □

Potential Function Argument. We use the potential function Φ to bound the total algorithm's cost. Let $T_\mathcal{K}$ be the hierarchical cover of C^* implied by Lemma 2.

$$\Phi = \sum_{K \in T_\mathcal{K}} \Phi_K \ , \quad \text{where } \Phi_K = \frac{(2x+1)(\lambda+1)}{x(\lambda-2)} \, \Xi_K - \frac{\lambda+1}{\lambda} \, \Upsilon_K \ .$$

The function Ξ_K is the sum of three components, $\Xi_K = \Xi_K^{(1)} + \Xi_K^{(2)} + \Xi_K^{(3)}$, where

$$\Xi_K^{(1)} = \sum_{j=0}^{\psi} \xi^{(1)}(K,j) , \quad \xi^{(1)}(K,j) = \begin{cases} f & \text{if } d(A,c_K^*) > R(j). \\ 0 & \text{if } d(A,c_K^*) \le R(j). \end{cases}$$

$$\Xi_K^{(2)} = \begin{cases} 0 & \text{if } K \text{ is the root of } T_K. \\ f \max\left\{ \ln\left(\dfrac{\min\{d(A,c_K^*),(\lambda+1)D(P_K)\}}{\lambda \frac{D(P_K)}{\gamma^2}} \right), 0 \right\} & \text{otherwise.} \end{cases}$$

$$\Xi_K^{(3)} = \begin{cases} 2f & \text{if } K \text{ is a typical coalition.} \\ f & \text{if } K \text{ is an isolated coalition.} \\ 0 & \text{if } K \text{ has become broken.} \end{cases}$$

The function Υ_K is defined as $\Upsilon_K = \begin{cases} \sum_{u \in \Lambda_K} d(A,c_K^*) & \text{if } K \in \mathsf{Act}(A). \\ 0 & \text{otherwise.} \end{cases}$

Let K be an active coalition. The function $\Xi_K^{(1)}$ compensates for the cost of opening the facility concluding the current phase of K. $\Xi_K^{(2)}$ compensates for the additional cost charged to the algorithm while K is typical active coalition (Lemma 7). $\Xi_K^{(3)}$ compensates for the cost of opening a facility which changes the status of K either from typical to isolated or from isolated to broken. The function Ξ_K never increases and can decrease only if a new facility closer to c_K^* is opened. The function Υ_K is equal to the potential accumulated by c_K^*. Υ_K increases when an inner demand of K is added to L and decreases when a new facility closer to c_K^* is opened.

In the following, $\Delta\Phi$ denotes the change in the potential function because of a demand w. More specifically, let Φ be the value of the potential function just before the arrival of w, and let Φ' be the value of the potential function just after the algorithm has finished processing w. Then, $\Delta\Phi = \Phi' - \Phi$. The same notation is used with any of the potential function components above.

We first prove that Φ_K remains non-negative (Lemma 5). If a demand w is added to L (i.e., no new facilities are opened), the algorithm incurs an assignment cost of $d(A,w)$, while if w is not added to L (i.e., a new facility at $\hat{w}$ is opened), the algorithm incurs a facility cost of f and an assignment cost of $d(\hat{w},w) < \frac{f}{x}$. In the former case, we show that $d(A,w) + \Delta\Phi \le (\lambda+1)\sigma^* + ((\lambda+1)\rho + 1)d_w^*$ (Lemma 6). In the latter case, we show that $f + d(\hat{w},w) + \Delta\Phi \le \frac{4(\lambda+1)}{\lambda-2}\left[(\lambda+1)\sigma^*|B_w| + ((\lambda+1)\rho+1)\sum_{u \in B_w} d_u^*\right]$ (Lemma 8).

Lemma 5. *For any coalition K, if $\ell_K \ge 0$, then $\Upsilon_K < \frac{\lambda}{\lambda-2}f$, while if $\ell_K = -1$, then $\Upsilon_K = 0$.*

Proof. In the last phase ($\ell_K = -1$), $\Upsilon_K = 0$ because there are no inner demands ($\mathsf{In}(K) = \emptyset$). Otherwise, DFL maintains the invariant that $\mathsf{Pot}(\Lambda_K) < f$ (Lemma 3). In addition, for any $u \in \Lambda_K$, $d(A,u) > \frac{\lambda-2}{\lambda}d(A,c_K^*)$, because of Ineq. (2). Therefore, $\Upsilon_K < \frac{\lambda}{\lambda-2}\mathsf{Pot}(\Lambda_K) < \frac{\lambda}{\lambda-2}f$. $\square$

Lemma 5 implies that Φ_K is non-negative, because if K is an active coalition and $\ell_K \geq 0$, then $\frac{\lambda+1}{\lambda}\Upsilon_K < \frac{\lambda+1}{\lambda-2}f \leq \frac{(2x+1)(\lambda+1)}{x(\lambda-2)}\Xi_K^{(1)}$. On the other hand, if either K is not an active coalition or $\ell_K = -1$, then $\Upsilon_K = 0$.

Lemma 6. *If the demand w is added to L, then $d(A, w) + \Delta\Phi \leq (\lambda+1)\sigma^* + ((\lambda+1)\rho+1)d_w^*$.*

Proof. Let K be the unique active coalition such that $w \in C_K$. If w is an inner demand of K, w is added to Λ_K, and $\Delta\Phi = -\frac{\lambda+1}{\lambda}\Delta\Upsilon_K = -\frac{\lambda+1}{\lambda}d(A, c_K^*)$. Using Ineq. (1), we conclude that $d(A, w) + \Delta\Phi \leq d_w^*$. If w is an outer demand of K, then $\Delta\Phi = 0$. Using Ineq. (3), we conclude that $d(A, w) + \Delta\Phi \leq (\lambda+1)\sigma^* + ((\lambda+1)\rho+1)d_w^*$. $\square$

We have also to consider demands w which are not added to L (i.e., a new facility at $\hat{w}$ is opened). Let A be the facility configuration just before the arrival of w, and let $A' = A \cup \{\hat{w}\}$. We observe that if either K is not an active coalition or $\ell_K = -1$, $\Upsilon_K = 0$ and Φ_K cannot increase due to the new facility at $\hat{w}$. Therefore, we focus on active coalitions K such that $\ell_K \geq 0$.

Lemma 7. *Let $\hat{w}$ be the facility opened when the demand w arrives. Then, for any typical active coalition K, the quantity $\frac{(2x+1)(\lambda+1)}{x(\lambda-2)}\Xi_K - \frac{(2x+1)(\lambda+1)}{x\lambda}\Upsilon_K$ cannot increase due to $\hat{w}$.*

Proof. If either the current phase ends ($d(\hat{w}, c_K^*) \leq R(\ell_K)$) or K stops being a typical active coalition due to $\hat{w}$, then $\Delta\Xi_K \leq -f$, and the lemma follows from $-\Delta\Upsilon_K < \frac{\lambda}{\lambda-2}f$.

If K remains a typical active coalition with respect to A' and the current phase does not end ($d(\hat{w}, c_K^*) > R(\ell_K)$), let $\tau_K^w = \frac{d(A, c_K^*)}{d(A', c_K^*)} \geq 1$ be factor by which $d(A, c_K^*)$ decreases because of the new facility at $\hat{w}$. K cannot be the root of T_K, which is an isolated coalition by definition. Moreover, since K is a typical active coalition with respect to both A and A', Lemma 2 implies that $(\lambda+1)D(P_K) > d(A, c_K^*) \geq d(A', c_K^*) > \lambda\frac{D(P_K)}{\gamma^2}$. Therefore,

$$\Delta\Xi_K^{(2)} = \left[\ln\left(\frac{d(A', c_K^*)}{\lambda\frac{D(P_K)}{\gamma^2}}\right) - \ln\left(\frac{d(A, c_K^*)}{\lambda\frac{D(P_K)}{\gamma^2}}\right)\right]f = \ln\left(\frac{d(A', c_K^*)}{d(A, c_K^*)}\right)f = -\ln(\tau_K^w)f$$

If $B_w \cap \ln(K) = \emptyset$, no demands are removed from Λ_K, and $-\Delta\Upsilon_K \leq (1 - \frac{1}{\tau_K^w})\Upsilon_K \leq \ln(\tau_K^w)\Upsilon_K$. Otherwise, we can show that $\tau_K^w > \frac{x}{3} > 3$, and $-\Delta\Upsilon_K \leq \Upsilon_K < \ln(\tau_K^w)\Upsilon_K$. In both cases, the lemma follows from $\Upsilon_K < \frac{\lambda}{\lambda-2}f$. $\square$

Lemma 8. *Let $\hat{w}$ be the facility opened when the demand w arrives. Then,*

$$f + d(\hat{w}, w) + \Delta\Phi \leq \frac{4(\lambda+1)}{\lambda-2}[(\lambda+1)\sigma^*|B_w| + ((\lambda+1)\rho+1)\sum_{u\in B_w} d_u^*].$$

Proof Sketch. Let Λ_w be the set of inner demands in B_w, and let $M_w = B_w \setminus \Lambda_w$ be the set of outer demands in B_w. We recall that $f + d(\hat{w}, w) \leq \frac{x+1}{x}f$.

Case Isolated. There exists an isolated active coalition K such that $d(\hat{w}, c_K^*) < d(A, c_K^*)$. Lemma 7 implies that for any typical active coalition K', $\Delta\Phi_{K'} \leq 0$. In addition, for $x \geq 10$, we can prove that (i) for any isolated active coalition K' different from K, $d(\hat{w}, c_{K'}^*) \geq d(A, c_{K'}^*)$, and (ii) for any active coalition K' different from K, $B_w \cap \ln(K') = \emptyset$. As a result, for any isolated active coalition K' different from K, $\Delta\Phi_{K'} = 0$. In addition, only inner demands of K are included in B_w ($\Lambda_w \subseteq \ln(K)$).

We have also to bound $\frac{x+1}{x} f + \Delta\Phi_K$. Since $-\Delta\Upsilon_K < \frac{\lambda}{\lambda-2} f$ and $\lambda = 3x + 2$, $\frac{x+1}{x} f + \Delta\Phi_K < \frac{2(\lambda+1)}{\lambda-2} f + \Delta\Xi_K$. We distinguish between two cases depending on the potential contributed by Λ_w.

Case Isolated.A. $\text{Pot}(\Lambda_w) \leq \frac{\text{Pot}(B_w)}{2}$. Then, $\frac{2(\lambda+1)}{\lambda-2} f$ cannot exceed $\frac{4(\lambda+1)}{\lambda-2} \text{Pot}(M_w)$. We also recall than $\Delta\Xi_K \leq 0$. Hence, both the algorithm's cost and the increase in the potential function can be charged to the potential of the outer demands in B_w. Using Ineq. (3), we conclude that

$$\tfrac{x+1}{x} f + \Delta\Phi_K < \tfrac{4(\lambda+1)}{\lambda-2} \text{Pot}(M_w) \leq \tfrac{4(\lambda+1)}{\lambda-2} [(\lambda+1)\sigma^* |B_w| + ((\lambda+1)\rho + 1) \textstyle\sum_{u \in B_w} d_u^*] .$$

Case Isolated.B. $\text{Pot}(\Lambda_w) > \frac{\text{Pot}(B_w)}{2}$. Since $\Lambda_w \subseteq \ln(K)$, Lemma 4 implies that $d(\hat{w}, c_K^*) < 8 \max\{D(K), r(\ell_K)\}$. Hence, either the current phase ends or the coalition K becomes broken. In both cases, $\Delta\Xi_K \leq -f$ and the decrease in Ξ_K compensates for both the algorithm's cost and the decrease in Υ_K.

Case Typical. For any isolated active coalition K, $d(\hat{w}, c_K^*) \geq d(A, c_K^*)$. Therefore, no inner demands of K are included in B_w, because it would be $d(\hat{w}, c_K^*) < \frac{x}{3} d(A, c_K^*)$ otherwise. As a result, $\Delta\Phi_K = \Delta\Upsilon_K = 0$.

If w is an inner demand, let K_w be the unique typical active coalition such that $w \in \ln(K_w)$. Similarly to the proof of Lemma 7, we can show that $\frac{x+1}{x} f + \Delta\Phi_{K_w} \leq 0$. In addition, for any typical active coalition K' different from K_w, Lemma 7 implies that $\Delta\Phi_{K'} \leq 0$.

If w is an outer demand, using the following upper bound on $\text{Pot}(B_w)$, we can charge the algorithm's cost to the potential of B_w.

$$\tfrac{x+1}{x} f \leq \text{Pot}(B_w) \leq \tfrac{x+1}{x} \left[(\lambda+1)\sigma^* |B_w| + ((\lambda+1)\rho+1) \sum_{u \in B_w} d_u^* \right] - \tfrac{x+1}{x} \tfrac{\lambda+1}{\lambda} \sum_{K \in \text{Act}(K)} \Delta\Upsilon_K$$

We conclude the proof by applying Lemma 7 for each typical active coalition. □

In addition to the initial credit provided by the potential function Φ, a demand's optimal assignment cost is considered at most once by Lemma 6 (i.e., when the demand is added to L) and at most once by Lemma 8 (i.e., when the demand is removed from L). Therefore, the algorithm's total cost cannot exceed $\frac{2(2x+1)(\lambda+1)}{x(\lambda-2)} \left[\psi + 3 + \ln\left(\frac{\lambda+1}{\lambda}\gamma^2\right) \right] F^* + \frac{5\lambda+2}{\lambda-2} [(\lambda+1)\rho + \lambda + 2] S^*$. Setting $\gamma = 8\lambda$ and $\psi = \rho = O(\frac{\log n}{\log\log n})$ yields the claimed competitive ratio. □

4 The Algorithm for Non-uniform Facility Costs

In this section, we outline the algorithm Non-Uniform Deterministic Facility Location – NDFL, which is a generalization of DFL and can handle non-uniform facility costs.

The algorithm first rounds down the facility costs to the nearest integral power of two. For each demand w, the algorithm computes r_w, B_w, $\mathsf{Pot}(B_w)$, and $\hat{w}$ as in Fig. 1. If $|B_w| > 1$, NDFL opens the cheapest facility in $B(w, r_w) \cup B(\hat{w}, r_w)$ if its cost does not exceed $\mathsf{Pot}(B_w)$. Ties are always broken in favour of $\hat{w}$. Namely, if there are many facilities of the same (cheapest) cost, the one nearest to $\hat{w}$ is opened. If a new facility is opened, the demands of B_w are removed from L. Otherwise, w is added to L. If $|B_w| = 1$, NDFL keeps opening the cheapest facility in $B(w, r_w)$ while there is a facility of cost no greater than $\mathsf{Pot}(B_w)$. In this case, $\hat{w}$ coincides with w and ties are broken in favour of w. After opening a new facility, the algorithm updates r_w and $\mathsf{Pot}(B_w)$ according to the new facility configuration and iterates. After the last iteration, w is added to L. As in Fig. 1, the algorithm finally assigns w to the nearest facility.

The following theorem can be proven by generalizing the techniques described in Section 3.

Theorem 3. *For any constant* $x \geq 12$, *the competitive ratio of* NDFL *is* $O(\frac{\log n}{\log \log n})$.

5 An Open Problem

In the framework of incremental clustering (e.g., [6,7]), an algorithm is also allowed to merge some of the existing clusters. On the other hand, the lower bound of Theorem 1 on the competitive ratio for Online Facility Location crucially depends on the restriction that facilities cannot be closed. A natural open question is how much the competitive ratio can be improved if the algorithm is also allowed to close a facility by re-assigning the demands to another facility (i.e., merge some of the existing clusters). This research direction is related to an open problem of [7] concerning the existence of an incremental algorithm for k-Median which achieves a constant performance ratio using $O(k)$ medians.

References

1. S. Albers and H. Koga. New online algorithms for the page replication problem. *J. of Algorithms*, 27(1):75–96, 1998.
2. B. Awerbuch, Y. Bartal, and A. Fiat. Competitive distributed file allocation. *Proc. of STOC '93*, pp. 164–173, 1993.
3. Y. Bartal. Probabilistic approximations of metric spaces and its algorithmic applications. *Proc. of FOCS '96*, pp. 184–193, 1996.
4. Y. Bartal, A. Fiat, and Y. Rabani. Competitive algorithms for distributed data management. *J. of Computer and System Sciences*, 51(3):341–358, 1995.

5. A. Borodin and R. El-Yaniv. *Online Computation and Competitive Analysis*. Cambridge University Press, 1998.
6. M. Charicar, C. Chekuri, T. Feder, and R. Motwani. Incremental clustering and dynamic information retrieval. *Proc. of STOC '97*, pages 626–635, 1997.
7. M. Charicar and R. Panigrahy. Clustering to minimize the sum of cluster diameters. *Proc. of STOC '01*, pages 1–10, 2001.
8. R. Fleischer and S. Seiden. New results for online page replication. *Proc. of APPROX '00, LNCS 1913*, pp. 144–154, 2000.
9. S. Guha. *Approximation Algorithms for Facility Location Problems*. PhD Thesis, Stanford University, 2000.
10. S. Guha and S. Khuller. Greedy strikes back: Improved facility location algorithms. *Proc. of SODA '98*, pp. 649–657, 1998.
11. M. Imase and B.M. Waxman. Dynamic Steiner tree problem. *SIAM J. on Discrete Mathematics*, 4(3):369–384, 1991.
12. K. Jain and V. Vazirani. Approximation algorithms for metric facility location and k-median problems using the primal-dual schema and Lagrangian relaxation. *J. of the ACM*, 48(2):274–296, 2001.
13. C. Lund, N. Reingold, J. Westbrook, and D.C.K. Yan. Competitive online algorithms for distributed data management. *SIAM J. on Computing*, 28(3):1086–1111, 1999.
14. M. Mahdian, Y. Ye, and J. Zhang. Improved approximation algorithms for metric facility location problems. *Proc. of APPROX '02, LNCS 2462*, pp. 229–242, 2002.
15. R.R. Mettu and C.G. Plaxton. The online median problem. *Proc. of FOCS '00*, pp. 339–348, 2000.
16. A. Meyerson. Online facility location. *Proc. of FOCS '01*, pp. 426–431, 2001.
17. D. Shmoys. Approximation algorithms for facility location problems. *Proc. of APPROX '00, LNCS 1913*, pp. 27–33, 2000.
18. D. Shmoys, E. Tardos, and K. Aardal. Approximation algorithms for facility location problems. *Proc. of STOC '97*, pp. 265–274, 1997.

A Study of Integrated Document and Connection Caching[*]

Susanne Albers[1] and Rob van Stee[2]

[1] Institut für Informatik, Albert-Ludwigs-Universität, Georges-Köhler-Allee, 79110 Freiburg, Germany. `salbers@informatik.uni-freiburg.de`.
[2] Centre for Mathematics and Computer Science (CWI), Kruislaan 413, NL-1098 SJ Amsterdam, The Netherlands. `Rob.van.Stee@cwi.nl`.

Abstract. Document caching and connection caching are extensively studied problems. In document caching, one has to maintain caches containing documents accessible in a network. In connection caching, one has to maintain a set of open network connections that handle data transfer. Previous work investigated these two problems separately while in practice the problems occur together: In order to load a document, one has to establish a connection between network nodes if the required connection is not already open.

In this paper we present the first study that integrates document and connection caching. We first consider a very basic model in which all documents have the same size and the cost of loading a document or establishing a connection is equal to 1. We present deterministic and randomized online algorithms that achieve nearly optimal competitive ratios unless the size of the connection cache is extremely small. We then consider general settings where documents have varying sizes. We investigate a FAULT model in which the loading cost of a document is 1 as well as a BIT model in which the loading cost is equal to the size of the document.

1 Introduction

Recently there has been considerable research interest in document caching [5, 7,8,9,10,11,12] and connection caching [2,3,4] in networks. In document caching, one has to maintain local caches containing documents available in the network. In connection caching, one has to maintain a set of open network connections that handle data transfer. However, previous work investigated these two problems separately, while in practice they are very closely related.

Consider a computer that is connected to a network. A user working at that computer wishes to access and download documents from other network sites. A downloaded document can be stored in local cache, so that it does not

[*] Work supported by the Deutsche Forschungsgemeinschaft, Project AL 464/3-1, and by the European Community, Projects APPOL and APPOL II. Work done while the second author was at the Institut für Informatik, Albert-Ludwigs-Universität, Freiburg, Germany.

J.C.M. Baeten et al. (Eds.): ICALP 2003, LNCS 2719, pp. 653–667, 2003.
© Springer-Verlag Berlin Heidelberg 2003

have to be retransmitted when the user wishes to access that document again. Serving requests to documents that are stored locally is much less expensive than transmitting requested documents over the network. Therefore, the local cache, which is of bounded capacity, should be maintained in a careful manner. The transmission of documents in a network is performed using protocols such as TCP (Transmission Control Protocol). If a network node v has to download a document available at node v', then there has to exist an open (TCP) connection between v and v'. If the connection is not already open, it has to be established at a cost. Most networks, such as the Web, today work with persistent connections, i.e. an established connection can be kept open and reused later. However, each network node can only maintain a limited number of open connections and the collection of open connections can be viewed as a *connection cache*. The goal is to maintain this cache so that the connection establishment cost is as small as possible.

Clearly, caching decisions made on the document and connection levels heavily affect each other. Evicting a document d from the document cache at node v has a very negative effect if the connection between node v and node v', where d is originally stored, is already closed. When d is requested again, one has to pay the connection establishment cost in addition to the necessary document transmission cost. A similar overhead occurs if a connection is closed that is frequently needed for data transfers. Therefore document and connection caching algorithms should coordinate their decisions. This can considerably improve the system's performance, i.e. the user perceived latency as well as the network congestion are reduced.

In this paper we present the first study of integrated document and connection caching. Formally, we consider a network node v. The node has two caches: one for the documents, also called *pages*, and one for the open connections currently maintained to other nodes. A sequence of *requests* must be served. Each request specifies a document d that the user at our network node wishes to access. If d resides in the document cache, then the request can be served at 0 cost. Otherwise a fault occurs and the request must be served by downloading d into the document cache at a cost of $cost(d) > 0$. Suppose that d is originally stored at network node v'. To load d into the document cache, an open connection must exist between v and v'. If the connection is already open, no cost is incurred. Otherwise the connection has to be established at a cost of $cost(v, v')$. The goal is to serve the request sequence so that the total cost is as small as possible.

The integrated document and connection caching problem is inherently online in that each request must be served without knowledge of future requests. We use competitive analysis to analyze the performance of online algorithms. We denote the cost of an algorithm A on a request sequence σ by $A(\sigma)$. The optimal cost to serve this sequence is denoted by $\text{OPT}(\sigma)$. The goal of an online algorithm A is to minimize the competitive ratio $R(A)$, which is defined as the smallest value R that satisfies $A(\sigma) \leq R \cdot \text{OPT}(\sigma) + a$, for any request sequence σ and some constant a independent of σ.

We remark here that a problem similar to that defined above arises in distributed databases. There, a user may have a file/page cache as well as a cache with pointers to files allowing fast access.

Previous work: As mentioned above document and connection caching have separately been the subjects of extensive research. There is a considerable body of work on document caching problems, see e.g [5,7,8,9,10,11,12]. However, the papers ignore that in a network setting, one may have to open a connection to load a document. If all documents have the same size and a loading cost of 1, which is the classical paging problem, the best competitive ratio of deterministic online algorithms is equal to k, where k is the number of documents that can be stored simultaneously in cache [11]. This competitiveness is achieved by the popular LRU (Least Recently Used) and FIFO (First-In First-Out) replacement strategies. On a fault, LRU evicts the page that was requested least recently and FIFO evicts the page that has been in cache longest. Fiat et al. [7] presented an elegant randomized paging algorithm called MARK that is $2H_k$-competitive against oblivious adversaries, where H_k is the k-th Harmonic number. More complicated algorithms that achieve an optimal competitiveness of H_k were given in [1,10]. Irani [9] initiated the algorithmic study of the document caching problem when documents have different sizes. She considered a FAULT model where the loading cost of each document is equal to 1 as well as a BIT model, where the loading cost is equal to the size of the document. She presented randomized $O(\log^2 k)$-competitive online algorithms for both settings. Young [12] gave a deterministic k-competitive online algorithm for a general cost model where the loading cost is an arbitrary non-negative value. Recently Feder et al. [5] studied a document caching problem where requests can be reordered. They concentrate on the case that the cache can hold one document. Gopalan et al. [8] study document caching in the Web when documents have expiration times. They assume all documents have the same size and a loading cost of 1.

Cohen et al. [3,4] introduced the connection caching problem. The input of the problem is a sequence of requests for TCP connections that must be established if not already open. Cohen et al. considered a distributed setting where requests occur at different network nodes. They gave deterministic k-competitive and randomized $O(H_k)$-competitive online algorithms if all connections incur the same establishment cost. Here k is the maximum number of connections that a network node can keep open simultaneously. The case that connections can have varying establishment costs was considered in [2].

Our contribution: We investigate document and connection caching in an integrated manner. In the following let k be the number of documents that can be stored in the document cache and k' be the number of connections that can be kept open. We start by studying a basic setting in which all documents have the same size and a loading cost of 1; the connections have an establishment cost of 1. We present a deterministic online algorithm that achieves a competitive ratio of $k+4$ if $k' \geq k$ and a ratio of $\min\{2k-k'+4, 2k\}$ if $k' < k$. Our algorithm uses LRU for the document cache and a phase based replacement strategy that tries to keep connections of documents that may be evicted soon. We develop a lower bound on the performance of any deterministic online algorithm which implies that our algorithm is nearly optimal if k' is not extremely small. We also consider randomized online algorithms and prove that by replacing LRU by a randomized Marking strategy we obtain a competitive ratio of $2H_k + \min\{2H_k, 2(k-k')+4\}$.

Additionally we investigate the problem that pages have varying sizes. If all documents have a loading cost of 1, which corresponds to Irani's FAULT model, we achieve a competitive ratio of $(4k + 14)/3$ if $k' \geq k$ and of $2k - 2k'/3 + 14/3$ if $k' < k$. Finally we consider a BIT model where the loading cost of a document is equal to the size of the document and the connection establishment cost is c, for some constant c. Here we prove a competitiveness of $(k + 5)(c' + 1)/2$ if $k' \geq k$, where $c' = c/s$ and s is the size of the smallest document ever requested. If $k' < k$, the competitiveness is $(2k - k' + 5)(c' + 1)/2$.

Finally we consider a distributed scenario, where requests can occur at different network nodes. We show that no deterministic online algorithm can in general be better than $2k$-competitive, where k is the maximum number of documents that can be stored at any network node. A competitive ratio of $2k$ is easily achieved by an online algorithm that uses a k-competitive paging algorithm for the document cache and any replacement strategy for the connection cache.

2   Algorithms for the Basic Model

In this section we study a very basic scenario where all documents have the same size. Loading a missing document costs 1 and establishing a connection also costs 1.

2.1   Deterministic Algorithms

We present a deterministic online algorithm ALG for our basic setting. ALG works in phases. Each phase is defined as a maximal subsequence of requests to k distinct pages, which starts after the previous phase finishes (the first phase starts with the first request). Within each phase ALG works as follows.

At the beginning of each phase, evict all connections that were not used in the previous phase.

On a page fault, use LRU to determine which page to evict from the page cache.

On a connection fault, if there is a free slot in the cache, use it;

    otherwise, use MRU (Most Recently Used) to determine which connection to evict.

For ease of exposition, we first consider the case where the size of the connection cache is at least the same size as the page cache, i.e. $k' \geq k$. We then extend our analysis to the case $k' < k$.

Theorem 1. *If $k' \geq k$, then $\mathcal{R}(\text{ALG}) \leq k + 4$.*

Proof. Consider a request sequence σ. We first study the case that $k' = k$. Suppose there are $N + 1$ phases, numbered $0, 1, \ldots, N$. For phase i, denote the number of page requests that cause a page fault by f_i; the number of page requests that do not cause a page fault by p_i (these pages were requested in the previous phase by definition of LRU); the number of MRU faults m_i, and the number of holes *created* by h_i (i. e. the number of connections evicted at the start of phase i). Define $F = \sum_{i=1}^{N} f_i$, $M = \sum_{i=1}^{N} m_i$, $H = \sum_{i=2}^{N} h_i$ and $P = \sum_{i=1}^{N} p_i$. (We ignore phase 0.) Note $h_1 = 0$ and $f_i + p_i = k$ for each phase i.

Each hole that is created, is filled at most once, and this happens on a connection fault. (It is possible that some holes are never filled.) Thus the number of connection faults that cause holes to be filled is at most H. Furthermore, the remaining connection faults are exactly the connection faults where MRU is applied; this happens M times. Thus

$$\text{ALG}(\sigma) \leq F + M + H = kN + M + H - P. \tag{1}$$

Note that our algorithm is defined in such a way that the number of page faults is independent of the number of connection faults or the decisions as to which connections are evicted. The page cache is simply maintained by LRU. By definition of LRU, there must be one OPT page fault in each phase. Thus

$$\text{OPT}(\sigma) \geq N. \tag{2}$$

Each phase can be visualized as follows. The connection cache is at all times divided into two sets, PREVIOUS and CURRENT. Here PREVIOUS contains the connection slots that were not (yet) used in this phase, while CURRENT contains the connection slots that were used in the current phase. At the start of each phase, CURRENT is empty and PREVIOUS contains all k slots. Note that some of these slots may contain holes, in case a connection was evicted that was not used in the previous phase.

For each page fault in a phase, there are two possibilities:

1. No connection fault:
    a) A not yet used connection slot is used for the first time in this phase (this connection was also used in the previous phase);
    b) A connection slot already used in the current phase is used again (two or more pages are at the same node).
2. Connection fault occurs:
    a) A hole is filled: a not yet used connection slot is used for the first time in this phase;
    b) A connection slot already used in the current phase is used again by MRU;
    c) (special case) A connection slot not yet used in the current phase is used by MRU.

Case 2.(c) can only occur if the very first page fault in a phase causes a connection fault; for a later page fault that also causes a connection fault, MRU always uses a slot that was already used in the current phase. From this list we have that only in cases 1.(a), 2.(a) and 2.(c), a connection slot moves from the set PREVIOUS to the set CURRENT.

Consider a phase $i > 0$. Suppose Case 2.(c) does not occur, and there are $m_i > 0$ MRU faults in phase i. Then at least m_i times, a connection slot already in CURRENT is used again. Hence at most $f_i - m_i$ times a connection slot moves from PREVIOUS to CURRENT. Therefore, at the end of phase i, there are at least $k - f_i + m_i$ connection slots still in PREVIOUS.

The pages requested in phase i can be divided into four groups:

1. pages that did not cause a page fault (p_i);
2. pages that caused a page fault, but no connection fault;
3. pages that caused a hole in the connection cache to be filled;
4. pages that caused a connection slot to be used again by MRU (m_i).

Every connection slot that at some point in phase i contains a connection to a page in group 2 or 3 (note that this may change later in the phase due to the use of MRU), is in CURRENT at the end of the phase. The other connection slots contain connections to pages that were either not requested in phase i (but were requested in phase $i - 1$, or they would have been evicted before), or that did not cause a fault. This last possibility occurs p_i times, so there are at least $k - f_i + m_i - p_i = m_i$ pages that are not requested again. This implies there are at least $k + m_i$ distinct pages requested in phase i and phase $i - 1$. Therefore OPT has at least m_i faults in phases $i - 1$ and i.

If Case 2.(c) does occur, then there were no holes at the start of phase i. Then the connections to the pages requested in phase $i - 1$ must all be distinct, $m_{i-1} = 0$ and $h_i = 0$. At the start of phase i, a connection slot moves from PREVIOUS to CURRENT using MRU. Case 2.(c) does not occur in the rest of the phase. Thus at the end of phase i, we have that there are at least $k - f_i + m_i - 1$ connection slots still in PREVIOUS. These slots correspond to connections that were used in the previous phase but not in this one, implying $k - f_i + m_i - p_i - 1 = m_i - 1$ pages that were requested in phase $i - 1$ but not in i. Then OPT has at least $m_i - 1$ faults in phases $i - 1$ and i. Moreover, it has at least one fault in phases $i - 2$ and $i - 1$, and $1 = m_{i-1} + 1$. By amortizing the cost, we find that OPT has at least m_i faults for every pair of phases $i - 1$ and i.

Thus $\text{OPT}(\sigma) \geq \sum_{i \text{ odd}} m_i$, and $\text{OPT}(\sigma) \geq \sum_{i \text{ even}} m_i$. This implies that

$$\text{OPT}(\sigma) \geq \frac{1}{2} \sum_{i>0} m_i = \frac{M}{2}. \tag{3}$$

The connections still in PREVIOUS at the end of phase i are evicted and become h_{i+1} holes. At most p_i of them lead to pages that were requested without a fault. Thus there are at least $k + h_{i+1} - p_i$ distinct pages requested in phases i and $i - 1$. This gives another bound for the cost of OPT:

$$\text{OPT}(\sigma) \geq \frac{1}{2} \sum_{i>0} (h_{i+1} - p_i) \geq \frac{H - P}{2}. \tag{4}$$

Combining (1), (2), (3) and (4) gives

$$\text{ALG}(\sigma) \leq kN + M + H - P \leq k \cdot \text{OPT}(\sigma) + 2\text{OPT}(\sigma) + 2\text{OPT}(\sigma) = (k+4)\text{OPT}(\sigma).$$

This proves the ratio. It can be seen that the proof also holds for $k' > k$.   □

Theorem 2. *If $k' < k$, then $\mathcal{R}(\text{ALG}) \leq \min(k + 4 + (k - k'), 2k)$.*

Proof. Clearly, $\mathcal{R}(\text{ALG}) \leq 2k$ since ALG has at most $2k$ faults per phase (k connection faults and k page faults). We still have (2) and (4) by the exact same reasoning as in the proof of Theorem 1.

For m_i, we have again that each time that MRU is applied, no connection moves from PREVIOUS to CURRENT (unless Case 2.(c) occurs). So at most $f_i - m_i$ times a connection moves from PREVIOUS to CURRENT. Therefore, at the end of the phase, at least $k' - f_i + m_i$ connections are still in PREVIOUS. At most p_i of them refer to pages requested without a fault in phase i, so at least $k' - f_i + m_i - p_i = k' - k + m_i$ pages are requested in phase $i - 1$ but not in phase i. Therefore there are at least $m_i + k'$ distinct pages requested in these two phases, and OPT has at least $m_i - (k - k')$ faults.

If Case 2.(c) occurs, there are only at least $k' - (f_i - (m_i - 1)) = m_i - (k - k') - 1$ connections still in PREVIOUS at the end. However, in that case we have $m_{i-1} \leq k - k'$ since there were no holes. Therefore $m_{i-1} - (k - k') \leq 0$ and we can amortize as before.

We therefore find

$$\text{OPT}(\sigma) \geq \frac{M - (k - k')N}{2}. \tag{5}$$

Using (2), this implies $M \leq 2\text{OPT}(\sigma) + (k - k')N \leq (k - k' + 2)\text{OPT}(\sigma)$. Therefore in this case

$$\text{ALG}(\sigma) \leq ((k + 2) + (k - k' + 2))\text{OPT}(\sigma) \leq (2k - k' + 4)\text{OPT}(\sigma).$$

This proves the lemma.   □

2.2 Randomized Algorithms

For the standard paging problem, the randomized algorithm MARK is $2H_k$-competitive, where H_k is the k-th Harmonic number [7]. Moreover, no randomized algorithm can have a competitive ratio less than H_k. The MARK algorithm processes a request sequence in phases. At the beginning of each phase, all pages in the memory system are unmarked. Whenever a page is requested, it is *marked*. On a fault, a page is chosen uniformly at random from among the unmarked pages in cache, and that page is evicted. A phase ends when all pages in cache are marked and a page fault occurs. Then, all marks are erased and a new phase is started.

In our algorithm ALG we substitute MARK for LRU to get a randomized algorithm. However, in this case it is also necessary to evict connections less greedily to get a good performance. In particular, at the start of a phase we will not evict any connections that are associated with pages requested in the previous phase. Note that some of these connections may not have been used in that phase, because the relevant page might not have caused a page fault.

Theorem 3. *For the randomized version of* ALG *and* $k' \geq k$, *we have* $\mathcal{R}(\text{ALG}) \leq 2H_k + 4$. *For* $k' < k$, *we have.* $\mathcal{R}(\text{ALG}) \leq 2H_k + \min(2H_k, 4 + 2(k - k'))$.

Proof. We analyze this algorithm very similarly to the original analysis of MARK [7] and to the analysis in Section 2.1. We define q_i as the number of *new* pages requested in phase i. A page is new if it is not in the cache at the start of the phase. We define h_i, m_i, H and M as before and write $Q = \sum q_i$. Then by [7],

$$\text{ALG}(\sigma) \leq H_k Q + H + M.$$

Moreover, $\text{OPT}(\sigma) \geq Q/2$.

Following the proof of the deterministic case, we now have that every connection slot that at some point in phase i contains a connection to a page in group 1, 2 or 3 (note that this may change later in the phase due to the use of MRU), is in CURRENT at the end of the phase. Therefore any connections that are still in PREVIOUS at that time (which get evicted and form holes) must be to pages not requested in the phase. Therefore $\text{OPT}(\sigma) \geq H/2$.

Suppose $k' \geq k$. Due to the randomization, we do not know whether or not Case 2.(c) occurs in a phase. However, as observed in the proof of the deterministic algorithm, we can amortize the offline faults if 2.(c) occurs to get the bound $\text{OPT}(\sigma) \geq M/2$. Therefore analogously to in the proof of Theorem 1, we have

$$\mathcal{R}(\text{ALG}) \leq 2H_k + 4.$$

We now consider the case $k' < k$. The only change is that the bound $\text{OPT}(\sigma) \geq M/2$ is replaced by

$$\text{OPT}(\sigma) \geq \frac{M - (k - k')N}{2} \geq \frac{M - (k - k')Q}{2},$$

where we have used $Q \geq N$, which follows from the fact that there must be at least one new page in every new phase by definition of the phases. This gives us

$$\mathcal{R}(\text{ALG}) \leq \frac{H_k Q + H + M}{\text{OPT}(\sigma)} \leq 2H_k + 4 + 2(k - k').$$

However, since the number of connection faults, $H + M$, is also upper bounded by the number of page faults $H_k Q$, we find

$$\mathcal{R}(\text{ALG}) \leq 2H_k + \min(2H_k, 4 + 2(k - k')).$$

$\square$

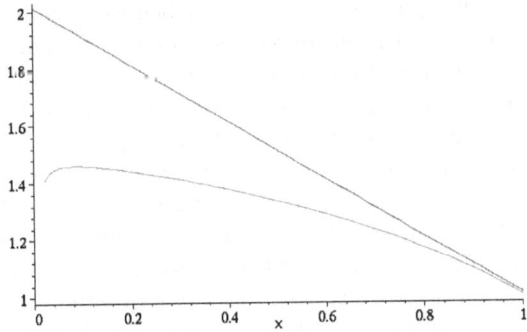

Fig. 1. The upper and lower bound: x-axis is k'/k, y-axis is $\mathcal{R}/k$

3   Lower Bounds

We present a lower bound on the performance of any deterministic online algorithm. The lower bound of Theorem 4 implies that if k' is not too small, our deterministic algorithm given in the last section is nearly optimal. Figure 1 depicts the lower as well as the upper bound.

Theorem 4. *Suppose $k' \geq 2$ and let $\alpha = k'/k$. Then for any online algorithm $\mathcal{A}$, we have*

$$\mathcal{R}(\mathcal{A}) \geq (k+1)\left(\frac{\alpha k - 1}{\alpha k} + \frac{1 - \alpha}{2 - \alpha + 3/k}\right).$$

Proof. We construct a lower bound as follows. We make use of $k+1$ pages that are stored at $k+1$ distinct nodes. Consider an online algorithm $\mathcal{A}$. Each page request in the sequence is to the (unique) page that $\mathcal{A}$ does not have in its cache. The sequence is divided into phases. In each phase, we count the number of distinct pages that have been requested in that phase; the first request to the $k+1$st distinct page is defined to be the start of the next phase. Since the connection cache has size k', $\mathcal{A}$ must have at least $k - k'$ connection faults in each phase. We define $\alpha = k'/k$, so that $k' = \alpha k$. We will write the average length of a phase as pk, where $p \geq 1$. The offline algorithm uses one of the following strategies depending on p.

Strategy 1. (For large p.) The first strategy is to always use LFD for the requested pages. We then count the number of offline page faults for each of the $k+1$ pages, and put $k' - 1$ connections to pages on which the most offline faults occur, in the connection cache. This part of the connection cache is fixed during the entire processing of the request sequence. The last slot is used for connection faults on the remaining $k + 1 - (k' - 1) = k - k' + 2$ pages.

Consider $k+1$ phases. There are at most $k+1$ offline faults, and on average at most $k - k' + 2$ of them are on pages of which the connections are not in the connection cache at all times. Thus there are on average at most $2k - k' + 3$ offline faults on $k+1$ phases.

Strategy 2. (For small p.) The second strategy begins by counting the number of requests to each page over the entire request sequence. Then, the $k - k' + 1$ pages that are requested the most often, are put in the page cache at the beginning, and the k' connections to the remaining pages are put in the connection cache. The entire connection cache is fixed throughout the sequence. The offline algorithm now uses LFD on the k' pages for which the connections are in the connection cache, and only uses the $k' - 1$ slots in the page cache that do not contain the $k - k' + 1$ most often requested pages. It has no connection faults at all.

Consider $(k + 1)(k' - 1)$ phases. These contain on average $(k + 1)(k' - 1)pk$ requests by definition of p. Thus, each page is requested on average $(k' - 1)pk$ times. The k' pages that are requested the least overall, must then be requested at most $k'(k' - 1)pk$ times on average at most. Since the offline algorithm has at most one fault every $k' - 1$ requests to this subset of pages, there are $k'pk$ offline faults.

Solving for p, we find that these two strategies have the same number of faults if

$$p = \frac{\alpha k - 1}{\alpha k}\left(2 - \alpha + \frac{3}{k}\right). \tag{6}$$

As long as this value is at least 1, we can use the first offline strategy if p is greater than the threshold, and the second strategy otherwise. The number of on-line faults in one phase must be at least $pk + (k - k')$ on average. This implies a competitive ratio of at least

$$\frac{(pk + k - k')(k + 1)(\alpha k - 1)}{k'pk} = (k + 1)\left(\frac{\alpha k - 1}{\alpha k} + \frac{1 - \alpha}{2 - \alpha + 3/k}\right).$$

Note that the threshold in (6) is greater than 1 for $k \geq k' \geq 2$.   □

We can show that the analysis of our algorithm ALG is asymptotically tight for $k' = 1$. Note that ALG behaves exactly like LRU in this case. This implies that even for $k' = 1$ it is nontrivial to find an algorithm with competitive ratio close to k.

Lemma 1. *For $k' = 1$, we have $\mathcal{R}(\mathrm{ALG}) \geq 2k - 2$.*

Proof. We use a set of pages numbered $1, 2, \ldots, k+1$ and request them cyclically. All the odd pages are at some node v_1 while the even pages are at another node v_2. It can be seen that our algorithm has a connection fault on every request, thus it has $2k$ faults per phase.

We now describe an off-line algorithm to serve this sequence. This algorithm only faults on pages in v_1, and each time evicts the page from that node that will be requested the furthest in the future. All pages in v_2 are in the cache at all times. Suppose k is even, then there are $k/2$ slots available in the cache for $k/2 + 1$ pages. Thus this off-line algorithm has a fault once every $k/2$ requests to pages in v_1.

Consider $k+1$ phases. It contains $k(k+1)$ requests, exactly k per page. Thus there are $2(k/2+1) = k+2$ offline faults in total, giving a competitive ratio of

$$\frac{2k(k+1)}{k+2} = 2k - \frac{2k}{k+2} \geq 2k - 2.$$

For odd k, there is one off-line fault per $(k-1)/2$ requests to pages in v_1. In $k-1$ phases there are $k(k-1)$ requests, thus $k(k-1)/2$ requests to pages in v_1 and in total k offline faults. This gives a ratio of exactly $2k - 2$. □

4   Generalized Models

In this section we study generalized problem settings in which the documents can have different sizes. For the standard multi-sized paging problem, the algorithm LRU is $(k+1)$-competitive in both the BIT and the FAULT model [6]. Here k is defined as the maximum number of pages that can fit in the cache, i.e. $k = K/s$ where K is the size of the cache (in bits) and s is the size of the smallest possible page. It is nontrivial to extend the analysis of our algorithm to these models.

In both models, a *phase* is now defined as a maximal subsequence of requests to a minimal volume of distinct pages that is larger than K. Thus there are at most $k+1$ page faults in a phase.

4.1   The Fault Model

For the FAULT model, we need to consider the number of pages requested in each phase, which can be less than k.

Theorem 5. *In the* FAULT *model,* $\mathcal{R}(\text{ALG}) \leq (4k+14)/3$ *for* $k' \geq k$ *and* $\mathcal{R}(\text{ALG}) \leq 2k - \frac{2}{3}k' + \frac{14}{3}$ *for* $k' < k$.

Proof. Suppose $k' = k$. Denote the number of pages requested in phase i by Φ_i. Write $\Delta_i = \Phi_i - \Phi_{i-1}$.

If there are m_i connection faults where MRU is applied, then m_i times a connection slot remains in CURRENT. Thus at most $k+1-m_i$ times a connection slot moves from PREVIOUS to CURRENT, and at least $m_i - 1$ connection slots are still in PREVIOUS at the end of the phase. These connections lead to at least $m_i - 1$ pages that were requested in phase $i-1$ but not in phase i.

Denote the set of pages requested in phase $i-1$ but not in phase i by F. Denote the set of pages requested in phase i by S. We partition F in two sets: F_1 contains the pages that OPT faults on, F_2 contains the rest. Consider the set F_2. OPT does not fault on these pages and thus has them in its cache at the start of phase $i-1$. This means that some pages in S are not yet in its cache and need to be loaded later.

Write the number of OPT faults in these two phases as $m_i - 1 - x$. If $x \leq 0$, we are done. Otherwise, F_2 contains $z \geq x > 0$ pages. OPT has exactly $z - x$ faults on the set S, that is, $z - x$ pages are loaded to come "in the place of" the z pages in F_2 (OPT does not necessarily replace exactly these pages in the cache). Since at most $k+1$ pages were requested in phase $i-1$, the set S then contains at most $k+1-x$ pages, i.e. our algorithm has at most $k+1-x$ page faults in phase $i+1$.

That is, if OPT has x faults less than $m_i - 1$ in phases $i-1$ and i, then our algorithm has (at least) x faults less than $k+1$ in phase i. Writing the number of OPT faults as $m_i - 1 - x_i$ in all cases where it is less than $m_i - 1$, this gives

$$\text{OPT}(\sigma) \geq \frac{M - N - X}{2},$$

where $X = \sum x_i$. (That is, all values x_i are positive.)

We can treat the holes that are created in the same way to find

$$\text{OPT}(\sigma) \geq \frac{H - P - X}{2}.$$

Finally we also still have $\text{OPT} \geq N$. We have

$$\text{ALG}(\sigma) \leq (k+1)N - X + M + H - P \quad \text{and} \quad \text{ALG} \leq 2((k+1)N - X),$$

where the second inequality follows since ALG has at most one connection fault for each page fault.

Thus if $X \geq \frac{kN-4}{3}$, we find that the competitive ratio is at most $4k/3 + 14/3$. On the other hand, if $X < \frac{kN-4}{3}$, then

$$\text{ALG}(\sigma) \leq (k+1)\text{OPT}(\sigma) + 4\text{OPT}(\sigma) + X \leq (k+5+k/3-4/3)\text{OPT}(\sigma) = \frac{4k+14}{3}\text{OPT}(\sigma).$$

This analysis can easily be extended to the case $k' < k$ as before, giving $\mathcal{R}(\text{ALG}) \leq 2k - \frac{2}{3}k' + \frac{14}{3}$. Details are omitted in this extended abstract. □

4.2   The Bit Model

In this section we investigate a BIT model in which the cost of loading a document is equal to the size of the document. We also assume that the cost of establishing a connection is equal to c, for some constant $c > 0$.

Theorem 6. *In the BIT model, $\mathcal{R}(\text{ALG}) \leq \frac{k+5}{2}(c'+1)$ for $k' \geq k$, where $c' = c/s$ is the cost of a connection fault divided by the size of the smallest possible page. For $k' < k$, $\mathcal{R}(\text{ALG}) \leq \frac{2k+5-k'}{2}(c'+1)$.*

Proof. Denote the average phase length by $K + \delta$ for some $\delta > 0$. Denote the average number of MRU faults in a phase by m and the average number of bits worth of old pages that are implied by m', then $m' \geq ms$. Denote the average number of pages on which there is no fault in a phase by p and the average number of bits that are requested without fault by p', then $p' \geq ps$. Finally, denote the average number of holes created in a phase by h. Denote the cost of a single connection fault by c and write $c' = c/s$. Similarly to in the previous section, it can be seen that for the average cost in a phase we have $\text{ALG}/s \leq k + \delta/s + (m+h)c/s - p'/s$ and $\text{OPT}/s \geq \max(\max(1, \delta/s), m/2, h - p/2)$. Here the first maximum in the second equation follows from $\sum_i \max(\delta_i, s)/Ns \geq \max(Ns, N\delta)/Ns = \max(1, \delta/s)$, where $K + \delta_i$ is the amount of bits from distinct requests requested in phase i.

Since the number of connection faults in a phase is bounded from above by the number of page faults, we have

$$m + h \leq \frac{K + \delta - p'}{s} \Rightarrow h \leq k + \frac{\delta}{s} - p - m \leq (k+1)\frac{\text{OPT}}{s} - p - m. \quad (7)$$

We also have $h \leq 2\frac{\text{OPT}}{s} + p$. Note that $2\frac{\text{OPT}}{s} + p = (k+1)\frac{\text{OPT}}{s} - p - m \Rightarrow 2p = (k-1)\text{OPT}/s - m$. Suppose $p \leq ((k-1)\text{OPT}/s - m)/2$. (The other case is handled similarly.) Then

$$\frac{\text{ALG}}{s} \leq (k+1)\frac{\text{OPT}}{s} + mc' + hc' - p \leq (k+1)\frac{\text{OPT}}{s} + mc' + 2\frac{\text{OPT}}{s}c' + p(c'-1)$$

$$\leq (k+1)\frac{\text{OPT}}{s} + mc' + 2\frac{\text{OPT}}{s}c' + (c'-1)((k-1)\frac{\text{OPT}}{s} - m)/2$$

$$\leq (k+1)\frac{\text{OPT}}{s} + (c'+1)m/2 + 2\frac{\text{OPT}}{s}c' + (c'-1)(k-1)\frac{\text{OPT}}{2s}$$

$$\leq (k + c' + 2 + 2c' + \frac{(c'-1)(k-1)}{2})\frac{\text{OPT}}{s} = \frac{(k+5)(c'+1)}{2} \cdot \frac{\text{OPT}}{s} \ .$$

For $k' < k$, we have $\text{OPT}(\sigma)/s \geq (m - (k - k'))/2$ and $\mathcal{R}(\text{ALG}) \leq \frac{(2k - k' + 5)(c' + 1)}{2}$; details are omitted in this extended abstract.   □

Hence the competitive ratio grows linearly with k and with c (c'). The reason for this is that we cannot identify connection faults by OPT; it is conceivable that OPT never has a connection fault.

5   The Distributed Setting

We finally study the distributed problem setting where requests can occur at various network nodes. Again, each node has a document cache and a connection cache. Here, a request is specified by a pair (v, d), indicating that document d is requested by the user at node v. The cost of serving requests is the same as before. The crucial difference is in the usage of connections. An open connection between nodes v and v' can be used for downloading documents from v to v' as well as from v' to v. However, if one of the nodes of the connection decides to close the connection, then the connection cannot be used by the other node either. Hence, the connection cache configurations affect each other.

Theorem 7. *In the distributed problem setting, no deterministic online algorithm can achieve a competitive ratio smaller than $2k/(1 + 1/k')$, where k is the size of the largest document cache and k' is the maximum number of connections that a network node can keep open.*

Proof. Consider a node v at which $k + 1$ documents are stored. Additionally we have $k' + 1$ nodes $v_i, 1 \leq i \leq k' + 1$, Each node in the network has a document cache of size k and a connection cache of size k'. Requests are generated as follows. At any time one of the connections (v, v_i) is closed in the configuration of an online algorithm A because v kan only maintain k' open connections and a connection is open only if it is cached by both of its endpoints. An adversary generates a request at this node v_i for the document that is currently not stored in A's document cache at v_i. Suppose that a request sequence consists of m requests and that m_i requests were generated at $v_i, 1 \leq i \leq k'+1$. The online cost is equal to $2m$. An optimal offline algorithm has at most $\lceil \frac{m_i}{k} \rceil$ document faults at v_i and hence no more than $\frac{m}{k} + k' + 1$ document faults in total. Furthermore an optimal algorithm can maintain the connection cache at v in such a way that at most $\lceil (\frac{m}{k} + k' + 1)/k' \rceil$ connection faults occur. Thus as $m \to \infty$, the ratio of the online to offline cost tends to $2/(\frac{1}{k} + \frac{1}{kk'}) = 2k(1 + 1/k')$. $\square$

Note that a competitive ratio of $2k$ is achieved by any caching algorithm that uses a k-competitive paging strategy for the document cache any replacement rule for the connection cache.

6   Conclusions

In this paper we studied integrated document and connection caching in a variety of problem settings. An open question left by our work is to find a better algorithm for the case where the connection cache is very small (relative to k). We conjecture that the true competitive ratio for this problem should be close to k.

References

1. D. Achlioptas, M. Chrobak, and J. Noga. Competitive analysis of randomized paging algorithms. *Theoretical Computer Science*, 234:203–218, 2000.
2. S. Albers. Generalized connection caching. In *Proceedings of the Twelfth ACM Symposium on Parallel Algorithms and Architectures*, pages 70–78. ACM, 2000.
3. E. Cohen, H. Kaplan, and U. Zwick. Connection caching. In *Proceedings of the 31st ACM Symposium on the Theory of Computing*, pages 612–621. ACM, 1999.
4. E. Cohen, H. Kaplan, and U. Zwick. Connection caching under various models of communication. In *Proceedings of the Twelfth ACM Symposium on Parallel Algorithms and Architectures*, pages 54–63. ACM, 2000.
5. T. Feder, R. Motwani, R. Panigraphy, and A. Zhu. Web caching with request reordering. In *Proceedings 13th ACM-SIAM Symposium on Discrete Algorithms*, pages 104–105, 2002.
6. A. Feldman, R. Karp, M. Luby, and L. A. McGeoch. Personal communication cited in [9].
7. A. Fiat, R.M. Karp, M. Luby, L.A. McGeoch, D.D. Sleator, and N.E. Young. Competitive paging algorithms. *Journal of Algorithms*, 12(4):685–699, Dec 1991.

8. P. Gopalan, H. Karloff, A. Mehta, M. Mihail, and N. Vishnoi. Caching with expiration times. In *Proceedings 13th ACM-SIAM Symposium on Discrete Algorithms*, pages 540–547, 2002.

9. S. Irani. Page replacement with multi-size pages and applications to web caching. In *Proceedings 29th ACM Symposium on Theory of Computing*, pages 701–710, 1997.

10. L. McGeoch and D. Sleator. A strongly competitive randomized paging algorithm. *J. Algorithms*, 6:816–825, 1991.

11. D. Sleator and R. E. Tarjan. Amortized efficiency of list update and paging rules. *Communications of the ACM*, 28:202–208, 1985.

12. N. Young. On-line file caching. In *Proceedings 9th ACM-SIAM Symposium on Discrete Algorithms*, pages 82–86, 1998.

A Solvable Class of Quadratic Diophantine Equations with Applications to Verification of Infinite-State Systems

Gaoyan Xie[1], Zhe Dang[1*], and Oscar H. Ibarra[2**]

[1] School of Electrical Engineering and Computer Science
Washington State University
Pullman, WA 99164, USA
[2] Department of Computer Science
University of California
Santa Barbara, CA 93106, USA

Abstract. A *k-system* consists of k quadratic Diophantine equations over nonnegative integer variables $s_1, ..., s_m, t_1, ..., t_n$ of the form:

$$\sum_{1 \le j \le l} B_{1j}(t_1, ..., t_n) A_{1j}(s_1, ..., s_m) = C_1(s_1, ..., s_m)$$

$$\vdots$$

$$\sum_{1 \le j \le l} B_{kj}(t_1, ..., t_n) A_{kj}(s_1, ..., s_m) = C_k(s_1, ..., s_m)$$

where l, n, m are positive integers, the B's are nonnegative linear polynomials over $t_1, ..., t_n$ (i.e., they are of the form $b_0 + b_1 t_1 + ... + b_n t_n$, where each b_i is a nonnegative integer), and the A's and C's are nonnegative linear polynomials over $s_1, ..., s_m$. We show that it is decidable to determine, given any 2-system, whether it has a solution in $s_1, ..., s_m, t_1, ..., t_n$, and give applications of this result to some interesting problems in verification of infinite-state systems. The general problem is undecidable; in fact, there is a fixed $k > 2$ for which the k-system problem is undecidable. However, certain special cases are decidable and these, too, have applications to verification.

1 Introduction

During the past decade, there has been significant progress in automated verification techniques for finite-state systems. One such technique is model-checking [5,19] that explores the state space of a finite-state system and checks that a desired temporal property is satisfied. Model-checkers like SMV [13] and SPIN [10] have been successful in many industrial-level applications. The successes have greatly inspired researchers to develop automatic techniques for analyzing infinite-state systems (such as systems that contain integer variables and parameters). However, in general, it is not possible to develop such techniques,

---

* Corresponding author (zdang@eecs.wsu.edu).
** The research of Oscar H. Ibarra has been supported in part by NSF Grants IIS-0101134 and CCR02-08595.

J.C.M. Baeten et al. (Eds.): ICALP 2003, LNCS 2719, pp. 668–680, 2003.

e.g., it is not possible to (automatically) verify whether an arithmetic program with two integer variables is going to halt [17]. Therefore, an important aspect of the research on infinite-state system verification is to identify what kinds of practically useful infinite-state models are decidable with respect to a particular form of properties (e.g., reachability).

In this paper, we look at a class of infinite-state systems that contain parameterized or unspecified constants. For instance, consider a nondeterministic finite state machine M. Each transition in M is assigned a label. On firing the transition $s \rightarrow^a s'$ from state s to state s' with label a, an *activity* a is performed. There are finitely many labels $a_1, \ldots, a_l$ in M. M can be used to model, among others, a finite state process where an execution of the process corresponds to an execution path (e.g., $s_0 \rightarrow^{a^0} s_1 \rightarrow^{a^1} \ldots \rightarrow^{a^r} s_{r+1}$, for some r) in M. On the path, a sequence of activities $a^0 \ldots a^r$ are performed. Let $\Sigma_1, \ldots, \Sigma_k$ be any k sets (not necessarily disjoint) of labels. An activity a is of type i if $a \in \Sigma_i$. An activity could have multiple types. Additionally, activities $a_1, \ldots, a_l$ are associated with *weights* $w_1, \ldots, w_l$ that are unspecified (or parameterized) constants in $\mathbf{N}$, respectively. Depending on the various application domains, the weight of an activity can be interpreted as, e.g., the time in seconds, the bytes of memory, or the budget in dollars, etc., needed to complete the activity. A type of activities is useful to model a "cluster" of activities. When executing M, we use nonnegative integer variables W_i to denote the accumulated weight of all the activities of type i performed so far, $1 \leq i \leq k$. One verification question concerns reachability:

(*) whether, for some values of the parameterized constants $w_1, \ldots, w_l$, there is an execution path from a given state to another on which $w_1, \ldots, w_l, W_1, \ldots, W_k$ satisfy a given Presburger formula P (a Boolean combination of linear constraints and congruences).

One can easily find applications for the verification problem. For instance, consider a packet-based network switch that uses a scheduling discipline to decide the order in which the packets from different incoming connections $c_1, \ldots, c_l$ are serviced (visited). Suppose that each connection c_i is assigned a weight w_i, $1 \leq i \leq l$, and each time when a connection is serviced (visited), the number of packets serviced from that connection is in proportion to its weight. But in this switch we have two outgoing connections (two servers in the "queue theory" jargon) o_1 and o_2 each of which serves a set of incoming connections C_1 and C_2 respectively ($C_1 \cup C_2 = \{c_1, \ldots, c_l\}$). The scheduling discipline for this switch can be modeled as a finite state system. If we take the event that an incoming connection is serviced by a specific server as an activity, then the weight of the activity could be the number of packets served that is in proportion to the weight of the incoming connection. Thus W_1 and W_2 could be used to denote the total amount (accumulated weights) of packets served by the two servers respectively. Later in the paper, we shall see how to model a fairness property using (*).

In this paper, we study the verification problem in (*) and its variants. First, we show that the problem is undecidable, in general. Then, we consider various restricted as well as modified cases in which the problem becomes decidable.

For instance, if P in (*) has only one linear constraint that contains some of $W_1, \ldots, W_k$, then the problem is decidable. Also, rather surprisingly, if in the problem in (*) we assume that the weight of each activity a_i can be nondeterministically chosen as any value between a concrete constant (such as 5) and a parameterized constant w_i, then it becomes decidable. We also consider cases when the transition system is augmented with other unbounded data structures, such as a pushdown stack, dense clocks, and other restricted counters.

In the heart of our decidability proofs, we first show that some special classes of systems of quadratic Diophantine equations/inequalities are decidable (though in general, these systems are undecidable [16]). This nonlinear Diophantine approach towards verification problems is significantly different from many existing techniques for analyzing infinite-state systems (e.g., automata-theoretic techniques in [14,3,7] , computing closures for Presburger transition systems [6, 4], etc.). Then, we study a more general version of the verification problem by considering weighted semilinear languages in which a symbol is associated with a weight. Using the decidability results on the restricted classes of quadratic Diophantine systems, we show that various verification problems concerning weighted semilinear languages are decidable. Finally, as applications, we "reinterpret" the decidability results for weighted semilinear languages into the results for some classes of machine models, whose behaviors (e.g., languages accepted, reachability sets, etc) are known to be semilinear, augmented with weighted activities.

Adding weighted activities to a transition system can be found, for instance, in [15]. In that paper, a "price" is associated with a control state in a timed automaton [2]. The price may be very complex; e.g., linear in other clock values etc. In general, the reachability problem for priced timed automata is undecidable [15]. Here, we are mainly interested in the decidable cases of the problem: what kind of "prices" (i.e., weights) can be placed such that some verification queries are still decidable, for transition systems like pushdown automata, restricted counter machines, etc., in addition to timed automata.

The paper is organized as follows. In the next section, we present the decidability results for the satisfiability problem of two special classes of quadratic Diophantine systems (Lemma 2 and Theorem 1). Then in Section 3, we generalize the verification problem in (*) in terms of weighted semilinear languages, and reduce the problem and its restricted versions to the classes of quadratic Diophantine systems studied in Section 2. In Section 4, we discuss the application aspects and extensions of the decidability results to other machine models. Due to space limitation, some of the proofs are omitted in the paper. The full version of the paper is accessible at www.eecs.wsu.edu/~zdang.

2 Preliminaries

Let $\mathbf{N}$ be the set of nonnegative integers and let $x_1, \ldots, x_n$ be n variables over $\mathbf{N}$. A *linear constraint* is defined as $a_1 x_1 + \ldots + a_n x_n > b$, where $a_1, \ldots, a_n$ and b are integers. A *congruence* is $x_i \equiv_b c$, where $1 \leq i \leq n$, and $b \neq 0, 0 \leq$

$c < b$. A Presburger formula is a Boolean combination of linear constraints and congruences using $\vee$ and $\neg$. Notice that, here, Presburger formulas are defined over nonnegative integer variables (instead of integer variables). It is well known that Presburger formulas are closed under quantifications ($\forall$ and $\exists$).

A subset S of $\mathbf{N}^n$ is a *linear set* if there exist vectors $v_0, v_1, \ldots, v_t$ in $\mathbf{N}^n$ such that $S = \{v|v = v_0 + b_1 v_1 + \ldots + b_t v_t, b_i \in \mathbf{N}\}$. The set S is a *semilinear set* if it is a finite union of linear sets. It is well known that S is semilinear iff S is Presburger definable (i.e., there is a Presburger formula P such that $P(v)$ iff $v \in S$).

A *linear polynomial* is a polynomial of the form $a_0 + a_1 x_1 + \ldots + a_n x_n$ where each coefficient a_i, $0 \le i \le n$, is an integer. The polynomial is *constant* if each $a_i = 0$, $1 \le i \le n$. The polynomial is *nonnegative* if each a_i, $0 \le i \le n$, is in $\mathbf{N}$. The polynomial is *positive* if it is nonnegative and $a_0 > 0$. A variable *appears* in a linear polynomial iff its coefficient in that polynomial is nonzero. The following result is needed in the paper.

Lemma 1. *It is decidable whether an equation of the following form has a solution in nonnegative integer variables $s_1, \ldots, s_m, t_1, \ldots, t_n$:*

$$L_0 + L_1 t_1 + \ldots + L_n t_n = 0 \qquad (1)$$

where $L_0, L_1, \ldots, L_n$ are linear polynomials over $s_1, \ldots, s_m$. The decidability remains even when the solution is restricted to satisfy a given Presburger formula P over $s_1, \ldots, s_m$.

Proof. The first part of the lemma has already been proved in [8], while the second part is shown below using a "semilinear transform". As we mentioned earlier, the set of all $(s_1, \ldots, s_m) \in \mathbf{N}^m$ satisfying P is a semilinear set (i.e., a finite union of linear sets). For each linear set of P, one can find nonnegative integer variables $u_1, \ldots, u_k$ for some k and a nonnegative linear polynomial $p_i(u_1, \ldots, u_k)$ for each $1 \le i \le m$ such that $(s_1, \ldots, s_m)$ is in the linear set iff each $s_i = p_i(u_1, \ldots, u_k)$, for some $u_1, \ldots, u_k$. The second part follows from the first part by substituting $p_i(u_1, \ldots, u_k)$ for s_i in $L_0, L_1, \ldots, L_n$. ∎

Let I, J and K be three pairwise disjoint subsets of $\{1, \ldots, n\}$. An *n-inequality* is an inequality over n nonnegative integer variables $t_1, \ldots, t_n$ and m (for some m) nonnegative integer variables $s_1, \ldots, s_m$ of the following form:

$$D_1 + a(\sum_{i \in I} L_{1i} t_i + \sum_{j \in J} L_{1j} t_j) \le D_2 + \sum_{i \in I} L_{2i} t_i + \sum_{k \in K} L_{2k} t_k$$

$$\le D_1' + a'(\sum_{i \in I} L_{1i} t_i + \sum_{j \in J} L_{1j} t_j), \qquad (2)$$

where $a < a' \in \mathbf{N}$, the D's (resp. the L's) are nonnegative (resp. positive) linear polynomials over $s_1, \ldots, s_m$, and $D_1 \le D_1'$ is always true (i.e., true for all $s_1, \ldots, s_m \in \mathbf{N}$).

Lemma 2. *For any n, it is decidable whether an n-inequality in (2) has a solution in nonnegative integer variables $s_1, \ldots, s_m, t_1, \ldots, t_n$. The decidability remains even when the solution is restricted to satisfy a given Presburger formula P over $s_1, \ldots, s_m$.*

Theorem 1. *It is decidable whether a system in the following form has a solution in nonnegative integer variables $s_1, \ldots, s_m$, $t_1, \ldots, t_n$: $P(D_1 + \sum_{1 \leq i \leq n} L_{1i} t_i, D_2 + \sum_{1 \leq i \leq n} L_{2i} t_i)$, where P is a Presburger formula over two nonnegative integer variables and the D's and the L's are nonnegative linear polynomials over $s_1, \ldots, s_m$.*

3   Semilinear Languages with Weights

We first recall the definition of semilinear languages. Let $\Sigma = \{a_1, \ldots, a_l\}$ be an alphabet. For each word α in Σ^*, the *Parikh* map of α is defined to be $\phi(\alpha) = (\phi_{a_1}(\alpha), \ldots, \phi_{a_l}(\alpha))$ where $\phi_{a_i}(\alpha)$ denotes the number of occurrences of symbol a_i in word α, $1 \leq i \leq l$. For a language $L \in \Sigma^*$, the Parikh map of L is $\phi(L) = \{\phi(\alpha) | \alpha \in L\}$. The language L is semilinear iff $\phi(L)$ is a semilinear set. L is effectively semilinear if the semilinear set $\phi(L)$ can be computed from the description of L.

Now, we add "weights" to a language L. A *weight measure* is a mapping that maps a symbol in Σ to a weight in $\mathbf{N}$. We shall use $w_1, \ldots, w_l$ to denote the weights for $a_1, \ldots, a_l$, respectively, under the measure. Let $\Sigma_1, \ldots, \Sigma_k$ be any k fixed subsets of Σ. For each $1 \leq i \leq k$, we use $W_i(\alpha)$ to denote the total weight of all the occurrences for symbols $a \in \Sigma_i$ in word α; i.e.,

$$W_i(\alpha) = \sum_{a_j \in \Sigma_i} w_j \cdot \phi_{a_j}(\alpha). \tag{3}$$

$W_i(\alpha)$ is called the *accumulated weight* of α wrt Σ_i. We are interested in the following *k-accumulated weight problem*:

- **Given:** An effectively semilinear language L, k subsets $\Sigma_1, \ldots, \Sigma_k$ of Σ, and a Presburger formula P over $l + k$ variables.
- **Question:** Is there a word α in L such that, for some $w_1, \ldots, w_l \in \mathbf{N}$,

$$P(w_1, \ldots, w_l, W_1(\alpha), \ldots, W_k(\alpha)) \tag{4}$$

holds?

In a later section, we shall look at the application side of the problem. The rest of this section investigates the decidability issues of the problem by transforming the problem and its restricted versions to a class of Diophantine equations.

A *k-system* is a quadratic Diophantine equation system that consists of k equations over nonnegative integer variables $s_1, \ldots, s_m, t_1, \ldots, t_n$ (for some m, n) in the following form:

$$\begin{cases} \sum_{1 \le j \le l} B_{1j}(t_1, ..., t_n) A_{1j}(s_1, ..., s_m) = C_1(s_1, ..., s_m) \\ \qquad \vdots \\ \sum_{1 \le j \le l} B_{kj}(t_1, ..., t_n) A_{kj}(s_1, ..., s_m) = C_k(s_1, ..., s_m) \end{cases} \quad (5)$$

where the A's, B's and C's are nonnegative linear polynomials, and l, n, m are positive integers.

Theorem 2. *For each k, the k-accumulated weight problem is decidable iff it is decidable whether a k-system has a solution.*

It is known [12] that there is a fixed k such that there is no algorithm to solve Diophantine systems in the following form: $t_1 F_1 = G_1$, $t_1 H_1 = I_1$, ..., $t_k F_k = G_k$, $t_k H_k = I_k$, where the F's, G's, H's, I's are nonnegative linear polynomials over nonnegative integer variables $s_1, ..., s_m$, for some m. Observe that the above systems are $2k$-systems. Therefore, from Theorem 2,

Theorem 3. *There is a fixed k such that the k-accumulated weight problem is undecidable.*

Currently, it is an open problem to find the maximal k such that the k-accumulated weight problem is decidable. Clearly, when $k = 1$, the problem is decidable. This is because 1-systems are decidable (Lemma 1). Below, using Theorem 1, we show that the problem is decidable when $k = 2$. Interestingly, it is still open whether the decidability remains for $k = 3$.

Theorem 4. *The 2-accumulated weight problem is decidable.*

In some restricted cases, the accumulated weight problem is decidable for a general k. We are now going to elaborate these cases. Consider a k-accumulated weight problem such that (4) is a disjunction of formulas in the following special form:

$$Q(w_1, ..., w_l) \ \wedge \ a_1 W_1(\alpha) + ... + a_k W_k(\alpha) + b_1 w_1 + ... + b_l w_l \sim a_0 \quad (6)$$

where Q is a Presburger formula over l variables, the a's and b's are integers, and $\sim \in \{=, \ne, >, <, \ge, \le\}$. Under this restriction, the k-accumulated weight problem is decidable.

Theorem 5. *For each k, the k-accumulated weight problem, in which (4) is a disjunction of formulas in the form of (6), is decidable.*

Currently we do not know whether Theorem 5 still holds if (6) is conjuncted with one additional inequality: $a_1' W_1(\alpha) + ... + a_k' W_k(\alpha) + b_1' w_1 + ... + b_l' w_l \sim a_0'$.

As in the statement of the problem at the beginning of this section, a weight measure assigns numbers $w_1, ..., w_l$ to symbols $a_1, ..., a_l$ respectively. Instead of a fixed one, suppose that the weight of a symbol a_i can take any value between a given number q_i and w_i. That is, the weight measure defines a possible weight range that a symbol can have, with the given number q_i being the lowest possible weight. Thus, in contrast to (3), $W_i(\alpha)$, $1 \leq i \leq l$, will be a set:

$$\{\hat{W}_i : \sum_{a_j \in \Sigma_i} q_j \cdot \phi_{a_j}(\alpha) \leq \hat{W}_i \leq \sum_{a_j \in \Sigma_i} w_j \cdot \phi_{a_j}(\alpha)\}. \qquad (7)$$

For instance, suppose $\Sigma_1 = \{a_1\}$, $q_1 = 2$, $w_1 = 7$, and a word $\alpha = a_1 a_1 a_1$. Clearly, 12 is a weight in $W_1(\alpha)$ according to (7).

With the new definition of $W_i(\alpha)$, the following *loose k-accumulated weight problem* can be formulated:

- **Given:** An effectively semilinear language L, numbers $q_1, \ldots, q_l \in \mathbf{N}$, k subsets $\Sigma_1, \ldots, \Sigma_k$ of Σ, and a Presburger formula P over $l + k$ variables.
- **Question:** Is there a word α in L such that, for some $w_1, \ldots, w_l \in \mathbf{N}$, and for some $\hat{W}_1, \ldots, \hat{W}_k$,

$$\hat{W}_1 \in W_1(\alpha) \wedge \ldots \wedge \hat{W}_k \in W_k(\alpha) \wedge P(w_1, \ldots, w_l, \hat{W}_1, \ldots, \hat{W}_k) \qquad (8)$$

holds?

Notice that the lower weight bounds $q_1, \ldots, q_l$ are in the **Given**-part, hence they are constants; while the upper bounds $w_1, \ldots, w_l$ in the **Question**-part, are essentially unspecified parameters. (Otherwise, if the lower bounds $q_1, \ldots, q_l$ are moved into the **Question**-part; i.e., both the lower and the upper bounds are parameterized constants, then the k-accumulated weight problem is a special case of the loose k-accumulated weight problem under this definition, by letting the lower bound and the upper bound be the same parameterized constant for each activity.)

The following result shows that the loose k-accumulated weight problem is decidable for each k. It is in contrast to Theorem 3 that the k-accumulated weight problem is undecidable for some large k.

Theorem 6. *For each k, the loose k-accumulated weight problem is decidable.*

4   Applications

In this section, we will apply the results presented in the previous section to some verification problems concerning infinite systems containing parameterized constants. We start with a general definition.

A transition system M can be described as a relation $T \subseteq S \times \Gamma^* \times \Sigma \times S \times \Gamma^*$, where S is a finite set of states, Γ is the configuration alphabet, and Σ is the activity alphabet. Obviously, we always assume that M can be effectively described; i.e., T is recursive. A configuration $\langle s, \beta \rangle$ of M is a pair of a state s in S and a word β in Γ^*. In the description of M, an initial configuration is also designated. According to the definition of T, an activity in Σ transforms one configuration to another. More precisely, we write $\langle s, \beta \rangle \xrightarrow{a} \langle s', \beta' \rangle$ if $T(s, \beta, a, s', \beta')$.

Let $\alpha \in \Sigma^*$ with $\alpha = a^1 \ldots a^m$ for some m. We say that $\langle s, \beta, \alpha \rangle$ is *reachable* if, for some configurations $\langle s_0, \beta_0 \rangle, \ldots, \langle s_m, \beta_m \rangle$, the following is satisfied

$$\langle s_0, \beta_0 \rangle \overset{u^1}{\rightarrow} \ldots \overset{u^m}{\rightarrow} \langle s_m, \beta_m \rangle, \tag{9}$$

where $\langle s_0, \beta_0 \rangle$ is the initial configuration, $s_m = s$ and $\beta_m = \beta$. We use L_s to denote the set $\{(\beta, \alpha) : \langle s, \beta, \alpha \rangle \text{ is reachable}\}$. M is a *semilinear system* if L_s is an effectively semilinear language for each $s \in S$ (i.e., the semilinear set of L_s is computable from the description of M). As before, we use $w_1, \ldots, w_l$ to denote a weight measure of $\Sigma = \{a_1, \ldots, a_l\}$, and use $\Sigma_1, \ldots, \Sigma_k$ to denote k subsets of Σ. We may introduce *weight counters* $W_1, \ldots, W_k$ into M to indicate that the accumulated weight on each Σ_i is incremented by w_i whenever an activity $a_j \in \Sigma_i$ is performed. That is, on a transition $\langle s, \beta \rangle \overset{a_j}{\rightarrow} \langle s', \beta' \rangle$ in M, the counters are updated as follows, for each $1 \leq i \leq k$, if $a_j \in \Sigma_i$ then $W_i := W_i + w_j$ else $W_i := W_i$. Similarly, for a loose weight measure $(q_1, w_1), \ldots, (q_l, w_l)$, the counters are updated on the transition as follows: for each $1 \leq i \leq k$, if $a_j \in \Sigma_i$ then $W_i := W_i + p_j$ else $W_i := W_i$, for some $q_j \leq p_j \leq w_j$ (i.e., p_j is nondeterministically chosen between q_j and w_j). Starting with 0, the weight counters are updated along an execution path in (9). We say that $\langle s, \beta, \alpha, W_1, \ldots, W_k \rangle$ is reachable (under the weight measure $w_1, \ldots, w_l$) if the weight counters have values $W_1, \ldots, W_k$ at the end of an execution path in (9) witnessing that $\langle s, \beta, \alpha \rangle$ is reachable.

Let $y_1, \ldots, y_u$ and $z_1, \ldots, z_v$ be distinct variables. A (u, v)-formula, denoted by $P([y_1, \ldots, y_u]; [z_1, \ldots, z_v])$, is a Presburger formula that is a Boolean combination (using $\wedge$ and $\neg$) of Presburger formulas over $y_1, \ldots, y_u$ and Presburger formulas over $z_1, \ldots, z_v$. For the M specified in above, we let $u = |\Gamma| + l$ and $v = l + k$. Now, we consider the k-*reachability problem* for M: given a state s and a (u, v)-formula P, are there $w_1, \ldots, w_l \in \mathbf{N}$ such that

$$P([\phi(\alpha), \phi(\beta)]; [w_1, \ldots, w_l, W_1, \ldots, W_k]) \tag{10}$$

holds for some reachable $\langle s, \beta, \alpha, W_1, \ldots, W_k \rangle$ (under the weight measure $w_1, \ldots, w_l$)? The *loose k-reachability problem* for M can be defined similarly where the lower weights $q_1, \ldots, q_l$ are given. Directly from Theorems 4, 5 and 6, one can show the following results.

Theorem 7. *The 2-reachability problem is decidable for semilinear systems.*

Theorem 8. *For each k, the k-reachability problem is decidable for semilinear systems, when P in (10) is a disjunction of formulas in the following form:*

$$Q([\phi(\alpha), \phi(\beta)]; [w_1, \ldots, w_l]) \wedge c_1 W_1 + \ldots + c_k W_k + d_1 w_1 + \ldots + d_l w_l \sim c_0,$$

where Q is a (u, l)-formula, the c's and d's are integers, and $\sim \in \{=, \neq, >, <, \geq, \leq\}$.

Theorem 9. *For each k, the loose k-reachability problem for semilinear systems is decidable.*

Many machine models are semilinear systems. We start with a simple model. Consider a nondeterministic finite state machine M, which is specified in Section 1 with a designated initial state. Notice that, in this case, $\Gamma = \emptyset$. Let s be a state. Clearly, L_s, the set of all the activity sequences when M moves from the initial state to s is a regular (and hence semilinear) language. Therefore, Theorems 7 and 8 hold for such M. Conversely, for any semilinear language L, one can construct, from the semilinear set of L, a regular language whose semilinear set is the same as the semilinear set of L [18]. From the regular language, one can easily construct a M and a state s such that the regular language is exactly L_s. It is routine to establish the fact that the k-reachability problem is decidable (for the M) iff the k-accumulated weight problem is decidable (for the L). From Theorem 3, one can show

Theorem 10. *There is a fixed k such that the k-reachability problem is undecidable for finite state machines M.*

In the definition of the k-reachability problem, the Presburger formula P in (10) is to specify the undesired values for the w's and the W's. When M is understood as a design of some system, a positive answer to the instance of the k-reachability problem indicates a design bug. In software engineering, it is highly desirable that a design bug is found as early as possible, since it is very costly to fix a bug once a system has already been implemented. It is noticed that in a specific implementation of the design, the parameterized constants are concrete, though the values differ from one implementation to another. Of course, one may test the specification by plugging in a particular choice for the concrete values. However, it is important to guarantee that for *any* concrete values for the parameterized constants, the design M is bug-free.

For instance, consider again the packet-based network switch example, where as we mentioned in Section 1, the switch is modeled as a finite state machine. Suppose the scheduling discipline is required to achieve such fairness property that no matter how the weights are assigned, the total packets serviced by o_1 must be greater than that of o_2 only if the summation of weights of connections in C_1 is greater than that of C_2 (we assume that all connections are nonempty at any time); i.e.,

$$\sum_{c_i \in C_1} w_i - \sum_{c_i \in C_2} w_i \geq 0 \rightarrow W_1 - W_2 \geq 0.$$

From Theorem 7, we know this fairness property can be automatically verified. When there are k servers involved in the example switch, a fairness property can be similarly formulated as a conjunction of the fairness between any two servers. In this case, the fairness property over k-servers is hard to be automatically verified, because of Theorem 10.

One may consider other variations on the model of M. For instance, an activity a_i is associated with, instead of one parameterized weight w_i, but two (or any fixed number of) parameterized weights w_i^1 and w_i^2, from which an instance of the activity can nondeterministically choose during execution. But this variation does not increase the expressive power of M, since "performing activity a_i" can be simulated by "performing activity a_i^1" or "performing activity a_i^2" (nondeterministically chosen) where activity a_i^1 (resp. a_i^2) has weight w_i^1 (resp. w_i^2). One may consider another variation on the model of M where an instance of activity a_i has a weight nondeterministically chosen in between some given number (such as 2) in $\mathbf{N}$ and a parameterized constant w_i. Clearly, from Theorem 9, the loose k-reachability problem is decidable for this model of M.

M can be further generalized; e.g., M is augmented with a pushdown stack. Each transition in M now is in the following form: $s \to^{a,b,\gamma} s'$, indicating that M moves from state s to state s' while performing an activity a and also updating the stack (replacing the top symbol b in the stack by a stack word γ). There are only finitely many transitions in the description of M. Initially, the stack contains a designated initial symbol (i.e., an *initialized stack*) and the machines stays at an initial state. Notice that, for this model of M, L_s is a permutation of a context-free (hence semilinear) language. Therefore, M is still a semilinear system. The results of Theorems 7, 8 and 9 apply for pushdown systems.

M can be further augmented with a finite number of reversal-bounded counters. A nonnegative integer counter is reversal-bounded [11] if it alternates between a nondecreasing mode and a nonincreasing mode (and vice versa) for a given finite number of times, independent of the computations. Hence, a transition in M, in addition to the stack operation, can increment/decrement a counter by one and test a counter against zero. When the counter values are encoded as unary strings, it is not hard to show that the language of L_s is a semilinear language [11]. Hence, this model of M is still a semilinear system, and hence, Theorems 7, 8 and 9 can be applied.

M can be further generalized by adding a number of dense clocks. A clock is a nonnegative real variable. Clock behavior in M includes progresses and resets. A clock progress makes all the clocks advance with the same rate for a nondeterministically chosen amount in positive reals. A clock reset brings a clock value to 0 while keeping all the other clocks unchanged. In M, a transition is either a *stay transition* or a *reset transition*. A stay transition makes M stay in the current state and not perform any stack and counter operations, but the clocks progress. A reset transition makes M move from a state to another while performing an activity, a stack operation, and/or a counter operation. In addition, the transition resets some clocks. A *clock constraint* is a Boolean combination of formulas $x \sim c$ and $x - y \sim c$ where x, y are clocks, and c is an integer, $\sim \in \{>, <, =, \geq, \leq\}$. A (stay/reset) transition in M is also associated with a clock constraint that must be satisfied in order for the transition to fire. The reader may have already noticed that, when M does not have reversal-bounded counters and the pushdown stack, and when each activity is understood as an "input symbol", M is simply equivalent to a timed automaton [2] that has been well studied in recent years for modeling and verifying real-time systems (see [1] for a survey). Here in this paper, an activity on a transition in M is associated with a weight. This weight can be understood as a special form of "prices" in the sense of [15] that tries to model some (e.g., linearly) time-dependent variables in a complex real-time systems. Though, in general, priced timed automata are undecidable for reachability [15], some restricted forms of prices should be decidable, as shown in below, when one understands a weight as a special form of prices.

Consider an execution of M that starts from the initial state and ends with state s. Initially, all the clocks and counters are 0 and the stack is initialized. At the end of the execution, we require that the clock values $(x_1, \ldots, x_t)$, the counter values $(y_1, \ldots, y_u)$, and the stack content (γ) satisfy a given formula $Q(x_1, \ldots, x_t, y_1, \ldots, y_u, z_1, \ldots, z_m)$ where z_i is the number of occurrences of stack symbol b_i in stack word γ. The form of the formula Q is a Boolean combination of $l(x_1, \ldots, x_t, y_1, \ldots, y_u, z_1, \ldots, z_m) \sim 0$ where l is a linear polynomial and $\sim \in \{>, <, =, \geq, \leq\}$. Notice that Q contains both dense variables and discrete variables. Here, we use L to denote the set of all activity sequences on all such executions. If M does not have counters and the stack, L is a regular language and Q is a clock constraint (i.e., as we defined earlier, comparing one clock or the difference of two clocks against a constant). The regularity can be shown using the classic region technique in [2]. In general, however, L is not regular. Using the main theorem in [9], one can show that L can be accepted by a nondeterministic pushdown automaton with reversal-bounded counters. Hence, L is still a semilinear language according to [11]. Associating an activity with a parameterized constant, one can formulate a k-reachability problem for M (similar to (10)): Is there an execution of M from the initial state to state s such that, at the end of the execution, the parameterized constants $w_1, \ldots, w_l$, the accumulated weights $W_1, \ldots, W_k$, the clocks values $x_1, \ldots, x_t$, the counter values $y_1, \ldots, y_u$, and the stack word counts $z_1, \ldots, z_m$, satisfy

$$P(w_1, \ldots, w_l, W_1, \ldots, W_k) \wedge Q(x_1, \ldots, x_t, y_1, \ldots, y_u, z_1, \ldots, z_m)?$$

Following the same proof ideas, one can show that the results of Theorems 7, 8 and 9 still hold for the M augmented with dense clocks, a pushdown stack and reversal-bounded counters.

As a final example, we use the decidability of 2-systems to strengthen recent results in [12]. Consider the model of a two-way deterministic finite automaton augmented with monotonic (i.e., nondecreasing) counters $C_1, \ldots, C_k$ operating on an input of the form $a_1^{i_1} \ldots a_n^{i_n}$ (for some fixed n), with left and right endmarkers. M starts in its initial state on the left end of the input with all counters initially zero. At each step, a counter can be incremented by 0 or 1, but the counters do not participate in the dynamics of the machine. An m-equality relation E over the counter values is a conjunction of m atomic relations of the form $c_i = c_j$. The m-equality relation problem is that of deciding, given a machine M, a state q, and an m-equality relation E, whether there is $(i_1, \ldots, i_n)$ such that M, on input $a_1^{i_1} \ldots a_n^{i_n}$, reaches some configuration where the state is q and the counter values satisfy E. Note that in dealing with the m-equality relation problem, we need only consider machines with at most $2m$ monotonic counters. It is open whether the m-equality relation problem is decidable. However, when $m = 1$, it was recently shown in [12] that the 1-equality relation problem is decidable. The proof of the decidability for $m = 1$ in [12] does not generalize to the case when the two counter values must satisfy an arbitrary Presburger formula E. We give a proof of this generalization below.

First we generalize the m-equality relation problem by allowing E to be an arbitrary Presburger relation $E(c_1, ..., c_k)$ over the counter values $c_1, ..., c_k$. Call this the *Presburger relation problem*. Note that the m-equality relation problem is a very special case of the Presburger relation problem. We can use the decidability of 2-systems to show that the Presburger relation problem for machines with only 2 monotonic counters is decidable. The idea is as follows. In [12], it was shown that the values c_1 and c_2 of the two counters at any time can effectively be represented by equations of the form:

$$c_1 = A_1 + yB_1 + C_1, c_2 = A_2 + yB_2 + C_2,$$

where y is a nonnegative integer variable, and $A_1, B_1, C_1, A_2, B_2, C_2$ are nonnegative linear polynomials in some nonnegative integer variables $x_1, ..., x_m$. (Even though C_1 and C_2 can be absorbed by A_1 and A_2, we use the formulation above to be consistent with the formulation in [12].) Since E (subset of $\mathbf{N}^2$) is Presburger, it is semilinear. First assume that E is a linear set. Then the two components of E can be represented by nonnegative linear polynomials $p_1(z_1, ..., z_r)$ and $p_2(z_1, ..., z_r)$ for some nonnegative integer variables $z_1, ..., z_r$. Thus, using the two equations above, we get: $A_1 + yB_1 + C_1 = p_1(z_1, ..., z_r)$, $A_2 + yB_2 + C_2 = p_2(z_2, ..., z_r)$. Rearranging terms, these two equations can be written as: $yB_1 = p_1 - A_1 - C_1$ and $yB_2 = p_2 - A_2 - C_2$. By semilinear transformation, we can reduce these equations to $yB_1 = D_1$ and $yB_2 = D_2$, where B_1, B_2, D_1, D_2 are nonnegative linear polynomials in some nonnegative integer variables $w_1, ..., w_t$. Since the above equations constitute a 2-system, it is solvable in $y, w_1, ..., w_t$. When E is a semilinear set, we just need to check if at least one of a finite number of equations of the form above has a solution.

It is open whether the Presburger relation problem is decidable when there are more than 2 monotonic counters (since the m-equality relation problem, which is a special case, is open). But suppose the Presburger relation E takes the following special form: $p_1(c_1, ..., c_k) \sim d_1 \wedge p_2(c_1, ..., c_k) \sim d_2 \wedge ... \wedge p_m(c_1, ..., c_k) \sim d_m$, where $d_1, ..., d_m$ are integers (positive, negative, zero) and each $p_i(c_1, ..., c_k)$ is a linear polynomial (not necessarily nonnegative), and each $\sim$ in $\{>, <, =, \geq, \leq\}$. It is easy to see that when $m = 2$, i.e., there are only two linear polynomials p_1 and p_2 involved in the conjunction above, then by adding "slack" variables and doing semilinear transformation, we can again reduce the problem to solving a system of the form: $yB_1 = D_1$, $yB_2 = D_2$, and, therefore, solvable. However, the case when $m > 2$ is open.

Acknowledgement. The authors would like to thank the anonymous referees for many valuable comments and suggestions.

References

1. R. Alur. Timed automata. In *CAV'99*, volume 1633 of *LNCS*, pages 8–22. Springer, 1999.
2. R. Alur and D. L. Dill. A theory of timed automata. *Theoretical Computer Science*, 126(2):183–235, April 1994.
3. A. Bouajjani, J. Esparza, and O. Maler. Reachability analysis of pushdown automata: application to model-checking. In *CONCUR'97*, volume 1243 of *LNCS*, pages 135–150. Springer, 1997.
4. T. Bultan, R. Gerber, and W. Pugh. Model-checking concurrent systems with unbounded integer variables: symbolic representations, approximations, and experimental results. *ACM Transactions on Programming Languages and Systems*, 21(4):747–789, July 1999.
5. E. M. Clarke, E. A. Emerson, and A. P. Sistla. Automatic verification of finite-state concurrent systems using temporal logic specifications. *ACM Transactions on Programming Languages and Systems*, 8(2):244–263, April 1986.
6. H. Comon and Y. Jurski. Multiple counters automata, safety analysis and Presburger arithmetic. In *CAV'98*, volume 1427 of *LNCS*, pages 268–279. Springer, 1998.
7. Z. Dang. Verifying and debugging real-time infinite state systems (PhD. Dissertation). *Department of Computer Science, University of California at Santa Barbara*, 2000.
8. Z. Dang, O. Ibarra, and Z. Sun. On the emptiness problems for two-way nondeterministic finite automata with one reversal-bounded counter. In *ISAAC'02*, volume 2518 of *LNCS*, pages 103–114. Springer, 2002.
9. Zhe Dang. Binary reachability analysis of pushdown timed automata with dense clocks. In *CAV'01*, volume 2102 of *LNCS*, pages 506–517. Springer, 2001.
10. G. J. Holzmann. The model checker SPIN. *IEEE Transactions on Software Engineering*, 23(5):279–295, May 1997. Special Issue: Formal Methods in Software Practice.
11. O. H. Ibarra. Reversal-bounded multicounter machines and their decision problems. *Journal of the ACM*, 25(1):116–133, January 1978.
12. O. H. Ibarra and Z. Dang. Deterministic two-way finite automata augmented with monotonic counters. 2002 (submitted).
13. K.L. McMillan. *Symbolic Model Checking*. Kluwer Academic Publishers, Norwell Massachusetts, 1993.
14. O. Kupferman and M.Y. Vardi. An automata-theoretic approach to reasoning about infinite-state systems. In *CAV'00*, volume 1855 of *LNCS*, pages 36–52. Springer, 2000.
15. K. Larsen, G. Behrmann, E. Brinksma, A. Fehnker, T. Hune, P. Pettersson, and J. Romijn. As cheap as possible: Efficient cost-optimal reachability for priced timed automata. In *CAV'01*, volume 2102 of *LNCS*, pages 493–505. Springer, 2001.
16. Y. V. Matiyasevich. *Hilbert's Tenth Problem*. MIT Press, 1993.
17. M. Minsky. Recursive unsolvability of Post's problem of Tag and other topics in the theory of Turing machines. *Ann. of Math.*, 74:437–455, 1961.
18. R. Parikh. On context-free languages. *Journal of the ACM*, 13:570–581, 1966.
19. M. Y. Vardi and P. Wolper. An automata-theoretic approach to automatic program verification. In *LICS'86*, pages 332–344. IEEE Computer Society Press, 1986.

Monadic Second-Order Logics with Cardinalities[*]

Felix Klaedtke[1] and Harald Rueß[2]

[1] Albert-Ludwigs-Universität Freiburg, Germany
[2] SRI International, CA, USA

Abstract. We delimit the boundary between decidability versus un-decidability of the weak monadic second-order logic of one successor (WS1S) extended with linear cardinality constraints of the form $|X_1| + \cdots + |X_r| < |Y_1| + \cdots + |Y_s|$, where the X_is and Y_js range over finite subsets of natural numbers. Our decidability and undecidability results are based on an extension of the classic logic-automata connection using a novel automaton model based on Parikh maps.

1 Introduction

In the automata-theoretic approach for solving the satisfiability problem of a logic one develops an appropriate notion of automata and establishes a translation from formulas to automata. The satisfiability problem for the logic then reduces to the automata emptiness problem. Most prominently, decidability of the (weak) monadic second-order logic of one successor (W)S1S is proved by a translation of formulas to word automata, see e.g. [27]. Despite the nonelementary worst-case complexity [19, 26], the automata-based decision procedure for WS1S, implemented in the Mona tool [10, 16], has been found to be effective for reasoning about a multitude of computation systems ranging from circuits [3, 2] to protocols [17, 25]. Furthermore, it has been integrated in theorem provers to decide well-defined fragments of higher-order logic [1, 21].

Many interesting verification problems, however, fall outside the scope of WS1S. For example, the verifications in WS1S for the sequential circuits considered in [3] are only with respect to concrete values of parameters such as setup time and minimum clock period since some linear arithmetic is used on these parameters. Also, certain distributed algorithms such as the Byzantine generals problem [18] of reaching distributed consensus in the presence of unreliable messengers and treacherous generals cannot be modeled in WS1S, since reasoning about the combination of (finite) sets and cardinality constraints on these sets is required here.

In order to support this kind of reasoning and to significantly extend the range of automated verification procedures we extend WS1S with atomic formulas of the form $|X_1| + \cdots + |X_r| < |Y_1| + \cdots + |Y_s|$, where the X_is and Y_js are

---

[*] This work was supported by SRI International internal research and development, and NASA through contract NAS1-00079.

J.C.M. Baeten et al. (Eds.): ICALP 2003, LNCS 2719, pp. 681–696, 2003.

monadic second-order (MSO) variables, and $|X|$ denotes the cardinality of the MSO variable X. The extension of WS1S with cardinality constraints is denoted by WS1S^{card}. Our main results are

(i) WS1S^{card} is undecidable. More precisely, (a) the fragment of WS1S^{card} consisting of the sentences of the form $\forall X \exists \overline{Y} \varphi$ is undecidable, where X is an MSO variable, $\overline{Y}$ is a vector of MSO variables ranging over finite sets of natural numbers, and all quantifiers in φ are first-order, that is, ranging over natural numbers. And, (b) the fragment of WS1S^{card} consisting of the sentences of the form $\exists X \forall y \exists \overline{Y} \varphi$, where y is a first-order variable and all quantifiers in φ are first-order.

(ii) The fragment consisting of the sentences of the form $Q_1 x_1 \ldots Q_\ell x_\ell Q \overline{Y} \varphi$ is decidable, where $Q_1, \ldots, Q_\ell \in \{\exists, \forall\}$ are first-order quantifiers and an MSO variable occurring in a cardinality constraint in φ is bound by $Q \in \{\exists, \forall\}$.

Together the results (i) and (ii) delimit the boundary between decidability versus undecidability of MSO logics with cardinality constraints.

We use an automata-theoretic approach for obtaining these results by defining a suitable extension of finite word automata. These extensions work over an extended alphabet in which a vector of natural numbers is attached to each letter of the input alphabet. An input is accepted if an input word is accepted in the traditional sense *and* if a projection of the word via a monoid homomorphism to a vector of natural numbers satisfies given arithmetic constraints. Since this monoid homomorphism generalizes Parikh's *commutative image* [23] on words, we call such an extended automaton a *Parikh finite word automaton* (PFWA). PFWAs characterize the expressiveness of the existential fragment of WS1S^{card}.

The undecidability results (i) follow from the undecidability of the universality problem for PFWAs and the undecidability of the halting problem for 2-register machines, whereas the decidability result (ii) is based on a two-step construction. First, we build a PFWA for the formula $\exists \overline{Y} \varphi$, and, second, we transform the PFWA into a corresponding Presburger arithmetic formula. This latter construction takes care of the quantification of the first-order variables $x_1, \ldots, x_\ell$. Compared to simply checking the emptiness problem for the PFWA associated with a formula, this two-stage translation yields a decision procedure for a much more expressive fragment of WS1S^{card}. These constructions can readily be extended to obtain corresponding results for cardinality constraints in second-order monadic logics over trees [15].

The paper is structured as follows. In §2 we introduce PFWAs. Then, in §3 we define WS1S^{card} and compare the expressiveness of the existential fragment of WS1S^{card} with PFWAs. In §4 we prove the results (i) and (ii), and illustrate applications of the decidability result (ii). Finally, in §5 we draw conclusions.

2   Parikh Automata

We introduce a framework that extends the acceptance condition of machines operating on words. In addition to the traditional acceptance condition of a machine, we require that an input satisfies arithmetic properties, where the input is associated with a vector of natural numbers. Parikh finite word automata are an instance of this framework.

Let $\Sigma = \{b_1, \ldots, b_n\}$ be a linearly ordered alphabet. Parikh's [23] *commutative image* $\Phi : \Sigma^* \to \mathbb{N}^{|\Sigma|}$ maps the elements of the free monoid Σ^*, so-called words, to vectors of natural numbers. The commutative image is defined by $\Phi(b_i) := \overline{e}_i$ and $\Phi(uv) := \Phi(u) + \Phi(v)$, where $\overline{e}_i \in \mathbb{N}^{|\Sigma|}$ is the unit vector with the ith coordinate equal to 1 and all other coordinates are 0. Intuitively, the ith position of $\Phi(w)$ counts how often b_i occurs in $w \in \Sigma^*$. We extend Parikh's commutative image by considering the Cartesian product of Σ and a nonempty set D of vectors of natural numbers.

Definition 1. *Let Γ be an alphabet of the form $\Sigma \times D$, where D is a nonempty subset of $\mathbb{N}^N$, for some $N \geq 1$. We define the **projection** $\Psi : \Gamma^* \to \Sigma^*$ and the **extended Parikh map** $\Phi : \Gamma^* \to \mathbb{N}^N$ as monoid homomorphisms.*

(i) $\Psi(b, \overline{d}) := b$, for $(b, \overline{d}) \in \Sigma \times D$, and $\Psi(uv) := \Psi(u)\Psi(v)$.
(ii) $\Phi(b, \overline{d}) := \overline{d}$, for $(b, \overline{d}) \in \Sigma \times D$, and $\Phi(uv) := \Phi(u) + \Phi(v)$.

Note that if we attach to each letter $b_i \in \Sigma$ the unit vector $\overline{e}_i \in \mathbb{N}^{|\Sigma|}$ in a word $w \in \Sigma^*$ then the extended Parikh map yields the commutative image of w.

We constrain a language by an arithmetic property given by a set of vectors of natural numbers.

Definition 2. *For a language $L \subseteq (\Sigma \times D)^*$ and $C \subseteq \mathbb{N}^N$, let*

$$L \restriction_C := \{\Psi(w) \mid w \in L \text{ and } \Phi(w) \in C\}$$

*be the **restriction** of L with respect to C.*

The acceptance condition of a machine operating on words can be extended in the following way. A word w over Σ is accepted if the machine accepts a word over $\Sigma \times D$ both in the traditional sense and if the sum of the attached vectors to the symbols in w is in a given subset of $\mathbb{N}^N$. Here, we are mainly concerned with finite word automata and arithmetic constraints restricted to *semilinear sets* $U \subseteq \mathbb{N}^s$. This means that there are linear polynomials $p_1, \ldots, p_m :$ $\mathbb{N}^r \to \mathbb{N}^s$ such that U is the union of the images of these polynomials, that is, $U = \bigcup_{1 \leq i \leq m} \{p_i(x_1, \ldots, x_r) \mid x_1, \ldots, x_r \in \mathbb{N}\}$.

Definition 3. *A **Parikh finite word automaton** (**PFWA**) of dimension $N \geq 1$ is a pair $(\mathcal{A}, C)$, where $\mathcal{A}$ is a finite word automaton with an alphabet of the form $\Sigma \times D$, D is a finite, nonempty subset of $\mathbb{N}^N$, and C is a semilinear subset of $\mathbb{N}^N$. The PFWA $(\mathcal{A}, C)$ **recognizes** the language $L(\mathcal{A}, C) := L(\mathcal{A}) \restriction_C$, where $L(\mathcal{A})$ is the language recognized by $\mathcal{A}$.*

*The PFWA $(\mathcal{A}, C)$ is **deterministic** if for the transition function δ of $\mathcal{A}$ it holds that for every state q and for every $(b, \overline{d}) \in \Sigma \times D$, $|\delta(q, (b, \overline{d}))| \leq 1$, and if $|\delta(q, (b, \overline{d}))| = 1$ then $|\delta(q, (b, \overline{d}'))| = 0$, for every $\overline{d} \neq \overline{d}'$.*

For example, the deterministic PFWA $(\mathcal{A}, \{\binom{z}{z} \mid z \in \mathbb{N}\})$, where $\mathcal{A}$ is given by the picture recognizes $\{a^{i+j}b^ic^j \mid i,j > 0\}$, which is context-sensitive but not context-free.

PFWAs are strictly more expressive than finite word automata, where the accepted words are constrained by their commutative images and semilinear sets. It is easy to define a deterministic PFWA automaton that recognizes the language $L := \{a^ib^ja^ib^j \mid i,j \geq 1\}$. But there does not exist a finite word automaton $\mathcal{A}$ with the alphabet $\{a,b\}$ and a set $C \subseteq \mathbb{N}^2$ such that $w \in L$ iff $w \in L(\mathcal{A})$ and $(k,\ell) \in C$, where (k,ℓ) is the commutative image of w.

A PFWA can be seen as a finite word automaton extended with counters, where a vector of natural numbers attached to a symbol is interpreted as an increment of the counters. In contrast to other counter automaton models in the literature, for example [4,7,11], we do not restrict the applicability of transitions in a run by additional guards on the values of the counters. Instead, a PFWA constrains the language of a finite word automaton over the extended alphabet by a semilinear set. It turns out (a) that PFWAs are equivalent to reversal-bounded multicounter machines [11,12] and (b) that PFWAs are equivalent to weighted finite automata over the groups $(\mathbb{Z}^k, +, 0)$ [6,20] with $k \geq 1$ in the sense that all these three different kinds of machines describe the same class of languages. We refer the reader to [15], for definitions and a detailed comparison of these automaton models, proofs of the equivalences, and a comparison of PFWAs to other automaton models.

We state some properties of PFWAs. The details can be found in [15].

Property 4. *(1) Deterministic PFWAs are closed under union, intersection, complement, and inverse homomorphisms, but not under homomorphisms.*
(2) PFWAs are closed under union, intersection, homomorphisms, inverse homomorphisms, concatenations, and left and right quotients, but not under complement.

The decidability of the emptiness problem relies on Parikh's result [23], which states that the commutative image of a context-free language is semilinear.

Lemma 5. *Let Γ be an alphabet of the form $\Sigma \times D$ with $D \subseteq \mathbb{N}^N$, for some $N \geq 1$. For every context-free language $L \subseteq \Gamma^*$, there are linear polynomials $q_1, \ldots, q_m : \mathbb{N}^r \to \mathbb{N}^N$, for some $r \geq 1$, such that*

$$\Phi(L) = \bigcup_{1 \leq i \leq m}\{q_i(x_1, \ldots, x_r) \mid x_1, \ldots, x_r \in \mathbb{N}\},$$

where Φ is the extended Parikh map of Γ. Moreover, the polynomials $q_1, \ldots, q_m$ are effectively constructible if L is given by a pushdown automaton.

For a PFWA $(\mathcal{A}, C)$, we know by Lemma 5 that the set $\Phi(L(\mathcal{A}))$ is semilinear and effectively constructible. The decidability of the emptiness problem follows from the facts that semilinear sets are effectively closed under intersection, and $L(\mathcal{A}, C) \neq \emptyset$ iff $\Phi(L(\mathcal{A})) \cap C \neq \emptyset$.

Property 6. *The emptiness problem for PFWAs is decidable.*

The undecidability of the universality problem for PFWAs can be shown by reduction from the word problem for Turing machines.

Property 7. *The universality problem for PFWAs is undecidable.*

Note that the universality problem for deterministic PFWAs is decidable since they are closed under complement and the emptiness problem is decidable.

3   WS1S with Cardinality Constraints

We extend WS1S in order to compare cardinalities of sets. We call this extension WS1S^{card}. The classic logic-automata connection of finite word automata and WS1S extends to PFWAs and the existential fragment of WS1S^{card}.

The Weak Monadic Second-Order Logic of One Successor. The atomic formulas of WS1S are membership Xx, and the successor relation $\text{succ}(x, y)$, where x, y are first-order (FO) variables, and X is a monadic second-order (MSO) variable. We adopt the following notation: lowercase letters $x, y, \ldots$ denote FO variables and uppercase letters $X, Y, \ldots$ denote MSO variables. Moreover, $\alpha, \beta, \ldots$ range over FO and MSO variables. Formulas are built from the atomic formulas and the connectives $\neg$ and $\vee$, and the existential quantifier $\exists$ for FO and MSO variables. We also use the connectives $\wedge$, $\rightarrow$ and $\leftrightarrow$, and the universal quantifiers $\forall$ for FO and MSO variables, and we use the standard conventions for omitting parentheses. A formula is **existential** if it is of the form $\exists \overline{X} \varphi$ where all bound variables in φ are FO.

Formulas are interpreted over the natural numbers with the successor relation, that is, the structure $(\mathbb{N}, \text{succ})$. An interpretation I maps FO variables to natural numbers and MSO variables are mapped to *finite* subsets of $\mathbb{N}$. The truth value of a formula φ in $(\mathbb{N}, \text{succ})$ with respect to an interpretation I, in symbols $(\mathbb{N}, \text{succ}), I \models \varphi$, is defined in the obvious way. Note that existential quantification for MSO variables only ranges over *finite* subsets of $\mathbb{N}$. We write $(\mathbb{N}, \text{succ}) \models \varphi$ if φ is a sentence, that is, φ does not have free variables.

Equality $x = y$ can be expressed by $\exists z (\text{succ}(x, z) \wedge \text{succ}(y, z))$. For a natural number $t \in \mathbb{N}$, we write $x = t$ for $\exists z_0 \ldots \exists z_t (x = z_t \wedge \bigwedge_{0 \le i < t} \text{succ}(z_i, z_{i+1}) \wedge \forall y \neg \text{succ}(y, z_0))$, and we write $Yx + 1$ for $\exists z (\text{succ}(x, z) \wedge Yz)$. The formula $part(U, X_1, \ldots, X_k)$ expresses that $X_1, \ldots, X_k$ is a partition of U, that is $\forall y (Uy \leftrightarrow \bigvee_{1 \le i \le k} X_i y) \wedge \neg \exists y (\bigvee_{1 \le i < j \le k} (X_i y \wedge X_j y))$. Note that in these formulas only FO variables are quantified.

A word $w = b_1 \ldots b_n \in (\{0, 1\}^k)^*$ determines the interpretation I_w for variables $\alpha_1, \ldots, \alpha_k$, where for all $1 \le j \le k$,

- $i \in I_w(\alpha_j)$ iff $\chi_j(b_i) = 1$, if α_j is an MSO variable, and
- $i = I_w(\alpha_j)$ iff $\chi_j(b_i) = 1$ and $\chi_j(b_{i'}) = 0$ for all $i' \ne i$, if α_j is an FO variable,

where χ_j projects a vector in $\{0, 1\}^k$ to its jth coordinate. We extend χ_j homomorphically to words. Note that we have implicitly assumed that the variables

$\alpha_1, \ldots, \alpha_k$ are ordered in the sense that the interpretation I_w of the variable α_j is determined by the jth projection of $b_1 \ldots b_n$, that is, $\chi_j(b_1 \ldots b_n)$. In the following, we write χ_{α_j} for χ_j. For a formula $\varphi(\alpha_1, \ldots, \alpha_k)$, we define

$$L(\varphi) := \{w \in (\{0,1\}^k)^* \mid (\mathbb{N}, \mathrm{succ}), I_w \models \varphi\}.$$

Later, we shall need the following facts that are due to Büchi, Elgot, and Trakhtenbrot. For more details, see, for example, [27].

Fact 8. $L(\varphi)$ is regular for every WS1S formula φ. Moreover, we can effectively build a finite word automaton recognizing $L(\varphi)$.

For the other direction, that is, describing regular languages by WS1S formulas, there is a subtlety that we want to point out. Note that natural numbers and finite subsets of $\mathbb{N}$ have several encodings, e.g., all the words in $\{0\}^*$ encode the empty set. It is easy to see that languages definable by WS1S formulas, are closed under $\bar{0}$-padding and $\bar{0}$-cutting, that is, $w \in L(\varphi)$ iff $w\bar{0} \in L(\varphi)$, where $\bar{0}$ is the letter $(0, \ldots, 0)$. We call a $\bar{0}$-padding and $\bar{0}$-cutting closed language **$\bar{0}$-closed**.

Fact 9. For every regular $\bar{0}$-closed language $L \subseteq (\{0,1\}^k)^*$ there is an existential WS1S formula $\varphi(X_1, \ldots, X_k)$ such that $L(\varphi) = L$.

To obtain an equivalence of the logic and the regular languages, one has to look at finite *word models* [27]. The main difference is that the universe of a finite word model is not $\mathbb{N}$, but $\{0, \ldots, n-1\}$ where n is given by the length of the word. The distinction between the different semantics is emphasized by using the name M2L(str) or MSO[+1] instead of WS1S. The results below carry over to finite word models. We use the WS1S semantics since it simplifies matters.

Cardinality Constraints. WS1S^{card} has in addition to the atomic formulas of WS1S the atomic formulas of the form $|X_1| + \cdots + |X_r| < |Y_1| + \cdots + |Y_s|$, where the truth value with respect to an interpretation I is defined as

$$(\mathbb{N}, \mathrm{succ}), I \models |X_1| + \cdots + |X_r| < |Y_1| + \cdots + |Y_s| \quad \text{iff} \quad \sum_{1 \le i \le r} |I(X_i)| < \sum_{1 \le i \le s} |I(Y_i)|.$$

Let C be the set of formulas of the form $|X_1| + \cdots + |X_r| < |Y_1| + \cdots + |Y_s|$ and their negations. We write formulas like $-2|X| = 3|Y| + |Z|$ which can be transformed to an equivalent Boolean combination of formulas in C by standard arithmetic. Moreover, we also use the summation symbol $\sum$ for a shorter representation.

Parikh Automata and WS1S^{card}. We carry over the Facts 8 and 9 to the existential fragment of WS1S^{card} and PFWAs. We start with the direction in Fact 9.

Theorem 10. *For every PFWA $(\mathcal{A}, C)$ where $L(\mathcal{A}, C) \subseteq (\{0,1\}^s)^*$ is $\bar{0}$-closed, there is an existential* WS1S^{card} *formula $\psi_{\mathcal{A},C}(U_1, \ldots, U_s)$ with $L(\varphi) = L(\mathcal{A}, C)$.*

Proof. Let $N \geq 1$ be the dimension of $(\mathcal{A}, C)$, and let $\mathcal{A} = (Q, \{0,1\}^s \times D, \delta, q_{\mathrm{I}}, F)$. Without loss of generality, we assume that $Q = \{1, \ldots, r\}$, for some $r \geq 1$. Let K be the maximal natural number occurring in a vector in D, that is $K := \max \left(\bigcup_{(d_1, \ldots, d_N) \in D} \{d_1, \ldots, d_N\} \right)$.

Let $b_0 \ldots b_{n-1} \in L(\mathcal{A}, C)$ with $n \geq 0$ and $b_{n-1} \neq \overline{0}$. The formula $\psi_{\mathcal{A}, C}$ describes the existence of an accepting run $\varrho = q_0 \ldots q_{n+m} \in Q^*$ on a word $(b_0, \overline{d}_0) \ldots (b_{n-1}, \overline{d}_{n-1})(\overline{0}, \overline{d}_n) \ldots (\overline{0}, \overline{d}_{n+m-1}) \in (\{0,1\}^s \times D)^*$, for some $m \geq 0$. Note that $L(\mathcal{A}, C)$ is $\overline{0}$-closed. It holds, $q_0 = q_{\mathrm{I}}$, $q_{i+1} \in \delta(q_i, (b_i, \overline{d}_i))$, for $0 \leq i < n+m$, and $q_{n+m} \in F$. We encode ϱ by pairwise disjoint sets $Y_1, \ldots, Y_r \subseteq \{0, \ldots, n+m\}$ such that Y_q contains those positions i with $q = q_i$. Moreover, we keep track of the numbers at the kth position of the vectors $\overline{d}_i$ with the sets $Z_k^0, \ldots, Z_k^K \subseteq \{0, \ldots, n+m\}$: it holds that $0 \in Z_k^0$ and $i \in Z_k^d$ iff the kth position of $\overline{d}_{i-1}$ is d, for $1 \leq i \leq n+m$. Therefore, the kth position of the vector $\overline{d}_0 + \cdots + \overline{d}_{n+m-1}$ is $\sum_{0 \leq d \leq K} d|Z_k^d|$. We have to check $\overline{d}_0 + \cdots + \overline{d}_{n+m-1} \in C$. Formally,

$$\exists Y_1 \ldots \exists Y_r \exists Z_1^0 \ldots \exists Z_1^K \ldots \exists Z_N^0 \ldots \exists Z_N^K \exists U \Big(domain(U, U_1, \ldots, U_s) \wedge$$
$$part(U, Y_1, \ldots, Y_r) \wedge \bigwedge_{1 \leq i \leq N} part(U, Z_i^0, \ldots, Z_i^K) \wedge$$
$$\forall x \Big((x = 0 \rightarrow Y_{q_1} x \wedge \bigwedge_{1 \leq i \leq N} Z_i^0 x) \wedge$$
$$(Ux \rightarrow \bigvee_{q \in \delta(p, (b, (d_1, \ldots, d_N)))} (Y_p x \wedge letter_b(x, U_1, \ldots, U_s) \wedge$$
$$Y_q x + 1 \wedge \bigwedge_{1 \leq i \leq N} Z_i^{d_i} x + 1)) \wedge$$
$$(Ux \wedge \neg Ux + 1 \rightarrow (\bigvee_{q \in F} Y_q x))) \Big) \wedge$$
$$\psi_C(Z_1^0, \ldots, Z_1^K, \ldots, Z_N^0, \ldots, Z_N^K) \Big),$$

where $domain(U, U_1, \ldots, U_s)$ is the formula $\forall x(U_1 x \vee \cdots \vee U_s x \rightarrow Ux + 1) \wedge \forall x(Ux + 1 \rightarrow Ux)$ and $letter_b(x, U_1, \ldots, Us) := (\bigwedge_{b_i = 0} \neg U_i x) \wedge (\bigwedge_{b_i = 1} U_i x)$, for $b = (b_1, \ldots, b_s) \in \{0,1\}^s$. It remains to define the formula ψ_C. Since C is semilinear we can assume that C is the union of the images of linear polynomials $p_1, \ldots, p_\ell : \mathbb{N}^k \rightarrow \mathbb{N}^N$, for some $k \geq 1$. For $1 \leq i \leq \ell$, let

$$\psi_{p_i} := \exists X_1 \ldots \exists X_k \Big(\bigwedge_{1 \leq j \leq N} \exists X \big(\forall y (Xy \leftrightarrow \bigvee_{0 \leq t < d_0^j} y = t) \wedge$$
$$\sum_{1 \leq d \leq K} d|Z_j^d| = |X| + d_1^j |X_1| + \cdots + d_k^j |X_k| \big) \Big),$$

where we assume that $p_i(x_1, \ldots, x_k)$ has the form $\overline{d}_0 + \overline{d}_1 x_1 + \cdots + \overline{d}_k x_k$ with $\overline{d}_j = (d_j^1, \ldots, d_j^N)$, for $0 \leq j \leq k$. Let $\psi_C := \psi_{p_1} \vee \cdots \vee \psi_{p_\ell}$. The quantification of the MSO variables in ψ_C can be pulled out to obtain an existential formula. $\dashv$

We give a set of formulas describing languages recognizable by PFWAs. This carries over Fact 8 to PFWAs and the existential fragment of WS1S^{card}. Let E be the set of WS1S^{card} formulas of the form $\exists X_1 \ldots \exists X_n \varphi$, where all MSO variables occurring in a subformula in C of φ are either free or bound by one of the existential quantifiers for $X_1, \ldots, X_n$. Note that an existential formula is in E, but E contains formulas that are not existential. A formula in E can contain MSO variables Y that are universally quantified if Y does not occur in subformulas in C and the quantification of Y happens below the existential quantification of $X_1, \ldots, X_n$.

Theorem 11. *For every $\varphi \in$ E, we can construct a PFWA recognizing $L(\varphi)$.*

Proof (Sketch). We can assume that $\varphi \in$ E is of the form $\exists X_1 \ldots \exists X_n (\bigvee_i \bigwedge_j \psi_{ij})$, where ψ_{ij} is either a WS1S formula or $\psi_{ij} \in$ C. By Fact 8 we can construct a finite word automaton $\mathcal{A}_{ij}$ with $L(\mathcal{A}_{ij}) = L(\psi_{ij})$ if ψ_{ij} is a WS1S formula, and for $\psi_{ij} \in$ C, it is straightforward to give a PFWA $(\mathcal{A}_{ij}, C_{ij})$ with $L(\mathcal{A}_{ij}, C_{ij}) = L(\psi_{ij})$. From Property 4 we can construct a PFWA recognizing $L(\varphi)$.   ⊣

Theorems 10 and 11 together reveal the following equivalence.

Corollary 12. *For a $\overline{0}$-closed language $L \subseteq (\{0, 1\}^s)^*$, the following two conditions are equivalent:*

(i) *L is recognizable by a PFWA, that is, there is a PFWA $(\mathcal{A}, C)$ with $L(\mathcal{A}, C) = L$.*

(ii) *L is definable in the existential fragment of $\text{WS1S}^{\text{card}}$, that is, there is an existential $\text{WS1S}^{\text{card}}$ formula φ with $L(\varphi) = L$.*

Another extension of the classical logic-automata connection with a similar flavor is given in [22] relating Petri net languages with the existential fragment of the MSO logic on words extended with partial orders $\leq_g$ and $=_g$ on subsets of $\{0, \ldots, n-1\}$ defined as $X \leq_g Y$ iff $|X \cap \{0, \ldots, m-1\}| \leq |Y \cap \{0, \ldots, m-1\}|$, for all $m \leq n$, and $X =_g Y$ iff $X \leq_g Y$ and $|X| = |Y|$. [22] does not investigate decidability problems about this logic as we will do in the next section for $\text{WS1S}^{\text{card}}$.

We want to point out that there is also a relationship between $\text{WS1S}^{\text{card}}$ and Petri nets. Petri net reachability is expressible in $\text{WS1S}^{\text{card}}$ [14]. From this it is not difficult to see that ($\overline{0}$-closed) Petri net languages can be described in $\text{WS1S}^{\text{card}}$. But the formulas for expressing the reachability problem (or describing Petri net languages) require a top-level quantification of the form $\exists x \exists \overline{X} \forall y \forall \overline{Y}$. In the next section we are going to show that the fragment with such a top-level quantification is undecidable. Note that the reachability problem for Petri nets is decidable.

4   Undecidability and Decidability Results

Decidability and undecidability results about MSO logics with cardinality constraints summarized in Figure 1 using the notation introduced below. Together, these results delimit the boundary between decidability versus undecidability in $\text{WS1S}^{\text{card}}$. Furthermore, we illustrate applications in hardware and software verification of a decidable fragment of $\text{WS1S}^{\text{card}}$.

We introduce the following notation to uniformly describe fragments of $\text{WS1S}^{\text{card}}$.

undecidable	$[\forall_{\mathrm{MSO}}\exists^*_{\mathrm{MSO}}\mathsf{FO};\mathrm{succ}]$	$[\exists_{\mathrm{MSO}}\forall_{\mathrm{FO}}\exists^5_{\mathrm{MSO}}\mathsf{FO};\mathrm{succ}]$
decidable	$[\mathsf{FO}(\exists^*_{\mathrm{MSO}} \cup \forall^*_{\mathrm{MSO}})\mathsf{FO};\mathsf{R}_{\mathrm{WS1S}}]$	

Fig. 1. Undecidable and decidable fragments of $\mathrm{WS1S}^{\mathrm{card}}$.

Definition 13. *Let* $\mathsf{Q} \subseteq \{\exists_{\mathrm{MSO}}, \forall_{\mathrm{MSO}}, \exists_{\mathrm{FO}}, \forall_{\mathrm{FO}}\}^*$ *and let* R *be a set of relations over the natural numbers and finite subsets of natural numbers. We write* $[\mathsf{Q}; \mathsf{R}]$ *for the set of sentences of the form* $Q_1\alpha_1 \ldots Q_n\alpha_n\varphi$, *where* $Q_1 \ldots Q_n \in \mathsf{Q}$ *and* φ *is a quantifier-free formula with relations in* R.

We write $[\mathsf{Q}; R_1, \ldots, R_n]$ for $[\mathsf{Q}; \{R_1, \ldots, R_n\}]$. Let $\mathsf{R}_{\mathrm{WS1S}}$ be the set of relations that are definable in WS1S. We will often give the set Q as a regular expression. For example, the set FO of arbitrary FO quantifier prefixes is $(\exists_{\mathrm{FO}} \cup \forall_{\mathrm{FO}})^*$, and we write, e.g., $\exists^2_{\mathrm{MSO}}$ for $\exists_{\mathrm{MSO}}\exists_{\mathrm{MSO}}$.

Undecidability Results.

Theorem 14. *The fragment* $[\forall_{\mathrm{MSO}}\exists^*_{\mathrm{MSO}}\mathsf{FO};\mathrm{succ}]$ *is undecidable.*

Proof. To prove this theorem we look at the universality problem for $\overline{0}$-closed PFWAs which is undecidable. This can be shown by adopting the proof that shows the undecidability for universality problem for PFWAs.

Let $(\mathcal{A}, C)$ be a $\overline{0}$-closed PFWA with $L(\mathcal{A}, C) \subseteq \{0,1\}^*$. For the formula $\psi_{\mathcal{A},C}(U_1)$ from Theorem 10, it holds, $(\mathbb{N}, \mathrm{succ}) \models \forall U_1 \psi_{\mathcal{A},C}$ iff $L(\mathcal{A}, C) = \{0,1\}^*$. Since $\psi_{\mathcal{A},C}$ is existential and the universality problem is undecidable, we have that the fragment $[\forall_{\mathrm{MSO}}\exists^*_{\mathrm{MSO}}\mathsf{FO};\mathrm{succ}]$ is undecidable. $\dashv$

Theorem 15. *The fragment* $[\exists_{\mathrm{MSO}}\forall_{\mathrm{FO}}\exists^5_{\mathrm{MSO}}\mathsf{FO};\mathrm{succ}]$ *is undecidable.*

Proof (Sketch). The undecidability is shown by encoding the halting problem for 2-register machines as a formula in $[\exists_{\mathrm{MSO}}\forall_{\mathrm{FO}}\exists^5_{\mathrm{MSO}}\mathsf{FO};\mathrm{succ}]$.

Let $\mathcal{C}$ be a 2-register machine. A computation of $\mathcal{C}$ can be encoded as a word $w \in \{0,1\}^*$ in the following way. The word w consists of segments of the form $110b_1 \ldots b_s 0 z_0 z_1 z'_0 z'_1$. The sequence $b_1 \ldots b_s$ encodes the state, namely $b_q = 1$ iff the state of the configuration is q. The sequence $z_0 z_1 z'_0 z'_1$ encodes the increment or decrement of a register: $z_i = 1$ iff the ith register is incremented, and $z'_i = 1$ iff the ith register is decremented. With the letters $110 \ldots 0 \ldots$ we can check whether a subword of w represents an encoding of a configuration.

We define a sentence of the form $\exists X \forall y \exists U \exists Z_0 \exists Z_1 \exists Z'_0 \exists Z'_1 \psi$, where ψ is FO. The details on this sentence are in [15]. Intuitively, X represents a word $w \in \{0,1\}^*$ that is an encoding of a computation of $\mathcal{C}$, where w is a concatenation of sequences of the form $110 \ldots 0 \ldots$ as explained above. The FO variable y intuitively ranges over all the configurations in X, and the MSO variables Z_i, Z'_i take care of the increments and decrements of the ith register up to the yth configuration. Therefore, $|Z_i| - |Z'_i|$ is the value of the ith counter in the yth configuration. The MSO variable U is used for some technical reasons; it represents the set $\{0, \ldots, y\}$. $\dashv$

Decidability Result. Since the emptiness problem for PFWAs is decidable and the construction of the PFWA in Theorem 11 for a given formula in E is constructive, we get a decision procedure for E: the formula is satisfiable iff the language of the constructed PFWA is nonempty. Here we show a stronger decidability result. Namely, we give a decision procedure for sentences that have an arbitrary prefix of FO quantifiers and the body of the sentence or its negation is in E. This is done by two constructions. We first construct a PFWA using Theorem 11, where we drop the prefix of FO quantifiers of the given sentence. Second, we construct from this PFWA a formula in Presburger arithmetic taking care of the quantification of the FO variables.

Theorem 16. *The fragment* $[\mathrm{FO}(\exists^*_{\mathrm{MSO}} \cup \forall^*_{\mathrm{MSO}})\mathrm{FO}; \mathrm{R}_{\mathrm{WS1S}}]$ *is decidable.*

Proof. Case I: $\varphi \in [\mathrm{FO}\exists^*_{\mathrm{MSO}}\mathrm{FO}; \mathrm{R}_{\mathrm{WS1S}}]$. Note that every relation R occurring in φ is expressible by a WS1S formula ψ_R. Therefore, we can assume that φ is of the form $Q_1 x_1 \ldots Q_m x_m \varphi'$ with $Q_1, \ldots, Q_m \in \{\exists, \forall\}$ and $\varphi' \in \mathrm{E}$ by substituting the relations R with ψ_R. By Theorem 11 we can construct a PFWA $(\mathcal{A}, C)$ with dimension $N \geq 1$ and $L(\varphi') = L(\mathcal{A}, C)$. Assume that $\mathcal{A} = (S, \Gamma, \delta, s_\mathrm{I}, F)$ with $\Gamma \subseteq \{0,1\}^m \times \mathbb{N}^N$. It holds, $(\mathbb{N}, \mathrm{succ}) \models Q_1 x_1 \ldots Q_m x_m \varphi'$ iff

$$Q_1 \tilde{x}_1 \in \mathbb{N}, \ldots, Q_m \tilde{x}_m \in \mathbb{N} \text{ there is a word } w \in L(\mathcal{A}, C) \text{ such that} \atop I_w(x_1) = \tilde{x}_1, \ldots, I_w(x_m) = \tilde{x}_m. \tag{1}$$

By definition, (1) is equivalent to

$$Q_1 \tilde{x}_1 \in \mathbb{N}, \ldots, Q_m \tilde{x}_m \in \mathbb{N} \text{ there is a word } w \in L(\mathcal{A}) \text{ such that} \atop I_{\Psi(w)}(x_1) = \tilde{x}_1, \ldots, I_{\Psi(w)}(x_m) = \tilde{x}_m \text{ and } \Phi(w) \in C, \tag{2}$$

where $\Psi : \Gamma \to \{0,1\}^m$ is the projection and Φ the extended Parikh map of Γ.

We extend the alphabet Γ to $\Gamma' := \{(b, \overline{v}, \overline{v'}) \mid (b, \overline{v}) \in \Gamma \text{ and } \overline{v'} \in \{0,1\}^m\}$, that is, we append the vectors in $\{0,1\}^m$ to each symbol $(b, \overline{v}) \in \Gamma$. Let Φ' be the extended Parikh map of Γ', and let $h : \Gamma'^* \to \Gamma^*$ be the homomorphism defined by $h(b, \overline{v}, \overline{v'}) := (b, \overline{v})$. We construct an automaton $\mathcal{A}'$ accepting $w \in \Gamma'^*$ iff $h(w) \in L(\mathcal{A})$ and $\Phi'(w) = (\Phi(h(w)), I_{\Psi(h(w))}(x_1), \ldots, I_{\Psi(h(w))}(x_m))$. Let $\mathcal{A}' := (S, \Gamma', \delta', s_\mathrm{I}, F)$, where the transition function δ' contains the same transitions as δ except that $\mathcal{A}'$ marks the positions in a word that determine the values of the interpretations for the FO variables $x_1, \ldots, x_m$. For each x_i, let $B_i \subseteq S$ be the set that contains all the states that are reachable *before* reading a symbol that determines the interpretation of x_i, that is $B_i := \bigcup_{j \in \mathbb{N}} B_i^j$, where $B_i^0 := \{s_\mathrm{I}\}$ and for $j > 0$, $B_i^{j+1} := \{s \in \delta(B_i^j, (b, \overline{v})) \mid (b, \overline{v}) \in \Gamma \text{ with } \chi_{x_i}(b) = 0\}$. Note that if a state s is in B_i and from s we can still reach an accepting state then for every word $w \in \Gamma^*$ with $\delta(s_\mathrm{I}, w) = s$ it holds that $\chi_{x_i}(w)$ is of the form $0 \ldots 0$. Otherwise, there would be a word in $L(\mathcal{A}, C)$ that is not an interpretation for the FO variable x_i. For $s \in S$, let $\overline{c}(s) \in \{0,1\}^m$ be the characteristic vector of s, that is $\overline{c}(s) := (c_1, \ldots, c_m)$, where $c_i = 1$ iff $s \in B_i$. Now, $\delta' : S \times \Gamma' \to \mathcal{P}(S)$ is defined by $\delta'(s, (b, \overline{v}, \overline{v'})) := \{s' \in \delta(s, (b, \overline{v})) \mid \overline{c}(s') = \overline{v'}\}$. By the construction of $\mathcal{A}'$, (2) is equivalent to

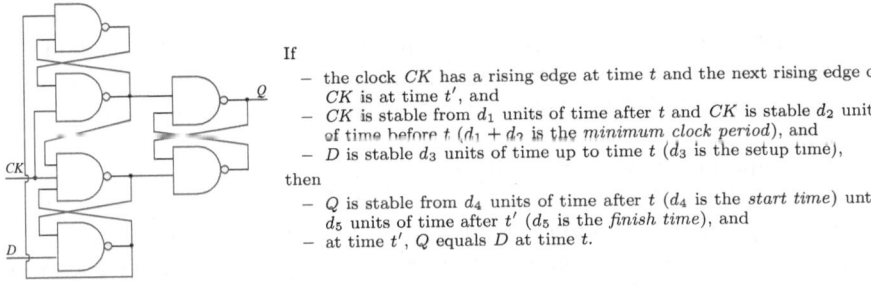

If

- the clock CK has a rising edge at time t and the next rising edge of CK is at time t', and
- CK is stable from d_1 units of time after t and CK is stable d_2 units of time before t ($d_1 + d_2$ is the *minimum clock period*), and
- D is stable d_3 units of time up to time t (d_3 is the *setup time*),

then

- Q is stable from d_4 units of time after t (d_4 is the *start time*) until d_5 units of time after t' (d_5 is the *finish time*), and
- at time t', Q equals D at time t.

Fig. 2. Circuit of an edge-triggered D-type flip-flop and its specification.

$$Q_1\tilde{x}_1 \in \mathbb{N}, \ldots, Q_m\tilde{x}_m \in \mathbb{N} \text{ there is a word } w \in L(\mathcal{A}') \text{ such that} \tag{3}$$
$$\Phi(h(w)) \in C \text{ and } \Phi'(w) = (\Phi(h(w)), \tilde{x}_1, \ldots, \tilde{x}_m).$$

From Lemma 5, we know that $\Phi'(L(\mathcal{A}'))$ is the union of the images of linear polynomials $q_1, \ldots, q_\ell : \mathbb{N}^r \to \mathbb{N}^N$, for some $r \geq 1$. Moreover, these polynomials are constructible from $\mathcal{A}'$. We conclude that (3) is equivalent to

$$Q_1\tilde{x}_1 \in \mathbb{N}, \ldots, Q_m\tilde{x}_m \in \mathbb{N} \text{ there are } y_1, \ldots, y_r \in \mathbb{N} \text{ and } \overline{v} \in \mathbb{N}^N \text{ such that}$$
$$\overline{v} \in C \text{ and } q_i(y_1, \ldots, y_r) = (\overline{v}, \tilde{x}_1, \ldots, \tilde{x}_m), \text{ for some } 1 \leq i \leq \ell. \tag{4}$$

Note that (4) can be expressed as a sentence in Presburger arithmetic. The claim follows from the decidability of Presburger arithmetic.

<u>Case II:</u> $\varphi \in [\mathsf{FO}\forall^*_{\mathsf{MSO}}\mathsf{FO}; \mathsf{R}_{\mathsf{WS1S}}]$. Follows from Case I by the duality of quantifiers.   ⊣

Applications. As an application, we sketch how this decidable fragment can be used to decide WS1S extended with some restricted linear arithmetic. Our example is the verification of an edge-triggered D-type flip-flop, taken from [3,8]. Although the circuit is built from only six nand-gates (left half of Figure 2), proving that the circuit meets its specification (right half of Figure 2) is "fairly complicated", as Gordon noted in [8]. The proof in [8] was done by paper and pencil, and contained a flaw, as reported in [3,28]. The correctness proof in [3] was done automatically by naturally expressing the higher-order logic formalization from [8] in WS1S and using the implementation of the automata-based decision procedure for WS1S in the Mona tool [10]. This verification technique works only if the parameters $d_1, \ldots, d_5$ are instantiated with concrete values because the specification contains some linear arithmetic, for example, "Q is stable from d_4 units of time after t until d_5 units of time after t'". Reusing most of the WS1S formalization from [3] we can formalize in the decidable fragment of WS1S^{card} whether the circuit meets its specification for all $d_1, \ldots, d_5 \in \mathbb{N}$ satisfying, for instance, the constraints $d_1 \geq 2$, $d_2 \geq 2$, $d_1 + d_2 \geq 5$, $d_3 \geq 3$, $d_4 \geq 3$, and $d_5 \leq 2$. Together with Theorem 16 this demonstrates that such parameterized verification problems are actually decidable problems.

We briefly recall the formalization in [3].[1] To keep the formulas readable, we use some syntactic sugar for WS1S. It will always be straightforward to translate the used notation to WS1S. Note that $x \leq y$ can be defined by $\forall Z(Zy \wedge \forall z(Zz + 1 \rightarrow Zz) \rightarrow Zx)$.

The temporal behavior of a unit-delay nand-gate with inputs X and Y, and output Z up to time \$ is described by $nand(\$, X, Y, Z) := \forall t(t < \$ \rightarrow (Zt+1 \leftrightarrow \neg(Xt \wedge Yt)))$, where \$ is an FO variable and X, Y, and Z are MSO variables. The temporal behavior of a nand-gate with three inputs can be described analogously. The circuit of the left half of Figure 2 implementing a D-type flip-flop can now be described by the following formula, where the internal wires are hidden by existential quantification.

$$imp(\$, D, CK, Q) := \exists W_1 \exists W_2 \exists W_3 \exists W_4 \exists W_5 \big(nand(\$, W_2, D, W_1) \wedge$$
$$nand3(\$, W_3, CK, W_1, W_2) \wedge nand(\$, W_4, CK, W_3) \wedge$$
$$nand(\$, W_1, W_3, W_4) \wedge nand(\$, W_3, W_5, Q) \wedge$$
$$nand(\$, Q, W_2, W_5)\big)$$

We recall the definitions [3] of the temporal concepts needed to formalize the flip-flop's specification.

- X is stable in the interval $[t, t')$:

$$stable(t, t', X) := \forall u\big(t \leq u < t' \rightarrow (Xu \leftrightarrow Xt)\big)$$

- X rises at t:

$$rise(t, X) := t > 0 \wedge \neg X(t-1) \wedge Xt$$

- t' is the next instance after t where X rises:

$$nextRise(t, t', X) := rise(t', X) \wedge \forall u\big(t < u < t' \rightarrow \neg rise(u, X)\big)$$

The flip-flop's specification given in the right half of Figure 2 can be formalized as

$$spec(\$, t, t', D, CK, Q) := \big(d_2 \leq t < t' \leq \$ - d_5 \wedge$$
$$rise(t, CK) \wedge nextRise(t, t', CK) \wedge$$
$$stable(t, t + d_1, CK) \wedge stable(t - d_2, t, CK) \wedge$$
$$stable(t + 1 - d_3, t + 1, D)\big) \rightarrow$$
$$stable(t + d_4, t' + d_5, Q) \wedge (Qt' \leftrightarrow Dt).$$

Note that this formula is a WS1S formula if $d_1, \ldots, d_5$ are not FO variables but natural numbers. For fixed values for $d_1, \ldots, d_5$, Mona checks automatically if the circuit meets its specification by computing the truth value of the formula

$$verify(\$, t, t', D, CK, Q) := imp(\$, D, CK, Q) \rightarrow spec(\$, t, t', D, CK, Q).$$

---

[1] Actually, Basin and Klarlund did not use WS1S but M2L(str). There are some technical differences between these logics as explained in §3. We have adopted their formalization to WS1S.

In the following, we show how the decidability result from Theorem 16 can be used to check whether the circuit is correct, for instance, for all $d_1, \ldots, d_5 \in \mathbb{N}$ with $d_1 > 2$, $d_2 \geq 2$, $d_1 + d_2 \geq 5$, $d_3 \geq 3$, $d_4 \geq 3$, and $d_5 \leq 2$. The constraints on the parameters can be expressed in WS1S by

$$constr(d_1, \ldots, d_5) := (d_1 \geq 2 \wedge d_2 \geq 3 \vee d_1 \geq 3 \wedge d_2 \geq 2) \wedge d_3 \geq 3 \wedge d_4 \geq 3 \wedge d_5 \leq 2.$$

Unfortunately, $\forall d_1 \ldots \forall d_5 \left(constr(d_1, \ldots, d_5) \rightarrow \forall \$ \forall t \forall t' \forall D \forall CK \forall Q \left(verify(\$, t, t', D, CK, Q) \right) \right)$ is not a WS1S^{card} formula, since $verify$ contains in the subformula $spec$ terms involving linear arithmetic, for example, $t + d_1$. But we can take a detour using MSO variables. For example, for the term $t + d_1$, we introduce an FO variable x_{t+d_1} and MSO variables T, D_1 with $x_{t+d_1} = |T| + |D_1|$, $T = \{0, \ldots, t-1\}$, and $D_1 = \{0, \ldots, d_1 - 1\}$. It holds $x_{t+d_1} = t + d_1$. Thus, the term $t + d_1$ can be substituted by x_{t+d_1}. Let $spec'$ be the formula, where we replace in $spec$ the terms τ involving linear arithmetic by fresh variables x_τ, that is,

$$spec'(\$, t, t', D, CK, Q, x_{\$-d_5}, x_{t+d_1}, x_{t-d_2}, x_{t-d_3}, x_{t+d_4}, x_{t'+d_5}) :=$$
$$\left(d_2 \leq t < t' \leq x_{\$-d_5} \wedge \right.$$
$$rise(t, CK) \wedge nextRise(t, t', CK) \wedge$$
$$stable(t, x_{t+d_1}, CK) \wedge stable(x_{t-d_2}, t, CK) \wedge$$
$$\left. stable(x_{t-d_3} + 1, t + 1, D)\right) \rightarrow$$
$$stable(x_{t+d_4}, x_{t'+d_5}, Q) \wedge (Qt' \leftrightarrow Dt).$$

We write $x = \pm|X_1| \pm \cdots \pm |X_r|$ for $\exists Z \left(\forall z (Zz \leftrightarrow z < x) \wedge |Z| = \pm|X_1| \pm \cdots \pm |X_r| \right)$, where x is an FO variable and the X_is are MSO variables. The formula aux ensures that the new variables in $spec'$ have the correct values, for example, x_{t+d_1} equals $t + d_1$.

$$aux(d_1, \ldots, d_5, \$, t, t', x_{\$-d_5}, x_{t+d_1}, x_{t-d_2}, x_{t-d_3}, x_{t+d_4}, x_{t'+d_5}) :=$$
$$\exists D_1 \ldots \exists D_5 \exists \pounds \exists T \exists T' \left(d_1 = |D_1| \wedge \cdots \wedge d_5 = |D_5| \wedge \right.$$
$$\$ = |\pounds| \wedge t = |T| \wedge t' = |T'| \wedge$$
$$x_{\$-d_5} = |\pounds| - |D_5| \wedge x_{t+d_1} = |T| + |D_1| \wedge$$
$$x_{t-d_2} = |T| - |D_2| \wedge x_{t-d_3} = |T| - |D_3| \wedge$$
$$\left. x_{t+d_4} = |T| + |D_4| \wedge x_{t'+d_5} = |T'| + |D_5|\right)$$

For proving the circuit correct, we have to check whether the formula

$$verify' := aux \wedge constr \rightarrow (imp \rightarrow spec')$$

is valid. This can be done automatically by Theorem 16, since $verify'$ can be transformed into a formula in $[\text{FO}\forall^*_{\text{MSO}}\text{FO}; \text{R}_{\text{WS1S}}]$ by universally quantifying over the FO variables and the MSO variables D, CK, Q, and by pulling out the existentially quantified MSO variables in aux. Note that the existential quantifiers become universal by this process. In addition to verification, our procedure may also be used for synthesizing sufficient parameter constraints if we do not restrict the parameters $d_1, \ldots, d_5$ by some constraints and do not universally quantify over them.

Although our decision procedure is built on top of a decision procedure for Presburger arithmetic and a translation from WS1S^{card} formulas to PFWAs and the worst-case complexity is in both cases very high we are encouraged by the outcomes with a prototype implementation. We tested our implementation on various case studies, such as the D-type flip-flop above and lemmas in a PVS theory about cardinalities of finite sets that were used in [24] to verify oral message algorithms. Such kinds of proofs are cumbersome and rather involved. Our decidability result opens up the possibility to effectively automate such kinds of verification problems.

5   Conclusions

We have extended WS1S with linear cardinality constraints, proved the undecidability of this extension, and identified decidable fragments (see Figure 1). These results were obtained by extending the logic-automata connection to fragments of WS1S with cardinality constraints and an appropriate automaton model that we call *Parikh finite word automata*. The resulting decision procedure has applications in both hardware and protocol verification [14,15], and initial experiments with an extension of the Mona tool with cardinality constraints are encouraging [13].

One advantage of our notion of Parikh word automata is that it easily generalizes to trees. A decidability result for a fragment of the weak monadic second-order logic of *two* successors with cardinality constraints using *Parikh finite tree automata* is included in [15]. Since monadic second-order logics on trees give a theoretical foundation of XML query languages [9], our results on trees may serve as a theoretical basis for extending current query languages as in [5].

The framework in §2 can also be generalized to infinite words and trees. A possible acceptance condition is in the spirit of the Büchi acceptance condition: one requires that the arithmetic constraints have to be satisfied for infinitely many prefixes in order to accept the input. Another extension that we want to look at is to generalize the framework to graphs with bounded tree-width.

Future work will include detailed complexity analyses, theoretically and practically, on Parikh automata and on the decision procedure for the decidable fragment of WS1S^{card}.

Acknowledgments. We thank J. Rushby for initiating and supporting this research, and the anonymous referees for their invaluable comments. The first author also thanks J. Meseguer.

References

1. D. BASIN AND S. FRIEDRICH, *Combining WS1S and HOL*, in FroCos'98, Applied Logic Series, 2000, pp. 39–56.

2. D. BASIN, S. FRIEDRICH, AND S. MÖDERSHEIM, *B2M: A semantic based tool for BLIF hardware descriptions*, in FMCAD'00, vol. 1954 of LNCS, 2000, pp. 91–107.

3. D. BASIN AND N. KLARLUND, *Automata based symbolic reasoning in hardware verification*, FM3D, 13 (1998), pp. 255–288.

4. H. COMON AND Y. JURSKI, *Multiple counters automata, safety analysis and Presburger arithmetic*, in CAV'98, vol. 1427 of LNCS, 1998, pp. 268–279.

5. S. DAL ZILIO AND D. LUGIEZ, *XML schema, tree logic and sheaves automata*, Research Report 4631, INRIA, 2002.

6. J. DASSOW AND V. MITRANA, *Finite automata over free groups*, International Journal of Algebra and Computation, 10 (2000), pp. 725–737.

7. A. FINKEL AND G. SUTRE, *Decidability of rechability problems for classes of two counter automata*, in STACS'00, vol. 1770 of LNCS, 2000, pp. 346–357.

8. M. GORDON, *Why higher-order logics is a good formalism for specifying and verifying hardware*, in Formal Aspects of VLSI Design, North-Holland, 1986, pp. 153–177.

9. G. GOTTLOB AND C. KOCH, *Monadic Datalog and the expressive power of languages for web information extraction*, in PODS'02, 2002, pp. 17–28.

10. J. HENRIKSEN, J. JENSEN, M. JORGENSEN, N. KLARLUND, B. PAIGE, T. RAUHE, AND A. SANDHOLM, *Mona: Monadic second-order logic in practice*, in TACAS'95, vol. 1019 of LNCS, 1995, pp. 89–110.

11. O. IBARRA, *Reversal-bounded multicounter machines and their decision problems*, JACM, 25 (1978), pp. 116–133.

12. O. IBARRA, J. SU, Z. DANG, T. BULTAN, AND R. KEMMERER, *Counter machines and verification problems*, TCS, 289 (2002), pp. 165–189.

13. F. KLAEDTKE, *CMona: Monadic second-order logics with linear cardinality constraints in practice.* in preparation, 2003.

14. F. KLAEDTKE AND H. RUESS, *WS1S with cardinality constraints*, Technical Report SRI-CSL-05-01, SRI International, 2001.

15. ———, *Parikh automata and monadic second-order logics with linear cardinality constraints*, Technical Report 177, Albert-Ludwigs-Universität Freiburg, 2002. (revised version).

16. N. KLARLUND, A. MØLLER, AND M. SCHWARTZBACH, *MONA implementation secrets*, in CIAA'00, vol. 2088 of LNCS, 2000, pp. 182–194.

17. N. KLARLUND, M. NIELSEN, AND K. SUNESEN, *Automated logical verification based on trace abstraction*, in PODC'96, 1996, pp. 101–110.

18. L. LAMPORT, R. SHOSTAK, AND M. PEASE, *The Byzantine Generals problem*, TOPLAS, 4 (1982), pp. 382–401.

19. A. MEYER, *Weak monadic second-order theory of successor is not elementary-recursive*, in Logic Colloquium, vol. 453 of LNM, 1975, pp. 132–154.

20. V. MITRANA AND R. STIEBE, *Extended finite automata over groups*, Discrete Applied Mathematics, 108 (2001), pp. 287–300.

21. S. OWRE AND H. RUESS, *Integrating WS1S with PVS*, in CAV'00, vol. 1855 of LNCS, 2000, pp. 548–551.

22. M. PARIGOT AND E. PELZ, *A logical approach of Petri net languages*, TCS, 39 (1985), pp. 155–169.

23. R. PARIKH, *On context-free languages*, JACM, 13 (1966), pp. 570–581.

24. J. RUSHBY, *Systematic formal verification for fault-tolerant time-triggered algorithms*, IEEE Trans. on Software Engineering, 2 (1999), pp. 651–660.

25. M. SMITH AND N. KLARLUND, *Verification of a sliding window protocol using IOA and MONA*, in FORTE/PSTV'00, vol. 183 of IFIP Conf. Proc., 2000, pp. 19–34.

26. L. STOCKMEYER, *The Complexity of Decision Problems in Automata Theory and Logic*, PhD thesis, Dept. of Electrical Engineering, MIT, Boston, Mass., 1974.

696   F. Klaedtke and H. Rueß

27. W. THOMAS, *Languages, automata, and logic*, in Handbook of Formal Languages, vol. 3, Springer-Verlag, 1997, pp. 389–455.

28. A. WILK AND A. PNUELI, *Specification and verification of VLSI systems*, in IC-CAD'89, 1989, pp. 460–463.

$\Pi_2 \cap \Sigma_2 \equiv AFMC$

Orna Kupferman[1*] and Moshe Y. Vardi[2**]

[1] Hebrew University, School of Engineering and Computer Science, Jerusalem 91904,
Israel
orna@cs.huji.ac.il, http://www.cs.huji.ac.il/~orna
[2] Rice University, Department of Computer Science, Houston, TX 77251-1892, U.S.A.
vardi@cs.rice.edu, http://www.cs.rice.edu/~vardi

Abstract. The *μ-calculus* is an expressive specification language in
which modal logic is extended with fixpoint operators, subsuming many
dynamic, temporal, and description logics. Formulas of μ-calculus are
classified according to their *alternation depth*, which is the maximal
length of a chain of nested alternating least and greatest fixpoint operators. Alternation depth is the major factor in the complexity of μ-calculus
model-checking algorithms. A refined classification of μ-calculus formulas
distinguishes between formulas in which the outermost fixpoint operator
in the nested chain is a least fixpoint operator (Σ_i formulas, where i is
the alternation depth) and formulas where it is a greatest fixpoint operator (Π_i formulas). The *alternation-free μ-calculus* (AFMC) consists
of μ-calculus formulas with no alternation between least and greatest
fixpoint operators. Thus, AFMC is a natural closure of $\Sigma_1 \cup \Pi_1$, which
is contained in both Σ_2 and Π_2. In this work we show that $\Sigma_2 \cap \Pi_2 \equiv$
AFMC. In other words, if we can express a property ξ both as a least fixpoint nested inside a greatest fixpoint and as a greatest fixpoint nested
inside a least fixpoint, then we can express ξ also with no alternation
between greatest and least fixpoints. Our result refers to μ-calculus over
arbitrary Kripke structures. A similar result, for directed μ-calculus formulas interpreted over trees with a fixed finite branching degree, follows
from results by Arnold and Niwinski. Their proofs there cannot be easily extended to Kripke structures, and our extension involves *symmetric
nondeterministic Büchi* tree automata, and new constructions for them.

1 Introduction

The *μ-calculus* is an expressive specification language in which formulas are built
from Boolean operators, existential ($\Diamond$) and universal ($\Box$) next-time modalities,
and least (μ) and greatest (ν) fixpoint operators [Koz83]. The discovery and use
of *symbolic model-checking* methods [McM93] for verification of large systems

* Supported in part by by NSF grant CCR-9988172 and by a research grant from the
Center for Pure and Applied Mathematics at the University of California, Berkeley
** Supported in part by NSF grants CCR-9988322, CCR-0124077, IIS-9908435, IIS-
9978135, and EIA-0086264, by BSF grant 9800096, and by a grant from the Intel
Corporation.

J.C.M. Baeten et al. (Eds.): ICALP 2003, LNCS 2719, pp. 697–713, 2003.
© Springer-Verlag Berlin Heidelberg 2003

has made the μ-calculus important also from a practical point of view: symbolic model-checking tools proceed by computing fixpoint expressions over the model's set of states. For example, to find the set of states from which a state satisfying some predicate p is reachable, the model checker starts with the set S of states in which p holds, and repeatedly add to S the set $\Diamond S$ of states that have a successor in S. Formally, the model checker calculates the set of states that satisfy the μ-calculus formula $\mu y.p \vee \Diamond y$.

Formulas of μ-calculus are classified according to their *alternation depth*, which is the maximal length of a chain of nested alternating least and greatest fixpoint operators. From a practical point of view, the classification is important, as the alternation depth is the major factor in the complexity of μ-calculus model-checking algorithms: the original algorithm for model checking a structure of size m with respect to a formula of length n and alternation depth d requires time $O(mn)^d$ [EL86], and more sophisticated algorithms can do the job in time roughly $O(mn)^{\lfloor \frac{d}{2} \rfloor + 1}$ [Jur00]. From a theoretical point of view, the classification naturally raises questions about the expressive power of the classes. In particular, the question whether the expressiveness hierarchy for the μ-calculus collapses (i.e., whether there is some $d \geq 1$ such that all μ-calculus formulas can be translated to formulas of alternation depth d) has been answered to the negative [Bra98]. The alternation-depth hierarchy of μ-calculus and the model-checking problem for the various classes in the hierarchy are strongly related to the index hierarchy in *parity games* and to the problem of deciding such games [Jur00].

A more refined classification of μ-calculus formulas distinguishes between formulas in which the outermost fixpoint operator in the nested chain is a least fixpoint operator (Σ_i formulas, where i is the alternation depth) and formulas where it is a greatest fixpoint operator (Π_i formulas). For example, the formula $\mu y.p \vee \Diamond y$ is a Σ_1 formula, as it has alternating depth 1 and its outermost fixpoint operator is μ. Similarly, the formula $\nu y.\mu z.\Box[(p \wedge y) \vee z]$ is a Π_2 formula[1]. By duality of the least and greatest fixpoint operators, the classes Π_i and Σ_i are complementary, in the sense that a formula ψ is in Π_i iff the formula $\neg\psi$ (in positive normal form, where negation is applied to atomic propositions only) is in Σ_i.

Some fragments of μ-calculus are of special interest in computer science: *Modal Logic* (ML) consists of μ-calculus formulas with no fixpoint operators (that is, ML $= \Sigma_0 \cup \Pi_0$). It is actually more correct to say that μ-calculus is the extension of ML with fixpoint operators. Extending ML with fixpoint operators still retain some of its basic semantic properties, in particular the property of being invariant under bisimulation [Ben91]. The *alternation-free μ-calculus* (AFMC) consists of μ-calculus formulas with no alternation between least and greatest fixpoint operators. Thus, AFMC is a natural closure of $\Sigma_1 \cup \Pi_1$, which is contained in both Σ_2 and Π_2. AFMC subsumes the branching temporal logic CTL and the dynamic logic PDL [FL79]. Formulas of AFMC

---

[1] An exact definition of the classes Σ_i and Π_i refers to the scope of the fixpoint operators. As we discuss in Section 4, several different definitions are studied in the literature, and we follow here the definition of [Niw86].

can be symbolically evaluated in time linear in the structure [CS91,KVW00]. While designers may prefer to use higher-level logics to specify properties, model-checking tools often proceed by evaluating the corresponding AFMC formulas [BRS99]. Finally, it is hard to produce an understandable formula with more than one alternation. Thus, $\Pi_2 \cup \Sigma_2$ subsumes almost all formulas one may wish to specify in practice. Formally, $\Pi_2 \cup \Sigma_2$ subsumes the branching temporal logic CTL*, and in fact, until [Bra98], the strictness of the expressiveness hierarchy of μ-calculus was known only for Π_i and Σ_i with $i \leq 2$ [AN90]. Also, the symbolic evaluation of linear properties is reduced to calculating a Π_2 formula [VW86, EL85].

For several hierarchies in computer science, even strict ones, it is possible to show local *coalescence*, where membership in some class of the hierarchy and in its complementary class implies membership in a lower class. For example, RE $\cap$ co-RE = Rec describes coalescence at the bottom of the arithmetical hierarchy [Rog67]. On the other hand, the analogous coalescence for the polynomial hierarchy is not known; it is a major open question whether NP $\cap$ co-NP = P [GJ79]. In [KV01], we showed that the bottom levels of the μ-calculus expressiveness hierarchy coalesce: $\Sigma_1 \cap \Pi_1 \equiv ML$. In other words, if we can express a property ξ both as a least fixpoint and as a greatest fixpoint, then we can express ξ without fixpoints. The proof uses the fact that μ-calculus formulas in $\Sigma_1 \cap \Pi_1$ correspond to languages that are both safety and co-safety. Consequently, for every property $\xi \in \Sigma_1 \cap \Pi_1$, we can construct two nondeterministic *looping* tree automata $\mathcal{U}$ and $\mathcal{U}'$ such that $\mathcal{U}$ and $\mathcal{U}'$ accept exactly all the trees that satisfy ξ and its complement, respectively (the fact that $\mathcal{U}$ and $\mathcal{U}'$ are looping means that they have trivial acceptance conditions – every infinite run is accepting). We showed in [KV01] how $\mathcal{U}$ and $\mathcal{U}'$ can be combined to a *cycle-free* automaton and then translate to an ML formula expressing ξ.

In this paper we show coalescence in higher classes of the hierarchy, namely $\Sigma_2 \cap \Pi_2 \equiv \text{AFMC}$.[2] In other words, if we can specify a property ξ both as a least fixpoint nested inside a greatest fixpoint and as a greatest fixpoint nested inside a least fixpoint, then we can express ξ also with no alternation between greatest and least fixpoints. Unfortunately, the technique of [KV01] is too weak to be helpful here. Indeed, formulas in Π_2 cannot be expressed by looping automata. As we explain below, the known automata-theoretic characterizations of Σ_2 and Π_2, and their relation to AFMC, cannot help us either.

One such known characterization [Niw86,AN92] refers to the expressive power of the μ-calculus over trees with fixed finite branching degrees. Over such trees, the existential next-time modality of the μ-calculus can be parameterized with *directions*. A modality parameterized with direction d means that the corresponding existential requirement should be satisfied in the d-th child of the current state. For example, for a binary tree in which each node has a left child and a right child, the formula $\Diamond_1 p$ means that the left child of the root satisfies

---

[2] The analogous complexity-theoretic result would be $\Sigma_2^p \cap \Pi_2^p = P^{NP}$, where Σ_2^p and Π_2^p form the second level of the polynomial hierarchy and P^{NP} is the polynomial closure of NP [GJ79].

p, and the formula $\mu y.p \vee \Diamond_r y$ means that some node in the rightmost path of the tree satisfies p. The ability of *directed μ-calculus* to distinguish between the various children of a node makes it convenient to translate formulas to tree automata and vice versa. In particular, it is known that directed-Π_2 is as expressive as *nondeterministic Büchi tree automata* [AN90,Kai95]. Our interest in this paper is in the expressive power of the μ-calculus over arbitrary *Kripke structures*, possibly with an infinite branching degree, which means that we cannot restrict attention to trees of fixed branching degrees.

An automata-theoretic framework for μ-calculus without directions is suggested in [JW95], by means of μ-*automata*, which are essentially symmetric alternating tree automata in a certain normal form. A related approach, in which alternation is more explicit, is presented in [Wil99]. Alternation allows the automaton to send several requirements to the same child. Symmetry means that the automaton does not distinguish between the different children of a node, and it sends copies to child nodes only in either a universal or an existential manner. It also means that the automaton can handle trees with a variable and even infinite branching degree. Formulas of μ-calculus in Π_i and Σ_i can be linearly translated to symmetric alternating parity/co-parity automata of index i. While it is possible to translate μ-calculus formulas to symmetric alternating automata, it is not immediately clear how such a translation can help in a translating of $\Sigma_2 \cap \Pi_2$ into the AFMC. By [AN92,KV99], formulas that are members of both directed-Π_2 and directed-Σ_2 can be translated to directed-AFMC. The proofs in [AN92,KV99] shows that given a formula $\psi \in \Sigma_2 \cap \Pi_2$, we can construct two nondeterministic Büchi tree automata $\mathcal{U}$ and $\mathcal{U}'$, for ψ and $\neg\psi$, and then combine the automata to a weak alternating automaton equivalent to ψ. The combination of $\mathcal{U}$ and $\mathcal{U}'$, however, crucially depends on the fact that the automata are nondeterministic (rather than alternating) and the fact that the automata can refer to particular directions in the tree.

The key to the results in [KV01] and here is a development of a theory of *symmetric nondeterministic* tree automata. In [KV01], we defined symmetric nondeterministic *looping* automata, and showed how to construct such automata for formulas in Π_1. In order to handle Σ_2 and Π_2, we define here symmetric nondeterministic *Büchi* automata, and translate Π_2 formulas to such automata. From a technical point of view, symmetric nondeterministic tree automata are essentially symmetric alternating automata with transitions in disjunctive normal form. Our main contribution is the development of various constructions for symmetric nondeterministic tree automata and their application to the study of the expressive power of the μ-calculus. Since removal of alternation in Büchi automata should take into an account the acceptance condition of the automaton and keep track of the states visited in each path of the run tree, the symmetry of the automaton poses real technical challenges. We then extend the construction in [KV99] to symmetric automata and combine the symmetric nondeterministic Büchi tree automata for ψ and $\neg\psi$ to a symmetric weak alternating automaton for ψ. Again, symmetry poses real technical challenges. (In fact, while the construction in [KV99] for the directed case is quadratic, here we end up with quadratically many states but exponentially many transitions.) Once we have a weak symmetric alternating automaton for ψ, it is possible to generate from it an equivalent AFMC formula [KV98].

2 Preliminaries

For a set $D \subseteq \mathbb{N}$ of directions, a D-tree is a nonempty set $T \subseteq D^*$, where for every $x \cdot d \in T$ with $x \in D^*$ and $d \in D$, we have $x \in T$. The elements of T are called *nodes*, and the empty word ε is the *root* of T. For every $x \in T$, the nodes $x \cdot d$, for $d \in D$, are the *children* of x. A node with no children is a *leaf*. The *degree* of a node x is the number of children x has. Note that the degree of x is bounded by $|D|$. For technical convenience, we assume that the set D is finite[3]. A D-tree is *leafless* if it has no leafs. Note that a leafless tree is infinite. A *path* π of a tree T is a set $\pi \subseteq T$ such that $\varepsilon \in \pi$ and for every $x \in \pi$, either x is a leaf or exactly one child of x is in π. For two nodes x_1 and x_2 of T, we say that $x_1 \le x_2$ iff x_1 is a prefix of x_2; i.e., there exists $z \in D^*$ such that $x_2 = x_1 \cdot z$. We say that $x_1 < x_2$ iff $x_1 \le x_2$ and $x_1 \ne x_2$. A *frontier* of a leafless tree is a set $E \subset T$ of nodes such that for every path $\pi \subseteq T$, we have $|\pi \cap E| = 1$. For example, the set $E = \{0, 100, 101, 11\}$ is a frontier of the $\{0, 1\}$-tree $\{0, 1\}^*$. For two frontiers E_1 and E_2, we say that $E_1 \le E_2$ iff for every node $x_2 \in E_2$, there exists a node $x_1 \in E_1$ such that $x_1 \le x_2$. We say that $E_1 < E_2$ iff for every node $x_2 \in E_2$, there exists a node $x_1 \in E_1$ such that $x_1 < x_2$. Note that while $E_1 < E_2$ implies that $E_1 \le E_2$ and $E_1 \ne E_2$, the other direction does not necessarily hold. Given an alphabet Σ, a Σ-*labeled D-tree* is a pair $\langle T, V \rangle$ where T is a D-tree and $V : T \to \Sigma$ maps each node of T to a letter in Σ. We extend V to paths in a straightforward way. For a Σ-labeled D-tree $\langle T, V \rangle$ and a set $A \subseteq \Sigma$, we say that E is an A-frontier iff E is a frontier and for every node $x \in E$, we have $V(x) \in A$. We denote by $trees(D, \Sigma)$ the set of all Σ-labeled D-trees, and denote by $trees(\Sigma)$ the set of all Σ-labeled D-trees, for some D. For a set $\mathcal{T} \subseteq trees(\Sigma)$, we denote by $comp(\mathcal{T})$ the set of Σ-labeled trees that are not in $\mathcal{T}$; thus $comp(\mathcal{T}) = trees(\Sigma) \setminus \mathcal{T}$.

Automata on infinite trees (tree automata, for short) run on leafless Σ-labeled trees. *Alternating tree automata* generalize nondeterministic tree automata and were first introduced in [MS87]. *Symmetric* alternating tree automata [JW95, Wil99] are capable of reading trees with variable branching degrees. When a symmetric automaton reads a node of the input tree it sends copies to all successors of that node or to some successor. Formally, for a given set X, let $\mathcal{B}^+(X)$ be the set of positive Boolean formulas over X. For a set $Y \subseteq X$ and a formula $\theta \in \mathcal{B}^+(X)$, we say that Y *satisfies* θ iff assigning **true** to elements in Y and assigning **false** to elements in $X \setminus Y$ satisfies θ. A symmetric alterating Büchi tree automaton (symmetric ABT, for short) is a tuple $\mathcal{A} = \langle \Sigma, Q, \delta, q_0, F \rangle$ where Σ is the input alphabet, Q is a finite set of states, $\delta : Q \times \Sigma \to \mathcal{B}^+(\{\Box, \Diamond\} \times Q)$ is

---

[3] As we detail in the proof of Theorem 6, due to the bounded-tree-model property for μ-calculus, this technical assumption does not prevent us from proving our main result also for general structures with an infinite branching degree.

a transition function, $q_0 \in Q$ is an initial state, and $F \subseteq Q$ is a Büchi acceptance condition. Intuitively, an atom $\langle \Box, q \rangle$ in $\delta(q, \sigma)$ denotes a universal requirement to send a copy of the automaton in state q to all the children of the current node. An atom $\langle \Diamond, q \rangle$ denotes an existential requirement to send a copy of the automaton in state q to some child of the current node. When, for instance, the automaton is in state q, reads a node x with k children $x \cdot 1, \ldots, x \cdot k$, and $\delta(q, V(x)) = (\Box, q_1) \wedge (\Diamond, q_2) \vee (\Diamond, q_3) \wedge (\Diamond, q_4)$, it can either send k copies in state q_1 to the nodes $x \cdot 1, \ldots, x \cdot k$ and send a copy in state q_2 to some node in $x \cdot 1, \ldots, x \cdot k$ or send one copy in state q_3 to some node in $x \cdot 1, \ldots, x \cdot k$ and send one copy in state q_4 to some node in $x \cdot 1, \ldots, x \cdot k$. So, while nondeterministic tree automata send exactly one copy to each child, symmetric alternating automata can send several copies to the same child. On the other hand, symmetric alternating automata cannot distinguish between the different successors and can send copies to child nodes only in either a universal or an existential manner. Formally, a *run* of $\mathcal{A}$ on an input Σ-labeled D-tree $\langle T, V \rangle$, for some set D of directions, is an $(D^* \times Q)$-labeled $\mathbb{N}$-tree $\langle T_r, r \rangle$ such that $\varepsilon \in T_r$ and $r(\varepsilon) = (\varepsilon, q_0)$, and for all $y \in T_r$ with $r(y) = (x, q)$ and $\delta(q, V(x)) = \theta$, there is a (possibly empty) set $S \subseteq \{\Box, \Diamond\} \times Q$, such that S satisfies θ, and for all $(c, s) \in S$, the following hold: (1) If $c = \Box$, then for each $d \in D$, there is $j \in \mathbb{N}$ such that $y \cdot j \in T_r$ and $r(y \cdot j) = (x \cdot d, s)$. (2) If $c = \Diamond$, then for some $d \in D$, there is $j \in \mathbb{N}$ such that $y \cdot j \in T_r$ and $r(y \cdot j) = (x \cdot d, s)$. Note that if $\theta = \textbf{true}$, then y need not have children. This is the reason why T_r may have leaves. Also, since there exists no set S as required for $\theta = \textbf{false}$, we cannot have a run that takes a transition with $\theta = \textbf{false}$. For a run $\langle T_r, r \rangle$ and an infinite path $\pi \subseteq T_r$, we define $inf(\pi)$ to be the set of states that are visited infinitely often in π, thus $q \in inf(\pi)$ if and only if there are infinitely many $y \in \pi$ for which $r(y) \in T \times \{q\}$. A run $\langle T_r, r \rangle$ is accepting if all its infinite paths satisfy the Büchi acceptance condition; thus $inf(\pi) \cap F \neq \emptyset$. A tree $\langle T, V \rangle$ is accepted by $\mathcal{A}$ iff there exists an accepting run of $\mathcal{A}$ on $\langle T, V \rangle$, in which case $\langle T, V \rangle$ belongs to $\mathcal{L}(\mathcal{A})$. A tree $\langle T, V \rangle$ is accepted by $\mathcal{U}$ iff there exists an accepting run of $\mathcal{A}$ on $\langle T, V \rangle$, in which case $\langle T, V \rangle$ belongs to the language, $\mathcal{L}(\mathcal{A})$, of $\mathcal{A}$.

The transition function of an ABT $\mathcal{A}$ induces a graph $G_{\mathcal{A}} = \langle Q, E \rangle$ where $E(q, q')$ if there is $\sigma \in \Sigma$ such that $(\Box, q')$ or $(\Diamond, q')$ appears in $\delta(q, \sigma)$. An ABT is a *weak alternating tree automaton* (AWT, for short) if for each strongly connected component $C \subseteq Q$ of $G_{\mathcal{A}}$, either $C \subseteq F$ or $C \cap F = \emptyset$ [MSS86]. Note that every infinite path of a run of an AWT ultimately gets "trapped" within some strongly connected component C of $G_{\mathcal{A}}$. The path then satisfies the acceptance condition if and only if $C \subseteq F$.

The symmetry condition can also be applied to nondeterministic tree automata. In a *symmetric nondeterministic Büchi tree automaton* (symmetric NBT, for short) $\mathcal{U} = \langle \Sigma, Q, \delta, q_0, F \rangle$, the state space is $Q = 2^S$ for some set S of *micro-states*, and the transition function $\delta : Q \times \Sigma \to 2^{2^S \times 2^S}$ maps a state and a letter to sets of pairs $\langle U, E \rangle$ of subsets of S. The set $U \subseteq S$ is the *universal set* and it describes the micro-states that should be members in all the child states. The set $E \subseteq S$ is the *existential set* and it describes micro-states each of which has to be a member in at least one child state. Formally, given $k \geq 1$, a k-tuple $\langle S_1, \ldots, S_k \rangle$ is induced by $\delta(q, \sigma)$ if there is $\langle U, E \rangle$ in $\delta(q, \sigma)$ such that for all $1 \leq i \leq k$ we have $U \subseteq S_i$, and for all $s \in E$ there is $1 \leq i \leq k$ such that $s \in S_i$. Intuitively, when the automaton reads a node x labeled σ that has k children, and it proceeds from the state q, it has to take two choices. First, the automaton chooses a pair $\langle U, E \rangle \in \delta(q, \sigma)$. Then, it chooses a way to deliver E among the k children. Thus, we can describe the two choices of the automaton by a pair $\langle U, \langle E_1, \ldots, E_k \rangle \rangle$, where $\langle U, \bigcup_{1 \leq i \leq k} E_z \rangle \in \delta(q, \sigma)$. Note that E_z may be empty. We denote by $\delta_k(q, \sigma)$ the set of such pairs. A *run* of $\mathcal{U}$ on an input tree $\langle T, V \rangle$ is a Q-labeled tree $\langle T, r \rangle$, such that $r(\varepsilon) = q_0$, and for every $x \in T$ with $r(x) = q$, there exists $\langle q_1, \ldots, q_k \rangle \in \delta_k(q, V(x))$ such that for all $1 \leq i \leq k$, we have $r(x \cdot i) = q_i$. Note that each node of the input tree corresponds to exactly one node in the run tree. A run $\langle T, r \rangle$ is accepting if all its paths satisfy the Büchi acceptance condition. Thus, for all paths π, we have $inf(\pi) \cap F = \emptyset$, where $q \in inf(\pi)$ if and only if there are infinitely many $x \in \pi$ for which $r(x) = q$. Equivalently, $\langle T, r \rangle$ is accepting iff $\langle T, r \rangle$ contains infinitely many F-frontiers $G_0 < G_1 < \ldots$. For a state $q \in Q$, let $\mathcal{U}^q$ be $\mathcal{U}$ with initial state q. We say that a symmetric NBT is *monotonic* if for every two states q and p such that $q \subseteq p$, we have that $\mathcal{L}(\mathcal{U}^p) \subseteq \mathcal{L}(\mathcal{U}^q)$, and $p \in F$ implies $q \in F$. In other words, the smaller the state is, the easier it is to accept from it. Note that symmetric nondeterministic tree automata are essentially symmetric alternating automata with transitions in disjunctive normal form (DNF); if we write the transition functions in DNF, then each disjunct is a conjunction of universal and existential requirements, corresponding to a pair $\langle U, E \rangle$.

3   From Symmetric NBT and Co-NBT to Symmetric AWT

Let $\mathcal{U} = \langle \Sigma, \mathcal{D}, Q, q_0, M, F \rangle$ and $\mathcal{U}' = \langle \Sigma, \mathcal{D}, Q', q'_0, M', F' \rangle$ be two NBT, and let $|Q| \cdot |Q'| = m$. In [Rab70], Rabin studies the joint behavior of a run of $\mathcal{U}$ with a run of $\mathcal{U}'$. Recall that an accepting run of $\mathcal{U}$ contains infinitely many F-frontiers $G_0 < G_1 < \ldots$, and an accepting run of $\mathcal{U}'$ contains infinitely many F'-frontiers $G'_0 < G'_1 < \ldots$. It follows that for every labeled tree $\langle T, V \rangle \in \mathcal{L}(\mathcal{U}) \cap \mathcal{L}(\mathcal{U}')$ and accepting runs $\langle T, r \rangle$ and $\langle T, r' \rangle$ of $\mathcal{U}$ and $\mathcal{U}'$ on $\langle T, V \rangle$, the joint behavior of $\langle T, r \rangle$ and $\langle T, r' \rangle$ contains infinitely many frontiers $E_i \subset T$, with $E_i < E_{i+1}$, such that $\langle T, r \rangle$ reaches an F-frontier and $\langle T, r' \rangle$ reaches an F'-frontier between E_i and E_{i+1}. Rabin shows that the existence of m such frontiers, in the joint behavior of some runs of $\mathcal{U}$ and $\mathcal{U}'$, is sufficient to imply that the intersection $\mathcal{L}(\mathcal{U}) \cap \mathcal{L}(\mathcal{U}')$ is not empty. We now extend Rabin's result to symmetric automata.

Assume that $\mathcal{U}$ and $\mathcal{U}'$ above are symmetric NBT. We say that a sequence $E_0, \ldots, E_m$ of frontiers of T is a *trap for $\mathcal{U}$ and $\mathcal{U}'$* iff $E_0 = \{\varepsilon\}$ and there exists a tree $\langle T, V \rangle$ and (not necessarily accepting) runs $\langle T, r \rangle$ and $\langle T, r' \rangle$ of $\mathcal{U}$ and $\mathcal{U}'$ on $\langle T, V \rangle$, such that for every $0 \leq i \leq m - 1$, we have that $\langle T, r \rangle$ contains an F-frontier G_i such that $E_i \leq G_i < E_{i+1}$, and $\langle T, r' \rangle$ contains an F'-frontier G'_i

such that $E_i \leq G'_i < E_{i+1}$. We say that $\langle T, r \rangle$ and $\langle T, r' \rangle$ *witness* the trap for $\mathcal{U}$ and $\mathcal{U}'$.

Theorem 1. *Consider two symmetric nondeterministic Büchi tree automata $\mathcal{U}$ and $\mathcal{U}'$. If there exists a trap for $\mathcal{U}$ and $\mathcal{U}'$, then $\mathcal{L}(\mathcal{U}) \cap \mathcal{L}(\mathcal{U}')$ is not empty.*

Proof. The proof follows the same line of reasoning as in [Rab70]. For a state $q \in Q$, let $\mathcal{U}^q$ be $\mathcal{U}$ with initial state q, and similarly for $q' \in Q'$ and $\mathcal{U}'^{q'}$. We define a sequence of relations over $Q \times Q'$. Let $H_0 = Q \times Q'$. Then, $\langle q, q' \rangle \in H_{i+1}$ iff $\langle q, q' \rangle \in H_i$ and there is a nonempty Σ-labeled D-tree $\langle T, V \rangle$, a frontier $E \subseteq T$, and runs $\langle T, r \rangle$ and $\langle T, r' \rangle$ of $\mathcal{U}^q$ and $\mathcal{U}'^{q'}$ on $\langle T, V \rangle$, such that there is an F-frontier $G < E$ and an F'-frontier $G' < E$, such that for all $x \in E$, we have $\langle r(x), r'(x) \rangle \in H_i$. It is easy to see that $H_0 \supseteq H_1 \supseteq H_2 \supseteq \dots$. Also, if $H_i = H_{i+1}$, then $H_i = H_{i+k}$ for all $k \geq 0$. In particular, since $|Q| \times |Q'| = m$, it must be that $H_m = H_{m+k}$ for all $k \geq 0$. As in [Rab70], it can now be shown that $\mathcal{L}(\mathcal{U}) \cap \mathcal{L}(\mathcal{U}') \neq \emptyset$ iff $H_m(q_0, q'_0)$, and the result follows.

Theorem 1 is the key to the construction described in Theorem 2 below.

Theorem 2. *Let $\mathcal{U}$ and $\mathcal{U}'$ be two symmetric monotonic NBT with $\mathcal{L}(\mathcal{U}') = comp(\mathcal{L}(\mathcal{U}))$. There exists a symmetric AWT $\mathcal{A}$ such that $\mathcal{L}(\mathcal{A}) = \mathcal{L}(\mathcal{U})$.*

Proof. Let $\mathcal{U} = \langle \Sigma, Q, q_0, M, F \rangle$ and $\mathcal{U}' = \langle \Sigma, Q', q'_0, M', F' \rangle$, and let $|Q| \cdot |Q'| = m$. Also, let S and S' be the micro-states of $\mathcal{U}$ and $\mathcal{U}'$, respectively, thus $Q = 2^S$ and $Q' = 2^{S'}$. We define the symmetric AWT $\mathcal{A} = \langle \Sigma, P, p_0, \delta, \alpha \rangle$ as follows.

- $P = Q \times Q' \times \{0, \dots, 2m-1\}$ and $p_0 = \langle q_0, q'_0, 0 \rangle$. Intuitively, a copy of $\mathcal{A}$ that visits the state $\langle q, q', i \rangle$ as it reads the node x of the input tree corresponds to runs r and r' of $\mathcal{U}$ and $\mathcal{U}'$ that visit the states q and q', respectively, as they read the node x of the input tree. Let $\rho = y_0, y_1, \dots, y_{|x|}$ be the path from ε to x. Consider the joint behavior of r and r' on ρ. We can represent this behavior by a sequence $\tau_\rho = \langle t_0, t'_0 \rangle, \langle t_1, t'_1 \rangle, \dots, \langle t_{|x|}, t'_{|x|} \rangle$ of pairs in $Q \times Q'$ where $t_j = r(y_j)$ and $t'_j = r'(y_j)$. We say that a pair $\langle t, t' \rangle \in Q \times Q'$ is an F-*pair* iff $t \in F$ and is an F'-*pair* iff $t' \in F'$. We can partition the sequence τ_ρ to blocks $\beta_0, \beta_1, \dots, \beta_i$ such that we close block β_b and open block β_{b+1} whenever we reach the first F'-pair that is preceded by an F-pair in β_b. In other words, whenever we open a block, we first look for an F-pair, ignoring F'-pairness. Once an F-pair is detected, we look for an F'-pair, ignoring F-pairness. Once an F'-pair is detected, we close the current block and we open a new block. Note that a block may contain a single pair that is both an F-pair and an F'-pair. The third element of a state keeps track of the visits to blocks. When we visit $\langle q, q', i \rangle$, the index of the last block in τ_ρ is $\lfloor \frac{i}{2} \rfloor$, and this block already contains an F-pair iff i is odd. We refer to i as the *status* of the state $\langle q, q, i \rangle$. For a status $i \in \{0, \dots, 2m-1\}$, let $P_i = Q \times Q \times \{i\}$ be the set of states with status i.

– In order to define the transition function δ, we first define a function $next$: $P \to \{0, \ldots, 2m - 1\}$ that updates the status of states. For that, we first define the function $next' : P \to \{0, \ldots, 2m\}$ as follows.

$$next'(\langle q, q', i \rangle) = \begin{bmatrix} i & \text{If } (i \text{ is even and } q' \notin F) \text{ or } (i \text{ is odd and } q' \notin F') \\ i+1 & \text{If } (i \text{ is even and } q \in F \text{ and } q' \notin F') \text{ or } (i \text{ is odd and } q' \in F') \\ i+2 & \text{If } i \text{ is even and } q \in F \text{ and } q' \in F'. \end{bmatrix}$$

Now, $next(\langle q, q', i \rangle) = \min\{next'(\langle q, q', i \rangle), 2m - 1\}$.

Intuitively, $next$ updates the status of states by recording and tracking of blocks. Recall that the status i indicates in which block we are and whether an F-pair in the current block has already been detected. The conditions for not changing i or for increasing it to $i+1$ and $i+2$ follow directly from the definition of the status. For example, the new status stays i if the current i is even and $\langle q, q' \rangle$ is not an F-block, or if i is odd and $\langle q, q' \rangle$ is not an F'-block. When i reaches or exceeds $2m - 1$, we no longer increase it, even if $q' \in F'$. The automaton $\mathcal{A}$ proceeds as follows. Essentially, for every run $\langle T, r' \rangle$ of $\mathcal{U}'$, the automaton $\mathcal{A}$ guesses a run $\langle T, r \rangle$ of $\mathcal{U}$ such that for every path ρ of T, the run $\langle T, r \rangle$ visits F along ρ at least as many times as $\langle T, r' \rangle$ visits F' along ρ. Thus, when we record blocks along ρ, we do not want to get stuck in an even status. Since $\mathcal{L}(\mathcal{U}) \cap \mathcal{L}(\mathcal{U}') = \emptyset$, then, by Theorem 1, no run $\langle T, r \rangle$ can witness with $\langle T, r' \rangle$ a trap for $\mathcal{U}$ and $\mathcal{U}'$. Consequently, recording of visits to F and F' along ρ can be completed once $\mathcal{A}$ detects that τ_ρ contains m blocks as above.

Recall that $Q = 2^S$ and $Q' = 2^{S'}$. For a set $E \subseteq S$, a *partition* of E is a set $\{E_1, \ldots, E_l\}$ with $E_i \subseteq E$ such that $E = \bigcup_{1 \leq i \leq l} E_i$, and for all $1 \leq i \neq j \leq n$, we have $E_i \cap E_j = \emptyset$. Let $par(E)$ be the set of partitions of E. Consider a set $E' \subseteq S'$ and a partition $\gamma' \in par(E')$. For a set $E \subseteq S$, we say that a partition η of $E \cup E'$ *agrees with* γ' if for all s'_1 and s'_2 in E', we have that s'_1 and s'_2 are in the same set in η iff they are in the same set in γ'. Let $agree(E, \gamma')$ be the set of partitions of $E \cup E'$ that agree with γ'. For example, if $E = \{s_1\}$ and $E' = \{s_2, s_3\}$, then the two possible partitions of E' are $\gamma'_1 = \{\{s_2, s_3\}\}$ and $\gamma'_2 = \{\{s_2\}, \{s_3\}\}$. Then, $agree(E, \gamma'_1)$ contains the two partitions $\{\{s_1, s_2, s_3\}\}$ and $\{\{s_1\}, \{s_2, s_3\}\}$, and $agree(E, \gamma'_2)$ contains the three partitions $\{\{s_1, s_2\}, \{s_3\}\}$, $\{\{s_1, s_3\}, \{s_2\}\}$, and $\{\{s_1\}, \{s_2\}, \{s_3\}\}$.

Now, let $p = \langle q, q', i \rangle$ be a state in P such that $M(q, \sigma) = \{\langle U_1, E_1 \rangle, \ldots, \langle U_n, E_n \rangle\}$ and $M'(q', \sigma) = \{\langle U'_1, E'_1 \rangle, \ldots, \langle U'_{n'}, E'_{n'} \rangle\}$. We distinguish between two cases.

- If $i < 2m - 1$ or $q \notin F$, then

$$\delta(p, \sigma) = \bigwedge_{1 \leq j' \leq n'} \bigwedge_{\gamma' \in par(E'_{j'})} \left(\bigvee_{1 \leq j \leq n} \bigvee_{\eta \in agree(E_j, \gamma')} go(j, j', \eta, next(p)) \right), \text{ where}$$

$$go(j, j', \eta, l) = \square \langle U_j, U'_{j'}, l \rangle \wedge \bigwedge_{X \in \eta} \lozenge \langle U_j \cup (X \cap E_j), U'_{j'} \cup (X \cap E'_{j'}), l \rangle.$$

That is, for every choice of $\mathcal{U}'$ for a $1 \leq j' \leq n'$ and for the way the existential requirements in $E'_{j'}$ are partitioned, there is a choice of $\mathcal{U}$

for a $1 \leq j \leq n$ and for the way the existential requirements in E_j are partitioned and combined with these in $E'_{j'}$ to a partition of $E_j \cup E'_{j'}$, such that the universal requirements in U_j and $U'_{j'}$ are sent to all directions, and existential requirements that are in the same set in the joint partition of $E_j \cup E'_{j'}$ are sent to the same direction. Note that the sets U_j and $U'_{j'}$ are sent along with the existential requirements. This guarantees that the states that are sent in the existential mode correspond to the states that $\mathcal{U}$ and $\mathcal{U}'$ visit, and not to subsets of such states.

- If $i = 2m - 1$ and $q \in F$, then $\delta(p, \sigma) = \textbf{true}$.

Note that $par(E')$ is exponential in $|E'|$, and the number of possible $\eta \in agree(E, \gamma')$ is exponential in $E \cup E'$. Thus, the size of δ is exponential in the sizes of M and M'.

- $\alpha = Q \times Q' \times \{i : i \text{ is odd}\}$. Thus, α makes sure that infinite paths of the run visits infinitely many states in which the status is odd, thus states in which we are in the second phase of blocks. moshe2:

Each set P_i is a strongly connected component, thus the automaton $\mathcal{A}$ is indeed an AWT. Note that, by the definition of α, a run is accepting iff no path of it gets trapped in a set of the form P_i, for an even i, namely a set in which $\mathcal{A}$ is waiting for a visit of $\mathcal{U}$ in a state in F. The number of states of $\mathcal{A}$ is $O(m^2)$. We prove that $\mathcal{L}(\mathcal{U}) = \mathcal{L}(\mathcal{A})$. We first prove that $\mathcal{L}(\mathcal{U}) \subseteq \mathcal{L}(\mathcal{A})$. Consider a D-tree $\langle T, V \rangle$. With every run $\langle T, r \rangle$ of $\mathcal{U}$ on $\langle T, V \rangle$ we can associate a run $\langle T_R, R \rangle$ of $\mathcal{A}$ on $\langle T, V \rangle$. Intuitively, the run $\langle T, r \rangle$ directs $\langle T_R, R \rangle$ in the nondeterminism in δ (that is, the choices of $1 \leq j \leq n$ and $\eta \in agree(E_j, \gamma')$). Formally, recall that a run of $\mathcal{A}$ on a D-tree $\langle T, V \rangle$ is a $(T \times P)$-labeled tree $\langle T_R, R \rangle$, where a node $y \in T_R$ with $R(y) = \langle x, p \rangle$ corresponds to a copy of $\mathcal{A}$ that reads the node $x \in T$ and visits the state p. We define $\langle T_R, R \rangle$ as follows.

- $\varepsilon \in T_R$ and $R(\varepsilon) = (\varepsilon, \langle q_0, q'_0, 0 \rangle)$.
- Consider a node $y \in T_R$ with $R(y) = (x, \langle q, q', i \rangle)$. By the definition of $\langle T_R, R \rangle$ so far, we have $r(x) = t$ for $q \subseteq t$. Consider first the case that $t = q$. Let $\{x \cdot 1, \ldots, x \cdot k\}$ be the children of x in T, and let $\langle U, \langle E_1, \ldots, E_k \rangle \rangle \in M_k(q, V(x))$ describe the choice $\mathcal{U}$ makes when it proceeds from the node x. Thus, for each $1 \leq z \leq k$, we have $r(x \cdot z) = U \cup E_z$. Let $j = next(\langle q, q', i \rangle)$. Consider the set

$$Y = \bigcup_{\substack{\langle U', \langle E'_1, \ldots, E'_k \rangle \rangle \in \\ M'_k(q', V(x))}} \{(1, \langle U, U', j \rangle), (1, \langle U \cup E_1, U' \cup E'_1, j \rangle), \ldots (k, \langle U, U', j \rangle), (k, \langle U \cup E_k, U' \cup E'_k, j \rangle)\}.$$

By the definition of δ, the set Y satisfies $\delta(\langle q, q', i \rangle, V(x))$ [4]. Let $l = |M'_k(q', V(x))|$, and let $\langle U'^w, E'^w_1, \ldots E'^w_k \rangle$, for $1 \leq w \leq l$, be the w-th pair in $M'_k(q', V(x))$. For all $1 \leq w \leq l$ and $1 \leq z \leq k$, we have $\{y \cdot (2k(w-1)+z-1), y \cdot (2k(w-1)+z)\} \subseteq T_R$, with $R(y \cdot (2k(w-1)+z-1)) = (x \cdot z, \langle U, U'^w, j \rangle)$ and $R(y \cdot (2k(w-1)+z)) = (x \cdot z, \langle U \cup E_z, U'^w \cup E'^w_z, j \rangle)$.

---

[4] Note that $\delta(\langle q, q', i \rangle, V(x))$ is a formula in $\mathcal{B}^+(\{\Box, \Diamond\} \times P)$, whereas $Y \subseteq \{1, \ldots, k\} \times P$, but the extension of the satisfaction relation to this setting is straightforward: an atom $(\Diamond, p)$ is satisfied in Y if there is $1 \leq z \leq k$ with $(z, p) \in Y$, and an atom $(\Box, p)$ is satisfied in Y if for all $1 \leq z \leq k$, we have $(z, p) \in Y$.

Note that the invariant that for all $y \in T_R$ with $R(y) = (x, \langle q, q', i \rangle)$, we have $r(x) = t$ for $q \subseteq t$, is maintained. If fact, we know that all the nodes $y \in T_R$ that correspond to copies of $\mathcal{A}$ that satisfy an existential requirement have $q = t$, and node $y \in T_R$ that correspond to copies of $\mathcal{A}$ that satisfy a universal requirement have $q = t$ iff the run r sends no existential requirement to the corresponding direction.

Consider now the case where $q \subset t$. Since $\mathcal{U}$ is monotonic, there is an accepting run $\langle T^x, r_q^x \rangle$ of $\mathcal{U}^q$ on the subtree of T with root x. We can proceed exactly as above, with $\langle T^x, r_q^x \rangle$ instead of $\langle T, r \rangle$.

Consider a tree $\langle T, V \rangle \in \mathcal{L}(\mathcal{U})$. Let $\langle T, r \rangle$ be an accepting run of $\mathcal{U}$ on $\langle T, V \rangle$, and let $\langle T_R, R \rangle$ be the run of $\mathcal{A}$ on $\langle T, V \rangle$ induced by $\langle T, r \rangle$ (and the "subtree runs", like $\langle T^x, r_q^x \rangle$ above). It can be shown that $\langle T_R, R \rangle$ is a legal accepting run. Indeed, since $\langle T, r \rangle$ and the subtree runs contains infinitely many F-frontiers, and since (by the definition of monotonic automaton) we do not lose visits to F when we switch to subset runs, no infinite paths of $\langle T_R, R \rangle$ can get trapped in a set P_i for an even i.

It is left to prove that $\mathcal{L}(\mathcal{A}) \subseteq \mathcal{L}(\mathcal{U})$. For that, we prove that $\mathcal{L}(\mathcal{A}) \cap \mathcal{L}(\mathcal{U}') = \emptyset$. Since $\mathcal{L}(\mathcal{U}) = comp(\mathcal{L}(\mathcal{U}'))$, it follows that every tree that is accepted by $\mathcal{A}$ is also accepted by $\mathcal{U}$. Consider a tree $\langle T, V \rangle$. With each run $\langle T_R, R \rangle$ of $\mathcal{A}$ on $\langle T, V \rangle$ and run $\langle T, r' \rangle$ of $\mathcal{U}'$ on $\langle T, V \rangle$, we associate a run $\langle T, r \rangle$ of $\mathcal{U}$ on $\langle T, V \rangle$. Intuitively, $\langle T, r \rangle$ makes the choices that $\langle T_R, R \rangle$ has made in its copies that correspond to the run $\langle T, r' \rangle$. Formally, $\langle T, r \rangle$ is such that $r(\varepsilon) = q_0$, and for all $x \in T$ with $r(x) = q$, we proceed as follows. Let $\{x \cdot 1, \ldots, x \cdot k\}$ be the children of x in T, and let $r'(x) = q'$. The run $\langle T, r' \rangle$ selects a pair $\langle U', \langle E_1', \ldots, E_k' \rangle \rangle \in M_k'(q', V(x))$ that $\mathcal{U}'$ proceeds with when it reads the node x. Formally, for all $1 \leq z \leq k$, we have $r'(x \cdot z) = U' \cup E_z'.$[5] By the definition of $r(x)$ so far, the run $\langle T_R, R \rangle$ contains a node $y \in T_R$ with $R(y) = \langle x, \langle q, q', i \rangle \rangle$ for some status i. If $\delta(\langle q, q', i \rangle, V(x)) = \mathbf{true}$, we define the reminder of $\langle T, r \rangle$ arbitrarily. Otherwise, let $1 \leq j' \leq n'$ and $\gamma' \in par(E_{j'}')$ be such that $\langle U', \langle E_1', \ldots, E_k' \rangle \rangle$ corresponds to j' and γ'. By the definition of δ, there are $1 \leq j \leq n$ and $\eta \in agree(E_j, \gamma')$ such that $go(j, j', \eta, next(\langle q, q', i \rangle))$ is satisfied and R proceeds according to j and η. Thus, if $\{E_j^1, \ldots, E_j^k\}$ is the partition of E_j that corresponds to η, then T_R contains at least k nodes $y \cdot c_z$, for $1 \leq z \leq k$, such that $R(y \cdot c_z) = \langle x \cdot z, \langle U_j \cup E_j^z, U' \cup E_z', next(\langle q, q', i \rangle) \rangle \rangle$. For all $1 \leq z \leq k$, we define $r(x \cdot z) = U_j \cup E_z^z$. Note that the invariant about the runs $\langle T, r \rangle$ and $\langle T_R, R \rangle$ is maintained. Note also that if $E_j^z \cup F_z' = \emptyset$, then the existence of a node $y \cdot c_z$ as above is guaranteed from universal part of δ, and if $E_j^z \cup E_z' \neq \emptyset$, its existence is guaranteed from the existential part (in which case it is crucial that we sent the universal requirements along with the existential ones).

We can now prove that $\mathcal{L}(\mathcal{A}) \cap \mathcal{L}(\mathcal{U}') = \emptyset$. Assume, by way of contradiction, that there exists a tree $\langle T, V \rangle$ such that $\langle T, V \rangle$ is accepted by both $\mathcal{A}$ and $\mathcal{U}'$. Let

---

[5] For a monotonic NBT, we assume that runs satisfy the requirements in transition function in an optimal way; thus when $\mathcal{A}$ chooses to proceed with $\langle U', \langle E_1', \ldots, E_k' \rangle \rangle \in M_k'(q', V(x))$, it is indeed the case that $r'(x \cdot z) = U' \cup E_z'$. If $r'(x \cdot z) \supset U' \cup E_z'$, we can replace r' with a run for which the equation holds.

$\langle T_R, R \rangle$ and $\langle T, r' \rangle$ be the accepting runs of $\mathcal{A}$ and $\mathcal{U}'$ on $\langle T, V \rangle$, respectively, and let $\langle T, r \rangle$ be the run of $\mathcal{U}$ on $\langle T, V \rangle$ induced by $\langle T_R, R \rangle$ and $\langle T, r' \rangle$. We claim that then, $\langle T, r \rangle$ and $\langle T, r' \rangle$ witness a trap for $\mathcal{U}$ and $\mathcal{U}'$. Since, however, $\mathcal{L}(\mathcal{U}) \cap \mathcal{L}(\mathcal{U}') = \emptyset$, it follows from Theorem 1, that no such trap exists, and we reach a contradiction. To see that $\langle T, r \rangle$ and $\langle T, r' \rangle$ indeed witness a trap, define $E_0 = \{\varepsilon\}$, and define, for $0 \leq i \leq m-1$, the set E_{i+1} to contain exactly all nodes x for which there exists $y \in T_R$ such that either $R(y) = \langle x, \langle (r(x), r'(x), 2i+1) \rangle \rangle$ and $r'(x) \in F'$ or $R(y) = \langle x, \langle (r(x), r'(x), 2i) \rangle \rangle$ and $r(x) \in F$ and $r'(x) \in F'$. That is, for every path ρ of T, the set E_{i+1} consists of the nodes in which the i'th block is closed in τ_ρ. By the definition of δ, for all $0 \leq i \leq m-1$, the run $\langle T, r \rangle$ contains an F-frontier G_i such that $E_i \leq G_i < E_{i+1}$ and the run $\langle T, r' \rangle$ contains an F'-frontier G_i' such that $E_i \leq G_i' < E_{i+1}$. Hence, $E_0, \ldots, E_m$ is a trap for $\mathcal{U}$ and $\mathcal{U}'$.

4   From $\Pi_2 \cap \Sigma_2$ to the Alternation-Free μ-Calculus

The μ-calculus is a propositional modal logic augmented with least and greatest fixpoint operators [Koz83]. Specifically, we consider a μ-calculus where formulas are constructed from Boolean propositions with Boolean connectives, the temporal operators $\exists\bigcirc$ ("exists next") and $\square$ ("for all next"), as well as least (μ) and greatest (ν) fixpoint operators. We assume that μ-calculus formulas are written in positive normal form (negation only applied to atomic propositions constants and variables). Formally, given a set AP of atomic proposition constants and a set APV of atomic proposition variables, a μ-calculus formula is either:

- **true, false**, p or $\neg p$ for all $p \in AP$.
- y for all $y \in APV$;
- $\varphi \wedge \psi$, $\varphi \vee \psi$, $\Diamond\varphi$, or $\square\varphi$, where φ and ψ are μ-calculus formulas;
- $\mu y.\varphi(y)$ or $\nu y.\varphi(y)$, where $y \in APV$ and $\varphi(y)$ is a μ-calculus formula containing y as a free variable.

We classify formulas to classes Σ_i and Π_i according to the nesting of fixpoint operators in them. Several versions to such a classification can be found in the literature [EL86,Niw86,Bra98]. We describe here the version defined in [Niw86]:

- A formula is in $\Sigma_0 = \Pi_0$ if it contains no fixpoint operators.
- A formula is in Σ_{i+1} if it is one of the following θ_i, $\theta_i \wedge \theta_i'$, $\theta_i \vee \theta_i'$, $\Diamond\theta_i$, $\square\theta_i$, $\mu y.\varphi_{i+1}(y)$, $\varphi_{i+1}(Y)[y \leftarrow \varphi_{i+1}']$, where θ_i and θ_i' are $\Sigma_i \cup \Pi_i$ formulas, φ_{i+1} and φ_{i+1}' are Σ_{i+1} formulas, $Y \subseteq APV$, $y \in Y$, and no free variable of φ_{i+1}' is in Y. In other words, to form Σ_{i+1}, we take $\Sigma_i \cup \Pi_i$ and close under Boolean and modal operations, $\mu y.\varphi(y)$ for $\varphi \in \Sigma_{i+1}$, and substitution of a free variable of $\varphi \in \Sigma_{i+1}$ by a formula $\varphi' \in \Sigma_{i+1}$ provided that no free variable of φ' is captured by φ.
- A formula is in Π_{i+1} if it is one of the following θ_i, $\theta_i \wedge \theta_i'$, $\theta_i \vee \theta_i'$, $\Diamond\theta_i$, $\square\theta_i$, $\nu y.\psi_{i+1}(y)$, $\psi_{i+1}(Y)[y \leftarrow \psi_{i+1}']$, where θ_i and θ_i' are $\Sigma_i \cup \Pi_i$ formulas, ψ_{i+1} and ψ_{i+1}' are Π_{i+1} formulas, $Y \subseteq APV$, $y \in Y$, and no free variable of ψ_{i+1}' is in Y.

Note that the "substitution step" suggests that the formula $\psi = \nu y.(\Diamond(y \wedge (\mu z.p \vee \Diamond z))$ is in both Π_2 and Σ_2. To see that ψ is in Σ_2 (it is easy to see that $\psi \in \Pi_2$), note that $\mu z.p \vee \Diamond z$ is in Σ_1, and hence also in Σ_2. In addition, the formula $\nu y.\Diamond(y \wedge x)$, for $x \in APV$, is in Π_1, and hence also in Σ_2. The formula $\mu z.p \vee \Diamond z$ has no free variables. Then, we can substitute x by it, get ψ, and stay in Σ_2. Note that for for classifications that do not allow such a substitution, the formula ψ is not in Σ_2. Note also that ψ is neither in Π_1 nor Σ_1.

Finally, we say that a formula is in Δ_i if it is one of the following θ_i, $\theta_i \wedge \theta_i'$, $\theta_i \vee \theta_i'$, $\Diamond\theta_i$, $\Box\theta_i$, $\theta(Y)[y \leftarrow \theta_i']$, where θ_i and θ_i' are $\Sigma_i \cup \Pi_i$ formulas, $Y \subseteq APV$, $y \in Y$, and no variable of θ_i' is in Y. In other words, to form Δ_i, we take $\Sigma_i \cup \Pi_i$ and close under Boolean and modal operations, and under substitution that does not increase the alternation depth. Note that Δ_0 is ML and Δ_1 is AFMC.

Essentially, Σ_i contains all Boolean and modal combinations of formulas in which there are at most $i - 1$ alternations of μ and ν, with the external fixpoint being a μ. Similarly, Π_i contains all Boolean and modal combinations of formulas in which there are at most i alternations of μ and ν, with the external fixpoint being a ν. A μ-calculus formula is *alternation free* if, for all atomic propositional variables y, there are no occurrences of ν (μ) on any syntactic path from an occurrence of μy (νy, respectively) to an occurrence of y. For example, the formula $\mu x.(p \vee \mu y.(x \vee EXy))$ is alternation free (and is in Σ_1) and the formula $\nu x.\mu y.((p \wedge x) \vee EXy)$ is not alternation free (and is in Π_2). The *alternation-free μ-calculus* is a subset of μ-calculus containing only alternation-free formulas. The alternation-free μ-calculus is a strict syntactic fragment of $\Pi_2 \cap \Sigma_2$. We now use Theorem 2 in order to show that $\Pi_2 \cap \Sigma_2$ is not more expressive than the alternation free μ-calculus. Thus, every formula in $\Pi_2 \cap \Sigma_2$ has an equivalent formula in AFMC.

For the alternation-free μ-calculus, an automata-theoretic characterization in terms of symmetric alternating weak automata is well known (a similar result is proven in [AN92] for directed trees):

Theorem 3. [KV98] *A set $\mathcal{T} \subseteq trees(\Sigma)$ can be expressed in AFMC iff $\mathcal{T}$ can be recognized by a symmetric weak alternating automaton.*

In [Kai95], Kaivola considered μ-calculus formulas in which the $\Diamond$ modality is parameterized with directions and translates Π_2 formulas to NBT. In order to apply Theorem 2, we should translate Π_2 formulas to symmetric monotonic NBT. For that, we first use a known translation of Π_2 formulas to symmetric ABT (Theorem 4; a similar translation for the directed case is described in [Niw86,Tak86]), and then remove alternation, with symmetry preserved (Theorem 5).

Theorem 4. [KVW00] *Given a Π_2 formula ψ, there is a symmetric alternating Büchi tree automaton $\mathcal{A}_\psi$ that accepts exactly all trees that satisfy ψ.*

Miyano and Hayashi described a translation of alternating Büchi word automata to equivalent nondeterministic Büchi word automata [MH84]. Mostowski extended the translation to tree automata [Mos84], and we extend it further to symmetric tree automata. Since the nondeterministic automaton needs to keep track of the states visited in each path of the run tree of the alternating automaton, the symmetry of the automaton poses real technical challenges.

Theorem 5. *Let $\mathcal{A}$ be a symmetric alternating Büchi tree automaton. There is a symmetric monotonic nondeterministic Büchi tree automaton $\mathcal{A}'$, with exponentially many states, such that $\mathcal{L}(\mathcal{A}') = \mathcal{L}(\mathcal{A})$.*

Proof. Let $\mathcal{A} = \langle \Sigma, S, s_{in}, \delta, \alpha \rangle$. Then $\mathcal{A}' = \langle \Sigma, Q, \{\langle s_{in}, 2 \rangle\}, \delta', \alpha' \rangle$, where

- $Q = 2^{S \times \{1,2\}}$. For a state $q \in Q$, let $q[1] = \{s : \langle s, 1 \rangle \in q\}$ and $q[2] = \{s : \langle s, 2 \rangle \in q\}$. Intuitively, the automaton $\mathcal{A}'$ guesses a run of $\mathcal{A}$. At a given node x of a run of $\mathcal{A}'$, it keeps in its memory the set of all the states of $\mathcal{A}$ that visit x in the guessed run. As it reads the next input letter, it guesses the way in which an accepting run of $\mathcal{A}$ proceeds from all of these states. This guess induces the states that the run of $\mathcal{A}'$ visit in the children of x. In order to make sure that every infinite path visits states in α infinitely often, the states are tagged by 1 or 2. States tagged by 1 correspond to copies that have already visit α, and states tagged by 2 correspond to copies that owe a visit to α. When all the copies visit α (that is, all the states are tagged by 1), we change the tag of all states to 2.

- Given $S' \subseteq S$, $\sigma \in \Sigma$, and a pair $\langle U, E \rangle$ of subsets of S, we say that $\langle U, E \rangle$ *covers* S' *and* σ if the set $\{\Box s : s \in U\} \cup \{\Diamond s : s \in E\}$ satisfies $\bigwedge_{s' \in S'} \delta(s', \sigma)$. Now, $\delta' : Q \times \Sigma \to 2^{Q \times Q}$ is defined, for all $q \in Q$ and $\sigma \in \Sigma$, as follows.
  - If $q[2] \neq \emptyset$, then $\delta'(q, \sigma)$ contains all pairs $\langle U, E \rangle$ such that there is $\langle U_1, E_1 \rangle$ that covers $q[1]$ and σ, and there is $\langle U_2, E_2 \rangle$ that covers $q[2]$ and σ, and the following hold.
    * $U = \{\langle s, 1 \rangle : s \in U_1 \cup (U_2 \cap \alpha)\} \cup \{\langle s, 2 \rangle : s \in U_2 \setminus \alpha\}$.
    * $E = \{\langle s, 1 \rangle : s \in E_1 \cup (E_2 \cap \alpha)\} \cup \{\langle s, 2 \rangle : s \in E_2 \setminus \alpha\}$.
  - If $q[2] = \emptyset$, then $\delta'(q, \sigma)$ contains all pairs $\langle U, E \rangle$ such that there is $\langle U_1, E_1 \rangle$ that covers $q[1]$ and σ and the following hold.
    * $U = \{\langle s, 1 \rangle : s \in U_1 \cap \alpha\} \cup \{\langle s, 2 \rangle : s \in U_1 \setminus \alpha\}$.
    * $E = \{\langle s, 1 \rangle : s \in E_1 \cap \alpha\} \cup \{\langle s, 2 \rangle : s \in E_1 \setminus \alpha\}$.

- $\alpha' = \{q : q[2] = \emptyset\}$. Note that a sequence of states of $\mathcal{A}$, which corresponds to the behavior of a copy of $\mathcal{A}$, changes the tag of its states from 2 to 1 when the copy visits a state in α. Also, once all the sequences change the tag of their states to 1, the attribution is changed back to 2. Thus, α' guarantees that all sequences visit α infinitely often.

It is easy to see that $\mathcal{A}$ is monotonic. Indeed, if $q \subseteq q'$, then $q[1] \subseteq q'[1]$ and $q[2] \subseteq q'[2]$. Thus, if a pair $\langle U, E \rangle$ covers $q'[1]$ and σ, then $\langle U, E \rangle$ also covers $q[1]$ and σ, and similarly for $q'[2]$ and $q[2]$. Hence, given an accepting run of $\mathcal{A}^{q'}$, we can make it an accepting run of $\mathcal{A}^q$ by changing the labels of the root from (ε, q') to (ε, q). In addition, if $q'[2]$ is empty, so is $q[2]$.

Remark 1. A related approach for translating μ-calculus formulas into symmetric automata is taken in [JW95] (see also [AN01]). First, μ-calculus formulas are transformed into a disjunctive form. The removal of conjunctions described there is similar to the removal of universal branches in alternating tree automata (and indeed it involves the same determinization construction that is present in the automata-theoretic approach [MS87]). It is then shown that disjunctive μ-calculus formulas correspond to μ-automata. Our focus here is on the translation of Π_2 formulas to symmetric monotonic nondeterministic Büchi tree automata. It is possible to recast our proof in an extension of the framework of μ-automata [Wal03], but we find our notion of symmetric nondeterministic automata more transparent.

Theorem 6. $\Pi_2 \cap \Sigma_2 \equiv AFMC$.

Proof. Since AFMC is a syntactic fragment of $\Pi_2 \cap \Sigma_2$, one direction is trivial. Let ξ be a property expressible in $\Pi_2 \cap \Sigma_2$. Given $\theta \in \Pi_2$ expressing ξ, we can construct, by Theorems 4 and 5, a symmetric monotonic NBT $\mathcal{U}_\theta$ that accepts exactly all trees that satisfy θ. Also, $\xi \in \Sigma_2$ implies that there is $\psi \in \Pi_2$ that is equivalent to $\neg\theta$, so we can also construct a symmetric monotonic NBT $\mathcal{U}_\psi$ that accepts exactly all trees that do not satisfy θ. Clearly, $\mathcal{L}(\mathcal{U}_\psi) = comp(\mathcal{L}(\mathcal{U}_\theta))$. Hence, by Theorem 2, there is a symmetric alternating weak automaton $\mathcal{A}_\theta$ that is equivalent to $\mathcal{U}_\theta$. By Theorem 3, the automaton $\mathcal{A}_\theta$ can be translated to a formula φ in AFMC such that a tree satisfies φ iff it is accepted by $\mathcal{U}_\theta$ iff it is not accepted by $\mathcal{U}_\psi$. We claim that φ is logically equivalent to θ over arbitrary structures (in particular, structures with an infinite branching degree). To see this, assume, by way of contradiction, that φ is not logically equivalent to θ. Then, either $\theta \wedge \neg\varphi$ or $\varphi \wedge \psi$ is satisfiable in some general structure. But then, either $\theta \wedge \neg\varphi$ or $\varphi \wedge \psi$ is satisfiable by a tree model [SE84] of a finite branching degree, contradicting the fact that a tree satisfies φ iff it is accepted by $\mathcal{U}_\theta$ iff it is not accepted by $\mathcal{U}_\psi$.

Remark 2. Since it is also known that the μ-calculus has the *finite-model property* [KP84], it follows that Theorem 6 can also be relativized to finite Kripke structures.

5   Concluding Remarks

We showed that $\Sigma_2 \cap \Pi_2 \equiv AFMC$. In other words, if we can specify a property ψ both as a least fixpoint nested inside a greatest fixpoint and as a greatest fixpoint nested inside a least fixpoint, we should be able to specify ψ also with no alternation between greatest and least fixpoints. This offers an elegant characterization of alternation freedom. The key to our results is a development of a theory of *symmetric nondeterministic Büchi* tree automata. A technical outcome of this theory is that the blow-up of our construction, i.e., going from formulas in $\Sigma_2 \cap \Pi_2$ to equivalent formulas in AFMC is doubly exponential. It would be interesting to try to improve this complexity or to prove its optimality.

Combining our result here with the result in [KV01] ($\Sigma_1 \cap \Pi_1 \equiv$ ML) suggests the possibility of a general coalescence result for the μ-calculus hierarchy. Recall the definition of Δ_i as the closure of $\Sigma_i \cap \Pi_i$ under Boolean and modal operations and under alternation-preserving substitutions. Then we have that $\Sigma_i \cap \Pi_i \equiv \Delta_{i-1}$ for $i = 1, 2$. It is tempting to conjecture that this holds for all $i > 0$, in analogy for such coalescence for the quantifier alternation hierarchy of first-order logic (cf. [Add62]). As is shown, however, in [AS03], this is not the case for $i > 2$.

Acknowledgements. We are grateful to J.W. Addison for valuable discussions regarding the first-order quantifier-alternation hierarchy and to I. Walukiewicz for discussions regarding μ-automata.

References

[Add62]   J.W. Addision, The theory of hierarchies. *Proc. Internat. Congr. Logic, Method. and Philos. Sci. 1960*, pages 26–37, Stanford University Press, 1962.

[AN90]   A. Arnold and D. Niwiński. Fixed point characterization of Büchi automata on infinite trees. *Information Processing and Cybernetics*, 8–9:451–459, 1990.

[AN92]   A. Arnold and D. Niwiński. Fixed point characterization of weak monadic logic definable sets of trees. In *Tree Automata and Languages*, pages 159–188, Elsevier, 1992.

[AN01]   A. Arnold and D. Niwiński. *Rediments of μ-calculus*. Elsevier, 2001.

[AS03]   A. Arnold and L. Santocanale, On ambiguous classes in the μ-calculus hierarchy of tree languages, *Proc. Workshop on Fixed Points in Computer Science*, Warsaw, Poland, 2003.

[Ben91]   J. Benthem. Languages in actions: categories, lambdas and dynamic logic. *Studies in Logic*, 130, 1991.

[Bra98]   J.C. Bradfield. The modal μ-calculus alternation hierarchy is strict. *TCS*, 195(2):133–153, March 1998.

[BRS99]   R. Bloem, K. Ravi, and F. Somenzi. Efficient decision procedures for model checking of linear time logic properties. In *Proc. 11th CAV*, LNCS 1633, pages 222–235. 1999.

[CS91]   R. Cleaveland and B. Steffen. A linear-time model-checking algorithm for the alternation-free modal μ-calculus. In *Proc. 3rd CAV, LNCS* 575, pages 48–58, 1991.

[EL85]   E.A. Emerson and C.-L. Lei. Temporal model checking under generalized fairness constraints. In *Proc. 18th Hawaii International Conference on System Sciences*, 1985.

[EL86]   E.A. Emerson and C.-L. Lei. Efficient model checking in fragments of the propositional μ-calculus. In *Proc. 1st LICS*, pages 267–278, 1986.

[FL79]   M.J. Fischer and R.E. Ladner. Propositional dynamic logic of regular programs. *Journal of Computer and Systems Sciences*, 18:194–211, 1979.

[GJ79]   M. Garey and D. S. Johnson. *Computers and Intractability: A Guide to the Theory of NP-completeness*. W. Freeman and Co., San Francisco, 1979.

[Jur00]   M. Jurdzinski. Small progress measures for solving parity games. In *17th STACS*, LNCS 1770, pages 290–301. 2000.

[JW95]   D. Janin and I. Walukiewicz. Automata for the modal μ-calculus and related results. In *Proc. 20th MFCS*, LNCS, pages 552–562. 1995.

[Kai95]   R. Kaivola. On modal μ-calculus and Büchi tree automata. *IPL*, 54:17–22, 1995.

[Koz83]   D. Kozen. Results on the propositional μ-calculus. *TCS*, 27:333–354, 1983.

[KP84]   D. Kozen and R. Parikh. A decision procedure for the propositional μ-calculus. In *Logics of Programs*, LNCS 164, pages 313–325, 1984.

[KV98]   O. Kupferman and M.Y. Vardi. Freedom, weakness, and determinism: from linear-time to branching-time. In *Proc. 13th LICS*, pages 81–92, June 1998.

[KV99]   O. Kupferman and M.Y. Vardi. The weakness of self-complementation. In *Proc. 16th STACS*, *LNCS* 1563, pages 455–466. 1999.

[KV01]   O. Kupferman and M.Y. Vardi. On clopen specifications. In *Proc. 8th LPAR*, LNCS 2250, pages 24–38. 2001.

[KVW00]   O. Kupferman, M.Y. Vardi, and P. Wolper. An automata-theoretic approach to branching-time model checking. *Journal of the ACM*, 47(2):312–360, March 2000.

[McM93]   K.L. McMillan. *Symbolic Model Checking*. Kluwer Academic Publishers, 1993.

[MH84]   S. Miyano and T. Hayashi. Alternating finite automata on ω-words. *TCS*, 32:321–330, 1984.

[Mos84]   A.W. Mostowski. Regular expressions for infinite trees and a standard form of automata. In *Computation Theory*, LNCS 208, pages 157–168. 1984.

[MS87]   D.E. Muller and P.E. Schupp. Alternating automata on infinite trees. *TCS*, 54:267–276, 1987.

[MSS86]   D.E. Muller, A. Saoudi, and P.E. Schupp. Alternating automata, the weak monadic theory of the tree and its complexity. In *Proc. 13th ICALP*, LNCS 226, 1986.

[Niw86]   D. Niwiński. On fixed point clones. In *Proc. 13th ICALP*, LNCS 226, pages 464–473. 1986.

[Rab70]   M.O. Rabin. Weakly definable relations and special automata. In *Proc. Symp. Math. Logic and Foundations of Set Theory*, pages 1–23, 1970.

[Rog67]   H. Rogers, Theory of recursive functions and effective computability. McGraw-Hill, 1967.

[SE84]   R.S. Street and E.A. Emerson. An elementary decision procedure for the μ-calculus. In *Proc. 11th ICALP*, LNCS 172, pages 465–472, 1984.

[Tak86]   M. Takahashi. The greatest fixed-points and rational ω-tree languages. TCS 44, pp. 259–274, 1986.

[VW86]   M.Y. Vardi and P. Wolper. An automata-theoretic approach to automatic program verification. In *Proc. 1st LICS*, pages 332–344, 1986.

[Wal03]   I. Walukiewicz. Private communication, 2003.

[Wil99]   T. Wilke. CTL$^+$ is exponentially more succinct than CTL. In *Proc. 19th FST& TCS*, LNCS 1738, pages 110–121, 1999.

Upper Bounds for a Theory of Queues

Tatiana Rybina and Andrei Voronkov

University of Manchester
{rybina,voronkov}@cs.man.ac.uk

Abstract. We prove an upper bound result for the first-order theory of a structure $\mathbf{W}$ of queues, i.e. words with two relations: addition of a letter on the left and on the right of a word. Using complexity-tailored Ehrenfeucht games we show that the witnesses for quantified variables in this theory can be bound by words of an exponential length. This result, together with a lower bound result for the first-order theory of two successors [6], proves that the first-order theory of $\mathbf{W}$ is complete in $\mathrm{LATIME}(2^{O(n)})$: the class of problems solvable by alternating Turing machines running in exponential time but only with a linear number of alternations.

1 Introduction

Theories of words are fundamental to computer science. Decision procedures for various theories of words are used in many areas of computing, for example in verification. Closely related to words are *queues* which can be regarded as words with two operations: deleting a letter on the left and adding a letter on the right. In this paper we prove upper bounds on the complexity of the first-order theory of queues. The upper bound is tight, i.e., it coincides with the respective lower bound up to a constant factor.

Denote by $\{0,1\}^*$ the set of all words over the finite alphabet $\{0,1\}$, by $ln(w)$ the length of the word w and by λ the empty word. We call the elements of $\{0,1\}^*$ simply *words*. By "$\cdot$" we denote concatenation of words. Define the following four relations on words:

$$l_0(a,b) \leftrightarrow b = 0 \cdot a; \qquad l_1(a,b) \leftrightarrow b = 1 \cdot a;$$
$$r_0(a,b) \leftrightarrow b = a \cdot 0; \qquad r_1(a,b) \leftrightarrow b = a \cdot 1.$$

The first-order structure $\mathbf{W} = \langle \{0,1\}^*, r_0, r_1, l_0, l_1 \rangle$ is called the *queue structure*. The *first-order theory of queues* is the first-order theory of $\mathbf{W}$.

Let us formulate the main result of this paper. See [11,10] for the precise definition of the *complexity class* $LATIME(2^{O(n)})$: it is the class of problems solvable by alternating Turing machines running in time $2^{O(n)}$ but only with a linear number of alternations. Of course, for this class polynomial-time or LOGSPACE reductions are too coarse, this class is closed with respect to LOGLIN-reductions [11], i.e., LOGSPACE reductions giving at most linear increase in length. The main result of this paper is the following.

J.C.M. Baeten et al. (Eds.): ICALP 2003, LNCS 2719, pp. 714–724, 2003.
© Springer-Verlag Berlin Heidelberg 2003

THEOREM 1 *The first-order theory of* **W** *is complete in* $LATIME(2^{O(n)})$ *with respect to LOGLIN-reductions.* ❏

This theorem will be proved using complexity tailored Ehrenfeucht games. We will show that in the first-order theory of **W** in every sentence witnesses for quantified variables can be bound by words of the size exponential in the size of the sentence.

The decidability of the first-order theory of this structure follows from the decidability of the first-order theory of two successors with the predicates of equal length and prefix [9]. It also immediately follows from the fact that this structure is automatic [7,4].

A lower bound on the first-order theory of **W** can be derived from the lower bound on the first-order theory of two successors, i.e., of the structure $\langle\{0,1\}^*, r_0, r_1\rangle$, proved in [11] based on [6] (a technique for proving lower bounds is also described in [5], some simple generalizations can be found in [12]). Expressive power of several theories of words, including the first-order theory of **W**, is discussed in [1].

For us the main motivation for these results was our case study of verification of a protocol with queues. Verification with queues was also extensively studied in [3,2]. In [8] we proved that first-order theories of some structures containing trees and queues are decidable. Our results were based on quantifier elimination and imply a non-elementary upper bound (a non-elementary lower bound also follows for the theory of trees [6]). However, if we consider a theory with queues only, it was clear that a non-elementary upper bound could be avoided. Indeed, the quantifier elimination arguments of [8] show that the main difference in expressive power between queues and stacks is periodicity constraints. However, these periodicity constraints, though they can express "deep" properties of queues (e.g., that all elements of a queue are 0's), still cannot distinguish queues which are indistinguishable by their "short" prefixes. Motivated by this observation we undertake a characterization of the exact complexity of the first-order theory of **W**.

In the proof of the upper bound for **W** we show, like [6], that all quantifiers in a formula can be replaced by quantifiers of an exponential size. However, our arguments are more technically involved. Moreover, some lemmas of [6] do not hold any more in this context.

2   Ehrenfeucht Games

$\mathbb{N}$ denotes the set of natural numbers. By $\bar{a}_k$ we denote the sequence $a_1, \ldots, a_k$ of k elements, and similar for other letters instead of **a**.

DEFINITION 2 (Norm) Let A be a structure. A *norm* on A, denoted $|| \cdot ||$, is a function from the domain of the structure A to $\mathbb{N}$. For an element a of A we write $||a||$ to denote the norm of a. ❏

The following definitions are similar to those of Ferrante and Rackoff [6].

DEFINITION 3 (Ehrenfeucht Equivalence) Let $n, k \in \mathbb{N}$, A be a structure, and $\bar{\mathbf{a}}_k$, $\bar{\mathbf{b}}_k$ be sequences of elements of A. Then we write $\bar{\mathbf{a}}_k \equiv_{n,k} \bar{\mathbf{b}}_k$ if for all formulas $F(x_1, \ldots, x_k)$ of quantifier depth at most n, $\bar{\mathbf{a}}_k$ satisfies $F(x_1, \ldots, x_k)$ in A if and only if $\bar{\mathbf{b}}_k$ satisfies $F(x_1, \ldots, x_k)$ in A. (In particular $\bar{\mathbf{a}}_k \equiv_{0,k} \bar{\mathbf{b}}_k$ means that $\bar{\mathbf{a}}_k$ and $\bar{\mathbf{b}}_k$ satisfy the same quantifier-free formulas.) $\qquad\square$

DEFINITION 4 (Boundedness) Let A be a structure with a norm $|| \cdot ||$ on it and $\mathcal{H} : \mathbb{N}^3 \to \mathbb{N}$ be a function. We say that A is $\mathcal{H}$-*bounded* if for all natural numbers k, n, m, a sequence $\bar{\mathbf{a}}_k$ of elements of A, and formula $F(x_1, \ldots, x_{k+1})$ of quantifier depth $\leq n$ the following property holds. If for all $i \leq k$ we have $||a_i|| \leq m$ and $A \models \exists x_{k+1} F(\bar{\mathbf{a}}_k, x_{k+1})$, then there exists $a_{k+1} \in A$ such that $||a_{k+1}|| \leq \mathcal{H}(n, k, m)$ and $A \models F(\bar{\mathbf{a}}_k, a_{k+1})$. $\qquad\square$

THEOREM 5 (Ferrante and Rackoff [6]) *Let A be a structure and $\mathcal{H} : \mathbb{N}^3 \to \mathbb{N}$ be a function. Let $\mathcal{E}_{n,k}$ be relations such that for all natural numbers n, k, m and sequences of elements $\bar{\mathbf{a}}_k$, $\bar{\mathbf{b}}_k$ of A the following properties are true:*

1. *if $\mathcal{E}_{0,k}(\bar{\mathbf{a}}_k, \bar{\mathbf{b}}_k)$ then $\bar{\mathbf{a}}_k \equiv_{0,k} \bar{\mathbf{b}}_k$;*
2. *if $\mathcal{E}_{n+1,k}(\bar{\mathbf{a}}_k, \bar{\mathbf{b}}_k)$ and for all $i \leq k$ we have $||b_i|| \leq m$ then for all $a_{k+1} \in A$ there exists $b_{k+1} \in A$ such that $\mathcal{E}_{n,k+1}(\bar{\mathbf{a}}_{k+1}, \bar{\mathbf{b}}_{k+1})$ and $||b_{k+1}|| \leq \mathcal{H}(n, k, m)$.*

Then:

1. *$\mathcal{E}_{n,k}(\bar{\mathbf{a}}_k, \bar{\mathbf{b}}_k) \Rightarrow \bar{\mathbf{a}}_k \equiv_{n,k} \bar{\mathbf{b}}_k$ for all $n, k \in \mathbb{N}$,*
2. *the structure A is $\mathcal{H}$-bounded.* $\qquad\square$

3   Main Argument

An Ehrenfeucht game decision procedure for **W** consists of defining a set of equivalence relations $\mathcal{E}_{n,k}$, which turn out to be refinements of the relations $\equiv_{n,k}$, defining a function $\mathcal{H}(n, k, m)$, and showing $\mathcal{H}$-boundedness of the structure **W**. Since the structure is $\mathcal{H}$-bounded, then the witnesses for quantifiers in formulas can be restricted by elements of a fixed depth. If the number of elements of every norm is finite, then we obtain a decision procedure.

Let a, v_1, v_2 be words. By $v_1[a]v_2$ we denote the word w, if it exists, such that $a = v_1 \cdot w \cdot v_2$.

In the sequel we will extensively use partial functions. Let us make a notational convention about their use.

CONVENTION 6 *Let e, e_1, e_2 be any expressions over words. We write $e \downarrow$ to denote that e exists. When we write $e_1 = e_2$, we mean that both e_1 and e_2 are defined and equal. If S is a set of words and we write $e \in S$, we mean that e is defined and e is a member of S.* $\qquad\square$

DEFINITION 7 (ε-word, ε-length, ε-correction) Let $\varepsilon \in \mathbb{N}$. A number $\ell \in \mathbb{N}$ is said to be an ε-*length* if $\ell \leq \varepsilon$. We will normally use this terminology when we speak about lengths of words. A word is an ε-*word* if its length is an ε-length. An ε-*correction* is a partial function α such that for some ε-words v_1, v_2, w_1, w_2 and for all words a we have $\alpha(a) = w_2 \cdot {}_{w_1}[a]_{v_1} \cdot v_2$. An ε-correction α is called *trivial* if for some words v, w and all words a we have $\alpha(a) = w \cdot {}_w[a]_v \cdot v$.   ❑

Let us note some useful properties of this definition.

LEMMA 8 *The following statements hold for ε-corrections.*

1. *If a is an ε_1-word and b is ε_2-word, then $a \cdot b$ is an $(\varepsilon_1 + \varepsilon_2)$-word.*
2. *If a is a ε_1-word and α is an ε_2-correction, then $\alpha(a)$ is an $(\varepsilon_1 + 2\varepsilon_2)$-word.*
3. *For every ε-correction α there exists an ε-correction inverse to α, denoted α^{-1}, such that for every word a we have*
   a) *if $\alpha(a) \downarrow$ then $\alpha^{-1}(\alpha(a)) = a$;*
   b) *if $\alpha(a) = b$ then $a = \alpha^{-1}(b)$ and $\alpha(\alpha^{-1}(b)) = b$;*
   c) *if $\alpha^{-1}(a) = b$ then $a = \alpha(b)$ and $\alpha^{-1}(\alpha(b)) = b$.*
4. *If α is an ε_1-correction, β is an ε_2-correction, and $\alpha(\beta(v))$ is defined for at least one word v, then their composition $\alpha\beta$ is an $(\varepsilon_1 + \varepsilon_2)$-correction.*   ❑

In the sequel we will often use this lemma implicitly.

For every word a, denote $a^* = \{a^n \mid n \in \mathbb{N}\}$.

LEMMA 9 (see [8]) *Let a, b, c be words such that $a \cdot b = b \cdot c$. Then*

1. *if $a \neq c$ then there exist words a_1 and a_2 such that $a = a_1 \cdot a_2$, $c = a_2 \cdot a_1$ and $b \in \{a^n \cdot a_1 \mid n \in \mathbb{N}\}$;*
2. *if $a = c$ then there exists a word s such that $c \in s^*$ and $b \in s^*$.*   ❑

LEMMA 10 *For every non-trivial ε-correction and word a, if $\alpha(a) = a$, then for some ε-word s and ε-correction γ we have $\gamma(a) \in s^*$.*

PROOF. By the definition of α there exist ε-words w_1, w_2, v_1, v_2 such that $\alpha(a) = w_2 \cdot {}_{w_1}[a]_{v_1} \cdot v_2$. On the other hand, we have $a = w_1 \cdot {}_{w_1}[a]_{v_1} \cdot v_1$. Thus $w_2 \cdot {}_{w_1}[a]_{v_1} \cdot v_2 = w_1 \cdot {}_{w_1}[a]_{v_1} \cdot v_1$. Since α is non-trivial, we have $w_1 \neq w_2$ and $v_1 \neq v_2$. The equality $\alpha(a) = a$ implies that either $ln(w_2) < ln(w_1)$, $ln(v_2) > ln(v_1)$, or $ln(w_2) > ln(w_1)$, $ln(v_2) < ln(v_1)$. Let us consider the case $ln(w_2) < ln(w_1)$, $ln(v_2) > ln(v_1)$, the other case is similar. In this case there exist words b and c such that $w_1 = w_2 \cdot b$ and $v_2 = c \cdot v_1$. Since w_1, v_2 are ε-words, b, c must be ε-words too. We have $w_2 \cdot {}_{w_1}[a]_{v_1} \cdot c \cdot v_1 = w_2 \cdot b \cdot {}_{w_1}[a]_{v_1} \cdot v_1$, hence $b \cdot {}_{w_1}[a]_{v_1} = {}_{w_1}[a]_{v_1} \cdot c$. By Lemma 9 there exist words s_1 and s_2 such that $b = s_1 \cdot s_2$, $c = s_2 \cdot s_1$ and ${}_{w_1}[a]_{v_1} \in \{(s_1 \cdot s_2)^n \cdot s_1 \mid n \in \mathbb{N}\}$. Evidently, s_1, s_2 are ε-words. Define $s = b$ and define γ as follows: for all v we have

$$\gamma(v) \overset{\text{def}}{=} \lambda \cdot {}_{w_2}[v]_{v_1} \cdot s_2.$$

The property $\gamma(a) \in s^*$ is not hard to check.   ❑

LEMMA 11 *Let b, c be ε-words, α, β be ε-corrections, and a be an arbitrary word.*

1. *If $\alpha(a) \in b^*$ then for all $w \in b^*$ such that $ln(w) \geq 2\varepsilon$ we have $\alpha^{-1}(w) \downarrow$.*
2. *If $ln(a) \geq 4\varepsilon$, $\alpha(a) \in b^*$, and $\beta(a) \in c^*$ then for all words $w \in b^*$ and $v \in c^*$ such that $ln(w), ln(v) \geq 4\varepsilon$ we have $\beta(\alpha^{-1}(w)) \in c^*$ and $\alpha(\beta^{-1}(v)) \in b^*$.* $\square$

The proof is straightforward but tedious.

The following definition of *indistinguishability* is the main technical notion of this paper. Define the following function L of two integer arguments: $L(n, k) = 2^{3n+k}$.

DEFINITION 12 (Indistinguishability) Let $\bar{a}_k$ and $\bar{b}_k$ be sequences of words and n be a natural number. We say that $\bar{a}_k$ and $\bar{b}_k$ are $\mathcal{E}_{n,k}$-*indistinguishable*, denoted $\bar{a}_k \, \mathcal{E}_{n,k} \, \bar{b}_k$, if the following conditions hold for all $i, j \in \{1, \ldots, k\}$. Let $\varepsilon = L(n, k)$.

1. For every ε-correction α we have $\alpha(a_i) = a_j$ if and only if $\alpha(b_i) = b_j$.
2. If either a_i or b_i is a 4ε-word, then $a_i = b_i$.
3. For every ε-correction α and ε-word a, $\alpha(a_i) \in a^*$ if and only if $\alpha(b_i) \in a^*$.

Prefix (respectively suffix) of the length ℓ of a word a, if it exists, is denoted $prefix(\ell, a)$ (respectively $suffix(\ell, a)$).

LEMMA 13 *Let $\bar{a}_k \mathcal{E}_{n,k} \bar{b}_k$. Define $\varepsilon = L(n, k)$. Then for every i*

1. *either $a_i = b_i$, or $prefix(\varepsilon, a_i) = prefix(\varepsilon, b_i)$ and $suffix(\varepsilon, a_i) = suffix(\varepsilon, b_i)$;*
2. *for every ε-correction α, $\alpha(a_i) \downarrow$ if and only if $\alpha(b_i) \downarrow$.*

PROOF. The second clause evidently follows from the first one, so we will only prove the first clause.

If $ln(a_i) \leq 4\varepsilon$ then, by Clause 2 of Definition 12, $a_i = b_i$. Otherwise we have $ln(a_i) > 4\varepsilon$. Define an ε-correction α by

$$\alpha(v) \overset{\text{def}}{=} prefix(\varepsilon, a_i) \cdot {}_{prefix(\varepsilon, a_i)}[v]_{suffix(\varepsilon, a_i)} \cdot suffix(\varepsilon, a_i).$$

It is easy to see that $\alpha(a_i) = a_i$, hence, by Clause 1 of Definition 12, $\alpha(b_i) = b_i$. Then $\alpha(b_i)$ is defined, hence $prefix(\varepsilon, b_i) = prefix(\varepsilon, a_i)$ and $suffix(\varepsilon, b_i) = suffix(\varepsilon, a_i)$. $\square$

By routine inspection of the definition of $\mathcal{E}_{n,k}$, we can also prove the following result.

COROLLARY 14 $\mathcal{E}_{n,k}$ *is an equivalence relation.* $\square$

The following lemma is a key to proving that **W** is $\mathcal{H}$-bounded.

LEMMA 15 *Let k, n be natural numbers and $\bar{a}_k$, $\bar{b}_k$ be sequences of words such that $\bar{a}_k \, \mathcal{E}_{n+1,k} \bar{b}_k$. Then for every word a_{k+1} there exists a word b_{k+1} such that $\bar{a}_{k+1} \, \mathcal{E}_{n,k+1} \, \bar{b}_{k+1}$.*

PROOF. Let $\varepsilon = L(n, k+1)$. In the proof we will construct the word b_{k+1} and prove $\mathcal{E}_{n,k+1}$-indistinguishability of $\bar{a}_{k+1}$ and $\bar{b}_{k+1}$ using the hypothesis about $\mathcal{E}_{n+1,k}$-indistinguishability of $\bar{a}_k$ and $\bar{b}_k$. In this respect note that $L(n+1, k) = 4 \cdot L(n, k+1)$. Therefore, in the proof we will use hypothesis about $4c$ words and prove statements about ε-words.

Let us note that while verifying Clauses 1–3 of Definition 12 for $\bar{a}_{k+1}$ and $\bar{b}_{k+1}$ we have to consider only the case $i = k+1$ or $j = k+1$ for Clause 1 and the case $i = k+1$ for Clauses 2–3. Moreover, for Clause 1 the proofs for the case $i = k+1$ are similar to the proofs for the case $j = k+1$, so we will only consider the case $i = k+1$.

Our choice of b_{k+1} depends on the properties of a_{k+1}, so we proceed by cases.

Case 1: a_{k+1} is a 4ε-word. We choose $b_{k+1} = a_{k+1}$.

Let us prove Clauses 1–3 of Definition 12 for $\bar{a}_{k+1}$ and $\bar{b}_{k+1}$.

1. Suppose α is an ε-correction $\alpha(a_i) = a_{k+1}$. We have to prove $\alpha(b_i) = b_{k+1}$. We only verify the case $i \leq k$ since the case $i = k+1$ is trivial. We know that a_{k+1} is 4ε-word and $\alpha^{-1}(a_{k+1}) = a_i$. Then a_i is 6ε-word. By the hypothesis, if a_i is 16ε-word, then $a_i = b_i$. Therefore, $\alpha(b_i) = b_{k+1}$.
2. We have to prove that if a_{k+1} is a 4ε-word or b_{k+1} is a 4ε-word, then $a_{k+1} = b_{k+1}$. But we have $a_{k+1} = b_{k+1}$ by our construction.
3. By our choice $a_{k+1} = b_{k+1}$, therefore for every ε-correction α and ε-word a : $\alpha(a_{k+1}) \in a^*$ if and only if $\alpha(b_{k+1}) \in a^*$.

Case 2: a_{k+1} is not a 4ε-word but there exist $j \leq k$ and ε-correction β such that $\beta(a_j) = a_{k+1}$. By Lemma 13, $\beta(b_j)$ is defined. We choose $b_{k+1} = \beta(b_j)$. Let us show that our choice of b_{k+1} satisfies the definition of $\mathcal{E}_{n,k+1}$-indistinguishability.

1. Suppose that α is an ε-correction and $i \leq k+1$. We need to verify that $\alpha(a_i) = a_{k+1}$ if and only if $\alpha(b_i) = b_{k+1}$.
   To prove the "only if" direction, suppose $\alpha(a_i) = a_{k+1}$. Since $a_{k+1} = \beta(a_j)$, we have $\beta^{-1}(a_{k+1}) = a_j$, hence $\beta^{-1}(\alpha(a_i)) = a_j$. Consider two cases: $i \neq k+1$ and $i = k+1$.
   Suppose $i \neq k+1$. Since $\beta^{-1}\alpha$ is a 2ε-correction, by the hypothesis we have $\beta^{-1}(\alpha(b_i)) = b_j$. This implies $\alpha(b_i) = \beta(b_j) = b_{k+1}$.
   Now suppose that $i = k+1$, then $\alpha(a_{k+1}) = a_{k+1}$, that is $\alpha(\beta(a_j)) = \beta(a_j)$, hence $\beta^{-1}(\alpha(\beta(a_j))) = a_j$. By the hypothesis, since $\beta^{-1}\alpha\beta$ is a 3ε-correction, $\beta^{-1}(\alpha(\beta(b_j))) = b_j$, hence $\alpha(\beta(b_j)) = \beta(b_j)$. But $\beta(b_j) = b_{k+1}$, so $\alpha(b_{k+1}) = b_{k+1}$.
   The "if" direction is similar.
2. Since a_{k+1} is not a 4ε-word to verify Clause 2 we have to show that b_{k+1} is not a 4ε-word. By our choice of b_{k+1}, $\beta^{-1}(b_{k+1}) = b_j$. Suppose that b_{k+1} is a 4ε-word, then b_j is a 6ε-word, so by our hypothesis $a_j = b_j$. Therefore, $\beta(a_j) = \beta(b_j)$, that is $a_{k+1} = b_{k+1}$. But then a_{k+1} would be a 4ε-word. Contradiction.

3. To verify Clause 3 we only have to show that for every ε-word a and every ε-correction α the following holds:

$$\alpha(a_{k+1}) \in a^* \leftrightarrow \alpha(b_{k+1}) \in a^*.$$

Suppose that $\alpha(a_{k+1}) \in a^*$, then $\alpha(\beta(a_j)) \in a^*$ and $\alpha\beta$ is a 2ε-correction. By the hypothesis, we have $\alpha(\beta(b_j)) \in a^*$, that is $\alpha(b_{k+1}) \in a^*$.

Case 3: a_{k+1} is not a 4ε-word and there are no $j \leq k$ and ε-correction α such that $\alpha(a_j) = a_{k+1}$ but there exist an ε-correction γ and an ε-word c such that $\gamma(a_{k+1}) \in c^*$.

If $\gamma(a_{k+1})$ is a 4ε-word, then a_{k+1} is a 5ε-word and we can choose $b_{k+1} = a_{k+1}$ and repeat the proof of Case 1.

Suppose that $\gamma(a_{k+1})$ is not a 4ε-word. Let ℓ be a natural number such that

$$ln(c^{\ell-1}) \leq \max(6\varepsilon, 4\varepsilon + \max_{i \leq k} ln(b_i)) < ln(c^\ell).$$

Then we choose $b_{k+1} = \gamma^{-1}(c^\ell)$ (notice that $ln(c^\ell) > 4\varepsilon$ and hence by Lemma 11 $\gamma^{-1}(c^\ell)$ is defined).

Let us prove some simple estimations on the length of b_{k+1}. Note that by our definition for all $i \leq k$ we have $ln(c^\ell) - ln(b_i) > 4\varepsilon$. Since b_{k+1} is an ε-correction of c^ℓ, this implies $ln(b_{k+1}) - ln(b_i) > 2\varepsilon$. In a similar way we can establish

$$ln(b_{k+1}) < \max(9\varepsilon, 7\varepsilon + \max_{i \leq k} ln(b_i)). \tag{1}$$

Let us prove Clauses 1–3 of Definition 12 for $\bar{a}_{k+1}$ and $\bar{b}_{k+1}$.

1. Let α be an ε-correction. We have to prove that $\alpha(a_i) = a_{k+1}$ if and only if $\alpha(b_i) = b_{k+1}$. Consider two cases: $i \leq k$ and $i = k + 1$.
   Let $i \leq k$. By the assumption $\alpha(a_i) \neq a_{k+1}$, so we have to prove $\alpha(b_i) \neq b_{k+1}$. Suppose, by contradiction, $\alpha(b_i) = b_{k+1}$. Then $ln(b_{k+1}) - ln(b_i) \leq 2\varepsilon$ which contradicts to $ln(b_{k+1}) - ln(b_i) > 2\varepsilon$.
   Let $i = k+1$. We have to prove that $\alpha(a_k+1) = a_{k+1}$ if and only if $\alpha(b_{k+1}) = b_{k+1}$. Suppose that $\alpha(a_{k+1}) = a_{k+1}$. By assumption, we have $\gamma(a_{k+1}) \in c^*$, i.e. there exists natural number z such that $\gamma(a_{k+1}) = c^z$. Without loss of generality we assume that c is non-periodic. Then $\alpha(\gamma^{-1}(c^z)) = \gamma^{-1}(c^z)$. This implies $\gamma(\alpha(\gamma^{-1}(c^z))) = c^z$. It is not hard to argue $\gamma\alpha\gamma^{-1}$ is a 2ε-correction. Since $\gamma(\alpha(\gamma^{-1}(c^z))) = c^z$, $\gamma\alpha\gamma^{-1}$ either is a trivial correction or for all $w \in c^*$ there exists $z_1 \in \mathbb{N}$ such that $ln(c^{z_1}) \leq 2\varepsilon$ and $\gamma\alpha\gamma^{-1}(w) = \lambda \cdot c^{z_1} [w]_\lambda \cdot c^{z_1}$. Thus $\gamma\alpha\gamma^{-1}(c^\ell) \downarrow$. It is now easy to see that $\gamma\alpha\gamma^{-1}(c^\ell) = c^\ell$. In the other direction the proof is similar.
2. Since $ln(a_{k+1}) > 4\varepsilon$ and $ln(b_{k+1}) > 4\varepsilon$ there is no need to verify Clause 2.
3. Suppose that $\alpha(a_{k+1}) \in a^*$ for some ε-correction α and ε-word a. We have to show $\alpha(b_{k+1}) \in a^*$. Since $\gamma(a_{k+1}) \in c^*$, by Lemma 11, we have $\alpha(\gamma^{-1}(c^\ell)) \in a^*$.

Case 4: a_{k+1} *is not a 4ε-word, there are no $j \leq k$ and ε-correction α such that* $\alpha(a_j) = a_{k+1}$ *and for every ε-correction α and ε-word a: $\alpha(a_{k+1}) \notin a^*$.* Define the set of words:

$$W = \{prefix(\varepsilon, a_{k+1}) \cdot q \cdot suffix(\varepsilon, a_{k+1}) \mid ln(q) = 2\varepsilon + 1\}.$$

Note that $|W| = 2^{2\varepsilon+1}$. It is not hard to argue that for all c, $d \in W$ and ε-corrections α, β the following holds:

$$\alpha(c) = \beta(d) \leftrightarrow c = d.$$

Therefore for every $i \leq k$ there exists at most one element $c \in W$ which can be obtained by an ε-correction from b_i.

Let us count the number of words $w \in W$ such that for some ε-correction α and ε-word a we have $\alpha(a) \in a^*$. It is not hard to argue that the number of such words is not greater than the number of ϵ-words, that is $2^{\varepsilon+1}$.

Now define the following set of words:

$$W' = \{d \in W \mid \text{ for all } i \leq k, \ \varepsilon\text{-words } a \text{ and } \varepsilon\text{-corrections } \beta :$$
$$\beta(b_i) \neq d \text{ and } \beta(d) \notin a^*\}.$$

Let us prove that W' is non-empty. Indeed, W' is be obtained from W by removing all ε-corrections of the words b_i and all ε-corrections of words belonging to some a^*, where a is an ϵ. Therefore, the cardinality of W' is at least $2^{2\varepsilon+1} - k - 2^{\varepsilon+1}$. We have

$$2^{2\varepsilon+1} - k - 2^{\varepsilon+1} > 2^{2\varepsilon+1} - 2^{\varepsilon+2} \geq 0,$$

so W' contains at least one element. Choose b_{k+1} to be any element of W'. Let us check that our choice of b_{k+1} satisfies the definition of $\mathcal{E}_{n,k+1}$-indistinguishability.

1. Let α be an ε-correction. By our assumption, for every $j \leq k$ we have $\alpha(a_j) \neq a_{k+1}$. By our construction of b_{k+1} we have $\alpha(b_j) \neq b_{k+1}$. So it remains to check that $\alpha(a_k + 1) = a_{k+1}$ if and only if $\alpha(b_k + 1) = b_{k+1}$. If α is trivial, then this property is straightforward, so assume that α is non-trivial. If $\alpha(a_{k+1}) = a_{k+1}$, then by Lemma 10 for some ε-word a and ε-correction β we would have $\beta(a_{k+1}) \in c^*$. This would contradict to our assumption, so we have $\alpha(a_{k+1}) \neq a_{k+1}$. Then we have to prove $\alpha(b_{k+1}) \neq b_{k+1}$. Suppose, by contradiction, $\alpha(b_{k+1}) = b_{k+1}$. Then by Lemma 10 for some ε-word a and ε-correction β we would have $\beta(b_{k+1}) \in c^*$. But this is impossible since $b_{k+1} \in W'$.
2. Since $ln(a_{k+1}) > 4\varepsilon$ and $ln(b_{k+1}) > 4\varepsilon$ there is no need to verify Clause 2.
3. We have to show that for every ε-correction α and ε-word a we have $\alpha(b_{k+1}) \notin a^*$. This is immediate by our choice of b_{k+1}.

The proof of Lemma 15 is completed.   □

LEMMA 16 *For all natural numbers k, n, all sequences of words $\bar{a}_k$ and $\bar{b}_k$, if $\bar{a}_k \, \mathcal{E}_{n+1,k} \bar{b}_k$ then for every word a_{k+1} there exists word b_{k+1} such that $\bar{a}_{k+1}$ and $\bar{b}_{k+1}$ are $\mathcal{E}_{n,k+1}$- indistinguishable and either*

1. $ln(b_{k+1}) \leq 9 * 2^{3n+k}$, or
2. for some $i \leq k$, $ln(b_{k+1}) \leq ln(b_i) + 7 * 2^{3n+k}$.

PROOF. By routine inspection of the proof of Lemma 15. These bounds appear from (1), other parts of the proof give lower bounds. □

For $w \in \{0,1\}^*$ and $n, k, m \in \mathbb{N}$, define $||w|| = ln(w)$ and $\mathcal{H}(n, k, m) = m + 9 * 2^{3n+k}$.

LEMMA 17 *For all natural numbers k, n, m, all sequences of words $\bar{a}_k$ and $\bar{b}_k$, if $\bar{a}_k \, \mathcal{E}_{n+1,k} \bar{b}_k$ and $||b_i|| \leq m$ for all $i \leq k$ then for every word a_{k+1} there exists word b_{k+1} such that $\bar{a}_{k+1}$ and $\bar{b}_{k+1}$ are $\mathcal{E}_{n,k+1}$-indistinguishable and $||b_{k+1}|| \leq \mathcal{H}(n, k, m)$.* □

This lemma proves the second conditions of Theorem 5, to prove the first condition note the following result.

LEMMA 18 *Let $\bar{a}_k, \bar{b}_k$ be sequences of words such that $\bar{a}_k \, \mathcal{E}_{0,k} \, \bar{b}_k$. Then $\bar{a}_k \, \overline{\overline{\equiv}}_{0,k} \, \bar{b}_k$.*

PROOF. Since $\bar{a}_k \, \mathcal{E}_{0,k} \, \bar{b}_k$, then for all $i, j \leq k$ the following equivalences hold:

$$\lambda[a_i]_0 = a_j \leftrightarrow \lambda[b_i]_0 = b_j; \, \lambda[a_i]_1 = a_j \leftrightarrow \lambda[b_i]_1 = b_j;$$
$$0[a_i]_\lambda = a_j \leftrightarrow 0[b_i]_\lambda = b_j; \, 1[a_i]_\lambda = a_j \leftrightarrow 1[b_i]_\lambda = b_j.$$

Thus

$$r_0(a_j, a_i) \leftrightarrow r_0(b_j, b_i); \, r_1(a_j, a_i) \leftrightarrow r_1(b_j, b_i);$$
$$l_0(a_j, a_i) \leftrightarrow l_0(b_j, b_i); \, l_1(a_j, a_i) \leftrightarrow l_1(b_j, b_i).$$

Using Definition 3, we conclude $\bar{a}_k \, \overline{\overline{\equiv}}_{0,k} \bar{b}_k$. □

4   Main Results

Lemma 17 and Lemma 18 prove the conditions for Theorem 5. Therefore, by this theorem we have the following key result.

THEOREM 19 *For all $n, k, m \in \mathbb{N}$:*

1. *for all sequences of words $\bar{a}_k$ and $\bar{b}_k$, if $\bar{a}_k$ and $\bar{b}_k$ are $\mathcal{E}_{n,k}$-indistinguishable then $\bar{a}_k \, \overline{\overline{\equiv}}_{n,k} \, \bar{b}_k$ for all $n, k \in \mathbb{N}$,*
2. *the structure $\mathbf{W}$ is $\mathcal{H}$-bounded.* □

Let us extend the first-order language by bounded quantifiers $(\exists v \preceq C)$ and $(\forall v \preceq C)$ for all natural numbers C with the following interpretation: $(\exists v \preceq C)A(v)$ holds if there exists a C-word such v that $A(v)$, and similar for $(\forall v \preceq C)$

LEMMA 20 Let $Q_1 x_1 \ldots Q_n x_n F(\bar{\mathbf{x}}_n)$ be a sentence such that $Q_i \in \{\forall, \exists\}$ and $F(\bar{\mathbf{x}}_n)$ is quantifier-free. Let $C = 9 * 2^{3n+1}$. Then

$$\mathbf{W} \models Q_1 x_1 \ldots Q_n x_n F(\bar{\mathbf{x}}_n) \ \leftrightarrow \ \mathbf{W} \models Q_1 x_1 \preceq C \ldots Q_n x_n \preceq C F(\bar{\mathbf{x}}_n). \qquad (2)$$

PROOF. Define $C_1 = 9 * 2^{3n}$ and for all $i > 1$, $C_{i+1} = C_i + 9 * 2^{3(n-i)+i}$. It follows from Theorem 19 that each of the quantifiers $Q_i x_i$ can be equivalently replaced by $(Q x_i \prec C_i)$. It is not hard to argue that $C_i < C$ for all i, which proves (2). □

Now we can prove our main result: Theorem 1.

PROOF (of Theorem 1). Recall that we have to prove that the first-order theory of $\mathbf{W}$ is complete in $\mathrm{LATIME}(2^{O(n)})$. It is known that the first-order theory of $\mathbf{W}$ is $\mathrm{LATIME}(2^{O(n)})$-hard already for formulas without the relations l_0, l_1 (see [6,11,10,12], so we should prove that the first-order theory of $\mathbf{W}$ belongs to the class $\mathrm{LATIME}(2^{O(n)})$. This can be proved by the following procedure running in exponential time by alternating Turing machines with a linear number of alternations: first, using Lemma 20, replace all quantifiers by quantifiers bound by words of length $2^{O(n)}$, and then "guess" the corresponding words using alternating Turing machines. The number of alternations is less than the number of quantifiers in the formula, and is therefore at most linear in n.   □

Acknowledgments. We thank Bakhadyr Khoussainov, Leonid Libkin, and Wolfgang Thomas for helpful remarks related to the first-order theory of $\mathbf{W}$.

References

1. M. Benedikt, L. Libkin, T. Schwentick, and L. Segoufin. A model-theoretic approach to regular string relations. In *Proc. 16th Annual IEEE Symposium on Logic in Computer Science, LICS 2001*, pages 431–440, 2001.

2. N.S. Bjørner. *Integrating Decision Procedures for Temporal Verification*. PhD thesis, Computer Science Department, Stanford University, 1998.

3. N.S. Bjørner. Reactive verification with queues. In *ARO/ONR/NSF/DARPA Workshop on Engineering Automation for Computer-Based Systems*, pages 1–8, Carmel, CA, 1998.

4. A. Blumensath and E. Grädel. Automatic structures. In *Proc. 15th Annual IEEE Symp. on Logic in Computer Science*, pages 51–62, Santa Barbara, California, June 2000.

5. K.J. Compton and C.W. Henson. A uniform method for proving lower bounds on the computational complexity of logical theories. *Annals of Pure and Applied Logic*, 48:1–79, 1990.

6. J. Ferrante and C.W. Rackoff. *The computational complexity of logical theories*, volume 718 of *Lecture Notes in Mathematics*. Springer-Verlag, 1979.

7. B. Khoussainov and A. Nerode.   Automatic presentations of structures.   In Daniel Leivant, editor, *Logic and Computational Complexity, International Workshop LCC '94*, volume 960 of *Lecture Notes in Computer Science*, pages 367–392. Springer Verlag, 1995.

8. T. Rybina and A. Voronkov. A decision procedure for term algebras with queues. *ACM Transactions on Computational Logic*, 2(2):155–181, 2001.

9. W. Thomas. Infinite trees and automaton definable relations over omega-words. *Theoretical Computer Science*, 103(1):143–159, 1992.

10. H. Volger. A new hierarchy of elementary recursive decision problems. *Methods of Operations Research*, 45:509–519, 1983.

11. H. Volger. Turing machines with linear alternation, theories of bounded concatenation and the decision problem of first order theories (Note). *Theoretical Computer Science*, 23:333–337, 1983.

12. S. Vorobyov and A. Voronkov. Complexity of nonrecursive logic programs with complex values. In *PODS'98*, pages 244–253, Seattle, Washington, 1998. ACM Press.

Degree Distribution of the FKP Network Model

Noam Berger[1], Béla Bollobás[2,3], Christian Borgs[4], Jennifer Chayes[4], and Oliver Riordan[3]

[1] Department of Statistics, University of California, Berkeley, CA 94720 [‡]
[2] Department of Mathematical Sciences, University of Memphis, Memphis TN 38152 [§]
[3] Trinity College, Cambridge CB2 1TQ, UK, and Royal Society research fellow Department of Pure Mathematics, Cambridge [¶]
[4] Microsoft Research, One Microsoft Way, Redmond, WA 98122.
noam@stat.berkeley.edu, {B.Bollobas,O.M.Riordan}@dpmms.cam.ac.uk, {borgs,jchayes}@microsoft.com

Abstract. Recently, Fabrikant, Koutsoupias and Papadimitriou [7] introduced a natural and beautifully simple model of network growth involving a trade-off between geometric and network objectives, with relative strength characterized by a single parameter which scales as a power of the number of nodes. In addition to giving experimental results, they proved a power-law lower bound on *part* of the degree sequence, for a wide range of scalings of the parameter. Here we prove that, despite the FKP results, the overall degree distribution is very far from satisfying a power law.

First, we establish that for almost all scalings of the parameter, either all but a vanishingly small fraction of the nodes have degree 1, or there is exponential decay of node degrees. In the former case, a power law can hold for only a vanishingly small fraction of the nodes. Furthermore, we show that in this case there is a large number of nodes with almost maximum degree. So a power law fails to hold even approximately at either end of the degree range. Thus the power laws found in [7] are very different from those given by other internet models or found experimentally [8].

1 Introduction

In the last few years there has been an explosion of interest in 'scale-free' random networks, based on measurements indicating that many large real-world networks have certain scale-free properties, for example power-law distributions of degrees and other parameters. The original observations of Faloutsos, Faloutsos and Faloutsos [8], and later many others, have led to a host of proposals for random graph models to explain these power laws, and to better understand the

[‡] Research undertaken during an internship at Microsoft Research.
[§] Research supported by NSF grant DSM 9971788 and DARPA grant F33615-01-C-1900.
[¶] Research undertaken while visiting Microsoft Research.

mechanisms at work in the growth of real-world networks such as the internet or web graphs; see [2,3,9] for a few examples. For extensive surveys of the huge amount of work in this area, see Albert and Barabási [1] and Dorogovtsev and Mendes [6]; for a survey of the rather smaller quantity of mathematical work see [4].

Most of the models introduced use a small number of basic mechanisms, mainly preferential attachment or copying, to produce power laws, and do not involve any reference to underlying geometry. Thus, while they may be appropriate for the web graph, for example, they do not seem to be suitable for the internet graph itself.

In [7], Fabrikant, Koutsoupias and Papadimitriou (FKP) proposed a new paradigm for power law behaviour, which they called 'heuristically optimized trade-offs': power laws may result from 'complicated optimization problems with multiple and conflicting objectives.' Their paradigm generalizes previous work of Carlson and Doyle [5] on 'highly optimized tolerance,' in which reliable design is one of the objectives.

In order to illustrate this paradigm, Fabrikant, Koutsoupias and Papadimitriou introduced a simple, natural network model with such a mechanism. As in many models, a network is grown one node at a time, and each node chooses a previous node to which it connects. However, in contrast to other network models, a key feature of the FKP model is the underlying geometry; the nodes are points chosen uniformly at random from some region, for example a unit square in the plane. The trade-off is between the geometric consideration that it is desirable to connect to a nearby point, and a networking consideration, that it is desirable to connect to a node which is 'central' in the network as a graph. Centrality may be measured by using, for example, the graph distance to the initial node.

Several variants of the basic model are considered by Fabrikant, Koutsoupias and Papadimitriou in [7]. The precise version we shall consider here is the principal version studied in [7]: fix a region $\mathcal{D}$ of area one in the plane, for example a disk or a unit square. The model is then determined by the number of nodes, $n + 1$, and a parameter, α. We start with a point x_0 of $\mathcal{D}$ chosen uniformly at random, and set $W(x_0) = 0$. For $i = 1, 2, \ldots, n$ we choose a new point x_i of $\mathcal{D}$ uniformly at random, and connect x_i to an earlier point x_j chosen to minimize

$$W(x_j) + \alpha d(x_i, x_j)$$

over $0 \le j < i$. Here $d(.,.)$ is the usual Euclidean distance. Having chosen x_j, we set $W(x_i) = W(x_j) + 1$. At the end we have a random tree $T = T(n, \alpha)$ on $n + 1$ nodes $x_0, \ldots, x_n$, where each node has a weight $W(x_i)$ which is just its graph distance in the tree from x_0.

As in [7], we consider $n \to \infty$ with α some function of n, typically a power.

One might think from the title or a first reading of [7] that the form of the degree sequence of this model has been essentially established. In fact, as we shall describe in the next section, this is not the case. Indeed, two of our results, while of course consistent with the actual results of [7], go against the impression given there that the entire degree sequence follows a power law.

2   Results

As in [7] we consider α in two ranges. Roughly speaking, *large* α will mean $\alpha > n^{1/2}$, and *small* α will mean $\alpha < n^{1/2}$. In fact, to keep things simple we will allow ourselves a logarithmic gap.

Most of the time we will work in terms of the tail of the distribution. Let $\alpha = \alpha(n)$ be given. For each $k = 1, 2, \ldots$, let $q_k(\alpha, n)$ be the expected number of nodes of $T(n, \alpha)$ with degree at least k, and let $\rho_k(\alpha) = \lim_{n \to \infty} q_k(\alpha, n)/n$ be the limiting proportion of nodes with degree at least k.

2.1   Small α

The impression given on first reading [7] is that for small α the whole degree distribution follows a power law. However, the experimental results of [7] strongly suggest that there is a new kind of power law, holding over a large range of degrees, from 2 up to a little below the maximum degree, but involving only a very small proportion of the vertices.

On a second look the situation is more confusing. Quoting the relevant part of the theorem (changing D to k for consistency with our notation):

> If $\alpha \geq 4$ and $\alpha = o(\sqrt{n})$, then the degree distribution of T is a power law; specifically, the expected number of nodes with degree at least k is greater than $c \cdot (k/n)^{-\beta}$ for some constants c and β (that may depend on α): $E[|\{i : \text{ degree of } i \geq k\}|] > c(k/n)^{-\beta}$. Specifically, for $\alpha = o(\sqrt[3]{n^{1-\epsilon}})$ the constants are: $\beta \geq 1/6$ and $c = O(\alpha^{-1/2})$.

The usual form of a power law would be that a proportion $k^{-\beta}$ of vertices have degree at least k, which is not what is claimed above. There are other problems: the constant c depends on α which depends on n, so c is not a constant. Allowing c to be variable, the claim may then become meaningless if c is very small.

Turning to the proof in [7], a nice geometric argument is given to show that, for $\alpha = o(n^{(1-\epsilon)/3})$ and $k \leq n^{1-\epsilon}/(C\alpha^3)$, which is far below the maximum degree, the expected number $q_k(\alpha, n)$ of vertices with degree at least k is at least $cn^{1/6}\alpha^{-1/2}k^{-1/6}$, where c and C are absolute constants. This supports the experimental results, showing that this interesting new model does indeed give power laws over a wide range; however, it tells us nothing about the vast majority of the vertices, namely all but $O(n^{1/6})$.

Now, in many examples of real-world networks, and in the preferential attachment and copying models of [2,9] and others, the power-law degree distribution involves almost all vertices, and, less clearly, holds very nearly up to the maximum degree. (In the latter case, the power law is often called a 'Zipf law', though in fact Zipf's law is a power law with a particular power.) Thus it is interesting to see whether this is the case for the FKP model.

Theorem 1. *Let $\alpha = o(n^{1/2}/(\log n)^2)$. Then,* **whp** *the tree $T(n, \alpha)$ has at least $n - O(\alpha^{1/2}n^{3/4}\log n) = n - o(n)$ leaves.*

In other words, almost all vertices of $T(n, \alpha)$ have degree 1; in particular, when $\alpha = n^a$ for some constant $a < 1/2$, the number of vertices with degree more than 1 is at most n^b for some constant $b < 1$. This contrasts strongly with the usual sense of power-law scaling, namely that the proportion of vertices with degree k converges to a function $f(k)$ which in turn decays like a power of k. This notion is implicit in [8] and [1], for example.

Our second result concerns the high degree vertices, showing that a 'Zipf-like' law does not hold. As usual, we write $O^*(\cdot)$ for $O((\log n)^C \cdot)$, suppressing constant powers of $\log n$, and similarly for $\Theta^*(\cdot)$. We write **whp** to mean *with high probability*, i.e., with probability $1 - o(1)$ as $n \to \infty$.

Theorem 2. *Suppose that* $(\log n)^7 \leq \alpha \leq n^{1/2}/(\log n)^4$. *Then there are constants* $c, C > 0$ *such that* **whp** *the maximum degree of* $T(n, \alpha)$ *is at most* Cn/α^2, *while* $T(n, \alpha)$ *has* $\Theta^*(\alpha^2)$ *nodes of degree at least* cn/α^2.

Taking $\alpha = n^a$ for a constant, $0 < a < 1/2$, for example, this says that there are many (a power of n) vertices with degree close to (within a constant factor of) the maximum degree. This contrasts sharply with a so-called Zipf distribution, where there would be a constant number of such vertices. In fact, our method will even show that there are many vertices with degree $(1 - o(1))$ times the maximum.

2.2   Large α

We now turn to the simpler case of large α. This case is interesting for three reasons: one is simply completeness. The second is that the case $\alpha = \infty$, while involving no trade-offs, is a very nice geometric model in its own right. Finally, the large α results will turn out to be useful in studying the small α case.

Theorem 3. *Suppose that* $\alpha = \alpha(n)$ *satisfies* $\alpha/(\sqrt{n} \log n) \to \infty$. *Then there are positive constants* A, A', C, C' *such that*

$$A'e^{-C'k} \leq \rho_k(\alpha) \leq Ae^{-Ck}$$

holds for every $k \geq 1$.

In other words, for large α the tail of the degree distribution decays exponentially, as for classical random graphs with constant average degree.

Our theorem strengthens the upper bound in [7], which says that $q_k(\alpha, n) \leq O(n^2)e^{-Ck}$, or, loosely speaking, that $\rho_k(\alpha) \leq O(n)e^{-Ck}$. Note that the upper bound of [7] gives information only for k larger than a constant times $\log n$, i.e., a vanishing fraction of the nodes. Furthermore, we complement our stronger upper bound with a matching lower bound.

We remark again that our results contain logarithmic factors that are presumably unnecessary; these help keep the proofs relatively simple.

3   The Pure Geometric Model

In this section we consider the case $\alpha = \infty$. In this case, each node x_i simply connects to the closest node among $x_0, \ldots, x_{i-1}$. Although this model is not our main focus, it is of interest in its own right, and it is somewhat surprising that it does not seem to have been extensively studied, unlike related objects such as the minimal spanning tree, for example (see [11,12]). We study this case for two reasons. First, for large α, $T(n, \alpha)$ approximates $T(n, \infty)$. Second, certain results about $T(n, \infty)$ will be useful to study $T(n, \alpha)$ even for very small α. We start with a simple but surprising exact result.

Lemma 1. *In the random tree $T(n, \infty)$, for $1 \leq t \leq n$ the probability that x_t is at graph distance r from x_0, i.e., has weight r, is exactly*

$$\sum_{1 \leq i_1 < i_2 < \ldots < i_{r-1} < t} \frac{1}{i_1 i_2 \ldots i_{r-1} t}$$

Proof. We write $i \rightarrow j$ if $j < i$ and x_i is adjacent (joined directly) to x_j. The key observation is as follows: suppose we fix the points $x_s, x_{s+1}, \ldots, x_n$, and also the *set* of points $S_{s-1} = \{x_0, x_1, \ldots x_{s-1}\}$, leaving undetermined the order of the points in S_{s-1}. Then x_s is joined to the closest point in S_{s-1}, which is a certain point x. When we choose the ordering of the points in S_{s-1}, the point x is equally likely to be x_0, x_1, or any other x_j, $j < s$. Taking $s = t$, it follows that the probability that $t \rightarrow j$ is exactly $1/t$. Using the same observation for $s = j$ we see that, given $t \rightarrow j$, the probability that $j \rightarrow k$ is $1/j$. Continuing, the probability that $t \rightarrow i_{r-1} \rightarrow i_{r-2} \rightarrow \cdots \rightarrow i_1 \rightarrow 0$ is $1/(t i_{r-1} i_{r-2} \cdots i_1)$. As these events are disjoint for different sequences, the lemma follows.

Another way of stating the lemma is that, for any fixed t, the distribution of the graph distance from t to 0 is the same as in a uniform random recursive tree. These are trees grown one node at a time, in which each new node is joined to an earlier node chosen uniformly at random. Such objects have been studied for some time; see, for example, the survey [10]. The radius (here, maximum node weight) of such a tree was shown by Pittel [13] to be $(c + o(1)) \log n$ for a certain constant $c = 1.79..$ given by a root of an equation. This result does not apply to $T(n, \alpha)$ because the dependence between nodes is different. We shall just give an upper bound.

Lemma 2. *Let $\alpha = \alpha(n)$ be arbitrary. Then as $n \rightarrow \infty$, **whp** every point in $T(n, \alpha)$ has weight at most $3 \log n$.*

Proof. For $\alpha = \infty$ this follows from Lemma 1 by straightforward calculation: the expected number of points with weight r is

$$\sum_{1 \leq i_1 < i_2 < \ldots < i_{r-1} < t \leq n} \frac{1}{i_1 i_2 \ldots i_{r-1} t} \leq \frac{1}{r!} \left(\sum_{i=1}^{n} \frac{1}{i} \right)^r \leq \frac{(1 + \log n)^r}{r!} \leq (e(1 + \log n)/r)^r .$$

Set $r = \lfloor 3 \log n \rfloor$. Then the expectation above tends to zero, so **whp** there are no points with weight r, and the radius, or maximum weight, is at most $r - 1$.

We can compare finite α with $\alpha = \infty$. Consider the sequence of points as fixed, let $W(x_i)$ be the weights for some finite $\alpha = \alpha(n)$, and let $W_\infty(x_i)$ be the weights obtained with $\alpha = \infty$. For any α, the weight of a point x_i is always at most one more than the weight of the nearest earlier point x_j: if we connect to a more distant point x_k it must have smaller weight than x_j. Since we have equality for $\alpha = \infty$, it follows that for any α we have $W(x_i) \leq W_\infty(x_i)$. As shown at the start of the proof, **whp** we have $W_\infty(x_i) \leq 3 \log n$ for every i, so we are done.

The lemma has a simple heuristic explanation: for $\alpha = \infty$ the closest earlier x_j to x_i will typically have index j around $i/2$, so it will take order $\log n$ steps to reach the origin. For finite α, any bias is towards earlier points. One might expect monotonicity of the weights as α decreases from one finite value to another, but this does not hold in general.

3.1   Degrees for $\alpha = \infty$

Here we are interested in the quantities $\rho_k(\infty)$ defined in section 2; our aim is to prove the $\alpha = \infty$ case of Theorem 3.

This result easy to see intuitively. As noted above, for $i < t \leq n$ the probability that $t \to i$ is exactly $1/t$. Thus the expected degree of node i in $T(n, \infty)$ is exactly

$$\frac{1}{i+1} + \frac{1}{i+2} + \cdots + \frac{1}{n} = \log(n/i) + O(i^{-1}).$$

If every degree were close to its expectation, this would give the result. In fact, it turns out that the probability of the degree of node i exceeding its expectation by some amount x decreases exponentially with x. To see this heuristically we use the notion of Voronoi cells: given a region $\mathcal{D}$ and a set of points X in $\mathcal{D}$, the region $\mathcal{D}$ is tiled by Voronoi cells V_x, one for each $x \in X$, defined as the set of points of $\mathcal{D}$ closer to x than to any other $y \in X$.

Here we consider $V_{i,t}$, the Voronoi cell of x_i with respect to $\{x_0, x_1, \ldots, x_t\}$. Note that $t \to i$ if and only if x_t is in $V_{i,t-1}$. Keeping i fixed, as t increases $V_{i,t}$ shrinks whenever x_t lands close enough to x_i. In particular, $V_{i,t}$ gets smaller whenever x_t lands in $V_{i,t-1}$ itself; the key point is that in this case the area of $V_{i,t}$ is on average less than that of $V_{i,t-1}$ by a factor f strictly less than 1. On average, $V_{i,i}$ has area $1/(i+1)$, and $V_{i,n}$ area $1/(n+1)$. Hence it is very unlikely that i has degree much bigger than $\log(n/i)$; otherwise the area of $V_{i,t}$ would decrease by too much as t increases from i to n.

Proof (of Theorem 3 for $\alpha = \infty$). We make the argument outlined above rigorous. The key observation is as follows: let V be a convex region and C a fixed point of V. Let X be a point of V chosen uniformly at random, and let V' be the set of points of V closer to C than to X. Then the expected area of V' is at most $15/16$ times the area of V. To see this, taking C as the origin divide

V into four parts Q_1, Q_2, Q_3, Q_4, the intersections of V with the four quadrants of $\mathbf{R}^2$. Suppose X falls in a certain Q_i. If Y is any other point of Q_i then $(X+Y)/2$ is closer to X than to C. This is easy to see geometrically: the vector $(X+Y)/2 - X = (Y-X)/2$ is shorter than $(Y+X)/2$, as the angle between X and Y is less than 90 degrees. Hence $V \setminus V'$ contains a copy of Q_i shrunk by a factor two in each direction, so in this case area$(V \setminus V') \geq$ area$(Q_i)/4$. Averaging, noting that the probability that X lies in Q_i is proportional to area(Q_i),

$$\mathbf{E}(\text{area}(V \setminus V')) \geq \sum_{i=1}^{4} \frac{\text{area}(Q_i)^2}{4\,\text{area}(V)} \geq \frac{\text{area}(V)}{16},$$

where the last step follows by convexity. Thus $\mathbf{E}(\text{area}(V')) \leq \frac{15}{16}\,\text{area}(V)$. Hence, fixing $x_0, \ldots, x_{t-1}$, conditional on $t \to i$, i.e., on $x_t \in V_{i,t-1}$, the expected area of $V_{i,t}$ is at most $\frac{15}{16}$ times the area of $V_{i,t-1}$.

Fix $0 \leq i \leq n$. Continuing the construction of $T(n, \infty)$ indefinitely, let $t_1 < t_2 < t_3 < \cdots$ be the points that send edges to i. Let $W_0 = V_{i,i}$ and $W_j = V_{i,t_j}$ be the Voronoi cells of i looked at at time i, and at each time when a new node joins to i. Note that $\mathbf{E}(\text{area}(W_0)) = 1/(i+1)$ as this is the cell corresponding to one of $i+1$ points chosen independently. It may be that the Voronoi cell containing i shrinks at intermediate times as well, but certainly given W_j, we have $\mathbf{E}(\text{area}(W_{j+1})) \leq \frac{15}{16}\,\text{area}(W_j)$. Hence

$$\mathbf{E}(\text{area}(W_k)) \leq \frac{1}{i+1}(15/16)^k. \tag{1}$$

We now consider time n: fix x_i and consider the n remaining points of $x_0, \ldots, x_n$ as random. Ignoring effects from the boundary of the region, if no other point lies within distance d of x_i, then the Voronoi cell $V_{i,n}$ contains a circle of radius $d/2$. In other words, for area$(V_{i,n})$ to be smaller than $\pi(d/2)^2$, one of the n points must lie in a disk of radius d, with area πd^2, an event with probability at most $n\pi d^2$. It turns out that boundary effects go the right way, so

$$\Pr(\text{area}(V_{i,n}) \leq x) \leq 4nx. \tag{2}$$

Finally, if i has degree at least $k+1$ in $T(n, \infty)$ then at least k of the first n points join to i, so $t_k \leq n$, and area$(V_{i,n}) \leq$ area(W_k). For any x, the probability of this is at most

$$\Pr(\text{area}(W_k) \geq x) + \Pr(\text{area}(V_{i,n}) \leq x),$$

which is at most

$$\frac{1}{x}\frac{1}{i+1}(15/16)^k + 4nx,$$

from (1), Markov's inequality and (2). The optimum choice

$$x = (15/16)^{k/2}/\sqrt{4n(i+1)}$$

yields

$$\Pr(\deg(i) \geq k+1) \leq 4\sqrt{\frac{n}{i+1}}(15/16)^{k/2}. \tag{3}$$

Summing over i by comparison with an integral, the expected number of nodes with degree at least $k+1$ is at most $(8+o(1))n(15/16)^{k/2}$, so $\rho_{k+1} \leq 8(15/16)^{k/2}$, proving the upper bound.

The lower bound also follows easily; the bound (3) shows that an individual degree is very unlikely to be much larger than its expectation. It follows that $\deg(i)$ has a significant (at least 1%, say) chance of being at least half its expectation, and the lower bound follows.

4   Observation

In the remaining proofs we will use again and again the following simple observation. At time t the points currently placed approximate a Poisson process with density $1/t$, so the closest earlier point x_j to x_t is 'typically' at distance $\Theta(1/\sqrt{t})$. In particular, for a fixed $t > 0$, if $\omega \to \infty$ then **whp** $\omega^{-1}t^{-1/2} \leq d(x_t, x_j) \leq \omega t^{-1/2}$.

Furthermore, for any positive constant c, **whp** at time t every disk of radius $c \log nt^{-1/2}$ contains a point already placed. (This is easy to check, and also follows from a more general and more precise result of Penrose [11].)

5   Large α

Proof (of Theorem 3). The case $\alpha = \infty$ was proved in section 3; to extend this result to α large requires only a little further work.

Suppose that $\alpha/(\sqrt{n}\log n) \to \infty$. Fix $\delta > 0$, and consider a point x_i with $i \geq \delta n$, and the nearest earlier point x_j. Since all weights are within $3\log n$ of one another, for x_i to join to some other point x_k we must have

$$d(x_i, x_k) \leq d(x_i, x_j) + 3\log n/\alpha = d(x_i, x_j) + o(n^{-1/2}). \tag{4}$$

As noted above, **whp** we have $d(x_i, x_j) \leq \omega i^{-1/2}$. Considering x_i and x_j as fixed, given that x_j is the closest earlier point to x_i, the other x_k, $k < i$, are distributed uniformly outside the circle centered at x_i with radius $d(x_i, x_j)$, and for a particular x_k to satisfy (4) it must lie in an annulus around this circle with thickness $o(n^{-1/2})$. This annulus has area $o(d(x_i, x_j)n^{-1/2}) = o((in)^{-1/2})$ (taking $\omega \to \infty$ slowly enough). Since there are $i-1$ points to consider, the probability that x_i does not join to the closest point x_j is at most $o(\sqrt{i/n}) = o(1)$. Thus, **whp** almost all points join to the nearest earlier point. In particular, the final tree $T(n, \alpha)$ differs in only $o(n)$ edges from $T(n, \infty)$, and hence the numbers ρ_k are the same as for $\alpha = \infty$.

The conclusion that $\rho_k(\alpha) = \rho_k(\infty)$ should hold provided only that $\alpha/\sqrt{n} \to \infty$; this is likely to be harder to show.

6 Critical α

If $\alpha - \Theta(\sqrt{n})$ then we expect the behaviour of the tree to be similar to that for $\alpha = \infty$. In particular, for $\alpha = cn^{1/2}$, $c > 0$, we expect limiting proportions $\rho_k = \rho_k(c)$ with $\rho_k(c) \to \rho_k(\infty)$ as $c \to \infty$ but $\rho_k(c)$ not in general equal to $\rho_k(\infty)$. Also, the radius, or maximum weight, should be $A(c) \log n$. We have not stated a result for this case, which is likely to be harder to analyze precisely.

Note that one might hope for a complete power law in the critical case, but this does not happen, as shown by, for example, the weak exponential upper bound in [7].

7 Small α

This case is the heart of our paper. Here *small* would ideally mean $o(n^{1/2})$; in fact, for simplicity we shall work with extra logarithmic factors. Throughout this section it will be convenient to re-scale by a factor of α: rather than choosing points in the unit square or disk, we choose points in a square $\mathcal{D}$ of side α; correspondingly, we join x_i to the earlier point x_j minimizing $W(x_j) + d(x_i, x_j)$. Note that the final density n/α^2 of points is high (compared to 1). The reason to consider this scaling is that differences in re-scaled distances of order 1 are what is relevant; in particular, as all weights are within $3 \log n$ of each other, no point ever connects to a point more than $3 \log n$ further away than its nearest point.

Considering the process defining $T(n, \alpha)$ as points arrive one by one, there is a transition in the behaviour around time $t = \alpha^2$. This is because in the re-scaled process, the density of points at time t is t/α^2. At times much earlier than α^2, this density is very small, so distances and their differences are typically large, and the process looks very much like the $\alpha = \infty$ case of connecting to the nearest point.

On the other hand, at times much later than α^2, the density of points is already very high. We expect that certain 'attractive' early points will have established 'regions of attraction' of order unit size; almost all later points then just join to the nearest attractive point by a short edge. In particular, almost all later points will themselves never be joined to.

7.1 Small Degrees

We now prove Theorem 1 from section 2, a precise version of the final observation from the paragraph above, that almost all points are leaves in $T(n, \alpha)$, i.e., have degree 1. In the proof we shall use the following simple geometric lemma.

Lemma 3. *Let $\mathcal{D}$ be a convex set in the plane, and let $X = \{x_0, \ldots, x_{k-1}\}$ be a set of points in $\mathcal{D}$. For $r > 0$ let $X(r)$ be the set of points in $\mathcal{D}$ at distance at most r from some x_i. For $0 < r_1 < r_2$ we have*

$$\text{area}(X(r_2)) \leq \frac{r_2^2}{r_1^2} \text{area}(X(r_1)).$$

Proof. A point $x \in \mathcal{D}$ lies in $X(r)$ if and only if $d(x, x_i) \leq r$ for x_i the closest point of X to x. Let us partition $\mathcal{D}$ into the Voronoi cells $V_i = \{x \in \mathcal{D} : d(x, x_i) = \min_j d(x, x_j)\}$. (We may ignore the boundaries.) Then, for any r, we have $\text{area}(X(r)) = \sum_i \text{area}(X(r) \cap V_i)$. But V_i is convex; thus if $X(r_2) \cap V_i$ is a certain region A, then $X(r_1) \cap V_i$ certainly contains the region obtained by shrinking A by a factor r_2/r_1 around the point x_i. Hence, $\text{area}(X(r_1) \cap V_i) \geq r_1^2/r_2^2 \, \text{area}(X(r_2) \cap V_i)$, and the lemma follows.

Of course, a corresponding result holds in any dimension, with exactly the same proof. Also, the result holds for an arbitrary (infinite) set X.

Proof (of Theorem 1). If x_i is joined to the earlier point x_j, we call $x_i x_j$ *the edge from* x_i. We consider edges with lengths in three ranges: writing γ for $\alpha^{1/2} n^{-1/4} = o(1/\log n)$, we call an edge of length ℓ *short* if $\ell < 1$, *long* if $\ell > 1 + \gamma$, and *medium* if $1 \leq \ell \leq 1 + \gamma$.

The key observation is that if the edge $x_i x_j$ from x_i is short, then x_i has degree 1 in the final graph $T(n, \alpha)$. To see this, note that no later point x_k can possibly join to x_i, since $W(x_i) = W(x_j) + 1$, while $d(x_k, x_j) < d(x_k, x_i) + 1$, so x_k would join to x_j in preference to x_i. To complete the proof we shall show that the number of medium and long edges is small.

Suppose that the edge $x_i x_j$ from x_i is medium. Writing w for $W(x_j)$, at time $i - 1$ there is no point with weight w within distance 1 of x_i, but there is such a point within distance $1 + \gamma$. Turning this around, let $X = \{x_j : W(x_j) = w, 0 \leq j \leq i - 1\}$. Then x_i lies in $X(1 + \gamma)$, but not in the interior of $X(1)$. By Lemma 3, $\text{area}(X(1 + \gamma)) \leq (1 + \gamma)^2 \, \text{area}(X(1))$. Hence, given $x_0, \ldots x_{i-1}$, the probability that x_i lies in $X(1 + \gamma) \setminus X(1)$ is at most $\frac{(1+\gamma)^2 - 1}{(1+\gamma)^2} \leq 2\gamma$. It follows from Lemma 2 that there are at most $\log n$ values of w to consider, so the probability that for a given i the edge $x_i x_j$ is medium is at most $2\gamma \log n = o(1)$. It follows that **whp** there are at most $2\gamma n \log n = 2\alpha^{1/2} n^{3/4} \log n = o(n)$ medium edges in the final tree.

We now consider long edges, i.e., edges of length at least $1 + \gamma$. The key observation is that when the edge from x_i is long, this edge provides a useful shortcut in future: new points near x_i have a better connection route than if x_i were deleted. To formalize this, given the final set of points $x_0, \ldots, x_n$ and their weights, for $1 \leq i \leq n$ let us define a function $c_i : \mathcal{D} \to \mathbf{R}$ by $c_i(x) = \min_{j<i} \{W(x_j) + d(x, x_j)\}$. Note that c_i only depends on the locations of $x_0, \ldots, x_{i-1}$, and that $c_i(x)$ is the 'cost' of connecting a potential new point at x to the existing tree on $x_0, \ldots, x_{i-1}$. In particular, x_i joins to the x_j attaining the minimum defining $c_i(x_i)$, and receives weight $W(x_j) + 1$. Suppose that $x_i x_j$ is long, i.e., has length at least $1 + \gamma$, and let $w = W(x_j)$. Then we have $c_i(x_i) = w + d(x_i, x_j) \geq w + 1 + \gamma$, but $c_{i+1}(x_i) = w + 1$. Hence

$$c_{i+1}(x_i) \leq c_i(x_i) - \gamma.$$

Our strategy is to consider the quantities $I_i = \int_{\mathcal{D}} c_i(x)$, $1 \leq i \leq n$. We shall show that I_i is positive, and decreases with i. Also, we shall show that **whp** I_{i_0} is not too large for some $i_0 = o(n)$, and that if the edge from i is long, then

$I_i - I_{i+1}$ is not too small; together these observations will give a bound on the number of long edges.

It is immediate from the definition that $c_i(x)$ and hence I_i are positive. Also, it is immediate that $c_{i+1}(x) \leq c_i(x)$—the minimum is taken over a larger set. Hence $I_{i+1} \leq I_i$ for each i.

Set $i_0 = \lfloor (\alpha \log n)^2 \rfloor = o(n)$. At time i_0 the overall density of points is at least $(\log n)^2$. Hence, **whp**, for every $x \in \mathcal{D}$ there is a $j < i_0$ with $d(x, x_j) < 1$. Since $W(x_j) \leq 3 \log n$ from section 5, we have $c_{i_0}(x) \leq 1 + 3 \log n$. Thus, **whp**,

$$I_{i_0} \leq (1 + 3 \log n)\text{area}(\mathcal{D}) = O(\alpha^2 \log n).$$

Finally, suppose that the edge from x_i is long. As shown above, we then have $c_{i+1}(x_i) \leq c_i(x_i) - \gamma$. Now each $c_k(x)$ is the minimum of a set of Lipschitz functions with constant 1, and is hence Lipschitz with constant 1. Thus for y at distance $\ell \leq \gamma/2$ from x_i we have $c_{i+1}(y) \leq c_i(y) - \gamma + 2\ell$. Integrating, we see that

$$I_{i+1} \leq I_i - \frac{1}{4} \int_{\ell=0}^{\gamma/2} (\gamma - 2\ell) 2\pi \ell \, d\ell = I_i - \frac{\pi}{48} \gamma^3.$$

(The initial factor of $1/4$ allows for the fact that the little disk we are integrating over may not lie entirely within $\mathcal{D}$.)

Since I_i is decreasing and positive, from the two equations above we see that **whp** the number of x_i, $i \geq i_0$, from which we have long edges is at most $O(\alpha^2 \log n/\gamma^3)$. Thus, **whp** we have $i_0 + O(\alpha^2 \log n/\gamma^3) = O(\alpha^{1/2} n^{3/4} \log n)$ long edges.

Combining the cases above completes the proof: we have shown that in total there are $O(\alpha^{1/2} n^{3/4} \log n) = o(n)$ medium and long edges, and hence $n - o(n)$ short edges. But every short edge gives rise to a leaf in T, so almost all nodes are leaves.

The above result shows that for small α the degree sequence of $T(n, \alpha)$ is not a power law in the usual sense, which is that for fixed k there is a limiting proportion p_k of nodes with degree k, which falls off as some power of k. In particular, here $p_1 = 1$, while $p_k = 0$ for all $k \neq 1$.

7.2   Large Degrees

We now turn to the opposite end of the degree sequence, showing that there is a bunching of degrees near the maximum, in the sense that for $\alpha = n^a$, $0 < a < 1/2$, a positive power of n nodes have degree within a constant factor of the maximum. This is easy to see heuristically: up to time α^2 the process looks like the $\alpha = \infty$ case, and all degrees are at most $O(\log n)$. Beyond this time, $\Theta(\alpha^2)$ attractive points will have become established, each of which will attract the $\Theta(n/\alpha^2)$ later points that fall in its zone of attraction, which will have re-scaled area $O(1)$, out of a total re-scaled area of α^2. Since no point can maintain a region of attraction much bigger than this for long, the maximum degree will also be of order $\Theta(n/\alpha^2)$.

As before, for simplicity we have allowed ourselves extra logarithmic factors when making this precise. In Theorem 2, which we now prove, the main case of interest is $\alpha = n^a$ for some constant a between 0 and $1/2$.

Proof (of Theorem 2). We start with the maximum degree, aiming to show that this is $O(n/\alpha^2)$. Let $t_0 = \alpha^2/(\log n)^2$. Arguing as in section 5 we see that **whp** at time t_0 the tree is essentially $T(t_0, \infty)$, and that all degrees are $O(\log n)$.

Fix a point x_i. To obtain the desired bound on the final degree of i we need only consider which x_j, $j > t_0$, join to x_i. Now at time t_0 the typical distance between points is $\log n$, and allowing for deviations no disk of radius $(\log n)^2$ is empty. (This is a rescaling of the final observation from section 4.) It follows that all later edges have length at most $2(\log n)^2$. Hence we need only consider a region R around x_i with radius $O((\log n)^2)$. We divide this into a 'good region', a disk of radius 1.1 around x_i, and a 'bad region', the rest of R. Note that $O(n/\alpha^2)$ points will fall into the good region, so we need only control the bad region. This is easy: the bad region can be covered by $O((\log n)^4)$ disks of radius 0.01. Within any such disk at most one point x_j, $j > i$, can join to i; a second point $x_{j'}$ landing in the same disk would rather join to x_j at distance < 0.01 than to x_i at distance at least 1.1, since the weight of x_j is only one larger than that of x_i. Hence the expected degree of x_i is at most

$$O(\log n) + O(n/\alpha^2) + O((\log n)^4) = O(n/\alpha^2).$$

Since the main term is at least $\Theta((\log n)^2)$ it is easy to check that large deviations are very unlikely, and hence that the maximum degree is $O(n/\alpha^2)$, as claimed.

Establishing the existence of 'attractive' points which remain attractive is not quite so easy, as the situation is not really as simple as the heuristic description suggests. However, with the flexibility allowed by logarithmic factors we can proceed as follows. Let us consider time $t_1 = \alpha^2/\omega$, where $\omega = (\log n)^7$. Set $S = \{x_0, \ldots, x_{t_1}\}$, noting that typical distances between nearest points of S are of order $\omega^{1/2}$. In fact, as S approximates a Poisson process with density ω^{-1}, one can check that **whp** every disk of radius $0.9\sqrt{\omega \log n}$ contains a point of S. (To see this, observe that S has very small probability of missing a given disk of radius $0.85\sqrt{\omega \log n}$.) For the moment we shall condition on $x_0, \ldots, x_{t_1}$, assuming that this property holds, and noting the consequence that all edges added after time t_1 have length at most $0.9\sqrt{\omega \log n} + 3\log n \leq (\log n)^4 - 1$; the nearest old point to any new point is within $0.9\sqrt{\omega \log n}$, and can have weight at most $3\log n$ more than the point actually joined to.

Let us say that a point of S is *isolated* if it is at distance at least 2 from every other point of S. Let us say that a point $x_i \in S$ of weight w is *good* if no other point $x_j \in S$ with smaller weight lies within distance $3(\log n)^5$ of x_i. Isolated good points are useful for the following reason: we claim that every later point x_k, $k > t_1$, within distance 1 of an isolated good point x_i will join to x_i. To see this, note that we have $x_k = x_{a_0} \to x_{a_1} \to x_{a_2} \to \cdots \to x_{a_{\ell-1}} \to x_{a_\ell}$ for some sequence $k = a_0 > a_1 \cdots > a_{\ell-1} > a_\ell$, with $a_{\ell-1} > t_1$, $a_\ell \leq t_1$. Suppose that $x_k \to x_i$ does not hold, i.e., $a_1 \neq i$. Then, as x_i is within distance one of x_k, we have $W(x_{a_1}) \leq W(x_i)$, and if equality holds, then x_{a_1} must also be within distance one of x_k. In the case of equality, since x_i is isolated it follows that $a_1 > t_1$, i.e., $\ell > 1$. Since $x \to y$ implies $W(x) = W(y) + 1$, it follows in either case that $W(x_{a_\ell}) < W(x_i)$. But then x_k is connected by a sequence of at most $3 \log n$ edges of length at most $(\log n)^4 - 1$ to a point $x_{a_\ell} \in S$ with smaller weight than x_i, contradicting that x_i is good, and establishing the claim.

Thus an isolated good point attracts all points after t_1 within distance 1, and will have final degree at least cn/α^2 **whp**. In fact, using only the Chernoff bounds, the deviation probability for one point is $o(n^{-1})$, so **whp** *every* isolated good point has final degree at least cn/α^2.

It remains to show that at time $t_1 = \alpha^2/(\log n)^7$ there are many isolated good points. We do this using a little trick. (We treat $3 \log n$ as an integer for notational convenience.)

Let $r_w = 3(\log n)^5(1 + 3 \log n - w)$, so $r_0 = O((\log n)^6)$, $r_{3 \log n} = 3(\log n)^5 \geq (\log n)^4$, and $r_w = r_{w-1} - 3(\log n)^5$. For $0 \leq w \leq 3 \log n$ let S_w be the set of points $x_i \in S$ with weight at most w, and let $T_w = S_w(r_w)$ be be the set of all points in $\mathcal{D}$ within distance r_w of some point in S_w. Note that T_0 has area $O((\log n)^{12})$, which is much less than α^2. On the other hand, $T_{3 \log n}$ is, **whp**, all of $\mathcal{D}$, since, as noted earlier, **whp** every point of $\mathcal{D}$ is within distance $(\log n)^4$ of some $x_i \in S$, which has weight at most $3 \log n$ by Lemma 2. Thus

$$\sum_w \text{area}(T_w \setminus T_{w-1}) \geq (1 - o(1))\alpha^2.$$

Suppose that $y \in T_w \setminus T_{w-1}$. Then there is some $x_i \in S$ with $W(x_i) \leq w$ and $d(y, x_i) \leq r_w$. On the other hand, there is no $x_j \in S$ with $W(x_j) \leq w - 1$ within distance $r_{w-1} = r_w + 3(\log n)^5$ of y. It follows that $W(x_i) = w$, and that x_i is good, so y is within distance r_w of a good x_i. As each such good x_i can only account for an area $\pi r_w^2 \leq \pi r_0^2 = O((\log n)^{12})$ of $T_w \setminus T_{w-1}$, it follows that **whp** the total number of good points in S is at least $g_0 = \Theta(\alpha^2(\log n)^{-12})$. On the other hand, since the density of points at time t_1 is $(\log n)^{-7}$, the probability that a given x_i, $i \leq t_1$, is not isolated is $\Theta((\log n)^{-7})$, and the expected number of non-isolated points in S is $\Theta(\alpha^2(\log n)^{-14})$. This is $o(g_0)$, so using Markov's inequality, **whp** almost all good points are isolated, completing the proof.

In fact, being a little more careful with the constants, we can show that both the maximum degree and the degrees of almost all isolated good points (those not too near the boundary of $\mathcal{D}$) are $(1 + o(1))\pi n/\alpha^2$. Thus there is a strong bunching of degrees near the maximum.

References

1. R. Albert and A.-L. Barabási, Statistical mechanics of complex networks, *Rev. Mod. Phys.* **74** (2002), 47–97.
2. A.-L. Barabási and R. Albert, Emergence of scaling in random networks, *Science* **286** (1999), 509–512.
3. B. Bollobás and O.M. Riordan, The diameter of a scale-free random graph, to appear in *Combinatorica*. (Preprint available from http://www.dpmms.cam.ac.uk/~omr10/.)

4. B. Bollobás and O. Riordan, Mathematical results on scale-free random graphs, in Handbook of Graphs and Networks, Stefan Bornholdt and Heinz Georg Schuster (eds.), Wiley-VCH, Weinheim (2002), 1–34.

5. J.M. Carlson and J. Doyle, Highly optimized tolerance: a mechanism for power laws in designed systems. *Phys. Rev. E* **60** (1999), 1412–1427.

6. S.N. Dorogovtsev and J.F.F. Mendes, Evolution of networks, *Adv. Phys.* **51** (2002), 1079.

7. A. Fabrikant, E. Koutsoupias and C.H. Papadimitriou, Heuristically optimized trade-offs: a new paradigm for power laws in the internet ICALP 2002, LNCS 2380, pp. 110–122.

8. M. Faloutsos, P. Faloutsos and C. Faloutsos, On power-law relationships of the internet topology, SIGCOMM 1999, *Comput. Commun. Rev.* **29** (1999), 251.

9. R. Kumar, P. Raghavan, S. Rajagopalan, D. Sivakumar, A. Tomkins and E. Upfal, Stochastic models for the web graph, FOCS 2000.

10. H.M. Mahmoud and R.T. Smythe, A survey of recursive trees, *Th. of Probability and Math. Statistics* **51** (1995), 1–27.

11. M.D. Penrose, A strong law for the largest nearest-neighbour link between random points, *J. London Math. Soc.* (2) **60** (1999), 951–960.

12. M.D. Penrose, A strong law for the longest edge of the minimal spanning tree. *Ann. Probab.* **27** (1999), 246–260.

13. B. Pittel, Note on the heights of random recursive trees and random m-ary search trees, *Random Struct. Alg.* **5** (1994), 337–347.

Similarity Matrices for Pairs of Graphs

Vincent D. Blondel and Paul Van Dooren

Division of Applied Mathematics, Université catholique de Louvain, 4 avenue Georges Lemaitre, B-1348 Louvain-la-Neuve, Belgium, blondel@inma.ucl.ac.be, http://www.inma.ucl.ac.be/~blondel/ vdooren@csam.ucl.ac.be

Abstract. We introduce a concept of similarity between vertices of directed graphs. Let G_A and G_B be two directed graphs with respectively n_A and n_B vertices. We define a $n_A \times n_B$ *similarity matrix* $\mathbf{S}$ whose real entry s_{ij} expresses how similar vertex i (in G_A) is to vertex j (in G_B) : we say that s_{ij} is their *similarity score*. In the special case where $G_A = G_B = G$, the score s_{ij} is the similarity score between the vertices i and j of G and the square similarity matrix $\mathbf{S}$ is the *self-similarity matrix* of the graph G. We point out that Kleinberg's "hub and authority" method to identify web-pages relevant to a given query can be viewed as a special case of our definition in the case where one of the graphs has two vertices and a unique directed edge between them. In analogy to Kleinberg, we show that our similarity scores are given by the components of a dominant vector of a non-negative matrix and we propose a simple iterative method to compute them.

Remark: Due to space limitations we have not been able to include proofs of the results presented in this paper. Interested readers are referred to the full version of the paper [1], and to [2] for a description of an application of our similarity concept to the automatic extraction of synonyms in a dictionary. Both references are available from the first authors web-site.

1 Generalizing Hubs and Authorities

Efficient web search engines such as Google are often based on the idea of characterizing the most important vertices in a graph representing the connections or links between pages on the web. One such method, proposed by Kleinberg [10], identifies in a set of pages relevant to a query search those that are good *hubs* or good *authorities*. For example, for the query "automobile makers", the home-pages of Ford, Toyota and other car makers are good authorities, whereas web pages that list these home-pages are good hubs. Good hubs are those that point to good authorities, and good authorities are those that are pointed to by good hubs. From these implicit relations, Kleinberg derives an iterative method that assigns an "authority score" and a "hub score" to every vertex of a given graph. These scores can be obtained as the limit of a converging iterative process which we now describe.

J.C.M. Baeten et al. (Eds.): ICALP 2003, LNCS 2719, pp. 739–750, 2003.

Let G be a graph with edge set E and let h_j and a_j be the hub and authority scores of the vertex j. We let these scores be initialized by some positive values and then update them simultaneously for all vertices according to the following *mutually reinforcing relation* : the hub score of vertex j is set equal to the sum of the authority scores of all vertices pointed to by j and, similarly, the authority score of vertex j is set equal to the sum of the hub scores of all vertices pointing to j :

$$\begin{cases} h_j \leftarrow \sum_{i:(j,i)\in E} a_i \\ a_j \leftarrow \sum_{i:(i,j)\in E} h_i \end{cases}$$

Let B be the adjacency matrix of G and let h and a be the vectors of hub and authority scores. The above updating equations take the simple form

$$\begin{bmatrix} h \\ a \end{bmatrix}_{k+1} = \begin{bmatrix} 0 & B \\ B^T & 0 \end{bmatrix} \begin{bmatrix} h \\ a \end{bmatrix}_k, \qquad k = 0,1,\ldots$$

which we denote in compact form by

$$x_{k+1} = M\, x_k, \qquad k = 0,1,\ldots$$

where

$$x_k = \begin{bmatrix} h \\ a \end{bmatrix}_k, \qquad M = \begin{bmatrix} 0 & B \\ B^T & 0 \end{bmatrix}.$$

We are only interested in the relative scores and we will therefore consider the *normalized* vector sequence

$$z_0 = x_0, \qquad z_{k+1} = \frac{M z_k}{\|M z_k\|_2}, \qquad k = 0,1,\ldots$$

Ideally, we would like to take the limit of the sequence z_k as a definition for the hub and authority scores. There are two difficulties with such a definition. Firstly, the sequence does not always converge. In fact, non-negative matrices M with the above block structure always have *two* real eigenvalue of largest magnitude and the resulting sequence z_k almost never converges. Notice however that the matrix M^2 is symmetric and non-negative definite and so, even though the sequence z_k may not converge, the even and odd sub-sequences do converge. Let us define

$$z_{even} = \lim_{k\to\infty} z_{2k} \quad \text{and} \quad z_{odd} = \lim_{k\to\infty} z_{2k+1}.$$

and let us consider both limits for the moment. The second difficulty is that the limit vectors z_{even} and z_{odd} do in general depend on the initial vector z_0 and there is no apparent natural choice for z_0. In Theorem 2, we define the set of all limit vectors obtained when starting from a positive initial vector

$$Z = \{z_{even}(z_0), z_{odd}(z_0) : z_0 > 0\}.$$

and prove that the vector z_{even} obtained for $z_0 = \mathbf{1}$ is the vector of largest possible 1-norm among all vectors in Z (throughout this paper we denote by $\mathbf{1}$

the vector, or matrix, whose entries are all equal to 1; the appropriate dimension of $\mathbf{1}$ is always clear from the context). Because of this extremal property, we take the two sub-vectors of $z_{even}(\mathbf{1})$ as definitions for the hub and authority scores. In the case of the above matrix M, we have

$$M^2 = \begin{bmatrix} BB^T & 0 \\ 0 & B^T B \end{bmatrix}$$

and from this it follows that, if the dominant invariant subspaces associated to $B^T B$ and BB^T have dimension one, then the normalized hub and authority scores are simply given by the normalized dominant eigenvectors of $B^T B$ and BB^T, respectively. This is the definition used in [10] for the authority and hub scores of the vertices of G. The arbitrary choice of $z_0 = \mathbf{1}$ made in [10] is given here an extremal norm justification. Notice that when the invariant subspace has dimension one, then there is nothing special about the starting vector $\mathbf{1}$ since any other positive vector z_0 would give the same result.

We now generalize this construction. The authority score of the vertex j of G can be seen as a similarity score between j and the vertex *authority* in the graph

$$hub \longrightarrow authority$$

and, similarly, the hub score of j can be seen as a similarity score between j and the vertex *hub*. The mutually reinforcing updating iteration used above can be generalized to graphs that are different from the hub-authority structure graph. The idea of this generalization is quite simple; we illustrate it in this introduction on the path graph with three vertices and provide a general definition for arbitrary graphs in Section 3. Let G be a graph with edge set E and adjacency matrix B and consider the *structure graph*

$$1 \longrightarrow 2 \longrightarrow 3.$$

To the vertex j of G we associate three scores x_{j1}, x_{j2} and x_{j3}; one for each vertex of the structure graph. We initialize these scores at some positive value and then update them according to the following mutually reinforcing relations

$$\begin{cases} x_{j1} \leftarrow & \sum_{i:(j,i)\in E} x_{i2} \\ x_{j2} \leftarrow \sum_{i:(i,j)\in E} x_{i1} & + \sum_{i:(j,i)\in E} x_{i3} \\ x_{j3} \leftarrow & \sum_{i:(i,j)\in E} x_{i2} \end{cases}$$

or, in matrix form (we denote by x_i the column vector with entries x_{ji}),

$$\begin{bmatrix} x_1 \\ x_2 \\ x_3 \end{bmatrix}_{k+1} = \begin{bmatrix} 0 & B & 0 \\ B^T & 0 & B \\ 0 & B^T & 0 \end{bmatrix} \begin{bmatrix} x_1 \\ x_2 \\ x_3 \end{bmatrix}_k , \qquad k = 0, 1, \dots$$

which we again denote by $x_{k+1} = M x_k$. The situation is now identical to that of the previous example and all convergence arguments given there apply here as well. The matrix M^2 is symmetric and non-negative definite, the normalized even and odd iterates converge, and the limit $z_{even}(1)$ is among all possible limits one that has largest possible 1-norm. We take the three components of this extremal limit $z_{even}(1)$ as definition for the similarity scores[1] s_1, s_2 and s_3 and define the similarity matrix by

$$\mathbf{S} = [s_1 \; s_2 \; s_3].$$

The rest of this paper is organized as follows. In Section 2, we describe some standard Perron-Frobenius results for non-negative matrices that will be useful in the rest of the paper. In Section 3, we give a precise definition of the similarity matrix together with different alternative definitions. The definition immediately translates into an approximation algorithm. In Section 4 we describe similarity matrices for the situation where one of the two graphs is a path graph; path graphs of lengths 2 and 3 are those that are discussed in this introduction. In Section 5, we consider the special case $G_A = G_B = G$ for which the score s_{ij} is the similarity between the vertices i and j in the graph G. Section 6 deals with graphs for which all vertices play the same rôle. We prove that, as expected, the similarity matrix in this case has rank one.

2   Graphs and Non-negative Matrices

With any directed graph $G = (V, E)$ one can associate a non-negative matrix via an indexation of its vertices. The so-called *adjacency matrix* of G is the matrix $B \in \mathbf{N}^{n \times n}$ whose entry d_{ij} equals the number of edges from vertex i to vertex j. Conversely, a square matrix B whose entries are non-negative integer numbers, defines a directed graph G with d_{ij} edges between i and j. Let B be the adjacency matrix of some graph G; the entry $(B^k)_{ij}$ is equal to the number of paths of length k from vertex i to vertex j. From this it follows that a graph is strongly connected if and only if for every pair of indices i and j there is an integer k such that $(B^k)_{ij} > 0$. Matrices that satisfy this property are said to be *irreducible*.

The Perron-Frobenius theory [8] establishes interesting properties about the eigenvectors and eigenvalues for non-negative and irreducible matrices. Let us denote the spectral radius[2] of the matrix C by $\rho(C)$. The following results follow from [8].

Theorem 1. *Let C be a non-negative matrix. Then*
(i) the spectral radius $\rho(C)$ is an eigenvalue of C – called the Perron root – and there exists an associated non-negative vector $x \geq 0$ ($x \neq 0$) – called the Perron vector – such that $Cx = \rho x$
(ii) if C is irreducible, then the algebraic multiplicity of the Perron root ρ is equal to one and there is a positive vector $x > 0$ such that $Cx = \rho x$

---

[1] In Section 4, we prove that the "central similarity score" s_2 can be obtained more directly from B by computing the dominating eigenvector of the matrix $BB^T + B^T B$.

[2] The spectral radius of a matrix is the largest magnitude of its eigenvalues.

(iii) if C is symmetric, then the algebraic and geometric multiplicity of the Perron root ρ are equal and there is a non-negative basis $X \geq 0$ associated to the invariant subspace associated to ρ, such that $CX = \rho X$.

In the sequel, we shall also need the notion of orthonormal projection on subspaces. Let $\mathcal{V}$ be a linear subspace of $\mathbf{R}^n$ and let $v \in \mathbf{R}^n$. The *orthogonal projection* of v on $\mathcal{V}$ is the vector in $\mathcal{V}$ with smallest distance to v. The matrix representation of this projection is obtained as follows. Let $\{v_1, \ldots, v_m\}$ be an orthogonal basis for $\mathcal{V}$ and arrange these column vectors in a matrix V. The projection of v on $\mathcal{V}$ is then given by $\Pi v = VV^T v$ and the matrix $\Pi = VV^T$ is the *orthogonal projector* on $\mathcal{V}$. From the previous theorem it follows that, if the matrix C is non-negative and symmetric, then the elements of the orthogonal projector Π on the vector space associated to the Perron root of C are all non-negative.

The next theorem will be used to justify our definition of similarity matrix between two graphs. The result describes the limit points of sequences associated with symmetric non-negative linear transformations.

Theorem 2. *Let M be a symmetric non-negative matrix of spectral radius ρ. Let $z_0 > 0$ and consider the sequence*

$$z_{k+1} = Mz_k/\|Mz_k\|_2, \quad k = 0, \ldots$$

Then the subsequences z_{2k} and z_{2k+1} converge to the limits

$$z_{even}(z_0) = \lim_{k \to \infty} z_{2k} = \frac{\Pi z_0}{\|\Pi z_0\|_2} \quad and \quad z_{odd}(z_0) = \lim_{k \to \infty} z_{2k+1} = \frac{\Pi M z_0}{\|\Pi M z_0\|_2},$$

where Π is the orthogonal projector on the invariant subspace of M^2 associated to its Perron root ρ^2. In addition to this, the set of all possible limits is given by:

$$Z = \{z_{even}(z_0), z_{odd}(z_0) : z_0 > 0\} = \{\Pi z/\|\Pi z\|_2 : z > 0\}$$

and the vector $z_{even}(\mathbf{1})$ is the unique vector of largest 1-norm in that set.

3   Similarity between Vertices in Graphs

We now introduce our definition of graph similarity for arbitrary graphs. Let G_A and G_B be two directed graphs with respectively n_A and n_B vertices. We think of G_A as a "structure graph" that plays the role of the graphs $hub \longrightarrow authority$ and $1 \longrightarrow 2 \longrightarrow 3$ in the introductory examples. Let $\text{pre}(v)$ (respectively $\text{post}(v)$) denote the set of ancestors (respectively descendants) of the vertex v. We consider real scores x_{ij} for $i = 1, \ldots, n_B$ and $j = 1, \ldots, n_A$ and simultaneously update all scores according to the following updating equations

$$[x_{ij}]_{k+1} = \sum_{r \in \text{pre}(i),\, s \in \text{pre}(j)} [x_{rs}]_k + \sum_{r \in \text{post}(i),\, s \in \text{post}(j)} [x_{rs}]_k \qquad (1)$$

These equations coincide with those given in the introduction. The equations can be written in more compact matrix form. Let X_k be the $n_B \times n_A$ matrix of entries $[x_{ij}]_k$. Then (1) takes the form

$$X_{k+1} = BX_k A^T + B^T X_k A, \qquad k = 0, 1, \ldots \qquad (2)$$

where A and B are the adjacency matrices of G_A and G_B. In this updating equation, the entries of X_{k+1} depend linearly on those of X_k. We can make this dependance more explicit by using the matrix-to-vector operator that develops a matrix into a vector by taking its columns one by one. This operator, denoted vec, satisfies the elementary property $\text{vec}(CXD) = (D^T \otimes C)\,\text{vec}(X)$ in which $\otimes$ denotes the Kronecker tensorial product (for a proof of this property, see Lemma 4.3.1 in [9]). Applying this property to (2) we immediately obtain

$$x_{k+1} = (A \otimes B + A^T \otimes B^T)\, x_k \qquad (3)$$

where $x_k = \text{vec}(X_k)$. This is the format used in the introduction. Combining this observation with Theorem 2 we deduce the following property for the normalized sequence Z_k.

Corollary 1. *Let G_A and G_B be two graphs with adjacency matrices A and B, select an initial positive matrix $Z_0 > 0$ and define*

$$Z_{k+1} = \frac{BZ_k A^T + B^T Z_k A}{\|BZ_k A^T + B^T Z_k A\|_2} \qquad k = 0, 1, \ldots.$$

Then, the matrix subsequences Z_{2k} and Z_{2k+1} converge to Z_{even} and Z_{odd}. Moreover, among all the matrices in the set

$$\{Z_{even}(Z_0), Z_{odd}(Z_0) : Z_0 > 0\}$$

the matrix $Z_{even}(1)$ is the unique matrix of largest 1-norm.

In order to be consistent with the vector norm appearing in Theorem 2, the matrix norm $\|.\|_2$ we use here is the square root of the sum of all squared entries (this norm is known as the Euclidean or Frobenius norm), and the 1-norm $\|.\|_1$ is the sum of all entries magnitudes. In view of this result, the next definition is now justified.

Definition 1. *Let G_A and G_B be two graphs with adjacency matrices A and B. The similarity matrix between G_A and G_B is the matrix*

$$\mathbf{S} = \lim_{k \to +\infty} Z_{2k}$$

obtained for $Z_0 = 1$ and

$$Z_{k+1} = \frac{BZ_k A^T + B^T Z_k A}{\|BZ_k A^T + B^T Z_k A\|_2}, \qquad k = 0, 1, \ldots$$

A direct algorithmic transcription of the definition leads to an approximation algorithm. An example of a pair of graphs and their corresponding similarity matrix is given in Figure 3. Notice that it follows from the definition that the similarity matrix between G_B and G_A is the transpose of the similarity matrix between G_A and G_B. Similarity matrices can alternatively be defined as the projection of the matrix $\mathbf{1}$ on an invariant subspace associated to the graphs and for particular classes of adjacency matrices, one can compute the similarity matrix $\mathbf{S}$ directly from the dominant invariant subspaces of matrices of the size of A or B; we provide explicit expressions for a few classes in the next sections. Similarity matrices can also be defined by their extremal property.

Corollary 2. *The similarity matrix of the graphs G_A and G_B of adjacency matrices A and B is the unique matrix of largest 1-norm among all matrices X that maximize the expression*

$$\frac{\|BXA^T + B^T XA\|_2}{\|X\|_2}.\tag{4}$$

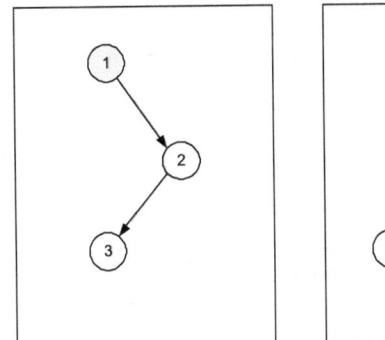 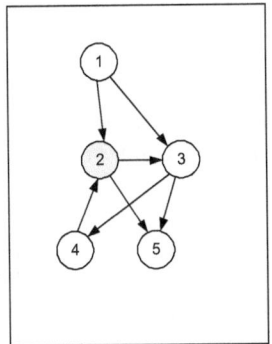

$$\mathbf{S} = \begin{bmatrix} 0.31 & 0.14 & 0 \\ 0.19 & 0.55 & 0.06 \\ 0.06 & 0.55 & 0.19 \\ 0.15 & 0.06 & 0.15 \\ 0 & 0.14 & 0.31 \end{bmatrix}$$

Fig. 1. Two graphs G_A and G_B and their corresponding similarity matrix $\mathbf{S}$. As an illustration, the similarity score between vertex 2 of graph G_A and vertex 3 of graph G_B is equal to 0.55.

4   Hubs, Authorities, Central Scores, and Path Graphs

As explained in the introduction, the hub and authority scores of a graph G_B can be expressed in terms of the adjacency matrix of G_B.

Theorem 3. *Let B be the adjacency matrix of the graph G_B. The normalized hub and authority scores of the vertices of G_B are given by the normalized dominant eigenvectors of the matrices $B^T B$ and BB^T, provided the corresponding Perron root is of multiplicity 1. Otherwise, it is the normalized projection of the vector $\mathbf{1}$ on the respective dominant invariant subspaces.*

The condition on the multiplicity of the Perron root is not superfluous. Indeed, even for strongly connected graphs, BB^T and B^TB may have multiple dominant roots: for cycle graph for example, both BB^T and B^TB are the identity matrix. Another interesting structure graph is the path graph of length three:

$$1 \longrightarrow 2 \longrightarrow 3$$

Similarly to the hub and authority scores, the resulting similarity score with vertex 2, a score that we call *central score*, can be given an explicit expression.

Theorem 4. *Let B be the adjacency matrix of the graph G_B. The normalized central scores of the vertices of G_B are given by the normalized dominant eigenvector of the matrix*

$$B^TB + BB^T,$$

provided the corresponding Perron root is of multiplicity 1. Otherwise, it is the normalized projection of the vector $\mathbf{1}$ on the dominant invariant subspace.

The above structure graphs are path graphs of length 2 and 3. For path graphs of arbitrary length ℓ we have:

Corollary 3. *Let B be the adjacency matrix of the graph G_B. Let G_A be the path graph of length ℓ:*

$$G_A \ : \quad 1 \longrightarrow 2 \longrightarrow \cdots \longrightarrow \ell.$$

Then the odd and even columns of the similarity matrix $\mathbf{S}$ can be computed independently as the projection of $\mathbf{1}$ on the dominant invariant subspaces of EE^T and E^TE where

$$E = \begin{bmatrix} B & & & \\ B^T & \ddots & & \\ & \ddots & B & \\ & & B^T & B \end{bmatrix} \quad \text{or} \quad E = \begin{bmatrix} B & & & \\ B^T & \ddots & & \\ & \ddots & B & \\ & & & B^T \end{bmatrix}$$

for ℓ even and ℓ odd, respectively.

5   Self-Similarity Matrix of a Graph

When we compare two equal graphs $G_A = G_B = G$, the similarity matrix $\mathbf{S}$ is a square matrix whose entries are similarity scores between vertices of G; this matrix is the *self-similarity matrix* of G. Various graphs and their corresponding self-similarity matrices are represented in Figure 2. In general, we expect vertices to have a high similarity score with themselves; that is, we expect the diagonal entries of self-similarity matrices to be large. We prove in the next theorem that the largest entry of a self-similarity matrix always appear on the diagonal and that, except for trivial cases, the diagonal elements of a self-similarity matrix are non-zero. As is shown with the last graph of Figure 2, it is however not true that diagonal elements dominate all elements on the same line and column.

Theorem 5. *The self-similarity matrix of a graph is positive semi-definite. In particular, the largest element of the matrix always appears on diagonal, and if a diagonal entry is equal to zero, then the corresponding line and column are equal to zero.*

For some classes of graphs, similarity matrices can be computed explicitly. We have for example:

Theorem 6. *The self-similarity matrix of the path graph of length ℓ is a diagonal matrix with diagonal elements equal to $\sin(j\pi/(\ell+1))$, $j = 1,\ldots,\ell$.*

When vertices of a graph are similar to each other, such as in cycle graphs, we expect to have a self-similarity matrix whose entries are all equal. This is indeed the case. Let us recall here that a graph is said to be *vertex-transitive* (or *vertex symmetric*) if all vertices play the same rôle in the graph. More formally, a graph G of adjacency matrix A is vertex-transitive if associated to any pair of vertices i, j, there is a permutation matrix T that satisfies $T(i) = j$ and $T^{-1}AT = A$.

Theorem 7. *All entries of the self-similarity matrix of a vertex-transitive graph are equal to $1/n$.*

6   Graphs Whose Vertices Are Symmetric to Each Other

We now analyze properties of the similarity matrix when one of the two graphs has all its vertices symmetric to each other, or has an adjacency matrix that is normal. We prove that in both cases the resulting similarity matrix has rank one.

Theorem 8. *Let G_A, G_B be two graphs and assume that G_A is vertex-transitive. Then the similarity matrix between G_A and G_B is a rank one matrix of the form*

$$\mathbf{S} = \alpha\, \mathbf{1}v^T$$

where $v = \Pi\mathbf{1}$ is the projection of $\mathbf{1}$ on the dominant invariant subspace of $(B + B^T)^2$ and α is the scaling factor $\alpha = 1/\|\mathbf{1}v^T\|$. In particular, if G_A and G_B are both vertex symmetric then the entries of their similarity matrix are all equal to $1/\sqrt{n_A n_B}$.

Cycle graphs have an adjacency matrix A that satisfies $AA^T = I$. This property corresponds to the fact that, in a cycle graph, all forward-backward paths from a vertex return to that vertex. More generally, we consider in the next theorem graphs that have an adjacency matrix A that is normal, i.e., such that $AA^T = A^T A$. In particular, graphs that have a symmetric adjacency matrix satisfy this property. We prove below that when one of the graphs has a normal adjacency matrix, then the similarity matrix has rank one and we provide an explicit expression for this matrix.

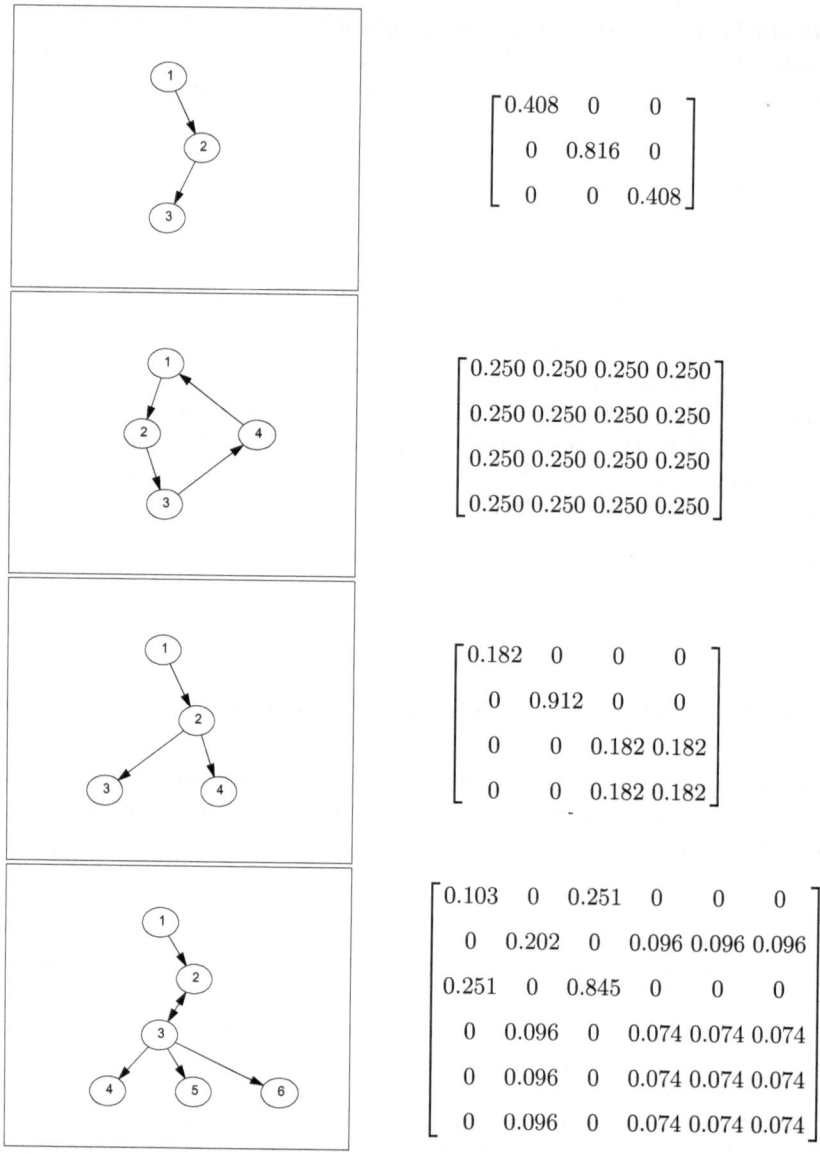

Fig. 2. Graphs and their corresponding self-similarity matrices.

Theorem 9. *Let G_A and G_B be two graphs and assume that A is a normal matrix. Then the similarity matrix between G_A and G_B is a rank one matrix $\mathbf{S} = uv^T$ where*

$$u = \frac{(\Pi_{+\alpha} + \Pi_{-\alpha})\mathbf{1}}{\|(\Pi_{+\alpha} + \Pi_{-\alpha})\mathbf{1}\|_2}, \quad v = \frac{\Pi_\beta \mathbf{1}}{\|\Pi_\beta \mathbf{1}\|_2}.$$

In this expression α is the Perron root of A, $\Pi_{+\alpha}$, $\Pi_{-\alpha}$ are the projectors on its invariant subspaces corresponding to the eigenvalues $+\alpha$ and $-\alpha$, β is the Perron root of $(B + B^T)$, and Π_β is the projector on the invariant subspace of $(B + B^T)^2$ corresponding to the eigenvalue β^2.

When one of the graphs G_A or G_B is vertex-transitive or has a normal adjacency matrix, the resulting similarity matrix $\mathbf{S}$ has rank one. Adjacency matrices of vertex-transitive graphs and normal matrices have the property that the projector $\Pi_{+\alpha}$ on the invariant subspace corresponding to the Perron root of A is also the projector on the subspace of A^T (and similarly for $-\alpha$). We conjecture here that the similarity matrix can only be of rank one if either A or B have this property.

7   Concluding Remarks

Investigations of properties and applications of the similarity matrix of graphs can be pursued in several directions. We outline here some possible research directions.

One natural extension of our concept is to consider networks rather than graphs; this amounts to consider adjacency matrices with arbitrary real entries and not just integers. The definitions and results presented in this paper use only the property that the adjacency matrices involved have non-negative entries, and so all results remain valid for networks with non-negative weights. The extension to networks makes a sensitivity analysis possible: How sensitive is the similarity matrix to the weights in the network? Experiments and qualitative arguments show that, for most networks, similarity scores are almost everywhere continuous functions of the network entries. Perhaps this can be analyzed for models for random graphs such as those that appear in [3]? These questions can probably also be related to the large literature on eigenvalues and eigenspaces of graphs; see, e.g., [4], [5] and [6].

More specific questions on the similarity matrix also arise. One open problem is to characterize the pairs of matrices that give rise to a rank one similarity matrix. The structure of these pairs is conjectured at the end of Section 6. Is this conjecture correct? A long-standing graph question also arise when trying to characterize the graphs whose similarity matrices have only positive entries. The positive entries of the similarity matrix between the graphs G_A and G_B can be obtained as follows. One construct the product graph, symmetrize it, and then identify in the resulting graph the connected component(s) of largest possible Perron root. The indices of the vertices in that graph correspond exactly to the nonzero entries in the similarity matrix of G_A and G_B. The entries of the similarity matrix will thus be all positive if and only if the product graph of G_A and G_B is weakly connected. The problem of characterizing all pairs of graphs that have a weakly connected product was introduced and analyzed in 1966 in [7]. The problem of efficiently characterizing all pairs of graphs that have a weakly connected product is a problem that is still open.

Another topic of interest is to investigate how the concepts proposed here can be used, possibly in modified form, for evaluating the similarity between two graphs, for clustering vertices or graphs, for pattern recognition in graphs or for data mining purposes.

Acknowledgment. Three of our students, Maureen Heymans, Anahí Gajardo and Pierre Senellart have provided inputs on several ideas developed in this paper. We are pleased to acknowledge the inputs of all these students.

This paper presents research supported by NSF under Grant No. CCR 99-12415 and by the Belgian Programme on Inter-university Poles of Attraction, initiated by the Belgian State, Prime Minister's Office for Science, Technology and Culture.

References

1. Vincent D. Blondel, Paul Van Dooren, A measure of similarity between graph vertices. With applications to synonym extraction and web searching. Technical Report UCL 02–50, submitted to journal, 2002.
2. Vincent D. Blondel, Pierre P. Senellart, Automatic extraction of synonyms in a dictionary, Technical report 2001–89, Université catholique de Louvain, Louvain-la-Neuve, Belgium. Also: Proceedings of the SIAM Text Mining Workshop, Arlington (Virginia, USA) April 11, 2002.
3. B. Bollobas, Random Graphs, Academic Press, 1985.
4. Fan R. K. Chung, Spectral Graph Theory, American Mathematical Society, 1997.
5. Dragoš Cvetković, Peter Rowlinson, Slobodan Simić, Eigenspaces of graphs, Cambridge University Press, 1997.
6. Dragoš Cvetković, M. Doob, H. Sachs, Spectra of graphs – Theory and applications (Third edition), Johann Ambrosius Barth Verlag, 1995.
7. Frank Harary, C. Trauth, Connectedness of products of two directed graphs, J. SIAM Appl. Math., 14, pp. 150–154, 1966.
8. R. A. Horn and C. R. Johnson, *Matrix Analysis*, Cambridge University Press, London, 1985.
9. R. A. Horn and C. R. Johnson, *Topics in Matrix Analysis*, Cambridge University Press, London, 1991.
10. Jon M. Kleinberg, Authoritative sources in a hyperlinked environment, Journal of the ACM, 46:5, pp. 604–632, 1999.
11. Pierre P. Senellart, Vincent D. Blondel, Automatic discovery of similar words. To appear in "A Comprehensive Survey of Text Mining", Springer-Verlag, 2003.

Algorithmic Aspects of Bandwidth Trading

Randeep Bhatia[1], Julia Chuzhoy[2], Ari Freund[2], and Joseph (Seffi) Naor[2]

[1] Bell Labs, 600 Mountain Ave, Murray Hill, NJ 07974
randeep@research.bell-labs.com
[2] Computer Science Dept., Technion, Haifa 32000, Israel
{cjulia,arief,naor}@cs.technion.ac.il

Abstract. We study algorithmic problems that are motivated by bandwidth trading in next generation networks. Typically, bandwidth trading involves sellers (e.g., network operators) interested in selling bandwidth pipes that offer to buyers a guaranteed level of service for a specified time interval. The buyers (e.g., bandwidth brokers) are looking to procure bandwidth pipes to satisfy the reservation requests of end-users (e.g., Internet subscribers). Depending on what is available in the bandwidth exchange, the goal of a buyer is to either spend the least amount of money to satisfy all the reservations made by its customers, or to maximize its revenue from whatever reservations can be satisfied.

We model the above as a real-time non-preemptive scheduling problem in which machine types correspond to bandwidth pipes and jobs correspond to the end-user reservation requests. Each job specifies a time interval during which it must be processed and a set of machine types on which it can be executed. If necessary, multiple machines of a given type may be allocated, but each must be paid for. Finally, each job has a revenue associated with it, which is realized if the job is scheduled on some machine.

There are two versions of the problem that we consider. In the cost minimization version, the goal is to minimize the total cost incurred for scheduling all jobs, and in the revenue maximization version the goal is to maximize the revenue of the jobs that are scheduled for processing on a given set of machines. We consider several variants of the problems that arise in practical scenarios, and provide constant factor approximations.

Keywords: Scheduling, bandwidth trading, approximation algorithms, primal-dual schema.

1 Introduction

We study algorithmic problems involving *bandwidth trading* in next generation networks. As network operators are building new high-speed networks, they look for new ways to sell or lease their plentiful bandwidth. At the same time there are emerging potential buyers of bandwidth, such as virtual network carriers, who would like to be able to expand capacity easily and rapidly to meet the ever-changing demands of their customers. Similarly, many companies are looking for ways to be able to reserve bandwidth for on-off events such as video-conferences.

J.C.M. Baeten et al. (Eds.): ICALP 2003, LNCS 2719, pp. 751–766, 2003.

Finally, there are network subscribers who would like to be able to buy bandwidth for the duration of a webcast, pay-per-movie on the web, etc. In this paper we consider some of the algorithmic problems arising in this context.

1.1   Bandwidth Trading in Practice

Our work is motivated by the emerging business model being discussed in the networking community, which we briefly describe here. (More details can be found, for example, at [19].) Although bandwidth exchange/trading is not new, the traditional methodology is often marred by long periods of binding contracts and slow provisioning time. With the recent advances in network technologies, however, there has been a tremendous leap forward both in network capacity and provisioning times. It is now possible to quickly provision end-to-end protocol-independent light paths with a specified Service Level Agreement (SLA) that takes into account QoS, bandwidth, restoration level, etc., in order to rapidly meet the changing bandwidth demands of network users. In addition, many identical low bandwidth data streams can be multiplexed over a single light path in the core network, thus enabling a core network operator to sell bandwidth at smaller granularities.

Driven to meet demands for high-speed data-centric applications, various up-start network carriers have been rolling out networks with vast amounts of excess capacity. With all this capacity up for grabs, a new generation of resellers, wholesalers, and on-line bandwidth brokers are poised to resell it to customers. Leading the pack in the bandwidth commodity effort are carriers such as Williams and a host of real-time on-line trading centers pioneered by the likes of Band-X and RateXchange.

Typically, bandwidth trading involves a *bandwidth exchange* which includes a marketplace for suppliers and buyers of bandwidth and a set of *pooling points* which are used for actually providing the bandwidth upon settlement.

Physically, a pooling point may be a fiber interconnection and switching site in a particular geographical location, with co-located points of presence for buyers (e.g. ISPs) and suppliers (e.g long haul carriers). In the pooling point, the buyer's network interfaces with the supplier's high-speed optical network, and data passing between the two is converted from electrical packets to optical signal and vice versa. It is assumed that a bandwidth exchange trades well-defined band-

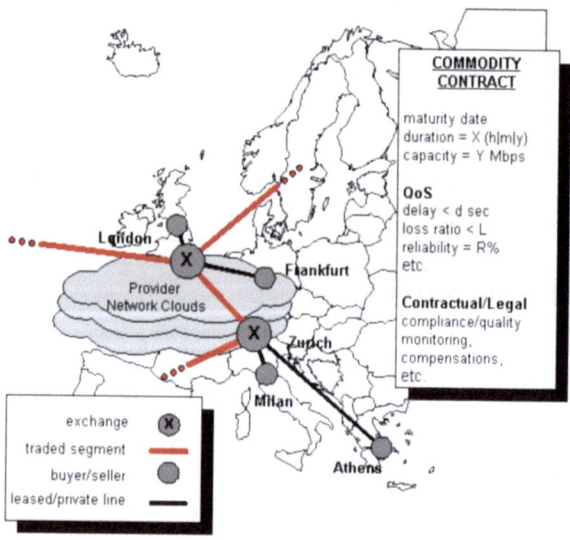

width contracts [6,12]. Each contract refers to a *bandwidth segment* between two pooling points, where a bandwidth segment is an abstraction of one or more high capacity networks providing connectivity between the two pooling points. Each bandwidth contract describes the duration for which connectivity will be made available as well as a Service Level Agreement (SLA) that takes into account QoS, bandwidth, restoration level, etc., The inset, taken from the web site of IBM Zürich Research Lab (`http://www.zurich.ibm.com/bandwidth/concepts.html`—the section titled QoS), summarizes the situation by showing an example of a bandwidth segment being offered between two pooling points connected to buyers' networks.

Optical technologies play a central role in next generation networks. A single strand of optical fiber is now capable of carrying a large number of high bandwidth data streams, each of which can be individually managed. Typically, a low bandwidth data stream is dedicated to a single end-user at any given time, but may be shared over time by multiple end-users, where the switch from one user to another is provisioned almost instantaneously. A core optical network operator is therefore able to trade a large number of identical bandwidth contracts (with same attributes), each corresponding to a low bandwidth data stream, all of which are multiplexed over a single high bandwidth light path. For practical purposes, one can therefore assume that an unlimited number of "copies" of each contract are available.

Finally, buyers are themselves service providers to their clients—the end users. End users generate bandwidth requests which are called *forward reservations*. Each forward reservation specifies the two endpoints (which translate into two pooling points) between which the bandwidth reservation is required, the time interval for which it is required, any other attributes of the connection (QoS, restoration level, etc.), and the revenue obtained by honoring the reservation. Buyers in the bandwidth exchange, who may be ISPs, bandwidth brokers, etc., are looking to procure bandwidth contracts to satisfy at the cheapest cost the forward reservation requests made by their clients (e.g., network subscribers, companies, or virtual network operators). A single procured bandwidth contract can be used to serve a set of "non-overlapping" forward reservation requests. Depending on what is available in the bandwidth exchange, the goal of a buyer is to either spend the least amount of money to buy enough bandwidth contracts to satisfy all the reservations, or, if the reservations cannot be all satisfied, to maximize its revenue from whatever reservations are possible.

Combining Contracts. Given a bandwidth exchange, a *contract graph* [6,12,19] is defined to be a graph whose nodes are the pooling points and edges represent the traded contracts. Several point-to-point segments (on a path in the graph) can be assembled to connect any two geographical locations. This leads to a new (path) contract whose attributes depend on the choice of the path in the contract graph. We stress that the new contract is *indivisible*. For example, consider three pooling points A, B, and C. Suppose that an (A, B) contract and a (B, C) contract are combined into an (A, C) contract. Then, the (A, C) contract cannot be used to also route traffic from A to B or from B to C, since this would require optical-to-electrical followed by an electrical-to-optical conversion at point B. Such conversions introduce substantial delays at the intermediate points and deteriorate end-to-end QoS, thus defeating the purpose of high-speed optical routing.

In general, we can assume that if a pair of pooling points have a point-to-point contract between them, then that is the cheapest way to connect the two points (for those attributes), since we consider a highly liquid bandwidth market in which arbitrage opportunities [7] are instantaneously removed. (A geographic arbitrage arises, for example, if the price of an indivisible point-to-point contract between New York and London is more than the price of a path contract with the same attributes that goes via Los Angeles.)

1.2   Wavelength Assignment in Optical Line Systems

Another motivation for the problems we consider comes from wavelength assignment in optical line systems [20] such as those involving DWDMs. An optical line system is a collection of nodes called Mesh Optical Add and Drop Multiplexers (MOADM) arranged in a line, with adjacent nodes connected by optical fibers. A demand enters the line system at one node and exits at some other node, and is routed on the same wavelength on the fibers connecting all the intermediate nodes. The set of wavelengths available on each fiber connecting two adjacent MOADM (say node i and $i+1$) may differ from fiber to fiber, and it is a function of the fiber characteristics, and also of the wavelengths which have been used up by previously provisioned demands. Given a set of demands, the problem is to assign wavelengths to them so that no two demands use the same wavelength on the same fiber.

We note that any optical line system as described above can be viewed as a set of windows, $\mathcal{I}$, where each $I \in \mathcal{I}$ is an interval corresponding to a single wavelength which is available between the two end-points of I. Given a set of demands D (i.e., D is a set of intervals), the wavelength assignment problem corresponds to packing the intervals belonging to D into $\mathcal{I}$ such that intervals packed into a window $I \in \mathcal{I}$ do not overlap.

1.3   Model Description, Notation, and Terminology

We model bandwidth trading as a real-time scheduling problem. As explained above, a bandwidth contract can only be used for routing traffic between its two end points. Therefore, we only need to consider the bandwidth trading problem for a single pair of pooling points. We view the bandwidth contracts as a set of machines of different types, where identical bandwidth contracts (with the same attributes) correspond to machines of the same type. Each machine type has a *cost* per machine. We can assume that an unlimited number of machines are available from each type, since the available bandwidth on the light paths in the core network is greater by many orders of magnitude than the end-user bandwidth requirement for any single data stream. The jobs correspond to reservation requests made by end users. Each job needs to be processed during a specified time interval, and it can only be processed on a machine of one of several specified types. Each job has a value associated with it corresponding to the revenue obtained for processing the job. At most one job can be scheduled on any machine at any given time (since no two overlapping reservation requests can be served by the same bandwidth contract).

In the *cost minimization* version of the problem, the goal is to find a set of machines of minimum cost so as to be able to schedule all the jobs on the machines. In this version of the problem we ignore the job revenues. In the *revenue maximization* version, the goal is to maximize the total revenue by selecting a subset of the jobs that can be scheduled using a given set of machines. In this version of the problem we ignore the machine costs.

Formally, we have a set of m machine types $\mathcal{T} = \{T_1, \ldots, T_m\}$. A *cost*, or *weight*, $w(T_i) \geq 0$, is incurred for allocating a machine of type T_i. There are n jobs belonging to an input set of jobs J, where each job j is associated with the following parameters: a revenue $w(j) \geq 0$, a set $S(j) \subseteq \mathcal{T}$ of machine types on which this job can be processed, and a time interval $I(j)$ during which the job is to be processed. We sometimes refer to a job and its interval interchangeably. At most one job can be processed on a given machine at any given moment.

The Problems. The general version of the cost minimization problem, where the sets $S(j)$ of machine types are arbitrary, is essentially equivalent (approximation-wise) to *set cover*. (Hardness can be shown by a simple reduction in which set elements become non-overlapping jobs and each set becomes a machine capable of processing only the jobs corresponding to the set's elements. Logarithmic approximability was shown by Jansen [11].) In practice, however, the definition of the sets $S(j)$ is usually based on properties of the machines and thus has a computationally more convenient structure. We consider two variants arising naturally in bandwidth trading and other real world applications.

Cost minimization with machine time intervals. Machine types are defined by time intervals. Each type T_i is associated with a time interval $I(T_i)$ during which machines of this type are available. A job can be processed by every machine that is available throughout its processing interval. Thus the sets $S(j)$ are defined by: $S(j) = \{T_i \in \mathcal{T} \mid I(j) \subseteq I(T_i)\}$. In the *unweighted* case all types have unit cost ($w(T_i) = 1$ for all i), and in the *weighted* case costs are arbitrary.

Cost minimization with machine strengths. There is a linear order of *strength* defined on the machine types, $T_1 \prec T_2 \prec \cdots \prec T_m$, such that a job that may be processed by a machine of a given type may also be processed on every stronger machine, i.e., for all jobs j, $S(j)$ has the form $\{T_{i_j}, T_{i_j+1}, \ldots, T_m\}$. We also assume that the stronger a machine is, the higher its cost (otherwise there is no point in using weaker machines). The linear order models a situation in which the SLA's are contained in each other in terms of the capabilities they specify.

We comment that no bounded approximation factors are possible for these problems (unless P = NP) if only a limited number of machines is available from each type. This follows by observing that it is NP-hard even to decide whether all the jobs can be scheduled on all available machines (by a simple reduction from the circular arc coloring problem which is NP-hard [9]).

We also consider the revenue maximization version.

Revenue maximization. We are given a collection $\mathcal{M}$ of machines (presumably, already paid for) and we wish to select a maximum revenue subset of the jobs that can be scheduled on these machines. Every job j specifies an arbitrary set, $S(j) \subseteq \mathcal{M}$, of machines on which it can be scheduled.

1.4   Our Contribution

The problems we consider are NP-hard. We present the first polynomial-time constant-factor approximation algorithms for both versions of the problem (cost minimization and revenue maximization).

In Section 2 we consider the cost minimization problem with machine time intervals. We describe a 3-approximation algorithm for the weighted case and a 2-approximation algorithm for the unweighted case. We remark that our 3-approximation algorithm for the weighted case can be extended to a *prize collecting* version of the problem, where it is not necessary to schedule all the jobs but there is a fine to be paid for each unscheduled job. We defer details to the full paper.

Our algorithm for the weighted case is based on a linear programming relaxation and it is a novel variant of the (combinatorial) primal-dual schema [18]. It is quite unique in that it copes with the difficulty posed by a constraint matrix containing both positive and negative entries. It is an interesting fact that the primal-dual schema is often incapable of dealing with both positive and negative coefficients. Our algorithm is also unconventional in that it departs from the common dual-ascent, reverse-delete paradigm. Rather than generating a minimal solution via a reverse-delete stage, we iteratively improve our schedule by a selective rescheduling of jobs which obviates some of the machines in our schedule. In addition, each iteration in our dual ascent stage increases several (but not all) of the dual variables. This contrasts with algorithms such as Goemans and Williamson's *network design* algorithm [10], where *all* dual variables are increased uniformly in each iteration, or Bar-Yehuda and Even's *vertex cover* algorithm [5], where a single dual variable is increased in each iteration.

In Section 3, we show a 2-approximation algorithm for the cost minimization problem with machine strengths. Both this algorithm and the algorithm for the unweighted version of the cost minimization problem with machine time intervals employ simple combinatorial lower bounds on the problems.

We conclude with the revenue maximization problem in Section 4. We present a $(1 - 1/e)$-approximation algorithm for this problem. This result improves on the approximation factor of $1/2$ implicit in [2] for this problem.

1.5   Related Work

Kolen and Kroon [13,15] and Jansen [11] considered the same scheduling model as ours in the context of aircraft maintenance. When aircraft arrive at the airport, they must be inspected by ground engineers between their arrivals and departures. There are different types of aircraft, and different types of engineer licenses, and there is an aircraft-type/engineer-license-type matrix specifying which license type permits an engineer to work on which aircraft type. In addition, the engineers work in shifts. The cost of assigning an engineer to an aircraft inspection job depends on the engineer's license type and the shift during which the inspection must be carried out. The goal is to enlist a minimum-cost multi-set of "engineer instances" to handle all aircraft inspections, where an engineer instance is defined by a (license,shift) pair. In our model, jobs correspond to inspections of aircraft, and machine types correspond to (license,shift) pairs.

Kolen and Kroon [13] study the computational complexity of this problem with respect to different aircraft/engineers matrices, when all the shifts are the same. In particular, their work implies that the cost minimization problem we consider in Section 3 is NP-hard. In [15], Kolen and Kroon study another version of this problem, where all the aircraft and license types are the same, and there are different time shifts. They show that the problem is NP-hard even for unit costs, implying that the problems we consider in Sections 2.1 and 2.2 are NP-hard as well. Jansen [11] gives an $O(\log n)$-approximation algorithm for the general problem, with both aircraft/license types and time shifts. When all the shifts are the same and all aircraft types are identical, the problem reduces to optimal coloring of interval graphs, and has a polynomial time algorithm [1].

Maximizing the throughput (revenue in our terminology) in real-time scheduling was studied extensively in [2,3,17,4,8]. They focused on the case where for each job, more than one time interval in which it can be performed is specified, while machines are available continuously. As here, jobs are scheduled non-preemptively and at each point of time only one job can be scheduled on a given machine. This model captures many applications, e.g., scheduling a space mission, bandwidth allocation, and communication in a linear network. The results of [2] on maximizing the throughput of unrelated parallel machines imply an approximation factor of 1/2 for our revenue maximization problem. This result was improved in [8] to $(1 - 1/e - \epsilon)$ (for any constant ϵ) for the unweighted version of the problem. The revenue maximization problem with machine time intervals was studied by Kolen and Kroon [14] (see also Kolen and Lenstra [16, pp. 1901-1903]). They solved the problem optimally with a dynamic programming algorithm whose running time is $O(n^m)$. This implies that the problem is polynomial-time solvable for a constant number of machines.

The wavelength assignment problem in optical line systems is studied in [20]. Their result implies that the resulting interval packing problem (which is a decision version of our revenue maximization problem) as described in Section 1.2 is NP-complete.

2   Cost Minimization with Machine Time Intervals

In this section we develop approximation algorithms for the special case of the cost minimization problem where each machine type T_i has a time interval $I(T_i)$ during which the machines of this type are available. The sets $S(j)$ of machine types allowable for job j are defined as follows: $S(j) = \{T_i \in \mathcal{T} \mid I(j) \subseteq I(T_i)\}$. We present a 3-approximation algorithm for the weighted case and a 2-approximation algorithm for the unweighted case.

2.1   The Weighted Case

Our algorithm for the weighted case is based on the primal-dual schema for approximation algorithms. The linear programming formulation of the problem contains two sets of variables: $\{x_i\}$ and $\{y_{ij}\}$. For each machine type T_i, variable x_i represents the number of machines allocated of type T_i, and for every pair of machine type T_i and job j such that $I(j) \subseteq I(T_i)$, variable y_{ij} indicates whether job j is assigned to a machine of type T_i. We also use the following notation: E is the set of endpoints of jobs and $J(t)$ is the set of jobs whose intervals contain time t. The linear program is:

$$\text{Min} \quad \sum_{i=1}^{m} w(T_i)x_i \qquad \text{s.t.}$$

$$\sum_{i=1}^{n} y_{ij} \geq 1, \qquad \forall j \in J; \tag{1}$$

$$x_i - \sum_{j \in J(t)} y_{ij} \geq 0, \qquad \forall 1 \leq i \leq m, \ \forall t \in E \cap I(T_i); \tag{2}$$

$$x, y \geq 0. \tag{3}$$

(The sums in Constraints (1) and (2) should be understood to include only variables y_{ij} that are defined.) The dual variables are $\{\alpha_j\}$ and $\{\beta_i^t\}$, corresponding to Constraints (1) and (2), respectively. The dual program is:

$$\text{Max} \quad \sum_{j \in J} \alpha_j \qquad \text{s.t.}$$

$$\sum_{t \in E \cap I(T_i)} \beta_i^t \leq w(T_i), \qquad \forall 1 \leq i \leq m; \tag{4}$$

$$\alpha_j - \sum_{t \in E \cap I(j)} \beta_i^t \leq 0, \qquad \forall 1 \leq i \leq m, \ \forall j \text{ s.t. } I(j) \subseteq I(T_i); \tag{5}$$

$$\alpha, \beta \geq 0. \tag{6}$$

Our algorithm proceeds in two phases. In the first phase it constructs a feasible schedule by iteratively allocating machines and scheduling jobs on them. In the second phase it improves the solution by considering the allocated machines in reverse order and (possibly) eliminates some of them, rescheduling jobs as necessary.

Phase 1: dual ascent. As mentioned, the first phase allocates machines and schedules jobs. Accordingly, at a given moment during this phase there are *scheduled* and *unscheduled* jobs, and *allocated* and *un-allocated* machines and machine types. Initially all jobs are unscheduled, all machines and machine types are unallocated, and all dual variables are set to 0.

The kth iteration in Phase 1 proceeds as follows. Let $t_k \in E$ be such that a maximum number of unscheduled jobs contain t_k, and let n_k denote the number of these jobs. Let T_k be the set of all un-allocated machine types whose intervals contain t_k. We increase $\beta_i^{t_k}$ for all i such that $T_i \in \mathcal{T}_k$ uniformly at the same rate until some constraint of type (4) becomes tight, i.e., we increase each of the βs in question by $\delta_k = \min\{w(T_i) - \sum_{t \in E \cap I(T_i)} \beta_i^t \mid T_i \in \mathcal{T}_k\}$. All the machine types that become tight are considered allocated from now on. For each currently unscheduled job j whose interval is contained in the interval one of these newly allocated machine types, we allocate a separate machine of the appropriate type, say T_i, schedule j on it, and set $\alpha_j = \sum_{t \in E \cap I(j)} \beta_i^t$.

We claim that the dual solution thus constructed is feasible. Clearly, the algorithm satisfies all constraints of type (4) at all times. To see that the solution satisfies all constraints of type (5) as well, consider any such constraint $\alpha_j - \sum_{t \in E \cap I(j)} \beta_i^t$. Suppose job j was scheduled in the kth iteration. Following the kth iteration, α_j remains unchanged and the sum of βs can only increase, so it suffices to show that the constraint is satisfied at the end of the kth iteration. Let $T_{i'}$ be the type of the machine on which job j was scheduled. If $i = i'$, the constraint is satisfied by equality. Otherwise, machine type T_i could not have been allocated prior to the kth iteration (for otherwise job j would have already been scheduled by the time the kth iteration commenced), and thus, for all $t \in E \cap I(j)$, the values of β_i^t and $\beta_{i'}^t$ must have increased identically during the first k iterations. Thus, $\sum_{t \in E \cap I(j)} \beta_i^t = \sum_{t \in E \cap I(j)} \beta_{i'}^t$ at the end of the kth iteration, and the claim follows.

Phase 2: reverse reschedule & delete. Let $\mathcal{M}$ be the set of machines allocated in the first phase. Later on we describe a reverse reschedule & delete procedure that returns a feasible schedule using a subset $\mathcal{M}' \subseteq \mathcal{M}$ of machines which has the property that for all k, the number of machines in $\mathcal{M}'$ of types from $\mathcal{T}_k$ is at most $3n_k$. We note that in standard primal-dual algorithms the second phase is a simple *reverse delete* phase, whose purpose is to yield a minimal solution. The approximation guarantee then follows from an upper bound on minimal solutions. In our case, we do not know how to find a minimal solution. In fact, even determining whether all jobs can be scheduled on a given set of machines is NP-hard. We therefore do not attempt to find a minimal solution, but instead, gradually discard some of the machines in a very special manner designed to achieve the above property.

Analysis. Let (α, β) be the dual solution constructed in Phase 1 and let (x, y) be the primal solution corresponding to the schedule generated in Phase 2. To show that this schedule is 3-approximate, it suffices to prove that $\sum_{i=1}^m w(T_i)x_i \le 3 \sum_{j \in J} \alpha_j$. This inequality from the next two claims.

Claim.
$$\sum_{i \in T} w(T_i)x_i \le 3 \sum_k n_k \delta_k.$$

Proof. For each allocated machine type T_i, $w(T_i)$ equals the sum of δ_k taken over all k such that machine type T_i was unallocated at the beginning of the kth iteration and $t_k \in I(T_i)$. Thus, $\sum_{i \in T} w(T_i)x_i = \sum_k \delta_k m_k$, where m_k is the number of machines of types in $\mathcal{T}_k$ used by the final schedule. The claim then follows, since $m_k \le 3n_k$. $\qquad \square$

Claim.
$$\sum_{j \in J} \alpha_j = \sum_k n_k \delta_k.$$

Proof. For each job j, α_j is the sum of all δ_k such that job j was still unscheduled at the beginning of the kth iteration and $t_k \in I(j)$. For each iteration k, the number of jobs that were unscheduled at the beginning of the kth iteration and contain t_k is exactly n_k. □

The reverse reschedule & delete procedure. Let $\mathcal{M}$ be the set of machines used in the schedule constructed in the first phase, and let $\mathcal{M}_k \subseteq \mathcal{M}$ be the subset of machines of types in $\mathcal{T}_k$. The purpose of the reverse reschedule & delete procedure is to prune each machine set $\mathcal{M}_k$, leaving only $3n_k$ (or less) of its members allocated, yet manage to feasibly schedule all of the jobs on the surviving machines. To achieve this we consider the sets $\mathcal{M}_k$ in reverse order (decreasing k), and prune each in turn.

The pruning procedure for $\mathcal{M}_k$ is the following. If $|\mathcal{M}_k| \leq 3n_k$, we do nothing. Otherwise, consider the jobs currently assigned to machines in $\mathcal{M}_k$. They are of three possible types: *left jobs*, which lie entirely to the left of t_k; *right jobs*, which lie entirely to the right of t_k; and *middle jobs*, which cross t_k.

The middle jobs are easiest. The number of middle jobs is exactly n_k (by definition), so they are currently scheduled on n_k different machines. We retain these machines, denoting them $\mathcal{M}_{\mathrm{mid}}$, and the scheduling of *all* jobs currently assigned to them (these may include some right jobs or left jobs in addition to all middle jobs).

The remaining left jobs are scheduled in the following manner. First note that $|\mathcal{M}_k \setminus \mathcal{M}_{\mathrm{mid}}| \geq 2n_k$, since $|\mathcal{M}_k| > 3n_k$. Denote by $\mathcal{M}_{\mathrm{left}}$ the set of n_k machines in $\mathcal{M}_k \setminus \mathcal{M}_{\mathrm{mid}}$ with leftmost left endpoints. Observe that the intervals of these machines all contain t_k by definition (as do the intervals of all machines in $\mathcal{M}_k$). Let t be the rightmost endpoint among the left endpoints of machines in $\mathcal{M}_{\mathrm{left}}$. All left jobs whose left endpoints are to the left of t must be currently scheduled on machines in $\mathcal{M}_{\mathrm{left}}$, so we leave them intact. We proceed to reschedule all remaining left jobs greedily in order of increasing left endpoint. Specifically, for each job j we select any machine in $\mathcal{M}_{\mathrm{left}}$ on which we have not already rescheduled a job that conflicts with j, and schedule j on it. To see that this is always possible, observe that all n_k machines are available between t and t_k, and thus if a job cannot be scheduled, then its left endpoint must be contained in n_k other jobs that were scheduled on machines in $\mathcal{M}_k$. These $n_k + 1$ jobs were therefore unscheduled at the beginning of the kth iteration (since we are pruning the sets $\mathcal{M}_k$ in reverse order), but this contradicts the definition of n_k, as these jobs all intersect at one time point.

The remaining right jobs are scheduled in a symmetric manner on the n_k machines in $\mathcal{M} \setminus \mathcal{M}_{\mathrm{mid}}$ with rightmost right endpoints. Some (or all) of these machines may belong to $\mathcal{M}_{\mathrm{left}}$, and therefore may already have left jobs scheduled on them, but that is not a problem because the intervals of left jobs and right jobs do not intersect.

2.2   The Unweighted Case

We present a 2-approximation algorithm for this case. Let L be the set of left endpoints of job intervals. For each point of time $t \in L$, let n_t be the number of jobs whose intervals contain t. The algorithm consists of two stages. In the first stage it solves the optimization problem of allocating a minimum number of machines such that for all $t \in L$, at least n_t of the allocated machines are available at time t. In the second stage it schedules the jobs using at most twice the number of machines allocated in the first stage.

Stage 1. Scan the points of time in L in left-to-right order. For each point $t \in L$, Let n'_t be the number of machines that are available at time t and have already been allocated. If $n'_t < n_t$, allocate another $n_t - n'_t$ machines of type T_t, where T_t is the machine type with the rightmost right endpoint among all machines available at time t.

Proposition 1. *The solution found in Stage 1 is optimal.*

Proof. We say that a time point $t \in L$ is *covered* by a set of machines $\mathcal{M}$ if $\mathcal{M}$ contains at least n_t machines that are available at time t. Let t_i be the time point considered in the ith iteration of Stage 1, and let $\mathcal{M}_i$ be the set of machines allocated in the first i iterations of Stage 1. We prove by induction that for all i, there exists an optimal solution that contains $\mathcal{M}_i$. For $i = 0$, $\mathcal{M}_i = \emptyset$ and the claim holds trivially. Consider $i > 0$. By the induction hypothesis there exists an optimal solution $\mathcal{M}^*$ such that $\mathcal{M}_{i-1} \subseteq \mathcal{M}^*$. If no new machines are allocated in the ith iteration, then $\mathcal{M}_i = \mathcal{M}_{i-1} \subseteq \mathcal{M}^*$. Otherwise, there are at least $n_{t_i} - n'_{t_i}$ machines in $\mathcal{M}^* \setminus \mathcal{M}_{i-1}$ that are available at time t_i. We remove any $n_{t_i} - n'_{t_i}$ of them from $\mathcal{M}^*$ and replace them by the newly allocated machines $\mathcal{M}_i \setminus \mathcal{M}_{i-1}$. This cannot affect the feasibility because all time points $t_j < t_i$ remain covered by $\mathcal{M}_{i-1}$, and the choice of $\mathcal{M}_i \setminus \mathcal{M}_{i-1}$ as machines with the rightmost right endpoints that are available at time t_i guarantees that they are all available at all times $t_j \geq t_i$ at which any of the machines they replace are available. □

Remark 1. A different approach to the solution of the optimization problem of Stage 1 is through the natural integer linear program for this problem. It is easy to see that the constraint matrix defining the linear program is totally unimodular (TUM) and thus the optimal solution to the linear program is always integral.

Stage 2. Let $\mathcal{M}$ be the set of machines allocated in Stage 1. Order the jobs by their left endpoint (from left to right) and schedule them in this order on machines in $\mathcal{M}$. Select for each job any machine on which no previously scheduled jobs intersect with the present job. The machine selected must also satisfy the condition that its time interval contains the job's left endpoint (though not necessarily the job's entire interval). The resultant schedule might, of course, be infeasible, due to jobs extending beyond the right endpoints of the machines on which they are scheduled, but at most one job per machine may do so. Fix the schedule by allocating new machines, one for each of these jobs. At most $|\mathcal{M}|$ new machines are added to the schedule.

Theorem 1. *Stage 2 returns a 2-approximate solution.*

Proof. The initial (infeasible) schedule constructed in Stage 2 contains all jobs, for if Stage 2 cannot schedule some job, then there are at least k other jobs containing its left endpoint $t_i \in L$, where k is the number of machines in $\mathcal{M}$ available at time t_i. This implies $n_{t_i} > k$, contradicting the fact that at each point of time $t \in L$, at least n_t machines are allocated in Stage 1. Thus, the final schedule constructed in Stage 2 is feasible and it uses at most $2|\mathcal{M}|$ machines. Since $|\mathcal{M}|$ is clearly a lower bound on the optimum, the solution is 2-approximate.

□

3   Cost Minimization with Machine Strengths

In this section we present a 2-approximation algorithm for the special case of the cost minimization problem where there is a linear order of *strength* on the machine types $T_1 \prec T_2 \prec \cdots \prec T_m$, such that a job that may be processed by a machine of a given type may also be processed on every stronger machine. In other words, $S(j)$ has the form $\{T_{i_j}, T_{i_j+1}, \ldots, T_m\}$ for all $j \in J$. We also assume that the stronger a machine, the higher its cost, i.e., $w(T_i) < w(T_{i+1})$ (otherwise there is no point in ever using weaker machines). We say that job j *exists* at time t if $t \in I(j)$.

For $1 \le i \le m$, let n_i be the maximum cardinality of a set of jobs that all exist simultaneously at some time point and all require machines of type T_i or stronger. Clearly, every feasible schedule requires at least n_i machines of type T_i or stronger, for all i. Thus, the cost of an optimal schedule is at least as high as the minimum cost of a set of machines with the property that for all i, the set contains at least n_i machines of type T_i or stronger. Define $n_{m+1} = 0$. Consider a set of machines $\mathcal{M}$ consisting of $n_i - n_{i+1}$ machines of type T_i, for all $1 \le i \le m$ (note that $n_i \ge n_{i+1}$ for all i, and the number of machines allocated in $\mathcal{M}$ of type T_i or stronger is n_i). Then $\mathcal{M}$ has the above property, and because stronger machines cost more than weaker ones, $\mathcal{M}$ is a minimum cost set with this property. Thus the cost of $\mathcal{M}$ is a lower bound on the cost of optimal solution. We show how to schedule all jobs on a set of machines containing at most two copies of each machine in $\mathcal{M}$. This schedule is therefore 2-approximate.

Let $M_1, \ldots, M_k$ (where $k = n_1$) be the machines in $\mathcal{M}$ ordered from weakest to strongest. Construct an *initial* infeasible schedule as follows. Consider the machines in order from M_1 to M_k. For each M_i, construct a schedule containing a subset of the jobs as follows. First, schedule on M_i all of the currently unscheduled jobs that can be processed on it. Ignore job overlap when constructing this schedule. Then, iterate: as long as there is a job j scheduled on M_i that is fully contained in the union of other jobs scheduled on M_i, un-schedule job j. Although the schedule thus constructed for M_i may contain overlapping jobs, it has the redeeming property that the interval graph it induces is 2-colorable, as it is an easy fact that if three intervals intersect, at least one of them must be contained in the union of the other two. Having constructed the initial schedule (on all machines), color the induced interval graph on each machine with two colors and create a feasible schedule by using two copies of $\mathcal{M}$, one for each color class.

It remains to show that the initial schedule contains all jobs. Restricting our attention to this schedule, we say that a time point t is *covered* on machine M_i if there is job containing t scheduled on M_i. By construction, the set of points covered on M_i is precisely the union of all jobs that could be processed on M_i and were still unscheduled when the algorithm reached M_i. It follows that if a time point t is not covered on M_i, then every job that contains it either cannot be processed on M_i, or is scheduled on some machine $M_{i'}$, $i' < i$. Thus, suppose the algorithm fails to schedule some job j. Let t be any time point in j. Then t must be covered on the strongest machine, i.e., M_k, since it is contained in an unscheduled job (namely j) that can be processed on it. Let i be minimal such that t is covered on machines $M_i, M_{i+1}, \ldots, M_k$. Let J' be the set of jobs scheduled on these machines that contain t. Assuming $i > 1$, point t is not covered on M_{i-1} by definition. Thus, by our previous observation, all of the jobs in $J' \cup \{j\}$ cannot be processed on M_{i-1}, and one (or two) of them are scheduled on M_i, so M_i is strictly stronger than M_{i-1}. Let T_{l-1} be the type of machine M_{i-1}. Then, all the jobs from $J' \cup \{j\}$ require machines of type at least as strong as T_l. Thus, $|J' \cup \{j\}| \le n_l$. On the other hand, the number of available machines of types T_l or stronger is $< |J' \cup \{j\}| \le n_l$, in contradiction with the fact the $\mathcal{M}$ contains n_l such machines. In the case $i = 1$ we get a contradiction directly: $n_1 \ge |J' \cup \{j\}| \ge k + 1 > n_1$.

Theorem 2. *Our algorithm returns a 2-approximate solution.*

4 Revenue Maximization

In the revenue maximization problem, we are given a set of machines $\mathcal{M} = \{M_1, \ldots, M_m\}$ (presumably, already paid for) and a set of jobs J. Since the set of machines is fixed here, we identify machines with machine types. For each job $j \in J$, there is a time interval $I(j)$ during which it should be processed and a non-negative profit (or *weight*) $w(j)$ associated with it. Every job j specifies an arbitrary set, $S(j) \subseteq \mathcal{M}$, of machines on which it can be scheduled. The goal is to find a feasible schedule of a subset of the jobs on the machines that maximizes the total profit of the jobs scheduled. We present a $(1 - 1/e)$-approximation algorithm for this problem. Our approach is to cast the problem as an integer problem and solve its linear programming (LP) relaxation. We then obtain an integral solution by randomly rounding the optimal fractional solution found for the LP relaxation.

The Linear Program: For each job j and for each machine M_i, there is a variable x_{ij} that indicates whether job j is scheduled on machine M_i.

$$\text{Max} \quad \sum_{j \in J} \sum_{M_i \in S(j)} w(j) x_{ij} \quad \text{s.t.}$$

$$\sum_{i=1}^{m} x_{ij} \le 1, \qquad \forall j \in J; \tag{7}$$

$$\sum_{j : t \in I(j)} x_{ij} \le 1, \qquad \forall i \in \{1, \ldots, m\}, \forall t; \tag{8}$$

$$x \ge 0. \tag{9}$$

Constraints (7) guarantee that each job is scheduled at most once. Constraints (8) guarantee that each machine executes at most one job at each time point.

Randomized Rounding: Let x be an optimal fractional solution. Choose N to be the smallest integer such that $N \cdot x_{ij}$ is integral for all i, j. We perform the randomized rounding on each machine separately. For each machine M_i, perform the following steps:

1. Construct an interval graph $\mathcal{I}$ as follows. For each job j, add $N \cdot x_{ij}$ copies of the time interval $I(j)$ to $\mathcal{I}$. Note that at each time point, the sum of the fractions of the jobs that are executed on machine M_i is at most 1. Thus the size of the maximum clique in the interval graph $\mathcal{I}$ is at most N.
2. Color $\mathcal{I}$ with N colors. Each color class induces a feasible schedule on machine M_i.
3. Choose one of the color classes uniformly at random. Schedule on M_i all the jobs that have time intervals belonging to this color class. If a job is scheduled on more than one machine, arbitrarily unassign it from all but one machine.

We remark that there is no need to build the interval graph $\mathcal{I}$ explicitly, i.e., to replicate intervals. In fact, a coloring satisfying the above can be computed in strongly polynomial time.

We now estimate the expected revenue of the schedule thus generated. For each job j, let $x_j = \Sigma_{i=1}^{m} x_{ij}$. For a job j, the probability of its being scheduled on a particular machine M_i is exactly x_{ij}. Therefore, the probability that it is not assigned to M_i is $1 - x_{ij}$. Thus, the probability that it is not assigned to any machine is

$$\prod_{i=1}^{m}(1 - x_{ij}) \leq \prod_{i=1}^{m}\left(1 - \frac{x_j}{m}\right) = \left(1 - \frac{x_j}{m}\right)^m < e^{-x_j}.$$

The probability that job j appears in the final schedule is therefore not less than $1 - e^{-x_j} \geq (1 - 1/e)x_j$, where the inequality follows from the fact that the real function $1 - e^{-x} - (1 - 1/e)x$ is non-negative in the range $0 \leq x \leq 1$ (as can be seen easily by differentiation). Thus the expected revenue is at least $(1 - 1/e)\sum_j w(j)x_j$. Using standard techniques, we can derandomize our algorithm without decreasing the approximation factor. The next theorem follows from the discussion above.

Theorem 3. *The algorithm yields a $(1 - 1/e)$-approximate solution.*

Remark 2. A similar algorithm can be used to obtain a $(1 - 1/e)$-approximation for the more general problem where each job j has a release date r_j, a deadline d_j and a processing time p_j, and $d_j - r_j < 2p_j$ for all j.

Acknowledgment. We thank Frits Spieksma for pointing out reference [13].

References

1. E. M. ARKIN AND E. B. SILVERBERG, Scheduling jobs with fixed start and end times. *Discrete Applied Mathematics,* Vol. 18, pp. 1–8, 1987.

2. A. BAR-NOY, S. GUHA, J. NAOR, AND B. SCHIEDER, Approximating the throughput of multiple machines in real-time scheduling, SIAM Journal on Computing, Vol. 31 (2001), pp. 331–352.

3. R. BAR-YEHUDA, A. BAR-NOY, A. FREUND, J. NAOR, AND B. SCHIEBER, A unified approach To approximating resource allocation and scheduling, *Proc. 32nd Annual ACM Symposium on Theory of Computing,* pp. 735–744, 2000.

4. P. BERMAN AND B. DASGUPTA, Multi-phase algorithms for throughput maximization for real-time scheduling. *Journal of Combinatorial Optimization,* Vol. 4, pp. 307–323, 2000.

5. R. BAR-YEHUDA AND S. EVEN, A linear time approximation algorithm for the weighted vertex cover problem. *Journal of Algorithms,* Vol. 2, pp. 198–203, 1981.

6. G. CHELIOTIS, Bandwidth Trading in the Real World: Findings and Implications for Commodities Brokerage. *3rd Berlin Internet Economics Workshop,* 26–27 May 2000, Berlin.

7. S. CHIU AND J. P. CRAMETZ, Surprising pricing relationships. *Bandwidth Special Report, RISK. ENERGY & POWER RISK MANAGEMENT* pages 12–14, July 2000. (http://www.riskwaters.com/bandwidth)

8. J. CHUZHOY, R. OSTROVSKY, Y. RABANI, Approximation algorithms for the job interval selection problem and related scheduling problems *Proc. 42nd Annual Symposium of Foundations of Computer Science,* pp. 348–356, 2001.

9. M.R. GAREY, D.S. JOHNSON, G.L. MILLER AND C.H. PAPADIMITRIOU, The complexity of coloring circular arcs and chords. *SIAM Journal on Algebraic and Discrete Methods,* Vol. 1, pp. 216–227, 1980.

10. M. X. GOEMANS AND D. P. WILLIAMSON, A general approximation technique for constrained forest problems. *SIAM J. on Computing,* Vol. 24, pp. 296–317, 1995.

11. K. JANSEN An approximation algorithm for the license and shift class design problem *European Journal of Operational Research 73* pp. 127–131, 1994.

12. C. KENYON AND G. CHELIOTIS, Stochastic Models for Telecom Commoditiy Prices *Computer Networks* 36(5–6):533–555, Theme Issue on Network Economics, Elsevier Science, 2001.

13. A. W. J. KOLEN AND L. G. KROON, On the computational complexity of (maximum) class scheduling, *European Journal of Operational Research,* Vol. 54, pp. 23–38, 1991.

14. A. W. J. KOLEN AND L. G. KROON, On the computational complexity of (maximum) shift scheduling, *European Journal of Operational Research,* Vol. 64, pp. 138–151, 1993.

15. A. W. J. KOLEN AND L. G. KROON, An analysis of shift class design problems, *European Journal of Operational Research,* Vol. 79, pp. 417–430, 1994.

16. A. W. J. KOLEN AND J. K. LENSTRA, Combinatorics in operations research, *Handbook of Combinatorics,* Eds: R. L. Graham, M. Grotschel, and L. Lovasz, North Holland, 1995.

17. F. C. R. SPIEKSMA. On the approximability of an interval scheduling problem. *Journal of Scheduling,* Vol. 2 pp. 215–227, 1999.

18. V. V. VAZIRANI. Approximation Algorithms. Springer-Verlag, 2001.
19. http://www.zurich.ibm.com/bandwidth/concepts.html
20. P. WINKLER AND L. ZHANG, Wavelength Assignment and Generalized Interval Graph Coloring, *Fourteenth Annual ACM-SIAM Symposium on Discrete Algorithms (SODA)*, 2003.

CTL$^+$ Is Complete for Double Exponential Time

Jan Johannsen and Martin Lange

Institut für Informatik
Ludwig-Maximilians-Universität München
Munich, Germany
{jjohanns,mlange}@informatik.uni-muenchen.de

Abstract. We show that the satisfiability problem for CTL$^+$, the branching time logic that allows boolean combinations of path formulas inside a path quantifier but no nesting of them, is 2-EXPTIME-hard. The construction is inspired by Vardi and Stockmeyer's 2-EXPTIME-hardness proof of CTL*'s satisfiability problem. As a consequence, there is no subexponential reduction from CTL$^+$ to CTL which preserves satisfiability.

1 Introduction

In the early 80s, a family of branching time logics was defined by Emerson and Halpern [3,4]. This included the commonly known logics CTL and CTL* as well as the less known logic CTL$^+$.

CTL formulas can only speak about states of a transition system, while CTL* allows properties of paths and states to be expressed. CTL$^+$ is the fragment of CTL* which does not allow temporal operators to be nested. It subsumes CTL syntactically.

Emerson and Halpern [3] already showed that every CTL$^+$ formula is equivalent to a CTL formula. The translation, however, yields formulas of exponential length. Recently, Wilke [10] and Adler and Immerman [1] have shown that this is unavoidable, i.e. that there are CTL$^+$ formulas of size n such that every equivalent CTL formula is of size $\Omega(n!)$.

This gap becomes apparent for example when the complexity of the model checking problem for these logics is considered. For CTL the problem is PTIME-complete, even in linear time, while the CTL$^+$ model checking problem is Δ_2-complete in the polynomial time hierarchy [8].

Kupferman and Grumberg [7] have shown that one can relax the syntactic restrictions CTL imposes on branching time formulas without having to give up linear time model checking. They define a logic CTL^2, which allows two temporal operators in the scope of a path quantifier - either nested or a boolean combination thereof. Syntactically, CTL$^+$ and CTL^2 are incomparable although semantically CTL^2 strictly subsumes CTL and therefore CTL$^+$ as well. To the best of our knowledge, no complexity bounds on CTL^2's satisfiability problem are given.

In contrast, CTL* which is known to be strictly more expressive than CTL, CTL$^+$ and even CTL^2, has a PSPACE-complete model checking problem [6].

J.C.M. Baeten et al. (Eds.): ICALP 2003, LNCS 2719, pp. 767–775, 2003.

Concerning the satisfiability checking problem, CTL is EXPTIME-complete while CTL* is 2-EXPTIME-complete. Inclusion in 2-EXPTIME was proved by Emerson and Jutla [5] after it had been shown to be contained in various deterministic and nondeterministic complexity classes between 2-EXPTIME and 4-EXPTIME. 2-EXPTIME-hardness was shown by Vardi and Stockmeyer [9] using a reduction from the word problem for an alternating exponential space bounded Turing Machine.

We use the basic ideas of their construction in order to prove 2-EXPTIME-hardness of CTL^+'s satisfiability checking problem. For instance, we also encode the computation tree of an alternating exponential space bounded Turing Machine on an input word by a tree model for a CTL^+ formula that describes the machine's behaviour. However, in order to overcome CTL^+'s weaknesses in expressivity compared to CTL* we need to make amendments to the models and the resulting formulas. Note that CTL^+ is, for example, not able to speak about the penultimate state on a finite path which is a crucial point in Vardi and Stockmeyer's reduction.

To overcome this problem we use a special type of alternating Turing Machine which is easily seen to be equivalent to a common one in terms of space complexity. This Turing Machine has states of three different types: those in which the tape head is deterministically moved, as well as existentially and universally branching states in which the symbol under the tape head is replaced and no movement takes place.

For this sort of alternating Turing Machine it becomes possible to describe the machine's behaviour by a CTL^+ formula. The distinction of Turing Machine states does not require formulas that speak about more than two consecutive states on a path of a transition system.

There are other CTL* formulas in Vardi and Stockmeyer's paper which cannot easily be transformed into CTL^+ because of CTL^+'s restriction regarding the nesting of path operators. E.g. the natural way of expressing that some event E happens at most once along a path uses two nested *until* formulas ("it is not the case that E happens at some point and at another point later on"). Formulas of this kind occur in properties like "there is exactly one tape head per configuration". To make the reduction work for CTL^+ too, we use additional atomic propositions in a model for the resulting CTL^+ formula.

Completeness follows from the fact that the satisfiability checking problem for CTL* is in 2-EXPTIME, but also because CTL^+ can be translated into CTL at the cost of an exponential blow-up. This does not only – to the best of our knowledge – provide the first complexity-theoretical completeness result for the CTL^+ satisfiability problem. It also shows the curious fact that concerning expressiveness CTL and CTL^+ fall into the same class different from CTL*. Concerning the model checking problem the three logics were shown to be complete for three (probably) different classes. But regarding satisfiability, CTL^+ and CTL* are complete for the same class which is different from the complexity of CTL satisfiability.

Finally, we present a consequence of CTL^+'s 2-EXPTIME-hardness. Wilke was the first to prove an exponential lower bound on the size of CTL formulas that arise under an equivalence preserving translation from CTL^+ [10]. This was improved by Adler and Immerman, who showed that there is indeed an $n!$ lower bound [1]. The 2-EXPTIME-hardness of the CTL$^+$ satisfiability problem strengthens Wilke's result in a different way: there is no subexponential reduction from CTL$^+$ to CTL that preserves satisfiability.

2   Preliminaries

The logic CTL$^+$. Let $\mathcal{P}$ be a finite set of propositional constants including tt and ff. A labelled transition system is a triple $\mathcal{T} = (\mathcal{S}, \rightarrow, L)$ s.t. $(\mathcal{S}, \rightarrow)$ is a directed graph, and $L : \mathcal{S} \rightarrow 2^{\mathcal{P}}$ labels the elements of $\mathcal{S}$, called *states*, with tt $\in L(s)$, ff $\notin L(s)$ for all $s \in \mathcal{S}$. $\mathcal{T}$ is called *total* if for all $s \in \mathcal{S}$ there is an $s' \in \mathcal{S}$ s.t. $s \rightarrow s'$.

A path in a total transition system $\mathcal{T}$ is an infinite sequence $\pi = s_0 s_1 \ldots$ of states s.t. $s_i \rightarrow s_{i+1}$ for all $i \in \mathbb{N}$. With π^i we denote the suffix of π starting with the i-th state.

Formulas of CTL$^+$ are given by the following grammar.

$$\varphi \quad ::= \quad q \mid \varphi \vee \varphi \mid \neg\varphi \mid \mathsf{E}\psi$$
$$\psi \quad ::= \quad q \mid \psi \vee \psi \mid \neg\psi \mid \mathsf{X}\varphi \mid \varphi \mathsf{U}\varphi$$

where q ranges over $\mathcal{P}$. The φ are often called *state formulas* while the ψ are *path formulas*. Only state formulas are CTL$^+$ formulas. Path formulas can only occur as subformulas of these.

We will use the standard abbreviations $\varphi \wedge \psi := \neg(\neg\varphi \vee \neg\psi)$, $\varphi \rightarrow \psi := \neg\varphi \vee \psi$, $\mathsf{A}\varphi := \neg\mathsf{E}\neg\varphi$, $\mathsf{F}\varphi := \mathsf{tt}\mathsf{U}\varphi$ and $\mathsf{G}\varphi := \neg\mathsf{F}\neg\varphi$. Furthermore, we will use a special until formula $\mathsf{F}_\psi\varphi := \neg\psi\mathsf{U}(\psi \wedge \varphi)$ which says that eventually φ holds in the first moment when ψ holds, too.

Formulas of CTL$^+$ are interpreted over paths $\pi = s_0 s_1 \ldots$ of a total transition system $\mathcal{T} = (\mathcal{S}, \rightarrow, L)$.

$$
\begin{aligned}
\pi &\models q &&\text{iff } q \in L(s_0) \\
\pi &\models \varphi \vee \psi &&\text{iff } \pi \models \varphi \text{ or } \pi \models \psi \\
\pi &\models \neg\varphi &&\text{iff } \pi \not\models \varphi \\
\pi &\models \mathsf{E}\varphi &&\text{iff } \exists\pi', \text{ s.t. } \pi' = s_0 \ldots \text{ and } \pi' \models \varphi \\
\pi &\models \mathsf{X}\varphi &&\text{iff } \pi^1 \models \varphi \\
\pi &\models \varphi\mathsf{U}\psi &&\text{iff } \exists k \in \mathbb{N} \text{ s.t. } \pi^k \models \psi \text{ and } \forall i < k : \pi^i \models \varphi
\end{aligned}
$$

Since the truth value of a state formula φ in a path $\pi = s_0 s_1 \ldots$ only depends on s_0, it is possible to write $s \models \varphi$ for a state s of a transition system and such a formula φ. A state formula φ is called *satisfiable* if there is a transition system $\mathcal{T}$ with a state s, s.t. $s \models \varphi$.

Alternating Turing Machines. We use the following model of alternating Turing Machine, which differs slightly from the standard model [2], but is easily seen to be equivalent w.r.t. space complexity. An alternating Turing Machine $\mathcal{M}$ is of the form $\mathcal{M} = (Q, \Sigma, q_0, q_a, q_r, \delta)$, where Q is the set of states, Σ is the alphabet, which contains a blank symbol $\square \in \Sigma$, and $q_0, q_a, q_r \in Q$.

The set Q of states is partitioned into $Q = Q_\exists \cup Q_\forall \cup Q_m \cup \{q_a, q_r\}$, where we write Q_b for $Q_\exists \cup Q_\forall$, these are the *branching* states. The transition relation δ is of the form

$$\delta \subseteq \left(Q_b \times \Sigma \times Q \times \Sigma\right) \cup \left(Q_m \times \Sigma \times Q \times \{L, R\}\right).$$

In a branching state $q \in Q_b$, the machine can act nondeterministically and writes on the tape, i.e., for each $a \in \Sigma$, there can be several transitions $(q, a, q', b) \in \delta$ for $q' \in Q$ and $b \in \Sigma$, meaning that the machine overwrites the a in the current tape cell with b, the machine enters state q', and the head does not move.

In a state $q \in Q_m$, the machine acts deterministically and moves its head, i.e., for each $a \in \Sigma$, there is exactly one transition $(q, a, q', D) \in \delta$, for $q' \in Q$ and $D \in \{L, R\}$, meaning that the head moves to the left (L) or right (R), and the machine enters state q'. For $q \in \{q_a, q_r\}$, there are no transitions in δ, and the machine halts.

We assume that the machine only halts when the state is q_a or q_r. A halting configuration is accepting iff the state is q_a. For the other configurations, the acceptance behaviour depends on the kind of state:

If the state is in Q_m, then the configuration is accepting iff its unique successor is accepting. If the state is in $Q_\exists$, then the configuration is accepting iff at least one of its successors is accepting. If the state is in $Q_\forall$, then the configuration is accepting iff all of its successors are accepting. The whole computation accepts if the initial configuration is accepting.

Double exponential time. The complexity class of double exponential time is defined as

$$\text{2-EXPTIME} = \bigcup_{k \in \mathbb{N}} \text{DTIME}(2^{2^{k \cdot n}})$$

where $\text{DTIME}(f(n))$ is the class of all languages which are accepted by a deterministic Turing Machine in time $f(n)$ where n is the length of the input word at hand.

It is well-known [2] that 2-EXPTIME coincides with

$$\text{AEXPSPACE} = \bigcup_{k \in \mathbb{N}} \text{ASPACE}(2^{k \cdot n})$$

the class of all languages accepted by an alternating Turing Machine using space which is at most exponential in the size of the input word.

3   The Reduction

Theorem 1. *Satisfiability of CTL$^+$ is 2-EXPTIME-hard.*

Proof. Suppose $\mathcal{M} = (Q, \Sigma, q_0, q_a, q_r, \delta)$ is an alternating exponential space bounded Turing Machine. Let $w = a_0 \ldots a_{n-1} \in \Sigma^*$ be an input for $\mathcal{M}$. W.l.o.g. we assume the space needed by $\mathcal{M}$ on input w to be bounded by $2^{kn} - 1$ for some $k \geq 1$. Let $N := 2^{kn} - 1$. Furthermore we assume that every computation ends in a configuration with the head on the rightmost tape cell while the machine is in either of he states q_a or q_r.

In the following we will construct a CTL$^+$ formula $\varphi_{M,w}$ s.t. $w \in L(M)$ iff $\varphi_{M,w}$ is satisfiable. Informally, an accepting computation of M on w will serve as a model for $\varphi_{M,w}$.

Like Vardi and Stockmeyer [9], we encode a configuration of M as a sequence of $2^{k \cdot n} - 1$ states in a possible model for $\varphi_{M,w}$. Successive configurations of the Turing Machine are modelled by concatenating these sequences, where we add one dummy state with index 0 between each pair of adjacent configurations.

The underlying set of propositions is $\mathcal{P} = Q \cup \Sigma \cup \{c_0, \dots, c_{k \cdot n-1}\} \cup \{x, z, e\}$.

- $q \in Q$ is true in a state of the model iff the head of the Turing Machine is on the corresponding tape cell in the corresponding configuration while the machine is in state q. The formula $h := \bigvee_{q \in Q} q$ says that the machine is in some state, i.e. the head is on that cell.
- $a \in \Sigma$ is true iff a is the symbol on the corresponding tape cell.
- $c_{k \cdot n-1}, \dots, c_0$ represent a counter in binary representation. The counter value in a state of the model is 0 at the dummy states and the number of the corresponding tape cell otherwise.
- x is used to denote that the corresponding configuration is accepting.
- z is used to mark the part of a tree model which corresponds to the computation. In order to be able to speak about a certain state somewhere on a path we let every state of the encoding have a successor which carries exatly the same amount of information except that it is labelled with $\neg z$. Thus, such a state can be seen as not belonging directly to the encoding of the computation tree but being a clone of a state in this tree.
- e indicates that the state at hand belongs to an "even" configuration, i.e. one with an even index in a sequence $C_0, C_1, \dots$ of configurations of the computation.

For every fixed m we can write a formula χ_m which says that the counter value is m in the current state, e.g.

$$\chi_0 := \bigwedge_{i=0}^{k \cdot n-1} \neg c_i \ , \quad \chi_1 := c_0 \wedge \bigwedge_{i=1}^{k \cdot n-1} \neg c_i \ \text{and} \ \chi_N := \bigwedge_{i=0}^{k \cdot n-1} c_i$$

for the dummy ($m = 0$), the leftmost ($m = 1$) and rightmost ($m = N$) position in a configuration.

In order to describe $M's$ behaviour on w we need to express several properties. The formula φ_0 says that there is always exactly one symbol on a tape cell, and M is never in two different states at the same time.

$$\varphi_0 := \text{AG}(\ (\neg \chi_0 \rightarrow \bigvee_{a \in \Sigma} a) \wedge (\chi_0 \rightarrow \neg h \wedge \bigwedge_{a \in \Sigma} \neg a) \wedge$$
$$\bigwedge_{a,b \in \Sigma, b \neq a} \neg(a \wedge b) \wedge \bigwedge_{q,q' \in Q, q \neq q'} \neg(q \wedge q') \ )$$

We can say that the counter value is not changed in the transition to the next state on a given path. This is used to clone states as indicated above. The value of e does not change in this case.

$$\psi_{rem} := (e \leftrightarrow \mathsf{X}e) \wedge \bigwedge_{j=0}^{k \cdot n - 1} (c_j \leftrightarrow \mathsf{X}c_j)$$

We can also say that the counter value is increased by 1 modulo $2^{k \cdot n}$. Then, a switch from e to $\neg e$ or vice versa occurs iff the counter is increased from $2^{k \cdot n} - 1$ to 0.

$$\psi_{inc} := ((e \leftrightarrow \mathsf{X}\neg e) \wedge \chi_N \wedge \mathsf{X}\chi_0) \vee$$
$$(e \leftrightarrow \mathsf{X}e) \wedge \bigvee_{j=0}^{k \cdot n - 1} (\neg c_j \wedge \mathsf{X}c_j \wedge \bigwedge_{i>j}(c_i \leftrightarrow \mathsf{X}c_i) \wedge \bigwedge_{i<j}(c_i \wedge \mathsf{X}\neg c_i))$$

The entire computation of $\mathcal{M}$ forms a tree. Each state is labelled with a symbol of Σ. Moreover, z holds on every state on the computation, and every state has at least one successor from which on z never holds. Furthermore, the subtree under this state reflects the labelling of its root's predecessor which still satisfies z. This idea is taken from Vardi and Stockmeyer's proof [9] and used to be able to speak about finite prefixes of infinite paths.

On all paths q_a or q_r is eventually reached and all following states do not satisfy z. The counter is only increased (modulo $2^{k \cdot n}$) in states satisfying z.

$$\psi_{eq} := \psi_{rem} \wedge \bigwedge_{q \in Q} q \leftrightarrow \mathsf{X}q \wedge \bigwedge_{a \in \Sigma} a \leftrightarrow \mathsf{X}a$$
$$\varphi_1 := \mathsf{AF}\neg z \wedge$$
$$\mathsf{AG}((z \wedge \neg q_a \wedge \neg q_r) \rightarrow (\mathsf{EX}z \wedge \mathsf{EX}\neg z) \wedge$$
$$\neg z \rightarrow \mathsf{A}(\mathsf{X}\neg z \wedge \psi_{eq}) \wedge$$
$$(q_a \vee q_r) \rightarrow \mathsf{AX}\neg z \wedge \chi_N) \wedge$$
$$\mathsf{AGA}((z \wedge \mathsf{X}z \leftrightarrow \psi_{inc}) \wedge (z \wedge \mathsf{X}\neg z \leftrightarrow \psi_{eq}))$$

There is at most one tape head in every configuration. (The fact that there is at least one will be guaranteed by φ_5 later on.) This is achieved by saying that there is no bit c_i which distinguishes two possible occurrences of an h in one configuration. To guarantee that one speaks about the same configuration for two such occurrences of h, we demand that the value of e never changes in between.

$$\varphi_2 := \mathsf{AGA}(\chi_0 \rightarrow (e \rightarrow \neg(\bigvee_{i=0}^{k \cdot n - 1} e\mathsf{U}(e \wedge h \wedge c_i) \wedge e\mathsf{U}(e \wedge h \wedge \neg c_i)) \wedge$$
$$\neg e \rightarrow \neg(\bigvee_{i=0}^{k \cdot n - 1} \neg e\mathsf{U}(\neg e \wedge h \wedge c_i) \wedge \neg e\mathsf{U}(\neg e \wedge h \wedge \neg c_i))))$$

The computation is accepting. Every q_a is marked with an x but no q_r is. Moreover, an x occurs together with an existential state only if there is a path along

z s.t. x holds together with the first occurrence of h. For universal or moving states all z-paths must satisfy x in their first occurrence of h.

$$\varphi_3 \; := \; x \wedge \text{AG}((q_a \to x) \wedge (q_r \to \neg x) \wedge$$
$$\bigwedge_{q \in Q_\exists} q \; \to \; (x \leftrightarrow \text{EXE}((z \wedge \neg h)\text{U}(z \wedge h \wedge x))) \wedge$$
$$\bigwedge_{q \in Q_\forall \cup Q_m} q \; \to \; (x \leftrightarrow \text{AXA}(z\text{U}(z \wedge h) \to F_h x)))$$

At the beginning, the tape contains $a_1 \ldots a_n \square \ldots \square$, the input word followed by $2^{k \cdot n} - n$ blank symbols. $\mathcal{M}$ is in state q_0 and the head is on the first symbol of w.

$$\varphi_4 \; := \; z \wedge e \wedge \chi_0 \wedge$$
$$\text{EX}(z \wedge q_0 \wedge a_1 \wedge$$
$$\text{EX}(z \wedge a_2 \wedge$$
$$\ldots \wedge$$
$$\text{EX}(z \wedge a_n \wedge$$
$$\text{EXE}((z \wedge \square)\text{U}(z \wedge \chi_0))) \ldots))$$

Now we have to say that two adjacent configurations comply with $\mathcal{M}$'s transition rules. In order to do so we need the following statements about a path. The counter value is 0 exactly once before $\neg z$ holds.

$$\psi_1 \; := \; e \to z\text{U}(z \wedge \neg e \wedge \chi_0) \wedge \neg e \to z\text{U}(z \wedge e \wedge \chi_0) \wedge$$
$$\neg (z\text{U}(e \wedge \chi_0) \wedge z\text{U}(\neg e \wedge \chi_0))$$

We need three formulas saying that the counter value in the first state not satisfying z is the same as the value of the first state on the path, resp. increased or decreased by 1. We explicitly forbid to increase a maximal value, resp. decrease a minimal one, i.e. do not calculate modulo $2^{k \cdot n}$, because these formulas are used to describe the tape head's moves. Note that it cannot go left at the right end of the tape and vice-versa.

$$\psi_= \; := \; \bigwedge_{i=0}^{k \cdot n - 1} c_i \leftrightarrow F_{\neg z} c_i$$

$$\psi_{+1} := \neg\chi_N \wedge \bigvee_{j=0}^{k \cdot n - 1} (\neg c_j \wedge F_{\neg z} c_j) \wedge \bigwedge_{i>j}(c_i \leftrightarrow F_{\neg z} c_i) \wedge \bigwedge_{i<j}(c_i \wedge F_{\neg z}\neg c_i)$$

$$\psi_{-1} := \neg\chi_1 \wedge \bigvee_{j=0}^{k \cdot n - 1} (c_j \wedge F_{\neg z}\neg c_j) \wedge \bigwedge_{i>j}(c_i \leftrightarrow F_{\neg z} c_i) \wedge \bigwedge_{i<j}(\neg c_i \wedge F_{\neg z} c_i)$$

Finally, we have to describe the machine's transition behaviour δ. On every state the following holds.

- If it is labelled with a $q \in Q_b$ then the actual symbol is replaced in every next configuration at the same position.

- If it is not labelled with a $q \in Q_b$, in particular no q at all, then the corresponding state of the next configuration carries the same symbol from Σ.
- If it is labelled with a $q \in Q_m$ then every next or previous state to the corresponding one in the next configuration is labelled with the machine state that is given by the transition relation.

Note that the second and third case do not exclude each other.

$$\varphi_5 := \mathtt{AG}\Big(\bigwedge_{q\in Q_b, a\in \Sigma}(q \wedge a \rightarrow \bigwedge_{(q,a,q',b)\in\delta} \mathtt{E}(\psi_1 \wedge \psi_= \wedge \mathtt{F}_{\neg z}(q' \wedge b)\,)\,) \wedge$$

$$\mathtt{A}(\psi_1 \wedge \psi_= \rightarrow \mathtt{F}_{\neg z}\bigvee_{(q,a,q',b)\in\delta}(q' \wedge b)\,)\,)$$

$$\wedge \bigwedge_{a\in\Sigma}\neg(\bigvee_{q\in Q_b}q)\wedge a \rightarrow \mathtt{A}(\psi_1 \wedge \psi_= \rightarrow \mathtt{F}_{\neg z}a\,)$$

$$\wedge \bigwedge_{(q,a,q',L)\in\delta}q\wedge a \rightarrow \mathtt{A}(\psi_1 \wedge \psi_{-1}\rightarrow \mathtt{F}_{\neg z}q'\,)$$

$$\wedge \bigwedge_{(q,a,q',R)\in\delta}q\wedge a \rightarrow \mathtt{A}(\psi_1 \wedge \psi_{+1}\rightarrow \mathtt{F}_{\neg z}q'\,)\,)$$

Altogether, the machine's behaviour is described by the formula

$$\varphi_{\mathcal{M},w} := \varphi_0 \wedge \varphi_1 \wedge \varphi_2 \wedge \varphi_3 \wedge \varphi_4 \wedge \varphi_5$$

Then, the part of a model for $\varphi_{\mathcal{M},w}$ that is marked with z corresponds to a successful computation tree of $\mathcal{M}$ on w. Conversely, such a tree can easily be extended to a model for $\varphi_{\mathcal{M},w}$.

Thus, $\mathcal{M}$ accepts w iff there exists a successful computation tree for $\mathcal{M}$ on w iff there exists a model for $\varphi_{\mathcal{M},w}$ iff $\varphi_{\mathcal{M},w}$ is satisfiable.

Finally, the size of $\varphi_{\mathcal{M},w}$ is quadratic in $|\Sigma|$ and $|Q|$ and linear in $|w|$ and $|\delta|$. □

Corollary 1. *There is no reduction* $r : CTL^+ \rightarrow CTL$ *s.t. for all* $\varphi \in CTL^+$:

- φ *is satisfiable iff* $r(\varphi)$ *is satisfiable, and*
- $|r(\varphi)| \leq f(|\varphi|)$ *for some* $f : \mathbb{N} \rightarrow \mathbb{N}$ *with* $f(n^2) = o(2^n)$.

Proof. Suppose there is a reduction from CTL^+ to CTL that preserves satisfiability and produces formulas of subexponential length $f(n)$. Then this reduction in conjunction with a satisfiability checker for CTL can be used to decide satisfiability of CTL^+ in asymptotically less time than $O(2^{f(n)})$. As a consequence of Theorem 1, every language in 2-EXPTIME can be decided in time $O(2^{f(n^2)})$ since it can be reduced to CTL^+ in quadratic time, and satisfiability for CTL can be decided in time $O(2^n)$. But according to the asymptotic restriction on f and the Time Hierarchy Theorem, there is a language in 2-EXPTIME which is not decidable in time $O(2^{f(n^2)})$. To see this note that

$$f(n^2) = o(2^n) \text{ iff } f(n^2) + \log f(n^2) = o(2^n) \text{ iff } 2^{f(n^2)} \cdot f(n^2) = o(2^{2^n})$$

□

References

1. M. Adler and N. Immerman. An $n!$ lower bound on formula size. In *Proc. 16th Symp. on Logic in Computer Science, LICS'01*, pages 197–208, Boston, MA, USA, June 2001. IEEE Computer Society.
2. A. K. Chandra, D. C. Kozen, and L. J. Stockmeyer. Alternation. *Journal of the ACM*, 28(1):114–133, January 1981.
3. E. A. Emerson and J. Y. Halpern. Decision procedures and expressiveness in the temporal logic of branching time. *Journal of Computer and System Sciences*, 30:1–24, 1985.
4. E. A. Emerson and J. Y. Halpern. "Sometimes" and "not never" revisited: On branching versus linear time temporal logic. *Journal of the ACM*, 33(1):151–178, January 1986.
5. E. A. Emerson and C. S. Jutla. The complexity of tree automata and logics of programs. *SIAM Journal on Computing*, 29(1):132–158, February 2000.
6. E. A. Emerson and C.-L. Lei. Modalities for model checking: Branching time logic strikes back. *Science of Computer Programming*, 8(3):275–306, 1987.
7. O. Kupferman and O. Grumberg. Buy one, get one free!!! *Journal of Logic and Computation*, 6(4):523–539, August 1996.
8. F. Laroussinie, N. Markey, and P. Schnoebelen. Model checking CTL^+ and $FCTL$ is hard. In *Proc. 4th Conf. Foundations of Software Science and Computation Structures, FOSSACS'01*, volume 2030 of *LNCS*, pages 318–331, Genova, Italy, April 2001. Springer.
9. M. Y. Vardi and L. Stockmeyer. Improved upper and lower bounds for modal logics of programs. In *Proc. 17th Symp. on Theory of Computing, STOC'85*, pages 240–251, Baltimore, USA, May 1985. ACM.
10. T. Wilke. CTL$^+$ is exponentially more succinct than CTL. In *Proc. 19th Conf. on Foundations of Software Technology and Theoretical Computer Science, FSTTCS'99*, volume 1738 of *LNCS*, pages 110–121. Springer, 1999.

Hierarchical and Recursive State Machines with Context-Dependent Properties*

Salvatore La Torre, Margherita Napoli, Mimmo Parente, and
Gennaro Parlato

Dipartimento di Informatica e Applicazioni
Università degli Studi di Salerno

Abstract. Hierarchical and recursive state machines are suitable abstract models for many software systems. In this paper we extend a model recently introduced in literature, by allowing atomic propositions to label all the kinds of vertices and not only basic nodes. We call the obtained models *context-dependent hierarchical/recursive state machines*. We study on such models cycle detection, reachability and LTL model-checking. Despite of a more succinct representation, we prove that LTL model-checking can be done in time linear in the size of the model and exponential in the size of the formula, as for standard LTL model-checking. Reachability and cycle detection become NP-complete, and if we place some restrictions on the representation of the target states, we can decide them in time linear in the size of the formula and the size of the model.

Keywords: Model Checking, Automata, Temporal Logic.

1 Introduction

Due to their complexity, the verification of the *correctness* of many modern digital systems is infeasible without suitable automated techniques. Formal verification has been very successful and recent results have led to the implementation of powerful design tools (see [CK96]). In this area one of the most successful techniques has been *model checking* [CE81]: a high-level specification is expressed by a formula of a logic and this is checked for fulfillment on an abstract model (state machine) of the system. Though model checking is linear in the size of the model, it is computationally hard since the model generally grows exponentially with the number of variables used to describe a state of the system (*state-space explosion*). As a consequence, an important part of the research on model checking has been concerned with handling this problem.

Complex systems are usually composed of relatively simple modules in a hierarchical manner. Hierarchical structures are also typical of object-oriented

---

* This research was partially supported by the MIUR in the framework of the project "Metodi Formali per la Sicurezza e il Tempo" (MEFISTO) and MIUR grant 60% 2002.

J.C.M. Baeten et al. (Eds.): ICALP 2003, LNCS 2719, pp. 776–789, 2003.

paradigms [BJR97,RBP+91,SGW94]. We consider systems modeled as *hierarchical finite state machines*, that is, finite state machines where a vertex can either expand to another hierarchical state machine or be a basic vertex (in the former case we call the vertex a *supernode*, in the latter simply a *node*).

The model we consider in this paper generalizes instead the model studied in [AY01]. There the authors consider the model checking on Hierarchical State Machines (HSM) where only the nodes are labeled with atomic propositions (AP). We relax this constraint and thus we allow to associate atomic propositions also with vertices that expand to a machine. Expanding a supernode v to a machine M, all vertices of M *inherit* the atomic propositions of v (*context*), so that different vertices expanding to M can place M into different contexts. For this reason, we call such a model a *hierarchical state machine with context-dependent properties* (in the following denoted by **Context-dependent Hierarchical State Machine**). The semantics of a **CHSM** is given by the corresponding natural flat model which is a Kripke structure.

By allowing this more general labeling, for a given system it is possible to obtain very succinct abstract models. In the following example, we show that the gain of succinctness can be exponential compared to the models used in [AY01]. Consider a digital clock with hours, minutes, and seconds. We can construct a hierarchical finite state machine $\mathcal{M}$ composed of three machines M_1, M_2, and M_3 such that the supernodes of M_3 expands to M_2 and the supernodes of M_2 expands to M_1. Machine M_1 is a chain of nodes. Machines M_2 and M_3 are chains of supernodes except for the initial and the output vertices that are nodes. In M_3 each supernode corresponds to a hour and they are linked accordingly to increasing time. Analogously, M_2 models minutes and M_1 seconds. A flat model for the digital clock has at least $24 \cdot 60 \cdot 60 = 86,400$ vertices, while the above hierarchical model has only $24 + 60 + 60 + 6 = 150$ vertices (6 are simply initial and output nodes). Assume that we are interested in checking properties that refer to a precise time expressed in hours, minutes and seconds. Clearly, it is not sufficient to label only the nodes (we would be able to capture only that an event happens at a certain second, but we would have no clue of the actual hour and minute). In the model defined in [AY01], at least $86,400$ nodes are needed, that is, there would be no gain with respect to a minimal flat model. In our model, we are able to label each supernode in M_3 with atomic propositions encoding the corresponding hour. Analogously we can use atomic propositions to encode minutes and seconds on M_2 and M_1, respectively. This way, each state of the corresponding flat model is labeled with the encoding of a hour, a minute and a second in a day and vertices are linked by increasing time.

A simple way of analyzing hierarchical systems is first to flatten them into equivalent non-hierarchical systems and then apply existing verification techniques on finite state systems. The drawback of such an approach is that the size of the flat system can be exponential in the hierarchical depth. In many recent papers, it has been shown that it is possible to reduce the complexity growth caused by handling large systems, by performing verification in a hierarchical manner [AGM00,AG00,BLA+99,AY01]. We follow this approach and study on

CHSMs standard decision problems which are related to system verification, such as reachability, cycle detection, and model checking. In this paper, we also consider **Context-dependent Recursive State Machines (CRSM)** which generalize **CHSMs** by allowing recursive expansions and we study on them the verification-related problems listed above. Recursive generalizations of the hierarchical model presented in [AY01] are studied in [AEY01,BGR01]. Recursive machines can be used to model the control flow of programs with recursive calls and thus are suitable for abstracting the behavior of reactive software systems.

Results. Given a transition system, a state s and a set of *target states* T, (usually expressed by a propositional boolean formula), the *reachability problem* is the problem of determining whether a state of T can be reached from s on a run of the system. In practice, this problem is relevant in the verification of systems, for example it is related to the verification of *safety requirements*: we want to check whether all the reachable states of the system belong to a given "safe" region (*invariant checking problem*).

We prove that reachability on **CRSMs** is NP-complete, and NP-hardness still holds if we restrict to **CHSMs**. We then give an algorithm to decide reachability on **CRSMs** that runs in time linear in the size of the model and exponential in the size of the formula. Finally, given a **CHSM** $\mathcal{M}$, we show effective sufficient conditions for solving reachability in time linear in both the size of the formula and the size of the model. Let us remark that these conditions are satisfied when we consider an instance of the reachability problem where the model is given by a Hierarchical State Machine (HSM) as defined in [AY01].

The *cycle detection problem* is the problem of verifying whether a given state can be reached repeatedly. Cycle detection is the basic problem for the verification of *liveness properties*: "some good thing will eventually happen".

We also consider the model checking of LTL formulas on **CRSMs**. Given a set of atomic propositions AP, a *linear temporal logic* (LTL) formula is built up in the usual way from atomic propositions, the boolean connectives, the temporal operators *next* and *until*. An LTL formula is interpreted over an infinite sequence over 2^{AP}. A **CRSM** satisfies a formula φ if every run in the corresponding flat model satisfies φ. Given an LTL formula φ and a **CRSM** $\mathcal{M}$, the *model checking problem* for $\mathcal{M}$ and φ is the problem to determine whether $\mathcal{M}$ satisfies φ. We give a decision algorithm that runs in $O(|\mathcal{M}| \cdot 8^{|\varphi|})$ time for **CHSMs** and an algorithm in $O(|\mathcal{M}| \cdot 16^{|\varphi|})$ time for **CRSMs**. Our algorithms do not need to flatten the system and mainly consist of reducing the model checking problem to the emptiness problem of recursive Büchi automata [AEY01].

The rest of the paper is organized as follows. In the next section definitions and notation are given. The NP-completeness of the cycle detection and of the reachability problems is shown in section 3 (actually the proofs for the cycle detection problems are omitted in this version, due to the lack of space). In section 4 we give the linear time algorithms for **CHSMs** and **CRSMs**. In Section 5, we discuss the model checking of LTL formulas. We conclude with few remarks in Section 6.

2   Context-Dependent State Machines

In this section we introduce the definitions and the notation we will use in the rest of the paper. We consider *Kripke structures*, that is, state transition graphs where each state is labeled by a subset of a finite set of atomic propositions (AP). A *Context-dependent Recursive State Machine* (**CRSM**) over AP is a tuple $\mathcal{M} = (M_1, \ldots, M_k)$ of Kripke structures with:

- a set of vertices N, split into disjoint sets $N_1, \ldots, N_k$; a set $IN = \{in_1, \ldots, in_k\}$ of initial vertices, where $in_i \in N_i$, and a set of output vertices OUT split into $OUT_1, \ldots, OUT_k$, with $OUT_i \subseteq N_i$;
- a mapping $expand : N \longrightarrow \{0, 1, \ldots, k\}$ such that $expand(u) = 0$, for each $u \in IN \cup OUT$. We define the closure of $expand$, $expand^+ : N \longrightarrow 2^{\{0,1,\ldots,k\}}$, as: $h \in expand^+(u)$ if either $h = expand(u)$ or $u' \in N_{expand(u)}$ exists such that $h \in expand^+(u')$.
- the sets of edges E_i, for $1 \leq i \leq k$, such that each edge in E_i is either a pair (u, v), with $u, v \in N_i$ and $expand(u) = 0$, or a triple $((u, z), v)$ with $z \in OUT_{expand(u)}$, and $u, v \in N_i$;
- a mapping $true : N \longrightarrow 2^{AP}$, such that $true(u) \cap true(v) = \emptyset$, for $v \in N_h, u \notin N_h$ and $h \in expand^+(u)$.

Informally, a **CRSM** is a collection of graphs which can call each other recursively. Each graph has an initial vertex and some output vertices. The mapping $expand$ gives the recursive-call structure. If $expand(u) = j > 0$, then the vertex u expands to the graph M_j and u is called a *supernode*; when $expand(u) = 0$ the vertex u is called a *node*. The mapping $true$ labels each vertex with a set of atomic propositions holding at that vertex. The *starting* node of a **CRSM** $\mathcal{M} = (M_1, \ldots, M_k)$ is the initial node in_k of M_k.

The Semantics of CRSMs. Every **CRSM** $\mathcal{M}$ corresponds to a flat model $\mathcal{M}^F$ which is a directed graph with (possibly infinite) vertices (*states*) labeled with atomic propositions. Informally speaking, the flat machine $\mathcal{M}^F$ is obtained starting from M_k and iteratively replacing every supernode u in it with the graph $M_{expand(u)}$. The flat machine $\mathcal{M}^F$ is defined as follows. A state of $\mathcal{M}^F$ is a tuple $X = [u_1, \ldots, u_m]$ where $u_1 \in N_k$, $u_{j+1} \in N_{expand(u_j)}$ for $j = 1, \ldots, m-1$, and $expand(u_m) = 0$. State X is labeled by a set of atomic proposition $true(X)$, consisting of the union of $true(u_j)$, for $j = 1, \ldots, m$. State $[in_k]$ is the initial state of $\mathcal{M}^F$. The set of transitions E is defined as follows. Let $X = [u_1, \ldots, u_m]$ be a state with $u_m \subset N_h$ and $u_{m-1} \in N_j$. Then, $(X, X') \in E$ provided that one of the following cases holds:

1. $(u_m, u') \in E_h$, $u' \in N_h$, and if $expand(u') = 0$ then $X' = [u_1, \ldots, u_{m-1}, u']$, otherwise $X' = [u_1, \ldots, u_{m-1}, u', in_l]$ for $l = expand(u')$.
2. $u_m \in OUT_h$, $((u_{m-1}, u_m), u') \in E_j$, $u' \in N_j$, and if $expand(u') = 0$ then $X' = [u_1, \ldots, u_{m-2}, u']$, otherwise $X' = [u_1, \ldots, u_{m-2}, u', in_l]$ for $l = expand(u')$.

Let $[u_1, \ldots, u_n]$ be a state of $\mathcal{M}^F$, a *prefix* of $[u_1, \ldots, u_n]$ is $u_1, \ldots, u_i$ for $i \leq n$.

A *Context-dependent Hierarchic State Machine* (**CHSM**) is a **CRSM** such that $expand(u) < i$, for every $u \in N_i$. A **CHSM** is a collection of graphs which are organized to form a hierarchy and $expand$ gives the hierarchical structure. The graph M_k is clearly the *top-level* graph of the hierarchy, i.e., no vertices expand to it and, as for **CRSM**s, its initial node in_k is the starting node of the **CHSM**.

3   Reachability and Cycle Detection Problems: Computational Complexity

In this section we discuss the computational complexity of the reachability and cycle detection problems for **CRSM**s and **CHSM**s. Given a **CRSM** $\mathcal{M} = (M_1, \ldots, M_k)$ and a propositional boolean formula φ, the *reachability problem* is the problem of deciding if a path in $\mathcal{M}^F$ exists from $[in_k]$ to a state X on which φ is satisfied. Analogously, the *cycle detection problem* is the problem of deciding if a cycle in $\mathcal{M}^F$ exists containing a reachable state X on which φ is satisfied.

We prove that for **CRSM**s and **CHSM**s these decision problems are NP-complete by showing NP-hardness for **CHSM**s and giving nondeterministic polynomial-time algorithms for **CRSM**s.

Lemma 1. *Reachability and cycle detection for* **CHSM**s *are NP-hard.*

Proof   We give a reduction in linear time with respect to the size of the formula from the satisfiability problem SAT. Given a boolean formula φ over the variables $x_1, \ldots, x_m$, we construct a **CHSM** $\mathcal{M} = (M_1, M_2, \ldots, M_m)$ over $AP = \{P_1, P_2, \ldots, P_m\}$, as follows. Each graph M_i has four vertices $in_i, p_i, notp_i, out_i$ forming a chain. Each vertex p_i is labeled by $\{P_i\}$ whereas the vertices $notp_i$, in_i and out_i are labeled by the empty set. Since an atomic proposition P_i does not label vertices in graphs other than M_i, this labeling implicitly corresponds to assigning $\neg P_i$ to $notp_i$. Vertices p_i and $notp_i$, for $i > 1$, are supernodes which expand into M_{i-1}, and p_1 and $notp_1$ are instead nodes.

Thus there are 2^m states of $\mathcal{M}^F$ of type $[u_1, \ldots, u_m]$ such that $u_{m-i+1} \in \{p_i, notp_i\}$ for $i = 1, \ldots, m$, and it is easy to verify that all these states are reachable from $[in_m]$. Clearly, given a truth assignment ν of $x_1, \ldots, x_m$, a state X of $\mathcal{M}^F$ exists such that ν assigns TRUE to x_i if and only if p_i occurs in X and, in turns, if and only if $P_i \in true(X)$. Thus a reachable state X of $\mathcal{M}^F$ exists whose labeling corresponds to a truth assignment fulfilling φ if and only if φ is satisfiable.

By definition of the cycle detection problem, checking for the existence of a cycle containing a state on which φ is satisfied requires to check for reachability first. Thus, NP-hardness is inherited from reachability.   □

To prove membership to NP of the reachability on **CRSM**s, we need to consider a notion of connectivity of vertices in a **CRSM**. We say that a vertex $u \in N$ is *connected* if a reachable state $[u_1, \ldots, u_m]$ of $\mathcal{M}^F$ exists, where $u = u_i$ for some $i = 1, \ldots, m$. Observe that the starting node in_k is clearly connected and a vertex $u \in N_j$ is connected if and only if in_j is connected and a path π in M_j from in_j to u exists, such that if π goes through an edge $((v, z), v') \in E_j$ then z is a connected vertex (recall that $z \in \mathrm{OUT}_{expand(v)}$). From this the following proposition holds.

Proposition 1. *A state* $[u_1, \ldots, u_m]$ *of* $\mathcal{M}^F$ *is reachable if and only if all the vertices* u_i, *for* $i = 1, \ldots, m$, *are connected.*

The above observation suggests also an algorithm to determine in linear time the connected vertices. We omit the proof of this result which is given by a rather simple modification of a depth-first search on a graph (see also [AEY01]).

Proposition 2. *Given a* **CRSM** $\mathcal{M}$, *the set of connected vertices of* $\mathcal{M}$ *can be determined in* $O(|\mathcal{M}|)$.

To prove membership to NP of the reachability on **CRSMs**, we need to prove the following technical lemma. Notice that this lemma is not needed for **CHSMs**, where the number of supernodes that compose a state of $\mathcal{M}^F$ is bounded from above by the number of component graphs.

Lemma 2. *Given a* **CRSM** $\mathcal{M}$, *for each state* $X = [u_1, \ldots, u_m]$ *such that* $m > n^2 + 1$, *where* n *is the number of supernodes of* $\mathcal{M}$, *a state* $X' = [u'_1, \ldots, u'_{m'}]$ *exists such that* $m' < m$ *and* $true(X) = true(X')$. *Moreover, if* X *is reachable then also* X' *is reachable.*

Proof Consider a sequence $v_1, \ldots, v_h \in N$. We say that a sub-sequence $v_i \ldots v_j$, $1 \leq i < j$, is a *cycle* if $v_i = v_j$. Moreover, we say that a cycle $v_i \ldots v_j$, is *erasable* if $\{v_{i+1}, \ldots, v_j\} \subseteq \{v_1, \ldots, v_i\}$. It is easy to verify that for a sequence $u_1 \ldots u_m$ such that $X = [u_1, \ldots, u_m]$ is a state of $\mathcal{M}^F$ and $u_i \ldots u_j$ is a cycle, we have that $X' = [u_1, \ldots, u_i, u_{j+1}, \ldots, u_m]$ is a state of $\mathcal{M}^F$ and if $u_i \ldots u_j$ is also erasable then $true(X) = true(X')$. Moreover, by Proposition 1, if X is reachable then also X' is reachable.

To conclude the proof we only need to show that for each state $X = [u_1, \ldots, u_m]$ such that $m > n^2 + 1$, where n is the number of supernodes in $\mathcal{M}$, $u_1 \ldots u_m$ contains an erasable cycle. Notice that $m > n^2 + 1$ implies that a supernode u exists, occuring at least $(n + 1)$ times in $u_1 \ldots u_m$. Suppose $u_1 \ldots u_m = \alpha_0 u \alpha_1 u \ldots \alpha_n u \beta$, where each α_i does not contain occurrences of u. A cycle $u \alpha_i u$ is not erasable only if it contains a supernode that is not in $\alpha_0 u \ldots \alpha_{i-1} u$. By a simple count, if $\alpha_0 u \ldots \alpha_{n-1}$ does not contain erasable cycles, then all supernodes occur in it. Thus, $u \alpha_n u$ is erasable. $\square$

Now, we can prove membership to NP of the reachability and the cycle detection problems on **CRSMs**.

Lemma 3. *Reachability and cycle detection for* **CRSMs** *are decidable in non-deterministic polynomial-time.*

Proof Consider the instance of the reachability problem given by a **CRSM** $\mathcal{M}$ and a propositional boolean formula φ. By Proposition 2 we can determine in

$O(|\mathcal{M}|)$ time the set of the connected vertices, and then, given a state X of $\mathcal{M}^F$, by Proposition 1 we can check if X is reachable in $O(|\mathcal{M}|+|X|)$ time. Verifying the fulfillment of φ on X takes $O(|\varphi|+|X|)$ time. Moreover, by Lemma 2 we need only to consider states $X = [u_1, \ldots, u_m]$ for $m \leq n^2 + 1$, where n is the number of supernodes of $\mathcal{M}$. Thus, we can conclude that the reachability problem on **CHSMs** is in NP.   □

By Lemmas 1 and 3 we have the following theorem.

Theorem 1. *Reachability and cycle detection for* **CRSMs** *(***CHSMs***) are NP-complete.*

4 Efficient Solutions to Reachability and Cycle Detection Problems

In this section, we give a linear time algorithm that solves reachability and cycle detection problems for **CHSMs** which are related to target sets by a particular condition (specified later). As a corollary we get three consequences: first the results regarding reachability and cycle detection for the model considered in [AY01] are obtained as particular cases, second we characterize a class of formulas guaranteeing that the algorithm works correctly and finally we show that the algorithm works also for DNF formulas, thus obtaining a general solution for any formula with a tight worst case running time of $O(|\mathcal{M}| \cdot 2^{|\varphi|})$.

Finally, we give a linear time reduction from the reachability problem on **CRSMs** for DNF formulas to the corresponding problem on **CHSMs**, thus the above general solution still holds for **CRSMs**.

Consider now **CHSMs**. Clearly a propositional formula φ can be evaluated in a state X of $\mathcal{M}^F$ by instantiating to TRUE the variables corresponding to the atomic propositions in $true(X)$ and to FALSE all the others. Now we wish to evaluate φ without constructing the graph $\mathcal{M}^F$, to this aim we use a greedy approach in a top-down fashion on the hierarchy: at each supernode we instantiate as many variables as possible. By traversing the hierarchy in a top-down fashion, once a node is reached, φ can only *partially* evaluated. On a supernode u of a **CHSM** all the variables instantiated to TRUE correspond to the atomic propositions in $true(u)$. Determining the variables to instantiate to FALSE is not so immediate. We define $AP(h)$ as the union of the sets labeling either the vertices in N_h or those having an ancestor in N_h, that is, $AP(h) = \bigcup_{v \in N_h} (true(v) \cup AP(expand(v)))$ where $AP(0) = \emptyset$. Moreover, for $u \in N_h$, we define the set $false(u)$ as $AP(h) \setminus (true(u) \cup AP(expand(u)))$. This set contains the atomic propositions that can be instantiated to FALSE at u, since a proposition $p \in false(u)$ if and only if $p \notin true(X)$, for every state X of $\mathcal{M}^F$ having the supernode u as a component. It is easy to see that the sets $false(u)$, $u \in N$, can be preprocessed in time $O(|\mathcal{M}|)$, by visiting $\mathcal{M}$ in a bottom-up way.

For a propositional boolean formula φ we denote by $\mathrm{Eval}(\varphi, u)$ the formula obtained by instantiating φ with $true(u)$ and $false(u)$. We generalize this notation to sequences of vertices defining $\mathrm{Eval}(\varphi, u_1, \cdots, u_i)$ as

---

Algorithm **Reachability**$(\mathcal{M}, \varphi)$
    return(**Reach**(M_k, φ)).

Function **Reach**(M_h, φ)

    VISITED$[h] \leftarrow MARK$;
    foreach $u \in N_h$ **do**
      $\varphi' = \text{Eval}(\varphi, u)$;
      if ($\varphi' ==$ **TRUE**) **then return TRUE**;
      if ($\varphi' ==$ **FALSE**) **then continue**;
      if ((expand(u)>0) AND ($VISITED[expand(u)]! = MARK$)) **then**
        if Reach$(M_{expand(u)}, \varphi')$ **then return TRUE**;
    endfor
    return FALSE;

Fig. 1. Algorithm Reachability.

Eval(Eval$(\varphi, u_1), u_2, \cdots, u_i)$. Finally, we will denote by $AP(\varphi)$ the set of atomic propositions corresponding to φ variables.

We consider a condition relating a **CHSM** $\mathcal{M}$ and a target set specified by a formula φ asserting that "when two supernodes expand to the same graph, then any partial evaluation of φ ending on them coincides". Formally, the condition is as follows:

Condition 1 *Let* $x_1, \cdots, x_i$ *and* $y_1, \cdots, y_j$ *be two prefixes of* $\mathcal{M}^F$ *states such that* $expand(x_i) = expand(y_j)$. *If neither* $Eval(\varphi, x_1, \cdots, x_i)$ *nor* $Eval(\varphi, y_1, \cdots, y_j)$ *is one of the constants* $\{TRUE, \ FALSE\}$, *then* $Eval(\varphi, x_1, \cdots, x_i) = Eval(\varphi, y_1, \cdots, y_j)$.

When reachability and cycle detection become tractable.

Theorem 2. *The reachability and cycle detection problems on a* **CHSM** $\mathcal{M}$ *and a formula* φ *satisfying Condition 1 are decidable in time* $O(|\mathcal{M}| \cdot |\varphi|)$.

Proof Consider a **CHSM** $\mathcal{M} = (M_1, \ldots, M_k)$ and without loss of generality assume that all the vertices of $\mathcal{M}$ are connected (see Proposition2). Algorithm **Reachability**$(\mathcal{M}, \varphi)$ (Figure 1), returns **TRUE** if and only if φ is evaluated to TRUE on a reachable state of $\mathcal{M}^F$. The function **Reach** uses a global array VISITED (initially unmarked in all the positions) to mark the visited graphs M_h. For each node u of M_h, φ is evaluated on it according to $true(u)$ and $false(u)$, call φ' the returned formula. If φ' evaluates TRUE on u, then **Reach** stops returning **TRUE**. (and the main algorithm stops too returning **TRUE**). If φ' evaluates FALSE, another vertex of M_h which has not yet been explored is processed. In case u is a supernode and $M_{expand(u)}$ has never been visited, then the function is called on the graph $M_{expand(u)}$ and φ'. Now note that Condition 1 assures that it is not necessary to visit a graph M_h more than once, thus the overall complexity of the algorithm is linear in $|\mathcal{M}|$ and $|\varphi|$ and clearly returns **TRUE** if and only if a node X in $\mathcal{M}^F$ exists on which φ is **TRUE**. $\square$

It is easy to see that given any formula φ and a Hierarchical State Machine (HSM) introduced in [AY01] (where only nodes are labeled with the mapping true, see the introduction), Condition 1 always holds, thus the linear time solutions for the reachability and cycle detection problems for HSM given in that paper are here obtained as particular cases.

Now we present a characterization of formulas for which Theorem 2 holds. A propositional boolean formula φ is said to be in $\mathcal{M}$-*normal form* if $\varphi = \varphi_1 \wedge \ldots \wedge \varphi_m$ and for every φ_i and for every vertex u of $\mathcal{M}$ it holds that either $AP(\varphi_i) \cap (true(u) \cup false(u)) = \emptyset$ or $AP(\varphi_i) \cap (true(u) \cup false(u)) = AP(\varphi_i)$. It is easy to see that also in this case Condition 1 holds.

Theorem 2 can be generalized for a finite disjunction of formulas satisfying Condition 1. Since a conjunction of literals is in $\mathcal{M}$-*normal form*, for all possible $\mathcal{M}$, then this generalization can be applied to DNF formulas. Thus, as any formula φ can always be transformed in a DNF formula, we have an algorithm for reachability and cycle detection problems whose worst case running time is $O(|\mathcal{M}| \cdot \mathrm{DNF}(\varphi))$, where $\mathrm{DNF}(\varphi)$ is the cost of the transformation of φ in Disjunctive Normal Form. All this yields a tight upper bound of $O(|\mathcal{M}| \cdot 2^{|\varphi|})$.

Reachability and cycle detection are also tractable on **CRSM**s if we restrict to formulas in disjunctive normal form as shown in the following theorem.

Theorem 3. *Reachability and cycle detection problems for a* **CRSM** $\mathcal{M}$ *and a formula* φ *in DNF are decidable in time* $O(|\mathcal{M}| \cdot |\varphi|)$.

Proof Consider a **CRSM** $\mathcal{M}$ and a DNF formula $\varphi = \psi_1 \vee \ldots \vee \psi_m$ where each ψ_i is a conjunction of literals. Our algorithm consists of reducing in $O(|\varphi| \cdot |\mathcal{M}|)$ time the reachability problem for $\mathcal{M}$ and ψ_i to the reachability problem for a **CHSM** $\bar{\mathcal{M}}$ and ψ_i, where size of $\bar{\mathcal{M}}$ is $O(|\mathcal{M}|)$. Then the result follows from Theorem 2.

Consider a disjunct clause ψ of φ. We simplify $\mathcal{M}$ using the following two steps.

1. for each graph M_i, delete all the existing edges and insert an edge from in_i to any other connected vertex of M_i;
2. if u is not an initial node and $true(u)$ contains an atomic proposition corresponding to a variable which is negated in ψ, then delete u from M_i.

This transformation can be performed in $O(|\psi| \cdot |\mathcal{M}|)$ time and preserves the reachability of the states of $\mathcal{M}^{\mathcal{F}}$ satisfying ψ, thanks to Proposition 2.

Now, define a supernode $u \in N_i$ as *recursively expansible* if $i \in expand^+(u)$ and a graph M_i as *recursively expansible* if it contains at least a recursively expansible supernode. We define the equivalence relation $\approx$ on the indices of recursively expansible graphs: $i \approx j$ if and only if vertices $u \in N_i$ and $v \in N_j$ exist such that $i \in expand^+(v)$ and $j \in expand^+(u)$. We want to define a **CHSM** $\bar{\mathcal{M}} = (\bar{M}_1, \bar{M}_2, \ldots, \bar{M}_{k'})$ such that $\bar{\mathcal{M}}$ has a component graph for each equivalence class of the relation $\approx$. Let $f : \{1, \ldots, k\} \longrightarrow \{1, \ldots, k'\}$ be the function that maps each i to j such that i is in the equivalence class corresponding to $\bar{M}_j$.

For a graph M_i which is not recursively expansible (i.e., $[i] = \{i\}$), we define $\bar{M}_{f(i)}$ as M_i except for the mapping *expand*, since $expand_{\bar{M}}(u) = f(expand_{\mathcal{M}}(u))$. For a recursively expansible graph M_i we define $\bar{M}_{f(i)}$ as follows. All vertices $u \in N_j$ which are not recursively expansible, with $j \approx i$, are vertices of $\bar{M}_{f(i)}$, the edges between them in M_i are edges of $\bar{M}_{f(i)}$ as well and $OUT_{f(i)} = \bigcup_{j,j \approx i} OUT_j$. Moreover, we add a new initial node $\bar{in}_{f(i)}$ and insert edges from $\bar{in}_{f(i)}$ to all vertices in_j, $j \approx i$. For each supernode u of $\bar{M}_{f(i)}$ we define $expand_{\bar{M}}(u) = f(expand_{\mathcal{M}}(u))$. Let $S_{\mathcal{M}}(i)$ be the set of all recursively expansible vertices belonging to all graphs M_j such that $j \approx i$. We define $true_{\bar{M}}(\bar{in}_{f(i)})$ as $true_{\bar{M}}(in_j)$ for an arbitrary $j \approx i$, and for each vertex u of $\bar{M}_{f(i)}$, $true_{\bar{M}}(u)$ as $\bigcup_{v \in S_{\mathcal{M}}(i)} true_{\mathcal{M}}(v) \cup true_{\mathcal{M}}(u)$ (note that no atomic proposition added in this way to the label of u corresponds to a variable which is negated in ψ).

Now observe that, by the above part 2 of the above simplification, if X is a state of $\mathcal{M}^F$ satisfying ψ and Y is a state of $\bar{\mathcal{M}}^F$ such that $true_{\mathcal{M}}(X) \subseteq true_{\bar{M}}(Y)$ and $true_{\bar{M}}(Y) \setminus true_{\mathcal{M}}(X)$ does not contain an atomic proposition corresponding to a variable which is negated in ψ, then Y satisfies ψ as well. Since the initial simplification also preserves reachability, we have that if a reachable state of $\mathcal{M}^F$ fulfilling ψ exists, then a state of $\bar{\mathcal{M}}^F$ fulfilling ψ also exists. Since by construction, states of $\bar{\mathcal{M}}^F$ corresponds to states of $\mathcal{M}^F$, the vice-versa also holds. □

As a consequence of Theorem 3 and the arguments for **CHSM**s and DNF formulas, the following theorem holds.

Theorem 4. *The reachability and cycle detection problems on a* **CRSM** *$\mathcal{M}$ and a propositional boolean formula φ are decidable in $O(|\mathcal{M}| \cdot 2^{|\varphi|})$ time.*

5 LTL Model Checking

Here we consider the verification problem of linear-time requirements, expressed by LTL-formulas [Pnu77]. We follow the automata theoretic approach to solving model checking [VW86]: given an LTL formula φ and a Kripke structure M, it is possible to reduce model checking to the emptiness problem of Büchi automata. To use this approach, we extend the Cartesian product between Kripke structures.

Given a transition graph with states labeled by subsets of atomic propositions and a state s, a *trace* is an infinite sequence $\alpha_1 \alpha_2 \ldots \alpha_i \ldots$ of labels of states occuring in a path starting from s. Moreover, given a **CRSM** $\mathcal{M}$, we define the language $\mathcal{L}(\mathcal{M})$ as the set of the traces of $\mathcal{M}^F$ starting from its initial state. A Büchi automaton $A = (Q, q_1, \Delta, L, T)$ is a Kripke structure (Q, Δ, L) together with a set of accepting states T and a starting state q_1. The language $\mathcal{L}(A)$ accepted by A is the set of the traces corresponding to paths visiting infinitely often a state of T.

Let $\mathcal{M} = (M_1, \ldots, M_k)$ be a **CRSM** and $A = (Q, q_1, \Delta, L, T)$, for $Q = \{q_1, \ldots, q_m\}$, be a Büchi automaton. Let $1 \leq i \leq k$, $1 \leq j \leq m$, and P be such that $P \subseteq AP$ and $P \cup true_{\mathcal{M}}(in_i) = L(q_j)$, we define the graphs $M_{(i,j,P)}$ as follows. Each $M_{(i,j,P)}$ contains vertices $[u, q, j, P]$ such that (u, q) belongs to the standard Cartesian product of M_i and A, and the labeling of q coincides with the labeling of u augmented with the atomic propositions that u inherits from its ancestors in a given context. The inherited set of atomic propositions is given by P. The property $P \cup true_{\mathcal{M}}(in_i) = L(q_j)$ assures that we consider only graphs $M_{(i,j,P)}$ whose initial vertex is compatible with the automaton state. Formally, we have:

- The set $N_{(i,j,P)}$ of the vertices of $M_{(i,j,P)}$ contains quadruples $[u, q, j, P]$, where $u \in N_i$, $q \in Q$, and
  - either $expand_{\mathcal{M}}(u) = 0$ and $L(q) = true_{\mathcal{M}}(u) \cup P$
  - or $expand_{\mathcal{M}}(u) = h > 0$ and $L(q) = true_{\mathcal{M}}(u) \cup true_{\mathcal{M}}(in_h) \cup P$.
- The initial vertex of $M_{(i,j,P)}$ is $[in_i, q_j, j, P]$ and the output nodes are $[u, q, j, P]$ for $u \in OUT_i$ and $q \in Q$;
- $M_{(i,j,P)}$ contains the following edges:
  - $([u, q', j, P], [v, q'', j, P])$, with $(q', q'') \in \Delta$ and $(u, v) \in E_i$,
  - $(([u, q_{j'}, j, P], [z, q', j', P \cup true_{\mathcal{M}}(u)]), [v, q'', j, P])$, with $(q', q'') \in \Delta$, $((u, z), v) \in E_i$, and $L(q) = true_{\mathcal{M}}(u) \cup true_{\mathcal{M}}(in_h) \cup P$ for $expand_{\mathcal{M}}(u) = h$.

From the above definition we observe that if u is a supernode then the labeling of q has to match also with the labeling of $in_{expand_{\mathcal{M}}(u)}$ since $[u, q, j, P]$ is a supernode of $\mathcal{M}'$ and one has to assure the correctness, with respect to the labeling, of its expansion. Note that when only the value of j varies, we have graphs which differ one each other only for the choice of the the initial vertex $[in_i, q_j, j, P]$. Moreover, the edges in $M_{(i,j,P)}$ are given by coupling the transitions (q', q'') of A with both kinds of edges (u, v) and $((u, z), v)$ in E_i. For $h = expand_{\mathcal{M}}(u)$, we have edges $(([u, q, j, P], [z, q', h, P \cup true_{\mathcal{M}}(u)]), [v, q'', j, P])$ for every $q \in Q$ such that $L(q) = true_{\mathcal{M}}(u) \cup true_{\mathcal{M}}(in_h) \cup P$. Thus, there might be as many as $|Q|$ edges, for every pair of edges $((u, z), v)$ and (q', q'').

We can now define $\mathcal{M}' = \mathcal{M} \bigotimes A$ as a **CRSM** constituted by some of the graphs $M_{(i,j,P)}$, and defined inductively as follows:

- $M_{(k,1,\emptyset)}$ is the graph containing the starting node of $\mathcal{M}'$;
- Let $M_{(i,j,P)}$ be a graph of $\mathcal{M}'$, and $[u, q_t, j, P]$ be a vertex of $M_{(i,j,P)}$.
  - If $expand_{\mathcal{M}}(u) = 0$ then $expand_{\mathcal{M}'}([u, q_t, j, P]) = 0$;
  - If $expand_{\mathcal{M}}(u) = h > 0$, and $P' = P \cup true_{\mathcal{M}}(u)$ then $M_{(h,t,P')}$ is a graph of $\mathcal{M}'$ and $expand_{\mathcal{M}'}([u, q_t, j, P]) = \langle h, t, P' \rangle$ where $\langle h, t, P' \rangle$ denotes the index of $M_{(h,t,P')}$;
- $true_{\mathcal{M}'}([u, q, j, P]) = true_{\mathcal{M}}(u)$, for every $[u, q, j, P]$.

Observe that $\mathcal{M}' = \mathcal{M} \bigotimes A$ is a **CRSM** and if $\mathcal{M}$ is a **CHSM**, then $\mathcal{M}'$ is a **CHSM**, as well. To determine the size of $\mathcal{M}'$, first consider the size of each graph $M_{(i,j,P)}$. The number of the edges is bounded by the product of the number of edges in M_i and the number of transitions in A multiplied at most by m, since we have at most $|Q|$ edges for any $(q', q'') \in \Delta$ and $((u, z), v) \in E_i$. Thus, an upper bound to the size of $M_{(i,j,P)}$ is given by $(m \cdot |E_i| \cdot |A|)$. The size of $\mathcal{M}'$ can be obtained now by counting the number of its component graphs.

Lemma 4. *Given a* **CRSM** $\mathcal{M}$, $\mathcal{M}' = \mathcal{M} \otimes A$ *is a* **CRSM** *that can be constructed in* $O(m^2 \cdot |\mathcal{M}| \cdot |A| \cdot |2^{AP}|)$ *time. Moreover, if* $\mathcal{M}$ *is a* **CHSM***, then* $\mathcal{M}'$ *is a* **CHSM** *that can be constructed in* $O(m^2 \cdot |\mathcal{M}| \cdot |A|)$ *time.*

Proof First recall that a graph $M_{(i,j,P)}$ of $\mathcal{M}'$ has the property that $P \cup true_{\mathcal{M}}(in_i) = L(q_j)$. Therefore, P is the union of two disjoint sets P_1 and P_2, such that P_1 is the set of the atomic propositions of $L(q_j)$ that do not belong to $true_{\mathcal{M}}(in_i)$, and $P_2 = P \cap true_{\mathcal{M}}(in_i)$ is a subset of $true_{\mathcal{M}}(in_i)$. Thus, for fixed values of i and j, P_1 is fixed and the number of different graphs $M_{(i,j,P)}$ is bounded above by the number of different subsets of $true_{\mathcal{M}}(in_i)$. Therefore, the size of $\mathcal{M}'$ is bounded above by $\sum_{j=1}^m \sum_{i=1}^k (2^{|AP|} \cdot m \cdot |M_i| \cdot |A|)$.

Now, let $\mathcal{M}$ be a **CHSM**. Given a graph $M_{(i,j,P)}$ of $\mathcal{M}'$, P is defined as the set of the propositions that the vertices of M_i inherit. Thus, $P \cap true_{\mathcal{M}}(u) = \emptyset$, for every vertex u of M_i and then $P \cap true_{\mathcal{M}}(in_i) = \emptyset$. Hence, in this case, P_2 is empty and then at most one graph $M_{(i,j,P)}$ exists for fixed values of i and j. Therefore, the size of $\mathcal{M}'$ is bounded above by $\sum_{j=1}^m \sum_{i=1}^k (m \cdot |M_i| \cdot |A|)$. $\square$

The **CRSM** $\mathcal{M} \otimes A$ can be used to check for the emptiness of the language given by the intersection of $\mathcal{L}(\mathcal{M})$ and $\mathcal{L}(A)$, as shown in the following lemma.

Lemma 5. *There exists an algorithm checking whether* $\mathcal{L}(\mathcal{M}) \cap \mathcal{L}(A) = \emptyset$ *in time linear in the size of* $\mathcal{M}' = \mathcal{M} \otimes A$.

Proof First, observe that if we consider as set of final states the vertices $[u, q, h, P]$ such that $q \in T$, the **CRSM** $\mathcal{M}'$ is a recursive Büchi automaton. Moreover, the set of the traces of $\mathcal{M}'^F$ is the same as the set of traces of the Cartesian product of $\mathcal{M}^F$ and A. Thus $\mathcal{L}(\mathcal{M}) \cap \mathcal{L}(A) \neq \emptyset$ if and only if $\mathcal{L}(\mathcal{M}') \neq \emptyset$. From [AEY01], for recursive Büchi automata with a single initial node for each graph, non-emptiness can be checked in linear time. $\square$

As a consequence of the above lemmas, we obtain an algorithm to solving the LTL model checking for **CRSMs**. Following the automata theoretic approach, one can construct a Büchi automaton $A_{\neg \varphi}$ of size $O(2^{|\varphi|})$ accepting the set $\mathcal{L}(A_{\neg \varphi})$ of the sequences which do not satisfy φ, and then φ is satisfied on all paths of $\mathcal{M}$ if and only if $\mathcal{L}(\mathcal{M}) \cap \mathcal{L}(A_{\neg \varphi})$ is empty. From Lemma 4, one can now construct $\mathcal{M} \otimes A_{\neg \varphi}$, whose size is $O(m^2 \cdot |\mathcal{M}| \cdot |A_{\neg \varphi}| \cdot 2^{|AP|}) = O(|\mathcal{M}| \cdot 16^{|\varphi|})$ (since $m = |A_{\neg \varphi}| = O(2^{|\varphi|})$ and $2^{|AP|} \leq 2^{|\varphi|}$). Moreover, this size reduces to $O(m^2 \cdot |\mathcal{M}| \cdot |A_{\neg \varphi}|) = O(|\mathcal{M}| \cdot 8^{|\varphi|})$, when $\mathcal{M}$ is a **CHSM**. Hence, by Lemma 5 we obtain the main result of this section.

Theorem 5. *The* LTL *model checking on a* **CRSM** $\mathcal{M}$ *and a formula* φ *can be solved in* $O(|\mathcal{M}| \cdot 16^{|\varphi|})$ *time. Moreover, if* M *is a* **CHSM** *the problem can be solved in* $O(|\mathcal{M}| \cdot 8^{|\varphi|})$ *time.*

6   Discussion

We have proposed new abstract models for sequential state machines: the context-dependent hierarchical and recursive state machines. On these models we have studied reachability, cycle detection and the more general problem of model checking with respect to linear-time specifications. An interesting feature of **CHSM**s is that they allow very succinct representations of systems, and this comes substantially at no cost if compared to analogous hierarchical models studied in the literature. Moreover, we prove that for some particular formulas we improve the complexity of previous approaches.

Several extensions of the introduced models can be considered.

Our models are sequential. If we add concurrency to **CHSM**s, the computational complexity of the considered decision problems grows significantly (we recall that reachability in communicating hierarchical state machines is EXPSPACE-complete [AKY99]). While for **CRSM**s with concurrency, reachability becomes undecidable since sequential **CRSM**s are as expressive as pushdown automata [AEY01,BGR01].

We have only considered models where a single entry node is allowed for each component machine. We can relax this limitation allowing multiple entry points. The semantics of this extension naturally follows from the semantics given for the single entry case. In the hierarchic setting, we can translate a multiple-entry **CHSM** $\mathcal{M}$ into an equivalent single-entry **CHSM** $\mathcal{M}'$ of size at most cubic in the size of $\mathcal{M}$. In fact, each component machine of $\mathcal{M}$ can be replaced in $\mathcal{M}'$ by multiple copies, each copy corresponding to an entry point and having as unique entry point the entry point itself. Expansions are redirected to the proper components in order to match the expansions in $\mathcal{M}$. Thus, supernodes may need to be replaced by multiple copies each pointing to the proper machine in $\mathcal{M}'$. If we apply this construction to a multiple-entry **CRSM**, the obtained single-entry **CRSM** does not satisfy the property $true(u) \cap true(v) = \emptyset$, for $v \in N_h, u \notin N_h$ and $h \in expand^+(u)$ (see definition of **CRSM**). This is a consequence of the fact that if a machine of the multiple-entry **CRSM** can directly or indirectly call itself, then there are two copies of this machine that may call each other recursively. We recall that the above property is sufficient to ensure that Condition 1 holds for conjunctions of literals, and thus is crucial to obtain the results given in Section 4. However, it is possible to prove that Theorem 3 also holds for multiple-entry **CRSM**s. We leave the details of this proof to the full paper.

For modeling purposes it is useful to have variables over a finite domain that can be passed from a component to another. We can extend our models to handle input, output and local variables. Consider a component machine M with h_e entry nodes, h_x exit nodes, and h_t internal vertices. If M is equipped also with k_i input boolean variables, k_o output boolean variables, and k_l local boolean variables, we can model by a machine having $2^{k_i} \cdot h_e$ entry nodes, $2^{k_o} \cdot h_x$ exit nodes, and $2^{k_i+k_l+k_o} \cdot h_t$ internal vertices.

References

[AEY01]   R. Alur, K. Etessami, and M. Yannakakis.   Analysis of recursive state machines.   In *Proc. of the 13th International Conference on Computer Aided Verification, CAV'01*, LNCS 2102, pages 207–220. Springer, 2001.

[AG00]   R. Alur and R. Grosu. Modular refinement of hierarchic reactive machines. In *Proc. of the 27th Annual ACM Symposium on Principles of Programming Languages*, pages 390–402, 2000.

[AGM00]   R. Alur, R. Grosu, and M. McDougall. Efficient reachability analysis of hierarchical reactive machines. In *Computer Aided Verification, 12th International Conference*, LNCS 1855, pages 280–295. Springer, 2000.

[AKY99]   R. Alur, S. Kannan, and M. Yannakakis. Communicating hierarchical state machines. In *Proc. of the 26-th International Colloquium on Automata, Languages and Programming, ICALP'99*, LNCS 1644, pages 169–178. Springer-Verlag, 1999.

[AY01]   R. Alur and M. Yannakakis. Model checking of hierarchical state machines. *ACM Transactions on Programming Languages and Systems (TOPLAS)*, 23(3):273–303, 2001.

[BGR01]   M. Benedikt, P. Godefroid, and T. W. Reps. Model checking of unrestricted hierarchical state machines. In *Proc. of the 28th International Colloquium Automata, Languages and Programming, ICALP'01*, LNCS 2076, pages 652–666. Springer, 2001.

[BJR97]   G. Booch, I. Jacobson, and J. Rumbaugh. *Unified Modeling Language User Guide*. Addison Wesley, 1997.

[BLA+99]   G. Behrmann, K.G. Larsen, H.R. Andersen, H. Hulgaard, and J. Lind-Nielsen. Verification of hierarchical state/event systems using reusability and compositonality. In *Proc. of the Tools and Algorithms for the Construction and Analysis of Systems, TACAS'99*, LNCS 1579, pages 163–177. Springer, 1999.

[CE81]   E.M. Clarke and E.A. Emerson. Design and synthesis of synchronization skeletons using branching time temporal logic. In *Proc. of Workshop on Logic of Programs*, LNCS 131, pages 52–71. Springer-Verlag, 1981.

[CK96]   E.M. Clarke and R.P. Kurshan. Computer-aided verification. *IEEE Spectrum*, 33(6):61–67, 1996.

[Pnu77]   A. Pnueli. The temporal logic of programs. In *Proc. of the 18th IEEE Symposium on Foundations of Computer Science*, pages 46–77, 1977.

[RBP+91]   J. Rumabaugh, M. Blaha, W. Premerlani, F. Eddy, and W. Lorensen. *Object-oriented Modeling and Design*. Prentice-Hall, 1991.

[SGW94]   B. Selic, G. Gullekson, and P.T. Ward. *Real-time object oriented modeling and design*. J. Wiley, 1994.

[VW86]   M.Y. Vardi and P. Wolper. Automata-theoretic techniques for modal logics of programs. *Journal of Computer and System Sciences*, 32:182–211, 1986.

Oracle Circuits for Branching-Time Model Checking

Philippe Schnoebelen

Lab. Spécification & Vérification
ENS de Cachan & CNRS UMR 8643
61, av. Pdt. Wilson, 94235 Cachan Cedex France
phs@lsv.ens-cachan.fr

Abstract. A special class of oracle circuits with tree-vector form is introduced. It is shown that they can be evaluated in deterministic polynomial-time with a polylog number of adaptive queries to an NP oracle. This framework allows us to evaluate the precise computational complexity of model checking for some branching-time logics where it was known that the problem is NP-hard and coNP-hard.

1 Introduction

Many different *temporal logics* have been proposed in the computer science literature [5]. Their main use is in the field of reactive systems, where *model checking* allows automated verification of correctness [3].

Comparing and classifying the different temporal logics is an important task. This is usually done along several axis, most notably *expressive power* and *computational complexity*. Regarding computational complexity, several open questions remain [16]. In particular, for several branching-time temporal logics, the complexity of model checking is not known. Advances in this domain are welcome since it is important to understand what ideas underly "optimal" algorithms, and what special cases may benefit from specialized methods.

Model checking in the polynomial-time hierarchy. There is a family of branching-time temporal logics for which the complexity of model checking is not known precisely. These logics can be described as branching-time logics where the underlying path properties are in NP or coNP (we give several examples in section 2). This leads to a P^{NP} (that is, Δ_2^p) upper bound for the full logic.

For such logics, the question of finding matching lower bounds saw no progress until recently, when Laroussinie, Markey, and the author managed to prove that some of them (including $B^*(\mathsf{F})$, CTL^+, and $FCTL$) have indeed a Δ_2^p-complete model checking problem [13,14].

However, for some remaining logics, the techniques used in [13,14] for proving Δ_2^p-hardness do not apply. The difficulty here is that, if these problems are *not* Δ_2^p-complete, we still lack methods for proving that a model checking problem has upper bounds higher than NP or coNP but lower than Δ_2^p.

J.C.M. Baeten et al. (Eds.): ICALP 2003, LNCS 2719, pp. 790–801, 2003.

Our contribution. In this paper we develop a framework that allows proving upper bounds below Δ_2^p and apply it to branching-time model checking problems. The approach is successful in that it allows us to prove model checking $B^*(\mathsf{X})$ is $\mathrm{P}^{\mathrm{NP}[O(\log^2 n)]}$-complete, and model checking *Timed* $B(\mathsf{F})$ is $\mathrm{P}^{\mathrm{NP}[O(\log n)]}$-complete.

Our framework is based on Boolean circuits with oracle queries (introduced in [21]). We identify two special classes of oracle circuits having *tree-vector* form (with special constraints on the oracle queries) for which we prove evaluation can be done in $\mathrm{P}^{\mathrm{NP}[O(\log n)]}$ and, respectively, $\mathrm{P}^{\mathrm{NP}[O(\log^2 n)]}$, i.e. they can be evaluated by a deterministic polynomial-time Turing machine that makes $O(\log n)$ (resp. $O(\log^2 n)$) adaptive queries to an NP-oracle (while Δ_2^p-complete problems require[1] polynomially-many adaptive queries).

Branching-time model checking problems lead naturally to tree-vector circuits, so that we obtain upper bounds directly by translations. The lower bounds are proved by ad-hoc reductions.

These results are important for several reasons:
1. The tree-vector oracle circuits may have more applications than just in model checking. In any case, they illuminate a structural feature of model checking where the formula is a modal expression *tree* evaluated over a *vector* of worlds.
2. The results help complete the picture in the classification of temporal logics. A logic like $B^*(\mathsf{X})$, the *full branching-time logic of "next"*, is perhaps not used in practice, but it is a fundamental fragment of CTL^*, for which we should be able to assess the complexity of model checking.
3. They provide examples of problems complete for $\mathrm{P}^{\mathrm{NP}[O(\log n)]}$ and $\mathrm{P}^{\mathrm{NP}[O(\log^2 n)]}$. Very few such examples are known. In particular, with the model checking of $B^*(\mathsf{X})$, we provide the first example (to the best of our knowledge) of a natural problem complete for $\mathrm{P}^{\mathrm{NP}[O(\log^2 n)]}$.

Related work. The best known framework for assessing the complexity of model checking problems is the automata-theoretic framework initiated by Vardi and Wolper [18]. By moving to tree-automata, this framework is able to deal with branching-time logics [12], where it has proved very successful. However, the tree-automata approach seems too coarse-grained for our problems where it seems we need a fine-grained look at the structure of the oracle calls.

Gottlob's work on NP trees [9] was an inspiration. His result prompted us to check whether certain tree-vectors of queries could be normalized.

Plan of the paper. We recall the necessary background in Section 2. Then Section 3 is devoted to tree-vector oracle circuits and flattening algorithms for evaluating them. This lays the ground for our proof that model checking $B^*(\mathsf{X})$ is $\mathrm{P}^{\mathrm{NP}[O(\log^2 n)]}$-complete (Section 4) and model checking *Timed* $B(\mathsf{F})$ is $\mathrm{P}^{\mathrm{NP}[O(\log n)]}$-complete (Section 5). The proofs that have been omitted for lack of space appear in the full version.

---

[1] That is, assuming Δ_2^p does not collapse to $\mathrm{P}^{\mathrm{NP}[O(\log^2 n)]}$ and $\mathrm{P}^{\mathrm{NP}[O(\log^2 n)]}$! We shall often write such sentences that implicitly assume the separation conjectures most complexity theorists believe are true.

2   Branching-Time Logics with Model Checking in Δ_2^p

2.1   Complexity Classes below Δ_2^p

We assume familiarity with computational complexity. The main definitions we need concern classes in the polynomial-time hierarchy (see [10,15]).

Δ_2^p is the class P^{NP} of problems solvable by deterministic polynomial-time Turing machines that may query an NP oracle. Some relevant subclasses of Δ_2^p have been identified:

- $P^{NP[O(\log n)]}$ only allows $O(\log n)$ oracle queries instead of polynomially-many. For example, PARITY-SAT (the problem where one is asked whether the number of satisfiable Boolean formulae from some input set $f_1, \ldots, f_n$ is odd or even) is $P^{NP[O(\log n)]}$-complete [19].
- $P_{\parallel}^{NP}$ only allows one round of *parallel* queries: the polynomially-many queries may not be adaptive (i.e., depend on the outcomes of earlier oracle queries) but must first be all formulated *before* the oracle is consulted on all queries. Then the computation proceeds normally, using the polynomially-many oracle answers.

$P^{NP[O(\log n)]}$ and $P_{\parallel}^{NP}$ coincide (and they further coincide with $P_{\parallel O(1)}^{NP}$, where a fixed number of parallel rounds is allowed). Wagner showed that many different and natural ways of restricting Δ_2^p all lead to the same $P^{NP[O(\log n)]}$ class (e.g. $P^{NP[O(\log n)]}$ coincide with L^{NP}), for which he introduced the name Θ_2^p [20]. Further variants were introduced by Castro and Seara, who proved that, for all $k \in \mathbb{N}$, $P^{NP[O(\log^k n)]}$ coincide with $P_{\parallel O(\log^{k-1} n)}^{NP}$ (where a succession of $O(\log^{k-1})$ parallel querying rounds are allowed) [1].

2.2   Branching-Time Logics and NP-Hard Path Modalities

We assume familiarity with temporal logic model checking [5,3,16]. Several branching-time logics combine the path quantifiers E and A with linear-time modalities whose path existence problem is in NP. Here are five examples:

- *FCTL* [8], or "Fair *CTL*", allows restricting to the fair paths of a Kripke structure, where the fair paths are defined by an arbitrary Boolean combination of $\overset{\infty}{F} \pm P_i$s. The existence of a fair path is NP-complete [8].
- *TCTL* [11], or "Timed *CTL*", allows adding timing subscripts to the usual modalities. In Timed KSs (i.e. Kripke structures where edges carry a discrete "duration" weight) the existence of a path of a given accumulated duration is NP-complete [14].
- *CTL*$^+$ [6] allows arbitrary Boolean combinations (not nesting) of the U and X modalities under a path quantifier. Thus *CTL*$^+$ is the branching-time extension of $L^1(U, X)$, the fragment of linear-time logic with modal depth one, for which the existence of a path is NP-complete [4].
- $B^*(F)$ and $B^*(X)$ are the branching-time extensions of $L(F)$ and $L(X)$ (resp.). $B^*(F)$ (called BT^* in [2]) is the *full branching-time logic of "eventually"*, while $B^*(X)$ is the *full branching-time logic of "next"*. The existence of a path satisfying an $L(F)$ or an $L(X)$ formula is NP-complete [17].

For these examples, NP-hardness is easy to prove by reduction from 3SAT. For example, consider an instance $\mathcal{I}$ of the form "$(x_1 \vee \overline{x_2} \vee \overline{x_4}) \wedge (\overline{x_1} \vee \cdots) \wedge \cdots$". With $\mathcal{I}$ we associate the following structure that applies to CTL^+, $B^*(\mathsf{F})$, and $B^*(\mathsf{X})$:

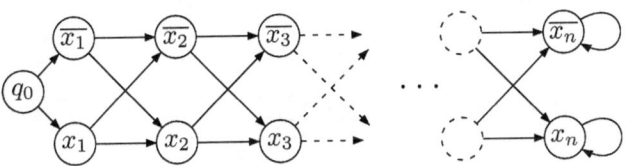

$$\mathcal{I} \text{ is satisfiable iff } q_0 \models \mathsf{E}\Big[(\mathsf{F}x_1 \vee \mathsf{F}\overline{x_2} \vee \mathsf{F}\overline{x_4}) \wedge (\mathsf{F}\overline{x_1} \vee \cdots) \wedge \cdots\Big] \qquad (1)$$

$$\text{iff } q_0 \models \mathsf{E}\Big[(\mathsf{X}x_1 \vee \mathsf{X}\mathsf{X}\overline{x_2} \vee \mathsf{X}\mathsf{X}\mathsf{X}\mathsf{X}\overline{x_4}) \wedge (\mathsf{X}\overline{x_1} \vee \cdots) \wedge \cdots\Big] \qquad (2)$$

For $FCTL$ we use a slight variant:

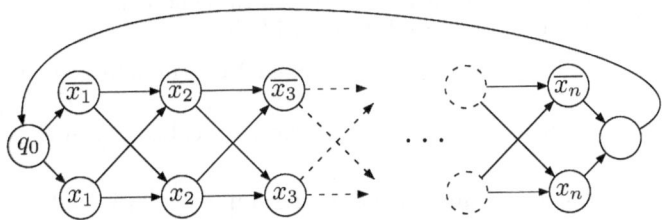

Here $\mathcal{I}$ is satisfiable iff

$$q_0 \models \mathsf{E}\left[\begin{array}{l}\neg(\overset{\infty}{\mathsf{F}}x_1 \wedge \overset{\infty}{\mathsf{F}}\overline{x_1}) \wedge \neg(\overset{\infty}{\mathsf{F}}x_2 \wedge \overset{\infty}{\mathsf{F}}\overline{x_2}) \wedge \cdots \wedge \neg(\overset{\infty}{\mathsf{F}}x_n \wedge \overset{\infty}{\mathsf{F}}\overline{x_n})] \\ \wedge (\overset{\infty}{\mathsf{F}}x_1 \vee \overset{\infty}{\mathsf{F}}\overline{x_2} \vee \overset{\infty}{\mathsf{F}}\overline{x_4}) \wedge (\overset{\infty}{\mathsf{F}}\overline{x_1} \vee \cdots) \wedge \cdots\end{array}\right] \qquad (3)$$

For $TCTL$ we reduce from SUBSET-SUM. With an instance $\mathcal{I}$ of the form "can one add numbers taken from $\{a_1, \ldots, a_n\}$ and obtain b?" we associate the following Timed KS:

Obviously $\mathcal{I}$ is solvable iff $q_0 \models \mathsf{EF}_{=b} q_n$.

2.3 Model Checking $B(L)$

Assume L is some linear-time logic, and write $B(L)$ for the associated branching-time logic. Emerson and Lei [8] observed that, from an algorithm for the existence

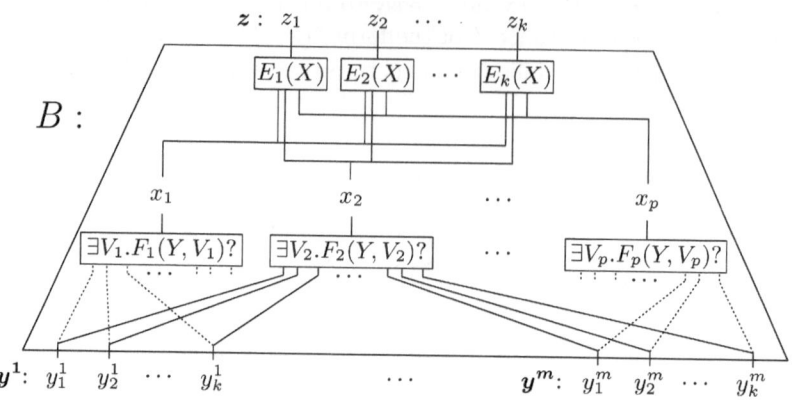

Fig. 1. General form of a "block" oracle circuit

of paths satisfying L properties, one easily derives a model checking algorithm for $B(L)$. Furthermore, this only needs a polynomial-time Turing reduction, so that if the existential problem for L belongs to some complexity class $\mathcal{C}$, then the model checking problem for $B(L)$ is in $\mathrm{P}^{\mathcal{C}}$ [8].

Example 2.1. With path modalities having an NP-complete existential problem, $B^*(\mathsf{F})$, $B^*(\mathsf{X})$, CTL^+, $ECTL^+$ (from [7]), $FCTL$, BTL_2 and $TCTL$ (over Timed KSs), all have a model checking in P^{NP}, the level called Δ_2^p in the polynomial-time hierarchy.[2]   □

Concerning the logics mentioned in Example 2.1, the only known lower bounds for their model checking problem were the obvious "NP-hard and coNP-hard" (or even DP-hard). However, all these logics have Θ_2^p-hard model checking (see Remark 5.3 below). Recently, Laroussinie, Markey and Schnoebelen showed Δ_2^p-hardness (hence Δ_2^p-completeness) for $FCTL$ and $B^+(\mathsf{F})$ in [13] (hence also for $B^*(\mathsf{F})$, CTL^+, $ECTL^+$, and BTL_2), and for $TCTL$ over Timed KSs in [14].

The techniques from [13,14] were not able to cope with $B^*(\mathsf{X})$, or with *Timed* $B(\mathsf{F})$ (the fragment of $TCTL$ where only the F modality may carry timing subscripts). This raises the question of whether these logics have Δ_2^p-hard model checking, and how to prove that. The Δ_2^p upper-bound is indeed too coarse: in the rest of the paper, we prove that model checking $B^*(\mathsf{X})$ is $\mathrm{P}^{\mathrm{NP}[O(\log^2 n)]}$-complete, and model checking *Timed* $B(\mathsf{F})$ is $\mathrm{P}^{\mathrm{NP}[O(\log n)]}$-complete.

3   Oracle Boolean Circuits and TB(SAT)

We consider special oracle Boolean circuits called *blocks*. As illustrated in Fig. 1, a block is a circuit B computing an output vector $\boldsymbol{z}$ of k bits from a set $\boldsymbol{y}^1, \dots, \boldsymbol{y}^m$ of m input vectors, again with k bits each. Inside the block, p internal gates $x_1, \dots, x_p$ query a SAT oracle: x_i evaluates to 1 iff $F_i(Y, V_i)$ is satisfiable, where

---

[2] For CTL^+ and $B^*(\mathsf{F})$, membership in Δ_2^p was observed as early as [2, Theo. 6.2].

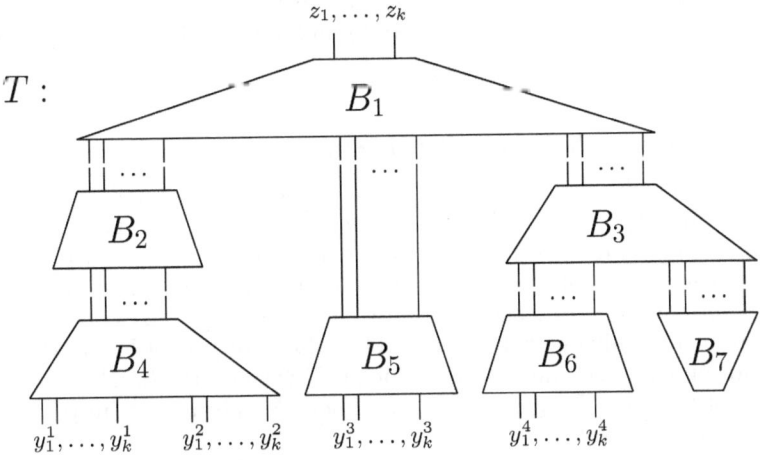

Fig. 2. A "tree of blocks" oracle circuit

F_i is a Boolean formula combining the km input bits $Y = \{y_l^j \mid j = 1, \ldots, m, \; l = 1, \ldots, k\}$ with some additional variables from some set V_i. Finally, the values of the output bits are computed from the x_i's by means of classical Boolean circuits (no oracles): z_i is some $E_i(X)$ where $X = \{x_1, \ldots, x_p\}$. We say m is the *degree* of the block, k is its *width*, and its size is the usual number of gates *augmented by the sizes of the F_i formulae.*

The obvious algorithm for computing the value of z for some km input bits is a typical instance of $P_{\|}^{NP}$: the p oracle queries are independent and can be asked in parallel. Building the queries and combining their answers to produce z is a simple polynomial-time computation.

Blocks are used to form more complex circuits: a *tree of blocks* is a circuit T obtained by connecting blocks having a same width k in a tree structure, as illustrated in Fig. 2 (where block B_7 has degree 0).

Every block in a tree has a level defined in the obvious way: in our example, $B_4, \ldots, B_7$ are at level 1, B_2, B_3 at level 2, and B_1, the *root*, at level 3. If the root of some tree is at level d, then the natural way of computing the value of the output z requires d rounds of parallel queries: in our example, the queries inside B_1 can only be formulated after the B_2 queries have been answered, and formulating these require that the B_4 queries have been answered before.

TB(SAT) is the decision problem where one is given a tree of blocks, Boolean values for its input bits, and is asked the value of (one of) the output bits. Compared to the more general problem of evaluating circuits with oracle queries (e.g. the Δ_2^p-complete **DAG**(SAT) problem of [9]), we impose the restriction of a tree-like structure, and compared to the more particular problem of evaluating Boolean trees with oracle queries (the Θ_2^p-complete **TREE**(SAT) problem of [9]), we allow each node of the tree to transmit a vector of bits to its parent node. Thus **TB**(SAT) is a restriction of **DAG**(SAT) and a generalization of **TREE**(SAT).

Fact 3.1 **TB**(SAT) *is* Δ_2^p*-complete.*

3.1   Circuits with Simple Oracle Queries

In a block of width k and degree m, we say a query $\exists V.F(Y,V)$? has *type* 1×M if it has the form $\exists l_1, \ldots, l_m \exists V'.F(y_{l_1}^1, \ldots, y_{l_m}^m, l_1, \ldots, l_m, V')$?, i.e. F only uses one bit from each input vector (but it can be any bit and this is existentially quantified upon). Our formulation quantifies upon indexes $l_1, \ldots, l_m$ in the $1, \ldots, k$ range but such a l_j is a shorthand for e.g. k bits "$l_j{=}1$", $\ldots$, "$l_j{=}k$" among which one and only one will be true. These bits are part of V and this is why F depends on $l_1, \ldots, l_m$ (and on V', which is V without the l_js). There is a similar notion of type $2 \times M$, type $3 \times M$, $\ldots$, where F only uses 2 (resp. 3, $\ldots$) bits from each input vector.

We say that a query has *type* 1×1 if it has the form $\exists j \exists l \exists V'.F(y_l^j, j, l, V')$?, i.e. F only uses one bit from one input vector (can be any bit from any vector and this is existentially quantified upon). Again, there is a similar notion of type 2×1, type 3×1, $\ldots$, where we only use 2 (resp. 3, $\ldots$) bits in total.

For a query type τ, we let **TB**(SAT)$_\tau$ denote the **TB**(SAT) problem restricted to trees of type τ (i.e. trees where all queries have type τ).

Before we see (in later sections) where such restricted queries appear, we show that they give rise to simpler oracle circuits:

Theorem 3.2. *For any $n > 0$*
1. **TB**(SAT)$_{n\times1}$ *is* $\mathrm{P}^{\mathrm{NP}[O(\log n)]}$*-complete,*
2. **TB**(SAT)$_{n\times\mathrm{M}}$ *is* $\mathrm{P}^{\mathrm{NP}[O(\log^2 n)]}$*-complete.*
We prove the upper bounds in the rest of this section. The lower bounds, Corollaries 4.6 and 5.6, are deduced from hardness results for model checking problems studied in the following sections.

3.2   Lowering TB(SAT)$_{1\times\mathrm{M}}$ Circuits

Assume block B is the parent of some B' inside a type $1 \times M$ tree T. Fig. 3 illustrates how one can merge B and B' into an equivalent single block B_{new}. Here B' is the leftmost child block of B, so that the input vector y^1 of B will play a special role, but the construction could have been applied with any other child.

The new block copies the u_i query gates and the G_i circuits from B' without modifying them. $2k$ new query gates $x_i^{s,b}$ are introduced for each x_i in B: $x_i^{s,b}$ is like x_i but it assumes $l_1 = s$ and $y_s^1 = b$ in F_i. The x_i query gates from B are replaced by new (non-query) circuits picking the best $x_i^{s,b}$ for which w_s agrees with the assumed value for y_s^1. The final B_{new} has type $1 \times M$ and degree $m' + m - 1$. $|B_{\mathrm{new}}|$ is $O(|B'| + 2k|B|)$: B was *expanded* but B' is unchanged.

The purpose of this merge operation is to lower the level of trees: we say a tree is *low* if its root has level at most $\log(1 + \text{number of blocks in the tree})$. The tree in Fig. 2 has 7 blocks and root at level 3, so it is (just barely) low.

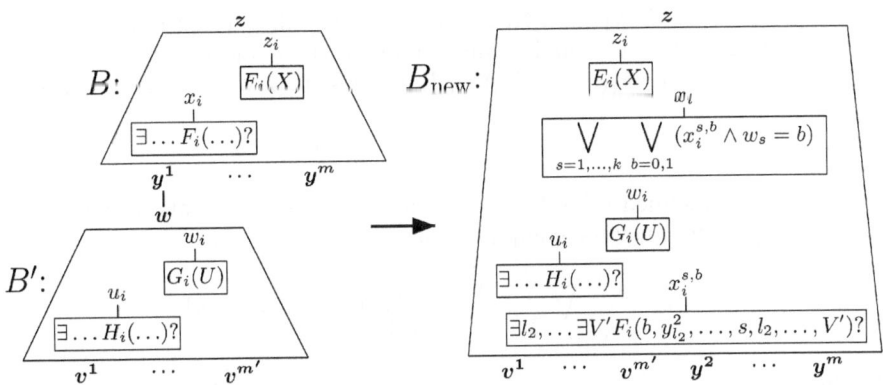

Fig. 3. Merging type $1 \times \mathrm{M}$ blocks

Lemma 3.3. *There is a logspace reduction that transforms type $1 \times \mathrm{M}$ trees of blocks into equivalent low trees.*

Proof. Consider a type $1 \times \mathrm{M}$ tree T. We say a block in T is *bad* if it is at some level $d > 1$ in T and has exactly one child at level $d - 1$ (called its *bad child*). For example B_2 is the only bad node in Fig. 2. If T has bad nodes, we pick a bad B of lowest level and merge it with its bad child. We repeat this until T has no bad node: the final tree T_{new} is low since any non-leaf block at level d must have at least two children at level $d - 1$ hence at least $2^d - 2$ descendants.

Observe that, when we merge a bad B at level d with its bad child B', the resulting B_{new} has level $d - 1$. Also, since we picked B lowest possible, B' was not bad, so B_{new} cannot be bad or have bad descendants. Thus B_{new} will never be bad again (though it can become a bad child) and will not be expanded a second time. Therefore T_{new} has size $O(k|T|)$ which is $O(|T|^2)$. $\qquad\square$

Observe that evaluating a low tree T only requires $O(\log|T|)$ rounds of parallel oracle queries. Therefore Lemma 3.3 provides a reduction from $\mathbf{TB}(\mathrm{SAT})_{1 \times \mathrm{M}}$ to $\mathrm{P}^{\mathrm{NP}}_{\|O(\log n)}$, a $\mathrm{P}^{\mathrm{NP}[O(\log^2 n)]}$-complete problem [1].

Corollary 3.4. $\mathbf{TB}(\mathrm{SAT})_{1 \times \mathrm{M}}$ *is in* $\mathrm{P}^{\mathrm{NP}[O(\log^2 n)]}$.

If now n is any fixed number, the obvious adaptation of the merging technique can lower trees of type $n \times \mathrm{M}$. Here the new block B_{new} uses $(2k)^n$ new query gates but since n is fixed, the transformation is logspace and the resulting T_{new} has size $O(|T|^{n+1})$.

Corollary 3.5. *For any $n \in \mathbb{N}$, $\mathbf{TB}(\mathrm{SAT})_{n \times \mathrm{M}}$ is in* $\mathrm{P}^{\mathrm{NP}[O(\log^2 n)]}$.

3.3   Flattening $\mathbf{TB}(\mathrm{SAT})_{1 \times 1}$ Circuits

Lemma 3.6. *For any $n \in \mathbb{N}$, there is a logspace reduction that transforms type $n \times 1$ trees of blocks into equivalent blocks.*

Proof (Sketch). With type 1×1 trees, one can merge all children $B_1, \dots, B_m$ with their parent B without incurring any combinatorial explosion. A query gate x_i of the form $\exists j \, \exists l \, \exists V'.F_i(y_l^j, j, l, V')$? will give rise to $2km$ new query gates

$$x_i^{r,s,b} := \exists V'.F_i(b, r, s, V')?$$

where r is the assumed value for j, s the assumed value for l and b the assumed value for y_s^r. x_i will now be computed via

$$x_i := \bigvee_{r=1,\dots,m} \bigvee_{s=1,\dots,k} \bigvee_{b=0,1} (x_i^{r,s,b} \wedge w_s^r = b).$$

We have $|B_{\text{new}}| = O(|B_1| + \dots + |B_m| + 2km|B|)$ so that a bottom-up repetitive application will transform a type 1×1 tree T into a single block of size $O(|T|^3)$.

For type n×1 trees, the obvious generalization introduces $(2km)^n$ new query gates when merging $B_1, \dots, B_m$ with their parent B, so that a tree T is flattened into a single block of size $O(|T|^{2n+1})$.   □

Lemma 3.6 reduces $\mathbf{TB(SAT)}_{n \times 1}$ to $\mathrm{P}_{\parallel}^{\mathrm{NP}}$, a $\mathrm{P}^{\mathrm{NP}[O(\log n)]}$-complete problem.

Corollary 3.7. *For any $n \in Nat$, $\mathbf{TB(SAT)}_{n \times 1}$ is in $\mathrm{P}^{\mathrm{NP}[O(\log n)]}$.*

4 Model Checking $B^*(\mathsf{X})$

In this section we show:

Theorem 4.1. *The model checking problem for $B^*(\mathsf{X})$ is $\mathrm{P}^{\mathrm{NP}[O(\log^2 n)]}$-complete.*

We start by introducing BX^*, a fragment of $B^*(\mathsf{X})$ where all occurrences of X are immediately over an atomic proposition, or an existential path quantifier (or an other X). Formally, BX^* is given by the following abstract syntax:

$$\varphi ::= \mathsf{E}f(\mathsf{X}^{n_1}\varphi_1, \dots, \mathsf{X}^{n_k}\varphi_k) \mid P_1 \mid P_2 \mid \dots$$

where $f(\dots)$ is any Boolean formula.

Lemma 4.2. *There exists a logspace transformation of $B^*(\mathsf{X})$ formulae into equivalent BX^* formulae.*

Proof (Idea). Bury the X's using $\mathsf{X}(\varphi \wedge \psi) \equiv \mathsf{X}\varphi \wedge \mathsf{X}\psi$ and $\mathsf{X}(\neg\varphi) \equiv \neg\mathsf{X}\varphi$.   □

Lemma 4.3. *There exists a logspace transformation from model checking for BX^* into $\mathbf{TB(SAT)}_{1 \times M}$.*

Proof. With a KS S and a BX^* formula φ we associate a tree of blocks where the width k is the number of states in S, and where there is a block B_ψ for every subformula ψ of φ (so that the structure of the tree mimics the structure of φ). The blocks are built in a way that ensures that the ith output bit of B_ψ is true iff q_i, the ith state in S, satisfies ψ. This only needs type $1 \times M$ blocks.

Assume ψ is some $\exists f(X^{n_1}\psi_1, \dots, X^{n_m}\psi_m)$ with $n_1 \leq n_2 \leq \dots \leq n_m$. Then, for $i = 1, \dots, k$, B_ψ computes whether $q_i \models \psi$ with a query gate x_i defined via

$$x_i := \exists l_1, \dots, l_m \left(f(y_{l_1}^1, \dots, y_{l_m}^m) \wedge \bigwedge_{j=1,\dots,m} Path(l_{j-1}, n_j - n_{j-1}, l_j) \right) ?$$

where $l_0 = i$, $n_0 = 0$ and $Path(l, n, l')$ (definition omitted) is a Boolean formula stating that S has an n-steps path from q_l to $q_{l'}$. $\qquad\square$

Corollary 4.4. *Model checking for $B^*(X)$ is in $P^{NP[O(\log^2 n)]}$.*

For Theorem 4.1, we need prove the corresponding lower bound:

Proposition 4.5. *Model checking for $B^*(X)$ is $P^{NP[O(\log^2 n)]}$-hard.*

We have to omit the proof of Proposition 4.5 for lack of space. The complete 3-pages proof can be found in the full version of this paper.

Corollary 4.6. $TB(SAT)_{1 \times M}$ *is $P^{NP[O(\log^2 n)]}$-hard.*

5 Model Checking *Timed B(F)*

In this section we show:

Theorem 5.1. *The model checking problem for Timed $B(F)$ over Timed KSs is $P^{NP[O(\log n)]}$-complete.*

This is obtained through the next two lemmas.

Lemma 5.2. *Model checking Timed $B(F)$ over Timed KSs is $P^{NP[O(\log n)]}$-hard.*

Proof. By reduction from PARITY-SAT. Assume we are given a set $\mathcal{I}_0, \dots, \mathcal{I}_{n-1}$ of SUBSET-SUM instances: we saw in section 2.2 how to associate with these a Kripke structure S and simple *Timed $B(F)$* formulae $\psi_0, \dots, \psi_{n-1}$ s.t. for every i, $\mathcal{I}_i$ is solvable iff $S \models \psi_i$. Assume w.l.o.g. that n is some power of 2: $n = 2^d$ and for every tuple $\langle b_1, \dots, b_k \rangle$ of $k \leq d$ bits define

$$\varphi_{\langle b_1,\dots,b_k \rangle} \overset{\text{def}}{=} \begin{cases} \neg\,\psi_{\sum_{j=1}^k b_j 2^{j-1}} & \text{if } k = d, \\ (\varphi_{\langle 0,b_1,\dots,b_k \rangle} \wedge \varphi_{\langle 1,b_1,\dots,b_k \rangle}) \vee (\neg\varphi_{\langle 0,b_1,\dots\rangle} \wedge \neg\varphi_{\langle 1,b_1,\dots\rangle}) & \text{otherwise.} \end{cases}$$

$S \models \varphi_{\langle b_1,\dots,b_k \rangle}$ iff there is an even number of solvable $\mathcal{I}_i$s among those whose index i has $b_1, \dots, b_k$ as last k bits. Therefore the total number of solvable $\mathcal{I}_i$ is even iff $S \models \varphi_{\langle\rangle}$. Since $d = \log n$, $|\varphi_{\langle\rangle}|$ is in $O(n\sum_i|\psi_i|)$ and the reduction is logspace. $\qquad\square$

We note that $P^{NP[O(\log n)]}$-hardness already occurs with a modal depth 1 formula.

Remark 5.3. Observe that this proof applies to all the logics we mentioned in section 2.2: it only requires that several SAT problems $f_1, \dots, f_n$ can be reduced to respective formulae $\psi_1, \dots, \psi_n$ other a same structure S. (This is always possible for logics having a reachability modality like EX or EF). $\qquad\square$

Lemma 5.4. *There exists a logspace transformation from model checking for Timed $B(F)$ over Timed KSs into* $\mathbf{TB}(SAT)_{1\times 1}$.

Proof (Sketch). We mimic the proof of Lemma 4.3: again we associate a block B_ψ for each subformula and k is the number of states of the Kripke structure S. Assume the edges $e_1, \ldots, e_r$ of S carry weights $d_1, \ldots, d_r$. Then, for ψ of the form $EF_{=c}\psi'$, block B_ψ will compute whether $q_i \models \psi$ by asking the query

$$x_i := \exists l \, \exists n_1, \ldots, n_r \left(y_l^1 \wedge c = \sum_{j=1,\ldots,r} n_j r_j \wedge Path'(i, n_1, \ldots, n_r, l) \right) ?$$

where $Path'(i, n_1, \ldots, n_r, l)$ (definition omitted) is a Boolean formula checking that there exists a path from q_i to q_l that uses exactly n_j times edge e_j for each $j = 1, \ldots, r$ (Euler's circuit theorem makes the check easy). We refer to [14, Lemma 4.5] for more details (e.g. how are the n_is polynomially bounded?) since here we only want to see that type 1×1 queries are sufficient for *Timed $B(F)$*. $AF_{=c}\psi$ is dealt with similarly. $\square$

Corollary 5.5. *Model checking Timed $B(F)$ over Timed KSs is in* $P^{NP[O(\log n)]}$.

Corollary 5.6. $\mathbf{TB}(SAT)_{1\times 1}$ *is* $P^{NP[O(\log n)]}$*-hard.*

6   Conclusion

We solved the model checking problems for $B^*(X)$ and *Timed $B(F)$*, two temporal logic problems where the precise computational complexity was left open.

For $B^*(X)$, the result is especially interesting because of the fundamental nature of this logic, but also because it provides the first example of a natural problem complete for $P^{NP[O(\log^2 n)]}$. Indeed, identifying the right complexity class for this problem was part of the difficulty.

Proving membership in $P^{NP[O(\log^2 n)]}$ required introducing a new family of oracle circuits. These circuits are characterized by their tree-vector form, and additional special logical conditions on the way an oracle query may depend on its inputs. The tree-vector form faithfully mimics branching-time model checking, while the special logical conditions originate from the modalities that appear in the path formulae. We expect our results on the evaluation of these circuits will be applied to other branching-time logics.

References

1. J. Castro and C. Seara. Complexity classes between Θ_k^p and Δ_k^p. *RAIRO Informatique Théorique et Applications*, 30(2):101–121, 1996.
2. E. M. Clarke, E. A. Emerson, and A. P. Sistla. Automatic verification of finite-state concurrent systems using temporal logic specifications. *ACM Trans. Programming Languages and Systems*, 8(2):244–263, 1986.
3. E. M. Clarke, O. Grumberg, and D. A. Peled. *Model Checking*. MIT Press, 1999.
4. S. Demri and Ph. Schnoebelen. The complexity of propositional linear temporal logics in simple cases. *Information and Computation*, 174(1):84–103, 2002.

5. E. A. Emerson. Temporal and modal logic. In J. van Leeuwen, editor, *Handbook of Theoretical Computer Science*, vol. B, chapter 16, pp 995–1072. Elsevier Science, 1990.

6. E. A. Emerson and J. Y. Halpern. Decision procedures and expressiveness in the temporal logic of branching time. *Journal of Computer and System Sciences*, 30(1):1–24, 1985.

7. E. A. Emerson and J. Y. Halpern. "Sometimes" and "Not Never" revisited: On branching versus linear time temporal logic. *J. ACM*, 33(1):151–178, 1986.

8. E. A. Emerson and Chin-Laung Lei. Modalities for model checking: Branching time logic strikes back. *Science of Computer Programming*, 8(3):275–306, 1987.

9. G. Gottlob. NP trees and Carnap's modal logic. *J. ACM*, 42(2):421–457, 1995.

10. D. S. Johnson. A catalog of complexity classes. In J. van Leeuwen, editor, *Handbook of Theoretical Computer Science*, vol. A, chapter 2, pp 67–161. Elsevier Science, 1990.

11. R. Koymans. Specifying real-time properties with metric temporal logic. *Real-Time Systems*, 2(4):255–299, 1990.

12. O. Kupferman, M. Y. Vardi, and P. Wolper. An automata-theoretic approach to branching-time model checking. *J. ACM*, 47(2):312–360, 2000.

13. F. Laroussinie, N. Markey, and Ph. Schnoebelen. Model checking CTL^+ and $FCTL$ is hard. In *Proc. 4th Int. Conf. Foundations of Software Science and Computation Structures (FOSSACS'2001)*, vol. 2030 of *Lect. Notes Comp. Sci.*, pp 318–331. Springer, 2001.

14. F. Laroussinie, N. Markey, and Ph. Schnoebelen. On model checking durational Kripke structures. In *Proc. 5th Int. Conf. Foundations of Software Science and Computation Structures (FOSSACS'2002)*, vol. 2303 of *Lect. Notes in Comp. Sci.*, pp 264–279. Springer, 2002.

15. C. H. Papadimitriou. *Computational Complexity*. Addison-Wesley, 1994.

16. Ph. Schnoebelen. The complexity of temporal logic model checking (invited lecture). In *Advances in Modal Logic, papers from 4th Int. Workshop on Advances in Modal Logic (AiML'2002)*. World Scientific, 2003. To appear.

17. A. P. Sistla and E. M. Clarke. The complexity of propositional linear temporal logics. *J. ACM*, 32(3):733–749, 1985.

18. M. Y. Vardi and P. Wolper. An automata-theoretic approach to automatic program verification. In *Proc. 1st IEEE Symp. Logic in Computer Science (LICS'86)*, pp 332–344. IEEE Comp. Soc. Press, 1986.

19. K. W. Wagner. More complicated questions about maxima and minima, and some closures of NP. *Theor. Comp. Sci.*, 51(1–2):53–80, 1987.

20. K. W. Wagner. Bounded query classes. *SIAM J. Computing*, 19(5):833–846, 1990.

21. C. B. Wilson. Relativized NC. *Mathematical Systems Theory*, 20(1):13–29, 1987.

There Are Spanning Spiders in Dense Graphs (and We Know How to Find Them)[*]

Luisa Gargano and Mikael Hammar

Dipartimento di Informatica ed Applicazioni
Università di Salerno, 84081 Baronissi (SA), Italy
fax:+39 089965272, {lg,hammar}@dia.unisa.it

Abstract. A spanning spider for a graph G is a spanning tree T of G with at most one vertex having degree three or more in T. In this paper we give density criteria for the existence of spanning spiders in graphs. We constructively prove the following result: Given a graph G with n vertices, if the degree sum of any independent triple of vertices is at least $n - 1$, then there exists a spanning spider in G. We also study the case of bipartite graphs and give density conditions for the existence of a spanning spider in a bipartite graph. All our proofs are constructive and imply the existence of polynomial time algorithms to construct the spanning spiders. The interest in the existence of spanning spiders originally arises in the realm of multicasting in optical networks. However, the graph theoretical problems discussed here are interesting in their own right.

Keywords: Graph theory, Graph and network algorithms.

1 Introduction

We consider the problem of constructing, for a given graph G, a spanning spider, that is, a spanning tree of G in which at most one vertex has degree larger than 2.

Much work has been devoted to the study of the existence of a Hamilton path in a given graph both from the algorithmic and the graph–theoretic point of view. Deciding if a graph admits a Hamilton path is a well known NP-complete problem, even in cubic graphs [11]. On the other hand, if the graph G satisfies any of a number of density conditions, a Hamilton path is guaranteed to exist. Dirac's classical theorem asserts that if G is a graph on n vertices and each vertex of G has degree at least $n/2$, then G has a Hamilton cycle. Dirac's proof also shows that if the sum of the degrees of any pair of independent vertices of G is at least $n - 1$, then G has a Hamilton path [5]. It is also well known that the

---

[*] This work is partially supported by the ministero dell'istruzione dell'università e della ricerca: the resource allocation in wireless networks project; and the European Union research training network: approximation and randomized algorithms in communication networks.

J.C.M. Baeten et al. (Eds.): ICALP 2003, LNCS 2719, pp. 802–816, 2003.

above density condition also provides an efficient algorithm to find the Hamilton path (start with any path and extend it by one edge at one of its endpoints; when this process cannot be iterated anymore, one gets the desired Hamilton path).

There are several natural generalizations of the Hamilton path problem. One may want for instance to minimize the maximum degree in a spanning tree of G — when asking for a spanning tree of maximum degree at most k, Dirac's density condition can be generalized to ask that the sum of the degrees of any k pairwise independent vertices is at least $n-1$ [17].

Another direction for generalizing the Hamilton path problem was considered in [7]: find a spanning tree T of a given graph G having the minimum possible number of branch vertices, where a branch vertex is a vertex of degree larger than two in T. The above minimum is zero if and only if the graph G has a Hamilton path. The interest in minimizing the number of branch vertices arises from a problem in optical networks [14,16]; it is motivated by an efficient use of new technologies (e.g., light splitting devices) in the realm of multicasting in optical networks; the interested reader is referred to [7,18].

Several algorithmic and graph–theoretic questions where studied in [7] concerning the construction of spanning trees with few branch vertices. In particular, density conditions that are sufficient to give upper bounds on the minimum number of branch vertices were studied; such conditions add to a degree bound the assumption that the graph is claw–free (e.g., it does not contain an induced $K_{1,3}$ (or $K_{1,4}$) subgraph). No non-trivial density condition is known to be sufficient without any additional assumption on the graph. The following conjecture was made in [7].

Conjecture 1. [**7**] Let G be a connected graph and k a nonnegative integer. If each vertex of G has degree at least $\frac{n-1}{k+2}$ (or more generally, if the sum of the degrees of any $k+2$ independent vertices is at least $n-1$) then there exists a spanning tree in G with at most k vertices of degree higher than 2.

For $k = 0$ the conjecture is true, being Dirac's condition for the existence of a Hamilton path.

A tree with at most one branch vertex is called a *spider*. Here we are interested in the existence (and construction) for a given graph G of a spanning spider — e.g. the case $k = 1$ in the above conjecture. We notice that the problem of deciding whether a given graph admits a spanning spider is computationally intractable in general.

Theorem 1. [**7**] *It is NP-complete to decide whether a graph G admits a spanning spider.*

1.1 Our Results

In this paper we study the problem of existence (and construction) of spanning spiders both in general and in bipartite graphs.

In case of general graphs we show that any graph G on n vertices in which the sum of the degrees of any three pairwise independent vertices is at least $n-1$ admits a spanning spider and this spider can be efficiently found. Namely, we prove the following Theorem.

Theorem 2. *Let G be a connected graph in which the sum of the degrees of any three independent vertices is at least $n - 1$. Then G contains a spanning spider. Furthermore, there is an $O(n^3)$ time algorithm to find a spanning spider in G.*

We also consider the case of bipartite graphs. It is well known that the degree bound for the existence of a Hamilton path can be improved when considering bipartite graphs [2]. The same holds for the existence of spanning spiders.

Theorem 3. *Let $G = (U, V, E)$ be a connected bipartite graph with $|U| \leq |V|$ such that for all $u \in U$ and $v \in V$, it holds that*

$$d(u) + d(v) \geq |U| \text{ and } d(v) \geq \frac{|V||U|}{|V| + |U|}.$$

Then G contains a spanning spider. Furthermore, there is an $O(n^3)$ time algorithm to find a spanning spider in G.

The stronger density condition given in the following theorem assures, for a bipartite graph $G = (U, V, E)$ with $|U| \leq |V|$, the existence of a spanning spider centered in u, for *each* vertex $u \in U$. Notice that if $|V| \geq |U|+2$ then G cannot contain a spanning spider centred at a vertex $v \in V$, even if G is the complete bipartite graph $K_{|U|,|V|}$.

Theorem 4. *Let $G = (U, V, E)$ be a connected bipartite graph with $|U| \leq |V|$ such that for all $u \in U$ and $v \in V$, it holds that $d(u) + d(v) \geq |V|$. Then G contains a spanning spider centred at any vertex $u \in U$. Furthermore, there is an $O(n^3)$ time algorithm to find a spanning spider in G.*

A basic tool in the construction of the desired spanning spiders is the construction of a sufficiently long path in the given graph. In Section 3 we give a local optimisation heuristic to find such paths. We define a set of *maximality criteria* and show how to find paths satisfying these criteria.

For the case of general graphs, our paths actually are Hamilton paths if the graph satisfies the density criterion of Dirac [5]. Indeed, our maximality criterion includes the simple one used in the original proof by Dirac (in that case a path is called maximal if it cannot be extended by adding one new vertex at one of its endpoints).

We also give a density criterion for the existence and construction of long paths in bipartite graphs — a generalization of our interest in the construction of paths that contain all vertices from the smaller partition of the vertex set. This criterion is a bit stronger than what is given by the more general theorem by Jackson [10], and we show a simple and efficient algorithm to generate such paths. In particular we prove the following result

Lemma 1. *Given a bipartite graph* $G = (U, V, E)$, *with* $|V| \geq |U|$. *If* $d(u)+d(v) \geq \delta$ *for any* $u \in U$ *and* $v \in V$, *then we can find, in time* $O(n^3)$, *a path in* G *that either spans all vertices in* U *or has size at least* 2δ.

1.2 Summary of the Paper

In Section 2 we state the notation used in the rest of the paper. In Section 3 we define maximal paths and show how to construct them. In Section 4 we show how to turn a maximal path into a spanning spider, in any graph satisfying the density condition of Theorem 2. In Section 5, we consider the construction of maximal paths and spanning spiders in bipartite graphs. In Section 6 we conclude and state some open problems. Please note that some proofs are omitted due to space limitations.

2 Notation

Let $G = (V, E)$ denote a connected graph on n vertices (in the rest of the paper any graph should be intended as a connected graph and we will reserve n to denote the number of vertices).

For a vertex $v \in V$ we let $d(v)$ denote the degree of v. For a subset $X \subseteq V$ we define $d_X(v)$ to be the number of vertices in X that are adjacent to v. We use $\delta(G) = \min_{v \in V} d(v)$ to denote the minimum degree of any vertex in G, and let $\delta_k(G)$ denote the minimum degree sum of any k pairwise independent vertices in G.

The neighborhood in G of a vertex x is denoted by $N(x)$, for a subset $X \subseteq V$ we define the neighbourhood of $v \in V$ with respect to X as

$$N_X(v) = \{u \in X \mid (u, v) \in E\}.$$

For sake of simplicity, whenever it is clear from the context, we will identify the vertex set of a (sub)graph H of G with H itself. Hence, we will use $|H|$ to indicate the number of vertices in the graph and $d_H(v)$ and $N_H(v)$ will represent, respectively, the degree and the neighborhood of v with respect to the vertex set of H.

Let $P = [v_0 v_1 \ldots v_t]$ denote a path in G. The *left neighbourhood* of $x \in V$ on P is the set

$$N_P^-(x) = \{v_i \mid (v_{i+1}, x) \in E\}.$$

The *right neighbourhood* of $x \in V$ on P is defined analogously as

$$N_P^+(x) = \{v_i \mid (v_{i-1}, x) \in E\}.$$

When the underlying path is evident from the context we write $N^-(x)$ and $N^+(x)$ for the left and right neighbourhoods, respectively.

Any left neighbour $v_i \in N^-(v_0)$ of v_0 is the end point of the path $P - (v_i, v_{i+1}) + (v_0, v_{i+1})$ containing the same set of vertices as P; by symmetry, the same holds for $N^+(v_t)$; see Figure 1. Therefore, we say that the elements in $N^-(v_0)$ and $N^+(v_t)$ are *potential endpoints* with respect to P.

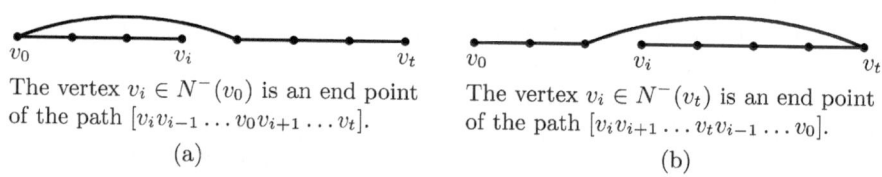

The vertex $v_i \in N^-(v_0)$ is an end point of the path $[v_i v_{i-1} \dots v_0 v_{i+1} \dots v_t]$.

(a)

The vertex $v_i \in N^-(v_t)$ is an end point of the path $[v_i v_{i+1} \dots v_t v_{i-1} \dots v_0]$.

(b)

Fig. 1. Potential end points in a path.

3   Maximal Paths in General Graphs

In order to construct a spanning spider in a dense graph, we first find a suitable long path in the graph. This path will then be turned into a spider that in a last step can be extended to span the whole graph. This section is devoted to finding the desired long paths in dense general graphs.

The following set of *maximality criteria* implicitly suggest a local optimisation heuristic to find suitably long paths in general dense graphs. We obtain this heuristic by showing how to find paths that satisfy the criteria.

Definition 1. *A path $P = [v_0 \dots v_t]$ is called* maximal *if either it is a Hamilton path or it satisfies each of the following conditions:*

i) $N(r) \cap N^-(v_0) = \emptyset = N(r) \cap N^+(v_t)$,   *for every $r \in V - P$.*
ii) $N(v_0) \cap N^+(v_t) = \emptyset$.
iii) $N^-(r)$ *is an independent set, for every $r \in V - P$.*
iv) *If $N^-(v_0) \cap N^+(v_t) \neq \emptyset$ then*
    (a) *no two consecutive vertices in P both have neighbours in $V - P$,*
    (b) *$V - P$ is an independent set.*

We show now that any non-maximal path $P = [v_0 \dots v_t]$ can be extended in polynomial time.

If condition **i)** is violated then there is a vertex r outside P that is adjacent to a potential end point of a path P'. Thus, we construct P' (if r is adjacent to v_0 or v_t then $P' = P$) as described in Figure 1 and add r to this path.

If condition **ii)** is violated then we can find a cycle in G that contains all the vertices of P; see Figure 2. Since G is connected and P is not a Hamilton path there is a vertex r outside P that is adjacent to a vertex v in P. Thus, we can

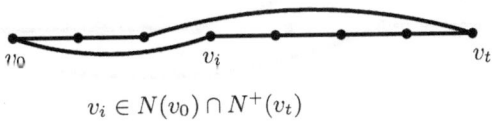

$$v_i \in N(v_0) \cap N^+(v_t)$$

Fig. 2. If $N(v_0) \cap N^+(v_t) \neq \emptyset$ then there is a cycle in G that contains all vertices in P.

extend P by constructing the path P' obtained by adding (r, v) to the cycle and removing any other edge incident to v.

If condition **iii)** is violated we find an edge between two vertices in $N^-(r)$ and extend P as described in Figure 3.

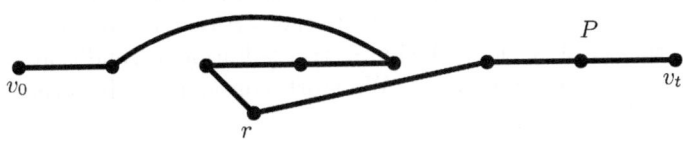

Fig. 3. If the left neighbourhood on P of a vertex r outside P is not an independent set then P can be extended to include r.

If condition **iv)** is violated then we have two cases to consider: either there are two consecutive vertices on P that are both adjacent to vertices in the subgraph $G-P$, or $V-P$ is not an independent set.

In the first case we identify the two vertices v_i and v_{i+1} that are both adjacent to vertices outside P. If they are both adjacent to the same vertex $r \in V-P$ then we directly add this new vertex to P obtaining the longer path $[v_0 \ldots v_i r v_{i+1} \ldots v_t]$. If they are adjacent to different vertices in $V-P$, we construct the cycle C containing all but one vertex of P as described in Figure 4. Let v denote the excluded vertex. Note that $(v_i, v_{i+1}) \in E(C)$ (otherwise either $v_i = v$ or $v_{i+1} = v$; but $v \in N^-(v_0)$ and such a vertex is not adjacent to vertices in $V-P$, by condition **i)**). Assume that $(v_i, r_1) \in E$ and $(v_{i+1}, r_2) \in E$, where $r_1, r_2 \notin P$. By removing (v_i, v_{i+1}) from C and adding (r_1, v_i) and (v_{i+1}, r_2) to C, we create a new path of size $|P|+1$, with end points r_1 and r_2.

For the second case, we observe that if $V-P$ is not an independent set then we can construct a path P' in $G-P$ containing at least two vertices, which by the connectivity of G can be connected to the cycle C described above. In this way we create a new path with size at least $|P|+1$.

A careful analysis of the violation checks above shows that an algorithm to find a maximal path can be implemented to run in $O(n^3)$ time.

Theorem 5. *A maximal path in a connected graph can be found in $O(n^3)$ time.*

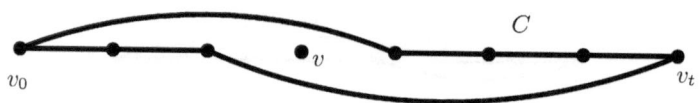

Fig. 4. The cycle C includes all vertices of the path $[v_0 \ldots v_t]$ except $v \in N^-(v_0) \cap N^+(v_t)$.

4 Spanning Spiders in General Graphs

In this section we give an algorithm to find spanning spiders in dense graphs, where dense in our case means graphs G for which $\delta_3(G) \geq n - 1$. We base our algorithm on the fact that we can compute maximal paths, as defined in Section 3, in $O(n^3)$ time. The given algorithm will prove Theorem 2.

Let P denote a maximal path in G according to Definition 1, with $P = [v_0 v_1 \ldots v_{t-1} v_t]$ and let $R = V - P$ denote the vertices of G outside P.

Recall that Definition 1 includes two additional conditions if the set $N^-(v_0) \cap N^+(v_t)$ is non-empty. We start considering the other case, i.e., $N^-(v_0) \cap N^+(v_t) = \emptyset$.

Lemma 2. *If P is maximal then either $N^-(v_0) \cap N^+(v_t) \neq \emptyset$ or there is a spanning spider in G whose centre is adjacent to all vertices outside P.*

Assume from now on that

$$N^-(v_0) \cap N^+(v_t) \neq \emptyset.$$

This implies that condition **iv (a)** and **iv (b)** of Definition 1 hold.

We will give an algorithm proving the following weaker theorem. Later we will extend it to the general case considered in Theorem 2.

Theorem 6. *Any connected graph G with $\delta(G) \geq (n-1)/3$ contains a spanning spider. Furthermore, there is an $O(n^3)$ time algorithm to find a spanning spider in G.*

The following lemma gives Theorem 6 when the size of R is small.

Lemma 3. *If $|R| \leq 2$ then G contains a spanning spider.*

Assume now that $|R| \geq 3$, with $R = \{r_1, r_2, \ldots, r_{|R|}\}$, and let r^* denote an arbitrary vertex in R. In order to prove Theorem 6, we construct a spanning spider out of the maximal path P. First we need to find a suitable centre for the spider. It turns out that a convenient property of such a centre is to be adjacent to many independent vertices which in turn are independent of R.

Lemma 4. *The set $N^-(r^*) \cup R$ is independent, with size $|R| + (n-1)/3$. Furthermore, there exists a vertex $v_i \in P - N^-(r^*)$ whose number of neighbours in $N^-(r^*) \cup R$ is at least*

$$\frac{(n-1)}{6} + \frac{3|R| - 1}{4}.$$

Proof. The independence is given by Definition 1 as follows. If $r \in R$ and $v \in N^-(r^*)$ then $(r, v) \notin E$ by condition **iv)** point **(a)**. R is an independent set by condition **iv)** point **(b)**. Left is to prove that $N^-(r^*)$ is independent, but this follows from condition **iii)**. The size of the union follows from the degree condition on r^*, and the fact that R and $N^-(r^*)$ are disjoint.

For the second part of the proof, consider the vertices in $N^-(r^*) \cup R$. Each of them is adjacent only to vertices in $P - N^-(r^*)$, since $N^-(r^*) \cup R$ is an independent set and $R \cap P = \emptyset$. By the pigeonhole principle there exists a vertex $v_i \in P - N^-(r^*)$ adjacent to at least

$$\frac{\frac{(n-1)}{3} \left| N^-(r^*) \cup R \right|}{|P - N^-(r^*)|} = \frac{\frac{n-1}{3} \left(\frac{n-1}{3} + |R| \right)}{n - \frac{n-1}{3} - |R|} = \frac{\frac{n-1}{3} \left(\frac{n-1}{3} - \frac{|R|-1}{2} + \frac{3|R|-1}{2} \right)}{2 \left(\frac{n-1}{3} - \frac{|R|-1}{2} \right)} \geq \frac{n-1}{6} + \frac{3|R|-1}{4}$$

vertices in $N^-(r^*) \cup R$. $\qquad\Box$

Let v_i be a vertex in $P - N^-(r^*)$ satisfying the condition given in Lemma 4.[1] Let Δ be the number of vertices in $N^-(r^*) \cup R$ adjacent to v_i, i.e.,

$$\Delta \geq \frac{n-1}{6} + \frac{3|R|-1}{4}. \tag{1}$$

Using the algorithm in Table 1 we construct a spider S, centred at v_i, with branches beginning at vertices in $N^-(r^*)$ and ending at vertices in $N(r^*)$. Note that S fails to include the tail of P. We let T denote this tail; see Figure 5.

Table 1. The spider construction algorithm for general graphs.

Algorithm   Spider construction in general graphs.
Input: A graph $G = (V, E)$, a maximal path P, and a vertex v_i satisfying the condition of Lemma 4.
Output: A spider S, centred at v_i, and a tail T, that collectively span P and a portion of R.
1 Initially let $S := P$.
2 **For each** $r \in R$ such that $(v_i, r) \in E$: add the edge (v_i, r) to S.
3 **If** all $r \in R$ are adjacent to v_i: **return** the spanning spider S. **Otherwise,**
4 **For each** $v_j \in P$ such that both (v_{j-1}, v_i) and (v_j, r^*) are in E: remove (v_{j-1}, v_j) from S, and add the edge (v_i, v_{j-1}) to S.
5 **If** there is an edge $(v_i, v_j) \in S$ with $j > i + 1$: remove the edge (v_i, v_{i+1}) from S (recall that v_i is the centre of the spider).
6 **Return** the spider S and the tail $T := P - S$.
End

Let L denote the leaves in S and let $R' = S - P - r^* \subset R$. We note that the number of leaves in S is at least $\Delta + 1$ but more importantly, the number of leaves adjacent to r^* is

$$d_L(r^*) \geq \Delta - |R'| - 2. \tag{2}$$

---

[1] Notice that $i < t$, since by condition **i)** of Definition 1, $(N^-(r^*) \cup R) \cap N(v_t) = \emptyset$.

To see this, note first of all that r^* is not adjacent to any leaf that belongs to R'. Secondly, the tail T is not in S, but contains exactly one vertex in $N(v_i)$ that also lies in $N^-(r^*)$. Finally, if v_i is adjacent to r^*, then r^* is itself a leaf in S, but is of course not adjacent to itself.

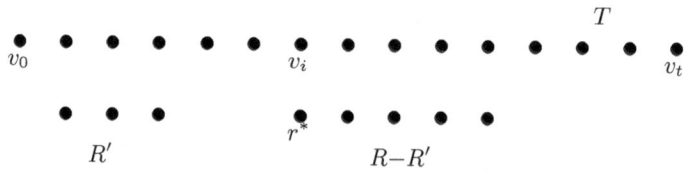

Fig. 5. The spider S, the tail T and the set $R - R'$, after the spider construction algorithm.

If there is a matching between the vertices in $R - R'$ and $L - R'$, then we can construct a spider covering G. Next we prove that there is such a matching. A vertex v in $S \cup T$ is called an internal vertex if $v \notin L$. We let I denote the set of internal vertices.

Lemma 5. *There exists a matching between $R - R'$ and $L - R'$.*

Proof. Since r^* is adjacent to more than $|R - R'|$ leaves in S, it suffices to show that there is a matching between $R - R' - \{r^*\}$ and $L - R'$. Let r denote an arbitrary vertex in $R - R' - \{r^*\}$. By definition,

$$d(r) = d_I(r) + d_L(r). \tag{3}$$

Since r is not adjacent to v_i, and v_{i+1} is a leaf by construction, neither v_i nor v_{i+1} is counted in $d_I(r)$. Neither are they counted in $d_L(r^*)$. This time, v_i is not counted, since it is not a leaf, and v_{i+1} is not counted because $v_i \notin N^-(r^*)$.

Therefore,

$$d_I(r) + d_L(r^*) \le (P - 2)/2,$$

since r and r^* cannot be adjacent to v_0 or v_t (Definition 1, condition **ii**)), nor to consecutive vertices on P (Definition 1, point **(a)** of condition **iv**)). Hence,

$$d_L(r) \ge d(r) + d_L(r^*) - (P - 2)/2. \tag{4}$$

Recalling that $d_L(r^*) \ge \Delta - |R'| - 2$ (by (2)) and that $|P| = n - |R|$, by using (4) we get

$$d_L(r) \ge \frac{n-1}{3} + (\Delta - |R'| - 2) - \frac{n - |R| - 2}{2}. \tag{5}$$

By using (1) we obtain

$$d_L(r) \ge \frac{n-1}{3} + \frac{n-1}{6} + \frac{3|R| - 1}{4} - |R'| - 2 - \frac{n - |R| - 2}{2}$$

$$= |R| - |R'| + (|R| - 7)/4$$

$$\geq |R - R'| - 1.$$

the last inequality holds since $|R| \geq 3$ by Lemma 3. Thus, each vertex in $R - R' - \{r^*\}$ is adjacent to at least $|R - R'| - 1$ leaves in S, so there exists a matching between $R - R' - \{r^*\}$ and $L - R'$.   □

Given the above guarantee of a matching we construct the spider as follows. Compute a matching between $R - R'$ and L. This gives us a new spider S' that contains all vertices except the tail T. The head of the tail is adjacent to r^*, and r^* is a leaf in S'. Add to S' the edge between r^* and the head of the tail to complete the spanning spider. This concludes the proof of Theorem 6.

Our main theorem follows easily from previous discussion.

Proof of Theorem 2 (sketch). We begin with the following observation. In any independent set, there can be at most two vertices with degree less than $(n-1)/3$. This follows directly from the degree sum criteria. Thus, in the set R there are at most two vertices with degree less than $(n-1)/3$, call these r and r'. It is easy to modify any maximal path so to contain the eventual low degree vertices, i.e., every vertex in R has at least $(n-1)/3$ neighbours.   □

5   Bipartite Graphs

In this section we consider bipartite graphs. As in the case of general graph, our spanning spider construction starts with the construction of a suitable long path.

5.1   Maximal Paths in Bipartite Graphs

In case of bipartite graphs the construction of a maximal path can be specialized to get Lemma 1, that is, to efficiently construct a path of size at least $2\min\{|U|, \delta\}$ in any bipartite graph $G = (U, V, E)$, with $|V| \geq |U|$ and $d(u) + d(v) \geq \delta$ for any $u \in U$ and $v \in V$.

The following special case will be used for the construction of spanning spiders in bipartite graphs.

Corollary 1. *Given a bipartite graph* $G = (U, V, E)$, *with* $|V| \geq |U|$. *If for any* $u \in U$ *and* $v \in V$, $d(u) + d(v) \geq |U|$ *then we can find, in time* $O(n^3)$, *a path in* G *that includes all vertices in* U.

Proof of Lemma 1. We first define a bipartite maximal path and then show that any such path has the desired property.

A path $P = [u_0 v_0 \ldots u_t v_t]$ in $G = (U, V, E)$ is called bipartite maximal if $|P| = 2|U|$ or it satisfies the following two conditions.

1) For any $u \in U$ and $v \in V$ the path P cannot be extended as either of the following

$$[uvu_0v_0 \ldots u_tv_t], \qquad [vu_0v_0 \ldots u_tv_tu], \qquad [u_0v_0 \ldots u_tv_tuv].$$

2) $N(u_0) \cap N^+(v_t) = \emptyset$.

It is possible to show that any bipartite maximal path has the desired length. To this aim, we show that any non maximal path P of size

$$|P| < 2\min\{|U|, \delta\} \tag{6}$$

can be extended to a path of size $|P| + 2$. This is obvious if condition **1)** does not hold. Consider then condition **2)** and let $v_i \in N(u_0) \cap N^+(v_t)$. Consider the cycle $P - \{(u_i, v_i)\} + \{(u_0, v_i), (u_i, v_t)\}$. W.l.o.g. denote it by

$$C = [u_0'v_0' \ldots u_t'v_t'].$$

We show that it is possible to obtain a path of size $|C| + 2 = |P| + 2$ from C. Let U' and V' denote the set of vertices outside C in U and V, respectively, i.e.,

$$U' = U - \{u_0', \ldots, u_t'\}, \text{ and } V' = V - \{v_0', \ldots, v_t'\}.$$

Since $|C| = |P| < 2|U|$ then $U' \neq \emptyset$.

Let us first assume that $R = U' \cup V'$ form an independent set. Let $u \in U'$ and $v \in V'$. If u and v are neighbours of two vertices that are adjacent on C then we can get a path of size $|C| + 2$ including all vertices in C together with u and v (see Figure 6).

Otherwise, for each pair u_i', v_i' of adjacent vertices in C it cannot hold that both $(u_i', v) \in E$ and $(u, v_i') \in E$. This implies that $d(u) + d(v) \leq |C|/2$, that is

$$|C| \geq 2(d(u) + d(v)) \geq 2\delta \geq 2\min\{|U|, \delta\}$$

contradicting (6).

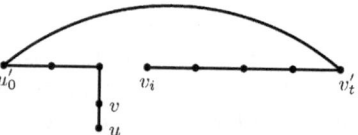

If R is independent then there exist vertices $u, v \in R$ adjacent to consecutive vertices on C.

(a)

If R is not independent then there exists an edge between vertices $u, v \in R$ adjacent to C.

(b)

Fig. 6. Extending paths in the bipartite case.

Suppose now that there exists an edge (u, v) between two vertices in R. Since G is connected, vertices u and v must be connected to C. This immediately implies that a path of at least 2 vertices in R can be ù added to C, thus giving a path of size at least $|C| + 2$. $\qquad \square$

5.2   Spanning Spiders in Bipartite Graphs

In this section we prove Theorem 3 and Theorem 4. We show how to construct the desired spanning spider starting from a bipartite maximal path as defined in Section 5.1.

Corollary 1 assures that if for each $u \in U$ and $v \in V$ it holds that $d(u)+d(v) \geq |U|$ then a bipartite maximal path includes all vertices of U (notice that this condition is satisfied by the hypothesis of both Theorem 4 and Theorem 3). Let $P = [u_0 v_0 \ldots u_{|U|-1} v_{|U|-1}]$ be such a path. Given a vertex $u_j \in U$, define the sets

$$R = V - P \quad \text{and} \quad R' = \{v \in R \,|\, (u_j, v) \in E\}. \tag{7}$$

We construct the spider S centred at u_j using the algorithm stated in Table 2.

Table 2. The spider construction algorithm for bipartite graphs.

Algorithm   Spider construction in bipartite graphs.
Input: A bipartite graph $G = (U, V, E)$, a maximal path P, and a vertex $u_j \in U$.
Output: A spider S, centred at u_j, and spanning P and R'.
1   Initially let $S := P$.
2   **For each** $v_i \in N_P(u_j)$ with $0 \leq i \leq |U| - 1,\ i \neq j - 1,\ i \neq j$: insert the edge (u_j, v) in S.
3   **For each** $v_i \in N_P(u_j)$ with $0 \leq i \leq |U| - 1,\ i \neq j - 1,\ i \neq j$: insert the edge (u_j, v_j) in S, remove from S the edge $\begin{cases} (u_i, v_i) & \text{if } i \geq j + 1, \\ (v_i, u_{i+1}) & \text{if } i \leq j - 2. \end{cases}$
End

This gives a spider S centred at u_j (see Figure 7) spanning all vertices in the path P and the set R'.

Let L and I denote, respectively, the leaves and the internal vertices of S that also belong to U. In order to obtain a spanning spider, one needs to include in S the vertices in $R - R'$. This will be done by finding a matching between the vertices in $R - R'$ and L.

Since the spider is centred in $u_j \in U$, it follows that all its leaves except $v_{|U|-1}$ and the elements in R' belong to U, i.e., $|L| = d(u_j) - 1 - |R'|$. Hence, the number of internal vertices of S that belong to U is

$$|I| = |U - L| = |U| - |L| = |U| - d(u_j) + |R'| + 1. \tag{8}$$

For each vertex $r \in R - R'$, we count the number of leaves in S that are adjacent to r, i.e., $d_L(r)$. Since, by definition, u_j is not adjacent to r, by (8)

$$d_L(r) = d(r) - d_I(r) \geq d(r) - (|I| - 1) = d(r) - |U| + d(u_j) - |R'|. \tag{9}$$

We need now to differentiate our reasoning in order to prove Theorem 3 and Theorem 4.

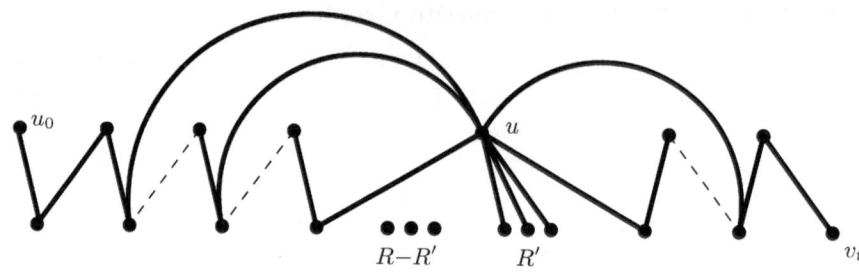

Fig. 7. Constructing a spider for a bipartite graph.

Proof of Theorem 3. Recall that we want to prove the existence of at least one spanning spider under the density criteria

(a) $d(u) + d(v) \geq |U|$,   (b) $d(v) \geq \alpha|U|$, where $\alpha = \frac{|V|}{|V|+|U|}$.

In order to conclude the proof of the theorem we need to choose a suitable center u_j for the above defined spider S. We choose u_j as the vertex of highest degree in U. By the pigeonhole principle

$$d(u_j) \geq \frac{\sum_{v \in V} d(v)}{|U|} \geq \frac{|V|(\alpha|U|)}{|U|} = \alpha|V|. \tag{10}$$

Take an arbitrary vertex $r \in R - R'$ (cfr. (7)). From (9) and (10) one has

$$d_L(r) \geq d(r) + d(u_j) - |R'| - |U| \geq d(r) + \alpha|V| - |R'| - |U|.$$

Using the density criteria **(b)**, we get

$$d_L(r) \geq \alpha|U| + \alpha|V| - |R'| - |U|.$$

Recalling that $\alpha = \frac{|V|}{|V|+|U|}$, and the equality $|R| = |V| - |U|$, we obtain

$$d_L(r) \geq \alpha(|U| + |V|) - |U| - |R'| = |V| - |U| - |R'| = |R| - |R'|.$$

By Hall's Theorem [9] there is a matching between $R - R'$ and the leaves of S and we can create a spanning spider by adding the vertices in $R - R'$ to the initial spider S centered in u_j.   □

Proof of Theorem 4. Recall that here u_j can be any vertex in U and that we are assuming that $d(u) + d(v) \geq |V|$, for any $u \in U$ and $v \in V$.

Using the fact that $d(r) + d(u_j) \geq |V|$, by (9) we get

$$d_L(r) \geq |V| - |U| - |R'| = |R - R'|.$$

By Hall's Theorem [9] there is a matching between the leaves of S and $R - R'$. Using this matching we create a spanning spider by adding the vertices in $R - R'$ to S.   □

6   Conclusions and Open Problems

We have considered the problem of constructing, for a given graph G, a spanning spider, that is, a spanning tree of G in which at most one vertex has degree larger than 2. We have considered both general and bipartite graphs. In particular, in the case of general graphs we have proved that given a graph G with n vertices, if the degree sum of any independent triple of vertices is at least $n - 1$, then there exists a spanning spider in G, thus proving a conjecture in [7].

The interest in the existence of spanning trees with limited number of vertices of degree larger than 2 (and of spanning spiders in particular) originally arises in the realm of multicasting in optical networks. However, the related algorithmic and graph theoretical problems are interesting in their own right.

The first obvious open question is whether Conjecture 1 holds for $k \geq 2$. Moreover, it would be interesting to extend the result presented in this paper in order to have non trivial density conditions for a graph to admit a spider covering at least a given fraction of its vertices.

References

1. C. Bazgan, M. Shanta, Z. Tuza, "On the Approximability of Finding A(nother) Hamiltonian Cycle in Cubic Hamiltonian Graphs", *Proc. 15th Annual Symposium on Theoretical Aspects of Computer Science*, LNCS, Vol. 1373, 276–286, Springer, (1998).
2. C. Berge. Graphs and Hypergraphs. North-Holland Publishing Company – Amsterdam and London, 1973.
3. J. A. Bondy. Properties of graphs with constraints on degrees. *Studia Sc. Math. Hung.*, 4:473–475, 1969.
4. V. Chvatal. On hamilton ideal's. *J. Comb. Theory*, 12 B:163–168, 1972.
5. G. A. Dirac. Some theorems on abstract graphs. *Proc. London Mathematical Society*, 2:69–81, 1952.
6. T. Feder, R. Motwani, C. Subi, "Finding Long Paths and Cycles in Sparse Hamiltonian Graphs", *Proc. Thirty second annual ACM Symposium on Theory of Computing (STOC'00)* Portland, Oregon, May 21–23, 524–529, ACM Press, 2000.
7. L. Gargano, P. Hell, L. Stacho, and U. Vaccaro. Spanning trees with bounded number of branch vertices. In *29-th International Colloquium on Automata, Languages, and Programming*, pages 355–365, 2002.
8. R. J. Gould. Updating the Hamiltonian problem—a survey. *J. Graph Theory*, 15(2):121–157, 1991.
9. P. Hall. On representatives of subsets. *Journal of London Mathematical Society*, pages 26–30, 1935.
10. B. Jackson. Long cycles in bipartite graphs. *Journal of Combinatorial Theory*, 38 B:118–131, 1985.
11. R. M. Karp. Reducibility among combinatorial problems. In *Complexity of Computer Applications*, pages 85–103. Plenum Press, New York, 1972.
12. S. Khuller, B. Raghavachari, N. Young, "Low degree spanning trees of small weight", *SIAM J. Comp.*, 25 (1996), 335–368.
13. J. Könemann, R. Ravi, "A Matter of Degree: Improved Approximation Algorithms for Degree-Bounded Minimum Spanning Trees", *Proc. Thirty second annual ACM Symp. on Theory of Computing (STOC'00)*, Portland, Oregon, 537–546, (2000).

14. B. Mukherjee, *Optical Communication Networks*, McGraw–Hill, New York, 1997.
15. L. Posa. A theorem concerning hamilton lines. *Magyar Tud. Akad. Mat. Kutato Int. Közl.*, 7:225–226, 1962.
16. T. E. Sterne, and K. Bala, *MultiWavelength Optical Networks*, Addison-Wesley, (1999).
17. S. Win, "Existenz von Gerüsten mit vorgeschriebenem Maximalgrad in Graphen", *Abh. Mat. Sem. Univ. Hamburg*, 43: 263–267, 1975.
18. X. Zhang, J. Wei, and C. Qiao Constrained Multicast Routing in WDM Networks with Sparse Light Splitting *Proc. of IEEE INFOCOM 2000*, vol. 3: 1781–1790, Mar. 2000.

The Computational Complexity of the Role Assignment Problem

Jiří Fiala[1]* and Daniël Paulusma[2]**

[1] Charles University, Faculty of Mathematics and Physics,
DIMATIA and Institute for Theoretical Computer Science (ITI)***,
Malostranské nám. 2/25, 118 00, Prague, Czech Republic.
fiala@kam.mff.cuni.cz
[2] University of Twente,
Faculty of Electrical Engineering, Mathematics and Computer Science,
Department of Applied Mathematics,
P.O. Box 217, 7500 AE Enschede, The Netherlands,
Phone: + 31 53 489 3421,
Fax: + 31 53 489 4858.
d.paulusma@math.utwente.nl

Abstract. A graph G is R-role assignable if there is a locally surjective homomorphism from G to R, i.e. a vertex mapping $r : V_G \to V_R$, such that the neighborhood relation is preserved: $r(N_G(u)) = N_R(r(u))$. Kristiansen and Telle conjectured that the decision problem whether such a mapping exists is an NP-complete problem for any connected graph R on at least three vertices. In this paper we prove this conjecture, i.e. we give a complete complexity classification of the role assignment problem for connected graphs. We show further corollaries for disconnected graphs and related problems.

Keywords: computational complexity, graph homomorphism, role assignment

2002 Mathematics Subject Classification: 05C15, 03D15.

1   Introduction

Given two graphs, say G and R, an *R-role assignment* for G is a vertex mapping $r : V_G \to V_R$, such that the neighborhood relation is maintained, i.e. all roles of the image of a vertex appear on the vertex's neighborhood. Such a condition can be formally expressed as

$$\text{for all } u \in V_G : r(N_G(u)) = N_R(r(u)),$$

where $N(u)$ denotes the set of neighbors of u in the corresponding graph.

---

* This author was partially supported by research grant GAUK 158/99.
** This author was partially supported by NWO grant R 61-507 and by Czech research grant GAČR 201/99/0242 during his stay at DIMATIA center in Prague.
*** Supported by the Ministry of Education of the Czech Republic as project LN00A056.

J.C.M. Baeten et al. (Eds.): ICALP 2003, LNCS 2719, pp. 817–828, 2003.

Such assignments have been introduced by Everett and Borgatti [6], who called them role colorings. They originated in the theory of social behavior. The graph R, i.e. the *role graph*, models roles and their relationships, and for a given society we can ask whether its individuals can be assigned roles such that the relationships are preserved: Each person playing a particular role has among its neighbors exactly all necessary roles as they are prescribed by the model.

From the computational complexity point of view it is interesting to know whether it is possible to decide quickly (i.e. in polynomial time) whether such assignment exists. This problem was considered by Roberts and Sheng [15], who focus on a more generalized problem called the 2-role assignment problem. If both graphs G and R are part of the input, the problem is NP-complete already for $R = K_3$ [12].

In order to make a more precise study we consider a class of R-role assignment problems, $RA(R)$, parameterized by the role graph R. Here the instance is formed only by the graph G, and we ask whether an R-role assignment of G exists.

The complexity study of this class of problems is closely related to a similar approach for locally constrained graph homomorphism problems [9]. A graph homomorphism from G to H is a vertex mapping $f : V_G \to V_H$ satisfying the property that whenever an edge (u, v) appears in E_G, then $(f(u), f(v))$ belongs to E_H as well.

The adjective "locally constrained" expresses the condition that the mapping f restricted to the neighborhood of any vertex u must satisfy further properties. (See [14,7] for a general model of such conditions.)

It may be required to be locally

- *bijective*, then the mapping is called a *full cover* of H, and the corresponding decision problem is called H-Cover [1,13],
- *injective*, then it is called a *partial cover* of H, and the problem H-PCover [8,9],
- *surjective*, then we get a *locally surjective cover* of H, and decision problem H-Colordomination [14].

All these problems are parameterized by a fixed graph H, and the instance is formed only by a graph G. The question is whether an appropriate graph homomorphism from G to H exists. Observe that the definition of a locally surjective cover is equivalent with the definition of an R-role assignment for $R = H$.

Full covers have important applications, for example in distributed computing [5], in recognizing graphs by networks of processors [2,3], or in constructing highly transitive regular graphs [4]. Similarly partial covers are used in distance constrained labelings of graphs [10].

Even if the first attempt to get some results on the computational complexity for the class of H-Cover problems was made a decade ago in [1], it is not fully classified yet neither for H-PCover nor for H-Colordomination ($RA(H)$) problems. However, several partial results are known. For example, if the H-Cover problem is NP-complete, then the corresponding H-PCover [9] and

H-COLORDOMINATION problems [14] are NP-complete as well. Moreover, the H-COVER problem is known to be NP-complete for all k-regular graphs H of valency $k \geq 3$ [9], and the NP-hardness hence propagates for partial and locally surjective covers of such graphs as well.

The H-COLORDOMINATION problem was proven to be NP-complete for paths, cycles and stars in [14]. It was conjectured there that for simple connected graphs the H-COLORDOMINATION problem is NP-complete if and only if H has at least three vertices.

Our Results

Our main result completely classifies the computational complexity of the H-COLORDOMINATION problem for all connected role graphs. This proves the conjecture made by Kristiansen and Telle [14]. We also fully determine the complexity of the problem for disconnected role graphs under the extra condition that each role must appear as the image of a vertex of the instance graph (cf. [15]). We finally generalize the result of Roberts and Sheng [15] on 2-role assignment problems by proving NP-completeness for the k-role assignment problem for any fixed $k \geq 2$.

The paper is organized as follows. The next section provides necessary definitions and basic observations. In the third section we show the construction of the main theorem, which proves the conjecture made in [14]. The fourth section describes the complexity of the role assignment problem for disconnected role graphs. We apply the main theorem to prove NP-completeness for the k-role assignment problem in the fifth section.

2   Preliminaries

Through the paper we use terminology stemming from the role assignment problems.

We consider simple graphs, denoted by $G = (V_G, E_G)$, where V_G is a finite vertex set of vertices and E_G is a set of unordered pairs of vertices, called edges. For a vertex $u \in V_G$ we denote its neighborhood, i.e. the set of adjacent vertices, by $N_G(u) = \{v \mid (u, v) \in E_G\}$.

The *degree* $\deg_G(u)$ of a vertex u is the number of edges incident with it, or equivalently the size of its neighborhood. The symbol $\delta(G)$ is the minimum degree among all vertices of G.

A graph G is called *connected* if for every pair of distinct vertices u and v, there exists a *path* connecting u and v, i.e. a sequence of distinct vertices starting by u and ending by v where each pair of consecutive vertices forms an edge of G. The *length* of the path is the number of its edges.

A graph that is not connected is called *disconnected*. Each maximal connected subgraph of a graph is called a *component*. A vertex whose removal causes a component of a graph to become disconnected is called a *cutvertex*. We say that a cutvertex u *separates* vertex v from w in G if v, w belong to different components of $G \setminus u$.

Two graphs G and $\tilde{G}$ are called *isomorphic*, denoted by $G \simeq \tilde{G}$, if there exists a one-to-one mapping f of vertices of G onto vertices of $\tilde{G}$ such that $(u, v) \in E_G$ if and only if $(f(u), f(v)) \in E_{\tilde{G}}$.

In the sequel the symbol G denotes the instance graph and R the so-called *role graph*.

Definition 1. *We say that G is R-role assignable if a mapping $r : V_G \to V_R$ exists satisfying:*

$$\text{for all } u \in V_G : r(N_G(u)) = N_R(r(u)),$$

where we use the notation $r(S) = \bigcup_{u \in S} r(u)$ for a set of vertices $S \subseteq V_G$. The function r is called an R-role assignment of G.

The goal of this paper is a full characterization of the computational complexity for the following class of problems:

R-ROLE ASSIGNMENT (RA(R))
Instance: A graph G.
Question: Does the graph G allow an R-role assignment?

We continue with some observations that we use later in the paper.

Observation 1 *If G is R-role assignable, then $\deg_G(u) \geq \deg_R(r(u))$ for all vertices $u \in V_G$.*

Proof. $\deg_G(u) = |N_G(u)| \geq |r(N_G(u))| = |N_R(r(u))| = \deg_R(r(u))$. $\qquad \square$

From this we easily derive that $\delta(G) \geq \delta(R)$, and moreover:

Lemma 1. *If G is R-role assignable and u is a vertex of G with $\deg_G(u) = \delta(R)$, then $\deg_R(r(u)) = \delta(R)$ and r restricted to $N_G(u)$ is an isomorphism between $N_G(u)$ and $N_R(r(u))$.*

Lemma 2. *Let G be R-role assignable and x, y be vertices of R connected by a path P_R. Then for each u with $r(u) = x$ a vertex $v \in V_G$ and a path P_G connecting u and v exist, such that r restricted to P_G is an isomorphism between P_G and P_R.*

Proof. We prove the statement by induction on the length of the path P_R. If x and y are adjacent, then the vertex u has a neighbor v mapping onto y, by the definition of the R-role assignment r.

Now assume that the path P_R is of length $k \geq 2$, and that the hypothesis is valid for all paths of length at most $k - 1$. Denote by y' the predecessor of y in P_R and by P'_R the truncation of P_R by the last edge, i.e. the path of length $k - 1$ connecting x and y'. By the induction hypothesis G contains a vertex v' and a path P'_G such that $P'_G \simeq P'_R$ under r. Then it is easy to find a neighbor v of v' satisfying $r(v) = y$ and tack it to P'_G to get the desired path P_G. $\qquad \square$

We get immediately the following:

Observation 2 *If G is R-role assignable and R is connected, then each vertex $v \in V_R$ appears as a role for some vertex $u \in V_G$.*

Lemma 3. *Let G be R-role assignable, u, u' be vertices of G such that $N_G(u) \subseteq N_G(u')$, and $\deg_G(u) = \delta(R)$. If all vertices of minimum degree in R are cutvertices then $r(u) = r(u')$.*

Proof. We denote $z = r(u)$. Since $\deg_R(z) \leq \deg_G(u) = \delta(R)$ we get that z is a vertex of minimum degree, and by our assumptions it is also a cutvertex in R. Let x, y be two of its neighbors that are separated by z and let $v, w \in N_G(u)$ be their preimages. (Their uniqueness is even guaranteed by Lemma 1.) The image of the path v, u', w is connected, hence it contains the vertex z as the role of u'. □

3 The Main Result

In this section we prove the conjecture of Kristiansen and Telle [14].

Theorem 1. *Let R be a connected role graph. Then the R-role assignment problem is polynomially solvable if $|V_R| \leq 2$ and it is NP-complete if $|V_R| \geq 3$.*

3.1 Sketch of the Proof

It is straightforward to see that the problem is polynomially solvable if the number of vertices of the role graph is at most two. For larger role graphs we prove NP-completeness by making a reduction from hypergraph 2-colorability.

The main idea is to split the problem in various cases depending on the number of cutvertices of minimum degree, the minimum degree and the second common neighborhood of a vertex of minimum degree of R.

For each case we construct an appropriate instance graph from an instance of the hypergraph 2-colorability problem. For this purpose we need several gadgets, which are explained in the next section.

3.2 Gadgets

For the garbage collection in our NP-completeness proof we need to construct a graph that allows two different role assignments.

Lemma 4. *Let R be a role graph. Then a graph H exists that has two R-role assignments r_1 and r_2, such that for any two roles v and w, a vertex u exists in H with $r_1(u) = v$, and $r_2(u) = w$. Moreover, H can be constructed in time being polynomial with respect to the size of R.*

Proof. Take H as the Cartesian product $R \times R$, defined by the vertex set $V_H = V_R \times V_R$, and edges $((a, b), (c, d)) \in E_H$ if and only if $(a, c), (b, d) \in E_R$.

The projections $r_1 : (a, b) \to a$ and $r_2 : (a, b) \to b$ are valid R-role assignments, and the vertex $u = (v, w)$ satisfies the statement of the Lemma.   □

Note that for our purposes, it is possible for any two roles v, w to construct a connected H with two role assignments — it is enough to select the component of $R \times R$ containing the vertex $u = (v, w)$.

Definition 2. *We say that a graph $\tilde{R}$ is glued in a graph G by a vertex $\tilde{v}$, if G can be obtained from $\tilde{R}$ and some other graph G' by identifying a vertex $x \in V_{G'}$ with the vertex $\tilde{v}$.*

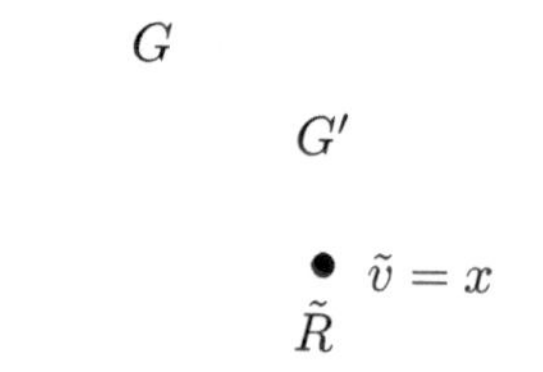

$$G$$

$$G'$$

$$\bullet \quad \tilde{v} = x$$
$$\tilde{R}$$

Fig. 1. A graph with a glued subgraph

As a convention we use letters x, y, z to denote roles, while u is reserved for vertices of the instance. The symbols v, w stand for roles, while $\tilde{v}$ or $\tilde{w}$ are vertices of the instance graph isomorphic to v, w. The proof of the following lemma is omitted in this extended abstract.

Lemma 5. *Let R be a connected role graph. Let G be an R-role assignable graph and $\tilde{R}$ be glued in G by a vertex $\tilde{v}$, where $\tilde{R}$ is isomorphic to R and v, the isomorphic copy of $\tilde{v}$ in $\tilde{R}$, is not a cutvertex of R. Then an R-role assignment r exists such that $r(\tilde{w}) = w$ for every $w \in V_R$.*

3.3   Proof of the Main Theorem

Proof. First we show that $\mathrm{RA}(R)$ is polynomially solvable for $|V_R| \leq 2$.

- $|V_R| = 1$. Clearly, a graph G is R-role assignable if and only if G contains only isolated vertices.
- $|V_R| = 2$. Clearly, a graph G is R-role assignable if and only if G is a bipartite graph that does not contain any isolated vertices.

Now let $|V_R| \geq 3$. Since we can guess a mapping $r : V_G \to V_R$ and check in polynomial time if r is an R-role assignment, the problem $\mathrm{RA}(R)$ is a member of NP. We prove NP-completeness by reduction from hypergraph 2-colorability. This is a well-known NP-complete problem (cf. [11]).

HYPERGRAPH 2-COLORABILITY (H2C)

Instance: A set $Q = \{q_1, \ldots, q_m\}$ and a set $\mathcal{S} = \{S_1, \ldots, S_n\}$ with $S_j \subseteq Q$ for $1 \leq j \leq n$.

Question: Is there a 2-coloring of $(Q, \mathcal{S})$, i.e., a partition of Q into $Q_1 \cup Q_2$ such that $Q_1 \cap S_j \neq \emptyset$ and $Q_2 \cap S_j \neq \emptyset$ for $1 \leq j \leq n$?

With such a hypergraph we associate its incidence graph I, which is a bipartite graph on $Q \cup \mathcal{S}$, where (q, S) forms an edge if and only if $q \in S$.

To prove the theorem we choose a vertex $v \in V_R$ of minimum degree. Because we cannot apply Lemma 5 if v is a cutvertex, we have to distinguish between the case, in which all vertices of minimum degree are cutvertices, and the case, in which a non-cutvertex of minimum degree exists.

Assume first that the vertex v is a vertex of minimum degree that *is not* a cutvertex. Denote the neighbors of v by $N_R(v) = \{w_1, \ldots, w_p\}$ and also the second common neighborhood as $M_R(v) = \bigcap_{u \in N_R(v)} N_R(u) = \{v, v_2, \ldots, v_l\}$. See Fig. 2 for a drawing of a possible situation.

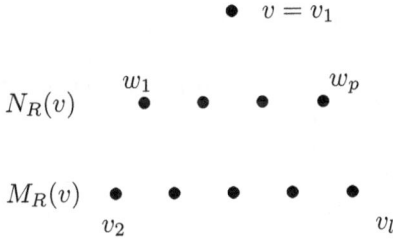

Fig. 2. Neighborhood of a vertex v in R.

We distinguish four cases according to possible values of p and l:

Case 1: $p = 1, l = 1$. Then $R = K_2$ and we have already discussed this case above.

Case 2: $p = 1, l \geq 3$. We extend the incidence graph I as follows: According to Lemma 4 we construct a graph H for which two role assignments exist mapping a particular vertex u to v_2 and v_3. We form an instance G as the union of the graph I and m disjoint copies of the graph H, where the vertex u of the i-th copy is identified with the vertex q_i of I. Finally we insert into G two extra copies $\tilde{R}, R'$ of the role graph R and add the following edges (cf. Fig 3):

- $(\tilde{v}, S_j)$ for all $S_j \in \mathcal{S}$,
- (v'_k, S_j) for all $S_j \in \mathcal{S}$ and all $4 \leq k \leq l$ (this set may be empty).

We show that the graph G formed in this way allows an R-role assignment if and only if $(Q, \mathcal{S})$ is 2-colorable.

Assume first that G is R-role assignable. Then according to Lemma 5 we assume that the vertex $\tilde{v}$ is assigned role v and all vertices S_j are mapped to role w_1. Since their neighborhoods are saturated by common $l - 3$ roles on

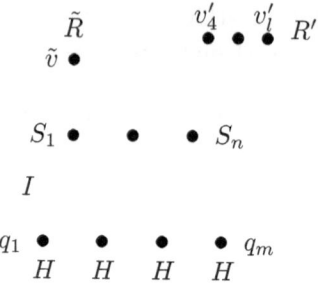

Fig. 3. Construction of the graph G in Case 2.

$v'_4, \ldots, v'_l$, at least two distinct roles $v_a, v_b \in M_R(v) \setminus r(\{v'_4, \ldots, v'_l\})$ exist that are used on some neighbors of each S_j in the set $\mathcal{S}$.

The partition $Q_1 = \{q_i \mid r(q_i) = v_a\}$ and $Q_2 = Q \setminus Q_1 \supseteq \{q_i \mid r(q_i) = v_b\}$ is the desired 2-coloring of $(Q, \mathcal{S})$.

In the opposite direction, any 2-coloring Q_1, Q_2 can be transformed into an R-role assignment r of G by letting $r(q_i) = v_a$ if $q_i \in Q_a$ for $a = 1, 2$ and by further extension according to the two projections of the graph H and graph isomorphisms $\tilde{R} \to R$, $R' \to R$.

Case 3: $p = 1, l = 2$. The case when R is isomorphic to the path P_4 was already shown to be NP-complete in [14]. If R is not isomorphic to a path on four vertices but v_2 is incident with a vertex v^* of degree one, then we can reduce this case to the previous case ($p = 1, l \geq 3$) by selecting v^* as the non-cutvertex of minimum degree. So without loss of generality we may assume that v_2 is not incident with a vertex of degree one.

We construct G from I as follows. First we insert n new vertices $S'_1, \ldots, S'_n$ and a copy $\tilde{R}$ of the role graph R. We identify each q_i with the vertex u of an extra copy of the graph H as in the previous case, but here H is constructed such that u can be assigned v or v_2.

These parts are linked as follows (cf. Fig. 4):

- $(\tilde{v}, S'_j) \in E_G$ for all $j \in \{1, \ldots, n\}$,
- $(q_i, S'_j) \in E_G$ if and only if $(q_i, S_j) \in E_I$.

If G is R-role assignable, then without loss of generality we may assume that $\tilde{v}$ has role v. Then all S'_j have role w_1 since w_1 is the only neighbor of v. The roles of all q_i hence belong to $N_R(w_1) = \{v, v_2\}$. Each S'_j requires the role v_2 to be present among its neighbors in Q. Moreover, if all neighbors of some S'_j in Q are assigned the role v_2, we get that S_j must be mapped to a neighbor of v_2 that is a leaf, but this is in contradiction with our assumptions. We conclude that each S_j is mapped to w_1. Hence both roles v, v_2 appear on its neighborhood and the partition $Q_1 = \{q_i \mid r(q_i) = v\}$ and $Q_2 = \{q_i \mid r(q_i) = v_2\}$ is a 2-coloring of $(Q, \mathcal{S})$.

In the opposite direction, an R-role assignment of G can be constructed from a 2-coloring of $(Q, \mathcal{S})$ in a straightforward way as in the previous case.

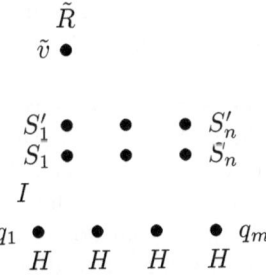

Fig. 4. Construction of the graph G in Case 3.

Case 4: $p \geq 2$. As above we first build the graph H which allows two R-role assignments mapping a vertex u either to w_1 or to w_2.

The graph G consists of the graph I, where each q_i is unified with the vertex u of an extra copy of H. We further include two copies of R denoted by $\tilde{R}$ and R'. Finally we extend the set of edges by (cf. Fig. 5):

- $(\tilde{v}, q_i)$ for all $q_i \in Q$,
- $(\tilde{v}, w'_k)$ for all $1 \leq k \leq p$,
- (S_j, w'_k) for all $3 \leq k \leq p$ (this set may be empty).

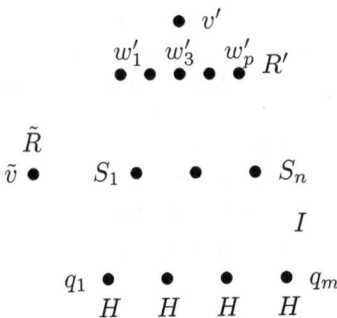

Fig. 5. Construction of the graph G in Case 4.

If an R-role assignment exists, then we assume that $r(\tilde{v}) = v$. For each S_j we have $N_G(S_j) \subseteq N_G(\tilde{v})$. So we know that S_j is assigned some role v_i for which $N_R(v_i) = N_R(v)$. However only $p - 2$ roles appear on vertices $w'_3, \ldots, w'_p$, so two distinct roles w_a and w_b are used on none of $w'_3, \ldots, w'_p$. Then we define a 2-coloring of $(Q, \mathcal{S})$ by selecting $Q_1 = \{q_i \mid r(q_i) = w_a\}$ and $Q_2 = Q \setminus Q_1 \supseteq \{q_i \mid r(q_i) = w_b\}$.

An R-role assignment can be derived from a 2-coloring of $(Q, \mathcal{S})$ as in the previous cases.

Finally, we return to the situation when all vertices of minimum degree in R are cutvertices. (Observe, that $\delta(R) \geq 2$ since vertices of degree one are not cutvertices.)

We construct the graph G as in Case 4 above (cf. Fig. 5). The argumentation goes in the same manner: Since $N_G(v') \subseteq N_G(\tilde{v})$, we get by Lemma 3 that $\tilde{v}$ is mapped to a role of minimum degree. For each S_j we have $N_G(S_j) \subseteq N_G(\tilde{v})$. So we know that S_j is assigned a role that has the same neighbors in R as role $r(\tilde{v})$. Each S_j then lacks two roles w_a, w_b that do not appear on $w'_3, \ldots, w'_p$. Hence we can define a valid 2-coloring of $(Q, \mathcal{S})$ according to the appearance of roles w_a and w_b on the set Q. $\qquad\square$

4   Disconnected Role Graphs

Up to now we have only considered role graphs that were connected. Due to this property we could easily derive that all roles appear as the image of a vertex in the instance graph (cf. Observation 2). We now focus our attention to the case of disconnected role graphs. Suppose R is a role graph with set of components $\mathcal{C} = \{C_1, \ldots C_m\}$. We order the components such that the latter have a higher number of vertices. (Formally, for all $i \leq j : |V_{C_i}| \leq |V_{C_j}|$.)

Note that the identity mapping $\pi : V_{C_1} \to V_R$ preserves the local constraint for role assignment, but Observation 2 is no longer valid here (take $G \simeq C_1$). Our argument guarantees that a locally surjective cover is globally surjective only for connected role graphs. Within some social network models it is natural to demand that all roles appear on the vertices of the instance graph. We show below that the computational complexity of the role assignment problem for disconnected role graphs depends whether such a property $r(V_G) = V_R$ is required or not.

We call an R-role assignment $r : V_G \to V_R$ a *globally R-role assignment* for G if r is an R-role assignment and $r(V_G) = V_R$ holds. Our generalized role assignment problem can now be formulated as

GLOBAL R-ROLE ASSIGNMENT (GRA(R))
Instance: A graph G.
Question: Is G globally R-role assignable?

With respect to the computational complexity we obtain the following result. (The proof is omitted in this abstract.)

Theorem 2. *Let R be a disconnected role graph. Then the GRA(R) problem is polynomially solvable if all components have at most two vertices and it is NP-complete otherwise.*

Now we show that without the condition of global surjectivity "$r(V_G) = V_R$", some polynomially solvable RA(R) problems exist for role graphs R with large components.

Take any role graph R with bipartite components (of arbitrary size) but assure that at least one of these components is isomorphic to K_2 (i.e. to a graph consisting of two vertices forming an edge). For simplicity assume that R has no isolated vertices. We claim that G is R-role assignable if and only if G is bipartite without isolated vertices. The necessity of such condition follows from the fact that non-bipartite graphs have no homomorphism to bipartite graphs. In the opposite direction, any homomorphism from G to K_2 can be viewed as an R-role assignment of G.

Our conjecture is that for all other simple role graphs the problem is NP-complete. Although we have shown above a proof of the polynomial part of the statement, we do not see a direct way for a possible NP-hardness construction.

5 k-Role Assignability

In this section we study a more general version of the role assignment problem. We call a graph G k-role assignable if there exists a role graph R on k vertices, such that G is globally R-role assignable.

k-ROLE ASSIGNMENT (k-RA)
Instance: A graph G.
Question: Is G k-role assignable?

This problem was studied by [15] and is of interest in social network theory where networks are modeled in which individuals of the same social role relate to other individuals in the same way. The networks of individuals are represented by simple graphs. Contrary to our previous results, in this new model two individuals that are related to each other may have the same role. Hence role graphs *that contain loops* are allowed.

Again our aim is to fully characterize the computational complexity of the k-RA problem. Clearly the 1-RA problem is solvable in linear time, since it is sufficient to check whether G has no edges ($R = K_1$) or whether all vertices in G have degree at least one (R consists of one vertex with a loop). The 2-RA problem is proven to be NP-complete in [15]. We generalize this result as follows. (The proof is omitted in this abstract.)

Corollary 1. *The k-RA problem is polynomially solvable for $k = 1$ and it is NP-complete for all $k \geq 2$.*

The computational complexity of the role assignment problem can be studied also for role graphs that contain some loops. If all components of R either consist of exactly one vertex or are isomorphic to K_2, the RA(R) problem is polynomially solvable. The conjecture is that in all other cases the problem is NP-complete, even if instances are restricted to simple graphs.

We expect that our constructions would work in a similar way. Instead of a graph isomorphic to the role graph an other appropriate graph should be glued in the instance graph to obtain a reduction from the H2C problem as we have used in the proof of Theorem 1.

References

1. ABELLO, J., FELLOWS, M. R., AND STILLWELL, J. C. On the complexity and combinatorics of covering finite complexes. *Australian Journal of Combinatorics 4* (1991), 103–112.
2. ANGLUIN, D. Local and global properties in networks of processors. In *Proceedings of the 12th ACM Symposium on Theory of Computing* (1980), 82–93.
3. ANGLUIN, D., AND GARDINER, A. Finite common coverings of pairs of regular graphs. *Journal of Combinatorial Theory B 30* (1981), 184–187.
4. BIGGS, N. Constructing 5-arc transitive cubic graphs. *Journal of London Mathematical Society II. 26* (1982), 193–200.
5. BODLAENDER, H. L. The classification of coverings of processor networks. *Journal of Parallel Distributed Computing 6* (1989), 166–182.
6. EVERETT, M. G., AND BORGATTI, S. Role coloring a graph. *Mathematical Social Sciences 21*, 2 (1991), 183–188.
7. FIALA, J., HEGGERNES, P., KRISTIANSEN, P., AND TELLE, J. A. Generalized *H*-coloring and *H*-covering of trees. In *Graph-Theoretical Concepts in Computer Science, 28th WG '02, Český Krumlov* (2002), no. 2573 in Lecture Notes in Computer Science, Springer Verlag, pp. 198–210.
8. FIALA, J., AND KRATOCHVÍL, J. Complexity of partial covers of graphs. In *Algorithms and Computation, 12th ISAAC '01, Christchurch, New Zealand* (2001), no. 2223 in Lecture Notes in Computer Science, Springer Verlag, pp. 537–549.
9. FIALA, J., AND KRATOCHVÍL, J. Partial covers of graphs. *Discussiones Mathematicae Graph Theory 22* (2002), 89–99.
10. FIALA, J., KRATOCHVÍL, J., AND KLOKS, T. Fixed-parameter complexity of λ-labelings. *Discrete Applied Mathematics 113*, 1 (2001), 59–72.
11. GAREY, M. R., AND JOHNSON, D. S. *Computers and Intractability*. W. H. Freeman and Co., New York, 1979.
12. HEGGERNES, P., AND TELLE, J. A. Partitioning graphs into generalized dominating sets. *Nordic Journal of Computing 5*, 2 (1998), 128–142.
13. KRATOCHVÍL, J., PROSKUROWSKI, A., AND TELLE, J. A. Covering regular graphs. *Journal of Combinatorial Theory B 71*, 1 (1997), 1–16.
14. KRISTIANSEN, P., AND TELLE, J. A. Generalized *H*-coloring of graphs. In *Algorithms and Computation, 11th ISAAC '01, Taipei, Taiwan* (2000), no. 1969 in Lecture Notes in Computer Science, Springer Verlag, pp. 456–466.
15. ROBERTS, F. S., AND SHENG, L. How hard is it to determine if a graph has a 2-role assignment? *Networks 37*, 2 (2001), 67–73.

Fixed-Parameter Algorithms for the
(k, r)-Center in Planar Graphs and Map Graphs

Erik D. Demaine[1], Fedor V. Fomin[2], Mohammad Taghi Hajiaghayi[1], and
Dimitrios M. Thilikos[3]

[1] MIT Laboratory for Computer Science, 200 Technology Square, Cambridge,
Massachusetts 02139, USA, {edemaine,hajiagha}@mit.edu
[2] Department of Informatics, University of Bergen, N-5020 Bergen, Norway,
fomin@ii.uib.no
[3] Departament de Llenguatges i Sistemes Informàtics, Universitat Politècnica de
Catalunya, Campus Nord – Mòdul C5, c/Jordi Girona Salgado 1-3, E-08034,
Barcelona, Spain, sedthilk@lsi.upc.es

Abstract. The (k, r)-*center problem* asks whether an input graph G has
$\leq k$ vertices (called *centers*) such that every vertex of G is within distance $\leq r$ from some center. In this paper we prove that the (k, r)-center
problem, parameterized by k and r, is fixed-parameter tractable (FPT)
on planar graphs, i.e., it admits an algorithm of complexity $f(k, r)n^{O(1)}$
where the function f is independent of n. In particular, we show that
$f(k, r) = 2^{O(r \log r)\sqrt{k}}$, where the exponent of the exponential term grows
sublinearly in the number of centers. Moreover, we prove that the same
type of FPT algorithms can be designed for the more general class of
map graphs introduced by Chen, Grigni, and Papadimitriou. Our results
combine dynamic-programming algorithms for graphs of small branchwidth and a graph-theoretic result bounding this parameter in terms of
k and r. Finally, a byproduct of our algorithm is the existence of a PTAS
for the r-domination problem in both planar graphs and map graphs.
Our approach builds on the seminal results of Robertson and Seymour on
Graph Minors, and as a result is much more powerful than the previous
machinery of Alber et al. for exponential speedup on planar graphs. To
demonstrate the versatility of our results, we show how our algorithms
can be extended to general parameters that are "large" on grids. In addition, our use of branchwidth instead of the usual treewidth allows us to
obtain much faster algorithms, and requires more complicated dynamic
programming than the standard leaf/introduce/forget/join structure of
nice tree decompositions. Our results are also unique in that they apply
to classes of graphs that are not minor-closed, namely, constant powers
of planar graphs and map graphs.

1 Introduction

Clustering is a key tool for solving a variety of application problems such as data
mining, data compression, pattern recognition and classification, learning, and

J.C.M. Baeten et al. (Eds.): ICALP 2003, LNCS 2719, pp. 829–844, 2003.
© Springer-Verlag Berlin Heidelberg 2003

facility location. Among the algorithmic problem formulations of clustering are k-means, k-medians, and k-center. In all of these problems, the goal is to partition n given points into k *clusters* so that some objective function is minimized.

In this paper, we concentrate on the (unweighted) (k,r)-*center* problem [7], in which the goal is to choose k *centers* from the given set of n points so that every point is within distance r from some center in the graph. In particular, the k-*center problem* [17] of minimizing the maximum distance to a center is exactly (k,r)-center when the goal is to minimize r subject to finding a feasible solution. In addition, the r-*domination problem* [7,16] of choosing the fewest vertices whose r-neighborhoods cover the whole graph is exactly (k,r)-center when the goal is to minimize k subject to finding a feasible solution.

A sample application of the (k,r)-center problem in the context of facility location is the installation of emergency service facilities such as fire stations. Here we suppose that we can afford to buy k fire stations to cover a city, and we require every building to be within r city blocks from the nearest fire station to ensure a reasonable response time. Given an algorithm for (k,r)-center, we can vary k and r to find the best bicriterion solution according to the needs of the application. In this scenario, we can afford high running time (e.g., several weeks of real time) if the resulting solution builds fewer fire stations (which are extremely expensive) or has faster response time; thus, we prefer fixed-parameter algorithms over approximation algorithms.

In this application, and many others, the graph is typically planar or nearly so. Chen, Grigni, and Papadimitriou [9] have introduced a generalized notion of planarity which allows local nonplanarity. In this generalization, two countries of a map are adjacent if they share at least one point, and the resulting graph of adjacencies is called a *map graph*. (See Section 2 for a precise definition.) Planar graphs are the special case of map graphs in which at most three countries intersect at a point.

Previous results. r-domination and k-center are NP-hard even for planar graphs. For r-domination, the current best approximation (for general graphs) is a $(\log n + 1)$-factor by phrasing the problem as an instance of set cover [7]. For k-center, there is a 2-approximation algorithm [17] which applies more generally to the case of weighted graphs satisfying the triangle inequality. Furthermore, no $(2 - \epsilon)$-approximation algorithm exists for any $\epsilon > 0$ even for unweighted planar graphs of maximum degree 3 [22]. For geometric k-center in which the weights are given by Euclidean distance in d-dimensional space, there is a PTAS whose running time is exponential in k [1]. Several relations between small r-domination sets for planar graphs and problems about organizing routing schemes with compact structures is discussed in [16].

The (k,r)-center problem can be considered as a generalization of the well-known dominating set problem. During the last two years in particular much attention has been paid to constructing fixed-parameter algorithms with exponential speedup for this problem. Alber et al. [2] were the first who demonstrated an algorithm checking whether a planar graph has a dominating set of size $\leq k$ in time $O(2^{70\sqrt{k}}n)$. This result was the first non-trivial result for the parameterized version of an NP-hard problem in which the exponent of the exponential term grows sublinearly in the parameter. Recently, the running time of this algorithm was further improved to $O(2^{27\sqrt{k}}n)$ [20] and $O(2^{15.13\sqrt{k}}k+n^3+k^4)$ [14]. Fixed-parameter algorithms for solving many different problems such as vertex cover, feedback vertex set, maximal clique transversal, and edge-dominating set on planar and related graphs such as single-crossing-minor-free graphs are considered in [11,21]. Most of these problems have reductions to the dominating set problem. Also, because all these problems are closed under taking minors or contractions, all classes of graphs considered so far are minor-closed.

Our results. In this paper, we focus on applying the tools of *parameterized complexity*, introduced by Downey and Fellows [12], to the (k, r)-center problem in planar and map graphs. We view both k and r as parameters to the problem. We introduce a new proof technique which allows us to extend known results on planar dominating set in two different aspects.

First, we extend the exponential speed-up for a generalization of dominating set, namely the (k, r)-center problem, on planar graphs. Specifically, the running time of our algorithm is $O((2r + 1)^{6(2r+1)\sqrt{k}+12r+3/2}n + n^4)$, where n is the number of vertices. Our proof technique is based on combinatorial bounds (Section 3) derived from the Robertson, Seymour, and Thomas theorem about quickly excluding planar graphs, and on a complicated dynamic program on graphs of bounded branchwidth (Section 4). Second, we extend our fixed-parameter algorithm to map graphs which is a class of graphs that is not minor-closed. In particular, the running time of the corresponding algorithm is $O((2r + 1)^{6(4r+1)\sqrt{k}+24r+3}n + n^4)$.

Notice that the exponential component of the running times of our algorithms depends only on the parameters, and is multiplicatively separated from the problem size n. Moreover, the contribution of k in the exponential part is sublinear. In particular, our algorithms have polynomial running time if $k = O(\log^2 n)$ and $r = O(1)$, or if $r = O(\log n/\log\log n)$ and $k = O(1)$. We stress the fact that we design our dynamic-programming algorithms using branchwidth instead of treewidth because this provides better running times.

Finally, in Section 6, we present several extensions of our results, including a PTAS for the r-dominating set problem and a generalization to a broad class of graph parameters.

2 Definitions and Preliminary Results

Let G be a graph with vertex set $V(G)$ and edge set $E(G)$. We let n denote the number of vertices of a graph when it is clear from context. For every nonempty $W \subseteq V(G)$, the subgraph of G induced by W is denoted by $G[W]$.

Given an edge $e = \{x, y\}$ of a graph G, the graph G/e is obtained from G by contracting the edge e; that is, to get G/e we identify the vertices x and y and remove all loops and duplicate edges. A graph H obtained by a sequence of edge contractions is said to be a *contraction* of G. A graph H is a *minor* of a graph G if H is a subgraph of a contraction of G. We use the notation $H \preceq G$ (resp. $H \preceq_c G$) for H is a minor (a contraction) of G.

(k, r)-**center.** We define the *r-neighborhood* of a set $S \subseteq V(G)$, denoted by $N_G^r(S)$, to be the set of vertices of G at distance at most r from at least one vertex of S; if $S = \{v\}$ we simply use the notation $N_G^r(v)$. We say a graph G has a (k, r)-*center* or interchangeably has an *r-dominating set of size k* if there exists a set S of centers (vertices) of size at most k such that $N_G^r(S) = V(G)$. We denote by $\gamma_r(G)$ the smallest k for which there exists a (k, r)-center in the graph. One can easily observe that for any r the problem of checking whether an input graph has a (k, r)-center, parameterized by k is $W[2]$-hard by a reduction from dominating set. (See Downey and Fellows [12] for the definition of the W Hierarchy.)

Map graphs. Let Σ be a sphere. A *Σ-plane graph* G is a planar graph G drawn in Σ. To simplify notation, we usually do not distinguish between a vertex of the graph and the point of Σ used in the drawing to represent the vertex, or between an edge and the open line segment representing it. We denote the set of regions (faces) in the drawing of G by $R(G)$. (Every region is an open set.) An edge e or a vertex v is *incident* to a region r if $e \subseteq \bar{r}$ or $v \subseteq \bar{r}$, respectively. ($\bar{r}$ denotes the *closure* of r.)

For a Σ-plane graph G, a *map* $\mathcal{M}$ is a pair (G, ϕ), where $\phi : R(G) \to \{0, 1\}$ is a two-coloring of the regions. A region $r \in R(G)$ is called a *nation* if $\phi(r) = 1$ and a *lake* otherwise.

Let $N(\mathcal{M})$ be the set of nations of a map $\mathcal{M}$. The graph F is defined on the vertex set $N(\mathcal{M})$, in which two vertices r_1, r_2 are adjacent precisely if $\bar{r}_1 \cap \bar{r}_2$ contains at least one edge of G. Because f is the subgraph of the dual graph G^* of G, it is planar. Chen, Grigni, and Papadimitriou [9] defined the following generalization of planar graphs. A *map graph $G_{\mathcal{M}}$ of a map $\mathcal{M}$* is the graph on the vertex set $N(\mathcal{M})$ in which two vertices r_1, r_2 are adjacent in $G_{\mathcal{M}}$ precisely if $\bar{r}_1 \cap \bar{r}_2$ contains at least one vertex of G.

For a graph G, we denote by G^k the *kth power of G*, i.e., the graph on the vertex set $V(G)$ such that two vertices in G^k are adjacent precisely if the distance in G between these vertices is at most k. Let G be a bipartite graph with a bipartition $U \cup W = V(G)$. The *half square $G^2[U]$* is the graph on the vertex set U and two vertices are adjacent in $G^2[U]$ precisely if the distance between these vertices in G is 2.

Theorem 1 ([9]). *A graph $G_{\mathcal{M}}$ is a map graph if and only if it is the half-square of some planar bipartite graph H.*

Here the graph H is called a *witness* for $G_{\mathcal{M}}$. Thus the question of finding a (k, r)-center in a map graph $G_{\mathcal{M}}$ is equivalent to finding in a witness H of $G_{\mathcal{M}}$ a set $S \subseteq V(G_{\mathcal{M}})$ of size k such that every vertex in $V(G_{\mathcal{M}}) - S$ has distance $\leq 2r$ in H from some vertex of S.

The proof of Theorem 1 is constructive, i.e., given a map graph $G_{\mathcal{M}}$ together with its map $\mathcal{M} = (G, \phi)$, one can construct a witness H for $G_{\mathcal{M}}$ in time $O(|V(G_\mathcal{M})| + |E(G_\mathcal{M})|)$. One color class $V(G_\mathcal{M})$ of the bipartite graph H corresponds to the set of nations of the map $\mathcal{M}$. Each vertex v of the second color class $V(H) - V(G_\mathcal{M})$ corresponds to an intersection point of boundaries of some nations, and v is adjacent (in H) to the vertices corresponding to the nations it belongs. What is important for our proofs are the facts that

1. in such a witness, every vertex of $V(H) - V(G_\mathcal{M})$ is adjacent to a vertex of $V(G_\mathcal{M})$, and
2. $|V(H)| = O(|V(G_\mathcal{M})| + |E(G_\mathcal{M})|)$.

Thorup [27] provided a polynomial-time algorithm for constructing a map of a given map graph in polynomial time. However, in Thorup's algorithm, the exponent in the polynomial time bound is about 120 [8]. So from practical point of view there is a big difference whether we are given a map in addition to the corresponding map graph. Below we suppose that we are always given the map.

Branchwidth. Branchwidth was introduced by Robertson and Seymour in their Graph Minors series of papers. A *branch decomposition* of a graph G is a pair (T, τ), where T is a tree with vertices of degree 1 or 3 and τ is a bijection from $E(G)$ to the set of leaves of T. The *order function* $\omega : E(T) \to 2^{V(G)}$ of a branch decomposition maps every edge e of T to a subset of vertices $\omega(e) \subseteq V(G)$ as follows. The set $\omega(e)$ consists of all vertices of $V(G)$ such that, for every vertex $v \in \omega(e)$, there exist edges $f_1, f_2 \in E(G)$ such that $v \in f_1 \cap f_2$ and the leaves $\tau(f_1), \tau(f_2)$ are in different components of $T - \{e\}$. The *width* of (T, τ) is equal to $\max_{e \in E(T)} |\omega(e)|$ and the *branchwidth* of G, $\mathbf{bw}(G)$, is the minimum width over all branch decompositions of G.

It is well-known that, if $H \preceq G$ or $H \preceq_c G$, then $\mathbf{bw}(H) \leq \mathbf{bw}(G)$.

The following deep result of Robertson, Seymour, and Thomas (Theorems (4.3) in [23] and (6.3) in [24]) plays an important role in our proofs.

Theorem 2 ([24]). *Let $k \geq 1$ be an integer. Every planar graph with no $(k \times k)$-grid as a minor has branchwidth $\leq 4k - 3$.*

Branchwidth is the main tool in this paper. All our proofs can be rewritten in terms of the related and better-known parameter *treewidth*, and indeed treewidth would be easier to handle in our dynamic program. However, branchwidth provides better combinatorial bounds resulting in exponential speed-up of our algorithms.

3   Combinatorial Bounds

Lemma 1. *Let $\rho, k, r \geq 1$ be integers and G be a planar graph having a (k, r)-center and with a $(\rho \times \rho)$-grid as a minor. Then $k \geq (\frac{\rho - 2r}{2r + 1})^2$.*

Proof. We set $V = \{1, \ldots, \rho\} \times \{1, \ldots, \rho\}$. Let

$$F = (V, \{((x, y), (x', y')) \mid |x - x'| + |y - y'| = 1\})$$

be a plane $(\rho \times \rho)$-grid that is a minor of some plane embedding of G. W.l.o.g. we assume that the external (infinite) face of this embedding of F is the one that is incident to the vertices of the set $V_{\text{ext}} = \{(x, y) \mid x = 1$ or $x = \rho$ or $y = 1$ or $y = \rho\}$, i.e., the vertices of F with degree < 4. We call the rest of the faces of F *internal faces*. We set $V_{\text{int}} = \{(x, y) \mid r + 1 \leq x \leq \rho - r, \ r + 1 \leq y \leq \rho - r\}$, i.e., V_{int} is the set of all vertices of F within distance $\geq r$ from all vertices in V_{ext}. Notice that $F[V_{\text{int}}]$ is a sub-grid of F and $|V_{\text{int}}| = (\rho - 2r)^2$. Given any pair of vertices $(x, y), (x', y') \in V$ we define $\delta((x, y), (x', y')) = \max\{|x - x'|, |y - y'|\}$.

We also define $d_F((x, y), (x', y'))$ to be the distance between any pair of vertices (x, y) and (x', y') in F. Finally we define J to be the graph occurring from F by adding in it the edges of the following sets:

$$\{((x, y), (x + 1, y + 1)) \mid 1 \leq x \leq \rho - 1, \ 1 \leq y \leq \rho - 1)\}$$

$$\{((x, y + 1), (x + 1, y)) \mid 1 \leq x \leq \rho - 1, \ 1 \leq y \leq \rho - 1)\}$$

(In other word we add all edges connecting pairs of non-adjacent vertices incident to its internal faces). It is easy to verify that $\forall (x, y), (x', y') \in V$ $\delta((x, y), (x', y')) = d_J((x, y), (x', y'))$. This implies the following.

If R is a subgraph of J, then $\forall (x, y), (x', y') \in V \delta((x, y), (x', y')) \leq d_R((x, y), (x', y'))$ (1)

For any $(x, y) \in V$ we define $B_r((x, y)) = \{(a, b) \in V \mid \delta((x, y), (a, b)) \leq r\}$ and we observe the following:

$$\forall_{(x,y) \in V} |V(B_r((x, y)))| \leq (2r + 1)^2. \tag{2}$$

Consider now the sequence of edge contractions/removals that transform G to F. If we apply on G only the contractions of this sequence we end up with a planar graph H that can obtained by the $(\rho \times \rho)$-grid F after adding edges to non-consecutive vertices of its faces. This makes it possible to partition the additional edges of H into two sets: a set denoted by E_1 whose edges connect non-adjacent vertices of some square face of F and another set E_2 whose edges connect pairs of vertices in V_{ext}. We denote by R the graph obtained by F if we add the edges of E_1 in F. As R is a subgraph of J, (1) implies that

$$\forall_{(x,y) \in V} N_R^r((x, y)) \subseteq B_r((x, y)) \tag{3}$$

We also claim that

$$\forall_{(x,y) \in V} N_H^r((x, y)) \subseteq B_r((x, y)) \cup (V - V_{\text{int}}) \tag{4}$$

To prove (4) we notice first that if we replace H by R in it then the resulting relation follows from (3). It remains to prove that the consecutive addition of edges of E_2 in R does not introduce in $N_R^r((x, y))$ any vertex of V_{int}. Indeed, this is correct because any vertex in V_{ext} is in distance $\geq r$ from any vertex in V_{int}. Notice now that (4) implies that $\forall_{(x,y) \in V} N_H^r((x, y)) \cap V_{\text{int}} \subseteq B_r((x, y)) \cap V_{\text{int}}$ and using (2) we conclude that

$$\forall_{(x,y) \in V} |N_H^r((x, y)) \cap V_{\text{int}}| \leq (2r + 1)^2 \tag{5}$$

Let S be a (k', r)-center in the graph H. Applying (5) on S we have that the r-neighborhood of any vertex in S contains at most $(2r+1)^2$ vertices from V_{int}. Moreover, any vertex in V_{int} should belong to the r-neighborhood of some vertex in S. Thus $k' \geq |V_{int}|/(2\rho+1)^2 = (\rho-2r)^2/(2\rho+1)^2$ and therefore $k' \geq (\frac{\rho-2r}{2r+1})^2$.

Clearly, the conditions that G has an r-dominating set of size k and $H \preceq_c G$ imply that H has an r-dominating set of size $k' \leq k$. (But this is not true for $H \preceq G$.) As H is a contraction of G and G has a (k, r)-center, we have that $k \geq k' \geq (\frac{\rho-2r}{2r+1})^2$ and lemma follows.

We are ready to prove the main combinatorial result of this paper:

Theorem 3. *For any planar graph G having a (k, r)-center,* $\mathbf{bw}(G) \leq 4(2r+1)\sqrt{k} + 8r + 1$.

Proof. Suppose that $\mathbf{bw}(G) > p = 4(2r+1)\sqrt{k} + 8r + \epsilon - 3$ for some ϵ, $0 < \epsilon \leq 4$, for which $p + 3 \equiv 0 \pmod 4$. By Theorem 2, G contains a $(\rho \times \rho)$-grid as a minor where $\rho = (2r+1)\sqrt{k} + 2r + \frac{\epsilon}{4}$. By Lemma 1, $k \geq (\frac{\rho-2r}{2r+1})^2 = (\frac{(2r+1)\sqrt{k}+\frac{\epsilon}{4}}{2r+1})^2$ which implies that $\sqrt{k} \geq \sqrt{k} + \frac{\epsilon}{8r+4}$, a contradiction.

Notice that the branchwidth of a map graph is unbounded in terms of k and r. For example, a clique of size n is a map graph and has a $(1, 1)$-center and branchwidth $\geq 2/3n$.

Theorem 4. *For any map graph $G_\mathcal{M}$ having a (k, r)-center and its witness H,* $\mathbf{bw}(H) \leq 4(4r+3)\sqrt{k} + 16r + 9$.

Proof. The question of finding a (k, r)-center in a map graph $G_\mathcal{M}$ is equivalent to finding in a witness H of $G_\mathcal{M}$ a set $S \subseteq V(G_\mathcal{M})$ of size k such that every vertex $V(G_\mathcal{M}) - S$ is at distance $\leq 2r$ in H from some vertex of S. By the construction of the witness graph, every vertex of $V(H) - V(G_\mathcal{M})$ is adjacent to some vertex of $V(G_\mathcal{M})$. Thus H has a $(k, 2r+1)$-center and by Theorem 3 the proof follows.

4 (k, r)-Centers in Graphs of Bounded Branchwidth

In this section, we present a dynamic-programming approach to solve the (k, r)-center problem on graphs of bounded branchwidth. It is easy to prove that, for a *fixed* r, the problem is in MSOL (monadic second-order logic) and thus can be solved in linear time on graphs of bounded treewidth (branchwidth). However, for r part of the input, the situation is more difficult. Additionally, we are interested in not just a linear-time algorithm but in an algorithm with running time $f(k, r)n$.

It is worth mentioning that our algorithm requires more than a simple extension of Alber et al.'s algorithm for dominating set in graphs of bounded treewidth [2], which corresponds to the case $r = 1$. In fact, finding a (k, r)-center is similar to finding homomorphic subgraphs, which has been solved only for special classes of graphs and even then only via complicated dynamic programs [18]. The main difficulty is that the path $v = v_0, v_1, v_2, \ldots, v_{\leq r} = c$ from a vertex v to its assigned center c may wander up and down the branch decomposition repeatedly, so that c and v may be in radically different 'cuts' induced by the branch decomposition. All we can guarantee is that the next vertex v_1 along the path from v to c is somewhere in a common 'cut' with v, and that vertex v_1 and v_2 are in a common 'cut', etc. In this way, we must propagate information through the v_i's about the remote location of c.

Let (T', τ) be a branch decomposition of a graph G with m edges and let $\omega' : E(T') \to 2^{V(G)}$ be the order function of (T', τ). We choose an edge $\{x, y\}$ in T', put a new vertex v of degree 2 on this edge, and make v adjacent to a new vertex r. By choosing r as a root in the new tree $T = T' \cup \{v, r\}$, we turn T into a rooted tree. For every edge of $f \in E(T) \cap E(T')$, we put $\omega(f) = \omega'(f)$. Also we put $\omega(\{x, v\}) = \omega(\{v, y\}) = \omega'(\{x, y\})$ and $\omega(\{r, v\}) = \emptyset$.

For an edge f of T we define E_f (V_f) as the set of edges (vertices) that are "below" f, i.e., the set of all edges (vertices) g such that every path containing g and $\{v, r\}$ in T contains f. With such a notation, $E(T) = E_{\{v,r\}}$ and $V(T) = V_{\{v,r\}}$. Every edge f of T that is not incident to a leaf has two children that are the edges of E_f incident to f. We denote by G_f the subgraph of G induced by the vertices incident to edges from the following set

$$\{\tau^{-1}(x) \mid x \in V_f \wedge (x \text{ is a leaf of } T')\}.$$

The subproblems in our dynamic program are defined by a coloring of the vertices in $\omega(f)$ for every edge f of T. Each vertex will be assigned one of $2r + 1$ colors

$$\{0, \uparrow 1, \uparrow 2, \ldots, \uparrow r, \downarrow 1, \downarrow 2, \ldots, \downarrow r\}.$$

The meaning of the color of a vertex v is as follows:

- 0 means that the vertex v is a chosen center.
- $\downarrow i$ means that vertex v has distance exactly i to the closest center c. Moreover, there is a neighbor $u \in V(G_f)$ of v that is at distance exactly $i - 1$ to the center c. We say that neighbor u *resolves* vertex v.
- $\uparrow i$ means that vertex v has distance exactly i to the closest center c. However, there is *no* neighbor of v in $V(G_f)$ resolving v. Thus we are guessing that any vertex resolving v is somewhere in $V(G) - V(G_f)$.

Intuitively, the vertices colored by $\downarrow i$ have already been resolved (though the vertex that resolves it may not itself be resolved), whereas the vertices colored by $\uparrow i$ still need to be assigned vertices that are closer to the center.

We use the notation $\updownarrow i$ to denote a color of either $\uparrow i$ or $\downarrow i$. Also we use $\updownarrow 0 = 0$.

For an edge f of T, a coloring of the vertices in $\omega(f)$ is called *locally valid* if the following property holds: for any two adjacent vertices v and w in $\omega(f)$, if v is colored $\updownarrow i$ and w is colored $\updownarrow j$, then $|i - j| \leq 1$. (If the distance from some vertex v to the closest center is i, then for every neighbor u of v the distance from u to the closest center can not be less than $i - 1$ or more than $i + 1$.)

For every edge f of T we define the mapping

$$A_f : \{0, \uparrow 1, \uparrow 2, \ldots, \uparrow r, \downarrow 1, \downarrow 2, \ldots, \downarrow r\}^{|\omega(f)|} \to \mathbb{N} \cup \{+\infty\}.$$

For a locally valid coloring $c \in \{0, \uparrow 1, \uparrow 2, \ldots, \uparrow r, \downarrow 1, \downarrow 2, \ldots, \downarrow r\}^{|\omega(f)|}$, the value $A_f(c)$ stores the size of the "minimum (k, r)-center restricted to G_f and coloring c". More precisely, $A_f(c)$ is the minimum cardinality of a set $D_f(c) \subseteq V(G_f)$ such that

- For every vertex $v \in \omega(f)$,
  - $c(v) = 0$ if and only if $v \in D_f(c)$, and
  - if $c(v) = \downarrow i$, $i \geq 1$, then $v \notin D_f(c)$ and either there is a vertex $u \in \omega(f)$ colored by $\updownarrow j$, $j < i$, at distance $i - j$ from v in G_f, or there is a path P of length i in G_f connecting v with some vertex of $D_f(c)$ such that no inner vertex of P is in $\omega(f)$.
- Every vertex $v \in V(G_f) - \omega(f)$ whose closest center is at distance $i \leq r$, either is at distance i in G_f from some center in $D_f(c)$, or is at distance j, $j < i$, in G_f from a vertex $u \in \omega(f)$ colored $\updownarrow(i - j)$.

We put $A_f(c) = +\infty$ if there is no such a set $D_f(c)$, or if c is not a locally valid coloring. Because $\omega(\{r, v\}) = \emptyset$ and $G_{\{r,v\}} = G$, we have that $A_{\{r,v\}}(c)$ is the smallest size of an r-dominating set in G.

We start computations of the functions A_f from leaves of T. Let x be a leaf of T and let f be the edge of T incident with x. Then G_f is the edge of G corresponding to x. We consider all locally valid colorings of $V(G_f)$ such that if a vertex $v \in V(G_f)$ is colored by $\downarrow i$ for $i > 0$ then there is an adjacent vertex w in $V(G_f)$ colored $\updownarrow i - 1$. For each such coloring c we define $A_f(c)$ to be the number of vertices colored 0 in $V(G_f)$. Otherwise, $A_f(c)$ is $+\infty$, meaning that this coloring c is infeasible. The brute-force algorithm takes $O(rm)$ time for this step.

Let f be a non-leaf edge of T and let f_1, f_2 be the children of f. Define $X_1 = \omega(f) - \omega(f_2)$, $X_2 = \omega(f) - \omega(f_1)$, $X_3 = \omega(f) \cap \omega(f_1) \cap \omega(f_2)$, and $X_4 = (\omega(f_1) \cup \omega(f_2)) - \omega(f)$.

Notice that

$$\omega(f) = X_1 \cup X_2 \cup X_3. \tag{6}$$

By the definition of ω, it is impossible that a vertex belongs to exactly one of $\omega(f), \omega(f_1), \omega(f_2)$. Therefore, condition $u \in X_4$ implies that $u \in \omega(f_1) \cap \omega(f_2)$ and we conclude that

$$\omega(f_1) = X_1 \cup X_3 \cup X_4, \tag{7}$$

and

$$\omega(f_2) = X_2 \cup X_3 \cup X_4. \tag{8}$$

We say that a coloring $c \in \{0, \uparrow 1, \uparrow 2, \ldots, \uparrow r, \downarrow 1, \downarrow 2, \ldots, \downarrow r\}^{|\omega(f)|}$ of $\omega(f)$ is *formed* from a coloring c_1 of $\omega(f_1)$ and a coloring c_2 of $\omega(f_2)$ if

1. For every $u \in X_1$, $c(u) = c_1(u)$;
2. For every $u \in X_2$, $c(u) = c_2(u)$;
3. For every $u \in X_3$,
   a) If $c(u) = \uparrow i$, $1 \leq i \leq r$, then $c(u) = c_1(u) = c_2(u)$. Intuitively, because vertex u is unresolved in $w(f)$, this vertex is also unresolved in $w(f_1)$ and in $w(f_2)$.
   b) If $c(u) = 0$ then $c_1(u) = c_2(u) = 0$.
   c) If $c(u) = \downarrow i$, $1 \leq i \leq r$, then $c_1(u), c_2(u) \in \{\downarrow i, \uparrow i\}$ and $c_1(u) \neq c_2(u)$. We avoid the case when both c_1 and c_2 are colored by $\downarrow i$ because it is sufficient to have the vertex u resolved in at least one coloring. This observation helps to decrease the number of colorings forming a coloring c. (Similar arguments using a so-called "monotonicity property" are made by Alber et al. [2] for computing the minimum dominating set on graphs of bounded treewidth.)
4. For every $u \in X_4$,
   a) either $c_1(u) = c_2(u) = 0$ (in this case we say that u is *formed by 0 colors*),
   b) or $c_1(u), c_2(u) \in \{\downarrow i, \uparrow i\}$ and $c_1(u) \neq c_2(u)$, $1 \leq i \leq r$ (in this case we say that u is *formed by $\{\downarrow i, \uparrow i\}$ colors*).
   This property says that every vertex u of $w(f_1)$ and $w(f_2)$ that does not appear in $w(f)$ (and hence does not appear further) should finally either be a center (if both colors of u in c_1 and c_2 were 0), or should be resolved by some vertex in $V(G_f)$ (if one of the colors $c_1(u), c_2(u)$ is $\downarrow i$ and one $\uparrow i$). Again, we avoid the case of $\downarrow i$ in both c_1 and c_2.

Notice that every coloring of $w(f)$ is formed from some colorings of $w(f_1)$ and $w(f_2)$. Moreover, if $D_f(c)$ is the restriction to G_f of some (k, r)-center and such a restriction corresponds to a coloring c of $w(f)$ then $D_f(c)$ is the union of the restrictions $D_{f_1}(c_1)$, $D_{f_2}(c_2)$ to G_{f_1}, G_{f_2} of two (k, r)-centers where these restrictions correspond to some colorings c_1, c_2 of $w(f_1)$ and $w(f_2)$ that form the coloring c.

We compute the values of the corresponding functions in a bottom-up fashion. The main observation here is that if f_1 and f_2 are the children of f, then the vertex sets $w(f_1)$ $w(f_2)$ "separate" subgraphs G_{f_1} and G_{f_2}, so the value $A_f(c)$ can be obtained from the information on colorings of $w(f_1)$ and $w(f_2)$.

More precisely, let c be a coloring of $w(f)$ formed by colorings c_1 and c_2 of f_1 and f_2. Let $\#_0(X_3, c)$ be the number of vertices in X_3 colored by color 0 in coloring c and and let $\#_0(X_4, c)$ be the number of vertices in X_4 formed by 0 colors. For a coloring c we assign

$$A_f(c) = \min\{A_{f_1}(c_1) + A_{f_2}(c_2) - \#_0(X_3, c_1) - \#_0(X_4, c_1) \mid c_1, c_2 \text{ form } c\}. \quad (9)$$

(Every 0 from X_3 and X_4 is counted in $A_{f_1}(c_1) + A_{f_2}(c_2)$ twice and $X_3 \cap X_4 = \emptyset$.) The time to compute the minimum in (9) is given by

$$O\left(\sum_c |\{\{c_1, c_2\} \mid c_1, c_2 \text{ form } c\}|\right).$$

Let $x_i = |X_i|$, $1 \le i \le 4$. For a coloring c let z_3 be the number of vertices colored by $\downarrow$ colors in X_3. Also we denote by z_4 the number of vertices in X_4 formed by $\{\downarrow i, \uparrow i\}$ colors, $1 < i \le r$. Thus the number of pairs forming c is $2^{z_3+z_4}$. The number of colorings of $w(f)$ such that exactly z_3 vertices of X_3 are colored by $\downarrow$ colors and such that exactly z_4 vertices of X_4 are formed by $\{\downarrow, \uparrow\}$ colors is

$$(2r+1)^{x_1}(2r+1)^{x_2}(r+1)^{x_3-z_3}\binom{x_3}{z_3}r^{z_3}\binom{x_4}{z_4}r^{z_4}.$$

Thus the number of operations needed to estimate (9) for all possible colorings of $w(f)$ is

$$\sum_{p=0}^{x_3}\sum_{q=0}^{x_4} 2^{p+q}(2r+1)^{x_1+x_2}(r+1)^{x_3-p}\binom{x_3}{p}r^p\binom{x_4}{q}r^q = (2r+1)^{x_1+x_2+x_4}(3r+1)^{x_3}.$$

Let ℓ be the branchwidth of G. By (6), (7) and (8),

$$
\begin{aligned}
x_1 + x_2 + x_3 &\le \ell \\
x_1 + x_3 + x_4 &\le \ell \\
x_2 + x_3 + x_4 &\le \ell.
\end{aligned}
\tag{10}
$$

The maximum value of the linear function $\log_{3r+1}(2r+1)\cdot(x_1+x_2+x_4)+x_3$ subject to the constraints in (10) is $\frac{3\log_{3r+1}(2r+1)}{2}\ell$. (This is because the value of the corresponding LP achieves maximum at $x_1 = x_2 = x_4 = 0.5\ell$, $x_3 = 0$.) Thus one can evaluate (9) in time

$$(2r+1)^{x_1+x_2+x_4}(3r+1)^{x_3} \le (3r+1)^{\frac{3\log_{3r+1}(2r+1)}{2}\ell} = (2r+1)^{\frac{3}{2}\cdot\ell}.$$

It is easy to check that the number of edges in T is $O(m)$ and the time needed to evaluate $A_{\{r,v\}}(c)$ is $O((2r+1)^{\frac{3}{2}\cdot\ell}m)$. Moreover, it is easy to modify the algorithm to obtain an optimal choice of centers by bookkeeping the colorings assigned to each set $w(f)$.

Summarizing, we obtain the following theorem:

Theorem 5. *For a graph G on m edges and with a given branch decomposition of width $\le \ell$, and integers k, r, the existence of a (k,r)-center in G can be checked in $O((2r+1)^{\frac{3}{2}\cdot\ell}m)$ time and, in case of a positive answer, constructs a (k,r)-center of G in the same time.*

Similar result can be obtained for map graphs.

Theorem 6. *Let H be a witness of a map graph $G_{\mathcal{M}}$ on n vertices and let k, r be integers. If a branch-decomposition of width $\le \ell$ of H is given, the existence of a (k,r)-center in $G_{\mathcal{M}}$ can be checked in $O((2r+1)^{\frac{3}{2}\cdot\ell}n)$ time and, in case of a positive answer, constructs a (k,r)-center of G in the same time.*

Proof. We give a sketch of the proof here. H is bipartite graph with a bipartition $(V(G_{\mathcal{M}}), V(H) - V(G_{\mathcal{M}}))$. There is a (k, r)-center in $G_{\mathcal{M}}$ if and only if H has a set $S \subseteq V(G_{\mathcal{M}})$ of size k such that every vertex $V(G_{\mathcal{M}}) - S$ is at distance $\leq 2r$ in H from some vertex of S. We check whether such a set S exists in H by applying arguments similar the proof of Theorem 5. The main differences in the proof are the following. Now we color vertices of the graph H by $\updownarrow i$, $0 \leq i \leq 2r$, where i is even. Thus we are using $2r + 1$ numbers. Because we are not interested whether the vertices of $V(H) - V(G_{\mathcal{M}})$ are dominated or not, for vertices of $V(H) - V(G_{\mathcal{M}})$ we keep the same number as for a vertex of $V(G_{\mathcal{M}})$ resolving this vertex. For a vertex in $V(G_{\mathcal{M}})$ we assign a number $\downarrow i$ if there is a resolving vertex from $V(H) - V(G_{\mathcal{M}})$ colored $\updownarrow (i - 2)$. Also we change the definition of locally valid colorings: for any two adjacent vertices v and w in $w(f)$, if v is colored $\updownarrow i$ and w is colored $\updownarrow j$, then $|i - j| \leq 2$.

Finally, H is planar, so $|E(H)| = O(|V(H)|) = O(n)$.

5   Algorithms for the (k, r)-Center Problem

For a planar graph G and integers k, r, we solve (k, r)-center problem on planar graphs in three steps.

Step 1: We check whether the branchwidth of G is at most $4(2r+1)\sqrt{k} + 8r + 1$. This step requires $O((|V(G)| + |E(G)|)^2)$ time according to the algorithm due to Seymour & Thomas (algorithm (7.3) of Section 7 of [25] — for an implementation, see the results of Hicks [19]). If the answer is negative then report that G has no any (k, r)-center and stop. Otherwise go to the next step.

Step 2: Compute an optimal branch-decomposition of a graph G. This can be done by the algorithm (9.1) in the Section 9 of [25] which requires $O((|V(G)| + |E(G)|)^4)$ steps.

Step 3: Compute, if it exists, a (k, r)-center of G using the dynamic-programming algorithm of Section 4.

It is crucial that, for practical applications, there are no *large hidden constants* in the running time of the algorithms in Steps 1 and 2 above. Because for planar graphs $|E(G)| = O(|V(G)|)$, we conclude with the following theorem:

Theorem 7. *There exists an algorithm finding, if it exists, a (k, r)-center of a planar graph in $O((2r + 1)^{6(2r+1)\sqrt{k}+12r+3/2} n + n^4)$ time.*

Similar arguments can be applied to solve the (k, r)-center problem on map graphs. Let $G_{\mathcal{M}}$ be a map graph. To check whether $G_{\mathcal{M}}$ has a (k, r)-center, we compute optimal branchwidth of its witness H. By Theorem 4, if $\mathbf{bw}(H) > 4(4r+3)\sqrt{k} + 16r + 9$, then $G_{\mathcal{M}}$ has no (k, r)-center. If $\mathbf{bw}(H) \leq 4(4r+3)\sqrt{k} + 16r + 9$, then by Theorem 6 we obtain the following result:

Theorem 8. *There exists an algorithm finding, if it exists, a (k, r)-center of a map graph in $O((2r + 1)^{6(4r+1)\sqrt{k}+24r+13.5} n + n^4)$ time.*

By a straightforward modification to the dynamic program, we obtain the same results for the *vertex-weighted* (k, r)-*center* problem, in which the vertices have real weights and the goal is to find a (k, r)-center of minimum total weight.

6   Concluding Remarks

In this paper, we presented fixed-parameter algorithms with exponential speed-up for the (k, r)-center problem on planar graphs and map graphs. Our methods for (k, r)-center can also be applied to algorithms on more general graph classes like constant powers of planar graphs, which are not minor-closed family of graphs. Extending these results to other non-minor-closed families of graphs would be instructive. Faster algorithms for (k, r)-center for planar graphs and map graphs can be obtained by adopting the proof techniques for planar dominating set from [14]. The disadvantage of this approach is that proofs (but not the algorithm itself) become much more difficult.

In addition, there are several interesting variations on the (k, r)-center problem. In *multiplicity-m* (k, r)-*center*, the k centers must satisfy the additional constraint that every vertex is within distance r of at least m centers. In *f-fault-tolerant* (k, r)-*center* [7], every non-center vertex must have f vertex-disjoint paths of length at most r to centers. (For this problem with $r = \infty$, [7] gives a polynomial-time $O(f \log |V|)$-approximation algorithm for k.) In *L-capacitated* (k, r)-*center* [7], each of the k centers can satisfy only L "customers", essentially forcing the assignment of vertices to centers to be load-balanced. (For this problem, [7] gives a polynomial-time $O(\log |V|)$-approximation algorithm for r.) In *connected* (k, r)-*center* [26], the k chosen centers must form a connected subgraph. In all these problems, the main challenge is to design the dynamic program on graphs of bounded treewidth/branchwidth. We believe that our approach can be used as the main guideline in this direction.

More generally, it seems that our approach should extend other graph algorithms (not just dominating-set-type problems) to apply to the rth power and/or half-square of a graph (and hence in particular map graphs). It would be interesting to explore to which other problems our approach can be applied. Also, obtaining "fast" algorithms for problems like feedback vertex set or vertex cover on constant powers of graphs of bounded branchwidth (treewidth), as we did for dominating set, would be interesting.

Map graphs can be seen as contact graphs of disc homeomorphs. A question is whether our results can be extended for another geometric classes of graphs. An interesting candidate is the class of unit-disk graphs. The current best algorithms for finding a vertex cover or a dominating set of size k on these graphs have $n^{O(\sqrt{k})}$ running time [4].

To demonstrate the versatility of our approach, notice that a direct consequence of our approach is the following theorem.

Theorem 9. *Let* **p** *be a function mapping graphs to non-negative integers such that the following conditions are satisfied:*

(1) *There exists an algorithm checking whether* $\mathbf{p}(G) \leq w$ *in* $f(\mathbf{bw}(G))n^{O(1)}$ *steps.*

(2) *For any* $k \geq 0$, *the class of graphs where* $\mathbf{p}(G) \leq k$ *is closed under taking of contractions.*

(3) *If* R *is any partially triangulated* $(j \times j)$-*grid*[1] *then* $\mathbf{p}(R) = \Omega(j^2)$.

Then there exists an algorithm checking whether $p(G) \leq k$ *on planar graphs in* $O(f(\sqrt{k}))n^{O(1)}$ *steps.*

For a wide source of parameters satisfying condition (1) we refer to the theory of Courcelle [10] (see also [5]). For parameters where $f(\mathbf{bw}(G)) = 2^{O(\mathbf{bw}(G))}$, this result is a strong generalization of Alber et al.'s approach which requires that the problem of checking whether $\mathbf{p}(G) \leq k$ should satisfy the "layerwise separation property" [3]. Moreover, the algorithms involved are expected to have better constants in their exponential part comparatively to the ones appearing in [3]. Similar results can also be obtained for constant powers of planar graphs and for map graphs.

Finally, let us note that combining Theorems 5 and 6 with Baker's approach [6] (see also [13] and [15]) adapted to branch decompositions instead of tree decompositions, we are able to obtain a PTAS for r-dominating set on planar and map graphs. We summarize these results in the following theorems:

Theorem 10. *For any integer* $p \geq 1$, *the* r-*dominating set problem on planar graphs has a* $(1 + 2r/p)$-*approximation algorithm with running time* $O(p(2r + 1)^{3(p+2r)}m))$.

Theorem 11. *For any integer* $p \geq 1$, *the* r-*dominating set problem on map graphs has a* $(1 + 4r/p)$-*approximation algorithm with running time* $O(p(4r + 3)^{3(p+4r)}m))$.

References

1. P. K. AGARWAL AND C. M. PROCOPIUC, *Exact and approximation algorithms for clustering*, Algorithmica, 33 (2002), pp. 201–226.
2. J. ALBER, H. L. BODLAENDER, H. FERNAU, T. KLOKS, AND R. NIEDERMEIER, *Fixed parameter algorithms for dominating set and related problems on planar graphs*, Algorithmica, 33 (2002), pp. 461–493.
3. J. ALBER, H. FERNAU, AND R. NIEDERMEIER, *Parameterized complexity: Exponential speed-up for planar graph problems*, in Electronic Colloquium on Computational Complexity (ECCC), Germany, 2001.
4. J. ALBER AND J. FIALA, *Geometric separation and exact solutions for the parameterized independent set problem on disk graphs*, in Foundations of Information Technology in the Era of Networking and Mobile Computing, IFIP 17th WCC/TCS'02, Montréal, Canada, vol. 223 of IFIP Conference Proceedings, Kluwer, 2002, pp. 26–37.

---

[1] A *partially triangulated* $(j \times j)$-*grid* is any graph obtained by adding noncrossing edges between pairs of nonconsecutive vertices on a common face of a planar embedding of an $(j \times j)$-grid.

5. S. ARNBORG, J. LAGERGREN, AND D. SEESE, *Problems easy for tree-decomposable graphs (extended abstract)*, in Automata, languages and programming (Tampere, 1988), Springer, Berlin, 1988, pp. 38–51.

6. B. S. BAKER, *Approximation algorithms for NP-complete problems on planar graphs*, J. Assoc. Comput. Mach., 41 (1994), pp. 153–180.

7. J. BAR-ILAN, G. KORTSARZ, AND D. PELEG, *How to allocate network centers*, J. Algorithms, 15 (1993), pp. 385–415.

8. Z.-Z. CHEN, *Approximation algorithms for independent sets in map graphs*, J. Algorithms, 41 (2001), pp. 20–40.

9. Z.-Z. CHEN, E. GRIGNI, AND C. H. PAPADIMITRIOU, *Map graphs*, J. ACM, 49 (2002), pp. 127–138.

10. B. COURCELLE, *Graph rewriting: an algebraic and logic approach*, in Handbook of theoretical computer science, Vol. B, Elsevier, Amsterdam, 1990, pp. 193–242.

11. E. D. DEMAINE, M. HAJIAGHAYI, AND D. M. THILIKOS, *Exponential speedup of fixed parameter algorithms on $K_{3,3}$-minor-free or K_5-minor-free graphs*, in The 13th Anual International Symposium on Algorithms and Computation—ISAAC 2002 (Vancouver, Canada), Springer, Lecture Notes in Computer Science, Berlin, vol.2518, 2002, pp. 262–273.

12. R. G. DOWNEY AND M. R. FELLOWS, *Parameterized complexity*, Springer-Verlag, New York, 1999.

13. D. EPPSTEIN, *Diameter and treewidth in minor-closed graph families*, Algorithmica, 27 (2000), pp. 275–291.

14. F. V. FOMIN AND D. M. THILIKOS, *Dominating sets in planar graphs: Branchwidth and exponential speed-up*, in Proceedings of the 14th Annual ACM-SIAM Symposium on Discrete Algorithms, 2003, pp. 168–177.

15. M. FRICK AND M. GROHE, *Deciding first-order properties of locally tree-decomposable structures*, J. Assoc. Comput. Mach., 48 (2001), pp. 1184–1206.

16. C. GAVOILLE, D. PELEG, A. RASPAUD, AND E. SOPENA, *Small k-dominating sets in planar graphs with applications*, in Graph-theoretic concepts in computer science (Boltenhagen, 2001), vol. 2204 of Lecture Notes in Comput. Sci., Springer, Berlin, 2001, pp. 201–216.

17. T. F. GONZALEZ, *Clustering to minimize the maximum intercluster distance*, Theoret. Comput. Sci., 38 (1985), pp. 293–306.

18. A. GUPTA AND N. NISHIMURA, *Sequential and parallel algorithms for embedding problems on classes of partial k-trees*, in Algorithm theory—SWAT '94 (Aarhus, 1994), vol. 824 of Lecture Notes in Comput. Sci., Springer, Berlin, 1994, pp. 172–182.

19. I. V. HICKS, *Branch Decompositions and their applications*, PhD thesis, Rice University, 2000.

20. I. KANJ AND L. PERKOVIĆ, *Improved parameterized algorithms for planar dominating set*, in Mathematical Foundations of Computer Science—MFCS 2002, Springer, Lecture Notes in Computer Science, Berlin, vol.2420, 2002, pp. 399–410.

21. T. KLOKS, C. M. LEE, AND J. LIU, *New algorithms for k-face cover, k-feedback vertex set, and k-disjoint set on plane and planar graphs*, in The 28th International Workshop on Graph-Theoretic Concepts in Computer Science(WG 2002), Springer, Lecture Notes in Computer Science, Berlin, vol. 2573, 2002, pp. 282–296.

22. J. PLESNÍK, *On the computational complexity of centers locating in a graph*, Apl. Mat., 25 (1980), pp. 445–452. With a loose Russian summary.

23. N. ROBERTSON AND P. D. SEYMOUR, *Graph minors. X. Obstructions to tree-decomposition*, J. Combin. Theory Ser. B, 52 (1991), pp. 153–190.

24. N. ROBERTSON, P. D. SEYMOUR, AND R. THOMAS, *Quickly excluding a planar graph*, J. Combin. Theory Ser. B, 62 (1994), pp. 323–348.

25. P. D. SEYMOUR AND R. THOMAS, *Call routing and the ratcatcher*, Combinatorica, 14 (1994), pp. 217–241.

26. C. SWAMY AND A. KUMAR, *Primal-dual algorithms for connected facility location problems*, in Proceedings of the 5th International Workshop on Approximation Algorithms for Combinatorial Optimization, vol. 2462 of Lecture Notes in Computer Science, Rome, Italy, September 2002, pp. 256–270.

27. M. THORUP, *Map graphs in polynomial time*, in The 39th Annual Symposium on Foundations of Computer Science (FOCS 1998), IEEE Computer Society, 1998, pp. 396–405.

Genus Characterizes the Complexity of Graph Problems: Some Tight Results

Jianer Chen[1], Iyad A. Kanj[2], Ljubomir Perković[2], Eric Sedgwick[2], and Ge Xia[1]

[1] Department of Computer Science, Texas A&M University, College Station, TX
77843-3112.*** {chen,gexia@cs.tamu.edu}
[2] School of Computer Science, Telecommunications and Information Systems,
DePaul University, 243 S. Wabash Avenue, Chicago, IL 60604-2301.
{ikanj,lperkovic,esedgwick@cs.depaul.edu}

Abstract. We study the fixed-parameter tractability, subexponential time computability, and approximability of the well-known NP-hard problems: INDEPENDENT SET, VERTEX COVER, and DOMINATING SET. We derive tight results and show that the computational complexity of these problems, with respect to the above complexity measures, is dependent on the genus of the underlying graph. For instance, we show that, under the widely-believed complexity assumption $W[1] \neq$ FPT, INDEPENDENT SET on graphs of genus bounded by $g_1(n)$ is fixed parameter tractable if and only if $g_1(n) = o(n^2)$, and DOMINATING SET on graphs of genus bounded by $g_2(n)$ is fixed parameter tractable if and only if $g_2(n) = n^{o(1)}$. Under the assumption that not all SNP problems are solvable in subexponential time, we show that the above three problems on graphs of genus bounded by $g_3(n)$ are solvable in subexponential time if and only if $g_3(n) = o(n)$.

1 Introduction

NP-completeness theory [13] serves as a foundation for the study of intractable computational problems. However, this theory does not obviate the need for solving these hard problems because of their practical importance. Many approaches have been proposed to solve these problems, including polynomial time approximation, fixed parameter tractable computation, and subexponential time algorithms. The INDEPENDENT SET, VERTEX COVER, and DOMINATING SET problems are among the celebrated examples of such problems. Unfortunately, these problems refuse to give in to most of these approaches. Recent research has shown [3] that none of them has a polynomial time approximation scheme unless P = NP. It is also unlikely that any of them is solvable in subexponential time [15]. In terms of fixed parameter tractability, INDEPENDENT SET and DOMINATING SET do not seem to have efficient algorithms even for small parameter values [11].

Variants of these problems where the input graph is constrained to have certain structural properties (bounded degree graphs, planar graphs, unit disk

*** This research is supported in part by the NSF under the grant CCR-0000206.

graphs, etc ...) were studied as well [1,2,4,12,13]. In particular, if we consider the above problems on the class of planar graphs (the problems remain NP-complete), they become more tractable in terms of the above three complexity measures. All of the three problems on planar graphs have polynomial time approximation schemes [5,16], and are solvable in subexponential time [16]. Recent research in fixed parameter tractability shows that all the three problems admit parameterized algorithms whose running time is subexponential in the parameter [2]. Very Recently, Ellis et al. showed that the DOMINATING SET problem on graphs of constant genus is fixed parameter tractable [12].

This raises an interesting question: What are the graph structures that determine the computational complexity of these important NP-hard problems?

In this paper, we demonstrate how the genus of the underlying graph plays an important role in characterizing the parameterized complexity, the subexponential time computability, and the approximability of the VERTEX COVER, INDEPENDENT SET, and DOMINATING SET problems. Our research shows that in most cases, graph genus is the sole factor that determines the complexity of the above problems. More precisely, in most cases, there is a precise genus threshold that determines the computational complexity of the problems in terms of the three complexity measures. For instance, we show that under the widely-believed complexity assumption $W[2] \neq$ FPT, DOMINATING SET is fixed parameter tractable if and only if the graph genus is $n^{o(1)}$. This result significantly extends both Alber et al. and Ellis et al.'s results for planar graphs and for constant genus graphs [1,12]. The proof is also simpler and more uniform. It is also shown that under the assumption $W[1] \neq$ FPT, INDEPENDENT SET is fixed parameter tractable if and only if the graph genus is $o(n^2)$. For the subexponential time computability, we show that under the assumption that not all SNP problems are solvable in subexponential time, VERTEX COVER, INDEPENDENT SET, and DOMINATING SET are solvable in subexponential time if and only if the genus of the graph is $o(n)$. In terms of approximability, we show that INDEPENDENT SET has a PTAS on graphs of genus $o(n/\lg n)$, but has no PTAS on graphs of genus $\Omega(n)$ unless P = NP. It is also shown that, unless P = NP, the VERTEX COVER and DOMINATING SET problems on graphs of genus $n^{\Omega(1)}$ have no PTAS. A summary of our main results and the previous known results is given in Table 1. Finally, we point out that our techniques can be extended to derive similar results for other NP-hard graph problems. Due to lack of space, the proofs of some results in the paper will be omitted.

We give a quick review on the terminologies related to this paper. Let G be a graph. A set of vertices C in the graph G is a *vertex cover* for G if every edge in G is incident to at least one vertex in C. An *independent set* I in G is a subset of vertices in G such that no two vertices in I are adjacent. A *dominating set* D in G is a set of vertices in G such that every vertex in G is either in D or adjacent to a vertex in D.

A *surface* of genus g is a sphere with g handles in the 3-space [14]. A graph G *embedded* in a surface S is a continuous one-to-one mapping from the graph into the surface. The embedding is *cellular* if each component of $S - G$, which

Table 1. Comparison between our results and the previous results

Prob.	FPT		Subexp. Time		Approximability	
	Ours	Previous	Ours	Previous	Ours	Previous
VC	–	FPT [11,16]	$2^{o(n)}$ iff $g=o(n)$	$2^{O(\sqrt{n})}$ if $g=c$ [2,16]	APX-C if $g=n^{\Omega(1)}$	PTAS if $g=c$ [5,16]
IS	FPT iff $g=o(n^2)$	FPT if $g=0$ [2]	$2^{o(n)}$ iff $g=o(n)$	$2^{O(\sqrt{n})}$ if $g=c$ [2,16]	PTAS if $g=o(\frac{n}{\log n})$ APX-H if $g=\Omega(n)$	PTAS if $g=c$ [5,16]
DS	FPT iff $g=n^{o(1)}$	FPT if $g=c$ [12]	$2^{o(n)}$ iff $g=o(n)$	$2^{O(\sqrt{n})}$ if $g=c$ [2,16]	APX-H if $g=n^{\Omega(1)}$	PTAS if $g=c$ [5,16]

is called a *face*, is homeomorphic to an open disk [14]. In this paper, we only consider cellular graph embeddings. The *size* of a face is the number of edge sides along the boundary of the face. The *(minimum) genus* $\gamma_{\min}(G)$ of a graph G is the smallest integer g such that G has an embedding on a surface of genus g. For more detailed discussions on data structures and algorithms for graph embedding on surfaces, the readers are referred to [7].

2 Genus and Parameterized Complexity

A *parameterized problem* consists of instances of the form (x, k), where x is the *problem description* and k is an integer called the *parameter*. For instance, the VERTEX COVER problem can be parameterized so that each instance of it is of the form (G, k), where G is a graph and k is the parameter, asking whether the graph G has a vertex cover of k vertices. Similarly, we can define the parameterized versions for INDEPENDENT SET and DOMINATING SET. A parameterized problem Q is *fixed parameter tractable* if it can be solved by an algorithm of running time $O(f(k)n^c)$, where f is a function independent of $n = |x|$ and c is a fixed constant [11]. Denote by FPT the class of all fixed parameter tractable problems. An example of the FPT problems is the VERTEX COVER problem that can be solved in time $O(1.285^k + kn)$ [9]. On the other hand, a large class of computational problems seems to not belong to FPT [11]. A hierarchy of parameterized intractability, the W-hierarchy, has been introduced. The 0th level of the hierarchy is the class FPT, and the ith level is denoted by $W[i]$ for $i > 0$ [11]. The hardness and completeness under a parameterized complexity preserving reduction (the *FPT-reduction*) have been defined for each level $W[i]$ in the W-hierarchy [11]. In particular, INDEPENDENT SET is $W[1]$-complete and DOMINATING SET is $W[2]$-complete. It is widely believed that no $W[i]$-hard problem is in the class FPT [11].

2.1 Genus and INDEPENDENT SET

We start by considering the parameterized complexity for the INDEPENDENT SET problem on graphs with genus constraints. A graph G is *p-colorable* if the vertices of G can be colored with p colors such that no two adjacent vertices are colored with the same color. The *chromatic number* $\chi(G)$ of G is the smallest integer p such that G is p-colorable.

Theorem 1. *The* INDEPENDENT SET *problem on graphs of genus bounded by* $g(n)$ *is fixed parameter tractable if* $g(n) = o(n^2)$.

Proof. Since $g(n) = o(n^2)$, there is a nondecreasing and unbounded function $r(n)$ such that $g(n) \leq n^2/r(n)$.[1] Without loss of generality, we can assume that $r(n) \leq n^2$ since otherwise $g(n) = 0$ and the theorem follows from [2]. Let G be a graph of n vertices and genus $g' \leq g(n)$. By Heawood's Theorem [14], the chromatic number $\chi(G)$ of the graph G is bounded by $(7 + \sqrt{1 + 48g'})/2$. From the definition, the chromatic number $\chi(G)$ of G implies an independent set of at least $n/\chi(G)$ vertices in G. Thus, the size $\alpha(G)$ of a maximum independent set in the graph G is at least $2n/(7 + \sqrt{1 + 48g'})$. Since $g' \leq g(n) \leq n^2/r(n)$, we get (note that $r(n) \leq n^2$)

$$\alpha(G) \geq \frac{2n}{7 + \sqrt{1 + 48n^2/r(n)}} \geq \frac{2n\sqrt{r(n)}}{7n + \sqrt{n^2 + 48n^2}} = \frac{\sqrt{r(n)}}{7} \tag{1}$$

Now we are ready for describing our parameterized algorithm. Note that one difficulty we must overcome is estimating the genus of the input graph. The graph minimum genus problem is NP-complete [18] and there is no known effective approximation algorithm for the problem. Therefore, some special tricks have to be used for this purpose. Here we will make use of the approximation algorithm for the graph minimum genus problem proposed in [8], which on an input graph G constructs an embedding of G whose genus is bounded by $\max\{4\gamma_{\min}(G), \gamma_{\min}(G) + 4n\}$. Consider the algorithm given in Figure 1.

ALGORITHM. IS-FPT
Input: a graph G of n vertices and an integer k
Output: decide if G has an independent set of k vertices
1. let $r_1(n) = \min\{r(n)/4, nr(n)/(n + 4r(n))\}$;
2. construct an embedding $\pi(G)$ of G using the algorithm in [8];
3. **if** the genus of $\pi(G)$ is larger than $n^2/r_1(n)$ **then**
    Stop ("the genus of G is larger than $g(n)$");
4. **if** $k \leq \sqrt{r_1(n)}/7$ **then**
    Stop ("the graph G has an independent set of k vertices")
    else try all vertex subsets of k vertices to derive a conclusion.

Fig. 1. A parameterized algorithm for INDEPENDENT SET

We analyze the complexity of the algorithm **IS-FPT**. First note that by our assumption on the function $r(n)$, the function $r_1(n)$ is also nondecreasing and

---

[1] In this paper, we only consider "simple" complexity functions whose value can be feasibly computed. Thus, in our discussion, the computational time for computing the values of complexity functions as such $g(n)$ and $r(n)$ will be neglected.

unbounded. The embedding $\pi(G)$ of the graph G in step 2 can be constructed in linear time [8], and the genus of the embedding $\pi(G)$ can also be computed in linear time [7].

Since $r_1(n) = \min\{r(n)/4, nr(n)/(n+4r(n))\}$, if the genus $\gamma(\pi(G))$ of the embedding $\pi(G)$ is larger than $n^2/r_1(n)$, then $\gamma(\pi(G))$ is larger than both $4n^2/r(n)$ and $n^2/r(n) + 4n$. According to [8], the genus $\gamma(\pi(G))$ of the embedding $\pi(G)$ is bounded by $\max\{4\gamma_{\min}(G), \gamma_{\min}(G) + 4n\}$. Thus, in case $\gamma(\pi(G)) \leq 4\gamma_{\min}(G)$, we have $4\gamma_{\min}(G) > 4n^2/r(n)$, and in case $\gamma(\pi(G)) \leq \gamma_{\min}(G) + 4n$, we have $\gamma_{\min}(G) + 4n > n^2/r(n) + 4n$. Thus, in all cases, we will have $\gamma_{\min}(G) > n^2/r(n) \geq g(n)$. In consequence, the algorithm **IS-FPT** concludes correctly if it stops at step 3.

If the algorithm **IS-FPT** reaches step 4, we know that the minimum genus of the graph G is bounded by $n^2/r_1(n)$. By the analysis above and the relation in (1), the size of a maximum independent set in G is at least $\sqrt{r_1(n)}/7$. Thus, in case $k \leq \sqrt{r_1(n)}/7$, there must be an independent set in G with k vertices. On the other hand, if $k > \sqrt{r_1(n)}/7$ then $\overline{r_1}(49k^2) \geq n$, where $\overline{r_1}$ is the inverse function of the function $r_1(n)$ defined by $\overline{r_1}(p) = \min\{ q \mid r_1(q) \geq p \}$. Since the function $r_1(n)$ is nondecreasing and unbounded, it is not difficult to see that the inverse function $\overline{r_1}(p)$ is also nondecreasing and unbounded. Since enumerating all vertex subsets of k vertices in the graph G can be done in $O(2^n)$ time, which is bounded by $O(2^{\overline{r_1}(49k^2)})$, we conclude that the total running time of the algorithm **IS-FPT** is bounded by $O(f(k) + n^2)$, where $f(k) = 2^{\overline{r_1}(49k^2)}$ is a function dependent only on k but not on n.

Thus, the algorithm **IS-FPT** solves the INDEPENDENT SET problem on graphs of genus bounded by $g(n)$ in time $O(f(k) + n^2)$, and the problem is fixed parameter tractable.

Remark. The algorithm **IS-FPT** does not have to know whether the input graph has its minimum genus bounded by $g(n)$. The point is, if the input graph has its minimum genus bounded by $g(n)$, then the algorithm **IS-FPT**, without needing to know this fact, will definitely and correctly decide whether it has an independent set of size k.

Theorem 2. *The* INDEPENDENT SET *problem on graphs of genus bounded by* $g(n)$ *is* $W[1]$*-complete if* $g(n) = \Omega(n^2)$.

Combining Theorem 1 and Theorem 2, and noting that the genus of a graph of n vertices is always bounded by $(n-3)(n-4)/12$ [14], we have the following tight result.

Corollary 1. *Assuming* $FPT \neq W[1]$, *the* INDEPENDENT SET *problem on graphs of genus bounded by* $g(n)$ *is not fixed parameter tractable if and only if* $g(n) = \Theta(n^2)$.

2.2 Genus and DOMINATING SET

We now discuss how graph genus affects the parameterized complexity of the DOMINATING SET problem. Efficient algorithms for DOMINATING SET on graphs of lower genus have been a recent focus in the study of parameterized computation. In particular, it is known that DOMINATING SET on planar graphs [1, 2] and on graphs of constant genus [12] is fixed parameter tractable. We will show a much stronger result: the DOMINATING SET problem on graphs of genus bounded by $g(n)$ is fixed-parameter tractable if and only if $g(n) = n^{o(1)}$.

For a given instance (G, k) of the DOMINATING SET problem, we apply the branch-and-bound search process to construct a dominating set D of k vertices in G. Initially, we have $D = \emptyset$, and all vertices of G are not yet dominated by vertices in D. In a more general form during the search process, we have included certain vertices in the dominating set D, and removed these vertices from the graph G. The remaining graph G' consists of white and black vertices, corresponding to the vertices that are dominated by vertices in D and the vertices that are still not yet dominated by vertices in D. The graph G' thus will be called a *BW-graph*. We call a set D' of vertices in the BW-graph G' a *B-dominating set* if every black vertex in G' is either in D' or is adjacent to a vertex in D'. Thus, our task is to construct a B-dominating set of k' vertices in the BW-graph G', where k' plus the number of vertices in D is equal to k.

Certain reduction rules can be applied to a BW-graph G':

    R1. Remove from G' all edges between white vertices;
    R2. Remove from G' all white vertices of degree 1;
    R3. If all neighbors of a white vertex u_1 are neighbors of another
          white vertex u_2, remove u_1 from G'.

It can be verified that there is a B-dominating set of k vertices in the graph before applying any of these rules if and only if there is a B-dominating set of k vertices in the graph after applying the rule [1,12]. A BW-graph G is called *reduced* if none of the above rules can be applied. According to rule **R1**, every edge in a reduced BW-graph either connects two black vertices or connects a black vertex and a white vertex (the edge will be called a *bb-edge* or a *bw-edge*, respectively).

Lemma 1. *Let G be a reduced BW-graph with n vertices, in which n_w are white and n_b are black, m edges, and minimum genus g, and suppose that G has neither multiple edges nor self-loops, then* (a) $m \leq 9n_b + 18g - 18$; *and* (b) $n \leq 4n_b + 6g - 6$.

Theorem 3. *The DOMINATING SET problem on graphs of genus bounded by $g(n)$ is fixed parameter tractable if $g(n) = n^{o(1)}$.*

Proof. Since $g(n) = n^{o(1)}$, we can write $g(n) \leq n^{1/r(n)}$ for some nondecreasing and unbounded function $r(n)$. For an instance (G, k) of the DOMINATING SET problem, where the graph G has n vertices and genus g', we apply the algorithm **DS-FPT** in Figure 2.

Let $\bar{r}$ be the inverse function of the function $r(n)$ defined by $\bar{r}(p) = \min\{ q \mid r(q) \geq p \}$. Then the function $\bar{r}$ is also nondecreasing and unbounded. In case $k \geq r(n)$, we have $\bar{r}(k) \geq n$. Thus, step 1 of the algorithm DS-FPT takes time $O(2^n) = O(2^{\bar{r}(k)})$.

ALGORITHM. DS-FPT

Input: a graph G of n vertices and an integer k

Output: decide if G has a dominating set of k vertices

1. **if** $k > r(n)$ **then**
   solve the problem by enumerating all subsets of k vertices in G; Stop;
2. $k_0 = k$; $D = \emptyset$; $G_0 = G$;   color all vertices of G_0 black;
3. **while** there is a black vertex u of degree $d \leq 19$ in G_0 **do**
   make a $(d+1)$-way branch each including either u or a neighbor of u in D;
   remove the new vertex in D from G_0 and color its neighbors in G_0 white;
   apply rules R1-R3 to make G_0 a reduced BW-graph;
   $k_0 = k_0 - 1$;
4. **if** the graph G_0 has at most $78n^{1/k}$ vertices **then**
   find a B-dominating set of k_0 vertices in G_0 by enumerating all vertex subsets of k_0 vertices in G_0
   else Stop ("the graph G has genus larger than $g(n)$");

Fig. 2. A parameterized algorithm for DOMINATING SET

Now suppose $k < r(n)$, step 3 repeatedly branches at a black vertex of degree bounded by 19 in the reduced BW-graph G_0. The search tree size $T(k)$ of step 3 thus satisfies the recurrence relation

$$T(k) \leq 20 \cdot T(k-1)$$

which has a solution $T(k) = O(20^k)$.

At the end of step 3, all black vertices in the reduced BW-graph G_0 have degree at least 20. Suppose at this point, the number of edges, the number of vertices, and the number of black vertices in G_0 are m_0, n_0 and n_b, respectively. Since $2m_0$ is equal to the sum of total vertex degrees in G_0, we have $2m_0 \geq 20n_b$. By Lemma 1, we also have $m_0 \leq 9n_b + 18g' - 18$ (note that the genus of the reduced BW-graph G_0 cannot be larger than the genus g' of the original graph G). Combining these two relations, we get $n_b \leq 18g' - 18$. Now again by Lemma 1, we have $n_0 \leq 4n_b + 6g' - 6$. Thus

$$n_0 \leq 4n_b + 6g' - 6 \leq 78g' - 78 < 78g'$$

Thus, if $g' \leq g(n) \leq n^{1/r(n)} < n^{1/k}$ (note $k < r(n)$), then the number n_0 of vertices in the graph G_0 must be bounded by $78n^{1/k}$. In this case, step 4 solves the problem in time $O(n_0^{k_0+1}) = O((n^{1/k})^k) = O(n)$. On the other hand, if G_0 has more than $78n^{1/k}$ vertices, then again step 4 concludes correctly that the genus of the input graph G is larger than $g(n)$.

In conclusion, the algorithm **DS-FPT** solves the DOMINATING SET problem on graphs of genus bounded by $g(n)$ in time $O(2^{\bar{r}(k)} + 20^k + n)$, and the problem is fixed parameter tractable.

We point out that the techniques used in Theorem 3 are simpler, more uniform, and derive much stronger results than those given in [1,12]. Also, similarly to the algorithm **IS-FPT**, the algorithm **DS-FPT** does not have to know whether the input graph has minimum genus bounded by $g(n)$. For any graph of minimum genus bounded by $g(n)$, the algorithm will definitely derive a correct conclusion.

Theorem 4. *The* DOMINATING SET *problem on graphs of genus bounded by* $g(n)$ *is* $W[2]$-*complete if* $g(n) = n^{\Omega(1)}$.

Combining Theorem 3 and Theorem 4, we derive the following tight result.

Corollary 2. *Assuming* $FPT \neq W[2]$, *the* DOMINATING SET *problem on graphs of genus bounded by* $g(n)$ *is fixed parameter tractable if and only if* $g(n) = n^{o(1)}$.

3   Genus and Subexponential Time Complexity

We say that a problem can be solved in *sublinear exponential time* (or *shortly subexponential time*) if it can be solved in time $O(2^{o(n)})$. Lipton and Tarjan used their planar graph separator theorem to show that a class of NP-hard planar graph problems, including VERTEX COVER, INDEPENDENT SET, and DOMINATING SET, are solvable in subexponential time [16]. They also described how their results can be extended to graphs of constant genus [16]. Recently, deriving lower bounds on the precise complexity of NP-hard problems has been attracting more and more attention [6,15]. In particular, Impagliazzo, Paturi, and Zane introduced the concept of SERF-reduction and showed that many well-known NP-hard problems are SERF-complete for the class SNP [15,17]. This implies that if any of these problems is solvable in subexponential time, then so are all problems in the class SNP, which seems quite unlikely.

In this section, we demonstrate how graph genus affects the subexponential time computability for the problems VERTEX COVER, INDEPENDENT SET, and DOMINATING SET. Our algorithmic results in this section extend Lipton and Tarjan's results on planar graphs and graphs of constant genus [16], and our lower bound results refine Impagliazzo, Paturi, and Zane's results on general graphs [15].

Proposition 1 ([10]). *Let* $G = (V, E)$ *be a graph of n vertices and genus g. There is a linear time algorithm that partitions V into three sets A, B, C, such that C separates A and B, $|A|, |B| \leq n/2$, $|C| \leq c\sqrt{(g'+1)n}$, where c is a fixed constant and $0 \leq g' \leq g$, and the graph induced by $A \cup B$ has genus bounded by* $g - g'$.

Theorem 5. *The problems* VERTEX COVER, INDEPENDENT SET, *and* DOMINATING SET *on graphs of genus bounded $g(n)$ are solvable in subexponential time if* $g(n) = o(n)$.

Proof. We first give a detailed description of our proof for the DOMINATING SET problem. Again, during the search for a minimum dominating set D in a graph G, we classify the vertices in G into five groups (instead of three groups as in Subsection 2.2):

(1) dominating vertices, which have been included in D;
(2) dominated vertices, which should not be in D and are dominated by vertices in D;
(3) white vertices, which are not in D but dominated by vertices in D;
(4) black vertices, which are not in D and neither yet dominated by vertices in D;
(5) red vertices, which should not be in D and are not yet dominated by vertices in D.

During our search process, the dominating vertices and dominated vertices are removed from the graph. Thus, the remaining graph consists of only black, red, and white vertices. Such a graph G will be called a *BWR-graph*. We will use Proposition 5 to partition the vertices of G into the three vertex subsets A, B, and C. Then we consider all possible assignments on the vertices in the set C. Each vertex u in C has the following possibilities:

- u is a white vertex. Then either u is in D or u is not in D;
- u is a red vertex. Then u must be dominated by either a vertex in C, or by a vertex in A, or by a vertex in B;
- u is a black vertex. Then either u is in D, or u is not in D, thus must be dominated by a vertex in C, by a vertex in A, or by a vertex in B.

Thus an assignment to the vertices in C can be as follows: each white vertex is assigned either "in-D" or "not-in-D", each red vertex is assigned either "in-A" or "in-B", and each black vertex is assigned either "in-D", "in-A", or "in-B". After this assignment, a white vertex will become either a dominating vertex (if it is "in-D") or a dominated vertex (if it is "not-in-D"), and thus will be removed from the graph; a red vertex adjacent to an "in-D" vertex in C will become a dominated vertex and will be removed from the graph (in this case, the assignment to the red vertex is ignored); a red vertex not adjacent to any "in-D" vertex in C will become a red vertex and will be added to the subgraph induced by either A or B (depending on whether it is an "in-A" or "in-B" vertex); a "in-D" black vertex will become a dominating vertex and will be removed from the graph G; a black vertex whose status is either "in-A" or "in-B" and is adjacent to a "in-D" vertex in C will become a dominated vertex, and will be removed from the graph; finally, an "in-A" black vertex (resp. an "in-B" black vertex) not adjacent to any "in-D" vertex in C will become a red vertex and will be added to the subgraph induced by A (resp. by B).

Since the set C separates the subgraphs induced by the sets A and B, it is not difficult to see that an assignment to the vertices in the set C will result in two separated BWR-subgraphs of G, one is induced by the set A plus certain vertices in C (we will call it the *A-subgraph*), and the other is induced by the set B plus some other vertices in C (we will call it the *B-subgraph*). Thus, the search process can be executed recursively on the A-subgraph and the B-subgraph.

We analyze the above algorithm. First note that the genus of a subgraph is always bounded by that of the original graph. Therefore, if the original graph has its genus bounded by $g(n)$, then all recursive calls of the algorithm are on graphs of genus bounded by $g(n)$. Thus, according to Proposition 5, the number of vertices in the set C constructed in each recursive call of the algorithm must be bounded by $c\sqrt{(g(n)+1)n}$.

The algorithm enumerates all possible assignments to the vertices in C. Since each vertex in C can get at most 3 different statuses, the total number of assignments to the vertices in C is bounded by $3^{|C|} \leq 3^{c\sqrt{(g(n)+1)n}}$. For each such assignment, the algorithm recursively works on the induced A-subgraph and B-subgraph. Since $|A|, |B| \leq n/2$, and $|C| \leq c\sqrt{(g(n)+1)n}$, the total number of vertices in each of the A-subgraph and the B-subgraph is bounded by $n/2 + c\sqrt{(g(n)+1)n}$, which is bounded by $2n/3$ when n is larger than a fixed constant. Therefore, the time complexity $T(n)$ of the algorithm is given by the recurrence relation

$$T(n) \leq 3^{c\sqrt{(g(n)+1)n}} \cdot 2T(2n/3) \leq 3^{c\sqrt{(g(n)+1)n}+1}T(2n/3)$$

From this and the fact that $g(n) = o(n)$, we can easily derive that $T(n) = 2^{o(n)}$, thus proving that for graphs of genus bounded by $g(n) = o(n)$, the DOMINATING SET problem can be solved in subexponential time.

The discussion on VERTEX COVER and INDEPENDENT SET is similar, and thus omitted.

Theorem 6. *For any function* $g(n) = \Omega(n)$, *if one of* VERTEX COVER, INDE-PENDENT SET, *and* DOMINATING SET *problems on graphs of genus bounded by* $g(n)$ *can be solved in subexponential time, then all problems in the class SNP can be solved in subexponential time.*

The class SNP contains many well-known NP-hard problems [15] including: k-SAT, k-COLORABILITY, k-SET COVER, VERTEX COVER, and INDEPENDENT SET. It is widely believed among researchers that it is quite unlikely that all problems in SNP are solvable in subexponential time. Based on this, and combining Theorem 5 and Theorem 6, we have the following tight results.

Corollary 3. *Assuming that not all the problems in SNP are solvable in subexponential time, the* VERTEX COVER, INDEPENDENT SET, *and* DOMINATING SET *problems on graphs of genus bounded by* $g(n)$ *are solvable in subexponential time if and only if* $g(n) = o(n)$.

4   Genus and Approximability

The reader is referred to [4,13] for the basic definitions and terminology of approximation algorithms.

Proposition 2 ([10]). *There is an* $O(n \log g)$ *time algorithm that for a given graph* G *of* n *vertices and genus* g *constructs a subset* P *of at most* $c \cdot \sqrt{gn \log g}$ *vertices, where* c *is a fixed constant, such that removing the vertices in* P *from* G *results in a planar graph.*

Theorem 7. *The* INDEPENDENT SET *problem on graphs of genus bounded by* $g(n)$ *has a PTAS if* $g(n) = o(n/\log n)$.

Proof. Let $g(n) \leq n/(r(n)\log n)$, where $r(n)$ is a nondecreasing and unbounded function. Our PTAS for the INDEPENDENT SET works as follows: for a given graph G of n vertices, we use the algorithm in Proposition 2 to construct the vertex subset P (this can be done in time $O(n\log n)$ even when the genus of G is larger than $g(n)$). If the number p_0 of vertices in P is larger than $c \cdot \sqrt{g(n)n\log g(n)}$, then we know that the input graph G has genus larger than $g(n)$ and we stop. Otherwise, the graph G_1 obtained by deleting the vertices in P from the graph G is a planar graph. We apply any known PTAS algorithm (e.g., those given in [5,16]) to construct an independent set I_1 for the graph G_1. The set I_1 is clearly an independent set in the original graph G. Thus, we simply output I_1 as a solution to the graph G.

It is obvious that the above algorithm runs in polynomial time and is an approximation algorithm for the INDEPENDENT SET problem on graphs of genus bounded by $g(n)$. What left is to analyze the approximation ratio of the algorithm. First note that because $g(n) \leq n/(r(n)\log n))$, the number p_0 of vertices in P is bounded by $p_0 \leq c \cdot \sqrt{g(n)n\log g(n)}) \leq cn/\sqrt{r(n)}$. Let n_1 be the number of vertices in the graph G_1, then $n_1 = n - p_0$. Let α and α_1 be the sizes of a maximum independent set in the graphs G and G_1, respectively. We have $\alpha_1 \leq \alpha \leq \alpha_1 + p_0$ (the second inequality is true because any maximum independent set in G minus the vertices in P makes an independent set in G_1). Moreover, because G_1 is a planar graph, by the Four-Color theorem [14], α_1 is at least $n_1/4$.

Let α'_1 be the number of vertices in the independent set I_1. Since the independent set I_1 is constructed by a PTAS on the planar graph G_1, we have $\alpha_1/\alpha'_1 \leq 1 + \epsilon$, where ϵ is the given error bound. Since the function $r(n)$ is nondecreasing and unbounded, there is a constant N_0 such that when $n \geq N_0$, we have

$$\frac{c}{4\sqrt{r(n)}} \leq \frac{1}{8} \quad \text{and} \quad \frac{8c(1+\epsilon)}{\sqrt{r(n)}} \leq \epsilon \qquad (2)$$

From the first inequality, we get

$$\alpha'_1 \geq \frac{\alpha_1}{1+\epsilon} \geq \frac{n_1}{4(1+\epsilon)} = \frac{n-p_0}{4(1+\epsilon)} \geq \frac{n - cn/\sqrt{r(n)}}{4(1+\epsilon)}$$

$$= n \cdot \left(\frac{1}{4(1+\epsilon)} - \frac{c}{4(1+\epsilon)\sqrt{r(n)}} \right) \geq \frac{n}{8(1+\epsilon)} \qquad (3)$$

Since $\alpha \leq \alpha_1 + p_0 \leq (1+\epsilon)\alpha'_1 + cn/\sqrt{r(n)}$, combining this with (2) and (3), we get

$$\frac{\alpha}{\alpha'_1} \leq 1 + \epsilon + \frac{cn}{\alpha'_1\sqrt{r(n)}} \leq 1 + \epsilon + \frac{8cn(1+\epsilon)}{n\sqrt{r(n)}} \leq 1 + 2\epsilon$$

This shows that our algorithm is a PTAS for the INDEPENDENT SET problem on graphs of genus bounded by $g(n)$.

Theorem 8. *Assuming $P \neq NP$, the* INDEPENDENT SET *problem on graphs of genus bounded by $g(n)$ has no PTAS if $g(n) = \Omega(n)$.*

Unfortunately, the analogous theorems to Theorem 7 do not hold for VERTEX COVER and DOMINATING SET.

Theorem 9. *Unless $P = NP$,* VERTEX COVER *and* DOMINATING SET *on graphs of genus bounded by $g(n)$ have no PTAS if $g(n) = n^{\Omega(1)}$.*

References

1. J. ALBER, H. FAN, M. FELLOWS, H. FERNAU, R. NIEDERMEIER, F. ROSAMOND, AND U. STEGE, Refined search tree technique for dominating set on planar graphs, *LNCS* **2136**, (2001), pp. 111–122.
2. J. ALBER, H. FERNAU, AND R. NIEDERMEIER, Parameterized complexity: exponential speedup for planar graph problems, *LNCS* **2076**, (2001), pp. 261–272.
3. S. ARORA, C. LUND, R. MOTWANI, M. SUDAN, AND M. SZEGEDY, Proof verification and hardness of approximation problems, *J. ACM* **45**, (1998), pp. 501–555.
4. G. AUSIELLO, P. CRESCENZI, G. GAMBOSI, V. KANN, A. MARCHETTI-SPACCAMELA, AND M. PROTASI, *Complexity and Approximation: Combinatorial Optimization Problems and Their Approximability Properties*, Springer-Verlag, Berlin Heidelberg, 1999.
5. B. BAKER, Approximation algorithms for NP-complete problems on planar graphs, *J. ACM* **41**, (1994), pp. 153–180.
6. L. CAI AND D. JUEDES, On the existence of subexponential-time parameterized algorithms, *JCSS*, to appear.
7. J. CHEN, Algorithmic graph embeddings, *TCS* **181**, (1987), pp. 247–266.
8. J. CHEN, S. KANCHI, AND A. KANEVSKY, A note on approximating graph genus, *Information Processing Letters* **61**, (1997), pp. 317–322.
9. J. CHEN, I. KANJ, AND W. JIA, Vertex cover: further observations and further improvements, *J. Algorithms* **41**, (2001), pp. 280–301.
10. H. DJIDJEV AND S. VENKATESAN, Planarization of graphs embedded on surfaces, *LNCS* **1017** (WG'95), (1995), pp. 62–72.
11. R. DOWNEY AND M. FELLOWS, *Parameterized Complexity*, Springer-Verlag, New York, 1999.
12. J. ELLIS, H. FAN, AND M. FELLOWS, The dominating set problem is fixed parameter tractable for graphs of bounded genus, *LNCS* **2368**, (2002), pp. 180–189.
13. M. GAREY AND D. JOHNSON, *Computers and Intractability: A Guide to the Theory of NP-completeness*, Freeman, San Francisco, 1979.
14. J. GROSS AND T. TUCKER, *Topological Graph Theory*, Wiley-Interscience, New York, 1987.
15. R. IMPAGLIAZZO, R. PATURI, AND F. ZANE, Which problems have strongly exponential complexity? *JCSS* **63**, (2001), pp. 512–530.
16. R. LIPTON AND R. TARJAN, Applications of a planar separator theorem, *SIAM Journal on Computing* **9**, (1980), pp. 615–627.
17. C. PAPADIMITRIOU AND M. YANNAKAKIS, Optimization, approximation and complexity classes, *JCSS* **43**, (1991), pp. 425–440.
18. C. THOMASSEN, The graph genus problem is NP-complete, *J. Algorithms* **10**, (1989), pp. 568–576.

The Definition of a Temporal Clock Operator

Cindy Eisner[1], Dana Fisman[1,2]*, John Havlicek[3], Anthony McIsaac[4], and
David Van Campenhout[5]**

[1] IBM Haifa Research Laboratory
[2] Weizmann Institute of Science
[3] Motorola, Inc.
[4] STMicroelectronics, Ltd.
[5] Verisity Design, Inc.

Abstract. Modern hardware designs are typically based on multiple
clocks. While a singly-clocked hardware design is easily described in
standard temporal logics, describing a multiply-clocked design is cumbersome. Thus it is desirable to have an easier way to formulate properties related to clocks in a temporal logic. We present a relatively simple
solution built on top of the traditional LTL-based semantics, study the
properties of the resulting logic, and compare it with previous solutions.

1 Introduction

Synchronous hardware designs are based on a notion of discrete time, in which
the flip-flop (or latch) takes the system from the current state to the next state.
The signal that causes the flip-flop (or latch) to transition is termed the *clock*.
In a singly-clocked hardware design, the behavior of hardware in terms of the
clock naturally maps to the notion of the next-time operator in temporal logics
such as LTL[10] and CTL[2], so that the following LTL formula:

$$G(p \rightarrow X\ q) \tag{1}$$

can be interpreted as "globally, if p then *at the next clock cycle, q*". Mapping
between a state of a model for the temporal logic and a clock cycle of hardware
can then be dealt with by the tool which builds a model from the source code
(written in some hardware description language, or HDL).

Modern hardware designs, however, are typically based on multiple clocks.
In such a design, for instance, some flip-flops may be clocked with *clka*, while
others are clocked with *clkb*. In this case, the mapping between states and clock
cycles cannot be done automatically; rather, the formula itself must contain some

---

* The work of this author was supported in part by the John Von Neumann Minerva
Center for the Verification of Reactive Systems.

** *E-mail addresses*: eisner@il.ibm.com (C. Eisner), dana@wisdom.weizmann.ac.il (D.
Fisman), john.havlicek@motorola.com (J. Havlicek), anthony.mcisaac@st.com (A.
McIsaac), dvc@verisity.com (D. Van Campenhout)

J.C.M. Baeten et al. (Eds.): ICALP 2003, LNCS 2719, pp. 857–870, 2003.

indication of which clock to use. For instance, a clocked version of Formula 1 might be:

$$(G(p \to X\ q))@clka \tag{2}$$

We would like to interpret Formula 2 as "globally, if p during a cycle of $clka$, then at the next cycle of $clka$, q". In LTL we can express this as:

$$G\ ((clka \wedge p) \to X[\neg clka\ W\ (clka \wedge q)]) \tag{3}$$

Thus, we would like to give semantics to a new operator @ such that Formula 2 is equivalent to Formula 3. The issue of defining what such a solution should be for LTL is the problem we explore in this paper.

We present a relatively simple solution built on top of the traditional LTL-based semantics. Our solution is based on the idea that the only role of the clock operator should be to define a projection of the path onto those states where the clock "ticks"[1]. Thus, $\neg(f@clk)$ should be equivalent to $(\neg f)@clk$, that is, the clock operator should be its own dual. Achieving this introduces a problem for paths on which the clock never ticks. We solve this problem by introducing a propositional strength operator that extends the semantics from non-empty paths to empty paths in the same way that the strong next operator [8] extends the semantics from infinite to finite paths. We present the resulting logic LTL@, and show that we meet the goal of the "projection view", as well as other design goals presented below. To show that the clock and propositional strength operators add no expressive power to LTL, we provide a set of rewrite rules to translate an LTL@ formula to an equivalent LTL formula.

The remainder of this paper is organized as follows. Section 2 describes related work. Section 3 defines hardware clocks. Section 4 discusses design requirements for the clock operator. Section 5 presents the definition of LTL@. In Section 6 we show that we have met the goals of Section 4. Section 7 discusses some additional properties of our logic. Section 8 concludes.

2   Related Work

Many modeling languages, such as Lustre [5] and Signal, incorporate the idea of a clock. However, in this paper we are interested in the addition of a clock operator to temporal logic. The work described in this paper is the result of discussions in the LRM sub-committee of the Accellera Formal Verification Technical Committee (FVTC). All four languages (Sugar2.0, ForSpec, Temporal e, CBV) examined by the committee enhance temporal logic with clock operators. Many of these languages distinguish between *strong* and *weak* clock operators, in a similar way as LTL distinguishes between strong and weak until.

---

[1] Actually, referring to a projection of the path is not precisely correct, as we allow access to states in between consecutive states of a projection in the event of a clock switch. However, the word "projection" conveys the intuitive function of the clock operator in the case that the formula is singly-clocked. Use of the word "projection" when describing the clocks of Sugar2.0 and ForSpec below is similarly imprecise.

Sugar2.0 supports both strong and weak versions of a clock operator. As originally proposed [3], a strongly clocked Sugar2.0 formula requires the clock to "tick long enough to ensure that the formula holds", while a weakly clocked formula allows it to stop ticking before then.

In ForSpec [1], which also supports strong and weak clocks, a strongly clocked formula requires only that the clock tick at least once, after which the only role of the clock is to define the projection of the path onto those states where the clock ticks. A weakly clocked formula, on the other hand, holds if the clock never ticks; if it does tick, then the role of the clock is the same as for a strongly clocked formula.

In Temporal e [9], which also supports multiple clocks, clocks are not attributed with strength. This is consistent with the use of Temporal e in simulation, in which behaviors are always finite in duration. Support for reasoning about infinite length behaviors is limited in Temporal e.

In CBV [6], clocking and alignment of formulas are supported by separate and independent sampling and alignment operators. The sampling operator is self-dual and determines the projection in the singly-clocked case. It is similar to the clock operator of LTL$^{@}$. The CBV alignment operators come in a strong/weak dual pair that take us to the first clock event, without affecting the sampling clock. The composition of the sampling operator with a strong/weak alignment operator on the same clock is provided by the CBV synchronization operators, which behave like the ForSpec strong/weak clock operators.

Clocked Temporal Logic [7], confusingly termed CTL by its authors, is another temporal logic that deals with multiple clocks. However, in their solution a clock is a pre-determined subset of the states on a path, and their approach is to associate a clock with each atomic proposition, rather than to clock formulas and sub-formulas.

Wang, Mok and Emerson have defined APTL [11], which enhances temporal logic with multiple real-time clocks. In this work, we are concerned with hardware clocks, which determine the granularity of time in a synchronous system, rather than with clocks in the sense of [11] that measure real time in an asynchronous system. Thus, for example [11] assumes the clock ticks infinitely often, while we are trying to face the problems arising when such an assumption is not adopted.

3   Hardware Clocks

A hardware clock is any signal connected to the *clock input* of a flip-flop or latch. A flip-flop or latch is a memory element, which passes on some function of its inputs to its outputs, but only when its clock input is active. At all other times, it remembers its previous input. A flip-flop responds only to a change in its clock input, while a latch will function continuously as long as the clock input is active.

There are many types of flip-flops and latches, each of which passes on different functions of its inputs to its outputs. Furthermore, real flip-flops and latches work in the real world, where time is continuous, and the amount of time during which a signal is asserted makes a difference. For the purposes of this paper, it is sufficient to examine one kind of flip-flop, working in an abstract world where time is discrete, defined as follows.

Definition 1 (Abstract flip-flop). *An abstract flip-flop is a hardware device with two inputs, d and c, and one output, o. Its functionality is described by the formula* $o' = (c \wedge d) \vee (\neg c \wedge o)$, *where* o' *is the value of o at the next point in time.*[2]

4   Issues in Defining the Clock Operator

We begin by trying to set the design requirements for the clock operator. What is the intuition it should capture? What are the problems involved?

The projection view. When only a single clock is involved we would like that a clocked formula $f@clk$ hold on a path π if and only if the unclocked formula f holds on a path π' where π' is π projected onto those states where clk holds.

Non-accumulation of clocks. In many hardware designs, large chunks of the design work on some main clock, while small pieces work on a secondary clock. Rather than require the user to specify a clock for each sub-formula, we would like to allow clocking of an entire formula on a main clock, and pieces of it on a secondary clock, in such a way that the outer clock (which is applied to the entire formula) does not affect the inner clock (which is applied to one or more sub-formulas). That is, we want a nested clock operator to have the effect of "changing the projection", rather than further projecting the projected path.

Finite and empty paths. The introduction of clocks requires us to deal with finite paths, since the projection of an infinite path may be finite. For LTL, this means that the single *next operator* X no longer suffices. To see why, consider an atomic proposition p and a path where the clock stops ticking. On the last state of the path, do we want $(X\ p)@clk$ to hold or not? Whatever we do, assuming we want to preserve the duality $\neg(X\ p) = X(\neg p)$ under clocks, and thus obtain a definition under which $\neg((X\ p)@clk)$ is equivalent to $(X(\neg p))@clk$, the result is unsatisfactory. For instance, if $(X\ p)@clk$ holds when the clock stops ticking, then $\neg((X\ p)@clk)$ does not. Letting $p = \neg q$, we get that $(X\ q)@clk$ does not hold if the clock stops ticking, which is a contradiction.

Thus, the addition of clocks to LTL-based semantics introduces problems similar to those of defining LTL semantics for finite paths. In particular, it requires us to make a decision as to the semantics of the next operator on the last clock tick of a path, with the result that the next operator is not dual to itself. Instead, we end up with two next operators, strong and weak, which are dual to each other [8].

---

[2] The value of the flip-flop's output is not defined at the first point in time.

Not only may the projection of an infinite path be finite, it may be empty as well. For LTL, this means that the duality problem exists not only for the next operator, but also for atomic propositions. Whatever choice we make for the semantics of $p@clk$ (where p is an atomic proposition) on an empty path, we cannot achieve the duality $\neg(p@clk) = (\neg p)@clk$ without adding something to the logic.

A natural solution for the semantics of a formula over a path where the clock does not tick is to take the strength from the temporal operator. Under this approach, for example, a clocked strong next does not hold on a path with no ticks, while a clocked weak next does hold on such a path. This solution breaks down in the case of a formula with no temporal operators. One way to deal with this is to make a decision as to the semantics of the clock operator on a path with no ticks, giving two clock operators which are dual to each other, rather than a single clock operator that is dual to itself. Below we discuss this issue in more detail.

Avoiding the problems of existing distinctions between strong and weak clocks. Three of the languages considered by the FVTC make a distinction between strong and weak clocks. However, each has significant drawbacks that we would like to avoid.

In Sugar2.0 as originally proposed [3], a strongly clocked formula requires the clock to "tick long enough to ensure that the formula holds", while a weakly clocked formula allows it to stop ticking before then. Thus, for instance, the formula $(F\ p)@clk!$ (where @ is the clock operator, clk is the clock, and the ! indicates that it is strong) requires there to be enough ticks of clk so that p eventually holds, whereas the formula $(F\ p)@clk$ (which is a weakly clocked formula, because there is no !) allows the case where p never occurs, if it "is the fault of the clock", i.e., if the clock ticks a finite number of times. Negation switches the clock strength, so that $\neg(f@clk) = (\neg f)@clk!$ and we get that $(G\ q)@clk!$ holds if the clock ticks an infinite number of times and q holds at every tick, while $(G\ q)@clk$ holds if q holds at every tick, no matter how many there are. Although initially pleasing, this semantics has the disadvantage that the formula $(F\ p) \wedge (G\ q)$ cannot be satisfactorily clocked for a finite path, because $((F\ p) \wedge (G\ q))@clk!$ does not hold on any finite path, while $((F\ p) \wedge (G\ q))@clk$ makes no requirement on p on such a path. Since our intent is to define a semantics that can be used in simulation (where every path is finite) as well as in model checking, this is unacceptable.

In ForSpec, a strongly clocked formula requires only that the clock tick at least once, after which the only role of the clock is to define the projection of the path onto those states where the clock ticks. A weakly clocked formula, on the other hand, holds if the clock never ticks; if it does tick, then the role of the clock is the same as for a strongly clocked formula. Thus, the only difference between strong and weak clocks in ForSpec is on paths whose projection is empty. This leads to the strange situation that a liveness formula may hold on some path π, but not on an extension of that path, $\pi\pi'$. For instance, if p is an atomic proposition, then $(\mathsf{F}\ p)@clk$ holds if there are no ticks of clk, but does not hold if there is just one tick, at which p does not hold.

In CBV, there is a self-dual clock operator, the sampling operator, according to which all temporal advances are aligned to the clock. However, the sampling operator causes no initial alignment. Therefore, sampled booleans are evaluated immediately; sampled next-times align to the next strictly future tick of the clock; and so forth. As a result, the projection defined by the CBV sampling operator includes the first state of a path, regardless of whether it is a tick of the clock. The CBV alignment and synchronization operators come in strong/weak dual pairs. The latter behave like the ForSpec strong/weak clock operators and therefore suffer from the same disadvantages.

Under the solutions described above, the clock or synchronization operator is given the role of determining the semantics in case the path is empty. As a result, the operator cannot be its own dual, resulting in two kinds of clocks. Our goal is to define a logic where the only role of the clock operator is to determine a projection. Thus, we seek a solution which solves the problem of an empty path in such a way that the clock operator is its own dual, eliminating the need for two kinds of clocks.

Equivalence and substitution. We would like the logic to adhere to an equivalence lemma as well as a substitution lemma. Loosely speaking, an equivalence lemma requires that two equivalent LTL formulas remain equivalent after the application of the clock operator. The substitution lemma guarantees that substituting sub-formula g for an equivalent sub-formula h does not change the truth value of the original formula.

Motivating example. We would like our original motivating example from the introduction to hold.

Goals

To summarize, our goals composed in light of the discussion above, are as follows:

1. When singly-clocked, the semantics should be that of the projection view.
2. Clocks should not accumulate.
3. The clock operator should be its own dual.
4. There should be a clocked version of $(\mathsf{F}\ p) \wedge (\mathsf{G}\ q)$ that is meaningful on paths with a finite number of clock ticks.
5. For any atomic proposition p, if $(\mathsf{F}\ p)@clk$ holds on a path, it should hold on any extension of that path.
6. For any clock c, two equivalent LTL formulas should remain equivalent when clocked with c.
7. Substituting sub-formula g for an equivalent sub-formula h should not change the truth value of the original formula.
8. The truth value of LTL$^@$ Formula 2 should be the same as the truth value of LTL Formula 3 for every path.

5   The Definition of LTL$^@$

We solve the problem of finite paths introduced by clocks in LTL based semantics by supplying both strong and weak versions of the next operator (X! and X).

We solve the problem of empty paths by introducing a new, propositional strength operator. Thus, if p is an atomic proposition, then $p!$ is as well. While p is a weak atomic proposition, and so holds on an empty path, $p!$ is a strong atomic proposition, and does not hold on such a path. The intuition behind this is that the role of the strength of a temporal operator is to tell us how far a finite path is required to extend. For strong until, as in $[f \cup g]$, we require that g hold somewhere on the path. For strong next, as in X! f, we require that there be a next state. Intuitively then, we get that a strong proposition, as in $p!$, requires that there be a current state.

Without clocks, there is never such a thing as not having a current state, so the problem of an empty path does not come up in traditional temporal logics. But for a clocked semantics, there may indeed not be a first state. In such a situation, putting the responsibility on the atomic proposition gives a natural extension to the idea of the formula itself telling us how far a finite path must extend. This leaves us with the desired situation that the sole responsibility of the clock operator will be to "light up" the states that are relevant for the current clock context, which is the intuitive notion of a clock.

5.1   Syntax

The syntax of LTL$^@$ is defined below, where we use the term *boolean expression* to refer to any application of the standard boolean operators to atomic propositions.

Definition 2 (Formulas of LTL$^@$).

- If p is an atomic proposition, then p and $p!$ are LTL$^@$ formulas.
- If *clk* is a boolean expression and f, f_1, and f_2 are LTL$^@$ formulas, then the following are LTL$^@$ formulas: $\neg f$, $f_1 \wedge f_2$, X! f, $[f_1 \cup f_2]$, $f@clk$.

Additional operators are derived from the basic operators defined above:[3]

- $f_1 \vee f_2 \overset{\text{def}}{=} \neg(\neg f_1 \wedge \neg f_2)$
- $f_1 \rightarrow f_2 \overset{\text{def}}{=} \neg f_1 \vee f_2$
- F $f \overset{\text{def}}{=} [\text{T} \cup f]$
- X $f \overset{\text{def}}{=} \neg$X! $\neg f$
- G $f \overset{\text{def}}{=} \neg$F $\neg f$
- $[f_1 \text{ W } f_2] \overset{\text{def}}{=} [f_1 \cup f_2] \vee$ G f_1

LTL is the subset of LTL$^@$ consisting of the formulas that have no clock operator and no sub-formulas of the form $p!$, for some atomic proposition p.

---

[3] Where T is an atomic proposition that holds on every state. In the sequel, we also use F, which is an atomic proposition that does not hold for any state.

5.2   Semantics

We define the semantics of LTL$^@$ formulas over words[4] from the alphabet 2^P. A letter is a subset of the set of atomic propositions P such that T belongs to the subset and F does not. We will denote a letter from 2^P by ℓ and an empty, finite, or infinite word from 2^P by w. We denote the length of word w as $|w|$. An empty word $w = \epsilon$ has length 0, a finite word $w = (\ell_0\ell_1\ell_2\cdots\ell_n)$ has length $n+1$, and an infinite word has length ∞. We denote the ith letter of w by w^i. We denote by $w^{i\cdots}$ the suffix of w starting at w^i. That is, $w^{i\cdots} = (w^iw^{i+1}\cdots w^n)$ or $w^{i\cdots} = (w^iw^{i+1}\cdots)$. We denote by $w^{i\cdots j}$ the finite sequence of letters starting from w^i and ending in w^j. That is, $w^{i\cdots j} = (w^iw^{i+1}\cdots w^j)$.

We first present the semantics of LTL$^@$ minus the clock operator over infinite, finite, and empty words (*unclocked semantics*). We then present the semantics of LTL$^@$ over infinite, finite, and empty words (*clocked semantics*). Later, we relate the two.

Unclocked semantics. We now present a semantics for LTL$^@$ minus the clock operator. The semantics is defined with respect to an infinite, finite, or empty word. The notation $w \models f$ means that formula f holds along the word w. The semantics is defined as follows, where p denotes an atomic proposition, f, f_1, and f_2 denote formulas, and j and k denote natural numbers (i.e., non-negative integers).

- $w \models p \Longleftrightarrow |w| = 0$ or $p \in w^0$
- $w \models p! \Longleftrightarrow |w| > 0$ and $p \in w^0$

- $w \models \neg f \Longleftrightarrow w \not\models f$
- $w \models f_1 \wedge f_2 \Longleftrightarrow w \models f_1$ and $w \models f_2$
- $w \models$ X! $f \Longleftrightarrow |w| > 1$ and $w^{1\cdots} \models f$
- $w \models [f_1 \cup f_2] \Longleftrightarrow$ there exists $k < |w|$ such that $w^{k\cdots} \models f_2$, and for every $j < k$ $w^{j\cdots} \models f_1$

Clocked semantics. We define the semantics of an LTL$^@$ formula with respect to an infinite, finite, or empty word w and a context c, where c is a boolean expression over P. For word w and boolean expression b, we say that $w^i \models b$ iff $w^{i\cdots i} \models b$. Second, we say that a finite word w *is a clock tick of* clock c if c holds at the last letter of w and does not hold at any previous letter of w. Formally,

Definition 3 (is a clock tick of). *We say that finite word w is a clock tick of c iff $|w| > 0$ and $w^{|w|-1} \models c$ and for every natural number $i < |w| - 1$, $w^i \not\models c$.*

The notation $w \models^c f$ means that formula f holds along the word w in the context of clock c. The semantics of an LTL$^@$ formula is defined as follows, where p denotes an atomic proposition, c, and c_1 denote boolean expressions, f, f_1, and f_2 denote LTL$^@$ formulas, and j and k denote natural numbers.

---

[4] Relating the semantics over words to semantics over models is done in the standard way. Due to lack of space, we omit the details.

- $w \models^c p \iff$ for all $j < |w|$ such that $w^{0..j}$ is a clock tick of c, $p \in w^j$
- $w \models^c p! \iff$ there exists $j < |w|$ such that $w^{0..j}$ is a clock tick of c and $p \in w^j$

- $w \models^c \neg f \iff w \not\models^c f$
- $w \models^c f_1 \wedge f_2 \iff w \models^c f_1$ and $w \models^c f_2$
- $w \models^c \mathsf{X}! \, f \iff$ there exist $j < k < |w|$ such that $w^{0..j}$ is a clock tick of c and $w^{j+1..k}$ is a clock tick of c and $w^{k..} \models^c f$
- $w \models^c [f_1 \cup f_2] \iff$ there exists $k < |w|$ such that $w^k \models c$ and $w^{k..} \models^c f_2$ and for every $j < k$ such that $w^j \models c$, $w^{j..} \models^c f_1$
- $w \models^c f @ c_1 \iff w \models^{c_1} f$

In LTL$^@$, every formula is evaluated in the context of a clock. The projection view requires that propositions are evaluated not at the first state of a path, but at the first state where the context clock ticks (if there is such a state). To be consistent with this, if the clock does not tick in the first state of a path, a formula $\mathsf{X}f$ or $\mathsf{X}!f$ must be evaluated in terms of the value of f at the second tick of the clock after the initial state.

6   Meeting the Goals

In this section, we analyze the logic LTL$^@$ with respect to the goals of Section 4. Due to lack of space all proofs are omitted; they can be found in the full version of the paper. The following definitions are needed for the sequel.

Definition 4 (Projection). *The projection of word w onto clock c, denoted $w|_c$, is the word obtained from w after leaving only the letters which satisfy c.*

Definition 5 (Unclocked equivalent). *Two LTL$^@$ formulas f and g with no clock operator are unclocked equivalent ($f \equiv g$) if for all words w, $w \models f$ if and only if $w \models g$.*

Definition 6 (Clocked equivalent). *Two LTL$^@$ formulas f and g are clocked equivalent ($f \stackrel{@}{\equiv} g$) if for all words w and all contexts c, $w \models^c f$ if and only if $w \models^c g$.*

Goal 1. The following theorem states that when a single clock is applied to a formula, the projection view is obtained.

Theorem 1 *Let f be an LTL$^@$ formula with no clock operator, c a boolean expression and w an infinite, finite, or empty word.*

$$w \models^c f \quad \text{if and only if} \quad w|_c \models f$$

It follows immediately that the clocked semantics is a generalization of the unclocked semantics - that is, that the clocked semantics reduces to the unclocked semantics when the context is T.

Corollary 1 *Let f be an* LTL$^@$ *formula with no clock operator, and w a word.*

$$w \models^T f \quad \text{if and only if} \quad w \models f$$

Goal 2. Looking at the semantics for $f@c_1$ in context c it is easy to see that $f@c_1@c_2 \stackrel{@}{\equiv} f@c_1$, and therefore clocks do not accumulate.

Goal 3. The following claim states that this goal is met.

Claim. $(\neg f)@b \stackrel{@}{\equiv} \neg(f@b)$

Goal 4. The clocked version of $(\mathsf{F}\ p) \wedge (\mathsf{G}\ q)$ is $((\mathsf{F}\ p) \wedge (\mathsf{G}\ q))@c$, and holds if p holds for some state and q holds for all states on the projected path.

Goal 5. The following claim states that Goal 5 is met.

Claim. Let b, clk and c be boolean expressions, w a finite word, and w' an infinite or finite word.

$$w \models^c (\mathsf{F}\ b)@clk \implies ww' \models^c (\mathsf{F}\ b)@clk$$

Goal 6. The following claim states that Goal 6 is met.

Claim. Let f and g be LTL$^@$ formulas with no clock operators, and let b be a boolean expression.

$$f \equiv g \quad \implies \quad f@b \stackrel{@}{\equiv} g@b$$

Note that if f and g are unclocked formulas then for some boolean expression c it may be that $f@c \stackrel{@}{\equiv} g@c$, even though $f \neq g$. For example, let $f = (\neg c) \to \mathsf{T}$ and let $g = (\neg c) \to \mathsf{F}$. Then $f@c \stackrel{@}{\equiv} g@c$, but $f \neq g$.

Goal 7. We use the notation $\varphi[\psi \leftarrow \psi']$ to denote the formula obtained from φ by replacing sub-formula ψ with ψ'. The following claim states that this goal is met.

Claim. Let g be a sub-formula of f, and let $g' \stackrel{@}{\equiv} g$. Then $f \stackrel{@}{\equiv} f[g \leftarrow g']$.

Goal 8. The following claim states that this goal is met.

Claim. For every word w,

$$w \models^T (\mathsf{G}(p \to \mathsf{X}\ q))@clka \iff w \models \mathsf{G}\ ((clka \wedge p) \to \mathsf{X}[\neg clka\ \mathsf{W}\ (clka \wedge q)])$$

7   Discussion

Looking backwards. In LTL, the evaluation of formula $G(p \rightarrow f)$, where p is a boolean expression, depends only on the evaluations of f starting at those points where p holds. In particular, satisfaction of $G(p \rightarrow f)$ on w is independent of the initial segment of w before the first occurrence of p. We might hope that satisfaction of $G(p@clkp \rightarrow f@clkf)$ on w will be independent of the initial segment of w before the first occurrence of p at a tick of $clkp$. This is not the case. For instance, consider the following formula:

$$G(p@clkp \rightarrow q@clkq) \tag{4}$$

which is a clocked version of the simple invariant $G(p \rightarrow q)$, where both p and q are boolean expressions. Formula 4 can be rewritten as

$$G([\neg clkp \; W \; (clkp \wedge p)] \rightarrow [\neg clkq \; W \; (clkq \wedge q)]) \tag{5}$$

by the rewrite rules in Theorem 2 below. The result is a dimension of temporality not present in the original, unclocked formula. For instance, for the behavior of p shown in Figure 1, Formula 5 requires that q hold at time 4 (because $[\neg clkp \; W \; (clkp \wedge p)]$ holds at time 3, and in order for $[\neg clkq \; W \; (clkq \wedge q)]$ to hold

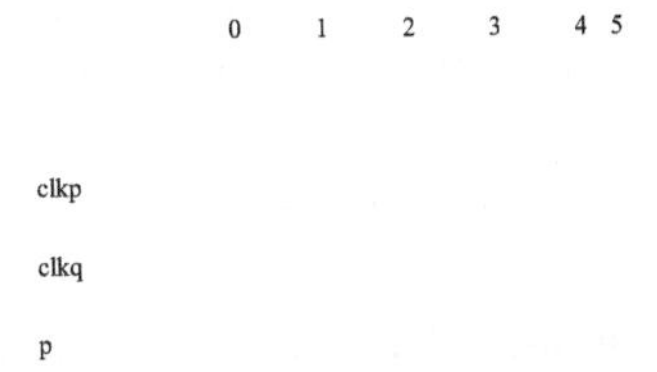

Fig. 1. Behavior of p illustrating a problem with Formula 5

at time 3, we need q to hold at time 4). Not only does Formula 5 require that q hold at time 4 for the behavior of p shown in Figure 1, it also requires that q hold at time 2 (because $[\neg clkp \; W \; (clkp \wedge p)]$ holds at time 2, and in order for $[\neg clkq \; W \; (clkq \wedge q)]$ to hold at time 2, we need q to hold at time 2). Thus, the direction of the additional dimension of temporality may be backwards as well as forwards.

To avoid the "looking backward" phenomenon the semantics of a boolean expression under clock operators should be non-temporal. For instance, we could define $p@clk = p$ and $p!@clk = p!$, or alternatively, $p@clk = clk \rightarrow p$ and $p!@clk = clk \wedge p!$. The disadvantage of these definitions is that the projection view is not preserved (because on a path such that p holds at the first clock but does not hold at the first state, $p@clk$ and/or $p!@clk$ do/does not hold).

We note that Formula 4 has the same backwards-looking feature in other semantics with strong and weak clocks [3,1], so the phenomenon does not arise purely from the design decisions we have taken here. Furthermore, if the multi-clocked version is taken as $(G(p \rightarrow (q@clkq)))@clkp$, then the phenomenon does not arise. Many properties of practical interest are of this form, for example a property asserting that the data is not corrupted between input and output interfaces clocked on different clocks:

$$(\mathsf{G}((receive \wedge (data\_in = d)) \rightarrow (\neg send \ \mathsf{U} \ (send \wedge (data\_out = d))))@clk\_out))@clk\_in \tag{6}$$

$[f \ \mathsf{U} \ g]$ as a fixed point. In standard LTL, $[f \ \mathsf{U} \ g]$ can be defined as a least solution of the equation $S = g \vee (f \wedge \mathsf{X}! \ S)$. In LTL$^@$, there is a fixed point characterization if f and g are themselves unclocked, because $[f \ \mathsf{U} \ g] \equiv (\mathsf{T}! \wedge g) \vee (f \wedge \mathsf{X}![f \ \mathsf{U} \ g])$ (the conjunction with T! is required in order to ensure equivalence on empty paths as well as the non-empty paths on which standard LTL formulas are interpreted). Thus by the claim of Goal 6 $[f \ \mathsf{U} \ g]@c \overset{@}{\equiv} ((\mathsf{T}!\wedge g) \vee (f \wedge \mathsf{X}![f \ \mathsf{U} \ g]))@c$ for any clock c and any formulas f and g containing no clock operators, and hence by the semantics, the truth value of $[f \ \mathsf{U} \ g]$ under context c is the same as the truth value of $(\mathsf{T}! \wedge g) \vee (f \wedge \mathsf{X}![f \ \mathsf{U} \ g])$ under context c, for any context c. If f and g contain clock operators, this equivalence no longer holds. Let p, q and d be atomic propositions, and let $f = q@d$. Consider a word w such that $w^0 \models d \wedge q$ and for all $i > 0$, $w^i \not\models d \wedge q$, and $w^0 \not\models c$. Then $w \not\models^c f$ hence $w \not\models^c (\mathsf{T}! \wedge f) \vee (p \wedge \mathsf{X}![p \ \mathsf{U} \ f])$. However, since $w^0 \not\models c$, and there is no state other than w^0 where $d \wedge q$ holds, $w \not\models^c_q [p \ \mathsf{U} \ f]$. Note that while of theoretical interest, the lack of a fixed point characterization of $[f \ \mathsf{U} \ g]$ is not an obstacle to model checking, since any LTL$^@$ formula can be translated to an equivalent LTL formula by the rewrite rules presented below.

$\mathsf{X}f$ and $\mathsf{X}!f$ on states where the clock does not hold. As mentioned earlier, another property of our definition is that on states where the clock does not hold, the next operators take us two clock cycles into the future, instead of the one clock cycle that we might expect. Further consideration shows that this is a direct result of the projection view: since $p@clk$ must mean that p holds at the next clock, it is clear that an application of a next operator (as in $(\mathsf{X}p)@clk$ or $(\mathsf{X}!p)@clk$) must mean that p holds at the one after that. This behavior of a clocked next operator is a consideration only in multi-clocked formulas, since in a singly-clocked formula, we are never "at" a state where the clock does not hold (except perhaps at the initial state).

Expressive power. The clock operator provides a concise way to express what would otherwise be cumbersome, but it does not add expressive power. Theorem 2 below states that the truth value of any LTL$^@$ formula under context clk is the same as that of the LTL formula $f' = \mathcal{T}^{clk}(f)$, where $\mathcal{T}^{clk}(f)$ is defined as follows:

- $\mathcal{T}^{clk}(p) = [\neg clk \ \mathsf{W} \ (clk \wedge p)]$
- $\mathcal{T}^{clk}(p!) = [\neg clk \ \mathsf{U} \ (clk \wedge p)]$
- $\mathcal{T}^{clk}(\neg f) = \neg \mathcal{T}^{clk}(f)$
- $\mathcal{T}^{clk}(f_1 \wedge f_2) = \mathcal{T}^{clk}(f_1) \wedge \mathcal{T}^{clk}(f_2)$

- $\mathcal{T}^{clk}(\mathsf{X!}\ f) = [\neg clk \cup (clk \wedge \mathsf{X!}[\neg clk \cup (clk \wedge \mathcal{T}^{clk}(f))])]$
- $\mathcal{T}^{clk}([f_1 \cup f_2]) = [(clk \rightarrow \mathcal{T}^{clk}(f_1)) \cup (clk \wedge \mathcal{T}^{clk}(f_2))]$
- $\mathcal{T}^{clk}(f @ clk_1) = \mathcal{T}^{clk_1}(f)$

Theorem 2 *Let f be any* LTL$^@$ *formula, c a boolean expression, and w a word.*

$$w \models^c f \quad \text{if and only if} \quad w \models \mathcal{T}^c(f)$$

Clearly $\mathcal{T}^{clk}()$ defines a recursive procedure whose application starting with $clk = \mathrm{T}$ results in an LTL formula with the same truth value in context T. Note that while we can rewrite a formula f into an LTL formula f' with the same truth value, we cannot use formulas f and f' interchangeably. For example, $p!@clk1$ translates to $[\neg clk1 \cup (clk1 \wedge p)]$, but these two are not clocked equivalent (because clocking each of them with $clk2$ will give different results).

8   Conclusion and Future Work

We have given a relatively simple definition of multiple clocking for LTL augmented with a clock operator that we believe captures the intuition behind hardware clocks, and have presented a set of rewrite rules that can be used as an implementation of the clock operator. In our definition, the only role of the clock operator is to define a projection of the path, and it is its own dual.

Our semantics, based on strong and weak propositions, achieves goals not achieved by semantics based on strong and weak clocks. In particular, it gives the projection view for singly-clocked formulas and a uniform treatment of empty and non-empty paths, including the interpretation of the operators G and F. It does not provide an easy solution to the question of how to define U as a fixed point operator for multi-clocked formulas. Future work should seek a way to resolve these issues without losing the advantages.

It may be noted that in the strong/weak clock semantics, alignment is always applied immediately after the setting of a clock context; while in the strong/weak proposition semantics, it is always applied immediately before an atomic proposition. Allowing more flexibility in where alignment (and strength) is applied may be a useful avenue for investigation.

Acknowledgements. We would like to thank Sharon Barner, Shoham Ben-David, Alan Hartman and Emmanuel Zarpas for their help with the formal definition of multiple clocks. We would also like to thank Mike Gordon, whose work on studying the formal semantics of Sugar2.0 with HOL [4] greatly contributed to our understanding of the problems discussed in this paper. Finally, thank you to Shoham Ben-David, Avigail Orni and Sitvanit Ruah for careful review and important comments.

References

1. R. Armoni, L. Fix, A. Flaisher, R. Gerth, B. Ginsburg, T. Kanza, A. Landver, S. Mador-Haim, E. Singerman, A. Tiemeyer, M. Y. Vardi, and Y. Zbar. The ForSpec temporal logic: A new temporal property-specification language. In J.-P. Katoen and P. Stevens, editors, *Proc. 8th International Conference on Tools and Algorithms for the Construction and Analysis of Systems (TACAS)*, volume 2280 of *Lecture Notes in Computer Science*. Springer, 2002.

2. E. Clarke and E. Emerson. Design and synthesis of synchronization skeletons using branching time temporal logic. In *Proc. Workshop on Logics of Programs*, LNCS 131, pages 52–71. Springer-Verlag, 1981.

3. C. Eisner and D. Fisman. Sugar 2.0 proposal presented to the Accellera Formal Verification Technical Committee, March 2002. At
   http://www.haifa.il.ibm.com/projects/verification/ sugar/Sugar_2.0_Accellera.ps.

4. M. J. C. Gordon. Using HOL to study Sugar 2.0 semantics. In *Proc. 15th International Conference on Theorem Proving in Higher Order Logics (TPHOLs)*, NASA Conference Proceedings CP-2002-211736, 2002.

5. N. Halbwachs, P. Caspi, P. Raymond, and D. Pilaud. The synchronous data-flow programming language LUSTRE. *Proceedings of the IEEE*, 79(9):1305–1320, 1991.

6. J. Havlicek, N. Levi, H. Miller, and K. Shultz. Extended CBV statement semantics, partial proposal presented to the Accellera Formal Verification Technical Committee, April 2002. At http://www.eda.org/vfv/hm/att-0772/01-ecbv_statement_semantics.ps.gz.

7. C. Liu and M. Orgun. Executing specifications of distributed computations with Chronologic(MC). In *Proceedings of the 1996 ACM Symposium on Applied Computing (SAC), February 17-19, 1996, Philadelphia, PA, USA*. ACM, 1996.

8. Z. Manna and A. Pnueli. *Temporal Verification of Reactive Systems: Safety*, pages 272–273. Springer-Verlag, New York, 1995.

9. M. Morley. Semantics of temporal e. In T. F. Melham and F. G. Moller, editors, *Proc. Banff'99 Higher Order Workshop (Formal Methods in Computation)*, 1999. University of Glasgow, Dept. of Computing Science Technical Report.

10. A. Pnueli. A temporal logic of concurrent programs. *Theoretical Computer Science*, 13:45–60, 1981.

11. F. Wang, A. K. Mok, and E. A. Emerson. Distributed real-time system specification and verification in APTL. *ACM Transactions on Software Engineering and Methodology*, 2(4):346–378, Oct. 1993.

Minimal Classical Logic and Control Operators

Zena M. Ariola[1] and Hugo Herbelin[2]

[1] University of Oregon, Eugene, OR 97403, USA
ariola@cs.uoregon.edu
[2] INRIA-Futurs, Parc Club Orsay Université, 91893 Orsay Cedex, France
Hugo.Herbelin@inria.fr

Abstract. We give an analysis of various classical axioms and character-ize a notion of minimal classical logic that enforces Peirce's law without enforcing Ex Falso Quodlibet. We show that a "natural" implementation of this logic is Parigot's classical natural deduction. We then move on to the computational side and emphasize that Parigot's $\lambda\mu$ corresponds to minimal classical logic. A continuation constant must be added to $\lambda\mu$ to get full classical logic. We then map the extended $\lambda\mu$ to a new theory of control, $\lambda\text{-}\mathcal{C}^-\text{-}top$, which extends Felleisen's reduction theory. $\lambda\text{-}\mathcal{C}^-\text{-}top$ allows one to distinguish between aborting and throwing to a continuation. It is also in correspondence with the proofs of a refinement of Prawitz's natural deduction.

1 Introduction

Traditionally, classical logic is defined by extending intuitionistic logic with ei-ther Pierce's law, excluded middle or the double negation law. We show that these laws are not equivalent and define *minimal classical logic*, which validates Peirce's law but not Ex Falso Quodlibet (EFQ), *i.e.* the law $\bot \to A$. The no-tion is interesting from a computational point of view since it corresponds to a calculus with a notion of control (such as `callcc`) which however does not allow one to abort a computation. We point out that closed typed terms of Parigot's $\lambda\mu$ [Par92] correspond to tautologies of minimal classical logic and not of (full) classical logic. We define a new calculus called $\lambda\mu\text{-}top$. Tautologies of classical natural deduction correspond to closed typed $\lambda\mu\text{-}top$ terms. We show the correspondence of $\lambda\mu\text{-}top$ with a new theory of control, $\lambda\text{-}\mathcal{C}^-\text{-}top$. The cal-culus $\lambda\text{-}\mathcal{C}^-\text{-}top$ is interesting in its own right, since it extends Felleisen's theory of control ($\lambda\text{-}\mathcal{C}$) [FH92]. The study of $\lambda\text{-}\mathcal{C}^-\text{-}top$ leads to the development of a refinement of Prawitz's natural deduction [Pra65] in which one can distinguish between aborting a computation and throwing to a continuation (aborting corre-sponds to throwing to the top-level continuation). This logic provides a solution to the mismatch between the operational and proof-theoretical interpretation of Felleisen's $\lambda\text{-}\mathcal{C}$ reduction theory.

We devote Section 2 to the definition of the various logics considered in this paper. Sections 3 through 5 explain their computational counterparts. We discuss related work in Section 6 and conclude in Section 7.

J.C.M. Baeten et al. (Eds.): ICALP 2003, LNCS 2719, pp. 871–885, 2003.

$$\frac{}{\Gamma, A \vdash_M A} \; Ax \qquad \frac{\Gamma, A \vdash_M B}{\Gamma \vdash_M A \to B} \to_i \qquad \frac{\Gamma \vdash_M A \to B \quad \Gamma \vdash_M A}{\Gamma \vdash_M B} \to_e$$

Fig. 1. Minimal Natural Deduction

2 Minimal, Intuitionistic, and Classical Logic

In this paper, we restrict our attention to propositional logic. We assume a set of *formulas*, denoted by roman uppercase letters A, B, *etc.*, which are built from an infinite set of *propositional atoms* (ranged over by X, Y, *etc.*), a distinguished formula $\bot$ denoting *false*, and *implication* written $\to$. We define *negation* as $\neg A \equiv A \to \bot$. A *named formula* is a pair of a formula and a name taken from an infinite set of *names*. We write A^x, B^α, *etc.* for named formulas. A *context* is a set of named formulas[1]. We use Greek uppercase letters Γ, Δ, *etc.* for contexts. We generally omit the names, unless there is an ambiguity. We will consider *sequents* of the form $\Gamma \vdash A$, $\Gamma \vdash$, $\Gamma \vdash; \Delta$, and $\Gamma \vdash A; \Delta$. The formulas in Γ are the *hypotheses* and the formulas on the right-hand side of the symbol $\vdash$ are the *conclusions*. In each case, the intuitive meaning is that the conjunction of the hypotheses implies the disjunction of the conclusions. A sequent with no conclusion means the negation of the conjunction of the hypotheses. As initially shown by Gentzen [Gen69] in his sequent calculus LK, classical logic can be obtained by considering sequents with several conclusions. Parigot extended this approach to natural deduction [Par92]. We will see that using sequents with several conclusions allows for a uniform presentation of different logics.

In the rest of the paper, we successively recall the definitions of minimal, intuitionistic and classical logic. We state simple facts about various classical axioms from which the definition of minimal classical logic emerges. Although we use natural deduction to formalize the various logics, we could have used sequent calculi instead (then the Curry-Howard correspondence would be with Herbelin's calculus [Her94]). If S is a schematic axiom or rule, we denote by $S, \Gamma \vdash A$ the fact that $\Gamma \vdash A$ is derivable using an arbitrary number of instances of S.

Minimal Logic. *Minimal natural deduction* implements minimal logic [Joh36]. It is defined by the set of (schematic) inference rules given in Figure 1. In minimal logic, $\bot$ is neutral and has no specific rule associated to it.

Normal proofs are an important tool for reasoning about provability in natural deduction. We say that an occurrence of $\to_e$ (also called Modus Ponens) is *normal* if its left premise is an axiom or another normal instance of Modus

---

[1] If interested only in provability, one could have defined contexts just as sets of formulas (not as sets of named formulas). But to assign terms to proofs, one needs to be able to distinguish between different occurrences of the same formula. This is the role of names. Otherwise, *e.g.* the two distinct normal proofs of $A, A \vdash A$ (representable by the λ-terms $\lambda x.\lambda y.x$ and $\lambda x.\lambda y.y$) would have been identified.

$$\frac{}{\Gamma, A \vdash_I A} \; Ax \qquad \frac{\Gamma \vdash_I}{\Gamma \vdash_I A} \; Activate$$

$$\frac{\Gamma \vdash_I \bot}{\Gamma \vdash_I} \; \bot_e \qquad \frac{\Gamma, A \vdash_I B}{\Gamma \vdash_I A \to B} \; \to_i \qquad \frac{\Gamma \vdash_I A \to B \quad \Gamma \vdash_I A}{\Gamma \vdash_I B} \; \to_e$$

Fig. 2. Intuitionistic Natural Deduction

Ponens. We say that a proof in minimal logic is *normal* if any occurrence of Modus Ponens in the proof is normal. As is well-known, a provable statement can be proved with a normal proof.

Theorem 1 (Prawitz). *If $\Gamma \vdash_M A$ is provable then there is a normal proof of $\Gamma \vdash_M A$.*

Intuitionistic logic. *Intuitionistic natural deduction* is described in Figure 2. The rule $\bot_e$ introduces a sequent with no conclusion, thus allowing the application of a weakening rule named *Activate*. Obviously, this presentation of intuitionistic logic is equivalent to minimal logic extended with the schematic axiom $\bot \to A$.

Proposition 1. $\Gamma \vdash_I A$ *iff* $EFQ, \Gamma \vdash_M A$.

In propositional or first-order predicate logic, there is no formula $\bot$ with the desired property, as stated by the following lemma which expresses that (propositional) intuitionistic logic is strictly stronger than minimal logic.

Proposition 2. $\nvdash_M EFQ$.

In contrast, in second-order logic, a formula having the property of $\bot$ is $\forall X.X$. However, the rule $\bot_e$ is still not valid for $\forall X.X$.

Classical axioms. We now give an analysis in minimal logic of different axiom schemes[2] leading to classical logic.

$(\neg A \to A) \to A$	Weak Peirce's law ($PL_\bot$)
$\neg A \vee A$	Excluded middle (EM)
$((A \to B) \to A) \to A$	Peirce's law (PL)
$(A \to B) \vee A$	Generalized excluded-middle (GEM)
$\neg\neg A \to A$	Double negation law (DN)

We classify the axioms in three categories: we call $PL_\bot$ and EM *weak classical axioms*, PL and GEM *minimal classical axioms*, and DN a *full classical axiom*. The main results of this section are that none of the classical axioms are indeed derivable in minimal logic and that the weak classical axioms are weaker in

---

[2] To reason about excluded-middle, we enrich the set of formulas with disjunction and the usual inference rules.

$$\frac{}{\Gamma, A \vdash_{MC} A; \Delta} \; Ax \qquad \frac{\Gamma \vdash_{MC}; A, \Delta}{\Gamma \vdash_{MC} A; \Delta} \; Activate \qquad \frac{\Gamma \vdash_{MC} A; A, \Delta}{\Gamma \vdash_{MC}; A, \Delta} \; Passivate$$

$$\frac{\Gamma, A \vdash_{MC} B; \Delta}{\Gamma \vdash_{MC} A \to B; \Delta} \to_i \qquad \frac{\Gamma \vdash_{MC} A \to B; \Delta \qquad \Gamma \vdash_{MC} A; \Delta}{\Gamma \vdash_{MC} B; \Delta} \to_e$$

Fig. 3. Minimal Classical Natural Deduction

minimal logic than the minimal classical axioms, which themselves are weaker than DN. Together with EFQ, weak and minimal classical axioms are however equivalent to DN.

Proposition 3. *In minimal logic, we have*

1. *neither $PL_\perp$, PL, EM, GEM nor DN is derivable.*
2. *$PL_\perp$ and EM are equivalent (as schemes).*
3. *GEM and PL are equivalent (as schemes).*
4. *GEM and PL imply EM and $PL_\perp$ but not conversely.*
5. *DN implies GEM and PL but not conversely.*
6. *DN, EM+EFQ, GEM+EFQ, $PL_\perp$+EFQ and PL+EFQ are all equivalent.*

The previous result suggests that there is space for a classical logic which does validate Peirce's law (or GEM) but not EFQ. Let us call this logic *minimal classical logic*. In contrast, EM and $PL_\perp$ without EFQ are weaker than PL, and their addition to minimal logic seems uninteresting. We will investigate a weaker form of EFQ at the end of this section.

Minimal Classical Logic. An axiom-free implementation of minimal classical logic is actually Parigot's classical natural deduction [Par92] (with no special rule for $\perp$). The inference rules are shown in Figure 3. Parigot's convention is to have two kinds of sequents, one with only named formulas on the right, written $\Gamma \vdash; \Delta$, and one with exactly one unnamed formula on the right, written $\Gamma \vdash A; \Delta$. We now state that minimal Parigot classical natural deduction is equivalent to minimal logic extended with Peirce's law, *i.e.* it implements minimal classical logic[3].

Proposition 4. $\Gamma \vdash_{MC} A$ *iff* $PL, \Gamma \vdash_M A$

Thanks to Proposition 3(4), we have, as a Corollary,

Corollary 1. *Minimal Parigot's classical natural deduction does not prove DN.*

Note however that $\vdash_{MC} \neg\neg A \to A; \perp$ is provable.

We now define the notion of *normal proof* for minimal Parigot classical natural deduction. We say that an occurrence of the rule *Passivate* is *normal* if its

---

[3] The proof involves replacing each instance of *Activate* on A by a number of instances of *PL* which is equal to the number of instances of *Passivate* on A.

premise is not an *Activate* rule. We say that a proof in minimal classical natural deduction is *normal* if any occurrence of Modus Ponens in the proof is normal (this is the same definition as for minimal non-classical natural deduction) and if any occurrence of *Passivate* is normal also.

Theorem 2 (Parigot). *If $\Gamma \vdash_{MC} A; \Delta$ is provable then there is a normal proof of $\Gamma \vdash_{MC} A; \Delta$*

$$\frac{}{\Gamma, A \vdash_C A; \Delta} \; Ax \qquad \frac{\Gamma \vdash_C; A, \Delta}{\Gamma \vdash_C A; \Delta} \; Activate \qquad \frac{\Gamma \vdash_C A; A, \Delta}{\Gamma \vdash_C; A, \Delta} \; Passivate$$

$$\frac{\Gamma \vdash_C \bot; \Delta}{\Gamma \vdash_C; \Delta} \; \bot_e \qquad \frac{\Gamma, A \vdash_C B; \Delta}{\Gamma \vdash_C A \to B; \Delta} \; \to_i \qquad \frac{\Gamma \vdash_C A \to B; \Delta \qquad \Gamma \vdash_C A; \Delta}{\Gamma \vdash_C B; \Delta} \; \to_e$$

Fig. 4. Classical Natural Deduction

Classical Logic. To obtain full classical logic from minimal Parigot's classical natural deduction[4] and thus derive DN, we explicitly add the elimination rule for $\bot$. The *(full) Parigot's classical natural deduction* is described in Figure 4. From Propositions 1, 3 and 4, we directly have:

Proposition 5. $\Gamma \vdash_C A$ *iff* $PL, \Gamma \vdash_I A$ *iff* $DN, \Gamma \vdash_M A$ *iff* $EFQ, \Gamma \vdash_{MC} A$.

We define *normal proofs* for classical natural deduction as for minimal classical natural deduction where the rule $\bot_e$ is normal if its premise is not an *Activate* rule (*i.e.* $\bot_e$ is considered at the same level as *Passivate*). Parigot's normalisation proof for minimal classical natural deduction applies also for full classical natural deduction.

Theorem 3 (Parigot). *If $\Gamma \vdash_C A; \Delta$ is provable then there is a normal proof of $\Gamma \vdash_C A; \Delta$.*

As expected, full classical logic is conservative over minimal classical logic for formulas not mentioning the $\bot$ formula, as stated by the following consequence of Theorem 3.

Proposition 6. *If $\bot$ does not occur in A then $\vdash_C A$ iff $\vdash_{MC} A$.*

---

[4] Parigot's original formulation of classical natural deduction [Par92] does not include the $\bot_e$-rule but gives direct rules for negation which are easily derivable from the elimination rule for $\bot$.

Remark 1. Minimal classical natural deduction without the *Passivate* rule yields minimal logic, since the context Δ is inert and can only remain empty in a derivation for which the end sequent has the form $\Gamma \vdash A$; (even the *Activate* rule cannot be applied). Similarly, classical natural deduction without the *Passivate* rule yields intuitionistic logic. As a consequence, minimal and intuitionistic natural deduction can both be seen as subsystems of classical natural deduction.

$$
\frac{}{\Gamma, A \vdash_{RAA} A} \; Ax
\qquad
\frac{\Gamma \vdash_{RAA} \bot^c}{\Gamma \vdash_{RAA} A} \; Activate
\qquad
\frac{\Gamma, \neg_c A \vdash_{RAA} \bot^c}{\Gamma \vdash_{RAA} A} \; RAA_c
$$

$$
\frac{\Gamma \vdash_{RAA} \bot}{\Gamma \vdash_{RAA} \bot^c} \; \bot_e^c
\qquad
\frac{\Gamma, A \vdash_{RAA} B}{\Gamma \vdash_{RAA} A \to B} \; \to_i
\qquad
\frac{\Gamma \vdash_{RAA} A \to B \quad \Gamma \vdash_{RAA} A}{\Gamma \vdash_{RAA} B} \; \to_e
$$

Fig. 5. Natural Deduction with RAA$_c$

Minimal Prawitz Classical Logic. Prawitz defines classical logic as minimal logic plus the Reductio Ad Absurdum rule (RAA) [Pra65]: from $\Gamma, \neg A \vdash \bot$ deduce $\Gamma \vdash A$. This rule implies EFQ (as DN implies EFQ) and hence yields full classical logic. In here we are interested in exploring the possibility of defining minimal classical logic from minimal logic and RAA but without deriving EFQ. Equivalently, we would like to devise a restricted version of EFQ that would allow one to prove PL from PL$_\bot$. This alternative formulation of (minimal) classical logic is obtained by distinguishing two different notions of $\bot$: $\bot$ for commands (written as $\bot^c$) and $\bot$ for terms (see Figure 5 where $\neg_c A$ stands for $A \to \bot^c$). If the context Δ is the set of formulas $A_1, \cdots, A_n$, then we write $\neg_c \Delta$ for the set $\neg_c A_1, \cdots, \neg_c A_n$. Sequents are of the form $\Gamma, \neg_c \Delta \vdash A$ or $\Gamma, \neg_c \Delta \vdash \bot_c$ and $\bot^c$ is not allowed to occur in Γ, Δ and A. The minimal subset does not contain the $\bot_e^c$ rule and is denoted by $\vdash_{MRAA}$.

Proposition 7. *Given a formula A and contexts Γ and Δ, all with no occurrences of $\bot^c$, we have*

1. $\Gamma \vdash_{MC} A; \Delta$ *iff* $\Gamma, \neg_c \Delta \vdash_{MRAA} A$.
2. $\Gamma \vdash_C A; \Delta$ *iff* $\Gamma, \neg_c \Delta \vdash_{RAA} A$.

3 Computational Content of Minimal Logic + Double Negation

To reason about Scheme programs, Felleisen *et al.* [FH92] introduced the λ-$\mathcal{C}$ calculus. $\mathcal{C}$ provides *abortive continuations*: the invocation of a continuation reinstates the captured context *in place* of the current one. Griffin was the first to observe that $\mathcal{C}$ is typable with $\neg\neg A \to A$. This extended the Curry-Howard

$$M ::= x \mid \lambda x.M \mid MM \mid (\mathcal{C}M)$$

$$\frac{}{\Gamma, x : A \vdash x : A} \; Ax \qquad \frac{\Gamma \vdash M : \neg\neg A}{\Gamma \vdash \mathcal{C}(M) : A} \; DN$$

$$\frac{\Gamma, x : A \vdash M : B}{\Gamma \vdash \lambda x.M : A \to B} \; \to_i \qquad \frac{\Gamma \vdash M : A \to B \quad \Gamma \vdash M' : A}{\Gamma \vdash MM' : B} \; \to_e$$

Fig. 6. The λ-$\mathcal{C}$ calculus

isomorphism to classical logic [Gri90]. The typing system for λ-$\mathcal{C}$ is given in Figure 6.

Proposition 8 (Griffin). *A formula A is provable in classical logic iff there exists a closed λ-$\mathcal{C}$ term M such that $\vdash M : A$ is provable.*

Felleisen also developed the λ-$\mathcal{K}$ calculus which axiomatizes the `callcc` (*i.e.* call-with-current-continuation) control operator. In contrast to $\mathcal{C}$, $\mathcal{K}$ leaves the current context intact as explicitly described in its usual encoding: $\mathcal{K}(M) = \mathcal{C}(\lambda k.k(Mk))$. $\mathcal{K}$ is not as powerful as $\mathcal{C}$ [Fel90]. In order to define $\mathcal{C}$ we need the abort primitive $\mathcal{A}$ (of type EFQ): $\mathcal{C}(M) = \mathcal{K}(\lambda k.\mathcal{A}(Mk))$. An alternative encoding, $\mathcal{K}(M) = \mathcal{C}(\lambda k.k(M\lambda x.\mathcal{A}(kx)))$, shows that $\mathcal{K}$ can be typed with PL. From Proposition 4, we have:

Proposition 9. *A formula A is provable in minimal classical logic iff there exists a closed λ-$\mathcal{K}$ term M such that $\vdash M : A$ is provable.*

The call-by-value and call-by-name reduction semantics of λ-$\mathcal{C}$ are presented in Figure 7. An important point to clarify is the presence of the abort operations in the right-hand sides of the reduction rules. As far as evaluation is concerned, they are not necessary. They are important in order to obtain a satisfying correspondence between the operational and reduction semantics. For example, the term $\mathcal{C}(\lambda k.(k\ \lambda x.x)N)$ evaluates to $\lambda x.x$. However, the absence of the abort from the reduction rules makes impossible to get rid of the control context $\lambda f.f\ N$. The abort steps signal that k is not a normal function but is an abortive continuation. As we explain in Section 5, these abort steps are different from the abort used in defining $\mathcal{C}$ in terms of $\mathcal{K}$. The aborts in the reduction rules correspond to throwing to a user defined continuation (*i.e.* a *Passivate* step), whereas the abort in the definition of $\mathcal{C}$ corresponds to throwing to the predefined top-level continuation (*i.e.* a $\perp_e$ step).

Remark 2. Parigot in [Par92] criticized Griffin's work because the proposed $\mathcal{C}$-typing did not fit the operational semantics. Actually, the only rule that breaks subject reduction is the top-level computation rule ($\mathcal{C}M \mapsto M(\lambda x.\mathcal{A}(x))$) which forces a conversion from $\perp$ to the top-level type. To solve the problem, instead of reducing M, Griffin proposed to reduce $\mathcal{C}(\lambda\alpha.\alpha M)$, where αM is of type $\perp$. As detailed in the next section, the classical version of Parigot's $\lambda\mu$ requires a similar intervention; a free continuation constant is needed which we call *top*.

$$\lambda_n\text{-}\mathcal{C} \quad \begin{cases} \beta: & (\lambda x.M)N \to M[x := N] \\ \mathcal{C}_L: & (\mathcal{C}M)N \to \mathcal{C}(\lambda k.M(\lambda f.\mathcal{A}(k(fN)))) \\ \mathcal{C}_{top}: & \mathcal{C}M \to \mathcal{C}(\lambda k.M(\lambda f.\mathcal{A}(kf))) \\ \mathcal{C}_{idem}: & \mathcal{C}(\lambda k.\mathcal{C}M) \to \mathcal{C}(\lambda k.M(\lambda x.\mathcal{A}(x))) \\ \mathcal{C}_{elim}: & \mathcal{C}(\lambda k.kM) \to M \qquad k \notin FV(M) \end{cases}$$

$$\begin{array}{c} \lambda_v\text{-}\mathcal{C} \\ V ::= x \mid \lambda x.M \end{array} \quad \begin{cases} \beta: & (\lambda x.M)V \to M[x := V] \\ \mathcal{C}_L: & (\mathcal{C}M)N \to \mathcal{C}(\lambda k.M(\lambda f.\mathcal{A}(k(fN)))) \cdot \\ \mathcal{C}_R: & V(\mathcal{C}M) \to \mathcal{C}(\lambda k.M(\lambda x.\mathcal{A}(k(Vx)))) \\ \mathcal{C}_{top}: & \mathcal{C}M \to \mathcal{C}(\lambda k.M(\lambda f.\mathcal{A}(kf))) \\ \mathcal{C}_{idem}: & \mathcal{C}(\lambda k.\mathcal{C}M) \to \mathcal{C}(\lambda k.M(\lambda x.\mathcal{A}(x))) \end{cases}$$

Fig. 7. $\lambda_n\text{-}\mathcal{C}$ and $\lambda_v\text{-}\mathcal{C}$ reduction rules

4 Computational Content of Classical Natural Deduction

Figure 8 describes Parigot's $\lambda\mu$ calculus [Par92] which is a term assignment for his classical natural deduction. The *Passivate* rule reads as follows: given a term producing a value of type A, if α is a continuation variable waiting for something of type A (*i.e. A cont*), then by invoking the continuation variable we leave the current context. Terms of the form $[\alpha]t$ are called *commands*. The *Activate* rule reads as follows: given a command (*i.e.* no formula is focused) we can select which result to get by capturing the associated continuation. If A^α is not present in the precondition then the rule corresponds to weakening. Note that the rule $\perp_e$ differs from Parigot's version. In [Par92], the elimination rule for $\perp$ is interpreted by an unnamed term $[\gamma]t$, where γ is any continuation variable (not always the same for every instance of the rule). In contrast, the rule is here systematically associated to the same primitive continuation variable, *top*, considered as a constant. This was also observed by Streicher *et al.* [SR98]. Parigot would represent DN as the term $\lambda y.\mu\alpha.[\gamma](y\lambda x.\mu\delta.[\alpha]x)$ whereas our representation is $\lambda y.\mu\alpha.[top](y\lambda x.\mu\delta.[\alpha]x)$. We use $\lambda\mu\text{-}top$ to denote the whole calculus with $\perp_e$ and $\lambda\mu$ to denote the calculus without $\perp_e$. The need for an

$$t :: x \mid \lambda x.t \mid tt \mid \mu\alpha.c$$
$$c ::= [\beta]t \mid [top]t$$

$$\frac{}{\Gamma, A^x \vdash x : A; \Delta} \; Ax \qquad \frac{c : \Gamma \vdash; A^\alpha, \Delta}{\Gamma \vdash \mu\alpha.c : A; \Delta} \; Activate \qquad \frac{\Gamma \vdash t : A; A^\alpha, \Delta}{[\alpha]t : \Gamma \vdash; A^\alpha, \Delta} \; Passivate$$

$$\frac{\Gamma \vdash t : \perp; \Delta}{[top]t : \Gamma \vdash; \Delta} \; \perp_e \qquad \frac{\Gamma \vdash t : A \to B; \Delta \quad \Gamma \vdash s : A; \Delta}{\Gamma \vdash ts : B; \Delta} \; \to_e \qquad \frac{\Gamma, A^x \vdash t : B; \Delta}{\Gamma \vdash \lambda x.t : A \to B; \Delta} \; \to_i$$

Fig. 8. $\lambda\mu$ and $\lambda\mu\text{-}top$ calculi

extra continuation constant to interpret the elimination of $\perp$ can be emphasized by the following statement:

Proposition 10. *A formula A is provable in minimal classical logic (resp. classical logic) iff there exists a closed $\lambda\mu$ term (resp. $\lambda\mu$-top term) t such that $\vdash t : A$ is provable.*

$$
\begin{array}{l}
\lambda\mu_n \\
\text{and} \\
\lambda\mu_n\text{-}top
\end{array}
\left\{
\begin{array}{lll}
\text{Logical rule:} & (\lambda x.t)s & \to t[x := s] \\
\text{Structural rule:} & (\mu\alpha.t)s & \to (\mu\alpha.t[[\alpha](ws)/[\alpha]w]) \\
\text{Renaming rule:} & \mu\alpha.[\beta]\mu\gamma.u \to \mu\alpha.u[\beta/\gamma] \\
\text{Simplification rule:} & \mu\alpha.[\alpha]u & \to u \quad \alpha \notin FV(u)
\end{array}
\right.
$$

$$
\begin{array}{l}
\lambda\mu_v \\
\text{and} \\
\lambda\mu_v\text{-}top \\
(v ::= x \mid \lambda x.t)
\end{array}
\left\{
\begin{array}{lll}
\text{Logical rule:} & (\lambda x.t)v & \to t[x := v] \\
\text{Left structural rule:} & (\mu\alpha.t)s & \to (\mu\alpha.t[[\alpha](ws)/[\alpha]w]) \\
\text{Right structural rule:} & v(\mu\alpha.t) & \to (\mu\alpha.t[[\alpha](vw)/[\alpha]w]) \\
\text{Renaming rule:} & \mu\alpha.[\beta]\mu\gamma.u \to \mu\alpha.u[\beta/\gamma] \\
\text{Simplification rule:} & \mu\alpha.[\alpha]u & \to u \quad \alpha \notin FV(u)
\end{array}
\right.
$$

Fig. 9. Call-by-name and call-by-value $\lambda\mu$ and $\lambda\mu$-*top* reduction rules

We write $\lambda\mu_n$ and $\lambda\mu_v$ (resp. $\lambda\mu_n$-*top* and $\lambda\mu_v$-*top*) for the $\lambda\mu$ calculus (resp. $\lambda\mu$-*top* calculus) equipped with call-by-name and call-by-value reduction rules, respectively. The reduction rules are given in Figure 9 (substitutions $[[\alpha](ws)/[\alpha]w]$ and $[[\alpha](sw)/[\alpha]w]$ are defined as in [Par92]). Note that the rules are the same for the $\lambda\mu$ and $\lambda\mu$-*top* calculi. $\lambda\mu_n$ is Parigot's original calculus, while our presentation of $\lambda\mu_v$ is similar to Ong and Stewart [OS97]. Both sets of reduction rules are well-typed and enjoy subject reduction.

Instead of showing a correspondence between the $\lambda\mu$-*top* calculi and the λ-$\mathcal{C}$ calculi as in [dG94], we have searched for an isomorphic calculus. This turns out to be interesting in its own right since it extends the expressive power of Felleisen λ-$\mathcal{C}$ and provides a term assignment for Prawitz's classical logic.

5 Computational Content of Prawitz's Classical Deduction

We consider a restricted form of λ-$\mathcal{C}$, called λ-$\mathcal{C}^-$-*top*. Its typing system is given in Figure 10. In λ-$\mathcal{C}^-$-*top*, we distinguish between capturing a continuation and expressing where to go next. We assume the existence of a top-level continuation called *top*. The control operator $\mathcal{C}^-$ can only be applied to a lambda abstraction. Moreover, the body of a $\mathcal{C}^-$-lambda abstraction is always of the form kM for a continuation variable k. In λ-$\mathcal{C}^-$-*top*, $\mathcal{K}$ and $\mathcal{C}$ are expressed as $\mathcal{C}^-(\lambda k.k\ M)$ and $\mathcal{C}^-(\lambda k.top\ M)$, respectively. In λ-$\mathcal{C}^-$-*top*, it is possible to distinguish between aborting a computation and throwing to a continuation. For example, one would write $\mathcal{C}^-(\lambda d.top\ M)$ to abort the computation M and $\mathcal{C}^-(\lambda d.k\ M)$ to invoke continuation k with M (d not free in M). Variables and continuation variables are kept distinct. The translation from $\lambda\text{-}\mathcal{C}$ to $\lambda\text{-}\mathcal{C}^-\text{-}top$ is is given in Figure 11. The call-by-name and call-by-value $\lambda\text{-}\mathcal{C}^-\text{-}top$ reduction rules are given in Figure 12. Note that one does not need the $\mathcal{C}_{top}$-rule, whose action is to wrap up an application of a continuation with a throw operation. $\mathcal{C}_{idem}^-$ is a generalization of $\mathcal{C}_{idem}$, which is obtained by instantiating the continuation variable k' to top (*i.e.* the continuation $\lambda x.\mathcal{A}(x)$): $\mathcal{C}^-(\lambda k.top\ \mathcal{C}(\lambda q.M)) \to \mathcal{C}^-(\lambda k.M[top/q])$. $\mathcal{C}_{idem}^-$ is similar to the rule proposed by Barbanera *et al.* [BB93]: $M(\mathcal{C}N) \to N(\lambda a.(Ma))$, where M has type $\neg A$. Felleisen proposed in [FH92] the following additional rules for $\lambda_v\text{-}\mathcal{C}$: $\mathcal{C}_E : E[\mathcal{C}M] \to \mathcal{C}(\lambda k.M(\lambda x.\mathcal{A}(k\ E[x])))$ (where E stands for a call-by-value evaluation context) and $\mathcal{C}_{elim} : \mathcal{C}(\lambda k.k\ M) \to M$, where k is not free in M. The first rule is a generalization of $\mathcal{C}_L$, $\mathcal{C}_R$, and $\mathcal{C}_{top}$ which adds expressive power to the calculus. The second rule, which also appears in [Hof95], leads to better simulation of evaluation. However, both rules destroy confluence of $\lambda_v\text{-}\mathcal{C}$. Felleisen left unresolved the problem of finding an extended theory that would include $\mathcal{C}_E$ or $\mathcal{C}_{elim}$ and still satisfy the classical properties of reduction theories. $\mathcal{C}_{elim}$ is already present in our calculi and $\mathcal{C}_E$ is derivable. Thus one may consider our calculi as a solution.

$$M ::= x \mid MM \mid \lambda x.M \mid \mathcal{C}^-(\lambda k.N)$$
$$N ::= k'M \mid topM$$

$$\frac{}{\Gamma, x : A \vdash x : A}\ Ax \qquad \frac{\Gamma \vdash N : \bot^c}{\Gamma \vdash \mathcal{C}^-(\lambda q.N) : A}\ Activate \qquad \frac{\Gamma, k : \neg_c A \vdash N : \bot^c}{\Gamma \vdash \mathcal{C}^-(\lambda k.N) : A}\ RAA_c$$

$$\frac{\Gamma \vdash M : \bot}{\Gamma \vdash topM : \bot^c}\ \bot_e^c \qquad \frac{\Gamma \vdash M : A \to B \quad \Gamma \vdash M' : A}{\Gamma \vdash MM' : B}\ \to_e \qquad \frac{\Gamma, x : A \vdash M : B}{\Gamma \vdash \lambda x.M : A \to B}\ \to_i$$

Fig. 10. $\lambda\text{-}\mathcal{C}^-$ and $\lambda\text{-}\mathcal{C}^-\text{-}top$ calculi

$$\overline{\overline{x}} = x \quad \overline{\overline{\lambda x.M}} = \lambda x.\overline{\overline{M}} \quad \overline{\overline{MN}} = \overline{\overline{M}}\ \overline{\overline{N}} \quad \overline{\overline{\mathcal{C}M}} = \mathcal{C}^-(\lambda k.top(\overline{\overline{M}}(\lambda x.\mathcal{C}^-(\lambda \delta.kx))))$$

Fig. 11. Translation from $\lambda\text{-}\mathcal{C}$ to $\lambda\text{-}\mathcal{C}^-\text{-}top$

Proposition 11.   *1. $\lambda_v\text{-}\mathcal{C}^-\text{-}top$ and $\lambda_n\text{-}\mathcal{C}^-\text{-}top$ are confluent and strongly normalizing.*

2. Subject reduction: Given $\lambda_v\text{-}\mathcal{C}^-\text{-}top$ ($\lambda_n\text{-}\mathcal{C}^-\text{-}top$) terms M, N, if $\Gamma \vdash M : A$ and $M \twoheadrightarrow N$ then $\Gamma \vdash N : A$.

Soundness and completeness properties for $\lambda_v\text{-}\mathcal{C}^-\text{-}top$ with respect to $\lambda_v\text{-}\mathcal{C}$ are stated below, where $\simeq_c$ denotes operational equivalence as defined in [FH92].

A λ_v-$\mathcal{C}^-$-*top* term M is translated into a λ_v-$\mathcal{C}$ term $\overline{M}$ by simply replacing $\mathcal{C}^-$ with $\mathcal{C}$ and by erasing the references to the *top* continuation.

Proposition 12. *1. Given λ_v-$\mathcal{C}$ terms M and N, if $M \twoheadrightarrow N$ then $\overline{M} \twoheadrightarrow \overline{N}$.*
2. Given λ_v-$\mathcal{C}^-$-top terms M and N, if $M \twoheadrightarrow N$ then $\overline{M} \simeq_c \overline{N}$.

Relation between the $\lambda\mu$-*top* and the λ-$\mathcal{C}^-$-*top* calculi. The λ-$\mathcal{C}^-$-*top* calculus has been designed in such a way that it is in one-to-one correspondence with the $\lambda\mu$-*top* calculus. The correspondence is given by $\overline{\lambda x.t} = \lambda x.\overline{t}$, $\overline{ts} = \overline{t}\overline{s}$, $\overline{\mu\alpha.[\gamma]t} = \mathcal{C}^-(\lambda\alpha.\gamma\overline{t})$. This correspondence extends to the reduction rules (Figure 12 matches Figure 9), as expressed by the following statement:

Proposition 13. *Let t, s be $\lambda\mu$-top-terms, then $t \to_{\lambda\mu_n\text{-top}} s$ iff $\overline{t} \to_{\lambda_n\text{-}\mathcal{C}^-\text{-top}} \overline{s}$ and $t \to_{\lambda\mu_v\text{-top}} s$ iff $\overline{t} \to_{\lambda_v\text{-}\mathcal{C}^-\text{-top}} \overline{s}$.*

$$
\boxed{
\begin{array}{ll}
\begin{array}{c}
\lambda_n\text{-}\mathcal{C}^- \\
\text{and} \\
\lambda_n\text{-}\mathcal{C}^-\text{-top}
\end{array}
\left\{
\begin{array}{lll}
\beta : & (\lambda x.M)N & \to M[x := N] \\
\mathcal{C}_L^- : & \mathcal{C}^-(\lambda k.M)N & \to \mathcal{C}^-(\lambda k.M[k\,(PN)/k\,P]) \\
\mathcal{C}_{idem}^- : & \mathcal{C}^-(\lambda k.k'\mathcal{C}^-(\lambda q.N)) & \to \mathcal{C}^-(\lambda k.N[k'/q]) \\
\mathcal{C}_{elim}^- : & \mathcal{C}^-(\lambda k.kM) & \to M \quad k \notin FV(M)
\end{array}
\right. \\[4em]
\begin{array}{c}
\lambda_v\text{-}\mathcal{C}^- \\
\text{and} \\
\lambda_v\text{-}\mathcal{C}^-\text{-top} \\
(V ::= x \mid \lambda x.M)
\end{array}
\left\{
\begin{array}{lll}
\beta : & (\lambda x.M)V & \to M[x := V] \\
\mathcal{C}_{elim}^- : & \mathcal{C}^-(\lambda k.kM) & \to M \quad k \notin FV(M) \\
\mathcal{C}_L^- : & \mathcal{C}^-(\lambda k.M)N & \to \mathcal{C}^-(\lambda k.M[k\,(PN)/k\,P]) \\
\mathcal{C}_R^- : & V\mathcal{C}^-(\lambda k.M) & \to \mathcal{C}^-(\lambda k.M[k\,(VP)/k\,P]) \\
\mathcal{C}_{idem}^- : & \mathcal{C}^-(\lambda k.k'\mathcal{C}^-(\lambda q.N)) & \to \mathcal{C}^-(\lambda k.N[k'/q])
\end{array}
\right.
\end{array}
}
$$

Fig. 12. Call-by-name and call-by-value λ-$\mathcal{C}^-$ and λ-$\mathcal{C}^-$-*top* reduction rules

Remark 3. Reducing the term corresponding to $\mathcal{C}(\lambda k.kIx)1$ we have:

$$(\mu\alpha.[top](\lambda k.kIx)(\lambda f.\mu\delta.[\alpha]f))1 \to (\mu\alpha.[top]((\lambda f.\mu\delta.[\alpha]f)I)x)1 \to (\mu\alpha.[top]((\mu\delta.[\alpha]I)x))1 \to$$
$$(\mu\alpha.[top](\mu\delta.[\alpha]I))1 \to (\mu\alpha.[top](\mu\delta.[\alpha](I1)))1 \to (\mu\alpha.[\alpha](I1))1 \to (\mu\alpha.[\alpha]1)1 \to 1$$

This reduction sequence is better than the corresponding sequence in λ-$\mathcal{C}$.

Proposition 14. *A formula A is provable in Prawitz's classical logic iff there exists a closed $\lambda\mathcal{C}^-$-top term M such that $\vdash M : A$ is provable.*

We define a subset of λ-$\mathcal{C}^-$-*top*, which does not allow one to abort a computation, *i.e.* terms of the form $\mathcal{C}^-(\lambda k.topM)$ are not allowed. We call this subset, which is isomorphic to $\lambda\mu$, λ-$\mathcal{C}^-$.

Proposition 15. *A formula A is provable in minimal Prawitz classical logic iff there exists a closed λ-$\mathcal{C}^-$ term M such that $\vdash M : A$ is provable.*

Remark 4. The λ-$\mathcal{C}^-$ term representing PL is $\lambda y.\mathcal{C}^-(\lambda k.k(y(\lambda x.\mathcal{C}^-(\lambda q.kx))))$, which can be written in ML as:

```
- fun PL y = callcc (fn k => (y  (fn x => throw k x)));
val PL = fn : (('a -> 'b) -> 'a) -> 'a
```

Notice how the throw construct corresponds to a weakening step. By Propositions 7, 9 and 15, λ-$\mathcal{C}^-$ is equivalent to λ-$\mathcal{K}$, assuming `callcc` is typed with PL, say `callcc`$_{\text{pl}}$. However, it might not be at all obvious how to use a continuation in different contexts, since we do not have weakening available. Consider for example the following ML term (with `callcc` and `throw` typable as in [DHM91]):

- `callcc (fn k => if (throw k 1) then 7 else (throw k 99));`

We use the continuation in both boolean and integer contexts. How can we write the above expression without making use of weakening or throw? The proof of Proposition 4 gives the answer:

- `callcc_pl (fn k => callcc_pl (fn q => if q 1 then 7 else k 99));`

We define a subset of λ-$\mathcal{C}^-$, called λ-$\mathcal{A}^-$, in which expressions of the form $\mathcal{C}^-(\lambda d.qM)$ are only allowed when d is not free in M and q is *top*, that is, we only allow throwing to the top-level continuation.

Proposition 16. *A formula A is provable in intuitionistic logic iff there exists a closed λ-$\mathcal{A}^-$ term M such that $\vdash M : A$ is provable.*

6   Related Work

The relation between Parigot $\lambda\mu$ and λ-$\mathcal{C}$ has been investigated by de Groote [dG94], who only considers the $\lambda\mu$ structural rule but not renaming and simplification. As for λ-$\mathcal{C}$, he only considers $\mathcal{C}_L$ and $\mathcal{C}_{top}$. However, these rules are not the original rules of Felleisen, since they do not contain abort. For example, $\mathcal{C}_{top}$ is $\mathcal{C}M \to \mathcal{C}(\lambda k.M(\lambda f.kf))$ which is in fact a reduction rule for λ-$\mathcal{F}$ [Fel88]. This work fails in relating $\lambda\mu$ to λ-$\mathcal{C}$ in an untyped framework, since it does not express continuations as abortive functions. It says in fact that $\mathcal{F}$ behaves as $\mathcal{C}$ in the simply-typed case. Ong and Stewart [OS97] also do not consider the abort step in Felleisen's rules. This could be justified because in a simply-typed setting these steps are of type $\bot \to \bot$. Therefore, it seems we have a mismatch. While the aborts are essential in the reduction semantics, they are irrelevant in the corresponding proof. We are the first to provide a proof theoretic justification for those abort steps, they correspond to the step $\bot \to \bot_c$. In addition to Ong and Stewart, Py [Py98] and Bierman [Bie98] have pointed out the peculiarity of having an open $\lambda\mu$ term corresponding to a tautology. Their solution is to abolish the distinction between commands and terms. A command is a term returning $\bot$. The body of a μ-abstraction is not restricted to a command, but can be of the form $\mu\alpha.t$, where t is of type $\bot$. Thus, one has $\lambda y.\mu\alpha.(y\ \lambda x.[\alpha]x) : \neg\neg A \to A$. We would then represent the term $\mathcal{C}(\lambda k.(kI)x)$ (where I is $\lambda x.x$) as $\mu\alpha.(\alpha I)x$. Whereas $\mathcal{C}(\lambda k.kIx)$ would reduce to $\mathcal{C}(\lambda k.kI)$ according to λ_n-$\mathcal{C}$ and to I in $\lambda\mu_n$-*top*, it would be in normal form in their calculus. Thus, their work in relating $\lambda\mu$ to λ-$\mathcal{C}$ only applies to typed λ-$\mathcal{C}$, whereas our work also applies to the untyped case. Crolard [Cro99] studied the relation between Parigot's $\lambda\mu$ and a calculus with a `catch` and `throw` mechanism. He showed that contraction corresponds to the `catch` operator ($\mu\alpha.[\alpha]t = $ `catch` $\alpha\ t$) and weakening corresponds to the `throw` operator ($\mu\delta.[\alpha]t = $ `throw` $\alpha\ t$ for δ not free in t). He only considers terms of the form $\mu\alpha.[\alpha]t$ and $\mu\beta.[\alpha]t$, where β does not occur free in t. This property is not preserved by the renaming rule, therefore reduction is restricted. We do not require such restrictions on reduction. We can simulate Ong and Stewart's $\lambda\mu$ and Crolard's calculus via this simple translation: $\mu\alpha.l$ becomes $\mu\alpha.[top]t$ and $[\beta]t$ becomes $\mu\delta.[\beta]t$, where δ is not free in t.

7   Conclusions

Our analysis of the logical strengths of EFQ, PL (or EM) and DN has led naturally to a restricted form of classical logic called *minimal classical logic*. Depending on whether EFQ, PL, or both are assumed in minimal logic, we get intuitionistic, minimal classical, or classical logic. Depending on whether or not we admit *Passivate* $(RAA_c{}^5)$ and $\perp_e$ $(\perp_e^c)$ in full classical natural deduction (on top of minimal natural deduction), we get the correspondences with the λ-calculi considered in this paper, as summarized above[6]. Among these systems, $\lambda\text{-}\mathcal{C}^-\text{-}top$ is a confluent extension of Felleisen's theory of control.

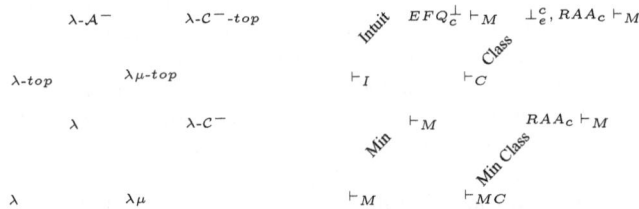

We also have some preliminary results regarding $\mathcal{F}$ [Fel88] which provides *functional continuations*, meaning that the invocation of a continuation reinstates the captured context *on top* of the current one. When a continuation is applied it acts like an ordinary function. We conjecture that $\mathcal{F}$ is still typable with DN. The difference with $\mathcal{C}$ is that, for $\mathcal{F}$, the $\perp$ type is equated to the top-level type. Therefore, we do not need the throw construct. As before, one could define a calculus $\lambda\text{-}\mathcal{F}^-\text{-}top$ with similar restrictions as for $\lambda\text{-}\mathcal{C}^-\text{-}top$, but without requiring the use of a throw construct to invoke a continuation. What is interesting is that the reduction rules for call-by-value and call-by-name $\lambda\text{-}\mathcal{F}^-\text{-}top$ would be the same as those given in Figure 12 with $\mathcal{F}^-$ replacing $\mathcal{C}^-$. In [Fel88], Felleisen also introduced the notion of a *prompt*, written #, with the reduction rules $\#_{\mathcal{F}} : \#(\mathcal{F}M) \rightarrow \#(M(\lambda x.x))$ and $\#_v : (\#V) \rightarrow V$. We can define prompt as $\#M = \mathcal{F}^-(\lambda top.top M)$. However, one would need one more reduction rule $(\mathcal{F}^-(\lambda top.M) \rightarrow \mathcal{F}^-(\lambda top.M[\lambda x.x/top]))$ and a proviso (the bound variable k cannot be top) on the lifting rules $\mathcal{F}_L^-$ and $\mathcal{F}_R^-$. To see why we need the proviso, consider the term $\mathcal{F}^-(\lambda top.3 * (top 2)) + 1$ which reduces to

---

[5] with restrictions on the use of $\perp_c$

[6] $\lambda\text{-}top$ is the subset of $\lambda\mu\text{-}top$ in which expressions of the form $\mu\delta.[\alpha]t$ are only allowed when δ is not free in t and α is top. $EFQ_c^\perp$ stands for $\perp_e^c$ and the rule $\Gamma \vdash \perp_c$ implies $\Gamma \vdash A$ (*i.e.* the restriction of RAA_c when $\neg_c A$ is not used in the proof).

7. If we had allowed the left lifting rule we would have obtained 9. We can also extend λ-$\mathcal{C}^-$-top with prompt, however, we do not need to extend the system with any additional reduction rules. With $\mathcal{C}^-$ and *prompt* one could also define shift and reset [DF89,DF90] as $(\text{shift } M) = \mathcal{C}^-(\lambda k.M(\lambda x.\#(kx)))$. This was also observed by Filinsky in [Fil94]. We plan to investigate these additional calculi and to extend our analysis to other control operators. The reader may refer to [Que93] for a complete list of them.

Acknowledgements. We thank Matthias Felleisen for numerous discussions about his theory of control. Miley Semmelroth helped us to improve the presentation of the paper. We also thank the anonymous referees for their comments. The first author has been supported by NSF grant 0204389.

References

[BB93]   F. Barbanera and S. Berardi. Extracting constructive content from classical logic via control-like reductions. In *TLCA'93, LNCS 664*, pages 45–59. 1993.

[Bie98]   G.M. Bierman. A computational interpretation of the lambda-mu calculus. In *MFCS'98, LNCS 1450*, pages 336–345, 1998.

[Cro99]   T. Crolard. A confluent lambda-calculus with a catch/throw mechanism. *Journal of Functional Programming*, 9(6):625–647, 1999.

[DF89]   Olivier Danvy and Andrzej Filinski. A functional abstraction of typed contexts. Technical Report 89/12, 1989.

[DF90]   O. Danvy and A. Filinski. Abstracting control. In *Proceedings of the 1990 ACM Conference on LISP and Functional Programming, Nice*, pages 151–160, New York, NY, 1990. ACM.

[dG94]   P. de Groote. On the relation between the lambda-mu calculus and the syntactic theory of sequential control. In *LPAR'94*, pages 31–43. 1994.

[DHM91]   B. F. Duba, R. Harper, and D. MacQueen. Typing first-class continuations in ML. In *POPL'91*, pages 163–173, 1991.

[Fel88]   M. Felleisen. The theory of practice of first-class prompt. In *POPL '88*, pages 180–190, 1988.

[Fel90]   M. Felleisen. On the expressive power of programming languages. In *ESOP '90, LNCS 432*, pages 134–151. 1990.

[FH92]   M. Felleisen and R. Hieb. A revised report on the syntactic theories of sequential control and state. *Theoretical Computer Science*, 103(2):235–271, 1992.

[Fil94]   Andrzej Filinski. Representing monads. In *Conf. Record 21st ACM SIGPLAN-SIGACT Symp. on Principles of Programming Languages, POPL'94, Portland, OR, USA, 17–21 Jan. 1994*, pages 446–457, New York, 1994. ACM Press.

[Gen69]   G. Gentzen. Investigations into logical deduction. In M.E. Szabo, editor, *Collected papers of Gerhard Gentzen*, pages 68–131. North-Holland, 1969.

[Gri90]   T. G. Griffin. The formulae-as-types notion of control. In *POPL'90*, pages 47–57, 1990.

[Her94]   H. Herbelin. A lambda-calculus structure isomorphic to Gentzen-style sequent calculus structure. In *CSL'94, LNCS 933*, 1994.

[Hof95]   M. Hofmann. Sound and complete axiomatization of call-by-value control operators. *Mathematical Structures in Computer Science*, 5:461–482, 1995.

[Joh36]   I. Johansson. Der Minimalkalkül, ein Reduzierter Intuitionistischer Formalismus. *Compositio Mathematica*, 4:119–136, 1936.

[OS97]   C.-H. Luke Ong and C. A. Stewart. A Curry-Howard foundation for functional computation with control. In *POPL'97*, pages 215–227. 1997.

[Par92]   M. Parigot. Lambda-mu-calculus: An algorithmic interpretation of classical natural deduction. In *LPAR '92*, pages 190–201, 1992.

[Pra65]   D. Prawitz. *Natural Deduction, a Proof-Theoretical Study*. Almquist and Wiksell, Stockholm, 1965.

[Py98]   W. Py. *Confluence en λμ-calcul*. PhD thesis, Université de Savoie, 1998.

[Que93]   Christian Queinnec. A library of high-level control operators. *ACM SIGPLAN Lisp Pointers*, 6(4):11–26, 1993.

[SR98]   T. Streicher and B. Reus. Classical logic: Continuation semantics and abstract machines. *Journal of Functional Programming*, 8(6):543–572, 1998.

Counterexample-Guided Control*

Thomas A. Henzinger, Ranjit Jhala, and Rupak Majumdar

EECS Department, University of California, Berkeley
{tah,jhala,rupak}@eecs.berkeley.edu

Abstract. A major hurdle in the algorithmic verification and control of systems is the need to find suitable abstract models, which omit enough details to overcome the state-explosion problem, but retain enough details to exhibit satisfaction or controllability with respect to the specification. The paradigm of counterexample-guided abstraction refinement suggests a fully automatic way of finding suitable abstract models: one starts with a coarse abstraction, attempts to verify or control the abstract model, and if this attempt fails and the abstract counterexample does not correspond to a concrete counterexample, then one uses the spurious counterexample to guide the refinement of the abstract model. We present a counterexample-guided refinement algorithm for solving ω-regular control objectives. The main difficulty is that in control, unlike in verification, counterexamples are strategies in a game between system and controller. In the case that the controller has no choices, our scheme subsumes known counterexample-guided refinement algorithms for the verification of ω-regular specifications. Our algorithm is useful in all situations where ω-regular games need to be solved, such as supervisory control, sequential and program synthesis, and modular verification. The algorithm is fully symbolic, and therefore applicable also to infinite-state systems.

1 Introduction

The key to the success of algorithmic methods for the verification (analysis) and control (synthesis) of complex systems is *abstraction*. Useful abstractions have two desirable properties. First, the abstraction should be *sound*, meaning that if a property (e.g., safety, controllability) is proved for the abstract model of a system, then the property holds also for the concrete system. Second, the abstraction should be *effective*, meaning that the abstract model is not too fine and can be handled by the tools at hand; for example, in order to use conventional model checkers, the abstraction must be both finite-state and of manageable size. Recent research has focused on a third desirable property of abstractions. A sound and effective abstraction (provided it exists) should be found *automatically*; otherwise, the labor-intensive process of constructing suitable abstract models often

---

* This research was supported in part by the DARPA SEC grant F33615-C-98-3614, the ONR grant N00014-02-1-0671, and the NSF grants CCR-9988172, CCR-0085949, and CCR-0225610.

J.C.M. Baeten et al. (Eds.): ICALP 2003, LNCS 2719, pp. 886–902, 2003.

negates the benefits of automatic methods for verification and control. The most successful paradigm in automatic abstraction is the method of *counterexample-guided abstraction refinement* [5,6,9]. According to that paradigm, one starts with a very coarse abstract model, which is effective but may not be informative, meaning that it may not exhibit the desired property even if the concrete system does. Then the abstract model is refined iteratively as follows: first, if the abstract model does not exhibit the desired property, then an abstract counterexample is constructed automatically; second, it can be checked automatically if the abstract counterexample corresponds to a concrete counterexample; if this is not the case, then, third, the abstract model is refined automatically in order to eliminate the spurious counterexample.

The method of counterexample-guided abstraction refinement has been developed for the *verification* of linear-time properties [9], and universal branching-time properties [10]. It has been applied successfully in both hardware [9] and software verification [6,18]. We develop the method of counterexample-guided abstraction refinement for the *control* of linear-time objectives. In verification, a counterexample to the satisfaction of a linear-time property is a *trace* that violates the property: for safety properties, a finite trace; for general ω-regular properties, an infinite, periodic (lasso-shaped) trace. In control, counterexamples are considerably more complicated: a counterexample to the controllability of a system with respect to a linear-time objective is a *tree* that represents a strategy of the system for violating the property no matter what the controller does. For safety objectives, finite trees are sufficient as counterexamples; for general ω-regular objectives on finite abstract models, infinite trees are necessary, but they can be finitely represented as graphs with cycles, because finite-state strategies are as powerful as infinite-state strategies [17].

In somewhat more detail, our method proceeds as follows. Given a two-player game structure (player 1 "controller" vs. player 2 "system"), we wish to check if player 1 has a strategy to achieve a given ω-regular winning condition. Solutions to this problem have applications in supervisory control [22], sequential hardware synthesis and program synthesis [8,7,21], modular verification [2,4,14], receptiveness checking [3,15], interface compatibility checking [12], and schedulability analysis [1]. We automatically construct an abstraction of the given game structure that is as coarse as possible and as fine as necessary in order for player 1 to have a winning strategy. We start with a very coarse abstract game structure and refine it iteratively. First, we check if player 1 has a winning strategy in the abstract game; if so, then the concrete system can be controlled; otherwise, we construct an abstract player-2 strategy that spoils against all abstract player-1 strategies. Second, we check if the abstract player-2 strategy corresponds to a spoiling strategy for player 2 in the concrete game; if so, then the concrete system cannot be controlled; otherwise, we refine the abstract game in order to eliminate the abstract player-2 strategy. In this way, we automatically synthesize "maximally abstract" controllers, which distinguish two states of the controlled system only if they need to be distinguished in order to achieve the control objective. It should be noted that ω-regular verification problems are but special cases of ω-regular control problems, where player 1 (the controller) has no choice of moves. Our method, therefore, includes as a special case counterexample-guided abstraction refinement for linear-time *verification*.

Furthermore, our method is fully *symbolic*: while traditional symbolic verification computes fixpoints on the iteration of a transition-precondition operator on regions (symbolic state sets), and traditional symbolic control computes fixpoints on the iteration of a more general, game-precondition operator *Cpre* (controllable *Pre*) [4,20], our counterexample-guided abstraction refinement also computes fixpoints on the iteration of *Cpre* and two additional region operators, called *Focus* and *Shatter*. The *Focus* operator is used to check if an abstract counterexample is genuine or spurious. The *Shatter* operator, which is used to refine an abstract model guided by a spurious counterexample, splits an abstract state into several states. Our top-level algorithm calls only these three system-specific operators: *Cpre*, *Focus*, and *Shatter*. It is therefore applicable not only to finite-state systems but also to infinite-state systems, such as hybrid systems, on which these three operators are computable (termination can be studied as an orthogonal issue along the lines of [13]; clearly, our abstraction-based algorithms terminate in all cases in which the standard, *Cpre*-based algorithms terminate, such as in the control of timed automata [20], and they may terminate in more cases).

In a previous paper, we improved the naive iteration of the "abstract-verify-refine" loop by integrating the construction of the abstract model and the verification process [18]. The improvement is called *lazy abstraction*, because the abstract model is constructed on demand during verification, which results in nonuniform abstractions, where some areas of the state space are abstracted more coarsely than others, and thus guarantees an abstract model that is as small as possible. The lazy-abstraction paradigm can be applied also to the algorithm presented here, which subsumes both verification and control. The details of this, however, need to be omitted for space reasons.

2   Games and Abstraction

Two-player games. Let Λ be a set of labels, and Φ a set of propositions. A (*two-player*) *game structure* $\mathcal{G} = (V_1, V_2, \delta, P)$ consists of two (possibly infinite) disjoint sets V_1 and V_2 of player-1 and player-2 states (let $V = V_1 \cup V_2$ denote the set of all states), a labeled transition relation $\delta \subseteq V \times \Lambda \times V$, and a function $P \colon V \to 2^{\Phi}$ that maps every state to a set of propositions. For every state $v \in V$, we call $L(v) = \{l \in \Lambda \mid \exists w. (v, l, w) \in \delta\}$ the set of *available moves*. In the sequel, i ranges over the set $\{1, 2\}$ of players. Intuitively, at state $v \in V_i$, player i chooses a move $l \in L(v)$, and the game proceeds nondeterministically to some state w satisfying $\delta(v, l, w)$.[1] We require that every player-2 state $v \in V_2$ has an available move, that is, $L(v) \neq \emptyset$. For a move $l \in \Lambda$,

---

[1] Even if the transition relation is deterministic, abstractions of the game may be nondeterministic.

let $Avl(l) = \{v \in V \mid l \in L(v)\}$ be the set of states in which move l is available. We extend the transition relation to sets via the operators $Apre, Epre$: $2^V \times \Lambda \to 2^V$ by defining $Apre(X, l) = \{v \in V \mid \forall w. \delta(v, l, w) \Rightarrow w \in X\}$ and $Epre(X, l) = \{v \in V \mid \exists w. \delta(v, l, w) \wedge w \in X\}$. For a proposition $p \in \Phi$, let $[p] = \{v \in V \mid p \in P(v)\}$ and $[\overline{p}] = V \setminus [p]$ be the sets of states in which p is true and false, respectively. We assume that Φ contains a special proposition $init$, which specifies a set $[init] \subseteq V$ containing the initial states. A run of the game structure $\mathcal{G}$ is a finite or infinite sequence $v_0 v_1 v_2 \ldots$ of states $v_j \in V$ such that for all $j \geq 0$, if v_j is not the last state of the run, then there is a move $l_j \in \Lambda$ with $\delta(v_j, l_j, v_{j+1})$. A $strategy\ of\ player\ i$ is a partial function $f_i: V^* \cdot V_i \to \Lambda$ such that for every state sequence $u \in V^*$ and every state $v \in V_i$, if $L(v) \neq \emptyset$, then $f_i(u \cdot v)$ is defined and $f_i(u \cdot v) \in L(v)$. Intuitively, a player-i strategy suggests, when possible, a move for player i given a sequence of states that end in a player-i state. Given two strategies f_1 and f_2 of players 1 and 2, the $possible$ $outcomes$ $\Omega_{f_1, f_2}(v)$ from a state $v \in V$ are runs: a run $v_0 v_1 v_2 \ldots$ belongs to $\Omega_{f_1, f_2}(v)$ iff $v = v_0$ and for all $j \geq 0$, either $L(v_j) = \emptyset$ and v_j is the last state of the run, or $v_j \in V_i$ and $\delta(v_j, f_i(v_0 \ldots v_j), v_{j+1})$. Note that the last state of a finite outcome is always a player-1 state.

Winning conditions. A $game$ $(\mathcal{G}, \Gamma)$ consists of a game structure $\mathcal{G}$ and an objective Γ for player 1. We focus on safety games, and briefly discuss games with more general ω-regular objectives at the very end of the paper. A $safety$ $game$ has an objective of the form $\Box \overline{err}$, where $err \in \Phi$ is a proposition which specifies a set $[err] \subseteq V$ of error states. Intuitively, the goal of player 1 is to keep the game in states in which err is false, and the goal of player 2 is to drive the game into a state in which err is true. Moreover, in all games we consider, whenever a dead-end state is encountered, player 1 loses. Formally, a run $v_0 v_1 v_2 \ldots$ is $winning\ for\ player\ 1$ if it is infinite and for all $j \geq 0$, we have $v_j \notin [err]$. Let Π_1 denote the set of runs that are winning for player 1. In general, an $objective$ for player 1 is a set $\Gamma \subseteq (2^\Phi)^\omega$ of infinite words over the alphabet 2^Φ, and Π_1 contains all infinite runs $v_0 v_1 \ldots$ such that $P(v_0), P(v_1), \ldots \in \Gamma$. The game starts from any initial state. A strategy f_1 is $winning\ for\ player\ 1$ if for all strategies f_2 of player 2 and all states $v \in [init]$, we have $\Omega_{f_1, f_2}(v) \subseteq \Pi_1$; that is, all possible outcomes are winning for player 1. Dually, a strategy f_2 is $spoiling\ for\ player\ 2$ if for all strategies f_1 of player 1, there is a state $v \in [init]$ such that $\Omega_{f_1, f_2}(v) \not\subseteq \Pi_1$. Note that in our setting, nondeterminism is always on the side of player 2. If the objective Γ is ω-regular, then either player 1 has a winning strategy or player 2 has a spoiling strategy [17]. We say that player 1 $wins$ the game if there is a player-1 winning strategy.

Example 1 [EXSAFETY]   Figure 1(a) shows an example of a safety game. The white states are player 1 states, and the black ones are player 2 states. The labels on the edges denote moves. The objective is $\Box \overline{p}$, that is, player 1 seeks to avoid the error states $[p]$. The player 1 states 1, 2, and 3 are the initial states, i.e., we wish player 1 to win from all three states. Note that in fact player 1 does win from the states 1, 2, and 3: at state 1, she plays the move C; at 2, she plays A;

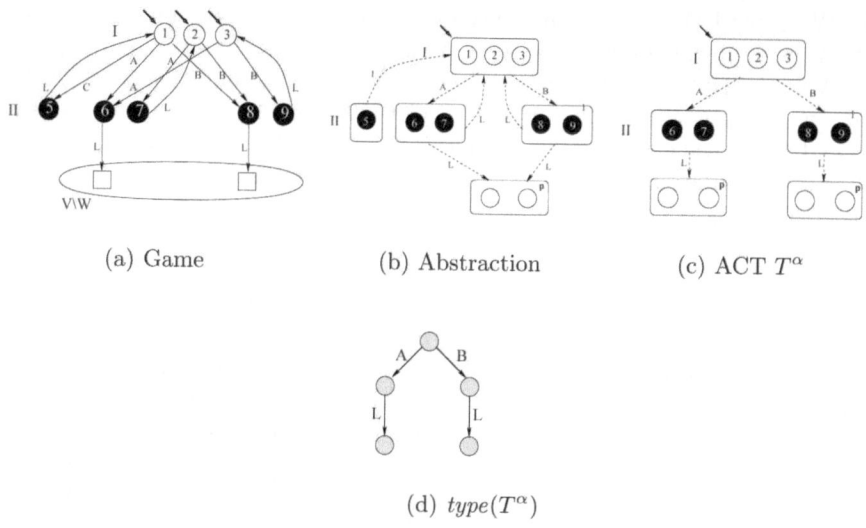

(a) Game   (b) Abstraction   (c) ACT T^α

(d) $type(T^\alpha)$

Fig. 1. Example ExSAFETY

and at 3, she plays B. In each case, the only move L available to player 2 brings the game back to the original state. This ensures that the game never reaches a state in $[p]$. ∎

The *(player-1) controllable predecessor* operator $Cpre_1 : 2^V \to 2^V$ denotes, for a set $X \subseteq V$ of states, the states from which player 1 can force the game into X in one step. Player 1 can force the game into X from a state $v \in V_1$ iff there is *some* available move l such that all l-successors of v are in X, and player 1 can force the game into X from a state $v \in V_2$ iff for *all* available moves l, all l-successors of v are in X. Formally:

$$Cpre_1(X) = (V_1 \cap \bigcup_{l \in \Lambda}(Avl(l) \cap Apre(X,l))) \cup (V_2 \cap \bigcap_{l \in \Lambda} Apre(X,l))$$

In particular, the set of states from which player 1 can keep the game away from *err* states is the greatest fixpoint $\nu X. [\overline{err}] \cap Cpre_1(X)$. Hence player 1 wins the safety game with objective $\Box \overline{err}$ iff $[init] \subseteq (\nu X. [\overline{err}] \cap Cpre_1(X))$.

Abstractions of games. Since solving a game may be expensive, we wish to construct sound abstractions of the game with smaller state spaces. Soundness means that if player 1 wins the abstract game, then she wins also the original, concrete game. To ensure soundness, we restrict the power of player 1 and increase the power of player 2 [19]. Therefore, we abstract the player-1 states so that *fewer* moves are available, and the player-2 states so that *more* moves are available. An *abstraction* $\mathcal{G}^\alpha$ for the game structure $\mathcal{G}$ is a game structure $(V_1^\alpha, V_2^\alpha, \delta^\alpha, P^\alpha)$ and a concretization function $[\![\cdot]\!]\colon V^\alpha \to 2^V$ (where $V^\alpha = V_1^\alpha \cup V_2^\alpha$ is the abstract state space) such that conditions (1)–(3) hold. (1) The abstraction preserves the player structure and propositions: for $i \subset [1,2]$ and all $v^\alpha \in V_i^\alpha$, we have $[\![v^\alpha]\!] \subseteq V_i$; for all $v^\alpha \in V^\alpha$, if $v, v' \in [\![v^\alpha]\!]$, then $P(v) = P(v')$ and $P^\alpha(v^\alpha) = P(v)$. (2) The abstract states cover the concrete state space: $\bigcup_{v^\alpha \in V^\alpha} [\![v^\alpha]\!] = V$. (3) For each player-1 abstract state $v^\alpha \in V_1^\alpha$, define $L^\alpha(v^\alpha) = \bigcap_{v \in [\![v^\alpha]\!]} L(v)$, and for each player-2 abstract state $v^\alpha \in V_2^\alpha$, define $L^\alpha(v^\alpha) = \bigcup_{v \in [\![v^\alpha]\!]} L(v)$. Then, for all $v^\alpha, w^\alpha \in V^\alpha$ and all $l \in \Lambda$, we have $\delta^\alpha(v^\alpha, l, w^\alpha)$ iff $l \in L^\alpha(v^\alpha)$ and there are states $v \in [\![v^\alpha]\!]$ and $w \in [\![w^\alpha]\!]$ with $\delta(v, l, w)$. Note that the abstract state space V^α and the concretization function $[\![\cdot]\!]$ uniquely determine the abstraction $\mathcal{G}^\alpha$. Intuitively, each abstract state $v^\alpha \in V^\alpha$ represents a set $[\![v^\alpha]\!] \subseteq V$ of concrete states. We will use only abstractions with finite state spaces. The controllable predecessor operator on the abstract game structure $\mathcal{G}^\alpha$ is denoted $Cpre_1^\alpha$.

Proposition 1 [Soundness of abstraction] *Let $\mathcal{G}^\alpha$ be an abstraction for a game structure $\mathcal{G}$, and let Γ be an objective for player 1. If player 1 wins the abstract game $(\mathcal{G}^\alpha, \Gamma)$, then player 1 also wins the concrete game $(\mathcal{G}, \Gamma)$.*

Example 2 [EXSAFETY] Figure 1(b) shows one particular abstraction for the game structure from Figure 1(a). The boxes denote abstract states with the states they represent drawn inside them. The dashed arrows are the abstract transitions. Note that from the starting player 1 box, the move C is not available, because it is not available at states 2 and 3. as not all the states in the box can do it. In the abstract game, player 2 has a spoiling strategy: after player 1 plays either move A or move B, player 2 can play move L and take the game to the error set $[p]$. ∎

3   Counterexample-Guided Abstraction Refinement

A *counterexample* to the claim that player 1 can win a game is a spoiling strategy for player 2. A counterexample for an abstract game $(\mathcal{G}^\alpha, \Gamma)$ may be either *genuine*, meaning that it corresponds to a counterexample for the concrete game $(\mathcal{G}, \Gamma)$, or *spurious*, meaning that it arises due to the coarseness of the abstraction. In the sequel, we check whether or not an abstract counterexample is genuine for a a fixed safety game $(\mathcal{G}, \Box \overline{err})$ and abstraction $\mathcal{G}^\alpha$. Moreover, if the counterexample is spurious, then we refine the abstraction in order to rule out that particular counterexample.

Abstract counterexample trees. Our abstract games are finite-state, and for safety games, memoryless spoiling strategies suffice for player 2. Finite trees are therefore a natural representation of counterexamples. We work with rooted, directed, finite trees with labels on both nodes and edges. Each node is labeled by an abstract state $v^\alpha \in V^\alpha$ or a concrete state $v \in V$, and possibly a set $r \subseteq V$ of concrete states. We write $\mathbf{n}{:}v^\alpha$ for node $\mathbf{n}$ labeled with v^α, and $\mathbf{n}{:}v^\alpha{:}r$ if $\mathbf{n}$ is labeled with both v^α and r. Each edge is labeled with a move $l \in \Lambda$. If $\mathbf{n} \xrightarrow{l} \mathbf{n}'$ is an edge labeled by l, then $\mathbf{n}'$ is called an l-child of $\mathbf{n}$. A leaf is a node without children. For two trees S and T, we write $S \preceq T$ iff S is a connected subgraph of T which contains the root of T. The *type* of a labeled tree T results from T by removing all node labels (but keeping all edge labels). Furthermore, $Subtypes(T) = \{type(S) \mid S \preceq T\}$. An *abstract counterexample tree* (ACT) T^α is a finite tree whose nodes are labeled by abstract states such that conditions (1)–(4) hold. (1) If the root is labeled by v^α, then $[\![v^\alpha]\!] \subseteq [init]$. (2) If $\mathbf{n}'{:}w^\alpha$ is an l-child of $\mathbf{n}{:}v^\alpha$, then $(v^\alpha, l, w^\alpha) \in \delta^\alpha$. (3) If node $\mathbf{n}{:}v^\alpha$ is a nonleaf player-1 node (that is, $v^\alpha \in V_1^\alpha$), then for *each* move $l \in L^\alpha(v^\alpha)$, the node $\mathbf{n}$ has at least one l-child. Note that if node $\mathbf{n}{:}v^\alpha$ is a nonleaf player-2 node ($v^\alpha \in V_2^\alpha$), then for *some* move $l \in L^\alpha(v^\alpha)$, the node $\mathbf{n}$ has at least one l-child. (4) If a leaf is labeled by v^α, then either $v^\alpha \in V_1^\alpha$ and $L^\alpha(v^\alpha) = \emptyset$, or $[\![v^\alpha]\!] \subseteq [err]$. Intuitively, T^α corresponds to a *set* of spoiling strategies for player 2 in the abstract safety game.

Example 3 [EXSAFETY]   Figure 1(c) shows an ACT T^α for the abstract game of Figure 1(b), and Figure 1(d) shows the type of T^α. After player 1 plays either move A or move B, player 2 plays L to take the game to the error set. ∎

Concretizing abstract counterexamples. A *concrete counterexample tree* (CCT) S is a finite tree whose nodes are labeled by concrete states such that conditions (1)–(4) hold. (1) If the root is labeled by v, then $v \in [init]$. (2) If $\mathbf{n}'{:}w$ is an l-child of $\mathbf{n}{:}v$, then $(v, l, w) \in \delta$. (3) If node $\mathbf{n}{:}v$ is a nonleaf player-1 node ($v \in V_1$), then for each move $l \in L(v)$, the node $\mathbf{n}$ has at least one l-child. (4) If a leaf is labeled by v, then either $v \in V_1$ and $L(v) = \emptyset$, or $v \in [err]$. The CCT S *realizes* the ACT T^α if $type(S) \in Subtypes(T^\alpha)$ and for each node $\mathbf{n}{:}w$ of S and corresponding node $\mathbf{n}{:}v^\alpha$ of T^α, we have $w \in [\![v^\alpha]\!]$. The ACT T^α is *genuine* if there is a CCT that realizes T^α, and otherwise T^α is *spurious*. To determine if the ACT T^α is genuine, we annotate every node $\mathbf{n}:v^\alpha$ of T^α, in addition, with a set $r \subseteq [\![v^\alpha]\!]$ of concrete states; that is, $\mathbf{n}:v^\alpha:r$. The result is called an *annotated* ACT. The set r represents an overapproximation for the set of states that can be part of a CCT with a type in $Subtypes(T^\alpha)$. Initially, $r = [\![v^\alpha]\!]$. The overapproximation r is sharpened repeatedly by application of a symbolic operator called *Focus*. For a node $\mathbf{n}$ of T^α, let $C(\mathbf{n}) = \{l \in \Lambda \mid \mathbf{n}$ has an l-child$\}$ be the set of moves that label the outgoing edges of $\mathbf{n}$. For each move $l \in C(\mathbf{n})$, let $\{\mathbf{n}_{l,j}{:}v_{l,j}^\alpha{:}r_{l,j}\}$ be the set of l-children of $\mathbf{n}$ (indexed by j). The operator *Focus*$(\mathbf{n}{:}v^\alpha{:}r)$ returns a subset of r:

$$
Focus(\mathbf{n}{:}v^\alpha{:}r) = \begin{cases} r & \text{if } \mathbf{n} \text{ leaf and } L^\alpha(v^\alpha) \neq \emptyset \\ r \cap \left(\bigcap_{l \in C(\mathbf{n})} Epre(\cup_j r_{l,j}, l) \right) \cap \left(\bigcap_{l \notin C(\mathbf{n})} \overline{Avl(l)} \right) & \text{if } \mathbf{n} \text{ other player-1 node} \\ r \cap \left(\bigcup_{l \in C(\mathbf{n})} Epre(\cup_j r_{l,j}, l) \right) & \text{if } \mathbf{n} \text{ player-2 node} \end{cases}
$$

Algorithm 1 AnalyzeCounterex(T^α)

Input: an abstract counterexample tree T^α with root $\mathbf{n}_0$.
Output: if T^α is spurious, then SPURIOUS and an annotation of T^α; otherwise GENUINE.
for each node $\mathbf{n}:v^\alpha$ of T^α **do** annotate $\mathbf{n}:v^\alpha$ by $[\![v^\alpha]\!]$
while there is some node $\mathbf{n}:v^\alpha:r$ with $r \neq Focus(\mathbf{n}:v^\alpha:r)$ **do**
   replace the annotation r of $\mathbf{n}:v^\alpha:r$ by $Focus(\mathbf{n}:v^\alpha:r)$
   if $r_0 = \emptyset$ for the annotated root $\mathbf{n}_0:\_:r_0$ **then return** (SPURIOUS, T^α with annotations)
   end while
return GENUINE

An application of $Focus(\mathbf{n}:v^\alpha:r)$ sharpens the set r by determining which of the states in r actually have successors that can be part of a spoiling strategy for player 2 in the concrete game. For leaves $\mathbf{n}:v^\alpha:r$ with $L^\alpha(v^\alpha) \neq \emptyset$, it must be that every state in r is an error state, and so can be part of a CCT. For all other player-1 nodes $\mathbf{n}:v^\alpha:r$, a state $v \in r$ can be part of a CCT only if (i) all moves available at v are contained in $C(\mathbf{n})$ and (ii) for every available move l, there is an l-child from which player 2 has a spoiling strategy; that is, for every available move l, the state v must have a successor in the union of all l-children's overapproximations. For player-2 nodes $\mathbf{n}:v^\alpha:r$, a state $v \in r$ can be part of a CCT only if there is some child from which player 2 has a spoiling strategy; that is, the state v must have a successor in the union of all children's overapproximations.

The procedure AnalyzeCounterex (Algorithm 1) iterates the *Focus* operator on the nodes of a given ACT T^α until there is no change. Let $Focus^*(\mathbf{n})$ denote the fixpoint value of the annotation for node $\mathbf{n}$ of T^α. For the root $\mathbf{n}_0$ of T^α, if $Focus^*(\mathbf{n}_0)$ is empty, then T^α is spurious. Otherwise, consider the annotated ACT that results from T^α by annotating each node $\mathbf{n}$ with $Focus^*(\mathbf{n})$, and removing all nodes $\mathbf{n}$ for which $Focus^*(\mathbf{n})$ is empty. This annotated ACT has a type in $Subtypes(T^\alpha)$, and moreover, its annotations contain exactly the states that can be part of a CCT that realizes T^α. Consequently, if $Focus^*(\mathbf{n}_0)$ is nonempty, then T^α is genuine, and the result of the procedure AnalyzeCounterex is a representation of the CCTs that realize T^α. The nondeterminism in the while loop of AnalyzeCounterex can be efficiently resolved by focusing each node after focusing all of its children. Since T^α is a finite tree, in this bottom-up way, each node is focused exactly once. Indeed, for finite-state game structures and nonsymbolic representations of ACTs, where all node annotations are stored as lists of concrete states, algorithm AnalyzeCounterex can be implemented in linear time.

Proposition 2 [Counterexample checking] *An ACT T^α for a safety game is spurious iff the procedure* AnalyzeCounterex(T^α) *returns* SPURIOUS. *Checking if an ACT for a safety game is spurious can be done in time linear in the size of the tree.*

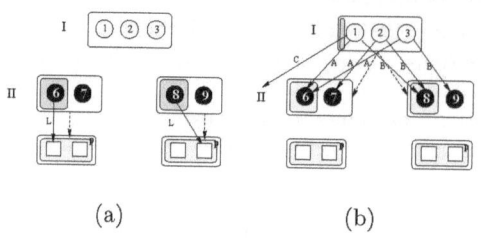

(a)   (b)

Fig. 2. Focusing T^α

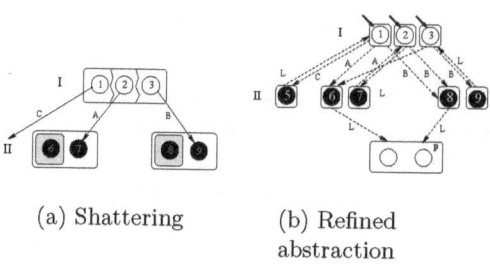

(a) Shattering   (b) Refined
                        abstraction

Fig. 3. Abstraction refinement

Example 4 [EXSAFETY] Figure 2 shows the result of running AnalyzeCounterex
on the ACT T^α from Figure 1(c). The shaded parts of the boxes denote the states
that may be a part of a CCT. The dashed arrows indicate abstract transitions,
and the solid arrows concrete transitions. Figure 2(a) shows the result of focusing
the player 2 nodes. All states in the leaves are error states, and therefore in the
shaded boxes. Only states 6 and 8 can go to the error region from the two
abstract player-2 states; hence only they are in the focused regions indicated by
shaded boxes. Figure 2(b) shows the result of a subsequent application of *Focus*
to the root. No state in the root can play only moves A and B and subsequently
go to states from which player 2 can spoil. Hence none of these states can serve
as the root of a CCT whose type is in $Subtypes(T^\alpha)$. Since the focused region of
the root is empty, we conclude that the ACT T^α is spurious.

Abstraction refinement. If we find an ACT T^α to be spurious, then we
must refine the abstraction $\mathcal{G}^\alpha$ in order to rule out T^α. Consider a state $\mathbf{n}:v^\alpha$
of T^α. Abstraction refinement may split the abstract state v^α into several states
$v_1^\alpha, \ldots, v_m^\alpha$, with $[\![v_k^\alpha]\!] = r_k$ for $1 \leq k \leq m$, such that $r_1 \cup \ldots \cup r_m = [\![v^\alpha]\!]$. For this
purpose, we define a symbolic *Shatter* operator, which takes, for a node $\mathbf{n}:v^\alpha:r$
of the annotated version of T^α generated by the procedure AnalyzeCounterex,
the triple $(\mathbf{n}, [\![v^\alpha]\!], r)$, and returns the set $\{r_1, \ldots, r_m\}$. The set r_1 is the "good"
set r (the annotation), from which player 2 does indeed have a spoiling strategy

---

Algorithm 2 RefineAbstraction($\mathcal{G}^\alpha, T^\alpha$)

Input: an abstraction $\mathcal{G}^\alpha$ and an abstract counterexample tree T^α.
Output: if T^α is spurious, then SPURIOUS and a refined abstraction; otherwise GENUINE.
if AnalyzeCounterex(T^α) = (SPURIOUS, S^α) **then**
$\quad R := \{[\![v^\alpha]\!] \mid v^\alpha \in V^\alpha\}$
$\quad$**for each** annotated node $\mathbf{n}:v^\alpha:r$ of S^α **do** $R := R \cup Shatter(\mathbf{n}, [\![v^\alpha]\!], r)$
$\quad$**return** (SPURIOUS, $Abstraction(R)$)
else return GENUINE

---

of a type in $Subtypes(T^\alpha)$. The sets $r_2, \ldots, r_m$ are "bad" subsets of $[\![v^\alpha]\!] \setminus r$, from which no such spoiling strategy exists. Each "bad" set r_k, for $2 \le k \le m$, is small enough that there is a simple single reason for the absence of a spoiling strategy. For player-1 nodes $\mathbf{n}$, a set r_k may be "bad" because every state $v \in r_k$ either (i) has a move available which is not in $C(\mathbf{n})$, or (ii) has a move l available such that none of the l-successors of v is in a "good" set, from which player 2 can spoil. For player-2 nodes $\mathbf{n}$, there is a single "bad" set, which contains the states that have no successor in a "good" set. Formally, the operator $Shatter(\mathbf{n}, q, r)$ is defined to take a node $\mathbf{n}$ of the ACT T^α, and two sets $q, r \subseteq V$ of concrete states such that $r \subseteq q$, and it returns a collection $R \subseteq 2^V$ of state sets $r_k \subseteq q$. For each move $l \in C(\mathbf{n})$, let $\{\mathbf{n}_{l,j}:v_{l,j}^\alpha:r_{l,j}\}$ again be the set of l-children of $\mathbf{n}$. Then:

$$Shatter(\mathbf{n}, q, r) = \begin{cases} \{r\} \cup \{(q \setminus r) \cap Avl(l) \mid l \notin C(\mathbf{n})\} \\ \quad \cup \{(q \setminus r) \cap \overline{Epre(\bigcup_j r_{l,j}, l)} \mid l \in C(\mathbf{n})\} & \text{if } \mathbf{n} \text{ is a player-1 node} \\ \{r, (q \setminus r)\} & \text{if } \mathbf{n} \text{ is a player-2 node} \end{cases}$$

Note that $\bigcup Shatter(\mathbf{n}, q, Focus(\mathbf{n} : v^\alpha : q)) = q$. The refinement of the given abstraction $\mathcal{G}^\alpha$ is achieved by the procedure RefineAbstraction (Algorithm 2). Given a collection $R \subseteq 2^V$ of state sets, define the equivalence relation $\equiv_R \subseteq V \times V$ by $v_1 \equiv_R v_2$ if for all sets $r \in R$, we have $v_1 \in r$ precisely when $v_2 \in r$. Let $Closure(R)$ denote the equivalence classes of $\equiv_R$. Given $V \subseteq \bigcup R$, the set $Closure(R) \subseteq 2^V$ of sets of concrete states uniquely specifies an abstraction for $\mathcal{G}$, denoted $Abstraction(R)$, which contains for each set $r \in Closure(R)$ an abstract state w_r^α with $[\![w_r^\alpha]\!] = r$ (from this the other components of the abstraction are determined). In particular, let $R_1 = \bigcup_{(\mathbf{n}:v^\alpha) \in T^\alpha} Shatter(\mathbf{n}, [\![v^\alpha]\!], Focus^*(\mathbf{n}))$ and $R_2 = \{[\![v^\alpha]\!] \mid v^\alpha \in V^\alpha\}$. Our refined abstraction is $Abstraction(R_1 \cup R_2)$. The new abstraction returned by the procedure RefineAbstraction($\mathcal{G}^\alpha, T^\alpha$) rules out ACTs that are similar to the spurious ACT T^α. Given two ACTs T^α and S^α, we say that T^α subsumes S^α if $type(S^\alpha) \in Subtypes(T^\alpha)$ and for each node $\mathbf{n}:w^\alpha$ of S^α and corresponding node $\mathbf{n}:v^\alpha$ of T^α, we have $[\![w^\alpha]\!] \subseteq [\![v^\alpha]\!]$.

Proposition 3 [Abstraction refinement] *If T^α is a spurious ACT for the abstraction $\mathcal{G}^\alpha$ of a safety game, then the abstraction returned by the procedure* RefineAbstraction($\mathcal{G}^\alpha, T^\alpha$) *has no ACT that is subsumed by T^α.*

Algorithm 3 CxSafetyControl($\mathcal{G}, \Box \overline{err}$)

Input: a game structure $\mathcal{G}$ and a safety objective $\Box \overline{err}$.
Output: either CONTROLLABLE and a player-1 winning strategy,
            or UNCONTROLLABLE and a player-2 spoiling strategy represented as ACT.
$\mathcal{G}^\alpha := InitialAbstraction(\mathcal{G}, \Box \overline{err})$
repeat
    $(winner, T^\alpha) := ModelCheck(\mathcal{G}^\alpha, \Box \overline{err})$
    if $winner = 2$ and RefineAbstraction$(\mathcal{G}^\alpha, T^\alpha) = (\text{SPURIOUS}, \mathcal{H}^\alpha)$ **then** $\mathcal{G}^\alpha := \mathcal{H}^\alpha$;
    $winner := \bot$
until $winner \neq \bot$
if $winner = 1$ **then return** (CONTROLLABLE, T^α)
return (UNCONTROLLABLE, T^α)

Example 5 [EXSAFETY]   Figure 3 shows the effect of the *Shatter* operator on the root of the ACT T^α from Figure 1(c), and the resulting refined abstract game for which T^α is no longer an ACT. For all nonroot nodes, shattering is trivial, namely, into the focused region and its complement. We break up the states in the root into (i) state 1, which can play the move C not available to the abstract state, (ii) state 2, which can proceed by move A to a state from which the abstract player-2 spoiling strategy fails (i.e., a state not inside a shaded box), and (iii) state 3, which can proceed by move B to a state from which the abstract player-2 spoiling strategy fails. ∎

4   Counterexample-Guided Controller Synthesis

Safety control. Given a game structure $\mathcal{G}$ and a safety objective $\Box \overline{err}$, we wish to determine if player 1 wins, and if so, construct a winning strategy ("synthesize a controller"). Our algorithm, which generalizes the "abstract-verify-refine" loop of [5,6,9], proceeds as follows:

Step 1 ("abstraction") We first construct an initial abstract game $(\mathcal{G}^\alpha, \Box \overline{err})$. This could be the trivial abstraction induced by the two propositions *init* and *err*, which has at most 8 abstract states (at most 4 for each player, depending on which of the two propositions are true).

Step 2 ("model checking") We symbolically model check the abstract game to find if player 1 can win, by iterating the $Cpre_1^\alpha$ operator. If so, then the model checker provides a winning player-1 strategy for the abstract game, from which a winning player-1 strategy in the concrete game can be constructed [13]. If not, then the model checker symbolically produces an ACT [11]. As the abstract state space is finite, the model checking is guaranteed to terminate.

Step 3 ("counterexample-guided abstraction refinement") If model checking returns an ACT T^α, then we use the procedure AnalyzeCounterex(T^α) to check if the ACT is genuine. If so, then player 2 has a spoiling strategy in the concrete game, and the system is not controllable. If the ACT is spurious, then we use the procedure RefineAbstraction($\mathcal{G}^\alpha, T^\alpha$) to refine the abstraction $\mathcal{G}^\alpha$, so that T^α (and similar counterexamples) cannot arise on subsequent invocations of the model checker. This step uses the operators *Focus* and *Shatter*, which are defined in terms of *Epre* and can therefore be implemented symbolically. Since T^α is a finite tree, also this step is guaranteed to terminate.

Goto step 2. The process is iterated until we find either a player-1 winning strategy in step 2, or a genuine counterexample in step 3.

The procedure is summarized in Algorithm 3. The function *InitialAbstraction*($\mathcal{G}, \Box\overline{err}$) returns a trivial abstraction for $\mathcal{G}$, which preserves *init* and *err*. The function *ModelCheck*($\mathcal{G}^\alpha, \Box\overline{err}$) returns a pair $(1, T^\alpha)$ if player 1 can win the abstract game, where T^α is a (memoryless) winning strategy for player 1, and otherwise it returns $(2, T^\alpha)$, where T^α is an ACT. From the soundness of abstraction, we get the soundness of the algorithm.

Proposition 4 [Partial correctness of CxSafetyControl]   *If the procedure* CxSafetyControl($\mathcal{G}, \Box\overline{err}$) *returns* CONTROLLABLE, *then player 1 wins the safety game* ($\mathcal{G}, \Box\overline{err}$). *If the procedure returns* UNCONTROLLABLE, *then player 1 does not win the game.*

In general, the procedure CxSafetyControl may not terminate for infinite-state games (it does terminate for finite-state games). However, one can prove sufficient conditions for termination provided certain state equivalences on the game structure have finite index [13]. For example, for timed games [3,20], where in the course of the procedure CxSafetyControl, the abstract state space always consists of blocks of clock regions, termination is guaranteed. Verification is the special case of control where all states are player-2 states. Hence our algorithm works also for verification, which is illustrated by the following example.

Example 6 [Safety verification]   Consider the transition system ExVERIF shown in Figure 4(a). All states are player-2 states. The initial states are 1 and 2, and we wish to check the safety property $\Box\overline{p}$, that the states 5 and 6 are not visited by any run. It is easy to see that the system satisfies this property. Figure 4(b) shows an abstraction for ExVERIF. This is a standard existential abstraction for transition systems. In verification, counterexaples are traces (trees without branches). Figure 4(c) shows a trace τ^α, which is an ACT for the abstraction (b). Figure 5 shows the result of running the algorithm AnalyzeCounterex on τ^α. Figure 5(a) shows the effect of applying *Focus* to the second abstract state in τ^α. All concrete states in the third abstract state are error states; hence they are all shaded. Only state 4 can go to one of the error states; hence it is the only state in the focused region of the second abstract state. Figure 5(b) shows the second application of *Focus*, to the root of the trace. As neither 1 nor 2 have 4 as a successor, the focused region of the root is empty. This implies that the counterexample is spurious. Figure 6(a) shows the effect of *Shatter* on the abstract trace τ^α. Since the shaded box of the second abstract state is $\{4\}$, this abstract state gets shattered into $\{3\}$ and $\{4\}$. No other abstract state is shattered. Figure 6(b) shows the refined abstraction, which is free of counterexamples. ∎

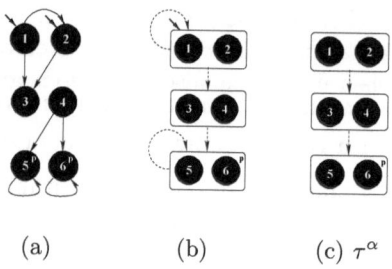

(a)   (b)   (c) τ^α

Fig. 4. Example EXVERIF

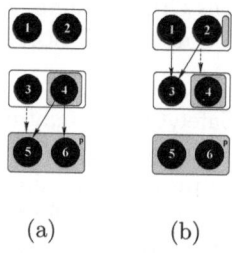

(a)   (b)

Fig. 5. Focusing τ^α

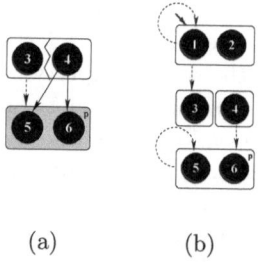

(a)   (b)

Fig. 6. Refinement

Omega-regular objectives. Counterexample-guided abstraction refinement can be generalized to games with arbitrary ω-regular objectives. To begin with, we must implement a symbolic model checker for solving ω-regular games: given a finite-state game structure $\mathcal{G}^\alpha$ and an ω-regular objective Γ, one can construct a fixpoint formula over the $Cpre_1^\alpha$ operator which characterizes the set of states from which player 1 can win [13]. Moreover, from the fixpoint computation, one

---

Algorithm 4 CombinedAnalyzeRefine($\mathcal{G}^\alpha, K^\alpha$)

Input: an abstraction $\mathcal{G}^\alpha$, and an abstract counterexample graph K^α with root n_0.
Output: if K^α is spurious, then SPURIOUS and a refined abstraction; otherwise GENUINE.
for each node $n : v^\alpha$ of K^α **do** annotate $n : v^\alpha$ by $[\![v^\alpha]\!]$
$R := \{[\![v^\alpha]\!] \mid v^\alpha \in V^\alpha\}$
while there is some node $n : v^\alpha : r$ with $r \neq Focus(n : v^\alpha : r)$ **do**
  $r' := Focus(n : v^\alpha : r)$
  $R := R \cup Shatter(n, r, r')$
  replace the annotation r of $n : v^\alpha : r$ by r'
  if $r_0 = \emptyset$ for the annotated root $n_0 : {}_- : r_0$ **then return** (SPURIOUS, $Abstraction(R)$)
end while
return GENUINE

---

can symbolically construct either a winning strategy for player 1 or a spoiling strategy for player 2 [13,20]. Counterexamples for finite-state ω-regular games are spoiling strategies with finite memory [17], which can be represented as finite graphs. Hence we generalize ACTs from trees to graphs as follows: an *abstract counterexample graph* (ACG) K^α is a rooted, directed, finite graph whose nodes are labeled by abstract states such that conditions (1)–(3) from the definition of ACT hold, and (4) if a leaf (a node with outdegree 0) is labeled by v^α, then $v^\alpha \in V_1^\alpha$ and $L^\alpha(v^\alpha) = \emptyset$. The definition of concrete counterexamples and of the operator *Subtypes* are generalized from trees to graphs in a similar, straightforward way, giving rise to the notion of whether an ACG is *genuine* or *spurious*. So suppose that the function $ModelCheck(\mathcal{G}^\alpha, \Gamma)$ returns a pair $(1, K^\alpha)$ if player 1 can win the abstract game, where K^α is a (finite-memory) winning strategy for player 1, and otherwise returns $(2, K^\alpha)$, where K^α is an ACG. In the latter case we must now check whether or not K^α is spurious, and if so, then refine the abstraction $\mathcal{G}^\alpha$.

While in the safety case, we analyzed counterexamples (Algorithm 1) before we refined the abstraction (Algorithm 2), for general ω-regular objectives, we combine both procedures (Algorithm 4). The algorithm CombinedAnalyzeRefine computes the fixpoint of the *Focus* operator on a given ACG K^α, and simultaneously refines the given abstraction $\mathcal{G}^\alpha$ by shattering an abstract state with each application of *Focus*. In contrast to the case of trees, for general graphs we cannot apply a bottom-up strategy for focusing. Indeed, in the presence of cycles, the computation of $Focus^*$ may require focusing a node several times before a fixpoint is reached, and CombinedAnalyzeRefine is not guaranteed to terminate (it does terminate for finite-state games). It is easy to see that the procedures AnalyzeCounterex and RefineAbstraction are a special case of CombinedAnalyzeRefine for the case that each node needs to be focused only once. In this case, all shattering can be delayed until focusing is complete, and thus repeated shattering while refocusing the same abstract state can be avoided. Suppose that the procedure CxControl is obtained from CxSafetyControl (Algorithm 3) by replacing the safety objective $\square \overline{err}$ with an arbitrary ω-

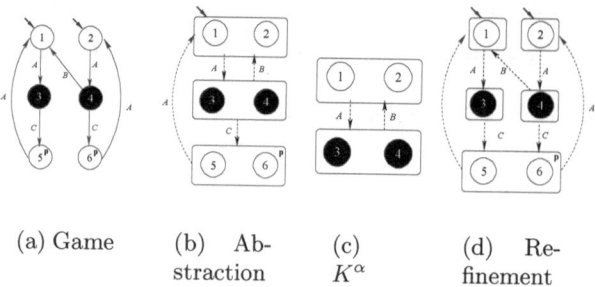

(a) Game (b) Ab- (c) (d) Re-
straction K^α finement

Fig. 7. Example ExBüchi

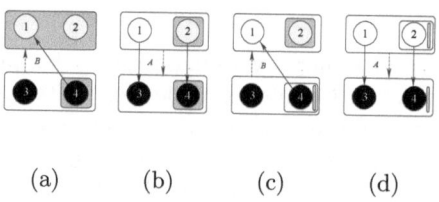

(a)   (b)   (c)   (d)

Fig. 8. CombinedAnalyzeRefine on K^α

regular objective Γ, and by calling the function CombinedAnalyzeRefine in place of RefineAbstraction. Then we have the following result.

Theorem 1. [Partial correctness of CxControl] *Let $\mathcal{G}$ be a game structure, and let Γ be an ω-regular objective. If the procedure* CxControl$(\mathcal{G}, \Gamma)$ *returns* Controllable, *then player 1 wins the game; if the procedure returns* Uncontrollable, *then player 1 does not win.*

Example 7 [Büchi game]   Figure 7(a) shows an example of a Büchi game. We wish to check if player 1 can force the game into a p-state infinitely often, i.e., the objective is $\Box\Diamond p$. Figure 7(b) shows an abstraction for the game. Figure 7(c) shows the result of solving the abstract game, namely, an ACG K^α that has player 2 force a loop not containing a p-state. Figure 8 shows how the ACG is analyzed and discovered to be spurious. Figure 8(a) shows the effect of *Focus* on the lower node of K^α. As only the state 4 has a move into $\{1, 2\}$, the shaded box for the lower node is $\{4\}$. Consequently, the abstract state $\{3, 4\}$ is shattered into $\{3\}$ and $\{4\}$. Figure 8(b) shows the effect of *Focus* on the upper node of K^α. Only state 2 has an A-successor in the shaded box of the lower node; hence the focused region for the upper node becomes $\{2\}$, and the upper node gets shattered into $\{1\}$ and $\{2\}$. In Figure 8(c) we again apply *Focus* to the lower node. Since no state has a B-move to the focused region of the upper node, the focused region of the lower node becomes empty. Figure 8(d) illustrates that after another *Focus* on the upper node, its focused region becomes empty as well. Figure 7(d) shows the resulting refined abstraction; it is easy to see that player 2 has no spoiling strategy. ∎

In [10], the authors consider counterexample-guided abstraction refinement for model checking universal CTL formulas. In this case (and for some more expressive logics considered in [10]), counterexamples are tree-like, and our algorithms for analyzing counterexamples and refining abstractions apply also (indeed, since in this case counterexamples are models of existential CTL formulas, abstract counterexample trees contain only player-2 nodes). More generally, the model-checking problem for the μ-calculus can be reduced to the problem of solving parity games [16]. Via this reduction, our method provides also a counterexample-guided abstraction refinement procedure for model checking the μ-calculus.

References

1. K. Altisen, G. Gössler, A. Pnueli, J. Sifakis, S. Tripakis, and S. Yovine. A framework for scheduler synthesis. In *RTSS: Real-Time Systems Symposium*, pages 154–163. IEEE, 1999.
2. R. Alur, L. de Alfaro, T.A. Henzinger, and F.Y.C. Mang. Automating modular verification. In *CONCUR: Concurrency Theory*, LNCS 1664, pages 82–97. Springer, 1999.
3. R. Alur and T.A. Henzinger. Modularity for timed and hybrid systems. In *CONCUR: Concurrency Theory*, pages 74–88. LNCS 1243, Springer, 2001.
4. R. Alur, T.A. Henzinger, and O. Kupferman. Alternating-time temporal logic. *Journal of the ACM*, 49:672–713, 2002.
5. R. Alur, A. Itai, R.P. Kurshan, and M. Yannakakis. Timing verification by successive approximation. *Information and Computation*, 118:142–157, 1995.
6. T. Ball and S.K. Rajamani. The SLAM project: debugging system software via static analysis. In *POPL: Principles of Programming Languages*, pages 1–3. ACM, 2002.
7. J.R. Büchi and L.H. Landweber. Solving sequential conditions by finite-state strategies. *Transactions of the AMS*, 138:295–311, 1969.
8. A. Church. Logic, arithmetic, and automata. In *International Congress of Mathematicians*, pages 23–35. Institut Mittag-Leffler, 1962.
9. E.M. Clarke, O. Grumberg, S. Jha, Y. Lu, and H. Veith. Counterexample-guided abstraction refinement. In *CAV: Computer-Aided Verification*, LNCS 1855, pages 154–169. Springer, 2000.
10. E.M. Clarke, S. Jha, Y. Lu, and H. Veith. Tree-like counterexamples in model checking. In *LICS: Logic in Computer Science*, pages 19–29. IEEE, 2002.
11. E.M. Clarke, O. Grumberg, K. McMillan, and X. Zhao. Efficient generation of counterexamples and witnesses in symbolic model checking. In *DAC: Design Automation Conference*, pages 427–432. ACM/IEEE, 1995.
12. L. de Alfaro and T.A. Henzinger. Interface automata. In *FSE: Foundations of Software Engineering*, pages 109–120. ACM, 2001.
13. L. de Alfaro, T.A. Henzinger, and R. Majumdar. Symbolic algorithms for infinite-state games. In *CONCUR: Concurrency Theory*, pages 536–550. LNCS 2154, Springer, 2001.

14. L. de Alfaro, T.A. Henzinger, and F.Y.C. Mang. Detecting errors before reaching them. In *CAV: Computer-Aided Verification*, LNCS 1855, pages 186–201. Springer, 2000.

15. D.L. Dill. *Trace Theory for Automatic Hierarchical Verification of Speed-independent Circuits.* MIT Press, 1989.

16. E.A. Emerson, C.S. Jutla, and A.P. Sistla. On model checking fragments of μ-calculus. In *CAV: Computer-Aided Verification*, LNCS 697, pages 385–396. Springer, 1993.

17. Y. Gurevich and L. Harrington. Trees, automata, and games. In *STOC: Symposium on Theory of Computing*, pages 60–65. ACM, 1982.

18. T.A. Henzinger, R. Jhala, R. Majumdar, and G. Sutre. Lazy abstraction. In *POPL: Principles of Programming Languages*, pages 58–70. ACM, 2002.

19. T.A. Henzinger, R. Majumdar, F.Y.C. Mang, and J.-F. Raskin. Abstract interpretation of game properties. In *SAS: Static-Analysis Symposium*, pages 220–239. LNCS 1824, Springer, 2000.

20. O. Maler, A. Pnueli, and J. Sifakis. On the synthesis of discrete controllers for timed systems. In *STACS: Theoretical Aspects of Computer Science*, LNCS 900, pages 229–242. Springer, 1995.

21. A. Pnueli and R. Rosner. On the synthesis of a reactive module. In *POPL: Principles of Programming Languages*, pages 179–190. ACM, 1989.

22. P.J. Ramadge and W.M. Wonham. Supervisory control of a class of discrete-event processes. *SIAM Journal of Control and Optimization*, 25:206–230, 1987.

Axiomatic Criteria for Quotients and Subobjects for Higher-Order Data Types

Jo Hannay

Department of Software Engineering,
Simula Research Laboratory, Pb. 134, NO-1325 Lysaker, Norway
johannay@simula.no

Abstract. Axiomatic criteria are given for the existence of higher-order maps over subobjects and quotients. These criteria are applied in showing the soundness of a method for proving specification refinement up to observational equivalence. This generalises the method to handle data types with higher-order operations, using standard simulation relations. We also give a direct setoid-based model satisfying the criteria. The setting is the second-order polymorphic lambda calculus and the assumption of relational parametricity.

1 Introduction

As a motivating framework for the results in this paper, we use specification refinement. We address specifications for data types whose operations may be higher order.

A *stepwise specification refinement* process transforms an abstract specification into one or more concrete specifications or program modules. If each step is proven correct, the resulting modules will be correct according to the initial abstract specification. This then describes a software development technique for producing small-scale certified components. Theoretical aspects to this idea have been researched thoroughly in the field of algebraic specification, see *e.g.,* [31,6].

When data types have higher-order operations, taking functions as arguments, several things in the refinement methodology break down. Most well-known perhaps, is the lack of correspondence between observational equivalence and the existence of simulation relations for data types, together with the lack of composability. The view is that standard notions of simulation relation are not adequate, and several remedies have been proposed; *pre-logical relations* [18,17], *lax logical relations* [28,20], *L-relations* [19], and *abstraction barrier-observing* simulation relations [11,12,13]. The latter, developed for System F in a logic [27] asserting *relational parametricity* [30], are directly motivated by the information-hiding mechanism in data types. Relational parametricity is in this context the logical assertion of the Basic Lemma [25,18] for simulation relations.

In this paper, we address a further issue. A general proof strategy for proving specification refinement up to observational equivalence is formalised in [4, 3]. For data types with first-order operations, the strategy is expressed in the

J.C.M. Baeten et al. (Eds.): ICALP 2003, LNCS 2719, pp. 903–917, 2003.

setting of System F and relational parametricity by axiomatising the existence of subobjects and quotients [29,36,9,12]. The axioms are sound w.r.t. the parametric *per* model of [1] which is a model for the logic in [27]. At higher order, more work is required, because in order to validate the axioms, one has to find a model which has higher-order operations over subobjects and quotients. Our solution is the core technical issue of this paper. First, we use a setoid-based semantics based on work on the syntactic level in [16]. Then we present general axiomatic criteria for the existence of higher-order functions over subobjects and quotients, and the setoid model is then an instance of this general schema. We think the axiomatic criteria are of general interest outside refinement issues. The results also answer the speculation in [36] about the soundness of similar axioms postulating quotients and subobjects at higher order.

Since simulation relations express observational equivalence, they play an integral part in the above proof strategy. At higher order, it is still possible to use standard simulation relations, because the strategy relies on establishing observational equivalence from the existence of simulation relations. In this paper, we exploit this fact and devise the axiomatic criteria for standard simulation relations. For the strategy to be complete however, one must utilise one of the above alternative notions of simulation relation, since there may not exist a standard simulation relation even in the presence of observational equivalence. To this end, abstraction barrier-observing (*abo*) simulation relations were used in [10,12], together with *abo*-relational parametricity, and a special *abo*-semantics. That approach does indeed yield higher-order operations over quotients and subobjects, but to devise general axiomatic criteria for the existence of higher-order functions over subobjects and quotients with alternative notions of simulation relations, is ongoing research.

2   Syntax

We review relevant formal aspects. For full accounts, see [2,25,8,27,1].

The second-order lambda-calculus F_2, or System F, has abstract syntax

$$\text{(types)} \ T ::= X \mid (T \to T) \mid (\forall X.T)$$
$$\text{(terms)} \ t \ ::= x \mid (\lambda x{:}T.t) \mid (tt) \mid (\Lambda X.t) \mid (tT)$$

where X and x range over type and term variables respectively. This provides polymorphic functionals and encodings of self-iterating inductive types [5], *e.g.*, $\text{Nat} \stackrel{\text{def}}{=} \forall X.X \to (X \to X) \to X$, with constructors, destructors and conditionals. Products $U_1 \times \cdots \times U_n$ encode as inductive types.

We use the logic for parametric polymorphism due to [27]; a second-order logic augmented with relation symbols, relation definition, and the axiomatic assertion of relational parametricity. See also [22,34]. Formulae now include relational statements as basic predicates and quantifiables,

$$\phi ::= (t =_A u) \mid t \ R \ u \mid \ \cdots \ \mid \forall R {\subset} A {\times} B \ . \ \phi \mid \exists R {\subset} A {\times} B \ . \ \phi$$

where R ranges over relation variables. Relation definition is accommodated by the syntax,

$$\Gamma \rhd (x\!:\!A, y\!:\!B) \, . \, \phi \quad \subset A \times B$$

where ϕ is a formula. For example $\mathrm{eq}_A \stackrel{def}{=} (x\!:\!A, y\!:\!A).(x =_A y)$.

We write $\alpha[\xi]$ to indicate possible occurrences of variable ξ in type, term or formula α, and write $\alpha[\beta]$ for the substitution $\alpha[\beta/\xi]$, following the appropriate rules regarding capture.

We get the *arrow-type relation* $\rho \to \rho' \subset (A \to A') \times (B \to B')$ from $\rho \subset A \times B$ and $\rho' \subset A' \times B'$ by

$$(\rho \to \rho') \stackrel{def}{=} (f\!:\!A \to A', g\!:\!B \to B') \, . \, (\forall x\!:\!A.\forall y\!:\!B \, . \, (x\rho y \;\Rightarrow\; (fx)\rho'(gy)))$$

The *universal-type relation* $\forall(Y, Z, R \subset Y \times Z)\rho[R] \;\subset\; (\forall Y.A[Y]) \times (\forall Z.B[Z])$ is defined from $\rho[R] \subset A[Y] \times B[Z]$, where Y, Z and $R \subset Y \times Z$ are free, by

$$\forall(Y, Z, R \subset Y \times Z)\rho[R] \stackrel{def}{=} (y\!:\!\forall Y.A[Y], z\!:\!\forall Z.B[Z]) \, . \, (\forall Y.\forall Z.\forall R \subset Y \times Z \, . \, ((yY)\rho[R](zZ)))$$

For n-ary $\boldsymbol{X}$, $\boldsymbol{A}$, $\boldsymbol{B}$, $\boldsymbol{\rho}$, where $\rho_i \subset A_i \times B_i$, we get $T[\boldsymbol{\rho}] \subset T[\boldsymbol{A}] \times T[\boldsymbol{B}]$, the *action of $T[\boldsymbol{X}]$ on $\boldsymbol{\rho}$*, by

$$\begin{aligned}
T[\boldsymbol{X}] = X_i : \qquad & T[\boldsymbol{\rho}] = \rho_i \\
T[\boldsymbol{X}] = T'[\boldsymbol{X}] \to T''[\boldsymbol{X}] : \; & T[\boldsymbol{\rho}] = T'[\boldsymbol{\rho}] \to T''[\boldsymbol{\rho}] \\
T[\boldsymbol{X}] = \forall X'.T'[\boldsymbol{X}, X'] : \; & T[\boldsymbol{\rho}] = \forall(Y, Z, R \subset Y \times Z)T'[\boldsymbol{\rho}, R]
\end{aligned}$$

The proof system is intuitionistic natural deduction, augmented with inference rules for relation symbols in the obvious way. There are standard axioms for equational reasoning implying extensionality for arrow and universal types.

Parametric polymorphism prompts all instances of a polymorphic functional to exhibit a uniform behaviour [33,1,30]. We adopt *relational parametricity* [30, 21]; a polymorphic functional instantiated at two related domains, should give related instances. This is asserted by the schema

$$\textsc{Param}: \; \forall \boldsymbol{Z}.\forall u\!:\!(\forall X.U[X, \boldsymbol{Z}]) \, . \, u \, (\forall X.U[X, \mathbf{eq}_{\boldsymbol{Z}}]) \, u$$

The logic with $\textsc{Param}$ is sound; we have the parametric *per*-model of [1] and the syntactic models of [14]. Relational parametricity yields the fundamental *Identity Extension Lemma*:

$$\forall \boldsymbol{Z}.\forall u, v\!:\!T[\boldsymbol{Z}] \, . \, (u \, T[\mathbf{eq}_{\boldsymbol{Z}}] \, v \;\Leftrightarrow\; (u =_{T[\boldsymbol{Z}]} v))$$

Constructs such as products, sums, initial and final (co-)algebras are encodable in System F [5]. With $\textsc{Param}$, these become provably universal constructions.

3   Specification Refinement

A specification determines a collection of data types realising the specification. A signature provides the desired namespace, and a set of formulae give properties to be fullfilled. Depending on refinement stage, these range from abstract to concrete implementational. A data type consists of a data representation and operations. In the logic, these are respectively a type A, and a term $\mathfrak{a}:T[A]$, where $T[X]$ plays the role of a signature. For instance, using a labeled product notation, $T_{\mathsf{STACK}_{\mathsf{Nat}}}[X] \overset{def}{=} (\mathsf{empty}:X, \mathsf{push}:\mathsf{Nat}\to X\to X, \mathsf{pop}:X\to X, \mathsf{top}:X\to\mathsf{Nat})$. Each $f_i : T_i[X]$ is a *profile* of the signature. Abstract properties are *e.g.*, $\forall x:\mathsf{Nat}, s:X \ . \ \mathfrak{x}.\mathsf{pop}(\mathfrak{x}.\mathsf{push}\, x\, s) = s \wedge \mathfrak{x}.\mathsf{top}(\mathfrak{x}.\mathsf{push}\, x\, s) = x$. A data type realising this stack specification, consists *e.g.*, of inductive type $\mathsf{List}_{\mathsf{Nat}}$ and $\mathfrak{l}$, where $\mathfrak{l}.\mathsf{empty} = \mathsf{nil}$, $\mathfrak{l}.\mathsf{push} = \mathsf{cons}$, $\mathfrak{l}.\mathsf{pop} = \lambda l:\mathsf{List}_{\mathsf{Nat}}.(\mathsf{cond}\,\mathsf{List}_{\mathsf{Nat}}\,(\mathsf{isnil}\,l)\,\mathsf{nil}\,(\mathsf{cdr}\,l))$, and $\mathfrak{l}.\mathsf{top} = \lambda l:\mathsf{List}_{\mathsf{Nat}}.(\mathsf{cond}\,\mathsf{Nat}\,(\mathsf{isnil}\,l)\,0\,(\mathsf{car}\,l))$. For encapsulation, data types would be given as packages of existential type, but our technical results are on the component level, so we omit this.

To each refinement stage, a set Obs of *observable types* is associated, containing inductive types, and also parameters. Two data types are interchangeable if it makes no difference which one is used in an observable computation. For example, an observable computation on natural-number stacks could be $\Lambda X.\lambda\mathfrak{x}:T_{\mathsf{STACK}_{\mathsf{Nat}}}[X] \ . \ \mathfrak{x}.\mathsf{top}(\mathfrak{x}.\mathsf{push}\, n\ \mathfrak{x}.\mathsf{empty})$. Thus, for $A, B, \mathfrak{a}:T[A], \mathfrak{b}:T[B]$, Obs,

Observational Equivalence: $\bigwedge_{D\in Obs} \forall f:\forall X.(T[X]\to D) \ . \ (fA\,\mathfrak{a}) = (fB\,\mathfrak{b})$

Observational equivalence can be hard to prove. A more manageable criterion for interchangeability lies in the concept of *data refinement* [15,7] and the use of relations to show *representation independence* [23,32,30], leading to *logical relations* for lambda calculus [24,25,35,26]. In the relational logic of [27] one uses the action of types on relations to express the above ideas. Two data types are related by a simulation relation if there exists a relation R on their respective data representations that is preserved by their corresponding operations:

Existence of Simulation Relation: $\exists R\subset A\times B \ . \ \mathfrak{a}\,(T[R])\,\mathfrak{b}$

With relational parametricity we get a connection to observational equivalence.

Theorem 1. *The following is derivable in the logic using* PARAM.

$$\forall A, B.\forall \mathfrak{a}:T[A], \mathfrak{b}:T[B] \ . \ \exists R\subset A\times B \ . \ \mathfrak{a}(T[R])\mathfrak{b}$$
$$\Rightarrow \quad \bigwedge_{D\in Obs} \forall f:\forall X.(T[X]\to D) \ . \ (fA\,\mathfrak{a}) = (fB\,\mathfrak{b})$$

Proof: This follows from the PARAM-instance $\forall Y.\forall f \ : \ \forall X.(T[X] \to Y) \ . \ f(\forall X.T[X]\to\mathsf{eq}_Y)f$. $\square$

Consider the assumption that $T[X]$ has only first-order function profiles:

FADT$_{Obs}$: Every profile $T_i[X] = T_{i1}[X] \to \cdots \to T_{n_i}[X] \to T_{c_i}[X]$ of $T[X]$ is first order, and such that $T_{c_i}[X]$ is either X or some $D \in Obs$.

Assuming *FADT$_{Obs}$* for $T[X]$, Theorem 1 becomes a two-way implication [11,12].

For data types with higher-order operations, we only have Theorem 1 in general. More apt relational notions for explaining interchangeability of data types have been found; prelogical relations [18,17], lax logical relations and L-relations [28,20,19], and *abo*-simulation relations [11,12,13].

For specification refinement one is interested in establishing observational equivalence. For this it suffices to find a simulation relation and then use Theorem 1. The problem at higher order is that there might not exist a simulation relation, even in the presence of observational equivalence.

Nonetheless, it is in many cases possible to find simulation relations at higher order. It is worthwhile to utilise this, since it is harder to deal with the alternative notions in practice; prelogical relations involve an infinite family of relations, *abo*-relations involve definability. Therefore, this paper establishes a proof strategy for refinement at higher order using standard simulation relations.

The strategy for proving observational refinement formalised by Bidoit *et al* [4,3], expresses observational abstraction in terms of a congruence. Using this congruence, one quotients over the data representation. Additionally, it may be necessary to restrict the data representation before quotienting, and in that case one also needs to construct subobjects. For example, sets might be implemented using lists for data representation, but the operations may be optimised, and otherwise fail, by assuming sorted lists. Since lists represent the same set up to duplication of elements, the list algebra is quotiented by a partial congruence that equates lists modulo duplicates, and which is defined only on sorted lists. This strategy is implemented in the type-theoretical setting by extending the logic with the following axiom schemata. They are tailored specifically for refinement.

Definition 1 (Existence of Subobjects (SUB) [9]).

$$\text{SUB} : \forall X . \forall \mathfrak{x} : T[X] . \forall R \subset X \times X . \ (\mathfrak{x} \ T[R] \ \mathfrak{x}) \ \wedge \ (\mathfrak{x} \ T[P_R] \ \mathfrak{x}) \ \Rightarrow$$
$$\exists S . \exists \mathfrak{s} : T[S] . \exists R' \subset S \times S . \exists \text{mono} : S \to X .$$
$$\forall s : S \ . \ s \ R' \ s \qquad\qquad\qquad \wedge$$
$$\forall s, s' : S \ . \ s \ R' \ s' \ \Leftrightarrow \ (\text{mono } s) \ R \ (\text{mono } s') \ \wedge$$
$$\mathfrak{x} \ (T[(x : X, s : S) \ . \ (x =_X (\text{mono } s))]) \ \mathfrak{s}$$

where $P_R \overset{def}{=} (x : X, y : X) \ . \ (x =_X y \wedge x \ R \ x)$. Intuitively, this essentially states that for any data type $\langle X, \mathfrak{x}\rangle$, if R is a relation that is compatible with the signature $T[X]$, then there exists a data type $\langle S, \mathfrak{s}\rangle$, a relation R', and a monomorphism from $\langle S, \mathfrak{s}\rangle$ to $\langle X, \mathfrak{x}\rangle$, such that R' is total on $\langle S, \mathfrak{s}\rangle$ and a restriction of R via mono, and such that $\langle S, \mathfrak{s}\rangle$ is a subalgebra of $\langle X, \mathfrak{x}\rangle$.

Definition 2 (Existence of Quotients (QUOT) [29]).

$$\text{QUOT} : \forall X . \forall \mathfrak{x} : T[X] . \forall R \subset X \times X . \qquad (\mathfrak{x} \ T[R] \ \mathfrak{x} \wedge equiv(R)) \ \Rightarrow$$
$$\exists Q . \exists \mathfrak{q} : T[Q] . \exists \text{epi} : X \to Q \ .$$
$$\forall x, y : X \ . \ xRy \ \Leftrightarrow \ (\text{epi } x) =_Q (\text{epi } y) \ \wedge$$
$$\forall q : Q . \exists x : X \ . \ q =_Q (\text{epi } x) \qquad\qquad \wedge$$
$$\mathfrak{x} \ (T[(x : X, q : Q).((\text{epi } x) =_Q q)]) \ \mathfrak{q}$$

where equiv(R) specifies R to be an equivalence relation.

Intuitively, this states that for any data type $\langle X, \mathfrak{x} \rangle$, if R is an equivalence relation on $\langle X, \mathfrak{x} \rangle$, then there exists a data type $\langle Q, \mathfrak{q} \rangle$ and an epimorphism from $\langle X, \mathfrak{x} \rangle$ to $\langle Q, \mathfrak{q} \rangle$, such that $\langle Q, \mathfrak{q} \rangle$ is a quotient algebra of $\langle X, \mathfrak{x} \rangle$.

Theorem 2. SUB, QUOT *hold in the parametric per-model of* [1], *under* $FADT_{Obs}$.

The proof of this theorem [12] relies on the model's ability to provide subobjects and quotients, and maps over these for any given morphism.

4   Higher-Order Quotient and Subobject Maps

In the *per*-model, first-order maps over subobjects and quotients are constructed from a given map by reusing the realiser. This does not work at higher order, since for functional arguments we have to contravariantly do this in reverse.

Consider *e.g.*, sequences over $\mathbb{N}$, whose encodings in $\mathbb{N}$ we write as the sequences themselves. Consider a function *rfi* on $\mathbb{N}$ that given a sequence, returns the sequence with the first item repeated. Define the *pers* $\mathcal{L}ist$, $\mathcal{B}ag$, and $\mathcal{S}et$ by

$$n \; \mathcal{L}ist \; m \;\; \Leftrightarrow \;\; n \text{ and } m \text{ encode the same list}$$
$$n \; \mathcal{B}ag \; m \;\; \Leftrightarrow \;\; n \text{ and } m \text{ encode the same list, modulo permutation}$$
$$n \; \mathcal{S}et \; m \;\; \Leftrightarrow \;\; n \text{ and } m \text{ encode the same list, modulo permutation and repetition}$$

Here, *rfi* is a realiser for a map $f_{rfi} : \mathcal{S}et \rightarrow \mathcal{S}et$, but is not a realiser for any map in $\mathcal{B}ag \rightarrow \mathcal{B}ag$, *i.e.*, we have *rfi* $(\mathcal{S}et \rightarrow \mathcal{S}et)$ *rfi* but not *rfi* $(\mathcal{B}ag \rightarrow \mathcal{B}ag)$ *rfi*.

In fact, the general problem is that there may not be a suitable function at all, let alone one sharing the same realiser.

In the following we sketch a setoid model based on ideas in [16]. This allows the construction of subobject and quotient maps by reusing realisers, also at higher order. Then we give axiomatic criteria for the construction of subobject and quotient maps at higher order. The setoid model fulfils these criteria.

We will work under the following reasonable assumption.

HADT$_{Obs}$: Every profile $T_i[X] = T_{i1}[X] \rightarrow \cdots \rightarrow T_{n_i}[X] \rightarrow T_{c_i}[X]$ of signature $T[X]$ is such that $T_{ij}[X]$ has no occurrences of universal types other than those in Obs, and $T_{c_i}[X]$ is either X or some $D \in Obs$.

4.1   A Setoid Model

Types are now interpreted as setoids, *i.e.*, pairs $\langle \mathcal{A}, \sim_{\mathcal{A}} \rangle$, consisting of a *per* $\mathcal{A}$ and a *per* $\sim_{\mathcal{A}}$ on $\mathcal{A}$, *i.e.*, a saturated *per* on $Dom(\mathcal{A}) \times Dom(\mathcal{A})$, giving the desired equality on the interpreted type. Given setoids $\langle \mathcal{A}, \sim_{\mathcal{A}} \rangle$ and $\langle \mathcal{B}, \sim_{\mathcal{B}} \rangle$, we form a setoid $\langle \mathcal{A}, \sim_{\mathcal{A}} \rangle \rightarrow \langle \mathcal{B}, \sim_{\mathcal{B}} \rangle \stackrel{def}{=} \langle \mathcal{A} \rightarrow \mathcal{B}, \sim_{\mathcal{A} \rightarrow \mathcal{B}} \rangle$, where $\sim_{\mathcal{A} \rightarrow \mathcal{B}}$ is the saturated relation $\sim_{\mathcal{A}} \rightarrow \sim_{\mathcal{B}} \subseteq Dom(\mathcal{A} \rightarrow \mathcal{B}) \times Dom(\mathcal{A} \rightarrow \mathcal{B})$. Saturation of $\sim$ is the condition $(m \; \mathcal{A} \; n \; \wedge \; n \sim n' \; \wedge \; n' \; \mathcal{B} \; m') \;\; \Rightarrow \;\; m \sim m'$.

A relation $\mathcal{R}$ between setoids $\langle \mathcal{A}, \sim_{\mathcal{A}} \rangle$ and $\langle \mathcal{B}, \sim_{\mathcal{B}} \rangle$ is now given by a saturated relation on $Dom(\sim_{\mathcal{A}}) \times Dom(\sim_{\mathcal{B}})$. Complex relations are defined as one would expect. The setoid definition of subobjects and quotients go as follows.

Definition 3 (Subobject Setoid). *Let $\mathcal{P}$ be a predicate on setoid $\langle \mathcal{X}, \sim_{\mathcal{X}} \rangle$, meaning that $\mathcal{P}$ fulfils the unary saturation condition $\mathcal{P}(x) \wedge x \sim_{\mathcal{X}} y \;\Rightarrow\; \mathcal{P}(y)$. Define the relation, also denoted $\mathcal{P}$, on $\langle \mathcal{X}, \sim_{\mathcal{X}} \rangle$ by $x \, \mathcal{P} \, y \overset{def}{\Leftrightarrow} x \sim_{\mathcal{X}} y \wedge \mathcal{P}(x)$. Then the subobject $\mathsf{R}_{\mathcal{P}}(\langle \mathcal{X}, \sim_{\mathcal{X}} \rangle)$ of $\langle \mathcal{X}, \sim_{\mathcal{X}} \rangle$ restricted on $\mathcal{P}$, is defined by $\langle \mathcal{X}, \mathcal{P} \rangle$.*

Definition 4 (Quotient Setoid). *Let $\mathcal{R}$ be a equivalence relation on setoid $\langle \mathcal{X}, \sim_{\mathcal{X}} \rangle$. Define the quotient $\langle \mathcal{X}, \sim_{\mathcal{X}} \rangle / \mathcal{R}$ of $\langle \mathcal{X}, \sim_{\mathcal{X}} \rangle$ w.r.t. $\mathcal{R}$ by $\langle \mathcal{X}, \mathcal{R} \rangle$.*

Theorem 3. *Suppose $T[X]$ adheres to HADT_{Obs}. Then SUB and QUOT hold in the setoid model indicated above.*

With setoids we may construct quotient maps from a given map and *vice versa* by reusing realisers, since the original domain inhabitation is preserved by subobjects and quotients. However, Theorem 3 is given as a corollary to a general result of the axiomatic criteria in the next section.

4.2   Axiomatic Criteria for Subobject and Quotient Maps

We now develop a general axiomatic scheme for obtaining subobject and quotient maps. The setoid approach in the previous section is an instance of this scheme.

For quotients, the general problem is that for a given map $f : X/R \to X/R$, there need not exist a map $g : X \to X$ such that for all $x : X$, $[g(x)] = f([x])$, i.e., $epi(g(x)) = f(epi(x))$, where $epi : X \to X/R$ maps an element to its equivalence class. This is the case for the *per*-model. The axiom of choice (AC) gives such a map g, because then epi has an inverse, and the desired g is given by $\lambda x : X.epi^{-1}(f epi(x))$. AC does not hold in the *per*-model, nor does it hold in the setoid model of the previous section. In this section, we develop both a weaker condition sufficient to give higher-order quotient maps, and a condition for obtaining higher-order subobject maps.

According to HADT_{Obs}, we consider arrow types over types $U_0, U_1, \ldots$, where any U_i is either X or some $D \in Obs$. For this, define families U^i by

$$U^0 = U_0$$
$$U^{i+1} = (U^i) \to U_{i+1}$$

For example, $U^2 = ((U_0 \to U_1) \to U_2)$.

Quotient Maps. For $U = U^n$, define $Q(U)^i$ for any equivalence relation R,

$$Q(U)^0 = U_0$$
$$Q(U)^1 = U_0/R \to U_1$$
$$Q(U)^{i+1} = (Q(U)^{i-1} \to U_i/R) \to U_{i+1}, \; 1 \le i \le n-1$$

where, $U_i/R = X/R$, if U_i is X, and $U_i/R = D$, if U_i is $D \in Obs$, e.g., $Q(U^2)^2 = ((U_0 \to U_1/R) \to U_2)$. In any $Q(U)^i$, quotients U_j/R occur only negatively.

Given $Q(U)^n$, we get derived relations, functions and types by the substitution operators $Q(U)^n[\xi]^+$ and $Q(U)^n[\xi]^-$, according to ξ being a relation, function or type; $Q(U)^n[\xi]^+$ substitutes ξ for positive occurrences of X in $Q(U)^n$, and $Q(U)^n[\xi]^-$ substitutes ξ for every (negative) occurrence of X/R in $Q(U)^n$. Relational and functional identities are then denoted by their domains.

Thus for $U = U^n$ and the equivalence relation R, we can define the relation

$$R(U)^n \overset{def}{=} Q(U)^n[R]^+$$

In any $R(U)^i$, R occurs positively, and identities U_j/R occur only negatively. The point of all this is that, if R is an equivalence relation on X, then $R(U)^i$ is an equivalence relation on $Q(U)^i$. This means that we may form the quotient $Q(U)^i/R(U)^i$. For example, consider $U = U^1 = X \to X$. Then $Q(U)^1 = X/R \to X$ and $R(U)^1 = X/R \to R$, and $X/R \to R$ is an equivalence relation on $X/R \to X$. In contrast, $R \to X/R$ is not necessarily an equivalence relation on $X \to X/R$. However, $(R \to X/R) \to R$ is an equivalence relation on $(X \to X/R) \to X$, that is, $R(U^2)^2$ is an equivalence relation on $Q(U^2)^2$, for $U^2 = (X \to X) \to X$.

Now consider the relation $graph(\mathsf{epi}) \overset{def}{=} (x : X, q : X/R) \, . \, ((\mathsf{epi}\, x) =_{X/R} q)$ where the map $\mathsf{epi} : X \to X/R$ maps elements to their R-equivalence class. A sufficient condition for obtaining higher-order functions over quotients is now

Quot-Arr: For R an equivalence relation on X, and any given $U = U^n$,

$$Q(U)^n/R(U)^n \cong Q(U)^n[X/R]^+$$

where the isomorphism $\mathsf{iso} : Q(U)^n[X/R]^+ \to Q(U)^n/R(U)^n$ is such that any f in the equivalence class $\mathsf{iso}(\beta)$ is such that $f \, (Q(U)^n[graph(\mathsf{epi})]^+) \, \beta$.

Note that **Quot-Arr** is not an extension to our logic; we do not have quotient types. Rather, **Quot-Arr** is a condition to check in any relevant model, in which the terminology concerning quotients is well defined. In [16], **Quot-Arr** is expressible in the logic, and **Quot-Arr** is shown strictly weaker than the axiom of choice.

Let us exemplify why **Quot-Arr** suffices. The challenge of this paper is higher-order operations in data types, and then the soundness of QUOT and SUB where $T[X]$ has higher-order operation profiles. To illustrate the use of **Quot-Arr** in semantically validating QUOT, suppose $T[X]$ has a profile $g : (X \to X) \to X$ and that $R \subset X \times X$ is an equivalence relation. Consider now any $\mathfrak{x} : T[X]$. Assuming $\mathfrak{x} \, (T[R]) \, \mathfrak{x}$, we must produce a $\mathfrak{q} : T[X/R]$, such that $\mathfrak{x} \, (T[graph(\mathsf{epi})]) \, \mathfrak{q}$. For $\mathfrak{x}.g : (X \to X) \to X$, this involves finding a $\mathfrak{q}.g : (X/R \to X/R) \to X/R$, such that

$$\mathfrak{x}.g \, ((graph(\mathsf{epi}) \to graph(\mathsf{epi})) \to graph(\mathsf{epi})) \, \mathfrak{q}.g \tag{1}$$

Consider now the following instance of **Quot-Arr**.

Quot-Arr$_1$: $(X/R \to X)/(X/R \to R) \cong (X/R \to X/R)$

With $Quot\text{-}Arr_1$ we can construct the following commuting diagram.

$$(X/R \to X) \xrightarrow{\ \text{epi} \to X\ } (X \to X) \xrightarrow{\ \mathfrak{r}.g\ } X \xrightarrow{\ \text{epi}\ } X/R$$

$$\text{epi}_{X/R \to X} \Big\downarrow$$

$$(X/R \to X)/(X/R \to R)$$

$$\text{iso} \Big\uparrow \qquad \xrightarrow{\ lift(\text{epi} \circ \mathfrak{r}.g \circ (\text{epi} \to X))\ }$$

$$X/R \to X/R$$

where epi $\to X$ maps any $f : X/R \to X$ to $\lambda x : X.f(\text{epi}\,x)$, and iso is so that any f in the equivalence class iso(β) satisfies $f\ (\text{eq}_{X/R} \to graph(\text{epi}))\ \beta$. The desired $\mathsf{q}.g : (X/R \to X/R) \to X/R$ is given by

$$lift(\text{epi} \circ \mathfrak{r}.g \circ (\text{epi} \to X)) \circ \text{iso}$$

Here *lift* is the operation that lifts any $\gamma : Z \to Y$ to $lift(\gamma) : Z/\!\!\sim \to Y$, given an equivalence relation $\sim$ on Z, provided that γ satisfies $x \sim y \Rightarrow \gamma x = \gamma y$ for all $x, y : Z$. Then, $lift(\gamma)$ is the map satisfying $lift(\gamma) \circ \text{epi} = \gamma$. To be able to lift epi $\circ \mathfrak{r}.g \circ (\text{epi} \to X)$ in this way, we must check that epi $\circ \mathfrak{r}.g \circ (\text{epi} \to X)$ satisfies $f\ (\text{eq}_{X/R} \to R)\ f' \Rightarrow (\text{epi} \circ \mathfrak{r}.g \circ (\text{epi} \to X))(f) =_{X/R} (\text{epi} \circ \mathfrak{r}.g \circ (\text{epi} \to X))(f')$, for all $f, f' : (X/R \to X)$. Assuming $f\ (\text{eq}_{X/R} \to R)\ f'$, we get $(\text{epi} \to X)(f)\ (R \to R)\ (\text{epi} \to X)(f')$. Then by $\mathfrak{r}\ T[R]\ \mathfrak{r}$, the result follows. This warrants the construction of $\mathsf{q}.g$.

To show that $\mathsf{q}.g$ is the desired function, we must check that it satisfies (1). This cannot be read directly from the above diagram; for instance, although $\mathsf{q}.g$ is constructed essentially in terms of $\mathfrak{r}.g$, it is clear that epi $\to X$ maps only to those α in $X \to X$ that do not discern between input of the same R-equivalence class, and these α might not cover the domain of inputs giving all possible outputs. Intuitively though, this suffices since R-equivalence is really all that matters.

More formally, suppose $\alpha : X \to X$ and $\beta : X/R \to X/R$ are such that

$$\alpha\ (graph(\text{epi}) \to graph(\text{epi}))\ \beta \tag{2}$$

We want $(\mathfrak{r}.g\,\alpha)\ graph(\text{epi})\ (\mathsf{q}.g\,\beta)$. First show for any $\alpha : X \to X$, there exists $f_\alpha : X/R \to X$, s.t. $(\text{epi} \to X)f_\alpha\ (R \to R)\ \alpha$ and iso$(\beta) = \text{epi}_{X/R \to X}(f_\alpha)$, *i.e.*,

$$\lambda x : X.f_\alpha(\text{epi}\,x))\ (R \to R)\ \alpha \tag{3}$$

$$\text{iso}(\beta) = [f_\alpha]_{X/R \to R} \tag{4}$$

The assumption on iso in $Quot\text{-}Arr_1$ is that any f in the equivalence class iso(β) is such that

$$f\ (\text{eq}_{X/R} \to graph(\text{epi}))\ \beta \tag{5}$$

so any of these f are candidates for f_α. For such an f we show $a\,R\,a' \Rightarrow (\lambda x:$ $X.f(\mathrm{epi}\,x)a)\,R\,\alpha a'$, i.e., $[a] = [a'] \Rightarrow [f[a]] = [\alpha a']$. We have from (5), $[a] =$ $[a'] \Rightarrow [f[a]] = \beta[a']$, and by (2), we have $[a] = [a'] \Rightarrow [\alpha a] = \beta[a']$. Together, this gives the desired property for f, so we have the existence of f_α satisfying (3) and (4). From (2) and (5) we also get

$$\lambda x:X.f_\alpha(\mathrm{epi}\,x)\quad (graph(\mathrm{epi}) \to graph(\mathrm{epi}))\quad \beta$$

From the above diagram, and (3) and (4), this gives $(\mathfrak{r}.g\,(\lambda x :$ $X.f_\alpha(\mathrm{epi}\,x)))\quad graph(\mathrm{epi})\quad (\mathfrak{q}.g\,\beta)$. By $\mathfrak{r}\,T[R]\,\mathfrak{r}$, and since we have $\alpha\quad (R \to R)\quad (\lambda x:X.f_\alpha(\mathrm{epi}\,x))$, we thus get $(\mathfrak{r}.g\,\alpha)\quad graph(\mathrm{epi})\quad (\mathfrak{q}.g\,\beta)$.

The general form of this diagram for any given $U = U^n$ and U_c, is

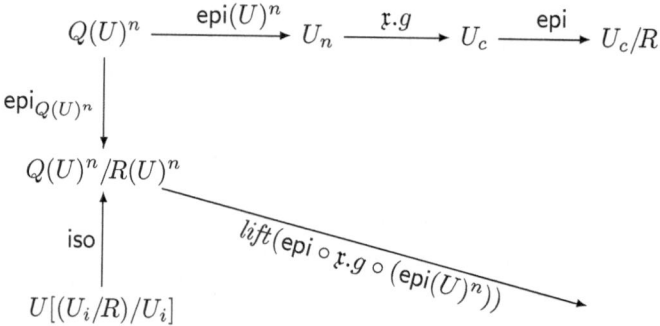

where for a given $U = U^n$, we define the function $\mathrm{epi}(U)^n \overset{def}{=} Q(U)^n[\mathrm{epi}]^-$.

Subobject Maps. A similar story applies to subobjects. For any predicate P on X, we write $R_P(X)$ for the subobject of X classified by those $x : X$ such that $P(x)$ holds. Let the monomorphism $\mathrm{mono} : R_P(X) \to X$ map elements to their correspondents in X. For use in arrow-type relations, we construct a binary relation from P, also denoted P, by

$$P \overset{def}{=} (x:X, y:X)\,.\,(x =_X y \,\wedge\, \exists y':R_P(X)\,.\,y = (\mathrm{mono}\,y'))$$

Now for a given $U = U^n$, define $S(U)^i$ for some P as follows

$$
\begin{aligned}
S(U)^0 &= R_P(U_0)\\
S(U)^1 &= U_0 \to R_P(U_1)\\
S(U)^{i+1} &= (S(U)^{i-1} \to U_i) \to R_P(U_{i+1}), 1 \le i \le n-1
\end{aligned}
$$

where, if U_i is X then $R_P(U_i) = R_P(X)$, and if U_i is $D \in Obs$ then $R_P(U_i) = D$. For example $S(U^2)^2 = ((R_P(U_0) \to U_1) \to R_P(U_2))$. In any $S(U)^i$, subobjects $R_P(U_j)$ occur only positively. For any $U = U^n$ and some P, define the relation

$$P(U)^n \overset{def}{=} S(U)^n[P]^-$$

The substitution operators $S(U)^n[\xi]^-$ and $S(U)^n[\xi]^+$ are analogues to $Q(U)^n[\xi]^-$ and $Q(U)^n[\xi]^+$. Identities are denoted by their domains.

Intuitively, one would think that for any given $U = U^n$, we should now postulate an isomorphism between $\mathsf{R}_{P(U)^n}(S(U)^n)$ and $S(U)^n[(\mathsf{R}_P(X))]^-$. This would be in dual analogy to *Quot-Arr*. However, this isomorphism does not exist even in the setoid model. For example, we will not be able to find an isomorphism between $\mathsf{R}_{P \to \mathsf{R}_P(X)}(X \to \mathsf{R}_P(X))$ and $\mathsf{R}_P(X) \to \mathsf{R}_P(X)$. However, it turns out that we can in fact use an outermost quotient instead of subobjects for the isomorphism, in the same way as we did for *Quot-Arr*.

Thus, if P is a predicate on X, then $P(U)^i$ is an equivalence relation on $S(U)^i$. This means that we may form the quotient $S(U)^i/P(U)^i$, e.g., if $U = U^1 = X \to X$, then $S(U)^1 = X \to \mathsf{R}_R(X)$ and $P(U)^1 = P \to \mathsf{R}_P(X)$, and $P \to \mathsf{R}_P(X)$ is an equivalence relation on $X \to \mathsf{R}_R(X)$. Again, in contrast, $\mathsf{R}_P(X) \to P$ is not necessarily an equivalence relation on $\mathsf{R}_P(X) \to X$. However, $(\mathsf{R}_P(X) \to P) \to \mathsf{R}_P(X)$ is an equivalence relation on $(\mathsf{R}_P(X) \to X) \to \mathsf{R}_P(X)$, that is, $P(U^2)^2$ is an equivalence relation on $S(U^2)^2$, for $U^2 = (X \to X) \to X$.

For the relation $graph(\mathsf{mono}) \stackrel{def}{=} (x : X, s : \mathsf{R}_P(X)) . (x =_X (\mathsf{mono}\, s))$, a sufficient condition for obtaining higher-order functions over subobjects is,

Sub-Arr: For P a predicate on X, and any given $U = U^n$,

$$S(U)^n/P(U)^n \cong S(U)^n[(\mathsf{R}_P(X))]^-$$

where the isomorphism $\mathsf{iso} : S(U)^n[(\mathsf{R}_P(X))]^- \to S(U)^n/P(U)^n$ is such that any f in the equivalence class $\mathsf{iso}(\beta)$ is such that $f\ (S(U)^n[graph(\mathsf{mono})]^-)\ \beta$.

Again, **Sub-Arr** is not an axiom in our logic, but is a condition that we can check for models in which the terminology in **Sub-Arr** has a well-defined meaning.

To illustrate **Sub-Arr**, suppose $T[X]$ has a profile $g : (X \to X) \to X$. For any $\mathfrak{x} : T[X]$, assume $\mathfrak{x}\ T[P]\ \mathfrak{x}$. We must exhibit a $\mathfrak{s} : T[\mathsf{R}_P(X)]$, such that $\mathfrak{x}\ T[graph(\mathsf{mono})]\ \mathfrak{s}$. For $\mathfrak{x}.g : (X \to X) \to X$, this means finding a $\mathfrak{s}.g : (\mathsf{R}_P(X) \to \mathsf{R}_P(X)) \to \mathsf{R}_P(X)$, s.t.

$$\mathfrak{x}.g\ ((graph(\mathsf{mono}) \to graph(\mathsf{mono})) \to graph(\mathsf{mono}))\ \mathfrak{s}.g \tag{6}$$

Consider now the following instance of **Sub-Arr**.

Sub-Arr₁: For a predicate P on X,

$$(X \to \mathsf{R}_P(X))/(P \to \mathsf{R}_P(X)) \cong \mathsf{R}_P(X) \to \mathsf{R}_P(X)$$

Using **Sub-Arr₁**, we can construct the following commuting diagram.

Then, $\mathfrak{s}.g : (\mathsf{R}_P(X) \to \mathsf{R}_P(X)) \to \mathsf{R}_P(X)$ is given by $lift(\mathfrak{x}.g \circ (X \to \mathsf{mono})) \circ \mathsf{iso}$. To justify the lifting of $\mathfrak{x}.g \circ (X \to \mathsf{mono})$, we must show for all $f, f' : X \to \mathsf{R}_P(X)$ satisfying $f \ (P \to \mathsf{R}_P(X)) \ f'$, that $\mathfrak{x}.g \circ (X \to \mathsf{mono})(f) =_X \mathfrak{x}.g \circ (X \to \mathsf{mono})(f')$. Note that $lift(\mathfrak{x}.g \circ (X \to \mathsf{mono}))$ then maps to X, so in addition we must show that $lift(\mathfrak{x}.g \circ (X \to \mathsf{mono}))$ in fact maps to $\mathsf{R}_P(X)$. Now, if $f \ (P \to \mathsf{R}_P(X)) \ f'$, we get $(X \to \mathsf{mono})(f) \ (P \to P) \ (X \to \mathsf{mono})(f')$. By assumption, we have $\mathfrak{x} \ T[P] \ \mathfrak{x}$, in particular $\mathfrak{x}.g \ ((P \to P) \to P) \ \mathfrak{x}.g$, and the result follows. If for some y, $\exists y' : \mathsf{R}_P(X) \ . \ \mathsf{mono} \, y' = y$, we assume that it is elementary to find such a y'. Thus, since mono is a monomorphism, we may map $lift(\mathfrak{x}.g \circ (X \to \mathsf{mono}))$ to $\mathsf{R}_P(X)$, and so we have a function $\mathfrak{s}.g : ((\mathsf{R}_P(X) \to \mathsf{R}_P(X)) \to \mathsf{R}_P(X))$.

To show that $\mathfrak{s}.g$ is the desired function, we must check that it satisfies (6). Suppose $\alpha : X \to X$ and $\beta : \mathsf{R}_P(X) \to \mathsf{R}_P(X)$ are such that

$$\alpha \ (graph(\mathsf{mono}) \to graph(\mathsf{mono})) \ \beta \tag{7}$$

We want $(\mathfrak{x}.g \, \alpha) \ graph(\mathsf{mono}) \ (\mathfrak{s}.g \, \beta)$. First show for any $\alpha : X \to X$, there exists $f_\alpha : X \to \mathsf{R}_P(X)$, such that $(X \to \mathsf{mono}) f_\alpha \ (P \to P) \ \alpha$ and $\mathsf{iso}(\beta) = \mathsf{epi}_{X \to \mathsf{R}_P(X)}(f_\alpha)$, i.e.,

$$\lambda x : X.\mathsf{mono}(f_\alpha x) \ (P \to P) \ \alpha \tag{8}$$

$$\mathsf{iso}(\beta) = [f_\alpha]_{P \to \mathsf{R}_P(X)} \tag{9}$$

The assumption on iso in $\textbf{\textit{Sub-Arr}}_1$ is that any f in the equivalence class $\mathsf{iso}(\beta)$ is such that

$$f \ (graph(\mathsf{mono}) \to \mathsf{eq}_{\mathsf{R}_P(X)}) \ \beta \tag{10}$$

so any of these f are candidates for f_α. For such an f, show (8), i.e., $a = a' \wedge \exists a''.\mathsf{mono} \, a'' = a' \Rightarrow \mathsf{mono}(fa) = \alpha a' \wedge \exists b \ . \ \mathsf{mono} \, b = \alpha a'$. We have from (10), $a = \mathsf{mono} \, a'' \Rightarrow fa = \beta \, a''$, and by assumption on α and β, we have $a' = \mathsf{mono} \, a'' \Rightarrow \alpha a' = \mathsf{mono}(\beta a'')$. This gives the desired property for f, so we have the existence of f_α satisfying (8) and (9). From (10) we also get

$$\lambda x : X.\mathsf{mono}(f_\alpha x) \ (graph(\mathsf{mono}) \to graph(\mathsf{mono})) \ \beta$$

From the above diagram, and (8) and (9), this gives $(\mathfrak{x}.g \, (\lambda x : X.\mathsf{mono}(f_\alpha x))) \ graph(\mathsf{mono}) \ (\mathfrak{s}.g \, \beta)$. By $\mathfrak{x} \ T[P] \ \mathfrak{x}$, and since $\alpha \ (P \to P) \ \lambda x : X.\mathsf{mono}(f_\alpha x)$, we thus get $(\mathfrak{x}.g \, \alpha) \ graph(\mathsf{mono}) \ (\mathfrak{s}.g \, \beta)$.

Here is the general form of this diagram for any given $U = U^n$ and U_c, is

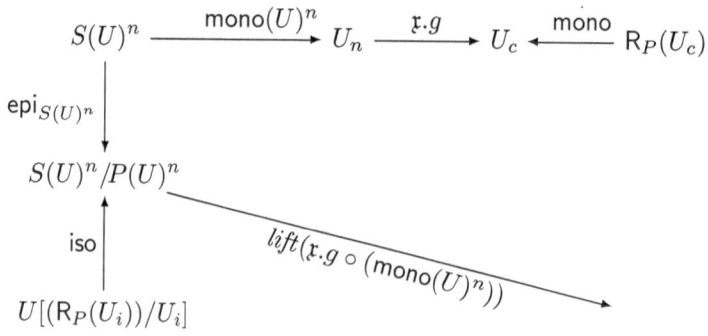

where for a given $U = U^n$, we define the function $\mathsf{mono}(U)^n \stackrel{\text{def}}{=} S(U)^n[\mathsf{mono}]^+$.

This schema is more general than what is called for in the refinement-specific SUB. In SUB, the starting point is a relation R, and the predicate with which one restricts the domain X, is $\Gamma_R(x) \stackrel{\text{def}}{=} x \mathrel{R} x$. The corresponding binary relation is then $P_R \stackrel{\text{def}}{=} (x{:}X, y{:}X) . (x =_X y \wedge x \mathrel{R} x))$.

In closing, we mention that the *per* model, parametric or not, does not satisfy *Quot-Arr* nor *Sub-Arr*. We summarise:

Theorem 4. *Suppose $T[X]$ adheres to $HADT_{Obs}$. Then* SUB *and* QUOT *hold in any model that satisfies* **Sub-Arr** *and* **Quot-Arr**.

Theorem 5. *The setoid model satisfies* **Quot-Arr** *and* **Sub-Arr**, *by the isomorphism being denotational equality.*

Proof: See [12].   □

Corollary 6. SUB *and* QUOT *hold in the setoid model indicated above.*

5   Final Remarks

We have devised and validated a method in logic for proving specification refinement for data types with higher-order operations. The method is based on standard simulation relations, accommodating the fact that these are easier to deal with than alternative notions when performing refinement. In general however, there may not exist standard simulation relations at higher order, even in the presence of observational equivalence. It is possible to devise specialised solutions to this using abstraction barrier-observing simulation relations [10,12], or pre-logical relations expressed in System F. Beyond that, it is desirable to find general axiomatic criteria analogous to *Sub-Arr* and *Quot-Arr*, using alternative notions of simulation relations. This is currently under investigation.

Acknowledgments. Martin Hofmann has contributed with essential input.

References

1. E.S. Bainbridge, P.J. Freyd, A. Scedrov, and P.J. Scott. Functorial polymorphism. *Theoretical Computer Science*, 70:35–64, 1990.
2. H.P. Barendregt. Lambda calculi with types. In S. Abramsky, D.M. Gabbay, and T.S.E. Maibaum, editors, *Handbook of Logic in Computer Science*, volume 2, pages 118–309. Oxford University Press, 1992.
3. M. Bidoit and R. Hennicker. Behavioural theories and the proof of behavioural properties. *Theoretical Computer Science*, 165:3–55, 1996.
4. M. Bidoit, R. Hennicker, M. Wirsing. Proof systems for structured specifications with observability operators. *Theoretical Computer Science*, 173:393–443, 1997.
5. C. Böhm and A. Berarducci. Automatic synthesis of typed λ-programs on term algebras. *Theoretical Computer Science*, 39:135–154, 1985.

6. M. Cerioli, M. Gogolla, H. Kirchner, B. Krieg-Brückner, Z. Qian, and M. Wolf, eds.. *Algebraic System Specification and Development. Survey and Annotated Bibliography, 2nd Ed.*, BISS Monographs, vol. 3. Shaker Verlag, 1997.

7. O.-J. Dahl. *Verifiable Programming, Revised version 1993.* Prentice Hall Int. Series in Computer Science; C.A.R. Hoare, Series Editor. Prentice-Hall, UK, 1992.

8. J.-Y. Girard, P. Taylor, and Y. Lafont. *Proofs and Types.* Cambridge Tracts in Theoretical Computer Science. Cambridge University Press, 1990.

9. J. Hannay. Specification refinement with System F. In *Computer Science Logic. Proc. of CSL'99*, vol. 1683 of *Lecture Notes in Comp. Sci.*, pages 530–545. Springer Verlag, 1999.

10. J. Hannay. Specification refinement with System F, the higher-order case. In *Recent Trends in Algebraic Development Techniques. Selected Papers from WADT'99*, volume 1827 of *Lecture Notes in Comp. Sci.*, pages 162–181. Springer Verlag, 1999.

11. J. Hannay. A higher-order simulation relation for System F. In *Foundations of Software Science and Computation Structures. Proc. of FOSSACS 2000*, vol. 1784 of *Lecture Notes in Comp. Sci.*, pages 130–145. Springer Verlag, 2000.

12. J. Hannay. *Abstraction Barriers and Refinement in the Polymorphic Lambda Calculus.* PhD thesis, Laboratory for Foundations of Computer Science (LFCS), University of Edinburgh, 2001.

13. J. Hannay. Abstraction barrier-observing relational parametricity. In *Typed Lambda Calculi and Applications. Proc. of TLCA 2002, Lecture Notes in Comp. Sci.*, Springer Verlag, 2002. To appear.

14. R. Hasegawa. Parametricity of extensionally collapsed term models of polymorphism and their categorical properties. In *Theoretical Aspects of Computer Software. Proc. of TACS'91*, vol. 526 of *Lecture Notes in Comp. Sci.*, pages 495–512. Springer Verlag, 1991.

15. C.A.R. Hoare. Proofs of correctness of data representations. *Acta Informatica*, 1:271–281, 1972.

16. M. Hofmann. *Extensional Concepts in Intensional Type Theory, Report CST-117-95 and Technical Report ECS-LFCS-95-327.* PhD thesis, Laboratory for Foundations of Computer Science (LFCS), University of Edinburgh, 1995.

17. F. Honsell, J. Longley, D. Sannella, and A. Tarlecki. Constructive data refinement in typed lambda calculus. In *Foundations of Software Science and Computation Structures. Proc. of FOSSACS 2000*, vol. 1784 of *Lecture Notes in Comp. Sci.*, pages 161–176. Springer Verlag, 2000.

18. F. Honsell and D. Sannella. Prelogical relations. *Information and Computation*, 178:23–43, 2002.

19. Y. Kinoshita, P.W. O'Hearn, J. Power, M. Takeyama, and R.D. Tennent. An axiomatic approach to binary logical relations with applications to data refinement. In *Theoretical Aspects of Computer Software. Proc. of TACS'97*, vol. 1281 of *Lecture Notes in Comp. Sci.*, pages 191–212. Springer Verlag, 1997.

20. Y. Kinoshita and J. Power. Data refinement for call-by-value programming languages. In *Computer Science Logic. Proc. of CSL'99*, vol. 1683 of *Lecture Notes in Comp. Sci.*, pages 562–576. Springer Verlag, 1999.

21. Q. Ma and J.C. Reynolds. Types, abstraction and parametric polymorphism, part 2. In *Mathematical Foundations of Programming Semantics, Proc. of MFPS*, vol. 598 of *Lecture Notes in Comp. Sci.*, pages 1–40. Springer Verlag, 1991.

22. H. Mairson. Outline of a proof theory of parametricity. In *Functional Programming and Computer Architecture. Proc. of the 5th acm Conf.*, vol. 523 of *Lecture Notes in Comp. Sci.*, pages 313–327. Springer Verlag, 1991.

23. R. Milner. An algebraic definition of simulation between programs. In *Joint Conferences on Artificial Intelligence, Proceedings of the 2nd International Conference, JCAI, London (UK)*, pages 481–489. Morgan Kaufman Publishers, 1971.

24. J.C. Mitchell. On the equivalence of data representations. In V. Lifschitz, editor, *Artificial Intelligence and Mathematical Theory of Computation: Papers in Honor of John McCarthy*, pages 305–330. Academic Press, 1991.

25. J.C. Mitchell. *Foundations for Programming Languages*. Foundations of Computing. MIT Press, 1996.

26. P.W. O'Hearn and R.D. Tennent. Relational parametricity and local variables. In *20th SIGPLAN-SIGACT Symposium on Principles of Programming Languages, Proceedings*, pages 171–184. ACM Press, 1993.

27. G.D. Plotkin and M. Abadi. A logic for parametric polymorphism. In *Typed Lambda Calculi and Applications. Proc. of TLCA'93*, vol. 664 of *Lecture Notes in Comp. Sci.*, pages 361–375. Springer Verlag, 1993.

28. G.D. Plotkin, J. Power, D. Sannella, and R.D. Tennent. Lax logical relations. In *Automata, Languages and Programming. Proc. of ICALP 2000*, vol. 1853 of *Lecture Notes in Comp. Sci.*, pages 85–102. Springer Verlag, 2000.

29. E. Poll and J. Zwanenburg. A logic for abstract data types as existential types. In *Typed Lambda Calculus and Applications. Proc. of TLCA'99*, vol. 1581 of *Lecture Notes in Comp. Sci.*, pages 310–324. Springer Verlag, 1999.

30. J.C. Reynolds. Types, abstraction and parametric polymorphism. In *Information Processing 83, Proc. of the IFIP 9th World Computer Congress*, pages 513–523. Elsevier Science Publishers B.V. (North-Holland), 1983.

31. D. Sannella and A. Tarlecki. Essential concepts of algebraic specification and program development. *Formal Aspects of Computing*, 9:229–269, 1997.

32. O. Schoett. Behavioural correctness of data representations. *Science of Computer Programming*, 14:43–57, 1990.

33. C. Strachey. Fundamental concepts in programming languages. Lecture notes from the International Summer School in Programming Languages, Copenhagen, 1967.

34. I. Takeuti. An axiomatic system of parametricity. *Fundamenta Informaticae*, 20:1–29, 1998.

35. R.D. Tennent. Correctness of data representations in Algol-like languages. In A.W. Roscoe, editor, *A Classical Mind: Essays in Honour of C.A.R. Hoare*. Prentice Hall International, 1997.

36. J. Zwanenburg. *Object-Oriented Concepts and Proof Rules: Formalization in Type Theory and Implementation in Yarrow*. PhD thesis, Tech. Univ. Eindhoven, 1999.

Efficient Pebbling for List Traversal Synopses

Yossi Matias[1]* and Ely Porat[2]

[1] School of Computer Science, Tel Aviv University, matias@cs.tau.ac.il
[2] Department of Mathematics and Computer Science, Bar-Ilan University, 52900
Ramat-Gan, Israel, (972-3)531-8407, porately@cs.biu.ac.il

Abstract. We show how to support efficient back traversal in a unidirectional list, using small memory and with essentially no slowdown in forward steps. Using $O(\lg n)$ memory for a list of size n, the i'th back-step from the farthest point reached so far takes $O(\lg i)$ time worst case, while the overhead per forward step is at most epsilon for arbitrary small constant $\epsilon > 0$. An arbitrary sequence of forward and back steps is allowed. A full trade-off between memory usage and time per back-step is presented: k vs. $kn^{1/k}$ and vice versa. Our algorithm is based on a novel pebbling technique which moves pebbles on a "virtual binary tree" that can only be traversed in a pre-order fashion.

The list traversal synopsis extends to general directed graphs, and has other interesting applications, including memory efficient hash-chain implementation. Perhaps the most surprising application is in showing that for any program, arbitrary rollback steps can be efficiently supported with small overhead in memory, and marginal overhead in its ordinary execution. More concretely: Let P be a program that runs for at most T steps, using memory of size M. Then, at the cost of recording the input used by the program, and increasing the memory by a factor of $O(\lg T)$ to $O(M \lg T)$, the program P can be extended to support an arbitrary sequence of forward execution and rollback steps, as follows. The i'th rollback step takes $O(\lg i)$ time in the worst case, while forward steps take $O(1)$ time in the worst case, and $1 + \epsilon$ amortized time per step.

1 Introduction

A unidirectional list enables easy forward traversal in constant time per step. However, getting from a certain object to its preceding object cannot be done effectively. It requires forward traversal from the beginning of the list and takes time proportional to the distance to the current object, using $O(1)$ additional memory. In order to support more effective back-steps on a unidirectional list, it is required to add auxiliary data structures.

* Research supported in part by the Israel Science Foundation.

J.C.M. Baeten et al. (Eds.): ICALP 2003, LNCS 2719, pp. 918–928, 2003.
© Springer-Verlag Berlin Heidelberg 2003

The goal of this work is to support memory- and time-efficient back traversal in unidirectional lists, without essentially increasing the time per forward traversal. In particular, under the constraint that forward steps should remain constant, we would like to minimize the number of pointers kept for the lists, the memory used by the algorithm, and the time per back-step, supporting an arbitrary sequence of forward and back steps.

Of particular interest are situations in which the unidirectional list is already given, and we have access to the list but no control over its implementation. The list may represent a data structure implemented in computer memory or in a database, it may reside on a separate computer system. The list may also represent a computational process, where the objects in the list are configurations in the computation and the next pointer represents a computational step.

1.1 Main Results

The main result of this paper is an algorithm that supports efficient back traversal in a unidirectional list, using small memory and with essentially no slowdown in forward steps: $1 + \epsilon$ amortized time per forward step for arbitrary small constant $\epsilon > 0$, and $O(1)$ time in the worst case. Using $O(\lg n)$ memory, back traversals can be supported in $O(\lg n)$ time per back-step, where n is the distance from the beginning of the list to farthest point reached so far. In fact, we show that a back traversal of limited scope can be executed more effectively: $O(\lg i)$ time for the i'th back-step, for any $i \leq n$, using $O(\lg n)$ memory.

The following trade-offs are obtained: $O(kn^{1/k})$ time per back-step, using k additional pointers, or $O(k)$ time per back-step, using $O(kn^{1/k})$ additional pointers; in both cases supporting $O(1)$ time per forward step (independent of k). Our results extend to general directed graphs, with additional memory of $\lg d_v$ bits for each node v along the backtrack path, where d_v is the outdegree of node v.

The crux of the list traversal algorithm is an efficient pebbling technique which moves pebbles on a virtual binary or k-ary trees that can only be traversed in a pre-order fashion. We introduce the *virtual pre-order tree data structure* which enables managing the pebbles positions in a concise and simple manner.

1.2 Applications

Consider a program P running running in time T. Then, using our list pebbling algorithm, the program can be extended to a program P' that supports rollback steps, where a rollback after step i means that the program returns to the configuration it had after step $i - 1$. Arbitrary rollback steps can be added to the execution of the program P' at a cost of increasing the memory requirement by a factor of $O(\lg T)$, and having the i'th rollback step supported in $O(\lg i)$ time.

The overhead for the forward execution of the program can be kept an arbitrary small constant.

Allowing effective rollback steps may have interesting applications. For instance, a desired functionality for debuggers is to allow pause and rollback during execution. Another implication is the ability to take programs that simulate processes and allow running them backward in arbitrary positions. Thus a program can be run with ϵ overhead and allow pausing at arbitrary points, and running backward an arbitrary number of steps with logarithmic time overhead. The memory required is keeping state configuration of $\lg T$ points, and additional $O(\lg T)$ memory. Often, debuggers and related applications avoid keeping full program states by keeping only differences between the program states. If this is allowed, then a more appropriate representation of the program would be a linked list in which every node represents a sequence of program states, such that the accumulated size of the differences is in the order of a single program state.

Our pebbling technique can be used to support backward computation of a hash-chain in time $O(kn^{1/k})$ using k hash values, or in time $O(k)$ using $O(kn^{1/k})$ hash values. A *hash-chain* is obtained by repeatedly applying a one way-hash function, starting with a secret seed. There are many cryptographic applications, including micro-payment, authentication, and session-key maintenance, which are based on rolling back a hash-chain. Our results enable effective implementation with arbitrary memory size.

The list pebbling algorithm extends to directed trees and general directed graphs. Applications include the effective implementation of the parent function ("..") for XML trees, and effective graph traversals with applications to "light-weight" Web crawling and garbage collection.

1.3 Related Work

The Schorr-Waite algorithm [9] has numerous applications; see e.g., [10,11,3]. It would be interesting to explore to what extent these applications could benefit from the non-intrusive nature of our algorithm. There is an extensive literature on graph traversal with bounded memory but for other problems than the one addressed in this paper; see, e.g., [5,2]. Pebbling models were extensively used for bounded space upper and lower bounds. See e.g., the seminal paper by Pippenger [8] and more recent papers such as [2].

The closest work to ours is the recent paper by Ben-Amram and Petersen [1]. They present a clever algorithm that, using memory of size $k \leq \lg n$, supports back-step in $O(kn^{1/k})$ time. However, in their algorithm forward steps take $O(k)$ time. Thus, their algorithm supports $O(\lg n)$ time per back-step, using $O(\lg n)$ memory but with $O(\lg n)$ time per forward step, which is unsatisfactory in our context. Ben-Amram and Petersen also prove a near-matching lower bound, implying that to support back traversal in $O(n^{1/k})$ time per back-step it is required to have $\Omega(k)$ pebbles. Our algorithm supports similar trade-off for back-steps as the Ben-Amram Petersen algorithm, while supporting simultaneously constant time per forward step. In addition, our algorithm extends to support $O(k)$ time per back-step, using memory of size $O(kn^{1/k})$.

Recently, and independently to our work, Jakobsson and Coppersmith [6,4] propose a so-called fractal-hashing technique that enables backtracking hash-chains in $O(\lg n)$ amortized time using $O(\lg n)$ memory. Thus, by keeping $O(\lg n)$ hash values along the hash-chain, their algorithms enables, starting at the end of the chain, to get repeatedly the preceding hash value in $O(\lg n)$ amortized time. Note that our pebbling algorithm enables a full memory-time trade-off for hash-chain execution, and can guarantee that the time per execution is bounded in the worst case.

The most challenging aspect of our algorithm is the proper management of the pointers positions under the restriction that forward steps have very little effect on their movement, to achieve ϵ-overhead per forward step. This is obtained by using the virtual pre-order tree data structure in conjunction with a so-called recycling-bin data structure and other techniques, to manage the positions of the back-pointers in a concise and simple manner.

Due to space limitations, many details are omitted from this extended abstract, and are given in the full paper [7].

2   The Virtual Pre-order Tree Data Structure

In this section we illustrate the basic idea of the list pebbling algorithm, and demonstrate it through a limited functionality of having a sequence of back-steps only. A full algorithm must support an arbitrary sequence of forward and backward steps, and we will also be interested in refinements, such as reducing to minimum the number of pebbles. Adapting the skeleton data structures to support the full algorithm and its refinements may be quite complicated, since controlling and handling the positions of the various pointers becomes a challenge. For the further restriction that forward steps do not incur more than constant overhead (independent of k), the problem becomes even more difficult and we are not aware of any previously known technique to handle this.

To have control over the pointers positioning, we present in Section 2.1 the *virtual pre-order tree* data structure, and show how it supports the sequence of back-steps similarly to the skeleton data structure. In the next sections, we will see how the virtual pre-order tree data structure is used to support the full algorithm as well as more advanced algorithms.

2.1   The Virtual Pre-order Tree Data Structure

The reader is reminded that in a pre-order traversal, the successor of an internal node in the tree is always its left child; the successor of a leaf that is a left child is its right sibling; and the successor of a leaf that is a right child is defined as the right sibling of the nearest ancestor that is a left child. An alternative description is as follows: consider the largest sub-tree of which this leaf is the right-most leaf, and let u be the root of that sub-tree. Then the successor is the right-sibling of u. Consequently, the backward traversal on the tree will be defined as follows. The successor of a node that is a left child is its parent. The successor of a node v that is a right child is the rightmost leaf of the left sub-tree of v's parent.

The *virtual pre-order tree* data structure consists of (1) an implicit binary tree, whose nodes correspond to the nodes of the linked list, in a pre-order fashion, and (2) an explicit sub-tree of the implicit tree, whose nodes are pebbled. For the basic algorithm, the pebbled sub-tree consists of the path from the root to the current position.

Each pebble represents a pointer; i.e., pebbled nodes can be accessed in constant time. We defer to later sections the issues of how to maintain the pebbles, and how to navigate within the implicit tree, without actually keeping it.

3   The List Pebbling Algorithm

In this section we describe the list pebbling algorithm, which supports an arbitrary sequence of forward and back steps. Each forward step takes $O(1)$ time, where each back-step takes $O(\lg n)$ amortized time, using $O(\lg n)$ pebbles. We will first present the basic algorithm which uses $O(\lg^2 n)$ pebbles, then describe the pebbling algorithm which uses $O(\lg n)$ pebbles without considerations such as pebble maintenance, and finally describe a full implementation using a so-called recycling bin data structure.

The list pebbling algorithm uses a new set of pebbles, denoted as *green pebbles*. The pebbles used as described in Section 2 are now called the *blue pebbles*. The purpose of the green pebbles is to be kept as placeholders behind the blue pebbles, as those are moved to new nodes in forward traversal. Thus, getting back into a position for which a green pebble is still in place takes $O(1)$ time.

3.1   The Basic List Pebbling Algorithm

Define a *left-subpath (right-subpath)* as a path consisting of nodes that are all left children (right children). Consider the (blue-pebbled) path p from the root to node i. We say that v is a left-child of p if it has a right sibling that is in p

(that is, v is not in p, it is a left child, and its parent is in p but not the node i). As we move forward, green pebbles are placed on right-subpaths that begin at left children of p. Since p consists of at most $\lg n$ nodes, the number of green pebbles is at most $\lg^2 n$.

When moving backward, green pebbles will become blue, and as a result, their left subpaths will not be pebbled. Re-pebbling these sub-paths will be done when needed. When moving forward, if current position is an internal node, then p is extended with a new node, and a new blue pebble is created. No change occurs with the green pebbles. If the current position is a leaf, then the pebbles at the entire right-subpath ending with that leaf are converted from blue to green. Consequently, all the green sub-paths that are connected to this right-subpath are un-pebbled. That is, their pebbles are released and can be used for new blue pebbles.

We consider three types of back-steps:

(i) *Current position is a left child:* The predecessor is the parent, which is on p, and hence pebbled. Moving takes $O(1)$ time; current position is to be un-pebbled.

(ii) *Current position is a right child, and a green sub-path is connected to its parent:* Move to the leaf of the green sub-path in $O(1)$ time, convert the pebbles on this sub-path to blue, and un-pebble the current position.

(iii) *Current position is a right child, and its parent's sub-path is not pebbled:* Reconstruct the green pebbles on the right sub-path connected to its parent v, and act as in the second case. This reconstruction is obtained by executing forward traversal of the left sub-tree of v. We amortize this cost against the sequence of back-steps starting at the right sibling of v and ending at the current position. This sequence includes all nodes in the right sub-tree of v. Hence, each back-step is charged with one reconstruction step in this sub-tree.

Consider a back step from a node u. Since such back step can only be charged once for each complete sub-tree that u belongs to, we have:

Claim. Each back step can be charged at most $\lg n$ times.

We can conclude that the basic list pebbling algorithm supports $O(\lg n)$ amortized list-steps per back-step, one list-step per forward step, using $O(\lg^2 n)$ pebbles.

3.2   The List Pebbling Algorithm with $O(\lg n)$ Pebbles

The basic list pebbling algorithm is improved by the following modification. Let v be a left child of p and let v' be the right sibling of v. Denote v to be the *last left child* of p if the left subpath starting at v' ends at the current position; let the right subpath starting at the last left child be the *last right subpath*. Then, if v is not the last left child of p, the number of pebbled nodes in the right subpath starting at v is at all time at most the length of the left subpath in p, starting at v'. If v is the last left child of p, the entire right subpath starting at v can be pebbled. We denote the (green) right subpath starting at v as the *mirror subpath* of the (blue) left subpath starting at v'. Nodes in the mirror subpath and the corresponding left subpath are said to be *mirrored* according to their order in the subpaths. The following clearly holds:

Claim. The number of green pebbles is at most $\lg n$.

When moving forward, there are two cases:

(1) *Current position is an internal node:* as before, p is extended with a new node, and a new blue pebble is created. No change occurs with the green pebbles (note the mirror subpath begins at the last left child of p).

(2) *Current position i is a leaf that is on a right subpath starting at v (which could be i, if i is a left child):* we pebble (blue) the new position, which is the right sibling of v, and the pebbles at the entire right subpath ending at i are converted from blue to green. Consequently, (1) all the green sub-paths that are connected to the right subpath starting at v are un-pebbled; and (2) the left subpath in p which ended at v now ends at the parent of v, so the mirror (green) node to v should now be un-pebbled. The released pebbles can be reused for new blue pebbles.

Moving backward is similar to the basic algorithm. There are three types of back-steps.

(1) *Current position is a left child:* predecessor is the parent, which is on p, and hence pebbled. Moving takes $O(1)$ time; current position is to be un-pebbled. No change occurs with green pebbles, since the last right subpath is unchanged.

(2) *Current position is a right child, and the (green) subpath connected to its parent is entirely pebbled:* Move to the leaf of the green subpath in $O(1)$ time, convert the pebbles on this subpath to blue, and un-pebble the current position. Since the new blue subpath is a left subpath, it does not have a mirror green subpath. However, if the subpath begins at v, then the left subpath in p ending at v is not extended, and its mirror green right subpath should be extended as well. This extension is deferred to the time the current position will become the end of this right subpath, addressed next.

(3) *Current position is a right child, and the (green) subpath connected to its parent is only partially pebbled:* Reconstruct the green pebbles on the right subpath connected to its parent v, and act as in the second case. This reconstruction is obtained by executing forward traversal of the sub-tree T_1 starting at v, where v is the last pebbled node on the last right subpath. We amortize this cost against

the back traversal starting at the right child of the mirror node of v and ending at the current position. This sequence includes back-steps to all nodes in the left sub-tree T_2 of the mirror of v. This amortization is valid since the right child of v was un-pebbled in a forward step in which the new position was the right child of the mirror of v. Since the size of T_1 is twice the size of T_2, each back-step is charged with at most two reconstruction steps.

As in Claim 3.1, we have that each back step can be charged at most $\lg n$ times, resulting with:

Theorem 1. *The list pebbling algorithm supports full traversal in at most* $\lg n$ *amortized list-steps per back-step, one list-step per forward step, using* $2\lg n$ *pebbles.*

3.3   Full Algorithm Implementation Using the Recycling Bin Data Structure

The allocation of pebbles is handled by an auxiliary data structure, denoted as the *recycling bin data structure*, or *RB*. The RB data structure supports the following operations:

Put pebble: put a released pebble in the RB for future use; this occurs in the simple back-step, in which the current position is a left child, and therefore its predecessor is its parent. (Back-step Case 1.)

Get pebble: get a pebble from the RB; this occurs in a simple forward step, in which the successor of the node of the current position is its left child. (Forward-step Case 1.)

Put list: put a released list of pebbles - given by a pair of pointers to its head and to its tail – in the RB for future use; this occurs in the non-simple forward step, in which the pebbles placed on a full right path should be released. (Forward-step Case 2.)

Get list: get the most recent list of pebbles that was put in the RB and is still there (i.e., it was not yet requested by a get list operation); this occurs in a non-simple back-step, in which the current position is a right child, and therefore its predecessor is a rightmost leaf, and it is necessary to reconstruct the right path from the left sibling of the current position to its rightmost leaf. It is easy to verify that the list that is to be reconstructed is indeed the last list to be released and put in the RB. (Back-step Cases 2 or 3.)

The RB data structure consists of a bag of pebbles, and a set of lists consisting of pebbles and organized in a double-ended queue of lists. The bag can be implemented as, e.g., a stack. For each list, we keep a pointer to its header and a pointer to its tail, and the pairs of pointers are kept in doubly linked list, sorted by the order in which they were inserted to RB. Initially, the bag includes $2\lg n$

pebbles and the lists queue is empty. Based on the claim, the $2 \lg n$ pebbles will suffice for all operations.

In situations in which we have a get pebble operation and an empty bag of pebbles, we take pebbles from one of the lists. For each list ℓ we keep a counter M_ℓ for the number of pebbles removed from the list.

The operations are implemented as follows:

Put pebble: Adding a pebble to the bag of pebbles (e.g., stack) is trivial; it takes $O(1)$ time.

Put list: a new list is added to the tail of the queue of lists in RB, to become the last list in the queue, and M_ℓ is set to 0. This takes $O(1)$ time.

Get pebble: If the bag of pebbles includes at least one pebble, return a pebble from the bag and remove it from there. If the bag is empty, then return and remove the last pebble from the list ℓ, which is the oldest among those having the minimum M, and increment its counter M_ℓ. This requires a priority queue according to the pairs $\langle M_\ell, R_\ell \rangle$ in lexicographic order, where R_ℓ is the rank of list ℓ in RB according to when it was put in it. We show below that such PQ can be supported in $O(1)$ time.

Get list: return the last list in the queue and remove it from RB. If pebbles were removed from this list (i.e., $M_\ell > 0$), then it should be reconstructed in $O(2^{M_\ell})$ time prior to returning it, as follows. Starting with the node v of the last pebble currently in the list, take 2^{M_ℓ} forward steps, and whenever reaching a node on the right path starting at node v place there a pebble obtained from RB using the get pebble operation. Note that this is Back-step Case 3, and according to the analysis and claim the amortized cost per back-step is $O(\lg n)$ time.

Claim. The priority queue can be implemented to support delmin operation in $O(1)$ time per retrieval.

We can conclude:

Theorem 2. *The list pebbling algorithm using the recycling bin data structure supports $O(\lg n)$ amortize time per back-step, $O(1)$ time per forward step, using $O(\lg n)$ pebbles.*

4   The Advanced List Pebbling Algorithm

The advanced list pebbling algorithm presented in this section supports back-steps in $O(\lg n)$ time per step in the worst case. Ensuring $O(\lg n)$ list steps per back-step in the worst case is obtained by processing the rebuilt of the missing green paths along the sequence of back traversal, using a new set of red pebbles. For each green path, there is one red pebble whose function is to progressively move forward from the deepest pebbled node in the path, to reach the next node to be pebbled. By synchronizing the progression of the red pebbles with the back-steps, we can guarantee that green paths will be appropriately pebbled whenever needed.

The number of pebbles used by the algorithm is bounded by $\lg n$ (rather than $O(\lg n)$). This is obtained by a careful implementation and a sequence of refinements, described in the appendix.

Theorem 3. *The list pebbling algorithm can be implemented on a RAM to support $O(\lg i)$ time in the worst-case per back-step, where i is the distance from the current position to the farthest point traversed so far. Forward steps are supported in $O(1)$ time in the worst case, $1+\epsilon$ amortized time per forward step, and no additional list-steps, using $\lg n$ pebbles. The memory used is at most $1.5(\lg n)$ words of $\lg n + O(\lg \lg n)$ bits each.*

5   Reversal of Program Execution

Uni-directional linked list can represent the execution of programs. A program state can be thought of as nodes of a list, and a program step is represented by a directed link between the nodes representing the appropriate program states. Since typically program states cannot be easily reversed, the list is in general uni-directional list.

Moving from a node in a linked list that represents a particular program back to its preceding node is equivalent to reversing the step represented by the link. Executing a back traversal on the linked list is hence equivalent to rolling back the program. Let the sequence of program states in a forward execution be $s_0, s_1, \ldots, s_T$. A *rollback* of a program of some state s_j is changing its state to the preceding state s_{j-1}. A rollback step from state s_j is said to be the i'th *rollback step* if state s_{j+i-1} is the farthest state that the program has reached so far.

We show how to efficiently support back traversal with negligible overhead to forward steps.

Theorem 4. *Let P be a program, using memory of size M and time T. Then, at the cost of recording the input used by the program, and increasing the memory by a factor of $O(\lg T)$ to $O(M \lg T)$, the program can be extended to support arbitrary rollback steps as follows. The i'th rollback step takes $O(\lg i)$ time in the worst case, while forward steps take $O(1)$ time in the worst case, and $1+\epsilon$ amortized time per step.*

The rolling method of Theorem 4 can be effectively combined with delta-encoding, which enables quick access to the last sequence of program states by encoding only the differences between them.

References

1. A. M. Ben-Amram and H. Petersen. Backing up in singly linked lists. In *Proceedings of the ACM STOC*, pages 780–786, 1999.
2. M. A. Bender, A. Fernandez, D. Ron, A. Sahai, and S. P. Vadhan. The power of a pebble: Exploring and mapping directed graphs. In *ACM Symposium on Theory of Computing*, pages 269–278, 1998.
3. Y. C. Chung, S.-M. Moon, K. Ebcioglu, and D. Sahlin. Reducing sweep time for a nearly empty heap. In *27th Annual ACM SIGPLAN-SIGACT Symposium on Principles of Programming Languages (POPL '00)*, Boston, MA, 2000. ACM Press.
4. D. Coppersmith and M. Jakobsson. Almost optimal hash sequence traversal. In *Proceedings of the Fifth Conference on Financial Cryptography (FC '02)*, 2002.
5. D. S. Hirschberg and S. S. Seiden. A bounded-space tree traversal algorithm. *Information Processing Letters*, 47(4):215–219, 1993.
6. M. Jakobsson. Fractal hash sequence representation and traversal. In *ISIT*, 2002.
7. Y. Matias and E. Porat. Efficient pebbling for list traversal synopses. Technical report, Tel Aviv University, 2002.
8. N. Pippenger. Advances in pebbling. In *In Proceedings of the International Colloquium on Automata, Languages and Programming*, pages 407–417, 1982.
9. H. Schorr and W. M. Waite. An efficient machineindependent procedure for garbage collection in various list structures. *Communications of the ACM*, 10(8):501–506, Aug. 1967.
10. J. Sobel and D. P. Friedman. Recycling continuations. In *Proceedings of the ACM SIGPLAN International Conference on Functional Programming (ICFP '98)*, volume 34(1), pages 251–260, 1999.
11. D. Walker and J. G. Morrisett. Alias types for recursive data structures. In *Types in Compilation*, pages 177–206, 2000.

Function Matching: Algorithms, Applications, and a Lower Bound

Amihood Amir[1][*], Yonatan Aumann[1], Richard Cole[2][**], Moshe Lewenstein[1], and Ely Porat[1]

[1] Bar-Ilan University
amir@cs.biu.ac.il
{aumann,moshe,porately}@cs.biu.ac.il
[2] New York University cole@cs.nyu.edu

Abstract. We introduce a new matching criterion – *function matching* – that captures several different applications. The *function matching problem* has as its input a text T of length n over alphabet Σ_T and a pattern $P = P[1]P[2] \cdots P[m]$ of length m over alphabet Σ_P. We seek all text locations i for which, for some function $f : \Sigma_P \to \Sigma_T$ (f may also depend on i), the m-length substring that starts at i is equal to $f(P[1])f(P[2]) \cdots f(P[m])$.

We give a randomized algorithm which, for any given constant k, solves the function matching problem in time $O(n \log n)$ with probability $\frac{1}{n^k}$ of declaring a false positive. We give a deterministic algorithm whose time is $O(n|\Sigma_P| \log m)$ and show that it is almost optimal in the newly formalized convolutions model. Finally, a variant of the third problem is solved by means of two-dimensional parameterized matching, for which we also give an efficient algorithm.

Keywords: Pattern matching, function matching, parameterized matching, color indexing, register allocation, protein folding.

1   Introduction

In the traditional pattern matching model, one seeks exact occurrences of a given pattern in a text, i.e. text locations where every text symbol is *equal* to its corresponding pattern symbol. In the *parameterized matching* problem, introduced by Baker [7], one seeks text locations where there exists a bijection f on the alphabet for which every text symbol is equal to *the image under f of* the corresponding pattern symbol. In the applications we will describe below, f cannot be a bijection. Rather, it should be simply a function. More precisely, P matches T at location i if for every element $a \in \Sigma$, *all* occurrences of a have the same corresponding symbol in T. In other words, unlike in parameterized

---

[*] Partially supported by a FIRST grant of the Israel Academy of Sciences and Humanities, and NSF grant CCR-01-04494.
[**] Partially supported by NSF grant CCR-01-05678.

J.C.M. Baeten et al. (Eds.): ICALP 2003, LNCS 2719, pp. 929–942, 2003.

matching, there may be a *several* different symbols in the pattern which are mapped to the same text symbol.

Consider the following problems where parameterized matching is insufficient and function matching is required.

Programming Languages: There is a growing class of real-time systems applications where software codes are embedded on small chips with limited memory, e.g. chips in appliances. In these applications it is important to have as small a number of memory variables as possible. A similar problem exists in compiler design, where it is desirable to minimize the register-memory traffic, and re-use global registers as much as possible. This need to compact code by global register allocation and spill code minimization is an active research topic in the programming languages community (see e.g. [13,12]).

Automatically identifying functionally equivalent pieces of such compact code would make it easier to reuse these pieces (and, for example, to replace multiple such pieces by one piece in embedded code). Baker's parameterized matching was a first step in this direction. It identified codes that are identical, up to a one-to-one mapping of the variable names. This paper considers a generalization that identifies codes in which the mapping of the variable names (or registers) is possibly a many-to-one mapping. This identifies a larger set of candidate code portions which might be functionally equivalent (the equivalence would depend on the interleaving of and updates to the variables and so would require further postprocessing for confirmation).

Computational Biology: The Grand Challenge *protein folding* problem is one of the most important problems in computational biology (see e.g. [14]). The goal is to determine a protein's tertiary structure (how it folds) from the linear arrangement of its peptide sequence.

This is an area of extremely active research and a myriad of methods have and are being considered in attempts to solve this problem. One possible technique that is being investigated is *threading* (e.g. [8,17]). The idea is to try to "thread" a given protein sequence into a known structure. A starting point is to consider peptide subsequences that are known to fold in a particular way. These subsequences can be used as patterns. Given a new sequence, with unknown tertiary structure, one can seek known patterns in its peptide sequence, and use the folding of the known subsequences as a starting point in determining the full structure. However, a subsequence of different peptides that bond in the same way as the pattern peptides may still fold in a similar way. Such functionally equivalent subsequences will not be detected by exact matching. Function matching can serve as a filter, that identifies a superset of possible choices whose bondings can then be more carefully examined.

Image Processing: One of the interesting problems in web searching is searching for color images (e.g. [16,6,3]). The simplest possible cases is searching for an icon in a screen, a task that the Human-Computer Interaction Lab at the University of Maryland was confronted with. If the colors are fixed, this is exact two-dimensional pattern matching [2]. However, if the color maps in pattern and text differ, the exact matching algorithm would not find the pattern. Parameterized two dimensional search is precisely what is needed. If, in addition, we are faced with a loss of resolution in the text, e.g. due to truncation, then we would need to use a two dimensional function matching search.

The above examples are a sample of diverse application areas encountering search problems that are not solved by state of the art methods in pattern matching. This need led us to introduce, in this paper, the *function matching* criterion, and to explore the *two dimensional parameterized matching* problem.

Function matching is a natural generalization of parameterized matching. However, relaxing the bijection restriction introduces non-trivial technical difficulties. Many powerful pattern matching techniques such as automata methods, subword trees, dueling and deterministic sampling assume transitivity of the matching relation (see [10] for techniques). For any pattern matching criteria where transitivity does not exist, the above methods do not help.

Examples of pattern matching with non-transitive matching relation are string matching with "don't cares", less-than matching, pattern matching with mismatches and swapped matching. It is interesting to note that the efficient algorithms for solving the above problems all used convolutions as their main tool. Convolutions were introduced by Fischer and Paterson [11] as a technique for solving pattern matching problems with wildcards, where indeed the match relation is not transitive.

It turns out that many such problems can be solved by a "standard" application of convolutions (e.g. matching with "don't cares", matching with mismatches in bounded finite alphabets, and swapped matching). Muthukrishnan and Palem were the first to explicitly identify this application method and introduced a *boolean convolutions model* [15] with locality restrictions and obtained several lower bounds in this model. Since the introduction of the boolean convolutions model, several papers appeared using general, rather than boolean convolutions. In this paper we provide a formal definition for a more general convolutions model that broadens the class of problems being considered.

The new convolutions model encapsulates the solution to many non-standard matching problems. Even more importantly, a rigorous formal definition of such a model is useful in proving lower bounds. While such bounds do not lower bound the solution complexity in a general RAM, they do help in understanding the limits of the convolution method, hitherto the only powerful tool for non-standard pattern matching.

There are three main contributions in this paper.

1. A solution to a number of search problems in diverse fields, achieved by the introduction of a new type of generalized pattern matching, that of function matching.

2. A formalization of a new general convolutions model. This leads to a deterministic solution. We prove that this solution is almost *tight* in the convolutions model. We also present an efficient randomized solution of the function matching problem.

3. Solutions to the problem of exact search in color images with different color maps. This is done via efficient randomized and deterministic algorithms for two-dimensional parameterized and function matching.

In section 2 we give the basic definitions and present progressively more efficient deterministic solutions, culminating in a $O(n|\Sigma_P| \log m)$ algorithm, where $|\Sigma_P|$ is the pattern alphabet size. We also present a Monte Carlo algorithm that solves the function matching problem in time $O(n \log m)$ time with failure probability no larger than $\frac{1}{n^k}$, where k is a given constant. In section 3 we formalize the new convolution model. We then show a lower bound proving that our deterministic algorithm is tight in the convolutions model and discuss the limitations of that model. Finally, in section 4 we present a randomized algorithm that solves the two-dimensional parameterized matching problem in time $O(n^2 \log n)$ with probability of false positives no larger than $\frac{1}{n^k}$, for given constant k. We also present a deterministic algorithm that solves the two-dimensional parameterized matching problem in time $O(n^2 \log^2 m)$.

2   Algorithms

The key notion is that of a *cover*.

Definition: Let U and V be equal length strings. Symbol τ in U is said to *cover* symbol σ in V if every occurrence of σ in V is aligned with an occurrence of τ in U (i.e. they occur in equal index locations). U is said to cover σ in V if there is some symbol τ in U covering σ. Finally, the cover is said to be an *exact* cover if every occurrence of τ in U is aligned with an occurrence of σ in V.

Definition: There is a *function match* of V with U if every symbol occurring in V is covered by U (but this relation need not be symmetric). If each of the covers is an exact cover the match is a *parameterized match* (and this relation is symmetric).

The term function match arises by considering the mapping from V's symbols to U's symbols specified by the match; it is a plain function in a function match and it is one-to-one in a parameterized match. In both cases the function is onto.

Definition: Given a text T (of length n) and a pattern P (of length m) the function matching problem is to find the alignments (positionings) of P such that P function matches the aligned portion of T. Note that every match may use a different function to associate the symbols of P with those in the aligned portion of T.

As is standard, we can limit T to have length at most $2m$, by breaking T into pieces of length $2m$, successive pieces overlapping by $m - 1$ symbols.

It is straightforward to give an $O(nm)$ time algorithm for function matching; it simply checks each possible alignment of the pattern in turn, each in time $O(m)$. This is left to the reader.

We start by outlining a simple $O(n|\Sigma_P||\Sigma_T| \log m)$ time algorithm, where Σ_P and Σ_T are the pattern and text alphabets, respectively. This algorithm finds, for each pair $\sigma \in \Sigma_P$ and $\tau \in \Sigma_T$, those alignments of the pattern with the text for which τ covers σ. This will take time $O(n \log m)$ for one pair. A function matching exists for an alignment exactly if every symbol occurring in P is covered.

Definition: The σ-*indicator* of string U, $\chi_\sigma(U)$ is a binary string of length U in which each occurrence of σ is replaced by a 1, and every other symbol occurrence is replaced by 0.

The procedure used the strings $\chi_\sigma(P)$ and $\chi_\sigma(T)$. For each alignment of $\chi_\sigma(P)$ with $\chi_\sigma(T)$ it computes the dot product of $\chi_\sigma(P)$ and the aligned portion of $\chi_\sigma(T)$. But the product is exactly the number of occurrences of σ in P aligned with occurrences of τ in T. This is a cover of σ by τ exactly if the dot product equals the number of occurrences of σ in P. The dot products, for each alignment of $\chi_\sigma(P)$ with $\chi_\sigma(T)$, are all readily computed in $O(n \log m)$ time by means of a convolution [11]. We have shown:

Theorem 1. *Function matching can be solved deterministically in time* $O(n|\Sigma_P||\Sigma_T| \log m)$.

We obtain a faster algorithm by determining simultaneously for all τ occurring in T and for one σ occurring in P, those alignments of P for which some τ covers σ. This is done in time $O(n \log m)$ and is repeated for each σ yielding an algorithm with running time $O(n|\Sigma_P| \log m)$. Our algorithm exploits the following observation.

Lemma 1. *Let* $a_1, ..., a_k$ *be natural numbers. Then* $k \sum_{h=1}^k (a_h)^2 = (\sum_{h=1}^k a_h)^2$ *iff* $a_i = a_j$, *for* $1 \le i < j \le k$.

The algorithm uses the strings T and T_2, where T_2 is defined by $T_2[i] = (T[i])^2$, $i = 0, ..., n - 1$. For each σ, for each alignment of P with each of T and T_2, the dot product of $\chi_\sigma(P)$ with the aligned portion of each of T and T_2 is computed. By Lemma 1 T covers σ in a given alignment exactly if the dot product of P with the aligned portion of T_2 is k times larger than the dot product of P with T, where k is the number of occurrences of σ in P. This yields:

Theorem 2. *The function matching problem can be solved deterministically in time* $O(n|\Sigma_P| \log m)$.

We seek further speedups via randomization. We give a Monte Carlo algorithm that, given a constant k, reports all function matches and with probability at most $\frac{1}{n^k}$ reports a non-match as a match. Our first step is to reduce function matching to *paired function matching*. In paired function matching the pattern is a *paired string*, a string in which each symbol appears at most twice. We then give a randomized algorithm for paired function matching.

For the reduction we create a new text T', whose length is $2n$, and a new pattern P', whose length is $2m$. There will be a match of P with T starting at location i in T exactly if there is a match of P' starting at location $2i - 1$ in T'. T' is obtained by replacing each symbol in T by two consecutive instances of the same symbol; e.g. if $T = abca$ then $T' = aabbccaa$. To define P', a little notation is helpful. Suppose symbol σ appears k times in P. Then new symbols $\sigma_1, \sigma_2, ..., \sigma_{k+1}$ are used in P'. The ith occurrence of σ is replaced by the pair of symbols σ_i, σ_{i+1}. e.g. if $P = aababca$ then $P' = a_1 a_2 a_2 a_3 b_1 b_2 a_3 a_4 b_2 b_3 c_1 c_2 a_4 a_5$. It is easy to see that function matches of P' in T' and of P in T correspond as described above. Thus it remains to give the algorithm for paired function matching.

This algorithm replaces the symbols of P' and T' by integers, chosen uniformly at random from the range $[1, 2n^{k+1}]$ as follows. For the text T', for each symbol σ, a random value v_σ is chosen, and each occurrence of σ is replaced by v_σ, forming a string T''. For the pattern P', for each symbol σ, occurring twice, a random value u_σ is chosen. The first occurrence of σ is replaced by u_σ and the second occurrence by $-u_\sigma$; if a symbol occurs once it is replaced by value 0. This forms string P''. Now, for each possible alignment of P'' with T'', the dot product of P'' with the aligned portion of T'' is computed. Clearly, if there is a function match of P with T, the corresponding dot product evaluates to 0. We show that when there is a function mismatch, the corresponding dot product is non-zero with high probability.

If there is a function mismatch then there is a symbol σ in P'' aligned with distinct symbols τ and ρ in T''. Imagine that the assignment of random values assigns values v_τ, v_ρ, u_σ last. Consider the dot product expressed as a function of v_τ, v_ρ, u_σ; it has the form $A + Bv_\tau + Cv_\rho + u_\sigma(v_\tau - v_\rho)$ (assuming the τ and ρ aligned with σ appear in left to right order), where $A, B,$ and C are the values obtained after making all the other random choices. It is easy to see that there is at most a $\frac{2}{2n^{k+1}} = \frac{1}{n^{k+1}}$ probability of this polynomial evaluating to 0. As there are $n - m + 1$ possible alignments of P with T, the overall failure probability is at most $\frac{1}{n^k}$. We have shown:

Theorem 3. *There is a randomized algorithm for function matching that, given a constant k, runs in time $O(kn \log m)$; it reports all function matches and, with probability at least $1 - \frac{1}{n^k}$ reports no mismatches as matches.*

3   Lower Bounds

The unfettered nature of the function matching problem is what makes it difficult. Traditional pattern matching methods such as automata, duels or witnesses, apparently are of no help since there is no transitivity in the function matching relation. Moreover, it is far from evident whether one can set rules during a pattern preprocessing stage that will allow text scanning, since the relationship between the text and pattern is quite loose. This is what pushed us to consider convolutions as the method for the upper bound. Unfortunately, our deterministic algorithm's complexity is no better than that of the naive for alphabets of unbounded size.

Whenever resorting to a randomized algorithm, it behooves the algorithm's developer to explain why they randomized. In this section we give evidence for the belief that an efficient deterministic solution to the problem, if such exists, may be very difficult. We do it by showing a lower bound of $\Omega(\frac{m}{b})$ convolutions with b-bit inputs and outputs for the function matching problem in the convolutions model. Convolutions, as a tool for string matching, were introduced by Fischer and Paterson [11]. Muthukrishnan and Palem [15] considered a Boolean convolutions model with locality restrictions for which they obtained a number of lower bounds. We did not find a formal definition of general convolutions as a resource in the literature. Recent uses of convolutions with non-Boolean inputs led us to broaden the class of convolutions being considered for lower bound proofs. In fact, Muthukrishnan and Palem proved a lower bound of $\Omega(\log \Sigma)$ boolean convolutions for string matching with wildcards with alphabet Σ; but their lower bound does not hold for more general convolutions as indicated by

Cole and Hariharan's recent two convolution algorithm [9]. Our model does not cover all conceivable convolutions-based methods. However, it broadens the class for which lower bounds can be proven. The next subsection formally defines the general convolutions model that we propose.

3.1   The Convolutions Model

We begin by defining the class of problems that are solved by the convolutions model.

Definition: A *pattern matching problem* is defined as follows:

MATCH RELATION: A binary relation $M(a, b))$, where $a = a_0...a_k, b = b_0...b_\ell$ and $a, b \in \Sigma^*$.

INPUT: A *text array*, $T = T[0], ..., T[n-1]$, and a *pattern array* $P = P[0], ..., P[m-1]$, $P[i], T[j] \in \Sigma$, $i = 0, ..., m-1$, $j = 0, ..., n-1$.

OUTPUT: The set of indices $S \subseteq \{0, ..., n-1\}$ where the pattern P *matches*, i.e. all indices i where $M(P, T_i)$, and T_i is the suffix of T starting at location i. We also call the output set of indices the *target elements*.

Example: *String Matching with Don't Cares*

The match relation is defined as follows. Let $\Sigma = \{0, 1\}$. Let $\phi \notin \Sigma$ (ϕ is the *don't care symbol*). Let $|a| = k$ and $|b| = \ell$. If $k > \ell$ then there is no match. Otherwise, a matches b iff $a_i = b_i$ or $a_i = \phi$ or $b_i = \phi$, $i = 0, ..., k-1$. The text and pattern arrays are $T = T[0], ..., T[n-1]$ and $P[0], ..., P[m-1]$, respectively. The target elements are all locations i in the text array T where there is an exact occurrence of P (where ϕ matches both 0 and 1).

As its name suggests, the convolutions model uses convolutions as basic operations on arrays. Another basic operation it uses is *preprocessing*. There is a difference, however, between pattern and text preprocessing. We place no restriction on the pattern preprocessing. The text preprocessing, however, must be *local*. When proving lower bounds in the convolutions model, we are mainly interested in the *number* of convolutions necessary to achieve the solution, rather than the time complexity of the solution (this is akin to counting the number of comparisons in the comparison model for sorting).

Definition: Let g be a pattern preprocessing function. A *g-local text preprocessing function* $f_g : N^n \to N^n$ is a function for which there exists n functions $f_g^j : N \to N$, such that $(f_g(T))[j] = f_g^j(T[j])$, $j = 0, ..., n-1$.

In words, the "locality" of function f_g is manifested by the fact that the value in index j of $f_g(T)$ is computed based solely on the pattern preprocessing (output $g(P)$), the index j, and the value of $T[j]$.

Examples:

1. Let T be an array. Then $\chi_a(T)$ is clearly a local array function, since the only index of T that participates in computing $\chi_a(T)[j]$ is j.

2. Let T be an array of numbers. The function f such that $f(T)[j] = T[j] - (\sum_{i=0}^{n-1} T[i])/n$ is not a local array function.

We now have all the building blocks of the convolutions model.

Definition: The *convolutions computation model* is a specialized model of computation that solves a subset of the pattern matching problems.

Given a pattern matching problem whose input is text T and pattern P, a solution in the convolutions model has the following form. Let g_i, $i = 1, ..., h(n)$ be pattern preprocessing functions, and let f_{g_i}, $i = 1, ..., h(n)$ be the corresponding local text preprocessing functions. The model also uses a parameter b.

1. Compute $h(n)$ convolutions $C_i \leftarrow f_{g_i}(T) \otimes g_i(P)$, $i = 1, ..., h(n)$, with b-bit inputs and outputs.
2. Compute the matches as follows. The decision of whether location j of the text is a match is decided by a computation whose inputs are a subset of $\{C_i[j] \mid i = 1, ..., h(n)\}$.

Examples:

1. *Exact String Matching with Don't Cares*
   This problem's solution was provided by Fischer and Paterson [11] is in the convolutions model. The two convolutions are:
   $C_1 \leftarrow \chi_0(T) \otimes \chi_1(P)$
   $C_2 \leftarrow \chi_1(T) \otimes \chi_0(P)$
   The text locations i for which $C_1[i] = C_2[i] = 0$ are precisely the match locations.
2. *Approximate Hamming Distance over a fixed bounded Alphabet*
   This problem was considered for unbounded alphabets in [1]. For bounded alphabets, the problem is defined in the convolutions model as follows. The matching relation $M_e(a, b)$ is all pairs of substrings over alphabet $\Sigma = \{1, ..., k\}$ for which $|a| \leq |b|$ and the number of *mismatches* between a and b (i.e. the indices j for which $a_j \neq b_j$) is no greater than e.
   the solution in the convolutions model is as follows. Compute the convolutions:
   $C_i \leftarrow \chi_i(T) \otimes \chi_{\bar{i}}(P)$, $i = 1, \cdots, k$, where

   $$\chi_{\bar{a}}(x) = \begin{cases} 1 \text{ if } x \neq a \\ 0 \text{ if } x = a \end{cases}$$

   The match locations are all indices j where $\sum_{i=1}^{k} C_i[j] \leq e$.
3. *Less-than Matching over a fixed bounded Alphabet*
   This problem was considered for unbounded alphabets in [4]. For bounded alphabets, the problem is defined in the convolutions model as follows. The matching relation $M(a, b)$ is all pairs of substrings over alphabet $\Sigma = \{1, ..., k\}$ for which $|a| \leq |b|$ and $a_j \leq b_j$ $\forall j = 0, ..., |a| - 1$.
   the solution in the convolutions model is as follows. Compute the convolutions:
   $C_i \leftarrow \chi_i(T) \otimes \chi_{<i}(P)$, i=1,..., k, where

   $$\chi_{<i}(x) = \begin{cases} 1 \text{ if } x < i \\ 0 \text{ if } x \geq i \end{cases}$$

   The match locations are all indices j where $\sum_{i=1}^{k} C_i[j] \neq 0$.
4. Our solutions in Section 2 are also algorithms in the convolutions model.

3.2   Lower Bounds

The solutions we presented in section 2 for the function matching problem were also in the convolutions model. The following theorem shows that our algorithm's complexity is almost tight in the convolutions model.

Theorem 4. *The function matching problem requires $\Omega(\frac{m}{b})$ convolutions in the convolutions model.*

Proof: We will show that the *word equality problem* can be linearly reduced to the function matching problem. The word equality problem is:

INPUT: Two m bit words, $W_1 = W_1[0], ..., W_1[m-1]$ and $W_2 = W_2[0], ..., W_2[m-1]$.

DECIDE: Whether $W_1 = W_2$ (i.e. $W_1[i] = W_2[i]$, $i = 0, ..., m-1$) or not.

The following communication complexity lower bound for the word equality problem is known. Suppose processor PA starts with word W_1 and processor PB with word W_2. Then to decide word equality they need to exchange $\Omega(m)$ bits [18].

We show that any algorithm for function matching in the convolutions model using $h(m)$ b-bit convolutions can be used to solve the word equality problem with a transmission of $b \cdot h(m)$ bits, implying $h(m) = \Omega(\frac{m}{b})$.

We consider the operation of the function matching algorithm on the following pattern and text. $T = W_1 W_2$, the concatenation of W_1 and W_2, and $P = 1, 2, \cdots, m, 1, 2, \cdots, m$. Note that T function matches P if and only if $W_1 = W_2$.

Now suppose that function matching is solved by some algorithm F in the convolutions model. F computes $h(m)$ convolutions $C_1, ..., C_{h(m)}$ and then uses the results of $C_i[1]$, $i = 1, ..., h(m)$ to decide whether there is a function-match. Note that for every convolution $C = A \otimes B$, $C[1] = \sum_{h=0}^{2m-1} A[h]B[h]$. However, this is equal to $\sum_{h=0}^{m-1} A[h]B[h] + \sum_{h=0}^{m-1} A[h+m]B[h+m]$. For each convolution, PA will compute $\sum_{h=0}^{m-1} A[h]B[h]$, which is based solely on $T[1], \cdots, T[m]$ and P, and PB will compute $\sum_{h=0}^{m-1} A[h+m]B[h+m]$, which is based solely on $T[m+1], \cdots, T[2m]$ and P. PA will then transmit its b-bit result to PB for each of the $h(m)$ convolutions used by F, and PB can at this point determine the result of the word equality problem. $\square$

It is important to be careful in interpreting the results of the convolutions model in a RAM complexity model since complexity in the convolutions model is measured by number of convolutions, rather than RAM operations. When evaluating the number of operations it takes to compute a convolution one must consider the number of bits in a RAM word. The standard in the pattern matching literature is an $O(\log m)$ bit word and the currently fastest known algorithm for computing convolutions is by using the FFT. Its time complexity is $O(n \log m)$ word operations. Thus, the number of RAM operations required to compute function matching in the convolutions model would appear to be $\Omega(nm)$. Of course, conceivably, by ingenious use of special case convolutions one might be able to evaluate the convolutions more quickly, though no such approach has occurred to us.

4   Two Dimensional Algorithms

The one dimensional parameterized matching problem was efficiently solved in [5]. However, as discussed in [3], the move to two dimensions implies a possible computational difficulty if no separable attributes exist. Parameterized matching is not separable – if all columns (or rows) have parameterized matches, it does not necessarily imply that the entire matrix has a parameterized match. Thus we are forced to seek other approaches.

Our Problem:

INPUT: Two dimensional text T of size $n \times n$, and two dimensional pattern P of size $m \times m$.

OUTPUT: All locations $[i, j]$ in T where there is a parameterized (function) occurrence of the pattern.

First we show how to reduce two-dimensional function matching to the one-dimensional case, yielding an $O(n^2 \log n)$ work randomized algorithm. We then show how, with an additional $O(n^2 \log n)$ work, to solve two-dimensional parameterized matching, again with a randomized algorithm. Finally, we give a deterministic algorithm for parameterized matching, which takes $O(n^2 \log^2 m)$ time.

The two-dimensional text T is written in row major order to give a one-dimensional text T'. The pattern is padded with wildcards (or equivalently, new characters, each appearing once) to produce m rows of length n; the padded pattern is written in row major order to give a one-dimensional pattern P'. Clearly, there is a match of P at location (i, j) in T exactly if there is a match of P' at location $n(i - 1) + j$ in T'. We have shown:

Corollary 1. *There is a randomized algorithm for two-dimensional function matching, which, when given a constant k, runs in time $O(kn^2 \log n)$, reports all function matches, and with probability at most $\frac{1}{n^k}$ falsely reports a mismatch as a match.*

In a parameterized match, the number of distinct symbols in the aligned portion of the text must equal the number of distinct symbols in the pattern. Amir, Church and Dar [3] gave an $O(n^2 \log n)$ time algorithm for this problem, the character count problem: determine, for each $m \times m$ subarray of an $n \times n$ array, the number of distinct characters appearing in the subarray. So we have shown:

Corollary 2. *There is a randomized algorithm for two-dimensional parameterized matching, which, when given a constant k, runs in time $O(kn^2 \log n)$, reports all parameterized matches, and with probability at most $\frac{1}{n^k}$ falsely reports a mismatch as a match.*

Next, we give an efficient deterministic algorithm for two-dimensional parameterized matching (time $O(n^2 \log^2 m)$). It uses one convolution on a size $O(n^2 \log m)$ vector, which can also be viewed as $O(\log m)$ convolutions on size n^2 vectors. As in one-dimension, this is considerably more efficient than what is known for function matching. Incidentally, we note that this convolution is outside the convolutions model.

It is helpful to recall the one-dimensional parameterized matching algorithm, due to Amir, Farach, and Muthukrishnan [7]. It is similar to the Knuth-Morris-Pratt string matching algorithm. The key idea is to recode the occurrences of each symbol in terms of their separation; namely, if symbol a occurs in the pattern (or text) at locations with indices $i_1 \leq i_2 \leq \ldots \leq i_k$, these occurrences of a are replaced by the numbers $0, i_2 - i_1, i_3 - i_2, \ldots, i_k - i_{k-1}$, respectively. For each symbol occurrence, this is simply the distance to the nearest occurrence of the same symbol to the left, if any. Except for the first occurrence of each symbol, a parameterized match in the original text and pattern corresponds to a standard match in the recoded text and pattern.

One perspective on this is that all occurrences of the same symbol in the pattern (and the text) have been connected into a structure; identifying a match becomes a question of finding the alignments for which the structures in the pattern and text match. We will seek a similar construction for the two-dimensional problem. However, this will now require creating connections to $O(\log n)$ neighbors of each symbol occurrence.

Our solution has the following form. For each occurrence I of a symbol in the pattern (resp. in the text) the relative locations of some $8 \log n$ instances $J_1, J_2, \ldots$ of the same symbol (resp. in the pattern and text) are recorded. We say that J_ℓ is a neighbor of I, for $\ell = 1, 2, \ldots$, and also that I and J_ℓ are *linked*. If I is in position (w, y) and J_ℓ is in position (x, z), their relative position is recorded as $(x - w, z - y)$. Each potential neighbor is selected according to a specific rule, which may or may not identify a neighbor (e.g., a rule could be: the next occurrence of the symbol to the right in the same row). The rules for pattern and text will be slightly different.

The collections of selected symbols have the following properties:

(i) For each alignment of the pattern with the text, for each symbol a, if the occurrences of a in the pattern and text match, then for each occurrence I_p of a in the pattern and the aligned occurrence I_t in the text, the neighbor of I_p selected by the ith rule for the pattern is aligned with the neighbor of I_t selected by the ith rule for the text (the converse need not be true, however).

(ii) All occurrences of a in the pattern are linked, for each symbol a.

To see why the converse need not hold in Property (i) consider the rule "the next instance of this symbol to the right in the same row"; since the text may extend further to the right than the pattern, for a given alignment this rule could yield a symbol occurrence in the text and not in the pattern (the text symbol in question would be to the right of the pattern) and of course this does not preclude a match.

The text and pattern are recoded using the following encoding for each symbol occurrence. Each symbol occurrence is encoded by an equal length sequence of $O(\log n)$ relative positions, ordered as follows: the relative position of the symbol occurrence yielded by rule (1), by rule (2), by rule (3), and so on. If a rule yields no occurrence this is recoded by the "relative position" $(0,0)$.

The matching problem is made one-dimensional by writing the recoded arrays in row major order using rows of length n (the missing entries in the pattern are replaced by sequences of $O(\log n)$ pairs $(0,0)$). Treating $(0,0)$ as a wildcard, it is easy to see that a parameterized match of the original pattern with the text corresponds exactly to a standard match with wildcards of the recoded pattern with the recoded text. The recoded text has length $O(n^2 \log m)$, thus this wildcard matching can be solved in time $O(n^2 \log m \log n)$ [9] (and by standard techniques this can be reduced to $O(n^2 \log^2 m)$ time).

We turn to the selection of neighbors. For the moment we suppose the pattern dimension $m = 2^i$ for some integer i. For each location (x, y) in the pattern, we divide most of the remainder of the pattern (excluding location (x, y)) into $8 \log n - 4$ disjoint rectangles. Each rectangle provides one neighbor.

The first four rectangles comprise row x and column y partitioned at location (x, y), i.e., (i) the points (x, z) with $z > y$, (ii) the points (x, w) with $w < x$, (iii) the points (z, y) with $z > x$, and (iv) the points (w, y) with $w < x$. Next, we describe how the quadrant below and to the right of (x, y) is divided into contiguous rectangles. Each rectangle comprises a distinct selection of contiguous rows, covering all columns from $y + 1$ to m, starting at row $x + 1$. From top to bottom, the sequence of rectangles have the following number of rows: $1, 2, 4, \ldots, m/4 = 2^{i-2}, m/4, m/8, \ldots, 2, 1, 1$, with the series stopping at the last rectangle that fits inside the pattern. This may mean that a portion of the quadrant is left uncovered.

Suppose a is the symbol in location (x, y). Each rectangle is traversed in column major order to find the first occurrence of an a, if any. These are the neighbors of the a in location (x, y). Analogous partitionings and traversals in directions away from location (x, y) are used for the other quadrants.

A very similar partitioning is used on the text, except that now each rectangle extends through $n - 1$ columns or to the right boundary of the text, whichever comes sooner. (This is for the SE quadrant; other quadrants are handled similarly.)

Clearly Property (i) above holds. It remains to show Property (ii).

Lemma 2. *Let (w, y) and (x, z) be two locations in the pattern both holding symbol a. Then they are linked.*

Proof: Clearly, if $w = x$ there is a series of links along row x connecting these two locations. Similarly if $y = z$.

So WLOG suppose that $w < x$ and $y < z$. We claim that either (x, z) lies in one of the rectangles defined for location (w, y) or (w, y) lies in one of the rectangles for (x, z) (or possibly both).

Suppose that $2^{k-1} < w \leq 2^k \leq m/2$. Then for (x, z) to lie outside one of (w, y)'s rectangles, $x > n - 2^{k-1}$ (for rows $w, w + 1, [w + 2, w + 3], \ldots, [w + m/4, w + m/2 - 1], [w + m - m/2, w + m - m/4 - 1], \ldots, [w + m - 2^{k+1}, w + m - 2^k - 1]$ are all included in (w, y)'s rectangles, and $w \geq 2^{k-1} + 1$). The symmetric argument for location (x, z) shows that (w, y) lies in one of z's rectangles if $x > n - 2^{k-1}$. This argument does not cover the case $w = 1$, but then (w, y)'s rectangles cover every row, nor the case $w > m/2$, but then (x, z)'s rectangles cover row w.

WLOG suppose that (x, z) lies in one of (w, y)'s rectangles. It need not be that (x, z) is a neighbor of (w, y), however. Nonetheless, by induction on $z - y$, we show they are linked. The base case, $y = z$, has already been demonstrated. Let (u, v) denote the neighbor of (w, y) in the rectangle containing (x, z). Then

$y < v \leq z$. By induction, (u, v) and (x, z) are linked and the inductive claim follows.   □

It remains to show how to identify the neighbors. This is readily done in $O(m^2 \log^2 m)$ time in the pattern and $O(n^2 \log n \log m)$ time in the text (and using standard additional techniques, in $O(n^2 \log^2 m))$ time). We describe the approach for the pattern.

The idea is to maintain, for each symbol a, a window of 2^i rows, for $i = 1, 2, \ldots, \log m - 2$, and in turn to slide each window down the pattern. In the window the occurrences of a are kept in a balanced tree in column major order. For each occurrence of a, its neighbors in the relevant window are found by means of $O(\log m)$ time searches. Thus, over all symbols and neighbors the searches take time $O(m^2 \log^2 m)$. To slide a window one row down requires deleting some symbol instances and adding others. This takes time $O(\log m)$ per change and as each symbol instance is added once and deleted once from a window of each size this takes time $O(m^2 \log^2 m)$ over all symbols and windows. (It is helpful to have a list of each character in row major order so as to be able to quickly decide which characters to add and to delete from the sliding window, but these lists take only $O(m^2)$ time to prepare for all the symbols.)

The preprocessing of the text is similar.

To extend this algorithm to arbitrary n, we simply expand the pattern to size $2^i \times 2^i$ by padding it with wildcards. We have shown:

Theorem 5. *There is an $O(n^2 \log^2 m)$ time algorithm for two-dimensional parameterized matching.*

References

1. K. Abrahamson. Generalized string matching. *SIAM J. Comp.*, 16(6):1039–1051, 1987.
2. A. Amir, G. Benson, and M. Farach. An alphabet independent approach to two dimensional pattern matching. *SIAM J. Comp.*, 23(2):313–323, 1994.
3. A. Amir, K. W. Church, and E. Dar. Separable attributes: a technique for solving the submatrices character count problem. In *Proc. 13th ACM-SIAM Symp. on Discrete Algorithms (SODA)*, pages 400–401, 2002.
4. A. Amir and M. Farach. Efficient 2-dimensional approximate matching of half-rectangular figures. *Information and Computation*, 118(1):1–11, April 1995.
5. A. Amir, M. Farach, and S. Muthukrishnan. Alphabet dependence in parameterized matching. *Information Processing Letters*, 49:111–115, 1994.
6. G.P. Babu, B.M. Mehtre, and M.S. Kankanhalli. Color indexing for efficient image retrieval. *Multimedia Tools and Applications*, 1(4):327–348, Nov. 1995.
7. B. S. Baker. A theory of parameterized pattern matching: algorithms and applications. In *Proc. 25th Annual ACM Symposium on the Theory of Computation*, pages 71–80, 1993.
8. J. H. Bowie, R. Luthy, and D. Eisenberg. A method to identify protein sequences that fold into a known three-dimensional structure. *Science*, (253):164–176, 1991.
9. R. Cole and R. Hariharan. Verifying candidate matches in sparse and wildcard matching. In *Proc. 34st Annual Symposium on the Theory of Computing (STOC)*, pages 592–601, 2002.
10. M. Crochemore and W. Rytter. *Text Algorithms*. Oxford University Press, 1994.

11. M.J. Fischer and M.S. Paterson. String matching and other products. *Complexity of Computation, R.M. Karp (editor), SIAM-AMS Proceedings*, 7:113–125, 1974.
12. W.C. Kreahling and C. Norris. Profile assisted register allocation. In *Proc. ACM Symp. on Applied Computing (SAC)*, pages 774–781, 2000.
13. G-Y. Lueh, T. Gross, and A-R. Adl-Tabatabai. Fusion-based register allocation. *ACM Transactions on Programming Languages and Sustems (TOPLAS)*, 22(3):431–470, 2000.
14. Jr. K. Merz and S. M. La Grand. *The Protein Folding Problem and Tertiary Structure Prediction*. Birkhauser, Boston, 1994.
15. S. Muthukrishnan and K. Palem. Non-standard stringology: Algorithms and complexity. In *Proc. 26th Annual Symposium on the Theory of Computing*, pages 770–779, 1994.
16. M. Swain and D. Ballard. Color indexing. *International Journal of Computer Vision*, 7(1):11–32, 1991.
17. J. Yadgari, Amihood Amir, and Ron Unger. Genetic algorithms for protein threading. In J. Glasgow, T. Littlejohn, F. Major, R. Lathrop, D. Sankoff, and C. Sensen, editors, *Proc. 6th Int'l Conference on Intellingent Systems for Molecular Biology (ISMB 98)*, pages 193–202. AAAI, AAAI Press, 1998.
18. A. C. C. Yao. Some complexity questions related to distributed computing. In *Proc. 11th Annual Symposium on the Theory of Computing (STOC)*, pages 209–213, 1979.

Simple Linear Work Suffix Array Construction*

Juha Kärkkäinen and Peter Sanders

Max-Planck-Institut für Informatik
Stuhlsatzenhausweg 85, 66123 Saarbrücken, Germany
[juha,sanders]@mpi-sb.mpg.de.

Abstract. A suffix array represents the suffixes of a string in sorted order. Being a simpler and more compact alternative to suffix trees, it is an important tool for full text indexing and other string processing tasks. We introduce the *skew algorithm* for suffix array construction over integer alphabets that can be implemented to run in linear time using integer sorting as its only nontrivial subroutine:
1. recursively sort suffixes beginning at positions i mod $3 \neq 0$.
2. sort the remaining suffixes using the information obtained in step one.
3. merge the two sorted sequences obtained in steps one and two.
The algorithm is much simpler than previous linear time algorithms that are all based on the more complicated suffix tree data structure. Since sorting is a well studied problem, we obtain optimal algorithms for several other models of computation, e.g. external memory with parallel disks, cache oblivious, and parallel. The adaptations for BSP and EREW-PRAM are asymptotically faster than the best previously known algorithms.

1 Introduction

The suffix *tree* [39] of a string is a compact trie of all the suffixes of the string. It is a powerful data structure with numerous applications in computational biology [21] and elsewhere [20]. One of the important properties of the suffix tree is that it can be constructed in linear time in the length of the string. The classical linear time algorithms [32,36,39] require a constant alphabet size, but Farach's algorithm [11,14] works also for integer alphabets, i.e., when characters are polynomially bounded integers. There are also efficient construction algorithms for many advanced models of computation (see Table 1).

The suffix *array* [18,31] is a lexicographically sorted array of the suffixes of a string. For several applications, the suffix array is a simpler and more compact alternative to the suffix tree [2,6,18,31]. The suffix array can be constructed in linear time by a lexicographic traversal of the suffix tree, but such a construction loses some of the advantage that the suffix array has over the suffix tree. The fastest *direct* suffix array construction algorithms that do not use suffix trees require $\mathcal{O}(n \log n)$ time [5,30,31]. Also under other models of computation, direct

---

* Partially supported by the Future and Emerging Technologies programme of the EU under contract number IST-1999-14186 (ALCOM-FT).

J.C.M. Baeten et al. (Eds.): ICALP 2003, LNCS 2719, pp. 943–955, 2003.
© Springer-Verlag Berlin Heidelberg 2003

algorithms cannot match suffix tree based algorithms [9,16]. The existence of an I/O-optimal direct algorithm is mentioned as an important open problem in [9].

We introduce the *skew algorithm*, the first linear-time direct suffix array construction algorithm for integer alphabets. The skew algorithm is simpler than any suffix tree construction algorithm. (In the appendix, we give a 50 line C++ implementation.) In particular, it is much simpler than linear time suffix tree construction for integer alphabets.

Independently of and in parallel with the present work, two other direct linear time suffix array construction algorithms have been introduced by Kim et al. [28], and Ko and Aluru [29]. The two algorithms are quite different from ours (and each other).

The skew algorithm.   Farach's linear-time suffix tree construction algorithm [11] as well as some parallel and external algorithms [12,13,14] are based on the following divide-and-conquer approach:

1. Construct the suffix tree of the suffixes starting at odd positions. This is done by reduction to the suffix tree construction of a string of half the length, which is solved recursively.
2. Construct the suffix tree of the remaining suffixes using the result of the first step.
3. Merge the two suffix trees into one.

The crux of the algorithm is the last step, merging, which is a complicated procedure and relies on structural properties of suffix trees that are not available in suffix arrays. In their recent direct linear time suffix array construction algorithm, Kim et al. [28] managed to perform the merging using suffix arrays, but the procedure is still very complicated.

The skew algorithm has a similar structure:

1. Construct the suffix array of the suffixes starting at positions $i \bmod 3 \neq 0$. This is done by reduction to the suffix array construction of a string of two thirds the length, which is solved recursively.
2. Construct the suffix array of the remaining suffixes using the result of the first step.
3. Merge the two suffix arrays into one.

Surprisingly, the use of two thirds instead of half of the suffixes in the first step makes the last step almost trivial: a simple comparison-based merging is sufficient. For example, to compare suffixes starting at i and j with $i \bmod 3 = 0$ and $j \bmod 3 = 1$, we first compare the initial characters, and if they are the same, we compare the suffixes starting at $i+1$ and $j+1$ whose relative order is already known from the first step.

Results.   The simplicity of the skew algorithm makes it easy to adapt to other models of computation. Table 1 summarizes our results together with the best previously known algorithms for a number of important models of computation. The column "alphabet" in Table 1 identifies the model for the alphabet Σ.

In a *constant* alphabet, we have $|\Sigma| = \mathcal{O}(1)$, an *integer* alphabet means that characters are integers in a range of size $n^{\mathcal{O}(1)}$, and *general* alphabet only assumes that characters can be compared in constant time.

Table 1. Suffix array construction algorithms. The algorithms in [11,12,13,14] are *indirect*, i.e., they actually construct a suffix *tree*, which can be then be transformed into a suffix array

model of computation	complexity	alphabet	source
RAM	$\mathcal{O}(n \log n)$ time	general	[31,30,5]
	$\mathcal{O}(n)$ time	integer	[11,28,29],skew
External Memory [38] D disks, block size B, fast memory of size M	$\mathcal{O}\left(\frac{n}{DB} \log_{\frac{M}{B}} \frac{n}{B} \log_2 n\right)$ I/Os $\mathcal{O}\left(n \log_{\frac{M}{B}} \frac{n}{B} \log_2 n\right)$ internal work	integer	[9]
	$\mathcal{O}\left(\frac{n}{DB} \log_{\frac{M}{B}} \frac{n}{B}\right)$ I/Os $\mathcal{O}\left(n \log_{\frac{M}{B}} \frac{n}{B}\right)$ internal work	integer	[14],skew
Cache Oblivious [15] M/B cache blocks of size B	$\mathcal{O}\left(\frac{n}{B} \log_{\frac{M}{B}} \frac{n}{B} \log_2 n\right)$ cache faults	general	[9]
	$\mathcal{O}\left(\frac{n}{B} \log_{\frac{M}{B}} \frac{n}{B}\right)$ cache faults	general	[14],skew
BSP [37] P processors h-relation in time $L + gh$ $P = \mathcal{O}(n^{1-\epsilon})$ processors	$\mathcal{O}\left(\frac{n \log n}{P} + (L + \frac{gn}{P}) \frac{\log^3 n \log P}{\log(n/P)}\right)$ time	general	[12]
	$\mathcal{O}\left(\frac{n \log n}{P} + L \log^2 P + \frac{gn \log n}{P \log(n/P)}\right)$ time	general	skew
	$\mathcal{O}(n/P + L \log^2 P + gn/P)$ time	integer	skew
EREW-PRAM [25]	$\mathcal{O}(\log^4 n)$ time, $\mathcal{O}(n \log n)$ work	general	[12]
	$\mathcal{O}(\log^2 n)$ time, $\mathcal{O}(n \log n)$ work	general	skew
arbitrary-CRCW-PRAM [25]	$\mathcal{O}(\log n)$ time, $\mathcal{O}(n)$ work (rand.)	constant	[13]
priority-CRCW-PRAM [25]	$\mathcal{O}(\log^2 n)$ time, $\mathcal{O}(n)$ work (rand.)	constant	skew

The skew algorithm for RAM, external memory and cache oblivious models is the first optimal *direct* algorithm. For BSP and EREW-PRAM models, we obtain an improvement over *all* previous results, including the first linear work BSP algorithm. On all the models, the skew algorithm is much simpler than the best previous algorithm.

In many applications, the suffix array needs to be augmented with additional data, the most important being the longest common prefix (lcp) array [1,2,26, 27,31]. In particular, the suffix tree can be constructed easily from the suffix and lcp arrays [11,13,14]. There is a linear time algorithm for computing the lcp array from the suffix array [27], but it does not appear to be suitable for parallel or external computation. We extend our algorithm to compute also the lcp array while retaining the complexities of Table 1. Hence, we also obtain improved suffix tree construction algorithms for the BSP and EREW-PRAM models.

The paper is organized as follows. In Section 2, we describe the basic skew algorithm, which is then adapted to different models of computation in Section 3. The algorithm is extended to compute the longest common prefixes in Section 4.

2   The Skew Algorithm

For compatibility with C and because we use many modulo operations we start arrays at position 0. We use the abbreviations $[a, b] = \{a, \dots, b\}$ and $s[a, b] = [s[a], \dots, s[b]]$ for a string or array s. Similarly, $[a, b) = [a, b - 1]$ and $s[a, b) = s[a, b - 1]$. The operator $\circ$ is used for the concatenation of strings. Consider a string $s = s[0, n)$ over the alphabet $\Sigma = [1, n]$. The suffix array SA contains the suffixes $S_i = s[i, n)$ in sorted order, i.e., if $\mathrm{SA}[i] = j$ then suffix S_j has rank $i + 1$ among the set of strings $\{S_0, \dots, S_{n-1}\}$. To avoid tedious special case treatments, we describe the algorithm for the case that n is a multiple of 3 and adopt the convention that all strings α considered have $\alpha[|\alpha|] = \alpha[|\alpha| + 1] = 0$. The implementation in the Appendix fills in the remaining details. Figure 1 gives an example.

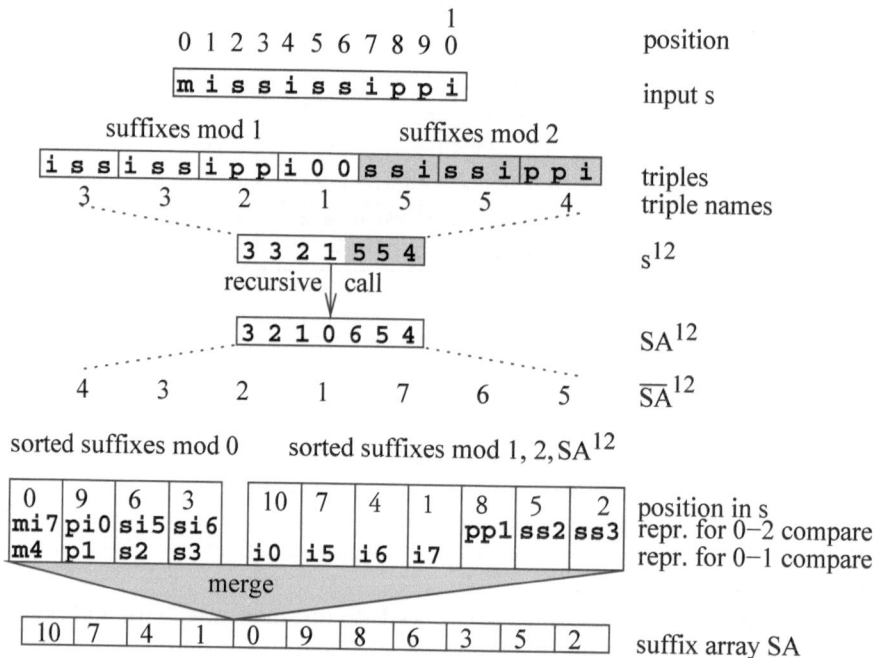

Fig. 1. The skew algorithm applied to $s = \mathtt{mississippi}$.

The first and most time consuming step of the skew algorithm sorts the suffixes S_i with $i \bmod 3 \neq 0$ among themselves. To this end, it first finds *lexicographic names* $s'_i \in [1, 2n/3]$ for the triples $s[i, i + 2]$ with $i \bmod 3 \neq 0$, i.e., numbers with the property that $s'_i \leq s'_j$ if and only if $s[i, i + 2] \leq s[j, j + 2]$. This can be done in linear time by radix sort and scanning the sorted sequence of triples — if triple $s[i, i+2]$ is the k-th different triple appearing in the sorted sequence, we set $s_i' = k$.

If all triples get different lexicographic names, we are done with step one. Otherwise, the suffix array SA^{12} of the string

$$s^{12} = [s_i' : i \bmod 3 = 1] \circ [s_i' : i \bmod 3 - 2]$$

is computed recursively. Note that there can be no more lexicographic names than characters in s^{12} so that the alphabet size in a recursive call never exceeds the size of the string. The recursively computed suffix array SA^{12} represents the desired order of the suffixes S_i with $i \bmod 3 \neq 0$. To see this, note that $s^{12}[\frac{i-1}{3}, \frac{n}{3})$ for $i \bmod 3 = 1$ represents the suffix $S_i = s[i, n) \circ [0]$ via lexicographic naming. The 0 characters at the end of s make sure that $s^{12}[n/3 - 1]$ is unique in s^{12} so that it does not matter that s^{12} has additional characters. Similarly, $s^{12}[\frac{n+i-2}{3}, \frac{2n}{3})$ for $i \bmod 3 = 2$ represents the suffix $S_i = s[i, n) \circ [0, 0]$.

The second step is easy. The suffixes S_i with $i \bmod 3 = 0$ are sorted by sorting the pairs $(s[i], S_{i+1})$. Since the order of the suffixes S_{i+1} is already implicit in SA^{12}, it suffices to stably sort those entries $SA^{12}[j]$ that represent suffixes S_{i+1}, $i \bmod 3 = 0$, with respect to $s[i]$. This is possible in linear time by a single pass of radix sort.

The skew algorithm is so simple because also the third step is quite easy. We have to merge the two suffix arrays to obtain the complete suffix array SA. To compare a suffix S_j with $j \bmod 3 = 0$ with a suffix S_i with $i \bmod 3 \neq 0$, we distinguish two cases:
If $i \bmod 3 = 1$, we write S_i as $(s[i], S_{i+1})$ and S_j as $(s[j], S_{j+1})$. Since $i + 1 \bmod 3 = 2$ and $j + 1 \bmod 3 = 1$, the relative order of S_{j+1} and S_{i+1} can be determined from their position in SA^{12}. This position can be determined in constant time by precomputing an array $\overline{SA}^{12}$ with $\overline{SA}^{12}[i] = j+1$ if $SA^{12}[j] = i$. This is nothing but a special case of lexicographic naming.[1]

Similarly, if $i \bmod 3 = 2$, we compare the triples $(s[i], s[i+1], S_{i+2})$ and $(s[j], s[j+1], S_{j+2})$ replacing S_{i+2} and S_{j+2} by their lexicographic names in $\overline{SA}^{12}$.

The running time of the skew algorithm is easy to establish.

Theorem 1. *The skew algorithm can be implemented to run in time* $\mathcal{O}(n)$.

Proof. The execution time obeys the recurrence $T(n) = \mathcal{O}(n) + T(\lceil 2n/3 \rceil)$, $T(n) = \mathcal{O}(1)$ for $n < 3$. This recurrence has the solution $T(n) = \mathcal{O}(n)$. □

3 Other Models of Computation

Theorem 2. *The skew algorithm can be implemented to achieve the following performance guarantees on advanced models of computation:*

---

[1] $\overline{SA}^{12} - 1$ is also known as the *inverse suffix array* of SA^{12}.

model of computation	complexity	alphabet
External Memory [38] D disks, block size B, fast memory of size M	$\mathcal{O}\left(\frac{n}{DB}\log_{\frac{M}{B}}\frac{n}{B}\right)$ I/Os $\mathcal{O}\left(n\log_{\frac{M}{B}}\frac{n}{B}\right)$ internal work	integer
Cache Oblivious [15]	$\mathcal{O}\left(\frac{n}{B}\log_{\frac{M}{B}}\frac{n}{B}\right)$ cache faults	general
BSP [37] P processors h-relation in time $L+gh$	$\mathcal{O}\left(\frac{n\log n}{P}+L\log^2 P+\frac{gn\log n}{P\log(n/P)}\right)$ time	general
$P=\mathcal{O}\left(n^{1-\epsilon}\right)$ processors	$\mathcal{O}(n/P+L\log^2 P+gn/P)$ time	integer
EREW-PRAM [25]	$\mathcal{O}\left(\log^2 n\right)$ time and $\mathcal{O}(n\log n)$ work	general
priority-CRCW-PRAM [25]	$\mathcal{O}\left(\log^2 n\right)$ time and $\mathcal{O}(n)$ work (rand.)	constant

Proof. **External Memory:** Sorting tuples and lexicographic naming is easily reduced to external memory integer sorting. I/O optimal deterministic[2] parallel disk sorting algorithms are well known [34,33]. We have to make a few remarks regarding internal work however. To achieve optimal internal work for all values of n, M, and B, we can use radix sort where the most significant digit has $\lfloor \log M \rfloor - 1$ bits and the remaining digits have $\lfloor \log M/B \rfloor$ bits. Sorting then starts with $\mathcal{O}\left(\log_{M/B} n/M\right)$ data distribution phases that need linear work each and can be implemented using $\mathcal{O}(n/DB)$ I/Os using the same I/O strategy as in [33]. It remains to stably sort the elements by their $\lfloor \log M \rfloor - 1$ most significant bits. For this we can use the distribution based algorithm from [33] directly. In the distribution phases, elements can be put into a bucket using a full lookup table mapping keys to buckets. Sorting buckets of size M can be done in linear time using a linear time internal algorithm.

Cache Oblivious: We use the comparison based model here since it is not known how to do cache oblivious integer sorting with $\mathcal{O}(\frac{n}{B}\log_{M/B}\frac{n}{B})$ cache faults and $o(n\log n)$ work. The result is an immediate corollary of the optimal comparison based sorting algorithm [15].

EREW PRAM: We can use Cole's merge sort [8] for sorting and merging. Lexicographic naming can be implemented using linear work and $\mathcal{O}(\log P)$ time using prefix sums. After $\Theta(\log P)$ levels of recursion, the problem size has reduced so far that the remaining subproblem can be solved on a single processor. We get an overall execution time of $\mathcal{O}\left(n\log n/P+\log^2 P\right)$.

BSP: For the case of many processors, we proceed as for the EREW-PRAM algorithm using the optimal comparison based sorting algorithm [19] that takes time $\mathcal{O}(n\log n/P+(gn/P+L)\frac{\log n}{\log(n/P)})$.

For the case of few processors, we can use a linear work sorting algorithm based on radix sort [7] and a linear work merging algorithm [17]. The integer

---

[2] Simpler randomized algorithms with favorable constant factors are also available [10].

sorting algorithm remains applicable at least during the first $\Theta(\log \log n)$ levels of recursion of the skew algorithm. Then we can afford to switch to a comparison based algorithm without increasing the overall amount of internal work.

CRCW PRAM: We employ the stable integer sorting algorithm [35] that works in $\mathcal{O}(\log n)$ time using linear work for keys with $\mathcal{O}(\log \log n)$ bits. This algorithm can be used for the first $\Theta(\log \log \log n)$ iterations. Then we can afford to switch to the algorithm [22] that works for polynomial size keys at the price of being inefficient by a factor $\mathcal{O}(\log \log n)$. Lexicographic naming can be implemented by computing prefix sums using linear work and logarithmic time. Comparison based merging can be implemented with linear work and $\mathcal{O}(\log n)$ time using [23]. □

The resulting algorithms are simple except that they may use complicated subroutines for sorting to obtain theoretically optimal results. There are usually much simpler implementations of sorting that work well in practice although they may sacrifice determinism or optimality for certain combinations of parameters.

4   Longest Common Prefixes

Let $\mathrm{lcp}(i, j)$ denote the length of the longest common prefix (lcp) of the suffixes S_i and S_j. The longest common prefix array LCP contains the lengths of the longest common prefixes of suffixes that are adjacent in the suffix array, i.e., $\mathrm{LCP}[i] = \mathrm{lcp}(SA[i], SA[i+1])$. A well-known property of lcps is that for any $0 \leq i < j < n$,

$$\mathrm{lcp}(i, j) = \min_{i \leq k < j} \mathrm{LCP}[k] \ .$$

Thus, if we preprocess LCP in linear time to answer range minimum queries in constant time [3,4,24], we can find the longest common prefix of any two suffixes in constant time.

We will show how the LCP array can be computed from the LCP^{12} array corresponding to SA^{12} in linear time. Let $j = SA[i]$ and $k = SA[i+1]$. We explain two cases; the others are similar.

First, assume that $j \bmod 3 = 1$ and $k \bmod 3 = 2$, and let $j' = (j-1)/3$ and $k' = (n+k-2)/3$ be the corresponding positions in s^{12}. Since j and k are adjacent in SA, so are j' and k' in SA^{12}, and thus $\ell = \mathrm{lcp}^{12}(j', k') = \mathrm{LCP}^{12}[\overline{\mathrm{SA}}^{12}[j'] - 1]$. Then $\mathrm{LCP}[i] = \mathrm{lcp}(j, k) = 3\ell + \mathrm{lcp}(j+3\ell, k+3\ell)$, where the last term is at most 2 and can be computed in constant time by character comparisons.

As the second case, assume $j \bmod 3 = 0$ and $k \bmod 3 = 1$. If $s[j] \neq s[k]$, $\mathrm{LCP}[i] = 0$ and we are done. Otherwise, $\mathrm{LCP}[i] = 1 + \mathrm{lcp}(j+1, k+1)$, and we can compute $\mathrm{lcp}(j+1, k+1)$ as above as $3\ell + \mathrm{lcp}(j+1+3\ell, k+1+3\ell)$, where $\ell = \mathrm{lcp}^{12}(j', k')$ with $j' = ((j+1)-1)/3$, $k' = (n+(k+1)-2)/3$. An additional complication is that, unlike in the first case, $j+1$ and $k+1$ may not be adjacent in SA, and consequently, j' and k' may not be adjacent in SA^{12}. Thus we have to compute ℓ by performing a range minimum query in LCP^{12} instead of a direct lookup. However, this is still constant time.

Theorem 3. *The extended skew algorithm computing both* SA *and* LCP *can be implemented to run in linear time.*

To obtain the same extension for other models of computation, we need to show how to answer $\mathcal{O}(n)$ range minimum queries on LCP^{12}. We can take advantage of the balanced distribution of the range minimum queries shown by the following property.

Lemma 1. *No suffix is involved in more than two lcp queries at the top level of the extended skew algorithm.*

Proof. Let S_i and S_j be two suffixes whose lcp $\text{lcp}(i, j)$ is computed to find the lcp of the suffixes S_{i-1} and S_{j-1}. (The other case that $\text{lcp}(i, j)$ is needed for the lcp of S_{i-2} and S_{j-2} is similar.) Then S_{i-1} and S_{j-1} are lexicographically adjacent suffixes and $s[i-1] = s[j-1]$. Thus, there cannot be another suffix S_k, $S_i < S_k < S_j$, with $s[k-1] = s[i-1]$. This shows that a suffix can be involved in lcp queries only with its two lexicographically nearest neighbors that have the same preceding character. □

We describe a simple algorithm for answering the range minimum queries that can be easily adapted to the models of Theorem 2. It is based on the ideas in [3,4] (which are themselves based on earlier results).

The LCP^{12} array is divided into blocks of size $\log n$. For each block $[a, b]$, precompute and store the following data:

- For all $i \in [a, b]$, a $\log n$-bit vector Q_i that identifies all $j \in [a, i]$ such that $\text{LCP}^{12}[j] < \min_{k \in [j+1, i]} \text{LCP}^{12}[k]$.
- For all $i \in [a, b]$, the minimum values over the ranges $[a, i]$ and $[i, b]$.
- The minimum for all ranges that end just before or begin just after $[a, b]$ and contain exactly a power of two full blocks.

If a range $[i, j]$ is completely inside a block, its minimum can be found with the help of Q_j in constant time (see [3] for details). Otherwise, $[i, j]$ can be covered with at most four of the ranges whose minimum is stored, and its minimum is the smallest of those minima.

Theorem 4. *The extended skew algorithm computing both* SA *and* LCP *can be implemented to achieve the complexities of Theorem 2.*

Proof. (Outline) **External Memory and Cache Oblivious:** The range minimum algorithm can be implemented with sorting and scanning.

Parallel models: The blocks in the range minima data structure are distributed over the processors in the obvious way. Preprocessing range minima data structures reduces to local operations and a straightforward computation proceeding from shorter to longer ranges. Lemma 1 ensures that queries are evenly balanced over the data structure. □

5   Discussion

The skew algorithm is a simple and asymptotically efficient direct algorithm for suffix array construction that is easy to adapt to various models of computation. We expect that it is a good starting point for actual implementations, in particular on parallel machines and for external memory.

The key to the algorithm is the use of suffixes S_i with $i \bmod 3 \in \{1, 2\}$ in the first, recursive step, which enables simple merging in the third step. There are other choices of suffixes that would work. An interesting possibility, for example, is to take suffixes S_i with $i \bmod 7 \in \{3, 5, 6\}$. Some adjustments to the algorithm are required (sorting the remaining suffixes in multiple groups and performing a multiway merge in the third step) but the main ideas still work. In general, a suitable choice is a periodic set of positions according to a *difference cover*. A difference cover D modulo v is a set of integers in the range $[0, v)$ such that, for all $i \in [0, v)$, there exist $j, k \in D$ such that $i \equiv k - j \pmod{v}$. For example $\{1, 2\}$ is a difference cover modulo 3 and $\{3, 5, 6\}$ is a difference cover modulo 7, but $\{1\}$ is not a difference cover modulo 2. Any nontrivial difference cover modulo a constant could be used to obtain a linear time algorithm. Difference covers and their properties play a more central role in the suffix array construction algorithm in [5], which runs in $\mathcal{O}(n \log n)$ time using sublinear extra space in addition to the string and the suffix array.

An interesting theoretical question is whether there are faster CRCW-PRAM algorithms for direct suffix array construction. For example, there are very fast algorithms for padded sorting, list sorting and approximate prefix sums [22] that could be used for sorting and lexicographic naming in the recursive calls. The result would be some kind of suffix list or padded suffix array that could be converted into a suffix array in logarithmic time.

References

1. M. I. Abouelhoda, S. Kurtz, and E. Ohlebusch. The enhanced suffix array and its applications to genome analysis. In *Proc. 2nd Workshop on Algorithms in Bioinformatics*, volume 2452 of *LNCS*, pages 449–463. Springer, 2002.
2. M. I. Abouelhoda, E. Ohlebusch, and S. Kurtz. Optimal exact string matching based on suffix arrays. In *Proc. 9th Symposium on String Processing and Information Retrieval*, volume 2476 of *LNCS*, pages 31–43. Springer, 2002.
3. S. Alstrup, C. Gavoille, H. Kaplan, and T. Rauhe. Nearest common ancestors: A survey and a new distributed algorithm. In *Proc. 14th Annual Symposium on Parallel Algorithms and Architectures*, pages 258–264. ACM, 2002.
4. M. A. Bender and M. Farach-Colton. The LCA problem revisited. In *Proc. 4th Latin American Symposium on Theoretical INformatics*, volume 1776 of *LNCS*, pages 88–94. Springer, 2000.
5. S. Burkhardt and J. Kärkkäinen. Fast lightweight suffix array construction and checking. In *Proc. 14th Annual Symposium on Combinatorial Pattern Matching*. Springer, June 2003. To appear.
6. M. Burrows and D. J. Wheeler. A block-sorting lossless data compression algorithm. Technical Report 124, SRC (digital, Palo Alto), May 1994.

7. A. Chan and F. Dehne. A note on coarse grained parallel integer sorting. *Parallel Processing Letters*, 9(4):533–538, 1999.
8. R. Cole. Parallel merge sort. *SIAM J. Comput.*, 17(4):770–785, 1988.
9. A. Crauser and P. Ferragina. Theoretical and experimental study on the construction of suffix arrays in external memory. *Algorithmica*, 32(1):1–35, 2002.
10. R. Dementiev and P. Sanders. Asynchronous parallel disk sorting. In *Proc. 15th Annual Symposium on Parallelism in Algorithms and Architectures*. ACM, 2003. To appear.
11. M. Farach. Optimal suffix tree construction with large alphabets. In *Proc. 38th Annual Symposium on Foundations of Computer Science*, pages 137–143. IEEE, 1997.
12. M. Farach, P. Ferragina, and S. Muthukrishnan. Overcoming the memory bottleneck in suffix tree construction. In *Proc. 39th Annual Symposium on Foundations of Computer Science*, pages 174–183. IEEE, 1998.
13. M. Farach and S. Muthukrishnan. Optimal logarithmic time randomized suffix tree construction. In *Proc. 23th International Conference on Automata, Languages and Programming*, pages 550–561. IEEE, 1996.
14. M. Farach-Colton, P. Ferragina, and S. Muthukrishnan. On the sorting-complexity of suffix tree construction. *J. ACM*, 47(6):987–1011, 2000.
15. M. Frigo, C. E. Leiserson, H. Prokop, and S. Ramachandran. Cache-oblivious algorithms. In *Proc. 40th Annual Symposium on Foundations of Computer Science*, pages 285–298. IEEE, 1999.
16. N. Futamura, S. Aluru, and S. Kurtz. Parallel suffix sorting. In *Proc. 9th International Conference on Advanced Computing and Communications*, pages 76–81. Tata McGraw-Hill, 2001.
17. A. V. Gerbessiotis and C. J. Siniolakis. Merging on the BSP model. *Parallel Computing*, 27:809–822, 2001.
18. G. Gonnet, R. Baeza-Yates, and T. Snider. New indices for text: PAT trees and PAT arrays. In W. B. Frakes and R. Baeza-Yates, editors, *Information Retrieval: Data Structures & Algorithms*. Prentice-Hall, 1992.
19. M. T. Goodrich. Communication-efficient parallel sorting. *SIAM J. Comput.*, 29(2):416–432, 1999.
20. R. Grossi and G. F. Italiano. Suffix trees and their applications in string algorithms. Rapporto di Ricerca CS-96-14, Università "Ca' Foscari" di Venezia, Italy, 1996.
21. D. Gusfield. *Algorithms on Strings, Trees, and Sequences: Computer Science and Computational Biology*. Cambridge University Press, 1997.
22. T. Hagerup and R. Raman. Waste makes haste: Tight bounds for loose parallel sorting. In *Proc. 33rd Annual Symposium on Foundations of Computer Science*, pages 628–637. IEEE, 1992.
23. T. Hagerup and C. Rüb. Optimal merging and sorting on the EREW-PRAM. *Information Processing Letters*, 33:181–185, 1989.
24. D. Harel and R. E. Tarjan. Fast algorithms for finding nearest common ancestors. *SIAM J. Comput.*, 13:338–355, 1984.
25. J. Jájá. *An Introduction to Parallel Algorithms*. Addison Wesley, 1992.
26. J. Kärkkäinen. Suffix cactus: A cross between suffix tree and suffix array. In Z. Galil and E. Ukkonen, editors, *Proc. 6th Annual Symposium on Combinatorial Pattern Matching*, volume 937 of *LNCS*, pages 191–204. Springer, 1995.
27. T. Kasai, G. Lee, H. Arimura, S. Arikawa, and K. Park. Linear-time longest-common-prefix computation in suffix arrays and its applications. In *Proc. 12th Annual Symposium on Combinatorial Pattern Matching*, volume 2089 of *LNCS*, pages 181–192. Springer, 2001.

28. D. K. Kim, J. S. Sim, H. Park, and K. Park. Linear-time construction of suffix arrays. In *Proc. 14th Annual Symposium on Combinatorial Pattern Matching*. Springer, June 2003. To appear.

29. P. Ko and S. Aluru. Space efficient linear time construction of suffix arrays. In *Proc. 14th Annual Symposium on Combinatorial Pattern Matching*. Springer, June 2003. To appear.

30. N. J. Larsson and K. Sadakane. Faster suffix sorting. Technical report LU-CS-TR:99-214, Dept. of Computer Science, Lund University, Sweden, 1999.

31. U. Manber and G. Myers. Suffix arrays: A new method for on-line string searches. *SIAM J. Comput.*, 22(5):935–948, Oct. 1993.

32. E. M. McCreight. A space-economic suffix tree construction algorithm. *J. ACM*, 23(2):262–272, 1976.

33. M. H. Nodine and J. S. Vitter. Deterministic distribution sort in shared and distributed memory multiprocessors. In *Proc. 5th Annual Symposium on Parallel Algorithms and Architectures*, pages 120–129. ACM, 1993.

34. M. H. Nodine and J. S. Vitter. Greed sort: An optimal sorting algorithm for multiple disks. *J. ACM*, 42(4):919–933, 1995.

35. S. Rajasekaran and J. H. Reif. Optimal and sublogarithmic time randomized parallel sorting algorithms. *SIAM J. Comput.*, 18(3):594–607, 1989.

36. E. Ukkonen. On-line construction of suffix trees. *Algorithmica*, 14(3):249–260, 1995.

37. L. G. Valiant. A bridging model for parallel computation. *Commun. ACM*, 22(8):103–111, Aug. 1990.

38. J. S. Vitter and E. A. M. Shriver. Algorithms for parallel memory, I: Two level memories. *Algorithmica*, 12(2/3):110–147, 1994.

39. P. Weiner. Linear pattern matching algorithm. In *Proc. 14th Symposium on Switching and Automata Theory*, pages 1–11. IEEE, 1973.

A   Source Code

The following C++ file contains a complete linear time implementation of suffix array construction. This code strives for conciseness rather than for speed — it has only 50 lines not counting comments, empty lines, and lines with a bracket only. A driver program can be found at
http://www.mpi-sb.mpg.de/~sanders/programs/suffix/.

```
inline bool leq(int a1, int a2,   int b1, int b2) // lexicographic order
{ return(a1 < b1 || a1 == b1 && a2 <= b2); }           // for pairs
inline bool leq(int a1, int a2, int a3,   int b1, int b2, int b3)
{ return(a1 < b1 || a1 == b1 && leq(a2,a3, b2,b3)); }    // and triples

// stably sort a[0..n-1] to b[0..n-1] with keys in 0..K from r
static void radixPass(int* a, int* b, int* r, int n, int K)
{ // count occurrences
  int* c = new int[K + 1];                          // counter array
  for (int i = 0;  i <= K;  i++) c[i] = 0;          // reset counters
  for (int i = 0;  i < n;  i++) c[r[a[i]]]++;     // count occurrences
  for (int i = 0, sum = 0;  i <= K;  i++)    // exclusive prefix sums
  { int t = c[i];  c[i] = sum;  sum += t; }
```

```
    for (int i = 0;  i < n;  i++) b[c[r[a[i]]]++] = a[i];      // sort
    delete [] c;
}

// find the suffix array SA of s[0..n-1] in {1..K}^n
// require s[n]=s[n+1]=s[n+2]=0, n>=2
void suffixArray(int* s, int* SA, int n, int K) {
    int n0=(n+2)/3, n1=(n+1)/3, n2=n/3, n02=n0+n2;
    int* s12  = new int[n02 + 3];  s12[n02]= s12[n02+1]= s12[n02+2]=0;
    int* SA12 = new int[n02 + 3];  SA12[n02]=SA12[n02+1]=SA12[n02+2]=0;
    int* s0   = new int[n0];
    int* SA0  = new int[n0];

    // generate positions of mod 1 and mod  2 suffixes
    // the "+(n0-n1)" adds a dummy mod 1 suffix if n%3 == 1
    for (int i=0, j=0;  i < n+(n0-n1);  i++) if (i%3 != 0) s12[j++] = i;

    // lsb radix sort the mod 1 and mod 2 triples
    radixPass(s12 , SA12, s+2, n02, K);
    radixPass(SA12, s12 , s+1, n02, K);
    radixPass(s12 , SA12, s  , n02, K);

    // find lexicographic names of triples
    int name = 0, c0 = -1, c1 = -1, c2 = -1;
    for (int i = 0;  i < n02;  i++) {
      if (s[SA12[i]] != c0 || s[SA12[i]+1] != c1 || s[SA12[i]+2] != c2)
      { name++; c0 = s[SA12[i]]; c1 = s[SA12[i]+1]; c2 = s[SA12[i]+2]; }
      if (SA12[i] % 3 == 1) { s12[SA12[i]/3]      = name; } // left half
      else                  { s12[SA12[i]/3 + n0] = name; } // right half
    }

    // recurse if names are not yet unique
    if (name < n02) {
      suffixArray(s12, SA12, n02, name);
      // store unique names in s12 using the suffix array
      for (int i = 0;  i < n02;  i++) s12[SA12[i]] = i + 1;
    } else // generate the suffix array of s12 directly
      for (int i = 0;  i < n02;  i++) SA12[s12[i] - 1] = i;

    // stably sort the mod 0 suffixes from SA12 by their first character
    for (int i=0, j=0; i < n02; i++) if (SA12[i] < n0) s0[j++] = 3*SA12[i];
    radixPass(s0, SA0, s, n0, K);

    // merge sorted SA0 suffixes and sorted SA12 suffixes
    for (int p=0, t=n0-n1, k=0;  k < n;  k++) {
#define GetI() (SA12[t] < n0 ? SA12[t] * 3 + 1 : (SA12[t] - n0) * 3 + 2)
      int i = GetI(); // pos of current offset 12 suffix
      int j = SA0[p]; // pos of current offset 0  suffix
      if (SA12[t] < n0 ? // different compares for mod 1 and mod 2 suffixes
        leq(s[i],        s12[SA12[t] + n0], s[j],        s12[j/3]) :
```

```
      leq(s[i],s[i+1],s12[SA12[t]-n0+1], s[j],s[j+1],s12[j/3+n0]))
  {          // suffix from SA12 is smaller
    SA[k] = i;  t++;
    if (t == n02) // done --- only SA0 suffixes left
       for (k++;  p < n0;  p++, k++) SA[k] = SA0[p];
  } else { // suffix from SA0  is smaller
    SA[k] = j;  p++;
    if (p == n0)  // done --- only SA12 suffixes left
       for (k++;  t < n02;  t++, k++) SA[k] = GetI();
  }
}
delete [] s12; delete [] SA12; delete [] SA0; delete [] s0;
}
```

Expansion Postponement via Cut Elimination in Sequent Calculi for Pure Type Systems[*]

Francisco Gutiérrez and Blas Ruiz

Departamento de Lenguajes y Ciencias de la Computación
Universidad de Málaga. Campus Teatinos 29071, Málaga. Spain
{pacog, blas}@lcc.uma.es

Abstract. The sequent calculus used in this paper is interesting because (1) it is equivalent to the standard formulation (*natural*) for Pure Type System (PTS), and (2) the corresponding cut-free subsystem makes it possible to introduce a notion of Cut Elimination (*CE*). This property has a deep impact on PTS and in logical frameworks based in PTS.
CE is an open problem for normalizing generic PTS. Likewise, other proposed versions of cut elimination have not been solved in dependent type systems.
Another interesting problem is *Expansion Postponement* (*EP*), posed by Henk Barendregt in August 1990. Except for PTS with important restrictions, *EP* is thus far an open problem, even for normalizing PTS. Surprisingly, in this paper we prove that *EP* is a consequence of *CE*.

Keywords: pure type systems, sequent calculi, cut elimination, expansion postponement.
Track: B.

1 Introduction

Pure Type Systems (PTSs) [1,2] provide a flexible and general framework to study dependent type system properties. These systems are the basis for logical frameworks and proof-assistants that heavily use dependent types [3,4].

In this paper we use the sequent calculi for PTS introduced in [5]. These sequent calculi are influenced by the correspondence between Gentzen's natural deduction NJ and the sequent calculus LJ for intuitionistic logics [6]. Recall that the natural system NJ uses rules to eliminate the connectives ($\rightarrow$, $\vee$, $\wedge$) on the right. An example of such rules is the rule ($\rightarrow E$) (or modus ponens). In the sequent calculus LJ, there is no rule that eliminates the connectives on the right; the rule ($\rightarrow E$) is replaced by the rules (*cut*) and ($\rightarrow L$):

$$(cut) \; \frac{\Gamma \vdash D \quad \Gamma, D \vdash C}{\Gamma \vdash C}, \qquad (\rightarrow L) \; \frac{\Gamma \vdash A \quad \Gamma, B \vdash C}{\Gamma, A \rightarrow B \vdash C}.$$

The standard (or natural) notion of derivation $\Gamma \vdash a : A$ for PTS is defined by the inductive system shown in Fig.1. By modifying the rules in Fig.1 we can

---

[*] This research was partially supported by the project TIC2001-2705-C03-02.

J.C.M. Baeten et al. (Eds.): ICALP 2003, LNCS 2719, pp. 956–968, 2003.
© Springer-Verlag Berlin Heidelberg 2003

obtain different systems. The standard PTS will be denoted by $\mathcal{N}$. In order to obtain a sequent calculus from the natural type inference relation, the (apl) rule or Π elimination rule

$$(apl) \quad \frac{\Gamma \vdash f : \Pi x : A.F \qquad \Gamma \vdash a : A}{\Gamma \vdash f\,a : F[x := a]}$$

has to be dispensed with, since it eliminates the Π constructor. Recall that Π is a generalization of the connective $\rightarrow$ for dependent types, and the (apl) rule corresponds to modus ponens. Influenced by the Howard-Curry-de Bruijn correspondence [7], an adaptation of Gentzen's $(\rightarrow L)$ rule will be used instead. In particular, we consider an adaptation of the left rule used by Kleene [8](page 481) in the G_3 system:

$$\frac{A \rightarrow B, \Gamma \vdash A \qquad A \rightarrow B, B, \Gamma \vdash C}{A \rightarrow B, \Gamma \vdash C}.$$

Hence, we consider the rules:

$$(\mathcal{K}) \quad \frac{\Gamma \vdash a : A \qquad \Gamma, x : S, \Delta \vdash c : C}{\Gamma, \Delta[x := y\,a] \vdash c[x := y\,a] : C[x := y\,a]} \quad \begin{cases} y : \Pi z : A.B \in \Gamma, \\ S =_\beta B[z := a], \end{cases}$$

$$(cut) \quad \frac{\Gamma \vdash d : D \qquad \Gamma, y : D \vdash c : C}{\Gamma \vdash c[y := d] : C[y := d]}.$$

$\mathcal{K}$ (for Kleene) denotes the system obtained by replacing the (apl) rule in the original system $\mathcal{N}$ (see Fig.1) by the $(\mathcal{K})$ and (cut) rules. Similarly, $\mathcal{K}^{cf}$ ($\mathcal{K}$–cutfree) denotes the system obtained by eliminating the (cut) rule.

The $\mathcal{K}$ system is equivalent to the natural system $\mathcal{N}$, and obviously, $\mathcal{K}^{cf}$ is also correct. A notion of completeness of $\mathcal{K}^{cf}$ generates the cut elimination property.

Recall Gentzen's *hauptsatz*: every LJ derivation can be obtained without using the cut rule, which is known as cut elimination. This result is an essential technique in proof theory [9]. Likewise, a similar notion of Cut Elimination (CE) can be formulated: every $\mathcal{K}$ *normalized* derivation can be obtained without using the cut rule. CE will have a deep impact on PTS. Thus, CE can be applied to develop proof-search strategies with dependent types, similar to those proposed in [10,11,3].

In [12] we prove that CE is equivalent to the admissibility of a rule to type applications in the system $\mathcal{K}^{cf}$. As a result, CE is obtained in two important families of systems.

CE is an open problem for generic normalized systems. This is not surprising, and in the present paper we prove that CE is actually sufficient to prove the *Expansion Postponement* (EP) problem [13] posed by Henk Barendregt in August 1990. If we consider the $\vdash_r$ system obtained when the (β) rule is substituted by the predicate β-reduction rule

$$\frac{\Gamma \vdash a : A \qquad A \twoheadrightarrow_\beta A'}{\Gamma \vdash a : A'},$$

then EP turns into the following conjecture: any judgement $\Gamma \vdash a : A$ can be obtained by first deriving $\Gamma \vdash_r a : A'$, for some redex A' $_\beta\!\leftarrow A$, and then possibly by applying β-expansion.

The relevance of EP stems from its application to the correctness proof of certain type checking systems ([14,13,15]). Bear in mind that except for PTS with important restrictions, EP is thus far an open problem, even for normalizing PTS [16].

Robert Pollack exposes in [17] the relation between EP and the problem of finding reasonable algorithms for type checking in PTS. In this sense, he proposes different ways to represent PTS in order to derive directly type checking algorithms: the syntax-directed systems.

[13] emphasizes that EP is a necessary condition for the completeness of the syntax-directed type checking algorithms proposed by Pollack, and therefore it is possible to find complete algorithms only for systems enjoying EP. Similarly, [14] also conjectures that the completeness of the algorithm is essentially equivalent to EP.

It is well-known that EP can be solved by the completeness of a certain system $\mathcal{N}_n$ that infers normal types only [13,18]. We will see that CE ensures that $\mathcal{K}$ is correct with respect to $\mathcal{N}_n$, and therefore EP is easily obtained.

The paper is organized as follows: in Section 2 we briefly describe PTS; in Section 3 we define the sequent calculus; Section 4 and Section 5 introduce cut elimination and expansion postponement properties; in Section 6 we prove the relation between cut elimination and expansion postponement, and finally, we present some conclusions and related works.

2   Pure Type Systems

In this section we review PTS and their main properties. For further details the reader is referred to [1,2,19,18].

Considering an infinite set of variables $\mathcal{V}$ $(x, y, \cdots \in \mathcal{V})$ and a set of constants or $sorts$ $\mathcal{S}$ $(s, s_1, \cdots \in \mathcal{S})$, the set $\mathcal{T}$ of terms for a PTS is inductively defined as:

$$a \in \mathcal{V} \cup \mathcal{S} \quad \Rightarrow \quad a \in \mathcal{T},$$
$$A, C, a, b \in \mathcal{T} \quad \Rightarrow \quad a\,b,\ \lambda x : A.b,\ \Pi x : A.C \in \mathcal{T}.$$

A PTS is defined by its specification, that is, a tuple $(\mathcal{S}, \mathcal{A}, \mathcal{R})$, where $\mathcal{A} \subseteq \mathcal{S}^2$ is a set of $axioms$, and $\mathcal{R} \subseteq \mathcal{S}^3$ a set of $rules$. Instances of this tuple embed important theories, such as $\lambda 2$, F^ω, and the Calculus of Constructions. The standard (or natural) notion of derivation $\Gamma \vdash a : A$ is defined by the inductive system shown in Fig.1, and the standard corresponding PTS will be denoted by $\mathcal{N}$.

We denote the β-reduction as $\twoheadrightarrow_\beta$ and the equality generated by $\twoheadrightarrow_\beta$ as $=_\beta$. The set of β-normal forms is denoted β-nf and a_β denotes the β-normal form of a; FV(a) denotes the set of free variables. $A[x := B]$ denotes, as usual, substitution. A $context$ Γ is a sequence (possibly empty) $\langle x_1 : A_1, \ldots, x_n : A_n \rangle$ of declarations $x_i : A_i$ where $x_i \in \mathcal{V}$ and $A_i \in \mathcal{T}$. We drop the $\langle \rangle$ symbols when there is no ambiguity. We write $x_i : A_i \in \Gamma$ when the declaration $x_i : A_i$ is in Γ, and by using the ($\doteq$) symbol to establish definitions, we have

$$
\begin{aligned}
\Gamma \subseteq \Gamma' &\doteq& \forall x \in \mathcal{V} \, [x : A \in \Gamma \Rightarrow x : A \in \Gamma'], \\
\text{Var}(\Gamma) &\doteq& \{x_1, \ldots, x_n\}, \\
\text{FV}(\Gamma) &\doteq& \text{FV}(A_1) \cup \cdots \cup \text{FV}(A_n).
\end{aligned}
$$

We can extend the β-reduction to contexts and therefore define the β-normal form for contexts:

$$
\langle \rangle_\beta \doteq \langle \rangle, \qquad\qquad \langle x : A, \Gamma \rangle_\beta \doteq \langle x : A_\beta, \Gamma_\beta \rangle.
$$

We say that Γ is a legal context (denoted with $\Gamma \vdash$) if $\exists c, C[\Gamma \vdash c : C]$. We recall elementary properties of PTS:

Lemma 1 (Elementary Properties) *If $\Gamma \vdash c : C$, then:*

(i) $\text{FV}(c : C) \subseteq \text{Var}(\Gamma)$, *and if* $x_i, x_j \in \text{Var}(\Gamma)$, *then* $i \neq j \Rightarrow x_i \neq x_j$.   *(FrVrs)*
(ii) $s_1 : s_2 \in \mathcal{A} \Rightarrow \Gamma \vdash s_1 : s_2$.   *(TypAx)*
(iii) $y : D \in \Gamma \Rightarrow \Gamma \vdash y : D$.   *(TypVr)*
(iv) *Type correctness:* $\Gamma \vdash c : C \wedge C \notin \mathcal{S} \Rightarrow \exists s \in \mathcal{S} \, [\Gamma \vdash C : s]$.   *(CrTyps)*
(v)   *Context correctness:* $\Gamma, x : A, \Delta \vdash d : D \Rightarrow \exists s \in \mathcal{S} \, [\Gamma \vdash A : s]$.   *(CrCtx)*

We also need typical properties of PTS: subject β-reduction $(S\beta)$, predicate β-reduction $(P\beta)$, substitution lemma (Sbs), and thinning lemma $(Thnng)$:

$$
\frac{\Gamma \vdash a : A \quad a \twoheadrightarrow_\beta a'}{\Gamma \vdash a' : A} \; (S\beta), \qquad \frac{\Gamma \vdash a : A \quad A \twoheadrightarrow_\beta A'}{\Gamma \vdash a : A'} \; (P\beta),
$$

$$
\frac{\Gamma \vdash d : D \quad \Gamma, y : D, \Delta \vdash c : C}{\Gamma, \Delta[y := d] \vdash c[y := d] : C[y := d]} \; (Sbs), \qquad \frac{\Gamma \vdash b : B \quad \Gamma \subseteq \Psi \vdash}{\Psi \vdash b : B} \; (Thnng).
$$

Let us recall that the natural system $\mathcal{N}$ satisfies a generation lemma (see Lemma 19 in [1]).

In this paper, every free *object* in the right hand side of an implication or in the conclusion of a rule is existentially quantified. For example, the *CrCtx* property of Lemma 1 can be enunciated as: $\Gamma, x : A, \Delta \vdash d : D \Rightarrow \Gamma \vdash A : s$.

The lemma below is rarely referred to in the literature; however, it will be used here to simplify some proofs. This lemma characterize the set of types for every term.

Lemma 2 (The Shape of Types) (van Benthem Jutting [19]) *The set of terms of a PTS can be divided in two disjoint classes T_v and T_s, inductively defined as:*

$$
\begin{aligned}
x &\in T_v & b \in T_v &\Rightarrow b\,c, \lambda x : A.b \in T_v, \\
s, \Pi x : A.B &\in T_s & b \in T_s &\Rightarrow b\,c, \lambda x : A.b \in T_s,
\end{aligned}
$$

so that

$$
\Gamma \vdash a : A, a : A' \quad \Rightarrow \quad \begin{cases} a \in T_v \Rightarrow A =_\beta A', \\ a \in T_s \Rightarrow A \twoheadrightarrow_\beta \Pi\Delta.s \wedge A' \twoheadrightarrow_\beta \Pi\Delta.s', \end{cases}
$$

where $\Pi\langle\rangle.M \doteq M$ and $\Pi\langle x : X, \Delta\rangle.M \doteq \Pi x : X.(\Pi\Delta.M)$.

$$(ax) \quad \frac{}{\vdash s_1 : s_2} \qquad s_1 : s_2 \in \mathcal{A}$$

$$(var) \quad \frac{\Gamma \vdash A : s}{\Gamma, x : A \vdash x : A} \qquad x \notin \mathsf{Var}(\Gamma)$$

$$(weak) \quad \frac{\Gamma \vdash b : B \qquad \Gamma \vdash A : s}{\Gamma, x : A \vdash b : B} \qquad b \in \mathcal{S} \cup \mathcal{V}, x \notin \mathsf{Var}(\Gamma)$$

$$(\Pi) \quad \frac{\Gamma \vdash A : s_1 \qquad \Gamma, x : A \vdash B : s_2}{\Gamma \vdash \Pi x : A.B \; : \; s_3} \qquad (s_1, s_2, s_3) \in \mathcal{R}$$

$$(apl) \quad \frac{\Gamma \vdash f : \Pi x : A.F \qquad \Gamma \vdash a : A}{\Gamma \vdash f\,a : F[x := a]}$$

$$(\lambda) \quad \frac{\Gamma \vdash \Pi x : A.B : s \qquad \Gamma, x : A \vdash b : B}{\Gamma \vdash \lambda x : A.b \; : \; \Pi x : A.B}$$

$$(\beta) \quad \frac{\Gamma \vdash a : A \qquad \Gamma \vdash A' : s}{\Gamma \vdash a : A'} \qquad A =_\beta A'$$

Fig. 1. Inference rules for PTS. For the sake of readability, $s_1 : s_2 \in \mathcal{A}$ stands for $(s_1, s_2) \in \mathcal{A}$.

3 Sequent Calculi for PTS

In order to obtain a sequent calculus from the natural type inference relation, the (apl) rule or Π elimination rule has to be dispensed with, since it eliminates the Π constructor.

Definition 3 *1. We consider the rules:*

$$(\mathcal{K}) \quad \frac{\Gamma \vdash a : A \qquad \Gamma, x : S, \Delta \vdash c : C}{\Gamma, \Delta[x := y\,a] \vdash c[x := y\,a] : C[x := y\,a]} \quad \begin{cases} y : \Pi z : A.B \in \Gamma, \\ S =_\beta B[z := a], \end{cases}$$

$$(cut) \quad \frac{\Gamma \vdash d : D \qquad \Gamma, y : D \vdash c : C}{\Gamma \vdash c[y := d] : C[y := d]} \; .$$

2. $\mathcal{K}$ *denotes the systems obtained by replacing the* (apl) *rule in the original system* $\mathcal{N}$ *(see Fig.1) by the* $(\mathcal{K})$ *and* (cut) *rules. The type inference relation of* $\mathcal{K}$ *will be denoted as* $\vdash_{\mathcal{K}}$.

3. *Similarly,* $\mathcal{K}^{cf}$ *denotes the systems obtained by eliminating the* (cut) *rule. Its type inference relation will be denoted as* $\vdash_{\mathcal{K}^{cf}}$.

Like PTS , $\mathcal{K}$ and $\mathcal{K}^{cf}$ denote many systems depending on the $(\mathcal{S}, \mathcal{A}, \mathcal{R})$ specification. Elementary properties of PTS hold for sequent calculi as well.

Lemma 4 *Lemma 1 holds for $\mathcal{K}$ and $\mathcal{K}^{cf}$ systems.*

Proof See Lemma 5 in [5]. ∎

Theorem 5 (Correctness and Completeness of Sequent Calculus)
$\mathcal{N} \equiv \mathcal{K}$.

Proof See [5]. ∎

Because of the form of the ($\mathcal{K}$) rule, every object (subject, context, and type) in each derivation in $\vdash_{\mathcal{K}^{cf}}$ is in β–normal form. In fact,

Lemma 6 (The Shape of Types in Cut Free Sequent Calculi) *In every* $\mathcal{K}^{cf}$ *system, we have that:*

(i) $\Gamma \vdash_{\mathcal{K}^{cf}} m : M \ \Rightarrow\ \Gamma, m, M \in \beta\text{-nf}.$
(ii) $a \in T_v \wedge \Gamma \vdash_{\mathcal{K}^{cf}} a : A, a : A' \ \Rightarrow\ A \equiv A'.$
(iii) $a \in T_s \wedge \Gamma \vdash_{\mathcal{K}^{cf}} a : A, a : A' \ \Rightarrow\ A \equiv \Pi\Delta.s \wedge A' \equiv \Pi\Delta.s'.$

Proof (i) it follows by IDs using the fact that the $[x := y\,a]$ operator preserves normal forms when $a \in \beta\text{-nf}$. In order to prove $(ii) - (iii)$, it suffices to apply Theorem 5 and then Lemma 2 and (i). ∎

Corollary 7 $\Gamma \vdash_{\mathcal{K}^{cf}} B : s \wedge \Gamma \vdash B : s' \ \Rightarrow\ \Gamma \vdash_{\mathcal{K}^{cf}} B : s'$ $s, s' \in \mathcal{S}.$

Proof. By induction on the structure of B. By the generation lemma, B cannot be an abstraction. If it is a constant, then again by the generation lemma in $\vdash$ we have $B : s' \in \mathcal{A}$ and apply $TypAx$ in $\mathcal{K}^{cf}$. If $B \in T_v$, we apply Lemma 6(ii) to get $s \equiv s'$. Should B be an application, it must have the form $y\,f_1 \ldots f_n$ that is in T_v, and the previous reasoning is applied again. If $B \equiv \Pi t : M.N$, we apply the generation lemma in both systems, followed by IH twice and (Π).∎

4   Cut Elimination

The notion of cut elimination for PTS is strongly influenced by the presence of the rule of β-conversion of types; therefore the system $\mathcal{K}$ can type objects (types, contexts, and terms) in not β–normal form. But from Lemma 6 we obtain that $\mathcal{K}^{cf}$ yields objects in β–normal form. Therefore, in $\mathcal{K}^{cf}$ system, we can dispense with the (β) rule since it does not yield different types. In a condensed form, *Cut Elimination* is enunciated as:

$$\Gamma, m, M \in \beta\text{-nf} \wedge \Gamma \vdash_{\mathcal{K}} m : M \ \Rightarrow\ \Gamma \vdash_{\mathcal{K}^{cf}} m : M. \qquad (CE)$$

This is the central property of the $\mathcal{K}$ sequent calculus.

By Theorem 5, $\vdash_{\mathcal{K}}$ can be taken as the standard relation (Fig.1). To prove CE, if we proceed by ID of $\Gamma \vdash m : M$ and the last rule applied is β-conversion, then IH cannot be applied on the premises since their types are not necessarily in β–normal form. We can then reformulate CE in the following equivalent way:

$$\Gamma, m \in \beta\text{-nf} \wedge \Gamma \vdash m : M \ \Rightarrow\ \Gamma \vdash_{\mathcal{K}^{cf}} m : M_\beta. \qquad (CE)$$

(where M_β denotes the β-normal form of M). However, the property of normalization must be imposed on the system[1]. Under these considerations, the problem above is avoided but a new problem arises when the last rule applied is the (apl) rule. Therefore, a new rule for typing applications in β–normal form is needed in the $\mathcal{K}^{cf}$ system.

Lemma 8 *If $\mathcal{K}$ is normalizing then CE is equivalent to the admissibility of the rule:*

$$\frac{\Gamma \vdash_{\mathcal{K}^{cf}} a : A \qquad \Gamma \vdash_{\mathcal{K}^{cf}} f : \Pi z : A.B}{\Gamma \vdash_{\mathcal{K}^{cf}} f a : B[z := a]_\beta} \; f a \in \beta\text{-nf.} \qquad (AplN)$$

Proof. ($\Rightarrow$) It follows from $\vdash_{\mathcal{K}^{cf}} \subseteq \vdash$, (apl), and CE.
($\Leftarrow$) Assume $AplN$. Then we prove that CE by ID $\Gamma \vdash m : M$. Only two cases are shown. When the last rule applied is (apl), we apply IH twice and then the $AplN$ property. When the last rule applied is

$$\frac{\Gamma \vdash A : s_1 \qquad \Gamma, x : A \vdash B : s_2 \qquad \Gamma, x : A \vdash b : B}{\Gamma \vdash \lambda x : A.b \; : \; \Pi x : A.B} \qquad (s_1, s_2, s_3) \in \mathcal{R},$$

by IH we have that $\Gamma \vdash_{\mathcal{K}^{cf}} A : s_1$ and $\Gamma, x : A \vdash_{\mathcal{K}^{cf}} b : B_\beta$, and then we have to prove that $\Gamma, x : A \vdash_{\mathcal{K}^{cf}} B_\beta : s_2$. We apply correctness of types in $\vdash_{\mathcal{K}^{cf}}$ to the last derivation:

— $B_\beta \equiv s \in \mathcal{S}$. Then, since $\Gamma, x : A \vdash s : s_2$, we have that $B_\beta : s_2 \in \mathcal{A}$, and then by Lemma 4($ii$) $(TypAx)$ we get $\Gamma, x : A \vdash_{cf} B_\beta : s_2$.
— $\Gamma, x : A \vdash_{cf} B_\beta : s$. By the second premise and $S\beta$ we obtain $\Gamma, x : A \vdash B_\beta : s_2$; then we apply Corollary 7.∎

5   Expansion Postponement

If we consider the $\vdash_r$ system obtained when the rule (β) is substituted by the predicate β-reduction rule ($P\beta$, see Section 2), then *Expansion Postponement* (EP) turns into the following conjecture: any judgement $\Gamma \vdash a : A$ can be obtained by first deriving $\Gamma \vdash_r a : A'$, for some redex A', and then possibly by applying β-expansion. Therefore EP is characterized by the following property:

$$\Gamma \vdash a : A \qquad \Rightarrow \qquad \Gamma \vdash_r a : A' \wedge A \twoheadrightarrow_\beta A'. \qquad (EP)$$

The EP formulation motivates the definition of the following reflexive and transitive relation $\sqsubseteq$:

$$\vdash_1 \sqsubseteq \vdash_2 \doteq \forall \Gamma, a, A \; [\; \Gamma \vdash_1 a : A \; \Rightarrow \; \Gamma \vdash_2 a : A' \;_{\beta\leftarrow} A \;].$$

Therefore, the property $\vdash \sqsubseteq \vdash_r$ captures EP.

An alternative to analyzing EP is to study the normalizing systems with types in β–normal form. In the sequel, we consider normalizing systems and let us then consider the system $\mathcal{N}_n$ obtained by the $\vdash_n$ relation, defined by the (ax),

$$(var_n) \quad \frac{\Gamma \vdash_n A : s}{\Gamma, x : A \vdash_n x : A_\beta} \quad x \notin \Gamma$$

$$(apl_n) \quad \frac{\Gamma \vdash_n f : \Pi x . A.\Gamma \qquad \Gamma \vdash_n a . A}{\Gamma \vdash_n f a : F[x := a]_\beta}$$

$$(\lambda_n) \quad \frac{\Gamma, x : A \vdash_n b : B \qquad \Gamma \vdash_n \Pi x : A.B : s}{\Gamma \vdash_n \lambda x : A.b : \Pi x : A_\beta.B}$$

Fig. 2. Additional rules for the $\vdash_n$ relation

(weak), (Π) rules (see Fig.1), and the rules of Fig.2. This system is considered in [13]. It is easy to prove by ID that $\Gamma \vdash_n a : A \Rightarrow A \in \beta$-nf and that the system is correct: $\vdash_n \subseteq \vdash$. On the other hand, the implication $\Gamma \vdash_n c : C \Rightarrow \Gamma \vdash_r c : C$ is easy by ID. Hence, EP is a consequence of $\vdash \subseteq \vdash_n$.

Except for PTS with important restrictions, EP and the $\sqsubseteq$-completeness of $\vdash_n$ are still open problems, but they admit solutions for particular PTS [18,16].

In order to study the implication $CE \Rightarrow EP$ we shall use two technical lemmas.

Lemma 9 (Semi–Commutation of Substitution and β–Reduction) *Let us assume that the β–normal form always exist. Then*

(i) $(A_\beta)^\circ_\beta \equiv A^\circ_\beta$.
(ii) $x \not\equiv y \wedge x \notin \mathrm{FV}(d) \Rightarrow (B^\circ_\beta[x := f^\circ])_\beta \equiv (B[x := f]_\beta)^\circ_\beta$.

where the priority of the operator $^\circ \equiv [y := d]$ is higher than that of $(\_)_\beta$.

Proof. (i). It suffices to apply substitutivity of $\twoheadrightarrow_\beta$ and Church-Rosser. (ii):
$(B^\circ_\beta[x := f^\circ])_\beta$
$\equiv \because$ (i) with $A := B^\circ$, $^\circ := [x := f^\circ]$
$(B^\circ[x := f^\circ])_\beta$
$\equiv \because$ untyped λ-calculus substitution lemma [2](Lemma 2.1.6):
$$x \not\equiv y \qquad x \notin \mathrm{FV}(d) \Rightarrow B^\circ[x := f^\circ] \equiv B[x := f]^\circ$$
$(B[x := f])^\circ_\beta$
$\equiv \because$ (i) with $A := B[x := f]$
$(B[x := f]_\beta)^\circ_\beta$ ∎

Lemma 10 (Context Substitution) *For any* PTS

$$\Gamma, y : Y, \Delta \vdash_n c : C \wedge \Gamma \vdash_n Y' : s \wedge Y' =_\beta Y \Rightarrow \Gamma, y : Y', \Delta \vdash_n c : C,$$

and we have the strong context substitution *property:*

$$\Gamma \vdash_n a : A \wedge \Gamma' \vdash_n \wedge \Gamma =_\beta \Gamma' \Rightarrow \Gamma' \vdash_n a : A.$$

---

[1] Recall that a PTS is normalizing if it verifies $\Gamma \vdash a : A \Rightarrow a$ is weak normalizing (also, by type correctness, A is weak normalizing).

Proof By ID of $\psi \equiv \Gamma, y : Y, \Delta \vdash_n c : C$. If ψ has been inferred using a rule whose premise includes the context $\Gamma' \supseteq \Gamma$, it suffices to apply IH and the same rule. Thus, we consider the (var_n) and $(weak)$ rules only. If $\Delta \equiv \langle \rangle$, we apply (var_n) or $(weak)$. The other cases follow by IH and the rule. The *strong context substitution* property follows by induction on the context using the first one. ∎

6   Expansion Postponement from Cut Elimination

In this section we prove that CE solves EP.

Lemma 11 (Correctness $\mathcal{K}^{cf}$ w.r.t. $\vdash_n$) *For every normalizing* PTS, *we have that* $\vdash_{\mathcal{K}^{cf}} \subseteq \vdash_n$.

Proof. In the first place, we prove that the $\vdash_n$ system satisfies the following restriction to the $(\mathcal{K})$ rule:

$$(n\mathcal{K}) \quad \frac{\Gamma \vdash_n a : A \quad \Gamma, x : S, \Delta \vdash_n c : C}{\Gamma, \Delta^\circ \vdash_n c^\circ : C^\circ} \quad \begin{cases} \Gamma, a, A, S, \Delta, c, C \in \beta\text{-nf}, \\ y : \Pi z : A.B \in \Gamma, \\ S =_\beta B[z := a], \end{cases}$$

where $^\circ \equiv [x := y\, a]$ and obviously Δ°, c°, and C° are in β–normal form too. We reason by ID of $\psi \equiv \Gamma, x : S, \Delta \vdash_n c : C$.

1. $\psi : -(var_n)^2$ with $\Delta \equiv \langle \rangle$, and $x : S \equiv c : C$, with $\Gamma \vdash_n S : s$ and $x \notin S$. Since Γ is a legal context, we have that $\Gamma \vdash_n y : \Pi z : A.B \in \Gamma$, and by applying the (apl_n) rule we get $\Gamma \vdash_n y\,a : B[z := a]_\beta (\equiv S \equiv S^\circ)$.
2. $\psi : -(var_n)$, with $\Delta \equiv \Delta_1, u : U$, and (because of $U \in \beta$-nf), $\Gamma, x : S, \Delta_1 \vdash_n U : s$, with $u : U \equiv c : C$; by IH we have $\Gamma, \Delta_1^\circ \vdash_n U^\circ : s$, and we finally apply the (var_n) rule.
3. $\psi : -(\lambda_n)$; because of $c \in \beta$-nf, we can assume:

    $$\frac{\Gamma, x : S, \Delta \vdash_n \Pi t : P.Q : s \quad \Gamma, x : S, \Delta, t : P \vdash_n q : Q}{\Gamma, x : S, \Delta \vdash_n \lambda t : P.q : \Pi t : P.Q},$$

    and then we apply IH on the premises, $P^\circ \in \beta$-nf, and the (λ_n) rule.
4. $\psi : -(apl_n)$

    $$\frac{\Gamma, x : S, \Delta \vdash_n f : \Pi t : G.F, \quad g : G}{\Gamma, x : S, \Delta \vdash_n f\,g : F[t := g]_\beta (\equiv c : C)}.$$

    By applying IH and (apl_n): $\Gamma, \Delta^\circ \vdash_n f^\circ g^\circ : (F^\circ[t := g^\circ])_\beta$. However, by Lemma 9 and $t \not\equiv x \land t \notin \mathrm{FV}(y\,a)$ we get $(F_\beta^\circ[t := g^\circ])_\beta \equiv (F[t := g]_\beta)_\beta^\circ$, and now it suffices to observe that the $^\circ$ operator preserves normal forms, and hence $(F^\circ[t := g^\circ])_\beta \equiv (F[t := g]_\beta)^\circ$.
5. The remaining cases $(weak, \Pi)$ follow by IH and the corresponding rule.

---

[2] $\psi : -(r)$ denotes that the last rule applied is (r).

Finally, we show that $\Gamma \vdash_{\mathcal{K}^{cf}} a : A \Rightarrow \Gamma \vdash_n a : A$ by ID. Since the system $\vdash_{\mathcal{K}^{cf}}$ can only infer objects in β–normal form, we only need to consider the case when the last rule applied is $(\mathcal{K})$. But in this case, we apply IH and the $(n\mathcal{K})$ rule. ■

To end this section, we state and prove the main result of this paper.

Theorem 12 *For every normalizing PTS, we have that $CE \Rightarrow\, \vdash\, \sqsubseteq\, \vdash_n$. Thus EP is a consequence of CE.*

Proof If CE holds, we have $\Gamma \vdash c : C \Rightarrow \Gamma_\beta \vdash_{\mathcal{K}^{cf}} c_\beta : C_\beta$ and apply Lemma 11 to obtain:
$$\Gamma \vdash a : A \Rightarrow \Gamma_\beta \vdash_n a_\beta : A_\beta.$$

Then, by applying the above property, $\vdash_n\, \sqsubseteq\, \vdash$ and context substitution, we have:
$$\Gamma \vdash_n A : s \Rightarrow \Gamma \vdash_n A_\beta : s. \qquad (KEY)$$

Now, we shall use the KEY property in order to prove $\Gamma \vdash c : C \Rightarrow \Gamma \vdash_n c : C_\beta$. We proceed by ID. The only interesting case is when the last rule applied is (λ):

$$\frac{\Gamma, x : A \vdash b : B \qquad \Gamma \vdash A : s_1 \qquad \Gamma, x : A \vdash B : s_2}{\Gamma \vdash \lambda x : A.b : \Pi x : A.B} \quad (s_1, s_2, s_3) \in \mathcal{R}.$$

Then we proceed in the following way:

$$\cfrac{\Gamma, x : A \vdash_n b : B_\beta \quad \text{IH} \qquad \cfrac{\cfrac{\Gamma \vdash_n A : s_1}{} \text{IH} \qquad \cfrac{\cfrac{\Gamma, x : A \vdash_n B : s_2}{\Gamma, x : A \vdash_n B_\beta : s_2} \text{IH} \atop KEY}{}}{\Gamma \vdash_n \Pi x : A.B_\beta : s_3} (\Pi)}{\Gamma \vdash_n \lambda x : A.b : \Pi x : A_\beta.B_\beta} (\lambda_n). \ ■$$

KEY property can also be proved by subject reduction. In order to prove subject reduction we try

$$\Gamma \vdash_n c : C \Rightarrow \begin{cases} c \to_\beta c' \Rightarrow \Gamma \vdash_n c' : C, \\ \Gamma \to_\beta \Gamma' \Rightarrow \Gamma' \vdash_n c : C, \end{cases}$$

by simultaneous ID. For the (apl_n) case, the following substitution lemma ($^\circ \equiv [y := d]$) is required:

$$\Gamma \vdash_n d : D_\beta \wedge \Gamma, y : D, \Delta \vdash_n c : C \Rightarrow \Gamma, \Delta^\circ \vdash_n c^\circ : C_\beta^\circ. \qquad (1)$$

[13] tries to obtain the substitution lemma (1) directly. Unfortunately their proof is not complete as indicated below. If we reason by induction on the derivation $\Gamma, y : D, \Delta \vdash_n c : C$, except for the (λ_n) case, all of them follow by IH and Lemma 9. The problem appears when the last rule applied is the (λ_n) rule

$$\frac{\Gamma, y : D, \Delta, x : A \vdash_n b : B \qquad \Gamma, y : D, \Delta \vdash_n \Pi x : A.B : s}{\Gamma, y : D, \Delta \vdash_n \lambda x : A.b \ : \ \Pi x : A_\beta.B}.$$

Then, by IH we get to $\Gamma, \Delta^\circ, x : A^\circ \vdash_n b^\circ : B_\beta^\circ$ and $\Gamma, \Delta^\circ \vdash_n \Pi x : A^\circ.B^\circ : s$. But we can not apply the (λ_n) rule again because we would need $\Gamma, \Delta^\circ \vdash_n \Pi x : A^\circ.B_\beta^\circ : s$. As consequence, EP is still an open problem.

Thus $\vdash \sqsubseteq \vdash_n$ is equivalent to the KEY property.

7   Conclusions and Related Works

In this paper we have proved that EP is a consequence of CE. The relevance of EP stems from its application to the correctness proof of numerous type checking systems.

Theoretical properties of sequent calculi presented in this paper have been studied in [5], but a general study of CE is very difficult due to the proviso $S =_\beta B[z := a]$ in the $(\mathcal{K})$ rule. This situation disappears by replacing $S \equiv B[z := a]$. This change will have a deep impact in the proof of cut elimination. In [12], we study the systems described by this new rule: $\mathcal{K}'$ and $\mathcal{K}'^{cf}$ (surprisingly, $\mathcal{K}' \equiv \mathcal{K}$, and trivially, $\mathcal{K}'^{cf} \subseteq \mathcal{K}^{cf}$). The cut elimination property obtained in these systems is the Strong Cut Elimination (SCE), stronger than the one presented in this paper.

While (weak) CE is an open problem for generic normalized systems, in [12] we have proven SCE (and also, CE) in two important families of systems characterized as follows. On the one hand, those PTSs where in every rule $(s_1, s_2, s_3) \in \mathcal{R}$, the constant s_2 does not occur in the right hand side of an axiom. Thus, we obtain proofs of SCE in the corners $\lambda \to$ and $\lambda 2$ of the λ-cube [2]. In addition, we have proven SCE for another class of systems, the Π-independent: the well-typed dependent products $\Pi z : A.B$ satisfy $z \notin \mathsf{FV}(B)$. This result yields SCE as a simple corollary, and corners $\lambda \to$ and $\lambda \underline{\omega}$ of Barendregt's λ-cube are particular cases. A generation lemma for the $\mathcal{K}'^{cf}$ system makes it possible to refute SCE for the remaining systems in the λ-cube, as well as in other interesting systems: $\lambda U, \lambda HOL, \lambda AUT\_QE, \lambda AUT - 68$, and λPAL, all of them described in [2]:216.

In summary, for a wide class of systems, the proof of CE is directly deduced from the axioms and rules of PTS, thus providing the proof of EP from the specification.

Recently, other authors [10,20] have introduced notions of CE for particular systems. Thus, our $(\mathcal{K})$ rule generalizes the rule used by Pym [10] in his proof of CE for the $\lambda \Pi$ system, a system with dependent types very similar to the λP PTS. Using Pym's rule, the generation lemma for applications cannot be proved. However, this lemma is essential in both our analysis of CE and in the proof of [10]; therefore, CE is an open problem in λP, but SCE does not hold in λP.

Acknowledgments. The authors are very grateful to Pablo López and the anonymous referees for comments on earlier versions of this paper.

References

1. H. Geuvers, M.-J. Nederhof, Modular proof of Strong Normalization for the Calculus of Constructions, Journal of Functional Programming 1 (1001) 15–189.
2. H. P. Barendregt, Lambda Calculi with Types, in: S. Abramsky, D. Gabbay, T. S. Maibaum (Eds.), Handbook of Logic in Computer Science, Oxford University Press, 1992, Ch. 2.2, pp. 117–309.
3. F. Pfenning, Logical frameworks, in: A. Robinson, A. Voronkov (Eds.), Handbook of Automated Reasoning, Vol. II, Elsevier Science, 2001, Ch. 17, pp. 1063–1147.
4. H. P. Barendregt, H. Geuvers, Proof-assistants using dependent type systems, in: A. Robinson, A. Voronkov (Eds.), Handbook of Automated Reasoning, Vol. II, Elsevier Science, 2001, Ch. 18, pp. 1149–1238.
5. F. Gutiérrez, B. C. Ruiz, A Cut Free Sequent Calculus for Pure Type Systems Verifying the Structural Rules of Gentzen/Kleene, in: International Workshop on Logic Based Program Development and Transformation (LOPSTR'02), September 17-20, Madrid, Spain, Vol. (to appear) of LNCS, Springer-Verlag, 2003, http://polaris.lcc.uma.es/~blas/publicaciones/.
6. G. Gentzen, Untersuchungen über das Logische Schliessen, Math. Zeitschrift 39 (1935) 176,–210,405–431, translation in [21].
7. H. Geuvers, Logics and type systems, Ph.D. thesis, Computer Science Institute, Katholieke Universiteit Nijmegen (1993).
8. S. C. Kleene, Introduction to Metamathematics, D. van Nostrand, Princeton, New Jersey, 1952.
9. M. Baaz, A. Leitsch, Cut-elimination and redundancy-elimination by resolution, Journal of Symbolic Computation 29 (2) (2000) 149–177.
10. D. Pym, A note on the proof theory of the $\lambda\Pi$–calculus, Studia Logica 54 (1995) 199–230.
11. D. Galmiche, D. J. Pym, Proof-search in type-theoretic languages: an introduction, Theoretical Computer Science 232 (1–2) (2000) 5–53.
12. F. Gutiérrez, B. C. Ruiz, Sequent Calculi for Pure Type Systems, Tech. Report 06/02, Dept. de Lenguajes y Ciencias de la Computación, Universidad de Málaga (Spain), http://polaris.lcc.uma.es/~blas/publicaciones/ (may 2002).
13. E. Poll, Expansion Postponement for Normalising Pure Type Systems, Journal of Functional Programming 8 (1) (1998) 89–96.
14. L. van Benthem Jutting, J. McKinna, R. Pollack, Checking Algorithms for Pure Type Systems, in: H. Barendregt, T. Nipkow (Eds.), Types for Proofs and Programs: International Workshop TYPES'93, no. 806 in Lecture Notes in Computer Science, Springer-Verlag, 1994, pp. 19–61.
15. G. Barthe, Type–checking injective pure type systems, Journal Functional Programming 9 (6) (1999) 675–698.
16. B. C. Ruiz, The Expansion Postponement Problem for Pure Type Systems with Universes, in: 9th International Workshop on Functional and Logic Programming (WFLP'2000), Dpto. de Sistemas Informáticos y Computación, Technical University of Valencia (Tech. Rep.), 2000, pp. 210–224, september 28-30, Benicassim, Spain.
17. R. Pollack, Typechecking in pure type systems, in: B. Nordström, K. Petersson, G. Plotkin (Eds.), Informal Proceedings of 1992 Workshop on Types for Proofs and Programs, Båstad, 1992, pp. 271–288, http://www.dcs.ed.ac.uk/lfcinfo/research/types-bra.

18. B. C. Ruiz, Sistemas de Tipos Puros con Universos, Ph.D. thesis, Universidad de Málaga (1999).
19. L. van Benthem Jutting, Typing in Pure Type Systems, Information and Computation 105 (1) (1993) 30–41.
20. M. Strecker, Construction and Deduction in Type Theories, Ph.D. thesis, Universität Ulm (1999).
21. G. Gentzen, Investigations into logical deductions, in: M. Szabo (Ed.), The Collected Papers of Gerhard Gentzen, North-Holland, 1969, pp. 68–131.

Secrecy in Untrusted Networks[*]

Michele Bugliesi[1], Silvia Crafa[1], Amela Prelic[2], and Vladimiro Sassone[3]

[1] Università "Ca' Foscari", Venezia;
[2] Max-Planck-Institut für Informatik;
[3] University of Sussex

Abstract. We investigate the protection of migrating agents against the untrusted sites they traverse. The resulting calculus provides a formal framework to reason about protection policies and security protocols over distributed, mobile infrastructures, and aims to stand to ambients as the spi calculus stands to π. We present a type system that separates trusted and untrusted data and code, while allowing safe interactions with untrusted sites. We prove that the type system enforces a privacy property, and show the expressiveness of the calculus via examples and an encoding of the spi calculus.

Introduction

Secure communication in the π-calculus relies on *private channels*. Process $(\boldsymbol{\nu}n)(\ \overline{n}\langle m\rangle \mid n(x).P\ )$ uses a private channel n to transmit message m. Intuitively, this guarantees the secrecy of m since no third process may interfere with n. In a distributed network, however, the subprocesses $\overline{n}\langle m\rangle$ and $n(x).P$ may be located at remote sites, and the link between them be physically insecure regardless of the privacy of n. It may therefore be desirable to implement a channel meant to deliver private information with lower level mechanisms, as for instance the *encrypted* connection over a *public* channel of the spi calculus [3]:

$$(\boldsymbol{\nu}n)(\ \overline{p}\langle\{m\}_n\rangle \mid p(y).case\ y\ of\ \{x\}_n\ in\ P\ ) \tag{1}$$

The knowledge of n is still confined here, but its role is different: n is an encryption key, rather than a channel. The message is encrypted and communicated along a public channel p; even though the encrypted packet is intercepted, only the intended receivers, which possess the key n, may decrypt it to obtain m (cf. [1] for a thorough discussion of the shortcomings of the scheme.)

Similar mechanisms for secrecy are available for Mobile Ambients, MA, [10]. The following process, for instance, provides for the exchange of messages between locations a and b.

$$(\boldsymbol{\nu}n)(a[\ n[\text{out } a.\text{in } b.\langle m\rangle]\ ] \mid b[\text{open } n.(x).P]) \tag{2}$$

[*] Research supported by EU FET-GC 'MyThS: Models and Types for Security in Mobile Distributed Systems' IST-2001-32617 and 'Mikado: Mobile Calculi based on Domains' IST-2001-32222, and by MIUR Project 'Modelli Formali per la Sicurezza'.

J.C.M. Baeten et al. (Eds.): ICALP 2003, LNCS 2719, pp. 969–983, 2003.

Ideally, no adversary can discover or seize m, or cause a different message to be delivered at b, as m is encapsulated into the secret ambient n.

The question we address in this paper is whether the abstract enveloping mechanism above can be turned into a realistic model of security for calculi of mobile agents that need to enforce protection policies and secrecy guarantees in untrusted environments. The answer we provide is articulated, and leads us to introduce new flexible, lower-level mechanisms. Our work is inspired by the development of spi from π in the ambition of identifying suitable such primitives.

Structure of the paper. §1 discusses how to achieve secrecy in (variants of) MA, and presents the motivations for introducing specific primitives; §2 provides a formal definition of the outcome, the SBA calculus, and illustrates it with a few examples. A key point of our work is the development in §3 of a type system which governs the interactions between trusted and untrusted (*opponents*) components travelling over open networks. Types split the world in two: the trusted *system* and the untrusted *context*. Relying on such intuition, data coming from the external environment is assigned an "unknown" type Public; Public values are handled with suspicion, since there is no saying what they are, or whom they are from. The type system guarantees a *secrecy* property proved in §4: a well-typed process does not disclose its secrets to any adversary, even though these may know its public names and traverse its sub-ambients. §5 presents an encoding of the spi calculus in SBA as a starting point for future comparisons between the two calculi.

1   A Sealing Mechanism for Ambients

The literature on mobile agent security focuses mainly on the dual problems of protecting a host from incoming agents and protecting a mobile agent from malicious hosts. Cryptography is used effectively in the latter case, by setting up a network of trusted sites and mechanisms of authentication between such sites and encrypted agents on the move (cf. [20,18]). The sealing mechanism we envision aims at protecting the secrecy of data inside ambient-like mobile agents which move freely in a network of possibly unreliable sites. The first question is whether these mechanisms are needed at all.

The security model of the Ambient Calculus [8] is centred around the idea that names provide the key to access the contents (data and code) encapsulated by ambients. Accordingly, as long an ambient name is secret, its content is protected from undesired access. The protocol for message exchange in (2), which here we question, is based on such secrecy assumption.

We start from the observation that ambient movement cannot occur without ambient revealing their names to some (not necessarily trusted) component of the underlying infrastructure. This happens in current implementations (cf. the ambient managers of [13] and the pointers-to-parent of [19]), and it is hard to conceive how it could possibly be otherwise. In the internetworking of the near future, crossing boundaries (routers, gateways, firewalls, ...) will involve running complex protocols. Travelling active packets will have to negotiate several conditions, such as QoS guarantees and bandwidth occupation, as well as paying for the service received. The principles of interoperability across different networks and of data encapsulation will require such protocols to work as direct dialogues between the interested parties. This can only rely on direct communication and, therefore, force agents to reveal their interfaces to the network. Thus, quite as secure remote communication cannot rely exclusively on private names, the security of a mobile ambient cannot be relegated to the confidentiality of its name. Back to our example, the encapsulation mechanism of (2) turns out not to be secure, as in a realistic scenario name n will have to be disclosed.

We may think of two ways to provide for stronger security guarantees. One possibility is to commit to agents their own security, by resorting to co-capabilities. For instance, process (1) can be recast in Safe Ambients, SA, [16] as shown below.

$$(\boldsymbol{\nu}n)(a[\,n[\,\text{out } a.\text{in } b.\overline{\text{open}}\ n.\langle m\rangle\,]\ |\ \overline{\text{out}}\ a\,]\ |\ b[\,\overline{\text{in}}\ b.\text{open } n.(x)P\,]) \qquad (3)$$

An alternative approach is to protect the secrecy of an ambient name by wrapping the ambient in a box that carries it to destination:

$$(\boldsymbol{\nu}n)(a[\,p[\,\text{out } a.\text{in } b\ |\ n[\,\langle m\rangle\,]\,]\,]\ |\ b[\,\text{open } p.\text{open } n.(x)P\,]) \qquad (4)$$

The first protocol guarantees that no one can enter n, or open it and read m before n reaches b, even though the name n is revealed while n is on the move. Notice that we ignore here the orthogonal issue of authenticating b against possible malicious impersonators. In (4), name n and message m are protected by the wrapper ambient p, to be opened at the target site b. Now n need not reveal its name – even though p is forcibly opened by an attacker – because it does not move.

Whether or not these protocols are satisfactory depends on the kind of agents and networks targeted. If we look at ambients as abstract physical devices, such as laptops or PDA's, then the first approach is likely to be all we need: physical devices can easily perform access control to protect their contents in ways similar to those encompassed by co-capabilities in (3). If instead, we think of ambients as representing "soft" agents, then (3) is only appropriate in "friendly" networks, where gateways respect the privacy of the code they route to the next hop to destination.

The second approach is more robust and applies well to the case of soft agents. In particular, in (4), we may think of $n[\,\langle m\rangle\,]$ as a piece of data encrypted under the key n: this is consistent with the structure of the protocol, as ambient n need not be active while inside p, since it is the thread out $a.$in b that routes p (hence n) to destination. On the other hand, this solution cannot be fruitfully applied to protect active agents, which cannot move autonomously when encrypted.

The solution we advocate combines the benefits of the two approaches just discussed, by introducing new abstract primitives (which can be read as) providing for subjective access control by ways of co-capabilities, and data encryption to preserve secrecy of data while allowing agents to move autonomously.

We develop our approach for the calculus NBA of [7], a calculus of (boxed) ambients based on two ideas: direct, named communication across parent-child boundaries, and dynamic learning of incoming ambients' names. An NBA ambient owns two channels, one for local, intra-ambient interactions, and one for hierarchical 'upward' communications. For instance $(x)^n.P \mid n[\langle m\rangle^\uparrow.Q]$ reduces to $P\{x := m\} \mid n[Q]$, and symmetrically with the roles of input and output swapped. Moreover, co-capabilities are binders, so that $a[\,\mathsf{in}\langle b,k\rangle.P\,] \mid b[\,\overline{\mathsf{in}}(x,k).Q\,]$ reduces to $b[\,a[P] \mid Q\{x := a\}\,]$, and similarly for the out capability. This means that Q inside b has learnt the name of the incoming agent a. Observe that k acts to control access to b, and must be matched by a for the move to take place. Actually, name binding and access control checking work in a way at all analogous to the exchange of names and credentials which occur when registering for a networked service (cf. [7] for a deeper discussion and for related work).

Following the intuitions highlighted above, on top of the communication and movement mechanisms à la NBA, we introduce a specific primitive to let an ambient 'seal' itself: $n[\,\mathsf{seal}\ k.P \mid Q] \longrightarrow n\{\!\!|\ P \mid Q\ |\!\!\}_k$. By exercising the capability seal k in one of its internal threads, ambient n blocks all its interactions with the outside and encrypts all its messages (to be exchanged either locally or across boundaries), included those in the thread Q. The flexibility of this mechanism derives from the fact that a "sealed" ambient $n\{\!\!|\ P \mid Q\ |\!\!\}_k$ is still (partially) active: in particular, it may still move over the network and perform limited forms of local synchronisation. On the contrary, its message exchanges are blocked and all its data encrypted, and so remain until it reaches a computational environment which knows k, the sealing key. The mechanism to unseal a sealed ambient is associated to movement and exercised through co-capabilities containing keys such as in the following process, where $n\{P'\}$ is an ambient that can be either sealed or not:

$$n\{\!\!|\ \mathsf{in}\ m.P \mid Q\ |\!\!\}_k \mid m\{\,\overline{\mathsf{in}}\ \{x\}_k.R \mid S\,\} \longrightarrow m\{\,n[P \mid Q] \mid R\{x := n\} \mid S\,\}$$

The resulting model can, in some respects, be viewed as a symmetric cryptosystem, with encryption associated with the sealing capability that secures the data inside an ambient, and decryption associated with the dual operation of unsealing performed at ambient boundaries.

2   Sealed Ambients

The syntax of the SBA calculus below is a proper extension of the syntax of Boxed Ambients, BA, [5], with movement co-capabilities and new 'sealing' primitives.

Expressions	M, N	$::= k \cdots q \mid x \cdots z \mid \mathsf{in}\ M \mid \mathsf{out}\ M \mid \overline{\mathsf{in}} \mid \overline{\mathsf{out}} \mid M.M$
Locations	η	$::= M \mid \uparrow \mid \star$
Prefixes	π	$::= M \mid (x_1,\dots,x_k)^\eta \mid \langle M_1,\dots,M_k\rangle^\eta \mid \overline{\mathsf{in}}\ \{x\}_M \mid \overline{\mathsf{out}}\ \{x\}_M \mid \mathsf{seal}\ M$
Processes	P	$::= 0 \mid \pi.P \mid (\nu n)P \mid P\mid P \mid !P \mid M[P] \mid M\{\!\!| P |\!\!\}_N$

Names $(k \cdots q)$ and variables $(x \cdots z)$ range over two disjoint sets; we use $a \cdots d$ to denote elements from either set, when the distinction is immaterial. Messages are formed as usual over names and (sequences of) capabilities. Locations indicate the target of a communication, i.e. a process in a child ambient M, in the parent ambient $(\uparrow)$, or a local process $(\star)$. The operators of inactivity, composition, restriction and replication are inherited from the π-calculus [17]. The process forms $(x_1, \ldots, x_k)^\eta.P$, $\langle M_1, \ldots, M_k \rangle^\eta.P$ and $M[P]$ denote directed (synchronous) input/output, as in BA, and ambients, as in MA. In addition, SBA provides a new construct for the formation of *sealed* ambients, noted $M\{\!| P |\!\}_N$, where M is the name and N is the sealing key. Three new prefix forms provide for the operations of unsealing, $\overline{\text{in}} \{x\}_k.P$, $\overline{\text{out}} \{x\}_k.P$, and sealing $\text{seal } k.P$.

We follow the usual conventions. Parallel composition has the lowest precedence among the operators, $\pi_1.\pi_2.P$ is read as $\pi_1.(\pi_2.P)$, while $\langle \tilde{M} \rangle^\eta$, $(\tilde{x})$ and $(\boldsymbol{\nu}\tilde{n})$ stand for $\langle M_1, \ldots, M_k \rangle^\eta$, $(x_1, \ldots, x_k)^\eta$, $(\boldsymbol{\nu} \, n_1, \ldots, n_k)$, respectively. We omit trailing and isolated dead processes, writing π for $\pi.0$, $\langle \tilde{M} \rangle$ for $\langle \tilde{M} \rangle.0$, and $n[\ ]$ for $n[0]$. The superscript $\star$ for local communication, is omitted. The operators $(\boldsymbol{\nu}n)P$, $\overline{\text{in}} \{x\}_a.P$, $\overline{\text{out}} \{x\}_a.P$, and $(\tilde{x})^\eta.P$ act as binders for the name n, and the variables x and $\tilde{x}$, respectively. The sets of *free names* and *free variables* of P, $fn(P)$ and $fv(P)$, are defined accordingly. A process is closed if it has no free variables (though it may have free names). In addition, we write $M\{P\}$ for $M\{\!| P |\!\}_N$ or $M[P]$ when the distinction may safely be disregarded; notice that in the following $M\{P\}$ always refers to the same kind of ambient on both the sides of a reduction rule.

Reduction. The operational semantics is defined as usual in terms of reduction and structural congruence. The definition of structural congruence is standard (cf. [10]). The basic idea behind the reduction relation is that ambients can be in two states, either *sealed* or *unsealed*. An ambient may be sealed at its formation, or become sealed as a result of one of its enclosed threads exercising a capability. When sealed, an ambient may move but not exchange any value, either locally or with the context. An unsealed ambient is fully operational and may move, as well as communicate. The two states for reductions are formalised by defining the reduction relation in terms of two, inter-dependent relations, formalised in Table 1.

The relation $\dashrightarrow$ (referred to as *silent reduction*) gives the semantics of mobility and sealing. Rules (*enter*) and (*exit*) allow any ambient, sealed or unsealed, to traverse any other ambient, sealed or unsealed: the move requires the target ambient to cooperate by offering a co-capability. Rules (*K-enter*) and (*K-exit*) provide an alternative mechanism for mobility, akin to that studied in [7]. As in *loc. cit.*, the incoming ambient is authenticated by a test on the sealing key k, and then its name registered by binding it to the variable x. In addition, the authentication mechanism of SBA has the effect of removing the seal on the incoming ambient, so as to enable it to interact with the accepting context. Rule *(seal)* shows the effect of sealing: the capability $\text{seal } k$ instructs a process to seal its enclosing ambient under a key k. Notice that encryption of individual messages remains indicated only implicitly by the $\{\!| \ldots |\!\}$ around the ambient; besides being a convenient notation, this abstracts away from irrelevant implementation details.

Silent reductions may occur within any context, except under prefixes. On the contrary, the reductions involving communication – which are exactly as in previous versions of (N)BA, viz. [11,7] – may only occur within unsealed ambients, as formalised by the relation $\longrightarrow$. This reflects the fact that semantically relevant local communications must involve clear-text messages and, therefore, be avoided in untrusted environments, i.e. when ambients are sealed. Finally, rule (*struct*) is standard, while rules (*silent*) and (*ambient*) guarantee that the two reduction relations are linked properly.

Remarks. For ease of presentation, the syntax and operational semantics are so defined as to guarantee that ambients cannot be sealed more than once. An alternative choice would be to separate the sealing primitives from those for mobility. Specifically, one could introduce an explicit unsealing prefix such as $\mathsf{unseal}\{x\}_k.P$, and define its semantics by the reduction $\mathsf{unseal}\{x\}_k.P \mid n\{\!\!\{\,Q\,\}\!\!\}_k \longrightarrow P\{x := n\} \mid n[\,Q\,]$. This, together with rules (*enter*) and (*exit*), would implement an unsealing mechanism similar to ours, albeit not atomic. However, our proposal appears to model faithfully the current practice in distributed and mobile systems, where the protocols for agent authentication and certification take place at domain (i.e. ambient) boundaries rather than after such boundaries have been crossed.

Examples. The kind of secure communication expected of the exchange of messages in (2) can now be achieved as in $(\nu n)a[\ p[\mathsf{seal}\ n.\mathsf{out}\ a.\mathsf{in}\ b.\langle m\rangle^{\uparrow}]\ ]$ $\mid \overline{\mathsf{out}} \mid\ b[\overline{\mathsf{in}}\ \{x\}_n.(y)^x.P \mid Q]$. The public ambient p seals itself with the private key n, shared by the sender and the intended receiver, moves over the network towards its destination, gets unsealed in the act of entering it, and becomes then ready to deliver its message. Like in the spi-calculus, it is the sealing key that is private, while the name of the ambient may be left public.

Incidentally, this formulation of the message exchange fixes a minor flaw in the protocols we discussed in §1 above. Namely, in the configuration $b[\mathsf{open}\ n.(x).Q \mid Q' \mid n[\overline{\mathsf{open}}\ n.\langle m\rangle]\,]$, as the opening of n and the delivery of its message are distinct steps, there is no guarantee that m will be received by the intended process when multiple threads are present inside b. In particular, m could end up in Q', even when it did not actually know the secret name n. Such behaviour is however inherent to the communication model of MA, and easily be avoided with the primitives for hierarchical communication of the present calculus

As a more realistic example, consider the case of an agent in search of vendors of a particular item over the network. The agent originates at a user site u, visits a collection s_i of network sites and reports the names of those which provide a specific item it. To protect the agent moving over the network, we use the SBA primitives as follows. Let k be a sealing key shared between user u and sites s_i. The user can be represented as the process $\mathsf{u}[\,(\nu a)a[P \mid R] \mid Q\,]$, where a is an agent with two threads: a router R, which controls movements, and a

Table 1. Reduction and Silent Reduction

Silent Reduction

Silent Reduction Context S ::= $-$ | $(\nu n)\mathsf{S}$ | $P|\mathsf{S}$ | $n[\mathsf{S}]$ | $n\{\!|\,\mathsf{S}\,|\!\}_k$

MOBILITY (I)

(enter)	$n\{\,\text{in } m.P \mid Q\,\} \mid m\{\,\overline{\text{in}}\,.R \mid S\,\} \dashrightarrow m\{\,n\{\,P \mid Q\,\} \mid R \mid S\,\}$
(exit)	$m\{\,n\{\,\text{out } m.P \mid Q\,\} \mid R\,\} \mid \overline{\text{out}}\,.S \dashrightarrow n\{\,P \mid Q\,\} \mid m\{\,R\,\} \mid S$

MOBILITY (II)

(K-enter)	$n\{\!|\,\text{in } m.P \mid Q\,|\!\}_k \mid m\{\,\overline{\text{in}}\,\{x\}_k.R \mid S\,\} \dashrightarrow m\{\,n[\,P \mid Q\,] \mid R\{x := n\} \mid S\,\}$
(K-exit)	$m\{\,P \mid n\{\!|\,\text{out } m.Q \mid R\,|\!\}_k\,\} \mid \overline{\text{out}}\,\{x\}_k.S \dashrightarrow m\{\,P\,\} \mid n[\,Q \mid R\,] \mid S\{x := n\}$
(seal)	$n[\,\text{seal } k.P \mid Q\,] \dashrightarrow n\{\!|\,P \mid Q\,|\!\}_k$

STRUCTURAL RULES

(struct)	$P \equiv Q, Q \dashrightarrow R,\ R \equiv S \ \Rightarrow\ P \dashrightarrow S$
(context)	$P \dashrightarrow Q \ \Rightarrow\ \mathsf{S}\{P\} \dashrightarrow \mathsf{S}\{Q\}$

Reduction

Reduction Context $\mathbf{E}$ ::= $-$ | $(\nu n)\mathbf{E}$ | $P|\mathbf{E}$ | $n[\mathbf{E}]$

COMMUNICATION

(local)	$(\tilde{x})P \mid \langle \tilde{M}\rangle Q \longrightarrow P\{\tilde{x} := \tilde{M}\} \mid Q$
(input n)	$(\tilde{x})^n P \mid n[\,\langle \tilde{M}\rangle^\uparrow Q \mid R\,] \longrightarrow P\{\tilde{x} := \tilde{M}\} \mid n[\,Q \mid R\,]$
(output n)	$\langle \tilde{M}\rangle^n P \mid n[\,(\tilde{x})^\uparrow Q \mid R\,] \longrightarrow P \mid n[\,Q\{\tilde{x} := \tilde{M}\} \mid R\,]$

STRUCTURAL RULES

(silent)	$P \dashrightarrow Q \ \Rightarrow\ P \longrightarrow Q$
(ambient)	$P \longrightarrow Q \ \Rightarrow\ n[\,P\,] \dashrightarrow n[\,Q\,]$
(struct)	$P \equiv Q, Q \longrightarrow R,\ R \equiv S \ \Rightarrow\ P \longrightarrow S$
(context)	$P \longrightarrow Q \ \Rightarrow\ \mathbf{E}\{P\} \longrightarrow \mathbf{E}\{Q\}$

---

communicator P, which interacts with the visited sites. We use two locks l and r to synchronise the two threads within a.

$$a[\,(\nu l, r)(\text{synch}(l) \mid \;!\,\overline{\text{synch}}(l).\text{seal } k.(\text{synch}(r) \mid \langle \text{it}\rangle^\uparrow.(x,y)^\uparrow([x = \text{it}]\langle y\rangle \mid \text{synch}(l)))$$
$$\mid \;\overline{\text{synch}}(r).\text{route}(u, s_1).\overline{\text{synch}}(r).\text{route}(s_1, s_2).\overline{\text{synch}}(r).\text{route}(s_2, u)\,]\,.$$

where $[a = b]P \triangleq (\nu c)\,(c[\,\langle\ \rangle^a \mid b[\,(\ )^\uparrow.\langle\ \rangle^\uparrow\,]\,] \mid (\ )^b.\langle\ \rangle^\uparrow\,] \mid (\ )^c.P), (c \notin fn(P))$
and $\text{synch}(n) \triangleq n[\,n\{\!|\,\text{out } n\,|\!\}_n\,]$, $\overline{\text{synch}}(n) \triangleq \overline{\text{out}}\,\{\_\}_n$

The first thread is a loop that, when activated, seals the agents under the key k, activates the router, and waits for a to be routed to the destination sites. Once there, it collects the name of the vendor, if this contains the desired item. The router thread, in turn, ships a across the network to visit the sites, in this case s_1 and s_2. However, before moving outside u or any of the s_i, it waits for the sibling thread to seal the agent using k. The reduction semantics guarantees that, whenever the ambient a is not inside a site which knows k, all data in a are sealed, hence kept secret.

To synchronise with each other, the router and the communicator use the process forms: $\text{synch}(n)$ and $\overline{\text{synch}}(n)$. Interestingly, local synchronisation between threads is available even though the ambient is sealed, since it does not rely on exchanges of messages. Finally, each of the visiting sites can be coded as $s_i[\overline{\text{in}}\ \{z\}_k.(x)^z.\langle f_i(x), s_i\rangle^z\ |\ \ldots]$. When agent a enters s_i it gets unsealed, so that it may hold exchanges with the site. Here the function f_i represents a lookup performed by the site searching for item x: the result is x if s_i has x on sale, or some different value otherwise. Of course, rather than total unsealing, a policy of selective decryption of sensitive data may be desirable when agents interact with sites only partially trusted; this variation of the example can easily be implemented in SBA.

3 A Type System

The type system separates trusted and untrusted data and code while allowing safe interactions with untrusted sites. In particular, a distinct type Un is used to type processes for which we cannot make any assumption on structure and/or behaviour. Correspondingly, we assign a 'default' type Public to data that comes from untyped processes, and we handle such data carefully. The structure of types is defined by the following productions:

$$
\begin{aligned}
\textit{Expression Types} \quad & W ::= \mathsf{Amb}[E]\ \mid\ \mathsf{Key}[E]\ \mid\ \mathsf{Public} \\
\textit{Exchange Types} \quad & E, F ::= \mathsf{shh}\ \mid\ (W_1, \ldots, W_k) \\
\textit{Process Types} \quad & T ::= [E, F]\ \mid\ \mathsf{Un}
\end{aligned}
$$

Untrusted processes are built upon expressions of type Public. In addition, the type Public is assigned to expressions that trusted processes may exchange with untrusted ones. Among such expressions, we include the movement (co-)capabilities, so as to enhance the flexibility of typing: there is no negative effect on safety (or security) in this choice, as the interaction among trusted components is enabled by the possession of shared keys, which are secret and hence protected from the untrusted components. The type $\mathsf{Key}[E]$ is the type of sealing keys: a key with this type may only be used to seal (trusted) ambients of type $\mathsf{Amb}[E]$. The latter, in turn, is the type of all the trusted ambients whose upward exchanges (if any) have type E. Notice that only ambients (not generic expressions) can be sealed. However, even untrusted ambients may be sealed, but in that case the sealing key is a generic expression of type Public and no security guarantee is made. As for process types, $[E, F]$ is the type of all processes that can be enclosed in ambients of type $\mathsf{Amb}[F]$, with E and F denoting the local and upward exchanges of the processes in question. Un is the type of the untrusted processes. In order to provide the intended privacy guarantees, the types of trusted and untrusted data and processes are kept separate (there is no subsumption rule, nor any common super-type). Nevertheless, the typing rules for processes allow non-trivial forms of interaction between trusted and untrusted processes. Specifically, ambients have full migration capabilities as the type system allow trusted ambients to traverse untrusted ones and vice versa (as in the example of §2). Instead, a trusted (resp. untrusted) sealed ambient may be unsealed only within trusted (resp. untrusted) contexts. As for communication, the following policy is adopted: (*i*) local exchanges are allowed everywhere except that at top level, where we disallow local exchanges between trusted and untrusted processes, and (*ii*) trusted and untrusted ambients may exchange values across boundaries, provided that such values have type Public. We proceed with the description of the typing rules, collected in Tables 2 and 3.

Typing Rules. Every (co-)capability is assigned type Public; accordingly, the rule (PREFIX) allows trusted ambients to traverse untrusted ones and vice versa without breaking the soundness of the type system. Note that ill-formed (paths of) capabilities, such as $a.b$ and in $(a.b)$ do type check in this system when a, b are Public. This is necessary to allow full flexibility in the typing of the opponent: on the other hand, we will prove that the type system providse the expected guarantees of secrecy and safety for any value exchange.

Each process form has two associated typing rules, depending on whether the process in question is to be considered trusted (Table 2) or deemed untrusted (Table 3): in the latter case, it could be an attacker or a trusted process tainted by an interaction with an untrusted component via its public names. For prefixes the two cases can be accounted for by a single rule, (PREFIX), where T stands for either $[E, F]$ or Un. For ambients we need four rules: rule (AMB SEAL) in Table 2, assigns a type to ambients formed with the 'right' key N, and enclosing a process P with the expected exchanges. Rule (AMB) in Table 2 is standard. Rules (UNTRUSTED AMB/AMB SEAL), in Table 3 are used to type untrusted, possibly ill-formed, ambients. In addition, observe that a trusted (sealed) ambient may be typed with type Un; this is perfectly correct and allows a trusted (sealed) ambient to traverse untrusted sites.

The same rationale applies to the prefix constructors for sealing and unsealing, as well as for local and upward communication. Three typing rules handle the case of input (output) from a sub-ambient M. As we noted above, we allow untrusted and trusted process to exchange values, as long as these have type Public, as required in rules (UNTRUSTED INPUT/OUTPUT M) in Table 3. Note also that in these rules we do not require that the arity of the downward communication matches that of the target ambient. This leaves full flexibility in the typing of opponent processes, as it is implied by the following proposition.

Proposition 1 (Typability). *Let P be a process with $fn(P) = \{a_1, \dots, a_n\}$ and $fv(P) = \{x_1, \dots, x_m\}$. Then a_1 : Public, $\dots, a_n$: Public, x_1 : Public, $\dots, x_m$: Public $\vdash P$: Un,*

In other words, no constraint is imposed on the structure of the opponent: only that it initially does not know any secret. In addition, one can easily prove the standard property of type preservation under reduction.

Proposition 2 (Subject Reduction). *If $\Gamma \vdash P : T$ and $P \longrightarrow Q$, then $\Gamma \vdash Q : T$.*

Table 2. Typing Rules: Trusted Processes

(EMPTY)

(ENV x)
$$\frac{\Gamma \vdash \diamond \quad x \notin Dom(\Gamma)}{\Gamma, x : W \vdash \diamond}$$

(PROJECTION)
$$\frac{\Gamma \vdash \diamond \quad \Gamma(M) = W}{\Gamma \vdash M : W}$$

(PATH)
$$\frac{\Gamma \vdash M_1 : \mathsf{Public} \quad \Gamma \vdash M_2 : \mathsf{Public}}{\Gamma \vdash M_1.M_2 : \mathsf{Public}}$$

$$\emptyset \vdash \diamond$$

(CO-IN)
$$\frac{\Gamma \vdash \diamond}{\Gamma \vdash \overline{\mathsf{in}} : \mathsf{Public}}$$

(CO-OUT)
$$\frac{\Gamma \vdash \diamond}{\Gamma \vdash \overline{\mathsf{out}} : \mathsf{Public}}$$

(IN M)
$$\frac{\Gamma \vdash M : W \quad W \in \{\mathsf{Amb}[E], \mathsf{Public}\}}{\Gamma \vdash \mathsf{in}\ M : \mathsf{Public}}$$

(OUT M)
$$\frac{\Gamma \vdash M : W \quad W \in \{\mathsf{Amb}[E], \mathsf{Public}\}}{\Gamma \vdash \mathsf{out}\ M : \mathsf{Public}}$$

(PREFIX)
$$\frac{\Gamma \vdash M : \mathsf{Public} \quad \Gamma \vdash P : T}{\Gamma \vdash M.P : T}$$

(SEAL)
$$\frac{\Gamma \vdash M : \mathsf{Key}[E] \quad \Gamma \vdash P : [F, E]}{\Gamma \vdash \mathsf{seal}\ M.P : [F, E]}$$

(AMB)
$$\frac{\Gamma \vdash M : \mathsf{Amb}[E] \quad \Gamma \vdash P : [F, E]}{\Gamma \vdash M[P] : T}$$

(AMB SEAL)
$$\frac{\Gamma \vdash N : \mathsf{Key}[E] \quad \Gamma \vdash M : \mathsf{Amb}[E] \quad \Gamma \vdash P : [F, E]}{\Gamma \vdash M\{\!| P |\!\}_N : T}$$

(CO-IN KEY)
$$\frac{\Gamma \vdash M : \mathsf{Key}[E] \quad \Gamma, x : \mathsf{Amb}[E] \vdash P : [G, H]}{\Gamma \vdash \overline{\mathsf{in}}\ \{x\}_M.P : [G, H]}$$

(CO-OUT KEY)
$$\frac{\Gamma \vdash M : \mathsf{Key}[E] \quad \Gamma, x : \mathsf{Amb}[E] \vdash P : [G, H]}{\Gamma \vdash \overline{\mathsf{out}}\ \{x\}_M.P : [G, H]}$$

(DEAD)
$$\frac{\Gamma \vdash \diamond}{\Gamma \vdash \mathbf{0} : T}$$

(PAR)
$$\frac{\Gamma \vdash P : T \quad \Gamma \vdash Q : T}{\Gamma \vdash P \mid Q : T}$$

(NEW)
$$\frac{\Gamma, n : W \vdash P : T}{\Gamma \vdash (\nu n)P : T}$$

(REPL)
$$\frac{\Gamma \vdash P : T}{\Gamma \vdash !P : T}$$

(LOCAL INPUT)
$$\frac{\Gamma, x_1 : W_1, \ldots, x_k : W_k \vdash P : [(W_1, \ldots, W_k), E]}{\Gamma \vdash (x_1, \ldots, x_k).P : [(W_1, \ldots, W_k), E]}$$

(LOCAL OUTPUT) $i = 1, \ldots, k$
$$\frac{\Gamma \vdash M_i : W_i \quad \Gamma \vdash P : [\tilde{W}, E]}{\Gamma \vdash \langle \tilde{M} \rangle.P : [\tilde{W}, E]}$$

(INPUT $\uparrow$)
$$\frac{\Gamma, x_1 : W_1, \ldots, x_k : W_k \vdash P : [E, (W_1, \ldots, W_k)]}{\Gamma \vdash (x_1, \ldots, x_k)^{\uparrow}.P : [E, (W_1, \ldots, W_k)]}$$

(OUTPUT $\uparrow$) $i = 1, \ldots, k$
$$\frac{\Gamma \vdash M_i : W_i \quad \Gamma \vdash P : [E, (W_1, \ldots, W_k)]}{\Gamma \vdash \langle M_1, \ldots, M_k \rangle^{\uparrow}.P : [E, (W_1, \ldots, W_k)]}$$

(INPUT M)
$$\frac{\Gamma \vdash M : \mathsf{Amb}[\tilde{W}] \quad \Gamma, \tilde{x} : \tilde{W} \vdash P : [E, F]}{\Gamma \vdash (\tilde{x})^M.P : [E, F]}$$

(OUTPUT M AMB)
$$\frac{\Gamma \vdash N : \mathsf{Amb}[\tilde{W}] \quad \Gamma \vdash \tilde{M} : \tilde{W} \quad \Gamma \vdash P : [E, F]}{\Gamma \vdash \langle \tilde{M} \rangle^N.P : [E, F]}$$

4   A Secrecy Theorem

We refer to a standard notion of secrecy in the literature of security protocols, namely: *a process preserves the secrecy of a piece of data M if it does not publish M, or anything that would permit the computation of M.* The formal definition is inspired by [2]. We adapt that definition to our framework by representing an attacker as a closed, but otherwise arbitrary, context. This leaves full power to an attacker, which can either take the role of an hostile context (or host) enclosing a trusted process, as in $a[Q \mid (-)]$, or the role of a malicious agent mounting an attack to a remote host, as in $a[\mathsf{in}\ p.\mathsf{in}\ q.Q \mid Q'] \mid (-)$. In addition, we characterise the initial knowledge of the attacker in terms of the names, the keys and the capabilities initially known to it. Interestingly, the knowledge of

Table 3. Typing Rules: Untrusted Processes

(UNTRUSTED AMB)

$\Gamma \vdash M : \mathsf{Public} \quad \Gamma \vdash P : \mathsf{Un}$

$$\Gamma \vdash M[\,P\,] : T$$

(UNTRUSTED AMB SEAL)

$\Gamma \vdash N : \mathsf{Public} \quad \Gamma \vdash M : \mathsf{Public} \quad \Gamma \vdash P : \mathsf{Un}$

$$\Gamma \vdash M \{\!| P |\!\}_N : T$$

(UNTRUSTED SEAL)

$\Gamma \vdash M : \mathsf{Public} \quad \Gamma \vdash P : \mathsf{Un}$

$$\Gamma \vdash \mathsf{seal}\ M.P : \mathsf{Un}$$

(UNTRUSTED CAP)

$\Gamma \vdash M : \mathsf{Public} \quad \Gamma \vdash P : \mathsf{Un}$

$$\Gamma \vdash M.P : \mathsf{Un}$$

(UNTRUSTED CO-IN)

$\Gamma \vdash M : \mathsf{Public} \quad \Gamma, x{:}\mathsf{Public} \vdash P : \mathsf{Un}$

$$\Gamma \vdash \overline{\mathsf{in}}\ \{x\}_M.P : \mathsf{Un}$$

(UNTRUSTED CO-OUT)

$\Gamma \vdash M : \mathsf{Public} \quad \Gamma, x{:}\mathsf{Public} \vdash P : \mathsf{Un}$

$$\Gamma \vdash \overline{\mathsf{out}}\ \{x\}_M.P : \mathsf{Un}$$

(UNTRUSTED LOCAL INPUT)

$\Gamma, x_1 : \mathsf{Public}, \dots x_k : \mathsf{Public} \vdash P : \mathsf{Un}$

$$\Gamma \vdash (x_1, \dots , x_k).P : \mathsf{Un}$$

(UNTRUSTED LOCAL OUTPUT)

$\Gamma \vdash M_i : \mathsf{Public} \quad i = 1, \dots, k \quad \Gamma \vdash P : \mathsf{Un}$

$$\Gamma \vdash \langle M_1, \dots , M_k \rangle.P : \mathsf{Un}$$

(UNTRUSTED INPUT ↑)

$\Gamma, x_1 : \mathsf{Public}, \dots x_k : \mathsf{Public} \vdash P : \mathsf{Un}$

$$\Gamma \vdash (x_1, \dots , x_k)^{\uparrow}.P : \mathsf{Un}$$

(UNTRUSTED OUTPUT ↑)

$\Gamma \vdash M_i : \mathsf{Public} \quad i = 1, \dots, k \quad \Gamma \vdash P : \mathsf{Un}$

$$\Gamma \vdash \langle M_1, \dots , M_k \rangle^{\uparrow}.P : \mathsf{Un}$$

(INPUT M UNTRUSTED)

$\Gamma \vdash M : \mathsf{Public} \quad \Gamma, x_1 : \mathsf{Public}, \dots x_k : \mathsf{Public} \vdash P : T$

$$\Gamma \vdash (x_1, \dots , x_k)^M.P : T$$

(OUTPUT M UNTRUSTED)

$\Gamma \vdash M : \mathsf{Public} \quad \Gamma \vdash M_i : \mathsf{Public} \quad i = 1, \dots, k \quad \Gamma \vdash P : T$

$$\Gamma \vdash \langle M_1, \dots , M_k \rangle^M.P : T$$

(UNTRUSTED INPUT M)

$\Gamma \vdash M : \mathsf{Amb}[\mathsf{Public}_1, \dots , \mathsf{Public}_n] \quad \Gamma, x_1 : \mathsf{Public}, \dots x_k : \mathsf{Public} \vdash P : \mathsf{Un}$

$$\Gamma \vdash (x_1, \dots , x_k)^M.P : \mathsf{Un}$$

(UNTRUSTED OUTPUT M)

$\Gamma \vdash M : \mathsf{Amb}[\mathsf{Public}_1, \dots , \mathsf{Public}_n] \quad \Gamma \vdash M_i : \mathsf{Public} \quad i = 1, .., k \quad \Gamma \vdash P : \mathsf{Un}$

$$\Gamma \vdash \langle M_1, \dots , M_k \rangle^M.P : \mathsf{Un}$$

capabilities is important here, since by exercising (a sequence of) capabilities an adversary may approach an agent and interact with it, even without knowing its name. As an example, if we take the process $b[\,\overline{\mathsf{in}}\ \{x\}_k.\langle a \rangle^x\,]$, an opponent may have access to the value a even without knowing the name b: knowing the capability 'in b' and the key 'k' is enough.

We define a context $\mathbf{A}(-)$ to be a process that contains exactly one variable $(-)$ (i.e. a hole). We denote with $\mathbf{A}(P)$ the process resulting from substituting the variable with P in A. Also, we denote with $fc(P)$ the set of capabilities formed over the free names of P: the inductive definition of this set is straightforward.

Definition 1 (S-adversary). *Let S be a finite set of names and capabilities. The closed context $\mathbf{A}(-)$ is an S-adversary if $fn(\mathbf{A}(-)) \cup fc(\mathbf{A}(-)) \subseteq S$.*

Next, we define what it means to preserve a secret: since capabilities are public, the definition of secrecy only applies to names. Let $\Longrightarrow$ be the reflexive and transitive closure of the reduction relation $\longrightarrow$.

Definition 2 (Revealing Names, Preserving their Secrecy). *Let P be a process, n a name free in P, and S a finite set of names and capabilities. P may reveal n to S iff there exists an S-adversary $\mathbf{A}(-)$, with $\mathbf{A}(P)$ closed, and a name $c \in S$ such that $\mathbf{A}(P) \Longrightarrow \mathbf{C}(c[\langle n\rangle^\uparrow \mid Q])$, for some context $\mathbf{C}(-)$ and process Q, with c not bound by $\mathbf{C}(-)$. Dually, P preserves the secrecy of n from S iff it does not reveal n to S.*

The definition extends readily to private names as follows (cf. [9]): $(\boldsymbol{\nu}n)P$ *may reveal n to S* if and only if there is a fresh name m such that $P\{n := m\}$ may reveal m to S, with $m \notin S \cup fn(P)$. Notice that an adversary may dynamically acquire new names and new capabilities (*i*) by creating its own fresh names, (*ii*) by receiving names over public channels, and (*iii*) by unsealing ambients sealed with a key it knows (thus learning the ambient's name). As an example, take $S = \{c\}$, and consider the process $P = c[\langle a\rangle^\uparrow] \mid a[\langle k\rangle^\uparrow]$. P does not preserve the secrecy of k from S, even though S does not include a. In fact, one can take the S-adversary $\mathbf{A}(-) = (x)^c.(y)^x.c[\langle y\rangle^\uparrow] \mid (-)$, and note that $\mathbf{A}(P) \Longrightarrow c[\langle k\rangle^\uparrow] \mid c[\,] \mid a[\,]$.

The secrecy theorem below states that a well-typed process P does not leak its secrets to any adversary that initially knows all the public names in P and has the capability to move in and out any ambient of P (included its secret ambients).

Theorem 1 (Secrecy). *Let P be a process such that $\Gamma \vdash P : \mathsf{Un}$ and $\Gamma \vdash s : W$ with $W \neq \mathsf{Public}$. Let $S = \{a \mid \Gamma \vdash a : \mathsf{Public}\} \cup \{\mathsf{in}\ a, \mathsf{out}\ a \mid a \in Dom(\Gamma)\}$. Then P preserves the secrecy of s from S.*

Notice that the theorem only holds for well-typed processes of type Un. This immediately rules out processes that exchange non-public data at top level. Indeed, for such processes no secrecy guarantee can be made, for adversaries always have free access to the anonymous top level channel of any process. On the other hand, the theorem captures precisely the security guarantees our approach was intended to provide. That follows by observing (*i*) that well-typed ambient processes can always be typed with type Un, and (*ii*) that ambients (i.e. agents) are indeed the objects of our security concerns.

5   Encoding of the Spi Calculus

We further illustrate the calculus with an encoding of spi-calculus [3]. To ease the presentation, we focus on the following fragment of the asynchronous spi-calculus, in which we disregard the construct for pairs, natural numbers and matching.

$$Expressions\ M, N ::= n \mid x \mid \{M_1, \ldots, M_n\}_N$$
$$Processes\quad P, Q \ ::= \mathbf{0} \mid \overline{M}\langle N_1, \ldots, N_n\rangle \mid M(x_1, \ldots, x_n)P$$
$$\mid\ P \mid Q \mid (\boldsymbol{\nu}n)P \mid case\ M\ of\ \{x_1, \ldots, x_n\}_N\ in\ P$$

Table 4. Encoding of the spi calculus

$\langle\, a\,\rangle_p = a$ (for a a name or a variable)

$\langle\, \{M_1, \ldots, M_n\}_k \,\rangle_p = p$

$[\![\, a\,]\!]_p = 0$ (for a a name or a variable)

$[\![\, \{M_1, \ldots, M_n\}_k \,]\!]_p = (\nu q_1, \ldots, q_n)$ $\{q_1, \ldots, q_n\} \cap (fn(M_1, \ldots, M_n) \cup \{k, p\}) = \emptyset$

$\qquad\qquad [\![\, M_1 \,]\!]_{q_1} \mid \ldots \mid [\![\, M_n \,]\!]_{q_n} \mid\, !\, p[\,(x)^\uparrow.\text{seal } k.\text{in } x.\langle\langle\, M_1 \,\rangle_{q_1}, \ldots, \langle\, M_n \,\rangle_{q_n})^\uparrow\,])$

$[\![\, 0 \,]\!] = 0, \qquad [\![\, (\nu n)P \,]\!] = (\nu n)[\![\, P \,]\!], \qquad [\![\, P \mid Q \,]\!] = [\![\, P \,]\!] \mid [\![\, Q \,]\!]$

$[\![\, \bar{b}\langle M_1, \ldots, M_n \rangle \,]\!] = (\nu q_1, \ldots, q_n)$ $\{q_1, \ldots, q_n\} \cap (fn(M_1, \ldots, M_n) \cup \{b\}) = \emptyset$

$\qquad\qquad [\![\, M_1 \,]\!]_{q_1} \mid \ldots \mid [\![\, M_n \,]\!]_{q_n} \mid b[\,\langle\langle\, M_1 \,\rangle_{q_1}, \ldots, \langle\, M_n \,\rangle_{q_n})^\uparrow\,]$

$[\![\, b(x_1, \ldots, x_n)P \,]\!] = (x_1, \ldots, x_n)^b.[\![\, P \,]\!]$

$[\![\, case\ M\ of\ \{x_1, \ldots, x_n\}_k\ in\ P \,]\!] = (\nu p)([\![\, M \,]\!]_p \mid (\nu c)(\langle c \rangle^{\langle\langle\, M \,\rangle\rangle_p} \mid$ $\{p, c\} \cap \{fn(P) \cup fn(M)\} = \emptyset$

$\qquad\qquad c[\,\overline{\text{in}}\ \{y\}_k.(x_1, \ldots, x_n)^y.(x_1, \ldots, x_n)^\uparrow\,] \mid (x_1, \ldots, x_n)^c.[\![\, P \,]\!]))$

The operational semantics of this fragment is standard (cf. [3]): in particular, decryption is governed by the following reduction: $case\ \{M_1, \ldots, M_n\}_k\ of\ \{x_1, \ldots, x_n\}_k\ in\ P \longrightarrow P\{x_i := M_i\}$.

The basic idea of the encoding is to represent an encrypted message with a sealed ambient that contains that message: communicating the encrypted message is then accounted for by communicating the name of the corresponding ambient. The formal definition is given in Table 4 in terms of three translation maps: $\langle\, \cdot \,\rangle_p :$ *Expressions* $\mapsto$ *Expressions*, $[\![\, \cdot \,]\!]_p :$ *Expressions* $\mapsto$ *Processes*, and $[\![\, \cdot \,]\!] :$ *Processes* $\mapsto$ *Processes*. In the first two (subsidiary) maps, p is the name of the ambient (if any) enclosing the message to be exchanged. In particular, if M is a name or a variable, then $\langle\, M \,\rangle_p$ returns M; if instead M is an encryption packet, $[\![\, M \,]\!]_p$ returns p, the name of the ambient that stores the packet. Correspondingly, $[\![\, M \,]\!]_p$ stores M into an ambient named p, if M is an encrypted message, and returns the inactive process otherwise. More precisely, if M is a message encrypted under a key k, the ambient generated by $[\![\, M \,]\!]_p$ first reads a name x, then gets sealed with k to move into x, where eventually gets unsealed and delivers its payload. The use of replication on the ambient encoding an encryption packet accounts for the possible non-linear usage of messages in spi.

The encoding can be shown to be sound with respect to appropriate choices of behavioural equivalences in the two calculi, noted $\cong_{spi}$ and $\cong_{SBA}$, respectively. In particular, we take $\cong_{spi}$ to be *testing* equivalence, the notion of equivalence for spi-calculus studied in [3], and for SBA, we define $\cong_{SBA}$ to be reduction barbed congruence, based on the following exhibition predicate: $P \downarrow_b^{SBA} \triangleq P \equiv (\nu\tilde{n})((\tilde{x})^b P_1 \mid P_2)$. Given these choices, one can prove that the encoding is equationally sound.

Theorem 2 (Soundness of the encoding). *If* $[\![\, P \,]\!] \cong_{SBA} [\![\, Q \,]\!]$ *then* $P \cong_{spi} Q$.

6 Conclusions

We have investigated new mechanisms to protect migrating agents against the untrusted networks they traverse. Our primitives are best understood as low-level primitives to be employed for a secure implementation of the abstract mechanisms for secrecy found in mainstream ambient calculi.

The resulting calculus, SBA, is derived as a natural extension of NBA, the variant of Boxed Ambients studied in [7]. In fact, NBA can be interpreted into SBA by defining the capability in$\langle n, k \rangle$ as seal k.in n, and similarly for out$\langle n, k \rangle$. (Observe though that this lacks the atomicity of movement and credential verification of NBA.) On the other hand, the sealing model of SBA appears to provide strictly more flexibility and expressiveness than the access control of NBA: a SBA agent can be sealed by any of its local threads. Hence, an agent can be sealed and protected from undesired interactions by firing an action in one of its local threads, and it is not clear that a corresponding mechanism can be recovered in NBA.

We have investigated the role of types in enforcing static guarantees of safety and secrecy in the presence of untyped opponents. It is worth remarking that even though our typing deals with untrusted networks, similar ideas can be used to generalise those presented in [6] and in [11] for access control and information flow security with untrusted components.

Similar studies have been conducted on other process calculi in the literature. In fact, our use of the trusted/untrusted rules is directly inspired by work on the π/spi calculus (Cardelli et al. [9], Gordon and Jeffrey [14], Abadi and Gordon [3]). Alternative approaches to the same problem have also been investigated. Among these, Hennessy and Riely [15] study an extension of the Dπ-calculus with a type system that labels some location as untrusted and relies on run-time type checking to enforce security restrictions for processes coming from untrusted locations. Similar approaches have also been advocated for Mobile Ambients [4], and other calculi (notably Klaim [12]).

Several questions remain to be explored, as for instance whether the data encryption underlying the sealing mechanisms we have introduced can be implemented effectively, and efficiently. Furthermore, in its current formulation, sealing an ambient has only the effect of guaranteeing the secrecy of data. More powerful mechanisms may be necessary to protect migrating agents by further hiding their structure or encrypting subcomponents consisting of data and code. Plans for future include work in both these directions.

Acknowledgements. We would like to thank Beppe Castagna for his suggestions, and the anonymous referees for their comments.

References

1. M. Abadi. Protection in programming-language translations. In *Proceedings of ICALP'98*, number 1443 in LNCS, pages 868–883. Springer-Verlag, 1998.

2. M. Abadi and B. Blanchet. Analyzing security protocols with secrecy types and logic programs. In *Proceedings of POPL'02*, pages 33–44. ACM Press, 2002.
3. M. Abadi and A. Gordon. A Calculus for Cryptographic Protocols: The Spi Calculus. *Information and Computation*, 148(1):1–70, 1999.
4. M. Bugliesi and G. Castagna. Secure safe ambients. In *Procedings of POPL'01*, pages 222–235. ACM Press, 2001.
5. M. Bugliesi, G. Castagna, and S. Crafa. Boxed ambients. In *Proceedings of TACS'01*, number 2215 in LNCS, pages 38–63. Springer-Verlag, 2001.
6. M. Bugliesi, G. Castagna, and S. Crafa. Reasoning about security in mobile ambients. In *Proceedings of CONCUR 2001*, number 2154 in LNCS, pages 102–120. Springer-Verlag, 2001.
7. M. Bugliesi, S. Crafa, M. Merro, and V. Sassone. Communication interference in mobile boxed ambients. In *FST&TCS 2002*, volume 2556 of *LNCS*, pages 71–84. Springer-Verlag, 2002.
8. L. Cardelli. Abstractions for mobile computations. In *Secure Internet Programming*, number 1603 in LNCS, pages 51–94. Springer-Verlag, 1999.
9. L. Cardelli, G. Ghelli, and A. D. Gordon. Secrecy and group creation. In *Proceedings of CONCUR'00*, number 1877 in LNCS, pages 365–379. Springer-Verlag, August 2000.
10. L. Cardelli and A. Gordon. Mobile ambients. In *FoSSaCS'98*, number 1378 in LNCS, pages 140–155. Springer-Verlag, 1998.
11. S. Crafa, M. Bugliesi, and G. Castagna. Information Flow Security for Boxed Ambients. *ENTCS*, 66(3), 2002.
12. R. De Nicola, G. Ferrari, and R. Pugliese. Klaim: a kernel language for agents interaction and mobility. *IEEE Transactions on Software Engeneering*, 24:315–330, 1998.
13. C. Fournet, J-J. Levy, and Schmitt. A. An asynchronous, distributed implementation of mobile ambients. In *Proceedings of IFIP TCS'00*, number 1872 in LNCS. Springer-Verlag, 2000.
14. A. D. Gordon and A. Jeffrey. Authenticity by typing for security protocols. In *Proceedings of CSFW 2001*, pages 145–159. IEEE Computer Society, 2001.
15. M. Hennesy and J. Riely. Type–safe execution of mobile agents in anonymous networks. In *Secure Internet Programming: Security Issues for Mobile and Distributed Objects*, number 1603 in LNCS, pages 95–115. Springer-Verlag, 1999.
16. F. Levi and D. Sangiorgi. Controlling interference in ambients. In *Proceedings of POPL'00*, pages 352–364. ACM Press, 2000.
17. R. Milner, J. Parrow, and D. Walker. A Calculus of Mobile Processes, Parts I and II. *Information and Computation*, 100:1–77, September 1992.
18. T. Sander and C. Tschudin. Towards mobile cryptography. In *Proceedings of the IEEE Symposium on Security and Privacy*. IEEE Computer Society Press, 1998.
19. D. Sangiorgi and A. Valente. A distributed abstract machine for safe ambients. In *Proc. of ICALP 2001*, pages 408–420, 2001.
20. U. G. Wilhelm, L. Buttyàn, and S. Staamann. On the problem of trust in mobile agent systems. In *Symposium on Network and Distributed System Security*. Internet Society, 1998.

Locally Commutative Categories

Arkadev Chattopadhyay and Denis Thérien[*]

School of Computer Science, McGill University, 3480 rue University, Montréal (PQ)
H3A 2A7, Canada
{achatt3,denis}@cs.mcgill.ca

Abstract. It is known that a finite category can have all its base monoids in a variety **V** (i.e. be locally **V**, denoted ℓ**V**) without itself dividing a monoid in **V** (i.e. be globally **V**, denoted **gV**). This is in particular the case when **V**=**Com**, the variety of commutative monoids. Our main result provides a combinatorial characterization of locally commutative categories. This is the first such theorem dealing with a variety for which local differs from global. As a consequence, we show that ℓ**Com** $\subset$ **gV** for every variety **V** that strictly contains the commutative monoids.

1 Introduction

In algebraic theory of automata, a language $L \subseteq A^*$ is said to be recognized by the finite monoid M if there exist a morphism $\phi : A^* \to M$ and a subset $F \subseteq M$ such that $L = \phi^{-1}(F)$. It is well-known that languages that can be so recognized are precisely the regular languages and that for each regular language there is a unique minimal monoid, called the *syntactic monoid of L* and denoted $M(L)$, that recognizes it. One expects that combinatorial properties of L would be reflected in the algebraic structure of $M(L)$: this intuition is completely valid and a driving theme of the field is to prove theorems of the following form:

"A language L belongs to the combinatorially-defined class $\mathcal{V}$ iff the syntactic monoid $M(L)$ belongs to the algebraically-defined class **V**."

For technical, but unavoidable, reasons, one sometimes has to deal with subsets of A^+ (instead of A^*) and semigroups (instead of monoids). Most often, "algebraically-defined" means that **V** is an M-variety, that is a class of finite monoids which is closed under division (i.e. morphic image and submonoid) and direct product. The notion of S-variety is similarly defined for finite semigroups. Books such as [1,2,4] offer a comprehensive treatment of this theory. One interesting by-product of results of the above form is that when membership in **V** is decidable, one gets a decision procedure to test if L is in $\mathcal{V}$, since the monoid $M(L)$ can be effectively computed from any of the common representations used for regular languages (automaton, regular expression, grammar, logical formula). Two classical theorems of that nature are the correspondence between star-free languages and aperiodic monoids [5] and the correspondence between piecewise-testable languages and $\mathcal{J}$-trivial monoids [6].

---

[*] Research supported in part by NSERC and FCAR grants.

J.C.M. Baeten et al. (Eds.): ICALP 2003, LNCS 2719, pp. 984–995, 2003.

Consider the situation where two automata are connected in series: for the second machine it is no longer the case that the space of inputs it can receive forms a free monoid, since the input sequence is mediated through the first machine and some combinations may never arise. Technically, the right point of view needed for analyzing the computations of the second automaton, is to view the machine as operating over a free category rather than over a free monoid. In order to understand the all-important case of serial connection of automata and its algebraic incarnation i.e the wreath product of monoids, it is essential to generalize the above setting to the level of categories, e.g. deciding if a monoid M divides a wreath product of the form $S \circ T$ amounts to decide if a certain category, constructible from M and T, divides S. In this framework, one considers languages as sets of finite-length paths in a directed multigraph (instead of finite-length sequences over a set) and such languages may be recognized by finite categories (instead of finite monoids). The notion of the syntactic category of a language appears naturally and so does the notion of a C-variety, i.e. a class of finite categories closed under division and direct product. Thus the manipulation and understanding of finite categories are essential ingredients in manipulation and understanding of regular languages as observed and formalised in the seminal work of [10].

Given a C-variety $\mathbf{W}$, it is easily seen that the monoids in $\mathbf{W}$ form an M-variety. It is thus natural to consider the following question: for a fixed M-variety $\mathbf{V}$, what are the C-varieties $\mathbf{W}$ for which the monoids in $\mathbf{W}$ are precisely those of $\mathbf{V}$? Two natural examples emerge readily: the variety $\mathbf{gV} = \{C : C$ divides M for some $M \in \mathbf{V}\}$, and the variety $\ell\mathbf{V} = \{C :$ every base monoid of C is in $\mathbf{V}\}$ are respectively the smallest and the largest C-variety with that property. It turns out that a combinatorial description of the languages recognized by monoids in $\mathbf{V}$ immediately implies a combinatorial description of the languages recognized by categories in $\mathbf{gV}$; similarly, an algebraic description of the monoids in $\mathbf{V}$ implies an algebraic description of the categories in $\ell\mathbf{V}$. Our understanding is thus complete whenever $\mathbf{gV} = \ell\mathbf{V}$; this happens in a number of interesting cases, e.g. for every non-trivial variety of groups, for semilattices, for aperiodic monoids. But there are also cases where $\mathbf{gV} \not\subseteq \ell\mathbf{V}$, e.g. for the trivial variety, for commutative monoids [9], for $\mathcal{J}$-trivial monoids [3]; apart from the case of the trivial variety, it becomes quite a challenge to find an algebraic description of $\mathbf{gV}$ or a combinatorial description of the languages recognized by members of $\ell\mathbf{V}$.

The main result of this paper is to provide a combinatorial description of the languages recognized by members of $\ell\mathbf{Com}$, the C-variety of locally commutative categories. This is the first instance of such result for a non-trivial variety $\mathbf{V}$ where $\mathbf{gV} \neq \ell\mathbf{V}$. We give our description via congruences of finite-index and some novel ideas have to be introduced. We also show that $\ell\mathbf{Com}$ is contained in $\mathbf{gV}$ for every M-variety $\mathbf{V}$ that strictly contains all commutative monoids. We then use known techniques to derive results about the S-variety $\mathbf{LCom} = \{S : eSe \in \mathbf{Com}$ for every $e = e^2\}$. The paper is organized as follows: section 2 presents the basic notions that are needed, section 3 proves the main theorem about locally commutative categories and section 4 describes the consequences of that result.

2   Basic Notions

A directed multigraph $G = (V, A, \alpha, \omega)$ consists of a set V of vertices, a set A of directed edges and two mappings $\alpha, \omega : A \to V$, which assigns to each edge a the start vertex $\alpha(a)$ and the end vertex $\omega(a)$ of that edge. Two edges a, b are consecutive iff $\omega(a) = \alpha(b)$. A path of length $n > 0$ is a sequence of n consecutive edges; we extend the mappings α and ω to paths in the natural way. For each vertex v we allow an empty path 1_v of length 0 for which $\alpha(1_v) = \omega(1_v) = v$. The length of a path x will be denoted by $|x|$, and the number of occurrences of an edge a in x by $|x|_a$. Two paths x, y are coterminal, denoted $x \sim y$ if $\alpha(x) = \alpha(y)$ and $\omega(x) = \omega(y)$. A loop is a path x such that $\alpha(x) = \omega(x)$ and a loop edge is a loop that consists of a single edge; we denote by $\overline{x}$ the path obtained from x by removing its loop edges. For a path x and a vertex v, let $x[v]$ stand for the subsequence of x consisting of all edges of the path that are incident on vertex v; note that $x[v]$ is not itself a path, and that when x is a loop $\overline{x}[v]$ has even length for each v.

An equivalence β on the set G^* of all paths in G is a graph congruence iff $x \, \beta \, y$ implies $x \sim y$ and $x_1 \, \beta \, y_1$, $x_2 \, \beta \, y_2$, $\omega(x_1) = \alpha(x_2)$ imply $x_1 x_2 \, \beta \, y_1 y_2$. The set of congruence classes, G^*/β, then forms a category. For each path x, we denote the corresponding congruence class containing x by $[x]_\beta$. We note that for every vertex v, the set $\{[x]_\beta : x \text{ is a loop on } v\}$ forms a monoid; we call these the base monoids of G^*/β. We refer the reader to [10] for the technical definition of division of categories and we define a C-variety to be a class of finite categories which is closed under division and direct product.

Monoids can be identified with 1-vertex categories in an obvious way. If we restrict a C-variety to its 1-vertex members, we then get an M-variety. In general, there may exist several C-varieties which coincide on the monoids they contain. For a given M-variety $\mathbf{V}$ it is always the case that $\mathbf{gV} = \{C : C \text{ divides } M \text{ for some } M \in \mathbf{V}\}$ is the smallest C-variety having $\mathbf{V}$ as its restriction to monoids: similarly $\ell\mathbf{V} = \{C : \text{every base monoid of } C \text{ is in } \mathbf{V}\}$ is always the largest C-variety with this property. Thus the C-variety corresponding to $\mathbf{V}$ is unique iff $\mathbf{gV} = \ell\mathbf{V}$. Although this holds in several instances, here are two examples where this is not the case.

Example 1. Let $\mathbf{V} = \mathbf{1}$ be the M-variety consisting of the 1-element monoid only. Then for every graph G, $G^*/\beta \in \mathbf{g1}$ iff β and $\sim$ coincide. On the other hand, let B be the subset of those edges of G for which the start and the end vertices belong to different strongly connected components. Define $x \, \gamma \, y$ iff $x \sim y$ and, for each $b \in B$, $x = x_0 b x_1$ iff $y = y_0 b y_1$. Clearly G^*/β is in $\ell\mathbf{1}$ but not in $\mathbf{g1}$ if B is non-empty. In fact, it is an exercise to show that $G^*/\beta \in \ell\mathbf{1}$ iff $\gamma \subseteq \beta$. An interesting consequence of this observation is that $\ell\mathbf{1} \subset \mathbf{gV}$ whenever $\mathbf{1} \subset \mathbf{V}$. Indeed an edge b of B can appear in a path zero or one time only: if M is a non-trivial monoid, i.e. M contains an element $m \neq 1$, it can be used to distinguish paths in which b occurs from paths in which it does not, by mapping b to m and every other edge of the graph to 1. Taking a direct product of $|B|$ copies of M

insures that we can recover the equivalence class (in γ) of a path from its value in $M^{|B|}$.

Example 2. Let $\mathbf{V} = \mathbf{Com}$, the variety of all commutative finite monoids. On any graph G, define $x\,\gamma_{t,q}\,y$ iff $x \sim y$ and for each $a \in \Lambda$ either ($|x|_a < t$ and $|x|_a = |y|_a$) or ($|x|_a \geq t, |y|_a \geq t$ and $|x|_a \equiv_q |y|_a$, where $\equiv_q$ denotes modulo q equality). It can be shown that $G^*/\beta \in \mathbf{gCom}$ iff $\gamma_{t,q} \subseteq \beta$ for some $t \geq 0, q \geq 1$. On the other hand, consider the following graph G:

a, c

1   2

b

define $x\,\beta\,y$ iff $x \sim y$ and ($|x| \leq 3$ and $x = y$) or ($|x| > 3$ and $x \sim y$). Then $G^*/\beta \in \ell\mathbf{Com}$ but not in $\mathbf{gCom}$.

This example is in some sense generic as [9] proves that a category C is in $\mathbf{gCom}$ iff it satisfies $xyz = zyx$ whenever x and z are coterminal; this result is combinatorially quite delicate to obtain. By definition, a category C is in $\ell\mathbf{Com}$ iff $xy = yx$ for every two loops x, y on the same vertex. The above example shows that knowing the number of occurrences of each edge in a path is not enough information to characterize the value of the path in a locally commutative category. Our present paper will provide a combinatorial description of the information that is missing in order to do so.

3 Combinatorial Characterization of Locally Commutative Categories

3.1 Free Locally Commutative Categories

Let G be a graph and define on G^* the congruence $x\,\gamma_\infty\,y$ iff $x \sim y$ and $|x|_a = |y|_a$ for every edge a. Let also θ_∞ be the coarsest congruence satisfying the equation $xyz\,\theta_\infty\,zyx$ whenever $x \sim z$. It was shown in [9] that $\gamma_\infty = \theta_\infty$.

The free locally commutative congruence on G^*, which we denote θ_∞^ℓ, is the coarsest congruence satisfying $xy\,\theta_\infty^\ell\,yx$ whenever x and y are loops on the same vertex. Obviously, θ_∞^ℓ refines $\theta_\infty = \gamma_\infty$. We also observe that $x\,\theta_\infty^\ell\,y$ iff $|x|_a = |y|_a$ for every loop edge a and $\overline{x}\,\theta_\infty^\ell\,\overline{y}$, i.e. the presence of loop edges cannot affect the congruence relation provided they are in equal number in both paths. There is another combinatorial property that is preserved by commutation of loops; let v be a vertex such that $|xy|_a = 0$ for each loop edge a on v and such that $|xy[v]|_a \leq 1$ for each a; then the subsequence $xy[v]$ is an even permutation of the subsequence $yx[v]$. We now proceed to show that these combinatorial properties, the last one suitably modified, characterize θ_∞^ℓ.

In general, it is not the case that every edge appears at most once in a path. Suppose $|x|_a = k$; we make the k occurrences of a in x formally distinct by labelling them, in the order they appear, as $a_{\lambda(1)}, \dots, a_{\lambda(k)}$, where λ is a permutation of $\{1, \dots, k\}$. A *labelling* $\Lambda(x)$ of a path x is the result of applying this process to each edge. Thus the edges forming $\Lambda(x)$ can be viewed as being distinct. We will write $I(x)$ when the labelling is based on identity permutations for each edge, i.e. for each a, if $|x|_a = k$, the occurrences of a in x are renamed $a_1, \dots, a_k$ in that order.

We define on G^* $x \, \gamma^\ell_\infty \, y$ iff $x \, \gamma_\infty \, y$ and there exists a labelling Λ for $\overline{y}$ such that for every vertex v the sequence $\Lambda(\overline{y})[v]$ is an even permutation of the sequence $I(\overline{x})[v]$. It can be checked that γ^ℓ_∞ is a congruence relation.

We state a useful property of γ^ℓ_∞

Proposition 1. *Let* $x = x_1 \rho x_2$ *and* $y = y_1 \rho y_2$ *be two paths in a graph such that* $x \, \gamma^\ell_\infty \, y$ *and* ρ *is a loop on some vertex* v *such that for each edge* a *in* ρ *we have* $|x|_a = |y|_a = 1$. *Then* $x_1 x_2 \, \gamma^\ell_\infty \, y_1 y_2$.

Proof. Clearly $x_1 x_2 \, \gamma_\infty \, y_1 y_2$. For the second property that we need to prove, we can assume that x and y do not contain any loop edge, since this property is dealing with $\overline{x}$ and $\overline{y}$. From the definition of γ^ℓ_∞ there exists a labelling function Λ such that for each vertex v, $\Lambda(y)[v]$ is an even permuation of $I(x)[v]$. But every edge that appears in ρ is unique and so Λ and I must have labelled ρ exactly the same way. Also for every vertex v, $\Lambda(\rho)[v]$ and $I(\rho)[v]$ have the same length, which is even since ρ is a loop. This implies that $\Lambda(y_1 y_2)[v]$ is an even permutation of $I(x_1 x_2)[v]$ for every vertex v, as required. $\square$

An immediate corollary follows

Corollary 1. *If two paths* x *and* y *satisfy* $x \, \gamma^\ell_\infty \, y$ *and there are* n *loops* $\rho_1, \dots, \rho_n$ *appearing in* x *and* y *where for each edge* a *in a loop* ρ_i *we have* $|x|_a = |y|_a = 1$ *then the paths obtained by deleting these loops from* x *and* y *(say* x' *and* y'*), satisfy* $x' \, \gamma^\ell_\infty \, y'$

Proof. This follows by repeatedly applying proposition 1 once for every loop ρ_i. $\square$

Lemma 1. *For two paths* x *and* y, $x \, \gamma^\ell_\infty \, y$ *iff* $x \, \theta^\ell_\infty \, y$.

Proof. The implication from right to left is easy and left to the reader.

Now for the other direction we assume $x \, \gamma^\ell_\infty \, y$. Since every loop edge appears the same number of times in the two paths, its suffices to show $\overline{x} \, \theta^\ell_\infty \, \overline{y}$, so we now suppose that x and y have no loop edges. Because of the labelling involved in the definition of γ^ℓ_∞, we can think of x and y as having at most one occurrence of any edge.

We will prove our claim by induction on the length of the paths. For the base case of $|x| = 1$ the lemma is trivially true. Also note that if x and y are two coterminal paths that start with the same edge a and $x = ax'$ and $y = ay'$, then $x \; \gamma_\infty^\ell \; y$ implies $x' \; \gamma_\infty^\ell \; y'$, since the occurrence of a is unique. Thus from the inductive hypothesis we obtain $x' \, \theta_\infty^\ell \, y'$ and this proves $x \, \theta_\infty^\ell \, y$.

Assume next that x and y start with different edges. Let $x = ax_0bx_1, y = by_0ay_1, v = \alpha(x) = \alpha(y)$; If v appears in y_1, i.e. $y_1 = y_{10}y_{11}$ with $\omega(y_{10}) = v$, then we can commute by_0 and ay_{10} and we are back at the previous case. A similar argument holds if x_1 contains v. Otherwise the vertex v, which is the common end vertex of x_0 and y_0, must appear at least once more in those two subpaths because $x[v]$ is an even permutation of $y[v]$. This implies the presence of an edge c in x_0 with start vertex v. This edge also appears in y, hence must appear in y_0. We can thus use loop commuting to bring the c as first edge in each path, and so $x \, \theta_\infty^\ell \, cx' \, \gamma_\infty^\ell \, cy' \, \theta_\infty^\ell \, y$ for some x' and y' (note that this follows from the already proven fact that θ_∞^ℓ refines γ_∞^ℓ). Now we are back to the case handled before. $\square$

The lemma above combinatorially captures the algebraic congruence θ_∞^ℓ and so provides a tool for describing the language recognized by free locally commutative categories. But it is impossible to work directly with the congruence γ_∞^ℓ for the case of finite categories since we have to deal with paths that are equivalent even though their lengths are different and so the concept of even permutations does not work anymore. This motivates us to find another way of characterising θ_∞^ℓ.

Consider the following special case. Let $x = ax_1yx_2zx_3a$ be a path where a is an edge which is coterminal with the subpaths y and z. One verifies that $x \, \theta_\infty^\ell \, zx_3yx_2ax_1a \, \theta_\infty^\ell \, zx_3yx_1ax_2a \, \theta_\infty^\ell \, zx_1ax_3yx_2a \, \theta_\infty^\ell \, ax_3zx_1yx_2a \, \theta_\infty^\ell \, ax_1yx_3zx_2a \, \theta_\infty^\ell \, ax_1zx_2yx_3a$. Thus we are able to interchange in x the coterminal subpaths y and z by using commutation of loops, because x contains an edge twice which is coterminal with these subpaths. The equivalence between exchange of coterminal paths and commutation of loops holds under a more general condition that we formalize below.

For a path x, define Γ_x^* as the reflexive and transitive closure of the relation Γ_x defined on the vertices by $v_1\Gamma_xv_2$ whenever there is an edge a such that $|x|_a \geq 2$ and $\alpha(a) = v_1, \omega(a) = v_2$ or $\alpha(a) = v_2, \omega(a) = v_1$.

Lemma 2. *For any path $x = x_1x_2x_3x_4x_5$ in G, if $\alpha(x_2) \, \Gamma_x^* \, \omega(x_2)$ and $x_2 \sim x_4$ then $x_1x_2x_3x_4x_5 \, \theta_\infty^\ell \, x_1x_4x_3x_2x_5$.*

Proof. Let $x = x_1x_2x_3x_4x_5$, $y = x_1x_4x_3x_2x_5$, $v_a = \alpha(x_2) = \alpha(x_4)$, $v_b = \omega(x_2) = \omega(x_4)$. If $v_a = v_b$, the result is immediate. Otherwise we prove the lemma by showing that the hypothesis implies $x \gamma_\infty^\ell y$. Clearly $|x|_a = |y|_a$ for each a. Consider now $\overline{x}$ and $\overline{y}$, or equivalently assume that x and y contain no loop edges. We have to show that there exists a labelling Λ which will make $\Lambda(y)[v]$ an even permutation of $I(x)[v]$ for every vertex v. Since y is obtained by interchanging subpaths of x, we get naturally from $I(x)$ a first labelling Λ for y. For each vertex $v \neq v_a, v_b$, we have that $x_2[v]$ and $x_4[v]$ have even length. Since $\Lambda(y)[v]$ is obtained from $I(x)[v]$ by interchanging two blocks of even length, it must be an even permutation. The problem is that $x_2[v_a]$ and $x_4[v_a]$ have odd length, hence the permutation $\Lambda(y)[v_a]$ is odd, and the same for v_b. Since $v_a \, \Gamma_x^* \, v_b$

there exists some $n > 0$ such that $v_a = v_0 \, \Gamma_x \, v_1 \, \Gamma_x \, v_2 \ldots \Gamma_x \, v_{n-1} \, \Gamma_x \, v_n = v_b$. Using the definition of Γ_x let e_i be the edge connecting v_{i-1} and v_i for $i > 0$. Each e_i is directed and its direction is arbitrary. Also there are at least two occurrences of e_i in both x and y. Let us create a new labelling Λ' that switches the labels (as given by Λ) of two arbitrarily chosen instances of e_i for each i. For all other edge occurrences, Λ' is the same as Λ. For each of $v_1, \ldots, v_{n-1}$, $\Lambda'(y)[v_i]$ differs from $\Lambda(y)[v_i]$ by two transpositions, hence it remains even. For v_0 and for v_n, the difference between Λ and Λ' is one transposition, hence these become even as well. $\square$

An edge e in a path x is called a *special edge for* x iff $\alpha(e)$ and $\omega(e)$ are not related by Γ_x^*. A maximal subpath in a path x that is completely contained inside an equivalence class of Γ_x^* is called a *component of* x. So special edges always connect components that are over different equivalence classes of Γ_x^*. Note that a component could consist of just the identity path in which case two special edges would be adjacent to each other. Clearly every special edge occurs exactly once in a path x. Every path x in G^* is thus now uniquely decomposed as $x_0 e_1 x_1 \ldots e_n x_n$, where the e_i's are the special edges for x and the x_i's are its components. The lemma above then gives the following result

Corollary 2. *If a path x has no special edges then for any path y, $x \, \theta_\infty^\ell \, y$ iff $x \, \theta_\infty \, y$ iff $x \, \gamma_\infty \, y$.*

In order to take into account the presence of special edges, we define, for each path x, a *reduced graph* $G_x = (V_x, A_x, \alpha_x, \omega_x)$ where $V_x = V/\Gamma_x^*$, A_x is the set of special edges for x, and α_x, ω_x are defined in the obvious way. The path x induces a path $Red(x)$ in the graph G_x by taking $Red(x)$ to be the sequence of special edges in the order they appear in x. Note that $Red(x)$ is a permutation of A_x and that $x \, \gamma_\infty \, y$ implies that $\Gamma_x^* = \Gamma_y^*$, hence that the graphs G_x and G_y are identical; furthermore we then have that $Red(x) \sim Red(y)$ in this graph.
  We now define a congruence on G^* by $x \, \delta_\infty^\ell \, y$ iff $x \, \gamma_\infty \, y$ and $Red(x) \, \gamma_\infty^\ell \, Red(y)$.

Lemma 3. *For two paths x and y in G if $x \, \delta_\infty^\ell \, y$ and $Red(x) = Red(y)$ then $x \, \theta_\infty^\ell \, y$.*

Proof. Let $x = x_0 e_1 x_1 \ldots e_n x_n$, $y = y_0 e_1 y_1 \ldots e_n y_n$. Observe that this forces $x_i \sim y_i$ for each i. Fix an equivalence class C in V/Γ_x^* and let $0 \le i_0 < i_1 < \ldots, i_t \le n$ be the indices for which x_{i_j} is a component of x over C; the same sequence of indices gives the components of y that are over C. For each j replace the subpath of x between $x_{i_{j-1}}$ and x_{i_j} by a "meta-edge" E_j that goes from $\omega(x_{i_{j-1}})$ to $\alpha(x_{i_j})$. Do the same for y. Consider the paths $X = x_{i_0} E_1 x_{i_1} \ldots E_t x_{i_t}$ and $Y = y_{i_0} E_1 y_{i_1} \ldots E_t y_{i_t}$. We have that $X \, \gamma_\infty \, Y$ and these two paths now have no special edges since the two endpoints of each E_j are in C. By corollary 2, X can be transformed into Y by commuting loops. The corresponding sequence of operations will transform x into a path $x' = x_0' e_1 x_1' \ldots e_n x_n'$ where $x_i' = y_i$ for $i \in \{i_0, \ldots, i_t\}$ and $x_i' = x_i$ otherwise. Doing this for each class of V/Γ_x^* in turn will transform x into y. $\square$

We are now in a position to prove the equivalence of δ_∞^ℓ and θ_∞^ℓ.

Lemma 4. *For any two paths x and y in G $x\,\delta_\infty^\ell\,y$ iff $x\,\theta_\infty^\ell\,y$.*

Proof. The implication from right to left is easy and left to the reader. We prove the second implication. Suppose $x = x_0 e_1 x_1 \ldots e_n x_n$; We fix in each equivalence class C of V/Γ_x^* a vertex v_C, and for each special edge e_i going from vertex v in C to a vertex v' in C', we augment the graph G by introducing four new edges: e_i^C going from v to v_C, f_i^C going from v_C to v, g_i^C going from v' to $v_{C'}$ and h_i^C going from $v_{C'}$ to v'. We create from x a new path x' in the augmented graph by the following process: if e_j is a special edge for x going from vertex v in C to a vertex v' in C', we replace e_j by $e_j^C f_j^C e_j g_j^{C'} h_j^{C'}$. If any loop edges have been added we remove them. We create y' from y similarly. $Red(x')\,\gamma_\infty^\ell\,Red(y')$ comes trivially from the fact that $x\,\delta_\infty^\ell\,y$ (since $Red(x) = Red(x')$ and $Red(y) = Red(y')$). Hence also $Red(x')\,\theta_\infty^\ell\,Red(y')$ by lemma 1. By construction, if there is a loop on vertex C appearing in $Red(x')$ in the reduced graph, there is a corresponding loop on vertex v_C appearing in x' in the augmented graph. Thus, corresponding to the sequence of loop commutations that transforms $Red(x')$ to $Red(y')$ in the reduced graph, there is a sequence of loop transformations, in the augmented graph, that transforms x' into a path (say w) in which the special edges appear in the same order as those of y'. Hence using lemma 3 it follows $x'\,\theta_\infty^\ell\,w\,\theta_\infty^\ell\,y'$. So $x'\,\gamma_\infty^\ell\,y'$ and by recalling that we obtained x' (y') from x (y) by adding a certain number of loops around every vertex v_C we apply proposition 1 and corollary 1 to get $x\,\gamma_\infty^\ell\,y$. Hence $x\,\theta_\infty^\ell\,y$. $\square$

Thus δ_∞^ℓ provides an alternative characterisation of locally commutative free categories. We will see in the next section that this characterisation can be naturally adapted to the case of finite categories.

3.2   Locally Commutative Finite Categories

We recall from [9] that the algebraic description of finite globally commutative categories is given by a path congruence $\theta_{t,q}$ generated by equations: $xyz\,\theta_{t,q}\,zyx$ for $x \sim z$ and $x^t\,\theta_{t,q}\,x^{t+q}$ where x is a loop.

The corresponding combinatorial congruence $\gamma_{r,q}$ is induced by relations: for $x \sim y$ we say $x\,\gamma_{r,q}\,y$ iff for all edge $a \in A$, either $(|x|_a, |y|_a < r$ and $|x|_a = |y|_a)$ or $(|x|_a, |y|_a \geq r$ and $|x|_a \equiv_q |y|_a)$

We summarize the main result from [9] in the lemma below:

Lemma 5. *For every $t \geq 0$ and graph G there exists s such that for two paths x and y, $x\,\gamma_{s,q}\,y$ implies $x\,\theta_{t,q}\,y$.*

Observe that $\theta_{t,q}$ can be thought of as a rewriting system. If a path y can be obtained from path x using only loop commuting $(uw \to wu)$ and loop replication $(u^t \to u^{t+q})$ <u>without</u> using loop deletion $(u^{t+q} \to u^t)$, then we write $x \leq_{t,q} y$. It is a trivial observation that $x \leq_{t,q} y$ implies $x\,\theta_{t,q}\,y$. Clearly $x \leq_{t,q} y$ implies for all $a \in A$, $|x|_a \leq |y|_a$.

We now state a result that follows from the argumentation given in [8] for proving his lemma B.3.10 in Appendix B.

Proposition 2. *For paths x and y and for $t > 0$ if $\forall a \in A(G)$, $|x|_a \leq |y|_a$ and $x \gamma_{t+1,q} y$ then $x \leq_{t,q} y$.*

Proof. This follows from the argument used in the proof of lemma B.3.10 given in [8] (and is left as an exercise for the reader). □

As an extension of ideas from free locally commutative categories we introduce $\theta_{t,q}^\ell$ to be the finite index path congruence generated by the conditions: $xy\,\theta_{t,q}^\ell\,yx$ where x and y are loops and $x^t\,\theta_{t,q}^\ell\,x^{t+q}$ where x is a loop. Analogous to the global case we write $x \leq_{t,q}^\ell y$ when $x\,\theta_{t,q}^\ell\,y$ and y can be obtained from x by just loop commuting and loop replication.

We also extend our combinatorial characterisation from the last section to $\delta_{t,q}^\ell$ meaning for two paths x and y $x\,\delta_{t,q}^\ell\,y$ iff $x\gamma_{t,q}y$ and $Red(x)\,\gamma_\infty^\ell\,Red(y)$, where γ_∞^ℓ gets defined on the reduced graph G_x. Note that this congruence only depends on the permutation of reduced paths which are of fixed length.

Using the definition of $\theta_{t,q}^\ell$ and the lemma 2 we can conclude the following:

Corollary 3. *For paths with no special edges, $\theta_{t,q}$ and $\theta_{t,q}^\ell$ are equivalent.*

This corollary along with lemma 5 gives us the intuition to expect the following result

Lemma 6. *If $x\,\delta_{s,q}^\ell\,y$ and $Red(x) = Red(y)$ then $x\,\theta_{t,q}^\ell\,y$ where s and t are related according to lemma 5.*

Proof. We direct the attention of the reader to the proof of lemma 3. Employing exactly the same technique as in that proof, fixing an equivalence class C in V/Γ_x^* we add "meta-edges" connecting two successive components of that class and obtain paths X and Y respectively from x and y. In our case here, $X\,\gamma_{s,q}\,Y$. Therefore using lemma 5 it follows $X\,\theta_{t,q}\,Y$ and since X and Y have no special edges from corollary 3 X can be transformed into Y by transformations preserving $\theta_{t,q}^\ell$. We apply the same operations on x to get a new path x' and then repeat the procedure with x' for each class of V/Γ_x^* to finally get y. □

We can now combine the lemma above and proposition 2 to obtain the following corollary

Corollary 4. *If $x\,\delta_{t+1,q}^\ell\,y$ and $Red(x) = Red(y)$ with $|x|_a \leq |y|_a$, then $x \leq_{t,q}^\ell y$.*

Lemma 7. *For every $t \geq 2$ and $q \geq 1$, there exists $R \geq t+1$ such that $x\,\delta_{R,q}^\ell\,y$ implies that there exists a path ρ satisfying $x\,\delta_{t+1,q}^\ell\,\rho$, where $\rho\,\theta_{t,q}^\ell\,y$ and for all edges $a \in A$, $|x|_a \leq |\rho|_a$.*

Proof. We will use lemmas 3.3 and 3.8 from [9] to prove this. Specifically let $R = m(G, t+1)(|E|+1) + 1$ where $m(G, t+1) = |V| + (t+1)(2^{|E|} - 1) + 2$ as defined in [9]. So for each edge a such that $|x|_a > |y|_a$ we have $|y|_a \geq R$ and since y can have at most $(|E|+1)$ components there is at least one component that has at least $m(G, t+1)$ occurences of a. We can now straight away apply the argument used to prove lemma 3.8 in [9] and obtain the result of the present lemma. □

Lemma 8. *If for two paths x and y, $|x|_a \leq |y|_a$ for all $a \in A$ and $x\,\delta^\ell_{t+1,q}\,y$, then $x\,\theta^\ell_{t,q}\,y$ for $t \geq 2$ and $q \geq 1$.*

Proof. We ask the reader to recall the technique used to prove lemma 4. We mimic the steps in that proof to augment the graph G by introducing four new edges for each special edge e_i and then modify the paths x and y to x' and y' respectively as prescribed there. (Note: we are using the same notation as in that proof.) Also let A' represent the set of edges of the augmented graph. The same argumentation of the earlier proof carries over to establish the existence of a path w such that $Red(w) = Red(y')$ and $x'\,\theta^\ell_\infty\,w\,\delta^\ell_{t+1,q}\,y'$. From corollary 4 it follows that $w \leq^\ell_{t,q} y'$ and hence $x' \leq^\ell_{t,q} y'$. This implies that there exists a series of loop commuting and loop duplicating transformations to obtain y' from x'. Let the loops that got duplicated, be called $\rho_1, \ldots, \rho_n$ and let them be around vertices $v_1, \ldots, v_n$ in G respectively. Also let n_i be the number of times ρ_i was duplicated. It is a trivial observation that every vertex v_i occurs somewhere in the path x and every loop ρ_i contains edges strictly from the unaugmented original graph G (since for each edge $a \in A' \backslash A$ we have $|x'|_a = |y'|_a$). Also no loop ρ_i contains any special edges as their count is one in both x' and y'. Hence every loop ρ_i could be added n_i times to path x to obtain a path u in G such that $u\,\delta^\ell_\infty\,y$ since $Red(x) = Red(u)$. This implies $x\,\delta^\ell_{t+1,q}\,y$ and hence from corollary 4 we have $x\,\theta^\ell_{t,q}\,u$. Now applying lemma 4 to u and y we get $u\,\theta^\ell_\infty\,y$ and hence $x\,\theta^\ell_{t,q}\,y$. $\square$

We now state the main result of this paper

Theorem 1. *β is a ℓ**Com**-congruence iff there exist $R \geq 2$, $q \geq 1$ such that $\delta^\ell_{R,q} \subseteq \beta$.*

Proof. The direction from right to left is trivial and left as an exercise for the reader (it can be verified that $\delta^\ell_{R,q}$ is a ℓ**Com**-congruence). For $t \geq 2$ we choose $R = m(G, t+1)(|E|+1)+1$ according to lemma 7. Then $x\,\delta^\ell_{R,q}\,y$ implies there exists a path z with $|x|_a \leq |z|_a$ for each edge $a \in A$ and $x\,\delta^\ell_{t+1,q}\,z\,\theta^\ell_{t,q}\,y$. Using lemma 8 we get $x\,\theta^\ell_{t,q}\,y$. We recall that for the cases $t = 0$ and $t = 1$, [9] proves ℓ**Com** coincides with **gCom**. $\square$

4   Consequences of the Main Result

In this section, we sketch some consequences of the combinatorial description obtained above. When an M-variety $\mathbf{V}$ is such that the C-varieties $\mathbf{gV}$ and $\ell\mathbf{V}$ differ, then $\ell\mathbf{V}$ cannot be equal to $\mathbf{gW}$ for any M-variety $\mathbf{W}$. How big should $\mathbf{W}$ be to insure $\ell\mathbf{V} \subset \mathbf{gW}$? In example 1 of section 2, we observed that for the trivial M-variety we have $\ell\mathbf{1} \subset \mathbf{gW}$ for every non-trivial $\mathbf{W}$. We now argue that a similar phenomenon occurs for **Com**.

Theorem 2. *ℓ**Com** $\subset \mathbf{gW}$ for every M-variety $\mathbf{W}$ that strictly contains **Com**.*

Proof. Our main result shows that in every locally commutative category, the value of a path is determined by the number of occurrences of each edge (threshold t, modulo q for some $t \geq 0, q \geq 1$) and the ordering of the so-called "special" edges. The first condition can be determined by using for each edge a cyclic counter of appropriate cardinality. For the second condition, let M be any non-commutative monoid, i.e. M contains two elements m and m' such that $mm' \neq m'm$. Fix two edges of the graph, a and b, map a to m, b to m' and every other edge to 1. If a path x contains at most one occurrence of each of a and b, which is necessarily the case when these two edges are special for x, the value of the path in M is in $\{1, m, m', mm', m'm\}$. In particular if both edges occur once, the order in which they appear can be recovered from the value in the monoid. If the graph has k edges, we can use the direct product of k cyclic counters to count occurrences of each edge, and $O(k^2)$ copies of M, one for each pair of edges. The value of the counters will determine the first condition and also which edges are special for a given path; we can then look up the appropriate copies of M to know in which order the special edges have appeared, hence recover the $\delta_{t,q}^{\ell}$-value of the path. □

Next, we transfer the last theorem to the S-variety $\mathbf{LCom} = \{S : eSe \in \mathbf{Com}$ for all $e = e^2\}$. For any semigroup S, consider the graph $G = (V, A, \alpha, \omega)$, where V is the set of idempotents of S, $A = V \times A \times V, \alpha(e, s, f) = e, \omega(e, s, f) = f$. Define the congruence β on G^* by identifying coterminal paths that multiply out to the same element in S. This construction trivially insures that $S \in \mathbf{LCom}$ iff $G^*/\beta \in \ell\mathbf{V}$. It follows from work of [7] that $S \in \mathbf{V} * \mathbf{D}$, where $\mathbf{D} = \{S : Se = e$ for all $e = e^2\}$ and $*$ denotes wreath product of varieties, iff $G^*/\beta \in \mathbf{gV}$. We thus get the following

Theorem 3. $\mathbf{LCom} \subset \mathbf{V} * \mathbf{D}$ *for every M-variety* $\mathbf{V}$ *that strictly includes the commutative monoids.*

5   Conclusion

In this paper we have proved a combinatorial description for the languages that can be recognized by finite locally commutative categories. This is the first result of that kind for a non-trivial M-variety for which the induced global and local C-varieties are different. We derived as a consequence the upper bound that for each M-variety $\mathbf{V}$ properly including the commutative monoids, the inclusion $\ell\mathbf{V} \subset \mathbf{gV}$ holds, which is similar to the situation for the trivial M-variety. It is easily checked that all these results can be proved, mutatis mutandis, for the C-variety of locally aperiodic commutative monoids. There is another famous case of an M-variety for which the induced global and local C-varieties are different, namely the variety $\mathbf{J}$ of $\mathcal{J}$-trivial monoids. However, Jorge Almeida has pointed out to us that there exists a C-variety $\mathbf{gV}$ where V is a M-variety of aperiodic monoids that strictly contains $\mathbf{J}$, such that $\mathbf{gV}$ does not contain $\ell\mathbf{J}$. It would be interesting to find an upper bound for $\ell\mathbf{J}$ in terms of globally defined C-varieties.

References

1. J. Almeida. *Finite Semigroups and Universal Algebra*. World Scientific, 1994.
2. S. Eilenberg. *Automata, Languages and Machines*, volume B. Academic Press, New York, 1976.
3. R. Knast. Some theorems on graph congruences. *RAIRO Inform. Théor.*, 17:331–342, 1983.
4. J. E. Pin. *Varieties of Formal Languages*. Plenum, London, 1986.
5. M. Schützenberger. On finite monoids having only trivial subgroups. *Inform. and Control*, 8:190–194, 1965.
6. I. Simon. Piecewise testable events. In 2nd *GI Conference*, volume 33 of *Lect.Notes in Comp.Sci*, pages 214–222, Berlin, 1975. Springer.
7. H. Straubing. Finite semigroup varieties of the form **V** * **D**. *Journal of Pure and Applied Algebra*, 36:53–94, 1985.
8. H. Straubing. *Finite Automata, Formal Logic and Circuit Complexity*. Birkhäuser, 1994.
9. D. Thérien and A. Weiss. Graph congruences and wreath products. *Journal of Pure and Applied Algebra*, 36:205–212, 1985.
10. B. Tilson. Categories as algebra: An essential ingredient in the theory of monoids. *Journal of Pure and Applied Algebra*, 48:83–198, 1987.

Semi-pullbacks and Bisimulations in Categories of Stochastic Relations

Ernst-Erich Doberkat

Chair for Software Technology
University of Dortmund
doberkat@acm.org

Abstract. The problem of constructing a semi-pullback in a category is intimately connected to the problem of establishing the transitivity of bisimulations. Edalat shows that a semi-pullback can be constructed in the category of Markov processes on Polish spaces, when the underlying transition probability functions are universally measurable, and the morphisms are measure preserving continuous maps. We demonstrate that the simpler assumption of Borel measurability suffices. Markov processes are in fact a special case: we consider the category of stochastic relations over Standard Borel spaces. At the core of the present solution lies a selection argument from stochastic dynamic optimization. An example demonstrates that (weak) pullbacks do not exist in the category of Markov processes.

Keywords: Bisimulation, semi-pullback, stochastic relations, labelled Markov processes, Hennessy-Milner logic.

1 Introduction

The existence of semi-pullbacks in a category makes sure that the bisimulation relation is transitive, provided bisimulation between objects is defined as a span of morphisms [10]. Edalat investigates this question for categories of Markov processes and shows that semi-pullbacks exist [6]. The category he focusses on has as objects universally measurable transition probability functions on Polish spaces, the morphisms are continuous, surjective, and probability preserving maps. His proof is constructive and makes essentially use of techniques of analytic spaces (which are continuous images of Polish spaces). The result implies that the semi-pullback of those transition probabilities which are measurable with respect to the Borel sets of the Polish spaces under consideration may in fact be universally measurable rather than simply Borel measurable. This then demands some unpleasant technical machinery when logically characterizing bisimulation for labelled Markov processes, cf. [2].

The distinction between measurability and universal measurability (both terms are defined in Sect. 2) may seem negligible at first. Measurability is the natural concept in measurable spaces (like continuity in topological spaces, or homomorphisms in groups), thus stochastic concepts are usually formulated in

J.C.M. Baeten et al. (Eds.): ICALP 2003, LNCS 2719, pp. 996–1007, 2003.

terms of it. Universal measurability requires a completion process using all (σ)-finite measures on the measure space under consideration. In a Polish space the Borel sets are generated by the open sets, so the generators are well known. Comparable generators for the universally measurable sets are not that easy identified, let alone put to use. Thus it appears to be sensible to search for solutions for the problem of constructing semi-pullbacks for stochastic relations or labelled Markov processes first within the realm of Borel sets.

We show that the semi-pullback of Borel Markov processes exists within the category of these processes, when the underlying space is Polish (like the real line). Edalat considers transition probability functions from one Polish space into itself, this paper considers the slightly more general notion of a stochastic relation, cf. [1,4], i.e., transition sub-probability functions from one Polish space to another one. Rather than constructing the function explicitly, we show that the problem can be formulated in terms of measurable set-valued maps for which a measurable selector exists.

The paper's contributions are as follows. First it is shown that one can in fact construct semi-pullbacks in a category of stochastic relations between Polish spaces (and, by the way, an example shows that weak pullbacks do not exist). The second contribution is the reduction of an existential argument to a selection argument, a technique borrowed from dynamic optimization. Third it is shown that the solution for characterizing bisimulations for labelled Markov processes proposed by Desharnais, Edalat and Panangaden [2] can be carried over to Standard Borel spaces with their simple Borel structure.

This note is organized as follows: Sect. 2 collects some basic facts from topology, and from measure theory. It is shown that assigning a Polish space its set of sub-probability measures is an endofunctor on this category. Sect. 3 defines the category of stochastic relations, shows how to formulate the problem in terms of a set-valued function, and proves that a selector exists. This also implies the existence of semi-pullbacks for some related categories. A counterexample destroys the hope for strengthening this results to weak pullbacks. Finally, we show in Sect. 4 that the bisimulation relation is transitive for the category of stochastic relations, and that bisimilar labelled Markov processes are characterized through a weak negation free logic. Sect. 5 wraps it all up by summarizing the results and indicating areas of further work.

2   A Small Dose Measure Theory

This Section collects some basic facts from topology and measure theory for the reader's convenience and for later reference.

A *Polish space* $(X, \mathcal{T})$ is a topological space which has a countable dense subset, and which is metrizable through a complete metric, a measurable space $(X, \mathcal{A})$ is a set X with a σ-algebra $\mathcal{A}$. The *Borel sets* $\mathcal{B}(X, \mathcal{T})$ for the topology $\mathcal{T}$ is the smallest σ-algebra on X which contains $\mathcal{T}$. A *Standard Borel* space $(X, \mathcal{A})$ is a measurable space such that the σ-algebra $\mathcal{A}$ equals $\mathcal{B}(X, \mathcal{T})$ for some Polish topology $\mathcal{T}$ on X. Although the Borel sets are determined uniquely through the topology, the converse does not hold, as we will see in a short while. Given two measurable spaces $(X, \mathcal{A})$ and $(Y, \mathcal{B})$, a map $f : X \to Y$ is $\mathcal{A} - \mathcal{B}$-*measurable* whenever $f^{-1}[\mathcal{B}] \subseteq \mathcal{A}$ holds, where $f^{-1}[\mathcal{B}] := \{f^{-1}[B] | B \in \mathcal{B}\}$ is the set of inverse images $f^{-1}[B] := \{x \in X | f(x) \in B\}$ of elements of $\mathcal{B}$. Note that $f^{-1}[\mathcal{B}]$ is in any case an σ-algebra. If the σ-algebras are the Borel sets of some topologies on X and Y, resp., then a measurable map is called *Borel measurable* or simply a *Borel* map. The real numbers $\mathbb{R}$ carry always the Borel structure induced by the usual topology.

A map $f : X \to Y$ between the topological spaces $(X, \mathcal{T})$ and $(Y, \mathcal{S})$ is $\mathcal{T} - \mathcal{S}$-*continuous* iff the inverse image of an open set from $\mathcal{S}$ is an open set in $\mathcal{T}$. Thus a continuous map is also measurable with respect to the Borel sets generated by the respective topologies.

When the context is clear, we will write down Polish spaces without their topologies, and the Borel sets are always understood with respect to the topology.

The following statement will be most helpful in the sequel. It states that, given a measurable map between Polish spaces, we can find a finer Polish topology on the domain, which has the same Borel sets, and which renders the map continuous; formally [11, Cor. 3.2.5, Cor. 3.2.6]:

Proposition 1. *Let $(X, \mathcal{T})$ and $(Y, \mathcal{S})$ be Polish spaces. If $f : X \to Y$ is a Borel measurable map, then there exists a Polish topology $\mathcal{T}'$ on X such that $\mathcal{T}'$ is finer than $\mathcal{T}$ (hence $\mathcal{T} \subseteq \mathcal{T}'$), $\mathcal{T}$ and $\mathcal{T}'$ have the same Borel sets, and f is $\mathcal{T}' - \mathcal{S}$ continuous.*

Given two measurable spaces X and Y, a *stochastic relation* $K : X \rightsquigarrow Y$ is a Borel map from X to the set $\mathbf{S}(Y)$, the latter denoting the set of all sub-probability measures on (the Borel sets of) Y. The latter set carries the *weak*-σ-algebra*. This is the smallest σ-algebra on $\mathbf{S}(Y)$ which renders all maps $\mu \mapsto \mu(B)$ measurable, where $B \subseteq Y$ is measurable. Hence $K : X \rightsquigarrow Y$ is a stochastic relation iff $K(x)$ is a sub-probability measure on (the Borel sets of) Y for all $x \in X$, such that $x \mapsto K(x)(B)$ is a measurable map for each Borel set $B \subseteq Y$.

Let Y be a Polish space, then $\mathbf{S}(Y)$ is usually equipped with the topology of weak convergence. This is the smallest topology on $\mathbf{S}(Y)$ which makes the map $\mu \mapsto \int_Y f \, d\mu$ continuous for each continuous and bounded $f : Y \to \mathbb{R}$. It is well known that this topology is Polish [9, Thm. II.6.5], and that its Borel sets is just the weak*-σ-algebra. If X is a Standard Borel space, then $\mathbf{S}(X)$ is also one: select a Polish topology $\mathcal{T}$ on X which induces the measurable structure, then $\mathcal{T}$ will give rise to the Polish topology of weak convergence on $\mathbf{S}(X)$ which in turn has the weak-*-σ-algebra as its Borel sets.

A Borel map $f : X \to Y$ between the Polish spaces X and Y induces a Borel map $\mathbf{S}(f) : \mathbf{S}(X) \to \mathbf{S}(Y)$ upon setting ($\mu \in \mathbf{S}(X), B \subseteq Y$ Borel) $\mathbf{S}(f)(\mu)(B) := \mu(f^{-1}[B])$

It is easy to see that a continuous map f induces a continuous map $\mathbf{S}(f)$, and we will see in a moment that $\mathbf{S}(f) : \mathbf{S}(X) \to \mathbf{S}(Y)$ is onto, provided $f : X \to Y$ is. Denote by $\mathbf{P}(X)$ the subspace of all probability measures on X.

Let $\mathcal{F}(X)$ be the set of all closed and non-empty subsets of the Polish space X, and call for Polish Y a relation, i.e., a set-valued map $F : X \to \mathcal{F}(Y)$ $\mathcal{C}$-

measurable iff, for any compact set $C \subseteq Y$, the weak inverse $\exists F(C) := \{x \in X | F(x) \cap C \neq \emptyset\}$ is measurable. A *selector* s for such a relation F is a single-valued map $s : X \to Y$ such that $\forall x \in X : s(x) \in F(x)$ holds. $\mathcal{C}$-measurable relations have Borel selectors:

Proposition 2. *Let X and Y be Polish spaces. Then each $\mathcal{C}$-measurable relation F has a measurable selector.*

Proof. Since closed subsets of Polish spaces are complete, the assertion follows from [8, Theorem 3]. □

As a first application it is shown that $\mathbf{S}$ actually constitutes an endofunctor on the category of Standard Borel spaces with surjective measurable map as morphisms. This implies that $\mathbf{S}$ is the functorial part of a monad similar to the one studied by Giry [7].

Lemma 1. $\mathbf{S}$ *is an endofunctor on the category $\mathfrak{SB}$ of Standard Borel spaces with surjective Borel maps as morphisms.*

Proof. 1. Let X and Y be Standard Borel spaces, and endow these spaces with a Polish topology the Borel sets of which form the respective σ-algebras. Since $\mathbf{S}(X)$ is a Polish space under the topology of weak convergence, and since a Borel map $f : X \to Y$ induces a Borel map $\mathbf{S}(f) : \mathbf{S}(X) \to \mathbf{S}(Y)$ with all the compositional properties a functor should have, only surjectivity of the induced map has to be shown.

2. In view of Prop. 1 it is no loss of generality to assume that f is continuous. Continuity and surjectivity together imply that $y \mapsto f^{-1}[\{y\}]$ has closed and non-empty values in X. It constitutes a $\mathcal{C}$-measurable relation, which has a measurable selector $g : Y \to X$ by Prop. 2, so that $f(g(y)) = y$ always holds. Let $\nu \in \mathbf{S}(Y)$, and define $\mu \in \mathbf{S}(X)$ as $\mu := \mathbf{S}(g)(\nu)$. Since $g^{-1}[f^{-1}[B]] = B$ for $B \subseteq Y$, it is now easy to establish that $\mathbf{S}(f)(\mu) = \nu$ holds. □

Finally, the concept of universal measurability is needed. Let $\mu \in \mathbf{S}(X, \mathcal{A})$ be a sub-probability on the measurable space $(X, \mathcal{A})$, then $A \subseteq X$ is called μ-*measurable* iff there exist $M_1, M_2 \in \mathcal{A}$ with $M_1 \subseteq A \subseteq M_2$ and $\mu(M_1) = \mu(M_2)$. The μ-measurable subsets of X form a σ-algebra $\mathcal{M}_\mu(\mathcal{A})$. The σ-algebra $\mathcal{U}(\mathcal{A})$ of universally measurable sets is defined by

$$\mathcal{U}(\mathcal{A}) := \bigcap \{\mathcal{M}_\mu(\mathcal{A}) | \mu \in \mathbf{S}(X, \mathcal{A})\}$$

(in fact, one considers usually all finite or σ-finite measures, but these definitions lead to the same universally measurable sets). If $f : X_1 \to X_2$ is an $\mathcal{A}_1$-$\mathcal{A}_2$-measurable map between the measurable spaces $(X_1, \mathcal{A}_1)$ and $(X_2, \mathcal{A}_2)$, then it is well known that f is also $\mathcal{U}(\mathcal{A}_1)$-$\mathcal{U}(\mathcal{A}_2)$-measurable; the converse does not hold, and one usually cannot conclude that a map $g : X_1 \to X_2$ which is $\mathcal{U}(\mathcal{A}_1)$-$\mathcal{A}_2$-measurable is also $\mathcal{A}_1$-$\mathcal{A}_2$-measurable.

3   Semi-pullbacks

The category $\mathfrak{SRel}$ of stochastic relations has as objects triplets $\langle X, Y, K \rangle$, where X and Y are Standard Borel spaces, and $K : X \rightsquigarrow Y$ is a stochastic relation. A morphism $\langle \varphi, \psi \rangle : \langle X, Y, K \rangle \to \langle X', Y', K' \rangle$ is a pair of surjective Borel maps $\varphi : X \to X'$ and $\psi : Y \to Y'$ such that $K' \circ \varphi = \mathbf{S}(\psi) \circ K$ holds. Thus we have for $x \in X, B' \subseteq Y'$ Borel the equality $K'(\varphi(x))(B') = K(x)(\psi^{-1}[B'])$, so that morphisms are in particular measure preserving. Morphisms compose componentwise.

The category of Markov processes is a subcategory of $\mathfrak{SRel}$: it has as objects pairs $\langle X, K \rangle$, where X is a Standard Borel space, and $K : X \rightsquigarrow X$ is a stochastic relation, i.e., a Borel measurable transition probability function. Morphisms are surjective and measurable measure preserving maps.

Edalat [6] investigates a similar category, called $\mathfrak{MProc}$ for easier reference: the objects are pairs $\langle X, K \rangle$ such that X is a Polish space, and K is a universally measurable transition sub-probability function. This requires that for each Borel set $A \subseteq X$ the map $x \mapsto K(x)(A)$ is $\mathcal{U}(\mathcal{B}(X))$-measurable, and that $K(x) \in \mathbf{S}(X, \mathcal{B}(X))$ for each $x \in X$. Morphisms in $\mathfrak{MProc}$ are surjective and continuous maps which are measure preserving. Note that an object $\langle X, K \rangle$ in $\mathfrak{MProc}$ has the property that the set $\{x \in X | K(x)(A) \leq r\}$ is universally measurable for each Borel set $A \subseteq X$ and for each $r \in \mathbb{R}$; since each Borel set is measurable, this is a weaker condition than the one we will be investigating.

Assume that $\langle \varphi_i, \psi_i \rangle : \langle X_i, Y_i, K_i \rangle \to \langle X, Y, K \rangle$ $(i = 1, 2)$ are morphisms in $\mathfrak{SRel}$, then a *semi-pullback* for this pair of morphisms is an object $\langle A, B, N \rangle$ together with morphisms $\langle \alpha_i, \beta_i \rangle : \langle A, B, N \rangle \to \langle X_i, Y_i, K_i \rangle$ $(i = 1, 2)$ so that this diagram is commutative in $\mathfrak{SRel}$:

$$
\begin{array}{ccc}
\langle A, B, N \rangle & \xrightarrow{\langle \alpha_1, \beta_1 \rangle} & \langle X_1, Y_1, K_1 \rangle \\
{\scriptstyle \langle \alpha_2, \beta_2 \rangle} \Big\downarrow & & \Big\downarrow {\scriptstyle \langle \varphi_1, \psi_1 \rangle} \\
\langle X_2, Y_2, K_2 \rangle & \xrightarrow[\langle \varphi_2, \psi_2 \rangle]{} & \langle X, Y, K \rangle
\end{array}
$$

This means in particular that $K_1 \circ \alpha_1 = \mathbf{S}(\beta_1) \circ N$ and $K_2 \circ \alpha_2 = \mathbf{S}(\beta_2) \circ N$ should hold, so that a bisimulation is to be constructed (cf. Def. 1). The condition that $\langle A, B, N \rangle$ is the object underlying a semi-pullback may be formulated in terms of measurable maps as follows: N is a map from the Standard Borel space A to the Standard Borel space $\mathbf{S}(B)$ so that N is also a measurable selector for the set-valued function

$$
b \mapsto \{\mu \in \mathbf{S}(B) | (K_1 \circ \alpha_1)(b) = \mathbf{S}(\beta_1)(\mu), (K_2 \circ \alpha_2)(b) = \mathbf{S}(\beta_2)(\mu)\}.
$$

This translates the problem of finding the object $\langle A, B, N \rangle$ of a semi-pullback to a selection problem for set-valued maps, provided the spaces A and B together with the morphisms are identified.

It should be noted that the notion of a semi-pullback depends only on the measurable structure of the Standard Borel spaces involved. The topological structure enters only through Borel sets, and Borel measurability. From Prop. 1 we see that there are certain degrees of freedom for selecting a Polish topology that generates the Borel sets. They will be capitalized upon in the sequel.

Our goal is to establish:

Theorem 1. $\mathfrak{SRel}$ *has semi-pullbacks for each pair of morphisms*

$$\langle X_1, Y_1, K_1 \rangle \xrightarrow{\langle \varphi_1, \psi_1 \rangle} \langle X, Y, K \rangle \xleftarrow{\langle \varphi_2, \psi_2 \rangle} \langle X_2, Y_2, K_2 \rangle$$

with a common range.

We begin with a rather technical measure-theoretic observation: in terms of probability theory, it states that there exists under certain conditions a common distribution for two random variables with values in a Polish space with preassigned marginal distributions. This is a cornerstone for the construction leading to the proof of Theorem 1, it shows in particular where Edalat's work enters the present discussion.

Proposition 3. *Let Z_1, Z_2, Z be Polish spaces, $\zeta_i : Z_i \to Z$ $(i = 1, 2)$ continuous and surjective maps, define $S := \{\langle x_1, x_2 \rangle \in Z_1 \times Z_2 | \zeta_1(x_1) = \zeta_2(x_2)\}$, and let $\nu_1 \in \mathbf{P}(Z_1), \nu_2 \in \mathbf{P}(Z_2), \nu \in \mathbf{P}(S)$ such that $\mathbf{P}(\pi_i)(\nu)(E_i) = \nu_i(E_i)$ holds for all $E_i \in \zeta_i^{-1}[\mathcal{B}(Z)]$ $(i = 1, 2)$, where $\pi_1 : S \to Z_1, \pi_2 : S \to Z_2$ are the projections; S carries the trace of the product topology. Then there exists $\mu \in \mathbf{P}(S)$ such that $\mathbf{P}(\pi_i)(\mu)(E_i) = \nu_i(E_i)$ is true for all $E_i \in \mathcal{B}(Z_i) : (i = 1, 2)$.*

Proof. $\zeta_i : Z_i \to Z$ are morphisms in Edalat's category of probability measures on Polish spaces. The assertion then follows from the proof of [6, Cor. 5.4]. $\square$

In important special cases, there are other ways of establishing the Proposition, as will be discussed briefly.

Remark 1. 1. If $\zeta_i : Z_i \to Z$ are bijections, then the Blackwell-Mackey Theorem [11, Thm. 4.5.7] shows that $\zeta_i^{-1}[\mathcal{B}(Z)] = \mathcal{B}(Z_i)$. In this case the given measure $\nu \in \mathbf{P}(S)$ is the desired one.

2. If Z_1, Z_2, Z are not only Polish but also locally compact (like the real line $\mathbb{R}$), then a combination of the Riesz Representation Theorem and the equally famous Hahn-Banach Theorem can be used to construct the desired measure directly. This is the line of attack in [5]. Consequently, the somewhat heavy machinery of regular conditional distributions on analytic spaces need not be used (on the other hand, the Hahn-Banach Theorem relies on the Axiom of Choice which is not listed among the light weight tools either). —

The preparations for establishing that $\mathfrak{SRel}$ has semi-pullbacks are complete.

Proof. **of Theorem 1**

1. In view of Prop. 1 we may assume that the respective σ-algebras on X_1 and X_2 are obtained from Polish topologies which render φ_1 and K_1 as well as φ_2 and K_2 continuous. These topologies are fixed for the proof. Put

$$A := \{\langle x_1, x_2 \rangle \in X_1 \times X_2 | \varphi_1(x_1) = \varphi_2(x_2)\},$$
$$B := \{\langle y_1, y_2 \rangle \in Y_1 \times Y_2 | \psi_1(y_1) = \psi_2(y_2)\},$$

then both A and B are closed, hence Polish. $\alpha_i : A \to X_i$ and $\beta_i : B \to Y_i$ are the projections, $i = 1, 2$. We know that for $x_i \in X_i$ the equalities $K(\varphi_1(x_1)) = \mathbf{S}(\psi_1)(K_1(x_1))$ and $K(\varphi_2(x_2)) = \mathbf{S}(\psi_2)(K_2(x_2))$ hold. The construction implies that $(\psi_1 \circ \beta_1)(y_1, y_2) = (\psi_2 \circ \beta_2)(y_1, y_2)$ is true for $\langle y_1, y_2 \rangle \in B$, and $\psi_1 \circ \beta_1 : B \to Y$ is surjective.

2. Fix $\langle x_1, x_2 \rangle \in A$. Lemma 1 shows that $\mathbf{S}$ is an endofunctor on $\mathfrak{SB}$, in particular that the image of a surjective map under $\mathbf{S}$ is onto again, so that there exists $\mu \in \mathbf{S}(S)$ with $\mathbf{S}(\psi_1 \circ \beta_1)(\mu) = K(\varphi_1(x_1))$, consequently, $\mathbf{S}(\psi_i \circ \beta_i)(\mu) = \mathbf{S}(\psi_i)(K_i(x_i))$ $(i = 1, 2)$. But this means that $\mathbf{S}(\beta_i)(\mu)(E_i) = K_i(x_i)(E_i)$ holds for all $E_i \in \psi_i^{-1}[\mathcal{B}(Y)]$. Put

$$\Gamma(x_1, x_2) := \{\mu \in \mathbf{S}(B) | \mathbf{S}(\beta_1)(\mu) = K_1(x_1) \wedge \mathbf{S}(\beta_2)(\mu) = K_2(x_2)\},$$

then Prop. 3 shows that $\Gamma(x_1, x_2) \neq \emptyset$.

3. Since K_1 and K_2 are continuous, $\Gamma(r_1, x_2) \subseteq \mathbf{S}(D)$ is closed, and the set $\exists \Gamma(C)$ is closed in A for compact $C \subseteq \mathbf{S}(B)$. In fact, let $(\langle x_1^{(n)}, x_2^{(n)} \rangle)_{n \in \mathbb{N}}$ be a sequence in this set with $x_i^{(n)} \to x_i$, as $n \to \infty$ for $i = 1, 2$, thus $\langle x_1, x_2 \rangle \in A$. There exists $\mu_n \in C$ such that $\mathbf{S}(\beta_i)(\mu_n) = K_i(x_i^{(n)})$. Because C is compact, there exists a converging subsequence $\mu_{s(n)}$ and $\mu \in C$ with $\mu = \lim_{n \to \infty} \mu_{s(n)}$ in the topology of weak convergence. Continuity of K_i implies that $\mathbf{S}(\beta_i)(\mu) = K_i(x_i)$, consequently $\langle x_1, x_2 \rangle \in \exists \Gamma(C)$, thus this set is closed, hence measurable. From Prop. 2 it can now be inferred that there exists a measurable map $N : A \to \mathbf{S}(B)$ such that $N(x_1, x_2) \in \Gamma(x_1, x_2)$ holds for every $\langle x_1, x_2 \rangle \in A$. Thus $N : A \rightsquigarrow B$ is a stochastic relation with $K_1 \circ \alpha_1 = \mathbf{S}(\beta_1) \circ N$, and $K_2 \circ \alpha_2 = \mathbf{S}(\beta_2) \circ N$. Hence $\langle A, B, N \rangle$ is the desired semi-pullback. $\square$

Specializing Theorem 1, we list some categories of stochastic relations which have semi-pullbacks. Whenever continuity enters the game, its proof shows that the semi-pullback has the continuity property, too.

Corollary 1. *The following categories have semi-pullbacks:*

1. *Objects are Standard Borel spaces with a sub-probability measure attached, morphisms are measure preserving and surjective Borel maps (continuous maps, resp.).*

2. *Objects are Markov processes over Standard Borel spaces (Polish spaces), morphisms are measure preserving and surjective Borel maps (continuous maps, resp.).*

3. *Objects are stochastic relations over Polish spaces, morphisms $\langle \varphi, \psi \rangle$ are as in $\mathfrak{SRel}$ with ψ continuous. In the subcategory in which φ is also continuous semi-pullbacks exists, too*

Hence we know that the semi-pullback $\langle X, K \rangle$ for morphisms involving Markov processes is a Markov process again (whereas Edalat's main result [6, Cor. 5.2] permits only to conclude that K is a universally measurable transition sub-probability function).

Remark 2. One might be tempted now and ask for pullbacks or at least for weak pullbacks in the categories involved, now that the upper left hand corner of a pullback diagram can be filled. Recall that in a category a pair $\langle \tau_1 : c \to a_1, \tau_2 : c \to a_2 \rangle$ is a weak pullback for the pair $\rho_1 : a_1 \to b, \rho_2 : a_2 \to b$ of morphisms iff it is a semi-pullback (so that $\rho_1 \circ \tau_1 = \rho_2 \circ \tau_2$ holds), and if $\langle \tau_1' : c' \to a_1, \tau_2' : c' \to a_2 \rangle$ is another semi-pullback for that pair, then there exists a morphism $\theta : c' \to c$ so that $\tau_i' = \tau_i \circ \theta$ $(i = 1, 2)$ holds. If the factor θ is unique, then the weak pullback is called a pullback.

The following example shows that even the category of Standard Borel spaces with probability measures where the morphisms are surjective and measure preserving measurable maps does not have always weak pullbacks: Let μ be the uniform distribution on $A := \{1, 2, 3\}$, put $B := \{a, b\}$ with $\nu(a) := \frac{2}{3}, \nu(b) := \frac{1}{3}$. Let $f : A \to B$ with $f(1) := f(2) := a, f(3) := b$. Then $f : \langle A, \mu \rangle \to \langle B, \nu \rangle$ is a morphism. Now compute the semi-pullback $\langle P, \gamma \rangle$ for the kernel pair represented by f. Then

$$P = \{\langle x, x' \rangle | f(x) = f(x')\} = \{\langle 1, 1 \rangle, \langle 1, 2 \rangle, \langle 2, 1 \rangle, \langle 2, 2 \rangle, \langle 3, 3 \rangle\},$$

and a suitable instance for γ is determined easily (e.g., $\gamma(\langle 3, 3 \rangle) = \frac{1}{3}$, all other pairs in P can be assigned $\frac{1}{6}$). The identity $\iota : \langle A, \mu \rangle \to \langle A, \mu \rangle$ has the property $f \circ \iota = f \circ \iota$. If a weak pullback exists, then we know about the factor ρ that $\rho(a) = \langle a, a \rangle$ holds for all $a \in A$; since f is not injective, ρ cannot be onto. This is a contradiction.

The reason for this is evidently that a weak pullback in e.g. $\mathfrak{SRel}$ would induce a weak pullback in the category of sets with ordinary maps as morphisms, but that it cannot be guaranteed there that the factor is onto, even if the morphisms for which the pullback is computed are.

Consequently, semi-pullbacks are the best we can do in $\mathfrak{SRel}$.

4   Bisimulation

This section demonstrates that the bisimulation relation on objects of $\mathfrak{SRel}$ is transitive, and serves as an application for the result that semi-pullbacks exist in this category. A final application is provided by proving the well known result due to Desharnais, Edalat and Panangaden that bisimilarity of labelled Markov processes may be characterized through a simple negation-free modal logic; the processes are based on Standard Borel spaces with measurable — rather than universally measurable — transition sub-probability functions.

We define a bisimulation for two objects in $\mathfrak{SRel}$ through a span of morphisms in that category [10]. This is similar to the notion of 1-bisimulation investigated in [4] for the comma category $\mathbb{1}_{\mathcal{M}} \downarrow \mathbf{S}$, were $\mathcal{M}$ is the category of all measurable spaces with measurable maps as morphisms.

Definition 1. *An object P in $\mathfrak{SRel}$ together with morphisms $\langle\sigma_1,\tau_1\rangle : P \to Q_1$ and $\langle\sigma_2,\tau_2\rangle : P \to Q_2$ is called a* bisimulation *of objects Q_1 and Q_2.*

We apply the semi-pullback for establishing the fact that the bisimulation relation is transitive in $\mathfrak{SRel}$.

Proposition 4. *The bisimulation relation between objects in the category $\mathfrak{SRel}$ of stochastic relations is transitive. The same is true for the subcategories of Markov processes introduced in Cor. 1.*

Finally the characterization of bisimulations for labelled Markov processes through a Hennessy-Milner logic will be discussed. This follows the lines of [2] (a completely different approach is pursued in [3]). We will capitalize on the possibility to construct semi-pullbacks in categories of Markov processes over Polish spaces with Borel (rather than universally) measurable transition sub-probabilities. Hence we can characterize bisimulation in what seems to be a much more natural way from a probabilistic point of view, albeit for a restricted class of Markov processes for which the argumentation can be kept within the realm of Standard Borel spaces.

Fix a countable set L of actions.

Definition 2. *Let S be a Standard Borel space, and assume that $k_a : S \rightsquigarrow S$ is a stochastic relation for each $a \in \mathsf{L}$. Then $(S,(k_a)_{a\in\mathsf{L}})$ is called a* labelled Markov process.

S serves as a state space for the process. If the process is in state $s \in S$, and action $a \in \mathsf{L}$ is taken, then $k_a(s,B)$ is the probability for the next state to be a member of Borel set $B \subseteq S$.

Before proceeding, recall that a subset $A \subseteq X$ of a Polish space X is called *analytic* iff there exists a Polish space P and a continuous map $f : P \to X$ such that $A = f[P]$ holds. If A is equipped with the trace of the Borel sets of X, viz., $\{A \cap B | B \in \mathcal{B}(X)\}$ then A together with this σ-algebra is called an *analytic space*. The definition of a labelled Markov process found in [2] resembles the one given above, but assumes that the state space is analytic; *generalized* labelled Markov processes are introduced in which the transition sub-probability is assumed to be universally measurable.

Returning to Def. 2, let $(S,(k_a)_{a\in\mathsf{L}})$ and $(S',(k'_a)_{a\in\mathsf{L}})$ be labelled Markov processes with the same set L of actions. A *morphism* $f : (S,(k_a)_{a\in\mathsf{L}}) \to (S',(k'_a)_{a\in\mathsf{L}})$ is a surjective Borel map $f : S \to S'$ such that $k'_a \circ f = \mathbf{S}(f) \circ k_a$ holds for all $a \in \mathsf{L}$. Hence f is probability preserving for each action. Thus we have for each action $a \in \mathsf{L}$ a morphism between the objects (S,k_a) and (S',k'_a) in the category described in Cor. 1.(2). Applying Cor. 1 for each action separately and collecting the results yields:

Corollary 2. *The category of labelled Markov processes with morphisms described above has semi-pullbacks.*

From now on we omit the set L of actions when writing down labelled Markov processes.

In essentially the same way bisimulations are introduced through a span of morphisms: the labelled Markov processes $(S, (k_a))$ and $(S', (k'_a))$ are called *bisimilar* iff there exists a labelled Markov process $(T, (\ell_a))$ and morphisms $(S, (k_a)) \leftarrow (T, (\ell_a)) \rightarrow (S', (k'_a))$.

We follow [2] in introducing syntax and semantics of the Hennessy-Milner logic $\mathcal{L}$. The syntax is given by

$$\top \mid \phi_1 \wedge \phi_2 \mid \langle a \rangle_q \phi$$

Here $a \in L$ is an action, and q is a rational number. Fix a labelled Markov process $(S, (k_a))$, then satisfaction of a state s for a formula ϕ is defined inductively. This is trivial for $\top$ and for formulas of the form $\phi_1 \wedge \phi_2$. The more complicated case is making an a-move: $s \models \langle a \rangle_q \phi$ holds iff we can find a measurable set $A \subseteq S$ such that $\forall s' \in A : s' \models \phi$ and $k_a(s, A) \geq q$ both hold. Intuitively, we can make an a-move in a state s to a state that satisfies ϕ with probability greater than q.

Denote by Φ the set of all formulas, and put $[\![\phi]\!]_S := [\![\phi]\!] := \{s \in S \mid s \models \phi\}$ as usual as the set of states that satisfy formula ϕ (we omit the subscript S if the context is clear). Let $(S', (k'_a))$ be another labelled Markov process, then define for $s \in S, s' \in S'$ the relation $s \approx s'$ iff s and s' satisfy all the same formulas. Formally, $s \approx s'$ holds iff $1_{[\![\phi]\!]}(s) = 1_{[\![\phi]\!]}(s')$ holds for all $\phi \in \Phi$, 1_A denoting the indicator function for the set A. Now define for labelled Markov processes the relation $\sim$ which indicates that two labelled Markov processes satisfy exactly the same formulas for logic $\mathcal{L}$: $(S, (k_a)) \sim (S', (k'_a))$ iff $[\forall s \in S \exists s' \in S' : s \approx s' \text{ and } \forall s' \in S' \exists s \in S : s' \approx s]$.

We will establish for labelled Markov processes the equivalence of bisimilarity and satisfying the same formulas, and we will follow essentially the line of attack pursued in [2]. But we want to stay within the realm of Standard Borel spaces. Working as in [2] with the set of equivalence classes with the final Borel structure for the quotient map for $\approx$ would bring us into the realm of analytic spaces. Instead we will work with a Borel set which intersects each equivalence class in exactly one element (what is usually called a *Borel cross section*, cf. [11, p. 186]). With T comes a surjective Borel map $f_T : S \rightarrow T$ which may stand in for the quotient map, so that we can construct from $(S, (k_a))$ another labelled Markov process $(T, (h_a))$ with f_T now acting as morphism. This is then applied to the case that $(S, (k_a)) \sim (S', (k'_a))$ by forming the sum of the processes and constructing from this sum through relation $\approx$ morphisms the semi-pullback of which will yield the desired bisimulation. So the plan is very similar to that in [2], but the terrain will be operated on in a slightly different manner.

We will restrict ourselves to processes for which the existence of a cross section is guaranteed:

Definition 3. *The $(S, (k_a))$ be a labelled Markov process is called* small *iff there exists a Borel cross section T for relation $\approx$, i.e., a Borel set $T \subseteq S$ which intersects each equivalence class in exactly one state.*

If S is locally compact, and each k_a is weakly continuous, then it can be shown that the process is small.

The cross section is a Standard Borel space in its own right, taking the Borel sets of S that are contained in T as its σ-algebra. Define $f_T : S \to T$ so that $f_T(x)$ is the unique element of $[x] \cap T$. Hence f_T picks from each class a representative, so that it may be interpreted as a selection map, because $s \approx f_T(s)$ always holds. f_T is a surjective Borel map [11, Prop. 5.1.9].

For the rest of the paper we assume that the processes involved are small.

Lemma 2. *Let $(S, (k_a))$ be small, and T, f_T as above. The σ-algebra on T is generated by the family $\mathcal{B}_0 := \{f_T[[\![\phi]\!]] \mid \phi \in \Phi\}$, and $\mathcal{B}_0$ is closed under finite intersections.*

From these data a labelled Markov process can be constructed:

Corollary 3. *Let S, T, f_T be as above, and put for $a \in \mathsf{L}, s \in S$ and the measurable $B \subseteq T$ $h_a(f_T(s))(B) := k_a(s)(f_T^{-1}[B])$. This defines a labelled Markov process $(T, (h_a))$ such that $f_T : (S, (k_a)) \to (T, (h_a))$ is a morphism.*

We can now prove that satisfying the same formulas and bisimilarity are equivalent, following the trail laid out in [2].

Theorem 2. *Two small labelled Markov processes are bisimilar iff they satisfy the same formulas.*

Proof. 1. The "only if"- part follows from [2, Cor. 9.3], so only the "if"-part needs to be established. We proceed as in the proof of [2, Theorem 9.10] by constructing from the labelled Markov processes $(S, (k_a))$ and $(S', (k'_a))$ a diagram of the form

$$(S, (k_a)) \longrightarrow (T, (h_a)) \longleftarrow (S', (k'_a))$$

2. Let S_0 be the sum of the Standard Borel spaces S and S', hence S_0 is a Standard Borel space again. Put for $a \in \mathsf{L}, s \in S_0$ and the Borel set $B \subseteq S_0$

$$\ell_a(s)(B) := \begin{cases} k_a(s)(S \cap B), & s \in S \\ k'_a(s)(S' \cap B), & s \in S' \end{cases}$$

Thus $(S_0, (\ell_a))$ is a labelled Markov process.

Since $(S, (k_a))$ is small, it has a Borel cross section T which is also a Borel cross section for S_0, since both processes satisfy the same formulas. Thus $(S_0, (\ell_a))$ is small. Construct the associated Borel map $f_T : S_0 \to T$ for T, and define the labelled Markov process $(T, (h_a))$ as in Lemma 3. Let $i : S \to S_0, i' : S' \to S_0$ be the embeddings of S resp. S' into S_0. Then $f_T \circ i : S \to T, f_T \circ i' : S' \to T$ are surjective. Both are morphisms. $\qquad \square$

Acknowledgements. The author wants to thank Georgios Lajios for his helpful and constructive comments. The referees' suggestions are gratefully acknowledged.

5   Conclusion

We show that one can construct a semi-pullback in the category of stochastic relations over Standard Borel spaces with Borel measurable and measure preserving maps as morphisms. The bisimulation relation is shown to be transitive. It is finally shown that the characterization of bisimulation through satisfiability in a simple logic may be derived in this conceptually simpler context, too.

We rely on selection arguments from the theory of set-valued relations. This technique permits drawing from the rich well of topology, in particular utilizing the weak topology on the space of all sub-probability measures. Selection arguments may be a helpful way of constructing objects; we illustrate this by showing that the map which assigns each Polish space its sub-probabilities and each surjective Borel measurable map the corresponding measure transform is actually a functor which may be difficult to establish otherwise.

Further work will investigate congruences on Markov processes. They arise in a natural fashion from morphisms and generalize relation $\approx$. Conditions on the smallness of these processes will be also of interest.

References

[1] S. Abramsky, R. Blute, and P. Panangaden. Nuclear and trace ideal in tensored *-categories. *Journal of Pure and Applied Algebra*, 143(1–3):3–47, 1999.

[2] J. Desharnais, A. Edalat, and P. Panangaden. Bisimulation of labelled markov-processes. *Information and Computation*, 179(2):163–193, 2002.

[3] J. Desharnais, V. Gupta, R. Jagadeesan, and P. Panangaden. Approximating labeled Markov processes. In *Proc. 15th Symposium on Logic in Computer Science*, pages 95–106. IEEE, 2000.

[4] E.-E. Doberkat. The demonic product of probabilistic relations. In Mogens Nielsen and Uffe Engberg, editors, *Proc. Foundations of Software Science and Computation Structures*, volume 2303 of *Lecture Notes in Computer Science*, pages 113–127, Berlin, 2002. Springer-Verlag.

[5] E.-E. Doberkat. A remark on A. Edalat's paper *Semi-Pullbacks and Bisimulations in Categories of Markov-Processes*. Technical Report 125, Chair for Software Technology, University of Dortmund, July 2002.

[6] A. Edalat. Semi-pullbacks and bisimulation in categories of Markov processes. *Math. Struct. in Comp. Science*, 9(5):523–543, 1999.

[7] M. Giry. A categorical approach to probability theory. In *Categorical Aspects of Topology and Analysis*, volume 915 of *Lecture Notes in Mathematics*, pages 68–85, Berlin, 1981. Springer-Verlag.

[8] C. J. Himmelberg and F. Van Vleck. Some selection theorems for measurable functions. *Can. J. Math.*, 21:394–399, 1969.

[9] K. R. Parthasarathy. *Probability Measures on Metric Spaces*. Academic Press, New York, 1967.

[10] J. J. M. M. Rutten. Universal coalgebra: a theory of systems. *Theoretical Computer Science*, 249(1):3–80, 2000. Special issue on modern algebra and its applications.

[11] S. M. Srivastava. *A Course on Borel Sets*. Number 180 in Graduate Texts in Mathematics. Springer-Verlag, Berlin, 1998.

Quantitative Analysis of Probabilistic Lossy Channel Systems

Alexander Rabinovich

School of Computer Science, Tel Aviv University, Israel
rabinoa@post.tau.ac.il

Abstract. Many protocols are designed to operate correctly even in the case where the underlying communication medium is faulty. To capture the behaviour of such protocols, *lossy channel systems (LCS)* [3] have been proposed. In an LCS the communication channels are modelled as FIFO buffers which are unbounded, but also unreliable in the sense that they can nondeterministically lose messages.

Recently, several attempts [5,1,4,6] have been made to study *Probabilistic Lossy Channel Systems (PLCS)* in which the probability of losing messages is taken into account and the following qualitative model checking problem is investigated: to verify whether a given property holds with probability one.

Here we consider a more challenging problem, namely to calculate the probability by which a certain property is satisfied. Our main result is an algorithm for the following *Quantitative model checking problem:*

Instance: A PLCS, its state s, a finite state ω-automaton $\mathcal{A}$, and a rational $\theta > 0$.

Task: Find a rational r such that the probability of the set of computations that start at s and are accepted by $\mathcal{A}$ is between r and $r + \theta$.

1 Introduction

Finite state machines which communicate through unbounded buffers (CFSM) have been popular in the modelling of communication protocols. A CFSM defines in a natural way an infinite state transition system. The fact that Turing machines can be simulated by CFSMs [7] implies that all the nontrivial verification problems are undecidable for CFSMs. Many protocols are designed to operate correctly even in the case where the underlying communication medium is faulty. To capture the behaviour of such protocols, *lossy channel systems (LCS)* [3] have been proposed as an alternative model. In an LCS the communication channels are modelled as FIFO buffers which are unbounded, but also unreliable in the sense that they can nondeterministically lose messages. Thought an LCS defines in a natural way an infinite state transition system, it has been shown that the reachability problem for LCS is decidable [3], while progress properties are undecidable [2].

Probabilistic Lossy Channel Systems. Since we are dealing with unreliable communication media, it is natural to deal with models in which the probability

J.C.M. Baeten et al. (Eds.): ICALP 2003, LNCS 2719, pp. 1008–1021, 2003.

of losing messages is taken into account. Recently, several attempts [10,5,1,4, 6] have been made to study *probabilistic Lossy Channel Systems (PLCS)* which introduce randomization into the behaviour of LCS. The works in [10,5,1,4, 6] define different semantics for PLCS, depending on the manner in which the messages may be lost inside the channels. All these models associate in a natural way a countable Markov Chain (M.C.) to a PLCS.

Baier and Engelen [5] consider a model which assumes that at most a single message may be lost during each step of the execution of the system. They showed decidability of the following problems under the assumption that the probability of losing messages is at least 0.5.

> *Qualitative Probabilistic Reachability*
> *Instance:* A PLCS M and its states s_1, s_2.
> *Question:* Is s_2 reached from s_1 with probability one?

> *Qualitative Probabilistic Model-checking*
> *Instance:* A PLCS M, its state s and a finite state ω-automaton $\mathcal{A}$.
> *Question:* Is the probability of the set of computations that start at s and are
> accepted by $\mathcal{A}$ equal to one?

The model in [1] assumes that messages can only be lost during send operations. Once a message is successfully sent to a channel, it continues to reside inside the channel until it is removed by a receive operation. Even the qualitative reachability problem was shown to be undecidable for this model of PLCS and losing probability $\lambda < 0.5$.

In [4,6] another semantics for PLCS was considered which is more realistic than that in [5,1]. More precisely, it was assumed that, during each step in the execution of the system, each message may be lost with a certain predefined probability. This means that the probability of losing a certain message will not decrease with the length of the channels (as it is the case with [5,1]).

For this model, the decidability of both the qualitative reachability and the qualitative model-checking problems was independently established in [4,6].

Our Contribution. All the above mentioned papers consider qualitative properties of PLCS. Here we consider a more challenging problem, namely to calculate the probability by which a certain property is satisfied.

Unfortunately, we were unable to prove that the probability of reaching a state s_2 from a state s_1 in PLCS is an algebraic number, or it is explicitly expressible by standard mathematical functions.

Therefore, we will approximate the probability by which a certain property is satisfied. Our main result is that the following two problems are computable.

> *Quantitative Probabilistic Reachability*
> *Instance:* A PLCS $\mathcal{L}$, its states s_1, s_2 and a rational $\theta > 0$.
> *Task:* Find a rational r such that s_2 is reached from s_1 with the probability
> between r and $r + \theta$.

> *Quantitative Probabilistic Model-checking*
> *Instance:* A PLCS $\mathcal{L}$, its state s, a finite state ω-automaton $\mathcal{A}$, and a rational $\theta > 0$.
> *Task:* Find a rational r such that the probability of the set of computations that
> start at s and are accepted by $\mathcal{A}$ is between r and $r + \theta$.

In order to approximate the probability p of the set of computations from a state s with a property φ in a PLCS $\mathcal{L}$ one can try to compute this probability p_n for the finite sub-chain $M_n = (S_n, P_n)$ of the countable Markov chain M generated by $\mathcal{L}$, where S_n is the set of states with at most n messages. There are two problems in this approach: (a) a state which was recurrent in M might become transient in M_n; (b) how to find n which will ensure that the result p_n approximates up to θ the probability p in M.

In order to overcome problem (a) we analyze the structure of the recurrent classes of the Markov chain generated by a PLCS $\mathcal{L}$. For problem (b) the value for n is computed from an appropriate reduction of the Markov chains generated by PLCSs to one dimensional random walks.

Outline. In the next two sections we give basics of transition systems and countable Markov chains respectively. In Sect. 4 the quantitative probabilistic reachability problem over countable Markov chains is considered. In Sections 5 and 6 the semantics of lossy channel systems and probabilistic lossy channel systems are described. In Sect. 7 the algorithm for the quantitative probabilistic reachability problem over PLCS is presented and its complexity is analyzed. In Sect. 8 we generalize our results to the verification of the properties definable by ω-behavior of finite state automata.

2   Transition Systems

In this section, we recall some basic concepts of transition systems.

A *transition system* T is a pair $(S, \longrightarrow)$ where S is a (potentially) infinite set of *states*, and $\longrightarrow$ is a binary relation on S. We write $s_1 \longrightarrow s_2$ to denote that $(s_1, s_2) \in \longrightarrow$ and use $\overset{*}{\longrightarrow}$ to denote the reflexive transitive closure of $\longrightarrow$. We say that s_2 is *reachable* from s_1 if $s_1 \overset{*}{\longrightarrow} s_2$. For sets $Q_1, Q_2 \subseteq S$, we say that Q_2 is *reachable* from Q_1, denoted $Q_1 \overset{*}{\longrightarrow} Q_2$, if there are $s_1 \in Q_1$ and $s_2 \in Q_2$ with $s_1 \overset{*}{\longrightarrow} s_2$. A *path* or *computation* π (from s_0) is a (finite or infinite) sequence $s_0, s_1, \ldots, s_n, \ldots$ of states with $s_i \longrightarrow s_{i+1}$ for $i \geq 0$. We use $\pi(i)$ to denote s_i, and write $s \in \pi$ to denote that there is an $i \geq 0$ with $\pi(i) = s$. For states s and s', we say that π *leads* from s to s', written, $s \overset{\pi}{\longrightarrow} s'$, if $s = s_0$ and $s' \in \pi$.

For $Q \subseteq S$, we define the *graph* of Q, denoted $Graph(Q)$, to be the transition system $(Q, \longrightarrow')$ where $s_1 \longrightarrow' s_2$ iff $s_1 \overset{*}{\longrightarrow} s_2$. A *strongly connected component (SCC)* in T is a maximal set $C \subseteq S$ such that $s_1 \overset{*}{\longrightarrow} s_2$ for each $s_1, s_2 \in C$.

We say that C is a *bottom SCC (BSCC)* if there is no other SCC C_1 in T with $C \xrightarrow{*} C_1$. In other words, the BSCCs are the leaves in the acyclic graph of SCCs (ordered by reachability).

3   Markov Chains

In this section, we recall basic properties of *Markov chains*. We also introduce attractors which play an important role in our analysis of recurrent classes of Markov chains.

A *Markov chain* M is a pair (S, P) where S is a (potentially infinite) set of states and P is a mapping from $S \times S$ to the set $[0, 1]$, such that $\sum_{s' \in S} P(s, s') = 1$, for each $s \in S$.

A Markov chain induces a transition system where the transition relation consists of pairs of states related by positive probabilities. Formally, the *underlying transition system* of M is $(S, \longrightarrow)$ where $s_1 \longrightarrow s_2$ iff $P(s_1, s_2) > 0$. In this manner, the concepts defined for transition systems can be lifted to Markov chains. For instance, an SCC in M is a SCC in the underlying transition system.

A Markov chain (S, P) induces a natural measure on the set of computations from every state s.

Let us recall some basic notions from probability theory. A *measurable space* is a pair (Ω, Δ) consisting of a non empty set Ω and a σ-algebra Δ of its subsets that are called *measurable sets* and represent random events. A *σ-algebra* over Ω contains Ω and is closed under complementation and countable union. Adding to a measurable space a *probability measure* $Prob : \Delta \rightarrow [0, 1]$ such that $Prob(\Omega) = 1$ and that is countably additive, we get a *probability space* $(\Omega, \Delta, Prob)$.

Consider a state s of a Markov chain (S, P). On the sets of computations that start at s, the probabilistic space $(\Omega, \Delta, Prob)$ is defined as follows (see [12]): $\Omega = sS^\omega$ is the set of all ω-sequences of states starting from s, Δ is the σ-algebra generated by the basic cylindric sets $D_u = uS^\omega$, for every $u \in sS^*$, and the probability measure $Prob$ is defined by $Prob(D_u) = \prod_{i=0,...,n-1} P(s_i, s_{i+1})$ where $u = s_0 s_1 ... s_n$; it is well-known that this measure is extended in a unique way to the elements of the σ-algebra generated by the basic cylindric sets.

Consider a set $Q \subseteq S$ of states and a path π. We say that π *reaches* Q if there is an $i \geq 0$ with $\pi(i) \in Q$. We say that π *repeatedly reaches* Q if there are infinitely many i with $\pi(i) \in Q$. Let s be a state in S. We say that a state s is *recurrent* if $Prob\{\pi : \pi$ is a path from s and π repeatedly reaches $s\} = 1$. We say that a state s is *transient* if $Prob\{\pi : \pi$ is a path from s and π repeatedly reaches $s\} = 0$.

The next theorem summarizes standard properties of countable Markov chains [13].

Theorem 3.1. *1. Every state is either transient or recurrent.*

2. If s is recurrent then all the states reachable from s are recurrent.

3. Let C be a strongly connected component of a Markov chain. Then, either all the states in C are transient or all the states in C are recurrent.

4. *Let C be a recurrent strongly connected component of a Markov chain and $s_1, \in C$. Then $Prob\{\pi : \pi$ starts at s_1 and repeatedly reaches every state of $C\} = 1$. For every state s and non-empty subset $B \subseteq C$ the probability to repeatedly reach every state of B from s is the same as the probability to reach B from s and is the same as the probability to reach s_1 from s.*
5. *A recurrent strongly connected component is always a bottom strongly connected component.*

A recurrent (transient) SCC is often called a recurrent (transient) class. We introduce a central concept which we use in our solution for the probabilistic reachability problem, namely that of *attractors*.

Definition 3.2 (attractor). *A set $A \subseteq S$ is said to be an attractor if, for each $s \in S$, the set A is reachable from s with probability one.*

In other words, regardless of the state in which we start, we are guaranteed that we will reach the attractor with probability one. It is clear that an attractor has a state in every recurrent class. The Lemma below follows from Theorem 3.1 and describes properties of the BSCCs of the graph of a finite attractor A.

Lemma 3.3. *Assume that a Markov chain M has a finite attractor A. Then (1) each BSCC C of $Graph(A)$ is a subset of a recurrent component in M. (2) A state is recurrent if and only if it is reachable from a BSCC C of $Graph(A)$. (3) For every s in M the set of recurrent states is reached from s with probability one.*

4   Approximating Probability for Countable Markov Chains

Let M be a M.C. and let s_1, s_2 be its states. We use $Prob_M(s_1 \xrightarrow{*} s_2)$ for the probability with which s_2 is reached from s_1 in M. Let $Comp_n(s_1)$ be the set of all the computations of length n in M from s_1. Partition $Comp_n(s_1)$ into three sets:

$$\text{Reach}_n(s_1, s_2) = \{\pi : \pi \in \text{Comp}_n(s_1) \wedge \exists i \leq n. \ \pi(i) = s_2\}$$

$$\text{Escape}_n(s_1, s_2) = \{\pi : \pi \in \text{Comp}_n(s_1) \setminus \text{Reach}_n(s_1, s_2) \wedge s_2$$

is unreachable from $\pi(n)$}

$$\text{Endecided}_n(s_1, s_2) = \text{Comp}_n(s_1) \setminus \text{Reach}_n(s_1, s_2) \setminus \text{Escape}_n(s_1, s_2)$$

All the computations in $\text{Reach}_n(s_1, s_2)$ reach s_2, and no computation in $\text{Escape}_n(s_1, s_2)$ extends to a computation that reaches s_2. Note that $Prob(\text{Comp}_n(s_1)) = 1$. Let $p_n^+ = Prob(\text{Reach}_n(s_1, s_2))$, $p_n^- = Prob(\text{Escape}_n(s_1, s_2))$ and $p_n^? = Prob(\text{Endecided}_n(s_1, s_2))$. Observe that p_n^+ and p_n^- are increasing sequences, while $p_n^?$ is decreasing and

$$p_n^+ \leq \lim p_n^+ = Prob_M(s_1 \xrightarrow{*} s_2) \leq p_n^+ + p_n^? \tag{1}$$

Path Enumeration (PE) scheme for approximation $Prob_M(s_1 \xrightarrow{*} s_2)$ is based on (1).

> **Path Enumeration Scheme for Approximating Probabilistic Reachability**
>
> *Instance:* A M.C. M, its states s_1, s_2 and a $\theta > 0$.
> *Task:* Find r such that s_2 is reached from s_1 with the probability
>   between r and $r + \theta$.
> **begin**
>   1. n:=0; $\Delta := 1$;
>   2. **while** $(\Delta > \theta)$ **do**
>   3. n:=n+1;   Compute $r := p_n^+$;   Compute $\Delta := p_n^?$
>   **end while**
>   4. return(r)
> **end**

In the above problem, we do not assume that M is finite. Hence, these are not instances of an algorithmic problem. In Sect. 7 we consider the quantitative reachability problem when countable Markov chains are described by probabilistic lossy channel systems. For such finite descriptions we investigate the corresponding algorithmic problem.

If M has finite branching and is presented effectively, then p_n^+ is computable. Moreover, if in addition the reachability problem for the transition system underlying M is decidable, then $\text{Escape}_n(s_1, s_2)$, $\text{Endecided}_n(s_1, s_2)$ and $p_n^?$ can be computed. Hence, in this case the scheme can be implemented. Observe that PE scheme terminates only if $\lim p_n^? < \theta$. Therefore,

Lemma 4.1. *If* $\lim p_n^? = 0$ *then PE scheme terminates.*

It is well-known that for a finite state Markov chains $\lim p_n^? = 0$. This property holds for Markov chains with finite attractors [15] as well.

Lemma 4.2. *If M has a finite attractor then* $\lim p_n^? = 0$.

Another class of Markov chains for which PE scheme terminates is the class of chains which satisfy the following property.

Definition 4.3. *A Markov chain $M = (S, P)$ has δ-reachability property for $\delta > 0$ if $\forall s_1, s_2 \in S($ s_2 is reachable from s_1 $) \Rightarrow Prob_M(s_1 \xrightarrow{*} s_2) > \delta$.*

Lemma 4.4. *If M has δ-reachability property then* $\lim p_n^? = 0$.

Theorem 4.5. *The PE scheme terminates over the class of Markov chains with finite attractor and over the class of Markov chains with δ-reachability property.*

A variant of PE scheme was suggested in [10] for the following decision problem.

> **A decision problem for Probabilistic Reachability**
>   *Instance:* A M.C. M, its states $s_1, s_2, \theta > 0$ and p.
>   *Question:* Is $p - \theta < Prob_M(s_1 \xrightarrow{*} s_2) < p + \theta$?

It was claimed in [10] that Eq. (1) implies that (a) when the scheme terminates it produces a correct answer (b) it terminates for the Markov chains defined by

PLCS under the semantics of [10]. However, assertion (a) was incorrect. Also, the Markov chains assigned to PLCSs in [10] do not have finite attractor property and the termination assertion (b) is unsound. It is an open question whether the above problem is decidable for the PLCSs defined in Sect. 6 (or considered in [5,4,6]) which have finite attractor property.

PE scheme is conceptually very simple, however, no information about the number of iterations before it terminates can be extracted from Theorem 4.5. For finite state M. C. standard algebraic methods allow to find the exact value of $Prob_M(s_1 \xrightarrow{*} s_2)$ in polynomial time; however, in this case PE scheme finds an approximation in time $|M|^{\Omega(\ln(\frac{1}{\theta}))}$. An alternative approach for approximation of $Prob_{M'}(s_1 \xrightarrow{*} s_2)$ is to "approximate" a countable M.C. M by a finite state M.C. M' and then to evaluate $Prob_{M'}(s_1 \xrightarrow{*} s_2)$ by standard algebraic methods. Below is a simple transformation which allows to reduce the size of Markov chains.

Let $M = (S, P)$ and let $U \subseteq S$ and let u be a new state. The chain $M' = (S', P')$ which is obtained from M by collapsing U into an absorbing state u is denoted by $M^{U,u}$ and is defined as follows: $S' = S \setminus U \cup \{u\}$ and

$$
P'(s, s') = \begin{cases} \sum_{d \in U} P(s, d) & \text{if } s \neq u \wedge s = u, \\ P(s, s') & \text{if } s \neq u \wedge s' \neq u, \\ 1 & \text{if } s = u = s', \\ 0 & \text{otherwise.} \end{cases}
$$

The following two lemmas are immediate, but useful for reductions of the size of M.C.

Lemma 4.6. *Let M be a M.C., let s_1, s_2 be states of M, let $u \notin S$, let C be a recurrent class such that $s_1 \notin C$ and let $M' = M^{C,u}$.*
1. If $s_2 \in C$ then $Prob_M(s_1 \xrightarrow{*} s_2) = Prob_{M'}(s_1 \xrightarrow{*} u)$.
2. If $s_2 \notin C$ then $Prob_M(s_1 \xrightarrow{*} s_2) = Prob_{M'}(s_1 \xrightarrow{*} s_2)$.

Lemma 4.7. *Let M be a M.C., let s_1, s_2 be states of M. Assume that $D \subseteq S \setminus \{s_1, s_2\}$ is such that either (1) $Prob\{\pi : \pi$ starts at s_1 and reaches $D\} \leq \theta$ or*
(2) $\forall s \in D. Prob\{\pi : \pi$ starts at s and reaches $s_2\} \leq \theta$. Let $M' = M^{D,d}$. Then $Prob_{M'}(s_1 \xrightarrow{*} s_2) \leq Prob_M(s_1 \xrightarrow{*} s_2) \leq Prob_{M'}(s_1 \xrightarrow{*} s_2) + \theta$.

In order to find D which satisfies the assumption of Lemma 4.7 we provide a reduction to a one-dimensional random walk. The following lemma is easily derived from standard properties of one dimensional random walks.

Lemma 4.8. *Let $M = (S, P)$ be a Markov chain where $S = \{0, 1, 2, 3, \ldots\}$, and*

– $P(0, 0) = 1$.
– $P(i, i+1) = \nu_i$, $P(i, i-1) = \mu_i$, and $P(i, i) = 1 - \mu_i - \nu_i$, for $i \geq 1$.
– There is $q > 0.5$ such that $\mu_i > q$ for all $i \geq 2$.

Let $N(\mu_1, q, \theta) = \lceil \frac{\ln(1-\mu_1) - \ln(\mu_1 \cdot \theta)}{\ln q - \ln(1-q)} \rceil + 1$, where $\lceil x \rceil$ stands for the smallest integer which is greater than or equal to x. Then, for each $\theta > 0$ and $n \geq N(\mu_1, q, \theta)$, the probability of reaching a state n from 1 is less than θ.

The main technical lemma for the correctness and the complexity analysis of the algorithm presented in Sect. 7 is a generalization of Lemma 4.8.

Lemma 4.9 (Main Lemma). *Consider a Markov chain $M = (S, P)$ such that*

1. *S is the union of disjoint sets S_i ($i \in \mathcal{N}$).*
2. *If $s \in S_i$, $s' \in S_j$, and $P(s, s') > 0$, then $j \leq i + 1$.*
3. *$S_0 = C \cup R$ and*
   - *For every state $s \in R$, only states in R are reachable from s.*
   - *For every state $s \in S_1$ there is a finite path to R with the probability $> \delta$ which is inside $C \cup R$.*
4. *There is $\alpha < \frac{1}{2}$ such that $\nu_i + \gamma_i < \alpha$ for each $i \geq 2$, where*

$$\nu_i = \sup_{s \in S_i} \left(\sum_{s' \in S_{i+1}} P(s, s') \right) \ \text{and} \ \gamma_i = \sup_{s \in S_i} \left(\sum_{s' \in S_i} P(s, s') \right).$$

Let $N_0 = N(\delta, 1 - \alpha, \theta)$, where N is defined as in Lemma 4.8. Then, for every $s \in S_0 \cup S_1$ the probability of reaching $\bigcup_{n \geq N_0} S_n$ from s is less than θ.

Hence,

Lemma 4.10. *Let M, S_i and N_0 be as in Lemma 4.9 and assume that $s_1, s_2 \in S_0$. Let $U = \bigcup_{n \geq N_0} S_n$. Let $M' = M^{U,u}$. Then $Prob_{M'}(s_1 \xrightarrow{*} s_2) \leq Prob_M(s_1 \xrightarrow{*} u) \leq Prob_{M'}(s_1 \xrightarrow{*} s_2) + \theta$.*

5   Lossy Channel Systems

In this section we consider lossy channel systems: processes with a finite set of local states operating on a number of unbounded and unreliable channels.

A *Lossy Channel System (LCS)* consists of a finite state process operating on a finite set of channels each of which behaves as a FIFO buffer which is unbounded and unreliable in the sense that it can nondeterministically lose messages. Formally, a *lossy channel system (LCS)* $\mathcal{L}$ is a tuple (S, C, M, T) where S is a finite set of *local states*, C is a finite set of *channels*, M is a finite *message alphabet*, and T is a set of *transitions* each of the form (s_1, op, s_2), where $s_1, s_2 \in S$, and op is an *operation* of one of the forms $c!m$ (sending message m to channel c), or $c?m$ (receiving message m from channel c). A *global state* s is of the form (s, w) where $s \in S$ and w is a mapping from C to M^*.

For words $x, y \in M^*$, we use $x \bullet y$ to denote the concatenation of x and y. We write $x \preceq y$ to denote that x is a (not necessarily contiguous) substring of y. We use $|x|$ to denote the length of x, and use $x(i)$ to denote the ith element of x where $i : 1 \leq i \leq |x|$. For $w_1, w_2 \in (C \mapsto M^*)$, we use $w_1 \preceq w_2$ to denote that $w_1(c) \preceq w_2(c)$ for each $c \in C$, and define $|w| = \sum_{c \in C} |w(c)|$. We also extend $\preceq$ to a relation on $S \times (C \mapsto M^*)$, where $(s_1, w_1) \preceq (s_2, w_2)$ iff $s_1 = s_2$ and $w_1 \preceq w_2$.

The LCS $\mathcal{L}$ induces a transition system $(S, \longrightarrow)$, where S is the set of global states, i.e., $S = (S \times (C \mapsto M^*))$, and $(s_1, w_1) \longrightarrow (s_2, w_2)$ iff one of the following conditions are satisfied:

 - There is a $\mathbf{t} \in \mathbf{T}$, where $\mathbf{t}$ is of the form $(\mathbf{s}_1, \mathbf{c!m}, \mathbf{s}_2)$ and $\mathbf{w}_2$ is the result of appending $\mathbf{m}$ to the end of $\mathbf{w}_1(\mathbf{c})$.
 - There is a $\mathbf{t} \in \mathbf{T}$, where $\mathbf{t}$ is of the form $(\mathbf{s}_1, \mathbf{c?m}, \mathbf{s}_2)$ and $\mathbf{w}_1$ is the result of removing $\mathbf{m}$ from the head of $\mathbf{w}_2(\mathbf{c})$.
 - Furthermore, if $(\mathbf{s}_1, \mathbf{w}_1) \longrightarrow (\mathbf{s}_2, \mathbf{w}_2)$ according to one of the previous two rules, then $(\mathbf{s}_1, \mathbf{w}_1) \longrightarrow (\mathbf{s}'_2, \mathbf{w}'_2)$ for each $(\mathbf{s}'_2, \mathbf{w}'_2) \preceq (\mathbf{s}_2, \mathbf{w}_2)$.

In the first two cases we define $\mathbf{t}(\mathbf{s}_1, \mathbf{w}_1) = (\mathbf{s}_2, \mathbf{w}_2)$.

A transition $(\mathbf{s}_1, \mathbf{op}, \mathbf{s}_2)$ is said to be *enabled* at $(\mathbf{s}, \mathbf{w})$ if $s = s_1$ and either $\mathbf{op}$ is of the form $\mathbf{c!m}$; or $\mathbf{op}$ is of the form $\mathbf{c?m}$ and $\mathbf{w}(\mathbf{c}) = \mathbf{m} \bullet x$, for some $x \in \mathbf{M}^*$. We define $enabled(\mathbf{s}, \mathbf{w}) = \{\mathbf{t} : \mathbf{t} \text{ is enabled at } (\mathbf{s}, \mathbf{w})\}$.

Remark on notation. We use $\mathbf{s}$ and $\mathbf{S}$ to range over local states and sets of local states respectively. On the other hand, s and S range over states and sets of states of the induced transition system (states of the transition system are global states of the LCS).

A set $Q \subseteq S$ is said to be *upward closed* if $s_1 \in Q$ and $s_1 \preceq s_2$ imply $s_2 \in Q$. The upward closure $Q \uparrow$ of a set Q is the set $\{s : \exists s' \in Q. s' \preceq s\}$.

Theorems in [3,9] imply the following decidability results for LCS:

Lemma 5.1. *(1) It is decidable whether a state s_2 is reachable from a state s_1. (2) It is decidable whether the upward closure of a finite set Q is reachable from a state s. (3) There is an algorithm Find-a-path($s_1, s_2, \mathcal{L}$) which returns a path from s_1 to s_2 in the lossy channel system $\mathcal{L}$ or returns "No" if s_2 is not reachable from s_1. (4) Graph(A) is computable for every finite set of global states A of an LCS.*

6   Probabilistic Lossy Channel Systems

We introduce a probabilistic behaviour into LCS obtaining *Probabilistic Lossy Channel Systems*. This semantics was considered in [4,6] and differs from that in [10,5]

A *probabilistic lossy channel system (PLCS)* $\mathcal{L}$ is of the form $(\mathbf{S}, \mathbf{C}, \mathbf{M}, \mathbf{T}, \lambda, w)$, where $(\mathbf{S}, \mathbf{C}, \mathbf{M}, \mathbf{T})$ is an LCS, $\lambda \in (0, 1)$, and w is a mapping from $\mathbf{T}$ to the natural numbers. Intuitively, we derive a Markov chain from the PLCS $\mathcal{L}$ by assigning probabilities to the transitions of the underlying transition system $(\mathbf{S}, \mathbf{C}, \mathbf{M}, \mathbf{T})$. The probability of performing a transition $\mathbf{t}$ from a global state $(\mathbf{s}, \mathbf{w})$ is determined by the weight $w(\mathbf{t})$ of $\mathbf{t}$ compared to the weights of the other transitions which are enabled at $(\mathbf{s}, \mathbf{w})$. Furthermore, after performing each transition, each message which resides inside one of the channels may be lost, independently of the other messages, with the probability λ. This means that the probability of the transition from $(\mathbf{s}_1, \mathbf{w}_1)$ to $(\mathbf{s}_2, \mathbf{w}_2)$ is equal to (the sum over all $(s_3, \mathbf{w}_3)$ of) the probability of reaching some $(s_3, \mathbf{w}_3)$ from $(\mathbf{s}_1, \mathbf{w}_1)$ through performing a transition t of the underlying LCS, multiplied by the probability of reaching $(\mathbf{s}_2, \mathbf{w}_2)$ from $(s_3, \mathbf{w}_3)$ through the loss of messages (see [4] for detailed calculations of the probabilities of the transitions).

To simplify the presentation, we assume from now on that PLCSs have no deadlock states, i.e., from every state a transition is enabled. The only probabilistic properties of PLCSs which we use are summarized in the next two lemmas from [4].

Lemma 6.1. *Let s be a state with m messages. The probability of the transitions from s to the set of states with $> m + 1$ messages is 0. The probability of the transitions from s to the set of states with $m + 1$ messages is $\leq (1 - \lambda)^{m+1}$. The probability of the transitions from s to the set of states with m messages is $\leq m\lambda(1 - \lambda)^m$.*

Lemma 6.2. *For each λ, w, and PLCS (S, C, M, T, λ, w) the set of the states with the empty set of messages is a finite attractor.*

The next lemma plays a key role in the algorithm presented in Sect. 7

Lemma 6.3. *For each PLCS $\mathcal{L} = (S, C, M, T, \lambda, w)$ there are $V_1, \ldots, V_k$ such that V_i are finite sets of global states and k is the number of the recurrent classes of $\mathcal{L}$ and for each state s: s is in the i-th recurrent class of $\mathcal{L}$ iff s is not in the upward closure of V_i. Moreover, $V_1, \ldots, V_k$ are computable from the underlying LCS (S, C, M, T).*

7   Algorithm for Approximating the Probability of Reachability

Lemmas 6.2, 5.1(1) and Theorem 4.5 imply that there is an algorithm based on PE scheme for the quantitative probabilistic reachability problem. However, no information about the complexity of this algorithm can be extracted from Theorem 4.5. In this section we provide an algorithm with a parametric complexity $f(\mathcal{L}, s_1, s_2) \times \frac{1}{\theta^3}$ for the quantitative probabilistic reachability problem.

The idea of the algorithm is to take the set $B_{\leq n}$ of states with at most n messages of the Markov chain M generated by PLCS $\mathcal{L}$. Construct a finite Markov chain $\widehat{M}$ by restricting the transition of M to $B_{\leq n}$, and then for each recurrence class D_i of M collapse the set of D_i states in $B_{\leq n}$ into one state of $\widehat{M}$. Finally, calculate the probability of reaching s_2 from s_1 in the finite M.C. $\widehat{M}$. The crucial fact in the correctness of our algorithm is that relying on Lemma 4.9 we can compute n big enough which will ensure that the probability of reaching s_2 from s_1 in the finite Markov chain $\widehat{M}$ approximates up to θ the probability of reaching s_2 from s_1 in the infinite Markov chain M.

In the rest of this section we describe the algorithm with a justification of its correctness and provide an analysis of its complexity.

Algorithm – for Quantitative Probabilistic Reachability Problem
Input PLCS $\mathcal{L} = (S, C, M, T, \lambda, w)$ with an underlying Markov chain $M = (S, P)$, states $s_1, s_2 \in S$, and a rational θ.
Output: a rational r such that $r \leq Prob_M(s_1 \xrightarrow{*} s_2) \leq r + \theta$.
Let A be the (finite) set of all states with 0 messages. A is an attractor by Lemma 6.2. By Lemma 5.1(4) we can construct $Graph(A)$. Then we can find the bottom

strongly connected components $C_1, \ldots, C_k$ in $Graph(A)$ and for $1 \leq i \leq k$ by Lemma 3.3 and by Lemma 6.3 we can compute finite sets of states V_i such that

$$\forall s \in S(s \text{ is in the recurrent class of } C_i) \text{ iff } s \text{ is not in the upward closure of } V_i \tag{2}$$

Hence, we can check whether s_1 (or any other state s) is in the i-th recurrent class.

In the case when s_1 is recurrent we proceed as follows: If s_2 is recurrent and in the same recurrent class as s_1 then output 1 else output 0. (The correctness of this answer follows by Lemma 3.1(4-5).)

Below we consider the case when s_1 is not recurrent. By Lemma 5.1(3) we can find l such that for every $u, v \in A \cup \{s_1, s_2\}$ if u is reachable from v then there is a path from v to u which passes only through nodes with at most l messages. Let m be such that $\forall n. \ m \leq n \rightarrow (1 - \lambda)^n (1 - \lambda + n\lambda) < \frac{1}{3}$ i.e., the probability to move from every state with $n \geq m$ messages to the set of states with at least n messages is less than $\frac{1}{3}$ (by Lemma 6.1). Let $h = max(l, m) + 1$.

Notations: Below we denote by B_i (respectively by $B_{\leq i}$) the set of states with i (respectively with at most i) messages.

For every state $s \in B_{<h}$ there is a path π_s which first chooses a lossy transition which leads to a state s' with 0 messages and then follows by a path from s' which is inside $B_{\leq l} \subseteq B_{\leq h}$ to a BSCC of $Graph(A)$. Let $\delta_s = Prob(\pi_s) > 0$ and let $0 < \delta = min(\delta_s : s \in B_{\leq h})$. Note that up to this point all our computations were independent of θ and their complexity depended only on $\mathcal{L}$, s_1 and s_2.

Observe that if we denote by R the set of recurrent states of M, by C the set of transient states with $< h$ messages; by S_0 the set $R \cup C$ and by S_i ($i > 0$) the set of transient states with $h + i - 1$ messages, then the assumptions of Lemma 4.9 are satisfied. Let $N_0 = N(\delta, \frac{2}{3}, \theta)$, where N is the function from Lemma 4.8 and let $n = h + N_0$. Note that n depends linearly on $\ln(\frac{1}{\theta})$. By Lemma 4.9 the probability to reach from s_1 the set $U = \bigcup_{n \geq N_0} S_n$ of transient states with $\geq n$ messages is at most θ. Therefore, by Lemma 4.10 we derive that $Prob_{M'}(s_1 \xrightarrow{*} s_2) \leq Prob_M(s_1 \xrightarrow{*} s_2) \leq Prob_{M'}(s_1 \xrightarrow{*} s_2) + \theta$ for $M' = M^{U, dead}$ obtained by collapsing U into a fresh state $dead$. The chain M' might be infinite. Below we are going to construct a finite state M.C. $\widehat{M}$ of size bounded by $|B_{\leq n}|$ such that $Prob_{M'}(s_1 \xrightarrow{*} s_2) = Prob_{\widehat{M}}(s_1 \xrightarrow{*} s_2)$. Hence, $Prob_{\widehat{M}}(s_1 \xrightarrow{*} s_2)$ will approximate up to θ the value of $Prob_M(s_1 \xrightarrow{*} s_2)$ which we are trying to compute.

The complexity of the construction of $\widehat{M}$ will be $O(|B_{\leq n}|^2)$ and by standard algebraic methods we can compute $Prob_{\widehat{M}}(s_1 \xrightarrow{*} s_2)$ in time $O(|B_{\leq n}|^3)$. Since n depends linearly on $\ln(\frac{1}{\theta})$, it follows that $|B_{\leq n}|$ depends linearly on $\frac{1}{\theta}$ and the complexity of the entire algorithm is $f(\mathcal{L}, s_1, s_2) \times \frac{1}{\theta^3}$.

We define $\widehat{M}$ by replacing every recurrent class of M' by an absorbing state. From Lemma 4.6 we will derive that this transformation preserves the probability of reaching s_2 from s_1. Formally a (finite) M. C. $\widehat{M} = (\widehat{S}, \widehat{P})$ is defined as follows.

Let D_i $(i = 1, \ldots, k)$ be the states with $\leq n$ messages, which are in the i-th recurrent class. (These sets can be computed by By Eq. (2) in time $O(|B_{\leq n}|)$. Let D be $B_{<n} \setminus \cup D_i$.

1. $\widehat{S} = D \cup \{d_1, \ldots, d_k\} \cup \{dead\}$. The states d_i correspond to the recurrent classes and the state $dead$ corresponds to the set of transient states of M with $\geq n$ messages.
2. The states $d_1, \ldots, d_k, dead$ are absorbing, i.e., for $d \in \{d_1, \ldots, d_k\} \cup \{dead\}$:

$$P'(d, d') = \begin{cases} 1 & \text{if } d' = d, \\ 0 & \text{otherwise.} \end{cases}$$

3. $\widehat{P}(d, d') = P(d, d')$ for $d, d' \in D$
4. $\widehat{P}(d, dead) = \sum_{s \in B_n \setminus \cup D_i} P(d, s)$ for $d \in D$.
5. $\widehat{P}(d, d_i) = \sum_{d' \in D_i} P(d, d')$, for $d \in D$ and $i : 1 \leq i \leq k$.

Recall that we treated already the case when s_1 is recurrent, hence s_1 is in D. Compute the output r which approximates $Prob_M(s_1 \xrightarrow{*} s_2)$ up to θ by the following cases:

1. if $s_2 \in D$ then compute by standard algebraic methods the probability r of reaching s_2 from s_1 in (a finite Markov chain) $\widehat{M}$.
2. if $s_2 \in D_i$ then compute the probability r of reaching d_i from s_1 in $\widehat{M}$.

We completed the presentation of the algorithm, established its correctness and proved that its complexity is $f(\mathcal{L}, s_1, s_2) \times \frac{1}{\theta^3}$. It was shown in [14] that the complexity of the reachability problem for LCSs is not bounded by any primitive recursive function in the size of LCS. Therefore, f is not primitive recursive in the size of PLCS.

8   Probability of Automata Definable Properties

In this section we consider more general properties than reachability. Let φ be a property of computations. We will be interested in approximating

$$Prob \{\pi \ : \ \pi \text{ is a computation from } s \text{ in PLCS } \mathcal{L} \text{ and } \pi \text{ satisfies } \varphi \}.$$

We show that if the properties of computations are specified by (the ω-behavior of) finite state automata or equivalently by formulas of the monadic second-order logic of order, then the above problem is computable.

We consider an extension of PLCS by adding a labeling function. A *state labeled* PLCS is an PLCS together with a finite alphabet Σ and a labeling function *lab* from the local states to Σ. We lift the labeling to the global states: the label of every global state is the same as the label of its local state component. Similarly, with a computation $s_0, s_1, \ldots$ we associate the ω-string $lab(s_1) lab(s_2) \ldots$ over the alphabet Σ.

The next lemma reduces the Quantitative Probabilistic Model-checking Problem to the Quantitative Probabilistic Reachability Problem for PLCSs (see Sect. 1).

Lemma 8.1. *There exists an algorithm which for a state labeled PLCS $\mathcal{L}$, a global state s in $\mathcal{L}$ and a finite state ω-automaton $\mathcal{A}$ produces a PLCS $\mathcal{L}'$, a global state s' in $\mathcal{L}'$ and a set $C_1, C_2, \ldots, C_p$ of BSCC for a finite attractor of $\mathcal{L}'$ such that the following are equivalent:*

1. *The probability that a computation of $\mathcal{L}$ that starts at s is accepted by $\mathcal{A}$ is r.*
2. *The probability that a computation of $\mathcal{L}'$ that starts at s' reaches $\bigcup_{i=1}^{p} C_i$ is r.*

Theorem 8.2. *The Quantitative Probabilistic Model-checking problem can be solved in time $g(\mathcal{L}, \mathcal{A}, s_1, s_2)) \times \frac{1}{\theta^3}$. .*

Proof. First apply the algorithm from Lemma 8.1. Observe that for $i \neq j$ no path reaches both C_i and C_j. For $i = 1, \ldots, p$ choose a state $s_i \in C_i$. Lemma 3.1(4) and Lemma 3.3 imply that the probability to reach s_i is the same as the probability to reach C_i. By the algorithm of Sect. 7 compute r_i which approximates up to $\frac{\theta}{p}$ the probability to reach s_i from s' in $\mathcal{L}'$. From Lemma 8.1 it follows that $r_1 + r_2 + \cdots + r_p$ approximates up to θ the probability that a computation of $\mathcal{L}$ that starts at s is accepted by $\mathcal{A}$.

Acknowledgments. We would like to thank Philippe Schnoebelen and an anonymous referee for pointing out that the path enumeration scheme terminates over Markov chains with finite attractors. We thank Parosh Abdulla, Danièle Beauquier, Philippe Schnoebelen and Anatol Slissenko for fruitful discussions and their useful comments.

References

1. P. A. Abdulla, C. Baier, P. Iyer, and B. Jonsson. Reasoning about probabilistic lossy channel systems. In *Proc. CONCUR 2000*, volume 1877 of *Lecture Notes in Computer Science*, 2000.
2. P. A. Abdulla and B. Jonsson. Undecidable verification problems for programs with unreliable channels. *Information and Computation*, 130(1):71–90, 1996.
3. P. A. Abdulla and B. Jonsson. Verifying programs with unreliable channels. *Information and Computation*, 127(2):91–101, 1996.
4. P. A. Abdulla and A. Rabinovich. Verification of probabilistic systems with faulty communication, 2003. In FOSSACS'03, volume 2620 of *LNCS*, pages 39–53. Springer Verlag, 2003.
5. C. Baier and B. Engelen. Establishing qualitative properties for probabilistic lossy channel systems. In *ARTS'99*, volume 1601 of *LNCS*, pages 34–52. Springer Verlag, 1999.
6. N. Bertrand and Ph. Schnoebelen. Model checking lossy channels systems is probably decidable. In FOSSACS'03, volume 2620 of *LNCS*, pages 120–135 Springer Verlag, 2003.
7. D. Brand and P. Zafiropulo. On communicating finite-state machines. *Journal of the ACM*, 2(5):323–342, 1983.
8. Gérard Cécé, Alain Finkel, and S. Purushothaman Iyer. Unreliable channels are easier to verify than perfect channels. *Information and Computation*, 124(1):20–31, 1996.

9. A. Finkel and Ph. Schnoebelen. Well structured transition systems everywhere!. *Theoretical Computer Science*, 256(1-2):63-92, 2001.
10. P. Iyer and M. Narasimha. Probabilistic Lossy Channel Systems. In Proc of *TAP-SOFT '97* LNCS 1214, 667-681 1997.
11. S. Karlin. *A First Course in Stochastic Processes*. Academic Press, 1966.
12. J. Kemeny, J. Snell, and Λ. Knapp. *Denumerable Markov Chains*. D Van Nostad Co., 1966.
13. J. R. Norris. *Markov Chains*. Cambridge University Press, 1997.
14. Ph. Schnoebelen. Verifying lossy channel systems has nonprimitive recursive complexity. *Information Processing Letters*, 83(5):251-261, 2002.
15. Ph. Schnoebelen. Personal communication, Jan. 2003.

Discounting the Future in Systems Theory[*]

Luca de Alfaro[1], Thomas A. Henzinger[2], and Rupak Majumdar[2]

[1] Department of Computer Engineering, UC Santa Cruz
luca@soe.ucsc.edu
[2] Department of Electrical Engineering and Computer Sciences, UC Berkeley
{tah,rupak}@eecs.berkeley.edu

Abstract. Discounting the future means that the value, today, of a unit payoff is 1 if the payoff occurs today, a if it occurs tomorrow, a^2 if it occurs the day after tomorrow, and so on, for some real-valued discount factor $0 < a < 1$. Discounting (or inflation) is a key paradigm in economics and has been studied in Markov decision processes as well as game theory. We submit that discounting also has a natural place in systems engineering: for nonterminating systems, a potential bug in the far-away future is less troubling than a potential bug today. We therefore develop a systems theory with discounting. Our theory includes several basic elements: discounted versions of system properties that correspond to the ω-regular properties, fixpoint-based algorithms for checking discounted properties, and a quantitative notion of bisimilarity for capturing the difference between two states with respect to discounted properties. We present the theory in a general form that applies to probabilistic systems as well as multicomponent systems (games), but it readily specializes to classical transition systems. We show that discounting, besides its natural practical appeal, has also several mathematical benefits. First, the resulting theory is robust, in that small perturbations of a system can cause only small changes in the properties of the system. Second, the theory is computational, in that the values of discounted properties, as well as the discounted bisimilarity distance between states, can be computed to any desired degree of precision.

1 Introduction

In systems theory, one models systems and analyzes their properties. Nonterminating discrete-time models, such as transition systems and games, are important in many computer science applications, and the ω-regular properties offer an accomplished theory for their analysis. The theory is expressive from a practical point of view [22,27], computational (algorithmic) [5,28], and abstract (language-independent) [21,34]. In its general setting, the theory considers games with ω-regular winning conditions [17,28], provides fixpoint-based algorithms for their solution [13,15], and property-preserving equivalence relations

---

[*] This research was supported in part by the NSF CAREER award CCR-0132780, the DARPA grant F33615-C-98-3614, the NSF grants CCR-9988172, CCR-0234690 and CCR-0225610, and the ONR grant N00014-02-1-0671.

J.C.M. Baeten et al. (Eds.): ICALP 2003, LNCS 2719, pp. 1022–1037, 2003.
© Springer-Verlag Berlin Heidelberg 2003

between structures [4,24]. From a systems engineering point of view, however, the theory has a significant drawback: it is too exact [1]. Since the ω-regular properties generalize finite behavior by considering behavior at infinity, they can distinguish behavior differences that occur arbitrarily late. This exactness becomes even more pronounced for probabilistic models [6,29,33], whose behaviors are specified using numerical quantities, because the theory can distinguish arbitrarily small perturbations of a system.

We propose an alternative formalism that is (in a certain sense) as expressive as the ω-regular properties, and yet achieves continuity in the Cantor topology by sacrificing exactness. In other words, we introduce an *approximate* theory of nonterminating discrete-time systems. The approximation is in two directions. First, instead of giving boolean answers to logical questions, we consider the value of a property to be a real number in the interval [0,1] [19]. Second, we generalize, as in [11,12,18], the classical notions of state equivalences to pseudometrics on states. Both are achieved by defining a *discounted* version of the classical theory. Discounting is inspired by similar ideas in Markov decision processes, economics, and game theory [16,31], and captures the natural engineering intuition that the far-away future is not as important as the near future. Consider, for example, the safety property that no unsafe state is visited. In the classical theory, this property is either true or false. In the discounted theory, its value is 1 if no unsafe state is visited ever, and $1 - a^k$, for some discount factor $0 < a < 1$, if no unsafe state is visited for k steps: the longer the system stays in safe states, the greater the value of the property. Our theory is *robust*, in that small perturbations of a system imply small differences in the numerical values of properties, and *computational*, in that numerical approximation schemes are available which converge geometrically to property values from both directions.

The key insight of this work is that discounting is most naturally and fundamentally applied not to properties, nor to state equivalences, but to the μ-calculus [20]. We introduce the *discounted μ-calculus*, a *quantitative* fixpoint calculus: rather than computing with sets of states, as the traditional μ-calculus does, we compute with functions that assign to each state a value between 0 and 1. A quantitative μ-calculus was introduced in [9] to compute the values of probabilistic ω-regular games by iterating a quantitative version of the predecessor (pre) operator. The discounted μ-calculus is obtained from the calculus of [9] by discounting the pre operator through multiplication with a discount factor $a < 1$. In the classical setting, there is a connection between (linear-time) ω-regular properties, (branching-time) μ-calculus, and games. By discounting the μ-calculus while maintaining this connection, we obtain a notion of discounted ω-regular properties, as well as algorithms for solving games with discounted ω-regular objectives. In the classical setting, the connection is as follows. The solution of a game with an ω-regular winning condition can be written as a μ-calculus formula [13,14]. The fixpoint formula defines the property: when evaluated on linear traces, it holds exactly on the initial states of the traces that satisfy the property. We extend this correspondence to the discounted setting by considering discounted versions of the μ-calculus formula: the discounted fixpoint formula, evaluated on linear traces, defines a discounted version of the original ω-regular property. At the same time, we show that the discounted formula, when evaluated on a game structure, computes the value of the game whose payoff is given by the discounted ω-regular property.

We develop our theory on the system model of concurrent probabilistic game structures [9,16]. These structures generalize several standard models of computation, including nondeterministic transition systems, Markov decision processes [10], and deterministic two-player games [2,32]. The use of discounting gives our theory two main features: *computationality* and *robustness*. Computationality is due to the fact that discount factors strictly less than 1 ensure the geometric convergence of each fixpoint computation by successive approximation (Picard iteration). This enables us to compute every fixpoint value to any desired degree of precision. Moreover, discounting entails the uniqueness of fixpoints. Together, the monotonicity of the μ-calculus operators, the geometric convergence of Picard iteration, and the uniqueness of fixpoints mean that we can iteratively compute geometrically converging lower and upper bounds for the value of every discounted μ-calculus formula. The existence of such approximation schemes is in sharp contrast to the situation for the undiscounted μ-calculus, where least and greatest fixpoints generally differ, where each (least or greatest) fixpoint can be approximated in one direction only (from below, or from above), and where in the quantitative case, no rate of convergence is known.

In the classical setting, the μ-calculus characterizes bisimilarity: two states are bisimilar iff they satisfy the same μ-calculus formulas. To extend this connection to the discounted setting, we define a quantitative, discounted notion of bisimilarity, which assigns a real-valued *distance* in the interval $[0,1]$ to every pair of states: the distance between two states is 1 if they satisfy different propositions, and otherwise it is coinductively computed from discounted distances between successor states. We show that in the discounted setting, the bisimilarity distance between two states is equal to the supremum, over all μ-calculus formulas, of the difference between the values of a formula at the two states. This is in fact the characterization of discounted bisimilarity from [11,12] extended to games. However, while in [11,12] the above characterization is taken to be the definition of discounted bisimilarity, in our case it is a theorem that can be proved from the coinductive definition. The theorem demonstrates the *robustness* of the theory: small perturbations in the numerical values of transition probabilities, as well as (small or large) perturbations that come far in the future, correspond to small bisimilarity distance, and hence to small differences in the numerical values of discounted properties. The numerical computation of discounted bisimilarity by successive approximation enjoys the same properties as the numerical evaluation of discounted μ-calculus formulas; in particular, geometrically-converging approximation schemes are available for computing both lower and upper bounds.

2   Systems: Concurrent Game Structures

For a countable set U, a *probability distribution* on U is a function $p: U \mapsto [0, 1]$ such that $\sum_{u \in U} p(u) = 1$. We write $\mathcal{D}(U)$ for the set of probability distributions on U. A two-player (*concurrent*) *game structure* [2,7] $\mathcal{G} = \langle Q, M, \Gamma_1, \Gamma_2, \delta \rangle$ consists of the following components:

- A finite set Q of states.
- A finite set M of moves.
- Two move assignments $\Gamma_1, \Gamma_2: Q \mapsto 2^M \setminus \emptyset$. For $i \in \{1, 2\}$, the assignment Γ_i associates with each state $s \in Q$ the nonempty set $\Gamma_i(s) \subseteq M$ of moves available to player i at state s.
- A probabilistic transition function $\delta: Q \times M^2 \mapsto \mathcal{D}(Q)$. For a state $s \in Q$ and moves $\gamma_1 \in \Gamma_1(s)$ and $\gamma_2 \in \Gamma_2(s)$, the function δ provides a probability distribution of successor states. We write $\delta(t \mid s, \gamma_1, \gamma_2)$ for the probability $\delta(s, \gamma_1, \gamma_2)(t)$ that the successor state is $t \in Q$.

At every state $s \in Q$, player 1 chooses a move $\gamma_1 \in \Gamma_1(s)$, and simultaneously and independently player 2 chooses a move $\gamma_2 \in \Gamma_2(s)$. The game then proceeds to the successor state $t \in Q$ with probability $\delta(t \mid s, \gamma_1, \gamma_2)$. The outcome of the game is a path. A *path* of $\mathcal{G}$ is an infinite sequence $s_0, s_1, s_2, \ldots$ of states in $s_k \in Q$ such that for all $k \geq 0$, there are moves $\gamma_1^k \in \Gamma_1(s_k)$ and $\gamma_2^k \in \Gamma_2(s_k)$ with $\delta(s_{k+1} \mid s_k, \gamma_1^k, \gamma_2^k) > 0$. We write Σ for the set of all paths.

The following are special cases of concurrent game structures. The structure $\mathcal{G}$ is *deterministic* if for all states $s \in Q$ and moves $\gamma_1 \in \Gamma_1(s)$, $\gamma_2 \in \Gamma_2(s)$, there is a state $t \in Q$ with $\delta(t \mid s, \gamma_1, \gamma_2) = 1$; in this case, with abuse of notation we write $\delta(s, \gamma_1, \gamma_2) = t$. The structure $\mathcal{G}$ is *turn-based* if at every state at most one player can choose among multiple moves; that is, for all states $s \in Q$, there exists at most one $i \in \{1, 2\}$ with $|\Gamma_i(s)| > 1$. The turn-based deterministic game structures coincide with the games of [32]. The structure $\mathcal{G}$ is *one-player* if at every state only player 1 can choose among multiple moves; that is, $|\Gamma_2(s)| = 1$ for all states $s \in Q$. The one-player game structures coincide with Markov decision processes (MDPs) [10]. The one-player deterministic game structures coincide with transition systems: in every state, each available move of player 1 determines a possible successor state.

A *strategy* for player $i \in \{1, 2\}$ is a function $\pi_i: Q^+ \mapsto \mathcal{D}(M)$ that associates with every nonempty finite sequence $\sigma \in Q^+$ of states, representing the history of the game, a probability distribution $\pi_i(\sigma)$, which is used to select the next move of player i. Thus, the choice of the next move can be history-dependent and randomized. We require that the strategy π_i can prescribe only moves that are available to player i; that is, for all sequences $\sigma \in Q^*$ and states $s \in Q$, if $\pi_i(\sigma s)(\gamma) > 0$, then $\gamma \in \Gamma_i(s)$. We write Π_i for the set of strategies for player i. The strategy π_i is *deterministic* if for all sequences $\sigma \in Q^+$, there exists a move $\gamma \in M$ such that $\pi(\sigma)(\gamma) = 1$. Thus, deterministic strategies are functions from Q^+ to M. The strategy π_i is *memoryless* if for all sequences $\sigma, \sigma' \in Q^*$ and states $s \in Q$, we have $\pi(\sigma s) = \pi(\sigma' s)$. Thus, the moves chosen by memoryless strategies depend only on the current state and not on the history of the game.

Given a starting state $s \in Q$ and two strategies π_1 and π_2 for the two players, the game is reduced to an ordinary stochastic process, denoted $\mathcal{G}_s^{\pi_1, \pi_2}$, which defines a probability distribution on the set Σ of paths. An *event* of $\mathcal{G}_s^{\pi_1, \pi_2}$ is a measurable set $A \subseteq \Sigma$ of paths. For an event $A \subseteq \Sigma$, we write $\mathrm{Pr}_s^{\pi_1, \pi_2}(A)$ for the probability that the outcome of the game belongs to A when the game starts from s and the players use the strategies π_1 and π_2. A *payoff function* v: $\Sigma \mapsto [0,1]$ is a measurable function that associates with every path a real in the interval $[0,1]$. Payoff functions define the rewards of the two players for each outcome of the game. For a payoff function v, we write $\mathrm{E}_s^{\pi_1, \pi_2}\{v\}$ for the expected value of v on the outcome when the game starts from s and the strategies π_1 and π_2 are used. If v defines the reward for player 1, then the *(player 1) value* of the game is a function that maps every state $s \in Q$ to the maximal expected reward $\sup_{\pi_1 \in \Pi_1} \inf_{\pi_2 \in \Pi_2} \mathrm{E}_s^{\pi_1, \pi_2}\{v\}$ that player 1 can achieve no matter which strategy player 2 chooses.

3   Algorithms: Discounted Fixpoint Expressions

Quantitative region algebra. The classical μ-calculus specifies algorithms for iterating boolean and predecessor (pre) operators on regions, where a region is a set of states. In our case a region is a function from states to reals. This notion of quantitative region admits the analysis both of probabilistic transitions and of real-valued discount factors. Consider a concurrent game structure $\mathcal{G} = \langle Q, M, \Gamma_1, \Gamma_2, \delta \rangle$. A *(quantitative) region* of $\mathcal{G}$ is a function $f \colon Q \mapsto [0,1]$ that maps every state to a real in the interval $[0,1]$. For example, for a given payoff function, the value of a game on the structure $\mathcal{G}$ is a quantitative region. We write $\mathcal{F}$ for the set of quantitative regions. By $\mathbf{0}$ and $\mathbf{1}$ we denote the constant functions in $\mathcal{F}$ that map all states in Q to 0 and 1, respectively. Given two regions $f, g \in \mathcal{F}$, define $f \leq g$ if $f(s) \leq g(s)$ for all states $s \in Q$, and define the regions $f \wedge g$ and $f \vee g$ by $(f \wedge g)(s) = \min\{f(s), g(s)\}$ and $(f \vee g)(s) = \max\{f(s), g(s)\}$, for all states $s \in Q$. The region $\mathbf{1} - f$ is defined by $(\mathbf{1} - f)(s) = 1 - f(s)$ for all $s \in Q$; this has the role of complementation. Given a set $T \subseteq Q$ of states, with abuse of notation we denote by T also the indicator function of T, defined by $T(s) = 1$ if $s \in Q$, and $T(s) = 0$ otherwise. Let $\mathcal{F}_{\mathbb{B}} \subseteq \mathcal{F}$ be the set of indicator functions (also called *boolean regions*). Note that in $\mathcal{F}_{\mathbb{B}}$, the operators $\wedge$, $\vee$, and $\leq$ correspond respectively to intersection, union, and set inclusion.

An operator $F \colon \mathcal{F} \mapsto \mathcal{F}$ is *monotonic* if for all regions $f, g \in \mathcal{F}$, if $f \leq g$, then $F(f) \leq F(g)$. The operator F is *Lipschitz continuous* if for all regions $f, g \in \mathcal{F}$, we have $|F(f) - F(g)| \leq |f - g|$, where $|\cdot|$ is the L_∞ norm. Note that Lipschitz continuity implies continuity: for all infinite increasing sequences $f_1 \leq f_2 \leq \cdots$ of regions in $\mathcal{F}$, we have $\lim_{n \to \infty} F(f_n) = F(\lim_{n \to \infty} f_n)$. The operator F is *contractive* if there exists a constant $0 < c < 1$ such that for all regions $f, g \in \mathcal{F}$, we have $|F(f) - F(g)| \leq c \cdot |f - g|$. For $i \in \{1, 2\}$, we consider so-called *pre operators* $\mathrm{Pre}_i \colon \mathcal{F} \mapsto \mathcal{F}$ with the following properties: (1) Pre_1 and Pre_2 are monotonic and Lipschitz continuous, and (2) for all regions $f \in \mathcal{F}$, we have $\mathrm{Pre}_1(f) = \mathbf{1} - \mathrm{Pre}_2(\mathbf{1} - f)$; that is, the operators Pre_1 and Pre_2 are dual. The following pre operators have natural interpretations on (subclasses of) concurrent game structures. The *quantitative pre operator* [9] $\text{Qpre}_1 \colon \mathcal{F} \mapsto \mathcal{F}$ is defined for every quantitative region $f \in \mathcal{F}$ and state $s \in Q$ by

$$\text{Qpre}_1(f)(s) \;=\; \sup_{\pi_1 \in \Pi_1} \inf_{\pi_2 \in \Pi_2} \text{E}_s^{\pi_1, \pi_2}\{v_{\bigcirc f}\},$$

where $v_{\bigcirc f} \colon \Sigma \mapsto [0,1]$ is the payoff function that maps every path $s_0, s_1, \ldots$ in Σ to the value $f(s_1)$ of f at the second state of the path. In words, $\text{Qpre}_1(f)(s)$ is the maximal expectation for the value of f that player 1 can achieve in a successor state of s. The value $\text{Qpre}_1(f)(s)$ can be computed by solving a matrix game:

$$\text{Qpre}_1(f)(s) \;=\; \text{val}_1 \!\left[\sum_{t \in Q} f(t) \cdot \delta(t \mid s, \gamma_1, \gamma_2)\right]_{\gamma_1 \in \Gamma_1(s), \gamma_2 \in \Gamma_2(s)}$$

where $\text{val}_1[\cdot]$ denotes the player 1 value (i.e., maximal expected reward for player 1) of a matrix game. The minmax theorem guarantees that this matrix game has optimal strategies for both players [35]. The matrix game can be solved by linear programming [9,26]. The player 2 operator Qpre_2 is defined symmetrically. The minmax theorem permits the exchange of the sup and inf in the definition, and thus ensures the duality of the two pre operators.

By specializing the quantitative pre operators Qpre_i to turn-based deterministic game structures, we obtain the *controllable pre operators* [2] $\text{Cpre}_i \colon \mathcal{F}_\mathbb{B} \mapsto \mathcal{F}_\mathbb{B}$, which are closed on boolean regions. In particular, for every boolean region $f \in \mathcal{F}_\mathbb{B}$ and state $s \in Q$, $\text{Cpre}_1(f)(s) = 1$ iff $\exists \gamma_1 \in \Gamma_1(s). \forall \gamma_2 \in \Gamma_2(s). f(\delta(s, \gamma_1, \gamma_2)) = 1$. In words, for a set $T \subseteq Q$ of states, $\text{Cpre}_1(T)$ is the set of states from which player 1 can ensure that the next state lies in T. For one-player game structures, this characterization further simplifies to $\text{Epre}_1(f)(s) = 1$ iff $\exists \gamma_1 \in \Gamma_1(s). f(\delta(s, \gamma_1, \cdot)) = 1$. This is the traditional definition of the existential pre operator on a transition system: for a set $T \subseteq Q$ of states, $\text{Epre}_1(T)$ is the set of predecessor states.

Discounted μ-calculus. We define a fixpoint calculus that permits the iteration of pre operators. The calculus is discounted, in that every occurrence of a pre operator is multiplied by a *discount factor* from [0,1]. If the discount factor of a pre operator is less than 1, this has the effect that each additional application of the operator in a fixpoint iteration carries less weight. We use a fixed set Θ of *propositions*; every proposition $T \in \Theta$ denotes a boolean region $[\![T]\!] \in \mathcal{F}_\mathbb{B}$. For a state $s \in Q$ with $[\![T]\!](s) = 1$, we write $s \models T$ and say that s is a *T-state*. The formulas of the *discounted μ-calculus* are generated by the grammar

$$\phi ::= T \mid \neg T \mid x \mid \phi \vee \phi \mid \phi \wedge \phi \mid \alpha \cdot \text{pre}_1(\phi) \mid \alpha \cdot \text{pre}_2(\phi) \mid$$
$$(1 - \alpha) + \alpha \cdot \text{pre}_1(\phi) \mid (1 - \alpha) + \alpha \cdot \text{pre}_2(\phi) \mid \mu x. \phi \mid \nu x. \phi$$

for propositions $T \in \Theta$, variables x from some fixed set X, and parameters α from some fixed set Λ. The syntax defines formulas in positive normal form. The definition of negation in the calculus, which is given below, makes it clear that we need two discounted pre modalities, $\alpha \cdot \mathrm{pre}_i(\cdot)$ and $(1 - \alpha) + \alpha \cdot \mathrm{pre}_i(\cdot)$, for each player $i \in \{1, 2\}$. A formula ϕ is *closed* if every variable x in ϕ occurs in the scope of a least-fixpoint quantifier μx or greatest-fixpoint quantifier νx.

A variable valuation $\mathcal{E} : X \mapsto \mathcal{F}$ is a function that maps every variable $x \in X$ to a quantitative region in $\mathcal{F}$. We write $\mathcal{E}[x \mapsto f]$ for the function that agrees with $\mathcal{E}$ on all variables, except that x is mapped to f. A formula may contain several different discount factors. A parameter valuation $\mathcal{P} : \Lambda \mapsto [0, 1]$ is a function that maps every parameter $\alpha \in \Lambda$ to a real-valued discount factor in the interval $[0, 1]$. Given a real $r \in [0, 1]$, the parameter valuation $\mathcal{P}$ is r-*bounded* if $\mathcal{P}(\alpha) \leq r$ for all parameters $\alpha \in \Lambda$. An *interpretation* is a pair that consists of a variable valuation and a parameter valuation. Given an interpretation $(\mathcal{E}, \mathcal{P})$, every formula ϕ of the discounted μ-calculus defines a quantitative region $[\![\phi]\!]^{\mathcal{G}}_{\mathcal{E}, \mathcal{P}} \in \mathcal{F}$ (the superscript $\mathcal{G}$ is omitted if the game structure is clear from the context):

$$[\![T]\!]_{\mathcal{E}, \mathcal{P}} = [\![T]\!]$$

$$[\![\neg T]\!]_{\mathcal{E}, \mathcal{P}} = 1 - [\![T]\!]$$

$$[\![x]\!]_{\mathcal{E}, \mathcal{P}} = \mathcal{E}(x)$$

$$[\![\alpha \cdot \mathrm{pre}_i(\phi)]\!]_{\mathcal{E}, \mathcal{P}} = \mathcal{P}(\alpha) \cdot \mathrm{Qpre}_i([\![\phi]\!]_{\mathcal{E}, \mathcal{P}})$$

$$[\![(1 - \alpha) + \alpha \cdot \mathrm{pre}_i(\phi)]\!]_{\mathcal{E}, \mathcal{P}} = (1 - \mathcal{P}(\alpha)) + \mathcal{P}(\alpha) \cdot \mathrm{Qpre}_i([\![\phi]\!]_{\mathcal{E}, \mathcal{P}})$$

$$[\![\phi_1 \genfrac{\{}{\}}{0pt}{}{\vee}{\wedge} \phi_2]\!]_{\mathcal{E}, \mathcal{P}} = [\![\phi_1]\!]_{\mathcal{E}, \mathcal{P}} \genfrac{\{}{\}}{0pt}{}{\vee}{\wedge} [\![\phi_2]\!]_{\mathcal{E}, \mathcal{P}}$$

$$[\![\genfrac{\{}{\}}{0pt}{}{\mu}{\nu} x. \phi]\!]_{\mathcal{E}, \mathcal{P}} = \genfrac{\{}{\}}{0pt}{}{\inf}{\sup} \{f \in \mathcal{F} \mid f = [\![\phi]\!]_{\mathcal{E}[x \mapsto f], \mathcal{P}}\}$$

The existence of the required fixpoints is guaranteed by the monotonicity and continuity of all operators. The region $[\![\phi]\!]_{\mathcal{E}, \mathcal{P}}$ is in general not boolean even if the game structure is turn-based deterministic, because the discount factors introduce real numbers. The discounted μ-calculus is closed under negation: if we define the negation of a formula ϕ inductively using $\neg(\alpha \cdot \mathrm{pre}_1(\phi')) = (1 - \alpha) + \alpha \cdot \mathrm{pre}_2(\neg \phi')$ and $\neg((1 - \alpha) + \alpha \cdot \mathrm{pre}_1(\phi')) = \alpha \cdot \mathrm{pre}_2(\neg \phi')$, then $[\![\neg \phi]\!]_{\mathcal{E}, \mathcal{P}} = 1 - [\![\phi]\!]_{\mathcal{E}, \mathcal{P}}$. This generalizes the duality $1 - \mathrm{Qpre}_1(f) = \mathrm{Qpre}_2(1 - f)$ of the undiscounted pre operators.

A parameter valuation $\mathcal{P}$ is *contractive* if $\mathcal{P}$ maps every parameter to a discount factor strictly less than 1. A fixpoint quantifier μx or νx occurs *syntactically contractive* in a formula ϕ if a pre modality occurs on every syntactic path from the quantifier to a quantified occurrence of the variable x. For example, in the formula $\mu x. (T \vee \alpha \cdot \mathrm{pre}_i(x))$ the fixpoint quantifier occurs syntactically contractive; in the formula $(1 - \alpha) + \alpha \cdot \mathrm{pre}_i(\mu x. (T \vee x))$ it does not. Under a contractive parameter valuation, every syntactically contractive occurrence of a fixpoint quantifier defines a contractive operator on the values of the free variables that are in the scope of the quantifier. Hence, by the Banach fixpoint theorem, the fixpoint is unique. In such cases, since there are unique fixpoints, we need not distinguish between μ and ν quantifiers, and we use a single (self-dual) fixpoint quantifier λ. Fixpoints can be computed by Picard iteration: $[\![\mu x. \phi]\!]_{\mathcal{E},\mathcal{P}} = \lim_{k\to\infty} f_k$ where $f_0 = \mathbf{0}$, and $f_{k+1} = [\![\phi]\!]_{\mathcal{E}[x\mapsto f_k],\mathcal{P}}$ for all $k \geq 0$; and $[\![\nu x. \phi]\!]_{\mathcal{E},\mathcal{P}} = \lim_{k\to\infty} f_k$ where $f_0 = \mathbf{1}$, and f_{k+1} is defined as in the μ case. If the fixpoint is unique, then both sequences converge to the same region $[\![\lambda x. \phi]\!]_{\mathcal{E},\mathcal{P}}$, one from below, and the other from above.

Approximating the undiscounted semantics. If $\mathcal{P}(\alpha) = 1$, then both discounted pre modalities $\alpha \cdot \mathrm{pre}_i(\cdot)$ and $(1-\alpha) + \alpha \cdot \mathrm{pre}_i(\cdot)$ collapse, and are interpreted as the quantitative pre operator $\mathrm{Qpre}_i(\cdot)$, for $i \in \{1,2\}$. In this case, we may omit the parameter α from formulas, writing instead the undiscounted modality $\mathrm{pre}_i(\cdot)$. The *undiscounted semantics* of a formula ϕ is the quantitative region $[\![\phi]\!]_{\mathcal{E},1}$ obtained from the parameter valuation 1 that maps every parameter in Λ to 1. The undiscounted semantics coincides with the quantitative μ-calculus of [9,23]. In the case of turn-based deterministic game structures, it coincides with the alternating-time μ-calculus of [2], and in the case of transition systems, with the classical μ-calculus of [19]. The following theorem justifies discounting as an approximation theory: the undiscounted semantics of a formula can be obtained as the limit of the discounted semantics as all discount factors tend to 1.[1]

Theorem 1. *Let $\phi(x)$ be a formula of the discounted μ-calculus with a free variable x and parameter α, which always occur in the context $\alpha \cdot \mathrm{pre}_i(x)$, for $i \in \{1,2\}$. Then*

$$\lim_{a\to 1} [\![\lambda x. \phi(\alpha \cdot \mathrm{pre}_i(x))]\!]_{\mathcal{E},\mathcal{P}[\alpha\mapsto a]} = [\![\mu x. \phi(\mathrm{pre}_i(x))]\!]_{\mathcal{E},\mathcal{P}}.$$

Furthermore, if x and α always occur in the context $(1-\alpha) + \alpha \cdot \mathrm{pre}_i(x)$, then

$$\lim_{a\to 1} [\![\lambda x. \phi((1-\alpha) + \alpha \cdot \mathrm{pre}_i(x))]\!]_{\mathcal{E},\mathcal{P}[\alpha\mapsto a]} = [\![\nu x. \phi(\mathrm{pre}_i(x))]\!]_{\mathcal{E},\mathcal{P}}.$$

Note that the assumption of the theorem ensures that the fixpoint quantifiers on x occur syntactically contractive on the discounted left-hand side, and therefore define unique fixpoints. Depending on the form of the discounted pre modality, the unique discounted fixpoints approximate either the least or the greatest undiscounted fixpoint. This implies that, in general, limits of discount factors are not interchangeable. Consider the formula

$$\varphi = \lambda y. \lambda x. ((\neg T \wedge \beta \cdot \mathrm{pre}_1(x)) \vee (T \wedge ((1-\alpha) + \alpha \cdot \mathrm{pre}_1(y)))).$$

Then $\lim_{\alpha\to 1} \lim_{\beta\to 1} \varphi$ is equivalent to $\nu y. \mu x. ((\neg T \wedge \mathrm{pre}_1(x)) \vee (T \wedge \mathrm{pre}_1(y)))$, which characterizes, in the turn-based deterministic case, the player 1 winning

---

[1] It may be noted that Picard iteration itself offers an approximation theory for fixpoint calculi: the longer the iteration sequence, the closer the approximation of the fixpoint. This approximation scheme, however, is neither syntactically robust nor compositional, because it is not closed under the unrolling of fixpoint quantifiers. By contrast, for every discounted μ-calculus formula $\kappa x. \phi(x)$, where $\kappa \in \{\mu,\nu\}$, we have $[\![\kappa x. \phi(x)]\!]_{\mathcal{E},\mathcal{P}} = [\![\phi(\kappa x. \phi(x))]\!]_{\mathcal{E},\mathcal{P}}$.

states of a Büchi game (infinitely many T-states must be visited). The inner (β) limit ensures that a T-state will be visited; the outer (α) limit ensures that this remains always the case. On the other hand, $\lim_{\beta \to 1} \lim_{\alpha \to 1} \varphi$ is equivalent to $\mu x. \nu y. ((\neg T \wedge \mathrm{pre}_1(x)) \vee (T \wedge \mathrm{pre}_1(y)))$, which characterizes the player 1 winning states of a coBüchi game (eventually only T-states must be visited). This is because the inner (α) limit ensures that only T-states are visited, and the outer (β) limit ensures that this will happen.

4   Properties: Discounted ω-Regular Winning Conditions

In the classical setting, the ω-regular languages can be used to specify system properties (or winning conditions of games), while the μ-calculus provides algorithms for verifying the properties (or computing the winning states). In our discounted approach, the discounted μ-calculus provides the algorithms; what, then, are the properties? We establish a connection between the semantics of a discounted fixpoint expression over a concurrent game structure, and the semantics of the same expression over the paths of the structure. This provides a trace semantics for the discounted μ-calculus, thus giving rise to a notion of "discounted ω-regular properties."

Reachability and safety conditions. A *discounted reachability game* consists of a concurrent game structure $\mathcal{G}$ (with state space Q) together with a winning condition $\Diamond_a T$, where $T \in \Theta$ is a proposition and $a \in [0, 1]$ is a discount factor. Starting from a state $s \in Q$, player 1 has the objective to reach a T-state as quickly as possible, while player 2 tries to prevent this. The reward for player 1 is a^k if a T-state is visited for the first time after k moves, and 0 if no T-state is ever visited. Formally, we define the payoff function $v_{\Diamond T}^a : \Sigma \mapsto [0, 1]$ on paths by $v_{\Diamond T}^a(s_0, s_1, \dots) = a^k$ for $k = \min\{i \mid s_i \models T\}$, and $v_{\Diamond T}^a(s_0, s_1, \dots) = 0$ if $s_k \not\models T$ for all $k \geq 0$. Then, for every state $s \in Q$, the *value* of the discounted reachability game at s is $(\langle\!\langle 1 \rangle\!\rangle \Diamond_a T)(s) = \sup_{\pi_1 \in \Pi_1} \inf_{\pi_2 \in \Pi_2} \mathrm{E}_s^{\pi_1, \pi_2}\{v_{\Diamond T}^a\}$. This defines a discounted stochastic game [31]. For $a = 1$, the value can be computed as a least fixpoint; for $a < 1$, as the unique fixpoint

$$\langle\!\langle 1 \rangle\!\rangle \Diamond_a T \;=\; [\![\lambda x. (T \vee \alpha \cdot \mathrm{pre}_1(x))]\!]_{\cdot, [\alpha \mapsto a]}.$$

Picard iteration yields $\langle\!\langle 1 \rangle\!\rangle \Diamond_a T = \lim_{k \to \infty} f_k$ where $f_0 = \mathbf{0}$, and $f_{k+1} = (T \vee a \cdot \mathrm{Qpre}_1(f_k))$ for all $k \geq 0$. This gives an approximation scheme from below to solve the discounted reachability game. The sequence converges geometrically in $a < 1$; more precisely, $(\langle\!\langle 1 \rangle\!\rangle \Diamond_a T)(s) - f_k(s) \leq a^k$ for all states $s \in Q$ and all $k \geq 0$. This permits the approximation of the value of the game for any desired precision. Furthermore, as the fixpoint is unique, an approximation scheme from above, which starts with $f_0 = \mathbf{1}$, also converges geometrically. For turn-based deterministic game structures, the value of the discounted reachability game $\langle\!\langle 1 \rangle\!\rangle \Diamond_a T$ at state s is a^k, where k is the length of the shortest path that player 1 can enforce to reach a T-state, if such a path exists (in the case of

one-player structures, k is the length of the shortest path from s to a T-state).
For general game structures and $a = 1$, the value $\langle\!\langle 1 \rangle\!\rangle \Diamond_1 T$ at s is the maximal
probability with which player 1 can achieve to reach a T-state [9]. A strategy
π_1 for player 1 is *optimal* (resp., ϵ-*optimal* for $\epsilon > 0$) for the reachability condi-
tion $\Diamond_a T$ if for all states $s \in Q$, we have $\inf_{\pi_2 \in \Pi_2} \mathrm{E}_s^{\pi_1,\pi_2}\{v_{\Diamond T}^a\} = (\langle\!\langle 1 \rangle\!\rangle \Diamond_a T)(s)$
(resp., $\inf_{\pi_2 \in \Pi_2} \mathrm{E}_s^{\pi_1,\pi_2}\{v_{\Diamond T}^a\} \geq (\langle\!\langle 1 \rangle\!\rangle \Diamond_a T)(s) - \epsilon$). While undiscounted ($a = 1$)
reachability games admit only ϵ-optimal strategies [16], discounted reachability
games have optimal memoryless strategies for both players [16,31].

The dual of reachability is safety. A *discounted safety game* consists of a
concurrent game structure $\mathcal{G}$ together with a winning condition $\Box_a T$, where
$T \in \Theta$ and $a \in [0,1]$. Starting from a state $s \in Q$, player 1 has the objective to
stay within the set of T-states for as long as possible. The payoff function $v_{\Box T}^a$:
$\Sigma \mapsto [0,1]$ is defined by $v_{\Box T}^a(s_0, s_1, \ldots) = 1 - a^k$ for $k = \min\{i \mid s_i \not\models T\}$, and
$v_{\Box T}^a(s_0, s_1, \ldots) = 1$ if $s_k \models T$ for all $k \geq 0$. For every state $s \in Q$, the *value* of the
discounted safety game at s is $(\langle\!\langle 1 \rangle\!\rangle \Box_a T)(s) = \sup_{\pi_1 \in \Pi_1} \inf_{\pi_2 \in \Pi_2} \mathrm{E}_s^{\pi_1,\pi_2}\{v_{\Box T}^a\}$.
For $a = 1$, the value can be computed as a greatest fixpoint; for $a < 1$, as the
unique fixpoint

$$\langle\!\langle 1 \rangle\!\rangle \Box_a T = [\![\lambda x.\,(T \wedge ((1-\alpha) + \alpha \cdot \mathrm{pre}_1(x)))]\!]_{\cdot,[\alpha \mapsto a]}.$$

For $a < 1$, the Picard iteration $\langle\!\langle 1 \rangle\!\rangle \Box_a T = \lim_{k \to \infty} f_k$ where $f_0 = \mathbf{0}$, and $f_{k+1} =$
$(T \wedge a \cdot \mathrm{Qpre}_1(f_k))$ for all $k \geq 0$, converges geometrically from below, and with
$f_0 = \mathbf{1}$, it converges geometrically from above. For turn-based deterministic
game structures and $a < 1$, the value $\langle\!\langle 1 \rangle\!\rangle \Box_a T$ at state s is $1 - a^k$, where k is
the length of the longest path that player 1 can enforce to stay in T-states. For
general game structures and $a = 1$, it is the maximal probability with which
player 1 can achieve to stay in T-states forever [9].

In summary, the mathematical appeal of discounting reachability and safety,
in addition to the practical appeal of emphasis on the near future, is threefold:
(1) geometric convergence from both below and above (no theorems on the rate
of convergence are known for $a = 1$); (2) the existence of optimal memoryless strategies (only ϵ-optimal strategies may exist for undiscounted reachability
games); (3) the continuous approximation property (Theorem 1), which shows
that for $a \to 1$, the values of discounted reachability and safety games converge
to the values of the corresponding undiscounted games.

Trace semantics of fixpoint expressions. Reachability and safety properties
are simple, and offer a natural discounted interpretation. For more general ω-
regular properties, however, there are often multiple candidates for a discounted
interpretation, as there are multiple algorithms for evaluating the property. Consider, for example, Büchi games. An undiscounted Büchi game consists of a con-
current game structure $\mathcal{G}$ together with a winning condition $\Box \Diamond T$, where $T \in \Theta$
specifies a set of Büchi states, which player 1 tries to visit infinitely often. The
value of the game at a state s, denoted $(\langle\!\langle 1 \rangle\!\rangle \Box \Diamond T)(s)$, is the maximal probability
with which player 1 can enforce that a T-state is visited infinitely often. The
value of an undiscounted Büchi game can be characterized as [9]

$$\langle\!\langle 1 \rangle\!\rangle \Box \Diamond T \;=\; \nu y.\,\mu x.\,((\neg T \wedge \mathrm{pre}_1(x)) \vee (T \wedge \mathrm{pre}_1(y))).$$

This fixpoint expression suggests several alternative ways of discounting the Büchi game. For example, one may require that the distances between the infinitely many visits to T-states are as small as possible, obtaining $\nu y.\,\lambda x.\,((\neg T \wedge \alpha \cdot \mathrm{pre}_1(x)) \vee (T \wedge \mathrm{pre}_1(y)))$. Alternatively, one may require that the number of visits to T-states is as large as possible, but arbitrarily spaced, obtaining $\lambda y.\,\mu x.\,((\neg T \wedge \mathrm{pre}_1(x)) \vee (T \wedge ((1 - \beta) + \beta \cdot \mathrm{pre}_1(y))))$. More generally, we can use both discount factors α and β, as in $\lambda y.\,\lambda x.\,((\neg T \wedge \alpha \cdot \mathrm{pre}_1(x)) \vee (T \wedge ((1 - \beta) + \beta \cdot \mathrm{pre}_1(y))))$, and study the effect of various relationships, such as $\alpha < \beta$, $\alpha = \beta$, and $\alpha > \beta$. All these discounted interpretations of Büchi games have two key properties: (1) the value of the game can be computed by algorithms that converge geometrically; and (2) if all discount factors tend to one, then the value of the discounted game tends to the value of the undiscounted game. So instead of defining a discounted Büchi (or more general ω-regular) winning condition, chosen arbitrarily from the alternatives, we take a discounted μ-calculus formula itself as specification of the game and show that, under each interpretation, the formula naturally induces a discounted property of paths.

We first define the semantics of a formula on a path. Consider a concurrent game structure $\mathcal{G}$. Every path $\sigma = s_0, s_1, \dots$ of $\mathcal{G}$ induces an infinite-state[2] game structure in a natural way: the set of states is $\{(k, s_k) \mid k \geq 0\}$, and at each state (k, s_k), both players have exactly one move available, whose combination takes the game deterministically to the successor state $(k+1, s_{k+1})$, for all $k \geq 0$. With abuse of notation, we write σ also for the game structure that is induced by the path σ. For this structure and $i \in \{1, 2\}$, $\mathrm{Qpre}_i(\{(k+1, s_{k+1})\})$ is the function that maps (k, s_k) to 1 and all other states to 0. For a closed discounted μ-calculus formula ϕ and parameter valuation $\mathcal{P}$, we define the *trace semantics* of ϕ under $\mathcal{P}$ to be the payoff function $[\phi]_{\mathcal{P}} : \Sigma \mapsto [0,1]$ that maps every path $\sigma \in \Sigma$ to the value $[\![\phi]\!]_{\mathcal{P}}^{\sigma}(s_0)$, where s_0 is the first state of the path σ (the superscript σ indicates that the formula is evaluated over the game structure induced by σ).

The Cantor metric d_C is defined on the set Σ of paths by $d_C(\sigma_1, \sigma_2) = \frac{1}{2^k}$, where k is the length of the maximal prefix that is common to the two paths σ_1 and σ_2. The following theorem shows that for discount factors strictly less than 1, the trace semantics of every discounted μ-calculus formula is a continuous function from this metric space to the interval $[0,1]$. This is in contrast to undiscounted ω-regular properties, which can distinguish between paths that are arbitrarily close.

Theorem 2. *Let ϕ be a closed discounted μ-calculus formula, and let $\mathcal{P}$ be a contractive parameter valuation. For every $\epsilon > 0$, there is a $\delta > 0$ such that for all paths $\sigma_1, \sigma_2 \in \Sigma$ with $d_C(\sigma_1, \sigma_2) < \delta$, we have $|[\phi]_{\mathcal{P}}(\sigma_1) - [\phi]_{\mathcal{P}}(\sigma_2)| < \epsilon$.*

A formula ϕ of the discounted μ-calculus is *player-1 strongly guarded* [8] if (1) ϕ is closed and consists of a string of fixpoint quantifiers followed by a

---

[2] The infiniteness is harmless, because we do not compute in this structure.

quantifier-free part, (2) ϕ contains no occurrences of pre_2, and (3) every conjunction in ϕ has at least one constant argument; that is, every conjunctive subformula of ϕ has the form $T \wedge \phi'$, where T is a boolean combination of propositions. In the classical μ-calculus, all ω-regular winning conditions of turn-based deterministic games can be expressed by strongly guarded (e.g., Rabin chain) formulas [13]. For player-1 strongly guarded formulas ϕ the following theorem gives the correspondence between the semantics of ϕ on structures and the semantics of ϕ on paths: the value of ϕ at a state s under parameter valuation $\mathcal{P}$ is the value of the game with start state s and payoff function $[\phi]_\mathcal{P}$.

Theorem 3. *Let $\mathcal{G}$ be a concurrent game structure, let ϕ be a player-1 strongly guarded formula of the discounted μ-calculus, and let $\mathcal{P}$ be a parameter valuation. For every state s of $\mathcal{G}$, we have $[\phi]^\mathcal{G}_{\cdot,\mathcal{P}}(s) = \sup_{\pi_1 \in \Pi_1} \inf_{\pi_2 \in \Pi_2} \mathrm{E}^{\pi_1,\pi_2}_s \{[\phi]_\mathcal{P}\}$.*

Rabin chain conditions. An undiscounted Rabin chain game [13,25] consists of a concurrent game structure $\mathcal{G}$ together with a winning condition $\bigvee_{i=0}^{n-1}(\Box\Diamond T_{2i} \wedge \neg\Box\Diamond T_{2i+1})$, where $n > 0$ and the T_j's are propositions with $\emptyset \subseteq [T_{2n}] \subseteq [T_{2n-1}] \subseteq \cdots \subseteq [T_0] = Q$. A more intuitive characterization of this winning condition can be obtained by defining, for all $0 \leq j \leq 2n - 1$, the set $C_j \subseteq Q$ of states of *color* j by $C_j = [T_j] \setminus [T_{j+1}]$. For a path $\sigma \in \Sigma$, let $MaxCol(\sigma)$ be the maximal j such that a state in C_j occurs infinitely often in σ. The winning condition for player 1 is that $MaxCol(\sigma)$ is even. The ability to solve games with Rabin chain conditions suffices for solving games with arbitrary ω-regular winning conditions, because every ω-regular property can be translated into a deterministic Rabin chain automaton [25,32].

As in the Büchi case, there are many ways to discount a Rabin chain game, so we use the corresponding fixpoint expression to explore various tradeoffs. Accordingly, for discount factors $a_0, \ldots, a_{2n-1} < 1$, we define the value of an $(a_0, \ldots, a_{2n-1})$-*discounted Rabin chain game* by

$$R(a_0, \ldots, a_{2n-1}) = [\lambda x_{2n-1} \ldots \lambda x_0. \bigvee_{0 \leq j < 2n} (C_j \wedge \mathrm{Rpre}(x_j))]_{\cdot,\{a_j \mapsto a_j \mid 0 \leq j < 2n\}},$$

where $\mathrm{Rpre}(x_j) = \alpha_j \cdot \text{pre}_1(x_j)$ if j is odd, and $\mathrm{Rpre}(x_j) = (1-\alpha_j) + \alpha_j \cdot \text{pre}_1(x_j)$ if j is even. Note that the fixpoint expression is player-1 strongly guarded. The value $R(a_0, \ldots, a_{2n-1})$ of the discounted Rabin chain game can be approximated monotonically by Picard iteration from below and above. Moreover, if the j-th fixpoint is computed for k_j steps, we can bound the cumulative error of the process. Let ε_j be the error in the value of the j-th fixpoint; then $\varepsilon_0 \leq a_0^{k_0}$, and $\varepsilon_j \leq \frac{\varepsilon_{j-1}}{1-a_j} + a_j^{k_j}$ for all $1 \leq j \leq 2n - 1$.

Theorem 4. *For a vector $(k_0, \ldots, k_{2n-1})$ of integers, let $R^\perp_{k_0,\ldots,k_{2n-1}}$ be the region obtained by approximating from below the j-th fixpoint of the discounted μ-calculus formula for $R(a_0, \ldots, a_{2n-1})$ for k_j iterations, and let $R^\top_{k_0,\ldots,k_{2n-1}}$ be the region obtained by approximating from above. If*

$a_0, \ldots, a_{2n-1} < 1$, then for each state s, we have $R(a_0, \ldots, a_{2n-1})(s) - \varepsilon_{2n-1} \leq R_{k_0, \ldots, k_{2n-1}}^\perp(s) \leq R_{k_0, \ldots, k_{2n-1}}^\top(s) \leq R(a_0, \ldots, a_{2n-1})(s) + \varepsilon_{2n-1}$. Moreover, if R is the value of the corresponding undiscounted Rabin chain game, then $\lim_{a_{2n-1} \to 1} \cdots \lim_{a_0 \to 1} R(a_0, \ldots, a_{2n-1}) = R$.

5   State Equivalences: Discounted Bisimilarity

Consider a concurrent game structure $\mathcal{G} = \langle Q, M, \Gamma_1, \Gamma_2, \delta \rangle$. A *distance function* $d: Q^2 \mapsto [0,1]$ is a pseudo-metric on the states with the range $[0,1]$. Distance functions provide a quantitative generalization for equivalence relations on states: distance 0 means "equivalent" in the boolean sense, and distance 1 means "different" in the boolean sense. For two distance functions d_1 and d_2, we write $d_1 \leq d_2$ if $d_1(s,t) \leq d_2(s,t)$ for all states $s,t \in Q$. Given a discount factor $a \in [0,1]$, we define the functor F_a mapping distance functions to distance functions: for every distance function d and all states $s,t \in Q$, we define $F_a(d)(s,t) = 1$ if there is a proposition $T \in \Theta$ such that $[\![T]\!](s) \neq [\![T]\!](t)$, and

$$F_a(d)(s,t) = a \cdot \max \left\{ \begin{array}{l} \sup_{\xi_1 \in \mathcal{D}_1(s)} \inf_{\hat{\xi}_1 \in \mathcal{D}_1(t)} \sup_{\xi_2 \in \mathcal{D}_2(t)} \inf_{\hat{\xi}_2 \in \mathcal{D}_2(s)} \mathrm{E}_{s,t}^{s':\xi_1\xi_2, t':\hat{\xi}_1,\hat{\xi}_2}\{d_a(s',t')\}, \\ \sup_{\hat{\xi}_1 \in \mathcal{D}_1(t)} \inf_{\xi_1 \in \mathcal{D}_1(s)} \sup_{\xi_2 \in \mathcal{D}_2(s)} \inf_{\hat{\xi}_2 \in \mathcal{D}_2(t)} \mathrm{E}_{s,t}^{s':\xi_1\xi_2, t':\hat{\xi}_1,\hat{\xi}_2}\{d_a(s',t')\} \end{array} \right\}$$

otherwise. In the above formula, $\mathcal{D}_i(u) = \mathcal{D}(\Gamma_i(u))$ is the set of probability distributions over the moves of player $i \in \{1,2\}$ at the state $u \in Q$. By $\mathrm{E}_{s,t}^{s':\xi_1\xi_2, t':\hat{\xi}_1,\hat{\xi}_2}\{d(s',t')\}$ we denote the expected value $d(s',t')$ of the distance function d when the state s' results from playing the distributions of moves ξ_1 and ξ_2 from s, and t' results from playing $\hat{\xi}_1$ and $\hat{\xi}_2$ from t. Formally,

$$\mathrm{E}_{s,t}^{s':\xi_1\xi_2, t':\hat{\xi}_1,\hat{\xi}_2}\{d(s',t')\} = \sum_{s',t' \in Q} \sum_{\gamma_1 \in \Gamma_1(s)} \sum_{\gamma_2 \in \Gamma_2(s)} \sum_{\hat{\gamma}_1 \in \Gamma_1(t)} \sum_{\hat{\gamma}_2 \in \Gamma_2(t)}$$
$$d(s',t') \cdot \delta(s' \mid s, \gamma_1, \gamma_2) \cdot \delta(t' \mid t, \hat{\gamma}_1, \hat{\gamma}_2) \cdot \xi_1(\gamma_1) \cdot \xi_2(\gamma_2) \cdot \hat{\xi}_1(\hat{\gamma}_1) \cdot \hat{\xi}_2(\hat{\gamma}_2).$$

The fixpoints of the functor F_a are called *a-discounted (game) bisimulations*. The least fixpoint of F_a is called *a-discounted (game) bisimilarity*, and denoted $B_a^{\mathcal{G}}$ (the superscript is omitted if the game structure $\mathcal{G}$ is clear from the context).[3] If $a < 1$, then F_a has a unique fixpoint; in this case, there is a unique a-discounted bisimulation, namely, B_a. If $a = 1$, instead of 1-discounted, we say *undiscounted*.

On MDPs (one-player game structures), for $a < 1$, discounted game bisimulation coincides with the discounted bisimulation of [12], and undiscounted game bisimulation coincides with the probabilistic bisimulation of [30]. On transition systems (one-player deterministic game structures), undiscounted game bisimulation coincides with classical bisimulation [24]. However, undiscounted game bisimulation is not equivalent to the alternating bisimulation of [3], which has been defined for deterministic game structures. By the minimax theorem [35],

---

[3] Bisimilarity is usually considered a greatest fixpoint, but in our setup, the distance function that considers all states to be equivalent in the boolean sense is the least distance function.

we can exchange the two middle sup and inf operators in the definition of F_a; that is, the roles of players 1 and 2 can be exchanged. Hence, there is only one version of (un)discounted game bisimulation, while there are distinct player 1 and player 2 alternating bisimulations. Alternating bisimulation corresponds to the case where the sets $\mathcal{D}_i(u)$, for $i \in \{1, 2\}$ and $u \in Q$, consist only of deterministic distributions, where each player must choose a specific move (indeed, the minimax theorem does not hold if the players are forced to use deterministic distributions). In the case of turn-based deterministic game structures, the two definitions collapse, but for concurrent game structures the sup-inf interpretation of winning is strictly weaker than the deterministic interpretation [7].

The a-discounted bisimilarity B_a can be computed using Picard iteration: starting from $d_a^{(0)}$, with $d_a^{(0)}(s, t) = 0$ for all states $s, t \in Q$, let $d_a^{(k+1)} = F_a(d_a^{(k)})$ for all $k \geq 0$. If $a < 1$, then we may start from any distance function $d_a^{(0)}$ (because the fixpoint is unique) and the convergence is geometric with rate a. The theorem below relates discounted and undiscounted game bisimulation.

Theorem 5. *On every concurrent game structure,* $\lim_{a \to 1} B_a = B_1$. *Moreover, for two states s and t, we have $B_1(s, t) = 0$ iff $B_a(s, t) = 0$ for any and all discount factors $a > 0$, and $B_1(s, t) = 1$ iff $B_a(s, t) = 1$ for any and all $a > 0$.*

Our main theorem on discounted game bisimulation states that for all states, closeness in discounted game bisimilarity corresponds to closeness in the value of discounted μ-calculus formulas. In other words, a small perturbation of a system can only cause a small change of its properties.

Theorem 6. *Consider two states s and t of a concurrent game structure, and a discount factor $a < 1$. For all closed discounted μ-calculus formulas ϕ and a-bounded parameter valuations $\mathcal{P}$, we have $|[\![\phi]\!]_{.,\mathcal{P}}(s) - [\![\phi]\!]_{.,\mathcal{P}}(t)| \leq B_a(s, t)$. Also, $\sup_\phi |[\![\phi]\!]_{.,\mathcal{P}}(s) - [\![\phi]\!]_{.,\mathcal{P}}(t)| = B_a(s, t)$.*

Let ϵ be a nonnegative real. A game structure $\mathcal{G}' = \langle Q, M, \Gamma_1, \Gamma_2, \delta' \rangle$ is an ϵ-perturbation of the game structure $\mathcal{G} = \langle Q, M, \Gamma_1, \Gamma_2, \delta \rangle$ if for all states $s \in Q$, all sets $X \subseteq Q$, and all moves $\gamma_1 \in \Gamma_1(s)$ and $\gamma_2 \in \Gamma_2(s)$, we have $|\sum_{t \in X} \delta(t \mid s, \gamma_1, \gamma_2) - \sum_{t \in X} \delta'(t \mid s, \gamma_1, \gamma_2)| \leq \epsilon$. We write $B_a^{\mathcal{G} \uplus \mathcal{G}'}$ for the a-discounted bisimililarity on the disjoint union of the game structures $\mathcal{G}$ and $\mathcal{G}'$. The following theorem, which generalizes a result of [11] from one-player structures to games, shows that discounted bisimilarity is robust under perturbations.

Theorem 7. *Let $\mathcal{G}'$ be an ϵ-perturbation of a concurrent game structure $\mathcal{G}$, and let $a < 1$ be a discount factor. For every state s of $\mathcal{G}$ and corresponding state s' of $\mathcal{G}'$, we have $B_a^{\mathcal{G} \uplus \mathcal{G}'}(s, s') \leq K \cdot \epsilon$, where $K = \sup_{k \geq 0} \{k \cdot a^k\}$.*

References

1. R. Alur and T.A. Henzinger. Finitary fairness. *ACM TOPLAS*, 20:1171–1194, 1994.

2. R. Alur, T.A. Henzinger, and O. Kupferman. Alternating-time temporal logic. *J. ACM*, 49:672–713, 2002.
3. R. Alur, T.A. Henzinger, O. Kupferman, and M.Y. Vardi. Alternating refinement relations. In *Concurrency Theory*, LNCS 1466, pp. 163–178. Springer, 1998.
4. M.C. Browne, E.M. Clarke, and O. Grumberg. Characterizing finite Kripke structures in propositional temporal logic. *Theoretical Computer Science*, 59:115–131, 1988.
5. J.R. Büchi. On a decision method in restricted second-order arithmetic. In *Congr. Logic, Methodology, and Philosophy of Science 1960*, pp. 1–12. Stanford University Press, 1962.
6. L. de Alfaro. Stochastic transition systems. In *Concurrency Theory*, LNCS 1466, pp. 423–438. Springer, 1998.
7. L. de Alfaro, T.A. Henzinger, and O. Kupferman. Concurrent reachability games. In *Symp. Foundations of Computer Science*, pp. 564–575. IEEE, 1998.
8. L. de Alfaro, T.A. Henzinger, and R. Majumdar. From verification to control: Dynamic programs for ω-regular objectives. In *Symp. Logic in Computer Science*, pp. 279–290. IEEE, 2001.
9. L. de Alfaro and R. Majumdar. Quantitative solution of ω-regular games. In *Symp. Theory of Computing*, pp. 675–683. ACM, 2001.
10. C. Derman. *Finite-State Markovian Decision Processes*. Academic Press, 1970.
11. J. Desharnais, V. Gupta, R. Jagadeesan, and P. Panangaden. Metrics for labeled Markov systems. In *Concurrency Theory*, LNCS 1664, pp. 258–273. Springer, 1999.
12. J. Desharnais, V. Gupta, R. Jagadeesan, and P. Panangaden. The metric analogue of weak bisimulation for probabilistic processes. In *Symp. Logic in Computer Science*, pp. 413–422. IEEE, 2002.
13. E.A. Emerson and C.S. Jutla. Tree automata, μ-calculus and determinacy. In *Symp. Foundations of Computer Science*, pp. 368–377. IEEE, 1991.
14. E.A. Emerson, C.S. Jutla, and A.P. Sistla. On model checking for fragments of μ-calculus. In *Computer-aided Verification*, LNCS 697, pp. 385–396. Springer, 1993.
15. E.A. Emerson and C.-L. Lei. Efficient model checking in fragments of the propositional μ-calculus. In *Symp. Logic in Computer Science*, pp. 267–278. IEEE, June 1986.
16. J. Filar and K. Vrieze. *Competitive Markov Decision Processes*. Springer, 1997.
17. Y. Gurevich and L. Harrington. Trees, automata, and games. In *Symp. Theory of Computing*, pp. 60–65. ACM, 1982.
18. C.-C. Jou and S.A. Smolka. Equivalences, congruences, and complete axiomatizations for probabilistic processes. In *Concurrency Theory*, LNCS 458, pp. 367–383. Springer, 1990.
19. D. Kozen. A probabilistic PDL. In *Symp. Theory of Computing*, pp. 291–297. ACM, 1983.
20. D. Kozen. Results on the propositional μ-calculus. *Theoretical Computer Science*, 27:333–354, 1983.
21. Z. Manna and A. Pnueli. A hierarchy of temporal properties. In *Symp. Principles of Distributed Computing*, pp. 377–408. ACM, 1990.
22. Z. Manna and A. Pnueli. *The Temporal Logic of Reactive and Concurrent Systems: Specification*. Springer, 1991.
23. A. McIver. Reasoning about efficiency within a probabilitic μ-calculus. *Electronic Notes in Theoretical Computer Science*, 22, 1999.
24. R. Milner. Operational and algebraic semantics of concurrent processes. In J. van Leeuwen, ed., *Handbook of Theoretical Computer Science*, vol. B, pp. 1202–1242. Elsevier, 1990.

25. A.W. Mostowski. Regular expressions for infinite trees and a standard form of automata. In *Computation Theory*, LNCS 208, pp. 157–168. Springer, 1984.

26. G. Owen. *Game Theory*. Academic Press, 1995.

27. A. Pnueli. The temporal logic of programs. In *Symp. Foundations of Computer Science*, pp. 46–57. IEEE, 1977.

28. M.O. Rabin. *Automata on Infinite Objects and Church's Problem*. Conference Series in Mathematics, vol. 13. AMS, 1969.

29. R. Segala. *Modeling and Verification of Randomized Distributed Real-Time Systems*. PhD thesis, MIT, 1995. Tech. Rep. MIT/LCS/TR-676.

30. R. Segala and N.A. Lynch. Probabilistic simulations for probabilistic processes. In *Concurrency Theory*, LNCS 836, pp. 481–496. Springer, 1994.

31. L.S. Shapley. Stochastic games. *Proc. National Academy of Sciences*, 39:1095–1100, 1953.

32. W. Thomas. On the synthesis of strategies in infinite games. In *Theoretical Aspects of Computer Science*, LNCS 900, pp. 1–13. Springer, 1995.

33. M.Y. Vardi. Automatic verification of probabilistic concurrent finite-state systems. In *Symp. Foundations of Computer Science*, pp. 327–338. IEEE, 1985.

34. M.Y. Vardi. A temporal fixpoint calculus. In *Symp. Principles of Programming Languages*, pp. 250–259. ACM, 1988.

35. J. von Neumann and O. Morgenstern. *Theory of Games and Economic Behavior*. Princeton University Press, 1947.

Information Flow in Concurrent Games*

Luca de Alfaro[1] and Marco Faella[1,2]

[1] Department of Computer Engineering, UC Santa Cruz, USA
[2] Dipartimento di Informatica ed Applicazioni, Università degli Studi di Salerno, Italy

Abstract. We consider games where the players have perfect information about the game's state and history, and we focus on the information exchange that takes place at each round as the players choose their moves. The ability of a player to gather information on the opponent's choice of move in a round determines her ability to counteract the move, and win the game. When the game is played between teams, rather than single players, the amount of intra-team communication determines the ability of the team members to coordinate their moves and win the game. We consider games with quantitative bounds on inter-team and intra-team information flow, and we provide algorithms and complexity bounds for their solution.

1   Introduction

We consider repeated games played for an infinite number of rounds on a finite state space [Sha53]. At each round of the game, each player selects a move; the selected moves jointly determine the next state of the game [Sha53,FV97, AHK97]. This process, repeated, gives rise to a play of the game, consisting in the infinite sequence of visited states. We consider safety games, where the goal consists in staying forever in a safe set of states, and reachability games, where the goal consists in reaching a desired subset of states [Tho95,Zie98]. The ability of a player to win such games depends on the information available to the player. In *partial information* games, players have incomplete information about the current state of the game and the past history; computing the sets of winning states for safety and reachability goals is EXPTIME complete [Rei84, KV97]. In this paper, we consider instead games where the players have perfect information about the game's current state and history, and we focus instead on the information exchange that takes place at each round, between players and within players, as the players choose their moves.

We first consider the distinction between *turn-based* and *concurrent* games. Usually, this distinction is defined structurally: a game is *concurrent* if at each state both players may have a choice of moves [FV97,AHK97]. and is turn-based if at each state at most one player has the choice among multiple moves [BL69,

---

* This research was supported in part by the NSF CAREER award CCR-0132780, the NSF grant CCR-0234690, the ONR grant N00014-02-1-0671, and the MIUR grant "Metodi Formali per la Sicurezza e il Tempo" (MEFISTO).

J.C.M. Baeten et al. (Eds.): ICALP 2003, LNCS 2719, pp. 1038–1053, 2003.

GH82,EJ91,Tho95]. We argue that the difference is best captured in terms of information: a game is concurrent if the two players must choose their moves independently, on the basis of the same information, and it is turn-based if one of the two players has full information about the opponent's choice of move when choosing her own move. Indeed, in a game where the players play simultaneously, if one player has full information about the other player's choice of move, the game is in effect turn-based, the player with full information playing second. While this may seem an odd way to play a game, it occurs in hardware design whenever a Moore machine, whose next outputs (the next move) can depend on the current inputs only, is composed with a Mealy machine, whose next outputs (move) can depend both on the current inputs, and on the *next* inputs from other machines. Effectively, the Moore machine chooses first, while the Mealy machine can look at the move chosen by the Moore machine before choosing its own move. Conversely, whenever in a turn-based game one of the players is prevented from observing the preceding opponent move (along with its effects), the game is effectively concurrent, even though the choice of moves is not simultaneous. Indeed, there would hardly be any concurrent game, if the distinction between concurrent and turn-based were based on truly simultaneous choice, rather than on independent choice under the same information.

Once the distinction between concurrent and turn-based games is phrased in terms of information, concurrent and turn-based games constitute the two extremes in a spectrum of games, which we call *semi-concurrent,* where one player is able to gather a bounded amount of information about the opponent's move before choosing her own. Games where players exchange information in a round have been considered in [dAHM00,dAHM01] to model the interaction of synchronous hardware; in those works the communication scheme is fixed, and is specified together with the game. We consider here games where the amount of information exchanged between players is bounded, but the information content, and the way it is gathered, is left to the discretion of the players. Semi-concurrent games have several applications. In the design of controllers for digital circuits, semi-concurrent games model the case where together with the controller, we can design combinatorial signals that provide information about the next state of the controlled system. Moreover, semi-concurrent games can be used to model games played with untrustworthy adversaries, who can exploit leaked information about our choice of move.

We provide algorithms for solving semi-concurrent games with respect to safety and reachability conditions. We consider both the case when the goal must be attained for all plays (*sure* winning), and the case when the goal must be attained with probability 1 (*almost* winning) [dAHK98], and we consider both the case when the player striving to achieve the goal can spy on the opponent, or is spied upon. We give tight bounds for the complexity of these algorithms, proving that for several combinations of goals, spying, and winning mode (sure or almost), deciding whether a player can win from a state is *NP*-complete. We also show that the larger the amount of information a player can gather about the opponent's choice of move, the more games the player can win; our results

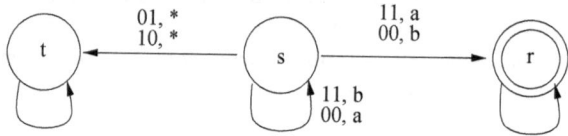

Fig. 1. A concurrent game. An edge label such as $11, a$ indicates that the edge is followed when player 1 chooses '11' and player 2 chooses 'a'. The game starts in s, and the goal is to reach r. States t and r are sink states, without outgoing transitions.

enable the determination of the minimum amount of information about the opponent's move that is required in order to win. Finally, investigate the need for randomization in winning reachability games. From [dAHK98] it is known that randomization is needed to win reachability games with probability 1 if the game is concurrent, but not if it is turn-based. We show that randomization in general is always needed to win semi-concurrent reachability games with probability 1, regardless of whether a player is spying or is being spied upon, as long as one of the players does not have perfect information about the other player's choice of move.

Concurrent games can also be seen as the extreme point of another spectrum, concerning the amount of communication *within* a player. While some players are single entities, others are internally composed of separate entities: such composite players are called *teams,* and the entities they comprise are called *team members.* We consider games where the move chosen by a team is a tuple, each team member choosing a component of the tuple. This problem was first studied in [PR79], where it was shown that team games where the players have incomplete information about the state of the game are in general undecidable. Later, [PR90] and [KV01] considered team games with linear or cyclic communication structure between team members, and showed that solving such games with respect to linear or branching time temporal logic conditions is decidable. These previous works considered games where each team member has a different, partial view of the state of the game. Here, we consider instead the situation in which the team members share complete information about the state of the game, but must coordinate their moves at each round.

A team can readily play any deterministic choice of move that can be played by a single-entity player: each team member simply chooses deterministically the desired move component. However, if the team members cannot communicate while choosing the move components, the team can only play *randomized* distributions of moves that result from the independent randomization of each member's choice. This limits the team's ability to win reachability games, as illustrated by the game of Figure 1. Player 1 can reach r from state s with probability 1 by choosing moves 00 and 11 with probability 1/2 each, and by choosing moves 01 and 10 with probability 0. However, assume player 1 consists in a two-member team, where each member chooses one of the bits. If the two team members cannot communicate while choosing the bits, the team can only play probability distributions $p(i,j)$ for $i,j \in \{0,1\}$ of the form $p(i,j) = q_1(i)q_2(j)$, where q_1 and q_2 are the distributions chosen by the team members. It is easy to see that team 1 cannot reach r with probability 1 from s using these distributions. If an arbitrary amount of communication can take place in a round between team members, the team can replicate the behavior of a single-entity player: thus, concurrent games constitute the limit case of team games for arbitrary communication.

Here, we study team games where a bounded (possibly 0) amount of communication can take place among team members in a round. Team games model controller-design problems where the controller consists of distributed sub-controllers that can observe the current state of the controlled system, but that have limited communication ability to coordinate their next move. For instance, in synchronous digital circuits, it may not be feasible for the sub-controllers to communicate their next state if they communicate through links that are slow compared to the system clock. Moreover, team games model the real-world situation when members of the same team must coordinate their next move covertly using limited bandwidth.

Team safety games can be solved in the same manner as concurrent safety games, since no randomization is required in the winning strategies. We present algorithms for solving team reachability games with communicating and non-communicating team members, and we provide tight bounds for their complexity, showing in particular that solving non-communicating reachability games is an *NP*-complete problem. While in the case of semi-concurrent games, the more the information is communicated, the more the games that can be won, we show that for team games, a single bit of information communicated between team members is as efficient as complete coordination. On the other hand, we show that probability 1 reachability team games with are in general not determined: if one team cannot win the game with probability 1, this does not imply the existence of a single adversary strategy that prevents the team from winning.

2   Games

For a finite set A, a *probability distribution* on A is a function $p : A \mapsto [0,1]$ such that $\sum_{a \in A} p(a) = 1$; we denote the set of probability distributions on A by $\mathcal{D}(A)$. A *game structure* is a tuple $G = (S, M, \Gamma_1, \Gamma_2, \tau)$, where:

- S is a finite set of *states*;
- M is a finite sets of *moves*;
- $\Gamma_1, \Gamma_2 : S \mapsto 2^M \setminus \{\emptyset\}$ are the *move assignments* of the players;
- $\tau : S \times M \times M \mapsto \mathcal{D}(S)$ is the *probabilistic transition function*.

At every state $s \in S$, player 1 chooses a move $a \in \Gamma_1(s)$, and player 2 chooses a move $b \in \Gamma_2(s)$; the game then proceeds to state $t \in S$ with probability $\tau(s, a, b)(t)$. For $s \in S$ and $a, b \in M$, we denote by $\delta(s, a, b) = \{t \in S \mid \tau(s, a, b)(t) > 0\}$ the set of possible successors of s when moves a and b are played. A *play* of G is an infinite sequence $s_0, \langle a_0, b_0 \rangle, s_1, \langle a_1, b_1 \rangle, \ldots$ such that for all $n \geq 0$, we have $a_n \in \Gamma_1(s_n)$, $b_n \in \Gamma_2(s_n)$, and $s_{n+1} \in \delta(s_n, a_n, b_n)$. We denote by $Plays_{s_0}$ the set of plays starting from $s_0 \in S$, and by $Plays = \bigcup_{s_0 \in S} Plays_{s_0}$ the set of all plays of G. A *history* σ is a finite play prefix $\sigma = s_0, \langle a_0, b_0 \rangle, s_1, \langle a_1, b_1 \rangle, \dots, s_n$ that terminates in a state; we denote by $last(\sigma)$ the last state s_n of σ. We denote by $Hist_{s_0}$ the set of histories of G starting from s_0, and by $Hist = \bigcup_{s_0 \in S} Hist_{s_0}$ the set of all histories of G. We define the *size* of G to be equal to the number of entries of the transition function δ; specifically, $|G| = \sum_{s \in S} \sum_{a \in \Gamma_1(s)} \sum_{b \in \Gamma_2(s)} |\delta(s, a, b)|$. Given $s \in S$, $Y \subseteq S$ and $B \subseteq M$, it is useful to define $Safe_1(s, Y, B)$ as the set of moves of player 1 that ensure staying in Y when the player 2 chooses moves from B. Formally, $Safe_1(s, Y, B) = \{a \in \Gamma_1(s) \mid \forall b \in B . \delta(s, a, b) \subseteq Y\}$. We define symmetrically $Safe_2(s, Y, A)$.

2.1   Strategies, Winning Conditions, and Games

Let Ω_s be the set of measurable subsets of $Plays_s$, defined as usual (see e.g. [Wil91]). A *family of strategies* $\langle \Upsilon_1, \Upsilon_2, \Pr \rangle$ consists of two sets Υ_1 and Υ_2 of strategies for players 1 and 2, together with a mapping $\Pr$ that associates a probability measure $\Pr_s^{\pi_1, \pi_2} : \Omega_s \mapsto [0, 1]$ with each initial state s and pair of strategies $\pi_1 \in \Upsilon_1$ and $\pi_2 \in \Upsilon_2$. Thus, $\Pr_s^{\pi_1, \pi_2}(\mathcal{E})$ is the probability that a game starting from $s \in S$ follows a play in $\mathcal{E} \in \Omega_s$ when players 1 and 2 play according to strategies π_1 and π_2, respectively. A probability measure $\Pr_s^{\pi_1, \pi_2}$ over Ω_s gives rise to a set $Outcomes(s, \Pr, \pi_1, \pi_2) \subseteq Plays_s$, of *outcome plays*, consisting of the plays whose finite prefixes can be followed with non-zero probability. Formally, for $\rho \in Plays_s$ and $n > 0$, let $\mathcal{E}(\rho, n) \in \Omega_s$ be the set of plays that agree with ρ up to round n: then, $\rho \in Outcomes(s, \Pr, \pi_1, \pi_2)$ if $\Pr_s^{\pi_1, \pi_2}(\mathcal{E}(\rho, n)) > 0$ for all $n > 0$.

We consider *safety games*, in which the winning condition $\Box R$ consists in remaining forever in a subset $R \subseteq S$ of states, and *reachability games*, in which the winning condition $\Diamond R$ consists in reaching a subset $R \subseteq S$ of states; we define $[\![\Box R]\!] = \{s_0, \langle a_0, b_0 \rangle, s_1, \langle a_1, b_1 \rangle, \dots \in Plays \mid \forall n \in \mathbb{N} . s_n \in R\}$ and $[\![\Diamond R]\!] = \{s_0, \langle a_0, b_0 \rangle, s_1, \langle a_1, b_1 \rangle, \dots \in Plays \mid \exists n \in \mathbb{N} . s_n \in R\}$. A *game* is thus a tuple $(G, \phi, i, \mathcal{M} \mid \Upsilon_1, \Upsilon_2, \Pr)$, composed of a game structure G, a winning condition $\phi \in \{\Box R, \Diamond R\}$, an integer $i \in \{1, 2\}$, a modality $\mathcal{M} \in \{sure, almost\}$, and a family of strategies. Given a family of strategies $\langle \Upsilon_1, \Upsilon_2, \Pr \rangle$ and $\phi \in \{\Box R, \Diamond R\}$, we define the sets $win_1(G, \phi, sure \mid \Upsilon_1, \Upsilon_2, \Pr)$ of player-1 *sure-winning* states and the set $win_1(G, \phi, almost \mid \Upsilon_1, \Upsilon_2, \Pr)$ of player-1 *almost-winning* states as follows [dAHK98]:

Sure-winning. For all $s \in S$, we have $s \in win_1(G, \phi, sure \mid \Upsilon_1, \Upsilon_2, \Pr)$ if there is $\pi_1 \in \Upsilon_1$ such that for all $\pi_2 \in \Upsilon_2$ we have $Outcomes(s, \Pr, \pi_1, \pi_2) \subseteq [\![\phi]\!]$.

Almost-winning. For all $s \in S$, we have $s \in win_1(G, \phi, almost \mid \Upsilon_1, \Upsilon_2, \Pr)$ if there is $\pi_1 \in \Upsilon_1$ such that for all $\pi_2 \in \Upsilon_2$ we have $\Pr_s^{\pi_1, \pi_2}([\![\phi]\!]) = 1$.

The sets of player-2 sure and almost-sure winning states are defined symmetrically. A *winning strategy* is a strategy that ensures victory to a player with the prescribed mode (sure, or almost), for all winning states. Precisely, for $\mathcal{M} \in \{sure, almost\}$, a winning strategy for $(G, \phi, 1, \mathcal{M} \mid \Upsilon_1, \Upsilon_2, \mathrm{Pr})$ is a strategy $\pi_1 \in \Upsilon_1$ such that, for all $s \in win_1(G, \phi, almost \mid \Upsilon_1, \Upsilon_2, \mathrm{Pr})$ and all $\pi_2 \in \Upsilon_2$, we have that $Outcomes(s, \mathrm{Pr}, \pi_1, \pi_2) \subseteq \phi$ if $\mathcal{M} = sure$, and $\mathrm{Pr}_s^{\pi_1, \pi_2}(\llbracket \phi \rrbracket) = 1$ if $\mathcal{M} = almost$. A *spoiling strategy* is an adversary strategy that prevents a player from winning whenever victory cannot be assured. Precisely, for $\mathcal{M} \in \{sure, almost\}$, a spoiling strategy for $(G, \phi, 1, \mathcal{M} \mid \Upsilon_1, \Upsilon_2, \mathrm{Pr})$ is a strategy $\pi_2 \in \Upsilon_2$ such that, for all $s \notin win_1(G, \phi, almost \mid \Upsilon_1, \Upsilon_2, \mathrm{Pr})$ and all $\pi_1 \in \Upsilon_1$, we have that $Outcomes(s, \mathrm{Pr}, \pi_1, \pi_2) \not\subseteq \phi$ if $\mathcal{M} = sure$, and $\mathrm{Pr}_s^{\pi_1, \pi_2}(\llbracket \phi \rrbracket) < 1$ if $\mathcal{M} = almost$. Analogous definitions hold for the winning problems that refer to player 2.

A *game type* is a tuple $(\triangle, i, \mathcal{M} \mid \Upsilon_1, \Upsilon_2, \mathrm{Pr})$ where $i \in 1, 2$, $\mathcal{M} \in \{sure, almost\}$, and $\triangle \in \{\square, \Diamond\}$. We say that a game type is *determined* iff, for all game structures G, player $i \in \{1, 2\}$, and $R \subseteq S$, both winning and spoiling strategies exist for $(G, \triangle R, i, \mathcal{M} \mid \Upsilon_1, \Upsilon_2, \mathrm{Pr})$.

2.2 Concurrent Games

In concurrent games, the players choose their moves simultaneously and independently. A *concurrent strategy* for player $i \in \{1, 2\}$ is a mapping $\pi_i \colon Hist \mapsto \mathcal{D}(M)$ that associates with every history σ of the game a probability distribution $\pi_i(\sigma)$ used to select the next move; for all $a \in M$, we require that $\pi_i(\sigma)(a) > 0$, implies $a \in \Gamma_i(last(\sigma))$, ensuring that the strategy selects only moves that are available to the players. We denote by Π_i^c the set of all concurrent strategies for player $i \in \{1, 2\}$. Given a strategy $\pi \in \Pi_1^c \cup \Pi_2^c$, we say that π is *memoryless* if for all $\sigma \in Hist$ we have $\pi(\sigma) = \pi(last(\sigma))$, and we say that π is *deterministic* if for all $\sigma \in Hist$ and all $a \in M$ we have $\pi(\sigma)(a) \in \{0, 1\}$. An initial state s_0 and a pair of strategies $\pi_1 \in \Pi_1^c$ and $\pi_2 \in \Pi_2^c$ give rise to a probability $\mathrm{Prb}_{s_0}^{\pi_1, \pi_2}$ on histories defined inductively by $\mathrm{PrbC}_{s_0}^{\pi_1, \pi_2}(s_0) = 1$ and, for $n \geq 0$, by $\mathrm{PrbC}_{s_0}^{\pi_1, \pi_2}(s_0, \ldots, s_n, \langle a_n, b_n \rangle, s_{n+1}) = \mathrm{PrbC}_{s_0}^{\pi_1, \pi_2}(s_0, \ldots, s_n) \cdot \pi_1(s_0, \ldots, s_n)(a_n) \cdot \pi_2(s_0, \ldots, s_n)(b_n) \cdot \tau(s_n, a_n, b_n)(s_{n+1})$. These probabilities on histories give rise to a probability measure $\mathrm{PrbC}_{s_0}^{\pi_1, \pi_2}$ on Ω_{s_0} [Wil91]. A *concurrent game* is a game in which the players use concurrent strategies. The winning states of concurrent safety and reachability games can be computed using μ-calculus; we briefly review the approach, as it will be the starting point of the algorithms we will present for other families of strategies.

Safety. The solution of concurrent safety games is entirely classical. The set of winning states can be computed using the *controllable predecessor* operator $CPre_1 : 2^S \mapsto 2^S$, defined for all $X \subseteq S$ by

$$CPre_1(X) = \{s \in S \mid \exists a \in \Gamma_1(s). \forall b \in \Gamma_2(s). \delta(s, a, b) \subseteq X\}.$$

Intuitively, $CPre_1(X)$ consists of all the states from which player 1 can force the game to X in one round; the operator $CPre_2$ for player 2 can be defined symmetrically. For $i \in \{1, 2\}$ and $R \subseteq S$ we have then:

$$win_i(G, \Box R, sure \mid \Pi_1^c, \Pi_2^c, \text{PrbC}) = win_i(G, \Box R, almost \mid \Pi_1^c, \Pi_2^c, \text{PrbC}) = \nu X.(R \cap CPre_i(X)), \tag{1}$$

where ν denotes the greatest fixpoint operator. The fixpoint can be computed by Picard iteration by letting $X_0 = S$ and, for $n \geq 0$, $X_{n+1} = R \cap CPre_i(X_n)$; the solution is then given by the limit $\lim_{n \to \infty} X_n$, which can be computed in at most $|S|$ iterations.

Reachability. For mode *sure*, the solution of concurrent reachability games is also classical: for player $i \in \{1, 2\}$ and target set $R \subseteq S$, we have

$$win_i(G, \Diamond R, sure \mid \Pi_1^c, \Pi_2^c, \text{PrbC}) = \mu X.(R \cup CPre_i(X)) \tag{2}$$

where μ denotes the least fixpoint operator. The solution can again be computed iteratively, as the limit $\lim_{n \to \infty} X_n$ of the sequence $X_0, X_1, X_2, \ldots$ defined by $X_0 = \emptyset$, and for $n \geq 0$, by $X_{n+1} = R \cup CPre_i(X_n)$. The solution for mode *almost* and player $i \in \{1, 2\}$ relies on the two-argument predecessor operator $APre_i : 2^S \times 2^S \mapsto 2^S$ [dAHK98,dAH00]. For $X, Y \subseteq S$, we have $s \in APre_1(Y, X)$ iff player i can force the game to stay in Y, while at the same time forcing a transition to X with positive probability:

$$APre_1(Y, X) = \{s \in S \mid \forall b \in \Gamma_2(s).\exists a \in Safe_1(s, Y, \Gamma_2(s)).\delta(s, a, b) \cap X \neq \emptyset\}.$$

The operator $APre_2$ for player 2 can be defined symmetrically. The set of states from which player $i \in \{1, 2\}$ wins with probability 1 with respect to the winning condition $\Diamond R$ can then be computed as a nested fixpoint [dAHK98,dAH00]:

$$win_i(G, \Diamond R, sure \mid \Pi_1^c, \Pi_2^c, \text{PrbC}) = \nu Y.\mu X.(R \cup APre_i(Y, X)). \tag{3}$$

To understand this algorithm, let Y^* set of winning states computed by (3). Since $Y^* = \mu X.(R \cup APre_i(Y^*, X))$, we can write $Y^* = \lim_{k \to \infty} X_k$, where $X_0 = R$, and for $k \geq 0$, where $X_{k+1} = (R \cup APre_i(Y^*, X_k))$. For $k \geq 0$, from $X_{k+1} \setminus X_k$ player i can ensure some probability of going to X_k, while never leaving Y^*. Hence, from any state in Y^*, player i can play a sequence of $|Y^*|$ rounds that ensure that (i) R is reached with positive probability, and (ii) Y^* is left only after R is reached. By repeating this $|Y^*|$-round sequence indefinitely, player i is able to reach R with probability 1. The following theorem summarizes the results on concurrent games.

Theorem 1 [dAHK98] *For all game structures G, players $i \in \{1, 2\}$, and sets $R \subseteq S$, the following assertions hold.*

Safety. *For $\mathcal{M} \in \{sure, almost\}$, the set $win_i(G, \Box R, \mathcal{M} \mid \Pi_1^c, \Pi_2^c, \text{PrbC})$ can be computed in time $\mathcal{O}(|G|)$ by (1). The game type $(\Box, i, \mathcal{M} \mid \Pi_1^c, \Pi_2^c, \text{PrbC})$ is determined, and there always exist winning strategies that are both deterministic and memoryless.*

Sure reachability. *The set $win_i(G, \Diamond R, sure \mid \Pi_1^c, \Pi_2^c, \text{PrbC})$ can be computed in time $\mathcal{O}(|G|)$ by (2). The game type $(\Diamond, i, sure \mid \Pi_1^c, \Pi_2^c, \text{PrbC})$ is determined, and there always exist winning strategies that are both deterministic and memoryless.*

Almost sure reachability. *The set* $win_i(G, \Diamond R, almost \mid \Pi_1^c, \Pi_2^c, PrbC)$ *can be computed in time* $\mathcal{O}(|G|^2)$ *by (3). The game type* $(\Diamond, i, almost \mid \Pi_1^c, \Pi_2^c, PrbC)$ *is determined, there always exist winning strategies that are memoryless, but the existence of deterministic winning strategies is not guaranteed.*

3   Semi-concurrent Games

In *semi-concurrent* games, one of the players, when choosing her move, has access to a bounded amount of information about the opponent's choice of move. To model inter-player communication within a round, we introduce semi-concurrent strategies. Let $\Sigma_k = \{1, 2, \ldots, k\}$. A *semi-concurrent strategy* of order $k > 0$ for player $i \in \{1, 2\}$ is a pair $\pi_i = \langle \pi_i^s, \pi_i^d \rangle$ consisting of a *spy strategy* $\pi_i^s :$ $Hist \times M \mapsto \mathcal{D}(\Sigma_k)$ and of a *decision strategy* $\pi_i^d : Hist \times \Sigma_k \mapsto \mathcal{D}(M)$, such that for all $\sigma \in Hist$, all $1 < j < k$ and all $a \in M$, we have that $\pi_i^d(\sigma, j)(a) > 0$ implies $a \in \Gamma_i(last(\sigma))$. The spy strategy represents the method used by player i to gather information about the opponent's move: after the history σ, if the opponent chooses move b, one of the integers in Σ_k is received by player i, according to the distribution $\pi_i^s(\sigma, b)$. Once player i receives an integer n, it chooses the move $a \in M$ with probability $\pi_i^d(\sigma, n)(a)$. Note that semi-concurrent strategies of order 1 are essentially concurrent strategies, as the only symbol carries no information, and semi-concurrent strategies of order $|M|$ give rise to turn-based games, since one of the players can obtain full information about the move of the other. For $k > 0$, we let $\vec{\Pi}_i^k$ be the set of all semi-concurrent strategies of order k for player $i \in \{1, 2\}$.

A *semi-concurrent game* is a game where one player uses semi-concurrent strategies, and the other uses concurrent strategies. We arbitrarily fix player 1 to be the player using semi-concurrent strategies. Hence, we consider the families of strategies $\langle \vec{\Pi}_1^k, \Pi_2^c, PrbS \rangle$ for $k > 0$, where for $\pi_1 = \langle \pi_1^s, \pi_1^d \rangle \in \vec{\Pi}_1^k$ and $\pi_2 \in \Pi_2^c$, $PrbS_{s_0}^{\pi_1, \pi_2}$ is defined inductively on histories by $PrbS_{s_0}^{\pi_1, \pi_2}(s_0) = 1$ and, for $n \geq 0$ and $\sigma \in Hist_{s_0}$, by

$$PrbS_{s_0}^{\pi_1, \pi_2}(\sigma, \langle a_n, b_n \rangle, s_{n+1}) = PrbS_{s_0}^{\pi_1, \pi_2}(\sigma) \cdot \pi_2(\sigma)(b_n) \cdot \tau(last(\sigma), a_n, b_n)(s_{n+1})$$
$$\cdot \sum_{j \in \Sigma_k} \pi_1^d(\sigma, j)(a_n) \cdot \pi_1^s(\sigma, b_n)(j).$$

Again, these probabilities on histories give rise to a probability measure $PrbS_{s_0}^{\pi_1, \pi_2}$ on Ω_{s_0}.

In general, both the spy strategy π_i^s and the decision strategy π_i^d can be history-dependent and randomized. For $i \in \{1, 2\}$, we say that a decision strategy π_i^d of order k is *memoryless* if $\pi_i^d(\sigma, j) = \pi_i^d(last(\sigma), j)$ for all $\sigma \in Hist$ and all $j \in \{1, 2, \ldots, k\}$, and we say that π_i^d is *deterministic* if $\pi_i^d(\sigma, j)(a) \in \{0, 1\}$ for all $\sigma \in Hist$, all $j \in \{1, 2, \ldots, k\}$, and all $a \in M$. Analogously, for $i \in \{1, 2\}$ and $k > 0$, we say that a spy strategy π_i^s is *memoryless* if $\pi_i^s(\sigma, b) = \pi_i^s(last(\sigma), b)$ for all $\sigma \in Hist$ and all $b \in M$, and we say that π_i^s is *deterministic* if $\pi_i^s(\sigma, b)(j) \in$

$\{0, 1\}$, for all $\sigma \in Hist$, all $j \in \{1, 2, \dots, k\}$, and all $b \in M$. We say that a semi-concurrent strategy $\pi_1 = \langle \pi_1^s, \pi_1^d \rangle$ is *memoryless* (respectively *deterministic*) if both π_1^s and π_1^d are memoryless (resp. *deterministic*).

3.1 Semi-concurrent Safety Games

Since the information in each round flows from player 2 to player 1, the solution of semi-concurrent safety games is not symmetrical with respect to players 1 and 2. In order to win a safety game, player 1 must be able at each round to issue a move that keeps the game into the safe region, regardless of the opponent's move. If player 1 can use an order-k semi-concurrent strategy, the best approach consists, at each round, in partitioning the moves of player 2 in k groups, and in using the spy strategy to communicate the group of the move chosen by player 2. If player 1 has a move for each of the k groups that ensures the game stays in the safe region, player 1 can win the game. Hence, we define the *order-k semi-concurrent* predecessor operator $SPre_1^k : 2^S \mapsto 2^S$ as follows. A *k-partition* of a set A consists in k subsets $A_1, \dots, A_k \subseteq A$ such that $A = \bigcup_{j=1}^k A_j$.

> For all $X \subseteq S$ and $s \in S$, we have $s \in SPre_1^k(X)$ iff there is a k-partition $B_1, \dots, B_k$ of $\Gamma_2(s)$ and $a_1, \dots, a_k \in \Gamma_1(s)$ (possibly not all distinct) such that, for all $b \in \Gamma_2(s)$, if $b \in B_j$ then $\delta(s, a_j, b) \subseteq X$.

Thus, when player 2 chooses move $b \in B_j$, player 1 can force the game to X by playing move a_j.

When player 2 tries to win a safety game using a concurrent strategy, the fact that player 1 uses a concurrent strategy, or a semi-concurrent strategy, is irrelevant: in fact, if player 2 had a move that guaranteed safety when not spied upon, the same move will guarantee safety also when spied upon by player 1. Thus, the game can be solved with the usual controllable predecessor operator $CPre_2$. The following theorem summarizes the results about semi-concurrent safety games.

Theorem 2   *For all game structures G, sets $R \subseteq S$, $k > 1$, and $\mathcal{M} \in \{sure, almost\}$, the following assertions hold:*

1. *We have $win_1(G, \square R, \mathcal{M} \mid \vec{\Pi}_1^k, \Pi_2^c, PrbS) = \nu X.(R \cap SPre_1^k(X))$; the set can be computed in time $\mathcal{O}(|G|^k)$. There are always winning strategies that are both deterministic and memoryless.*

2. *We have $win_2(G, \square R, \mathcal{M} \mid \vec{\Pi}_1^k, \Pi_2^c, PrbS) = win_2(G, \square R, \mathcal{M} \mid \Pi_1^c, \Pi_2^c, Prb)$, and as in the case of concurrent games, the above sets can be computed in time $\mathcal{O}(|G|)$ by (1). There are always winning strategies that are both deterministic and memoryless.*

As for determinacy, the following theorem holds.

Theorem 3   *For $i \in \{1, 2\}$, the game type $(\square, i, \mathcal{M} \mid \vec{\Pi}_1^k, \Pi_2^c, PrbS)$ is determined.*

If we consider the order $k > 0$ to be part of the input, we obtain the following *NP*-completeness result. The result is proved by reducing Vertex Cover [GJ79] to the problem of computing $SPre_1^k(\cdot)$.

Theorem 4   *Given input (k, G, R), the membership problem in $win_1(G, \Box R, \mathcal{M} \mid \vec{\Pi}_1^k, \Pi_2^c, PrbS)$ is NP-complete.*

3.2   Player One Reachability Games

In order to win a reachability game with mode *sure*, player 1 must guarantee that, at each round, deterministic progress is made toward the goal. Hence, the solution of player 1 reachability games for mode *sure* uses again the controllable predecessor operator $SPre_1^k$.

When the desired winning mode is *almost*, rather than *sure*, the game is solved using a semi-concurrent version $SAPre_1^k$ of the operator $APre_1$ for concurrent games, for $k > 0$. Again, the best approach for player 1 consists in partitioning the adversary's moves into k subsets $B_1, \ldots, B_k$, using the spy strategy to learn the subset in which the move played by player 2 lies. Thus, if the conditions of operator $APre_1$ hold for each subset $B_1, \ldots, B_k$ of moves, then player 1 is able to ensure probabilistic progress toward the goal. The definition is as follows.

> Given two sets $X, Y \subseteq S$ and a state $s \in S$, we say that $s \in SAPre_1^k(Y, X)$ if and only if there exist a k-partition $B_1, \ldots, B_k$ of $\Gamma_2(s)$ such that, for all $b \in \Gamma_2(s)$, if $b \in B_j$ then there is $a \in Safe_1(s, Y, B_j)$ such that $\delta(s, a, b) \cap X \neq \emptyset$.

Theorem 5   *For all game structures G, $R \subseteq S$, and $k > 1$, the following assertions hold:*

1. *We have $win_1(G, \Diamond R, sure \mid \vec{\Pi}_1^k, \Pi_2^c, PrbS) = \mu X.(R \cup SPre_1^k(X))$; the fixpoint can be computed in time $\mathcal{O}(|G|^k)$. There always exist deterministic and memoryless winning strategies.*

2. *We have $win_1(G, \Diamond R, almost \mid \vec{\Pi}_1^k, \Pi_2^c, PrbS) = \nu Y.\mu X.(R \cup SAPre_1^k(Y, X))$. Deciding whether a state belongs to the above fixpoint is NP-complete in $|G|$. There always exist memoryless winning strategies, but there may not be deterministic winning strategies.*

The theorem states, in particular, that computing the set of winning states of a player 1 semi-concurrent reachability game is an *NP*-complete problem even when $k = 2$, i.e., when the spy strategies can communicate at most 1 bit of information about player 2's choice of move. The *NP*-completeness result is proved by reducing 3-SAT to the problem of deciding $s \in SAPre_1^2(\cdot, \cdot)$. Then, it is shown that the result for $k = 2$ implies the result for all $k > 1$.

Theorem 6   *The following assertions hold:*

1. *The game type* $(\Diamond, 1, sure \mid \vec{\Pi}_1^k, \Pi_2^c, PrbS)$ *is determined.*
2. *The game type* $(\Diamond, 1, almost \mid \vec{\Pi}_1^k, \Pi_2^c, PrbS)$ *is not determined. On the other hand, if only memoryless spy strategies are considered, the latter game type is determined.*

The following theorem states that the more information is available to player 1, the more games player 1 can win.

Theorem 7   *For* $\triangle \in \{\Diamond, \square\}$ *and* $\mathcal{M} \in \{sure, almost\}$, *if* $k_1 > k_2 > 0$ *then there is a game structure* G *and a subset of states* $R \subseteq S$ *such that*

$$win_1(G, \triangle R, \mathcal{M} \mid \vec{\Pi}_1^{k_2}, \Pi_2^c, PrbS) \subsetneq win_1(G, \triangle R, \mathcal{M} \mid \vec{\Pi}_1^{k_1}, \Pi_2^c, PrbS).$$

The theorem is proved by constructing a game where player 1 has moves $a_1, \ldots, a_m$, and player 2 has moves $b_1, \ldots, b_m$. In order to win, player 1 must match each move b_j, for $1 \leq j \leq m$, with move a_j. Obviously, player 1 can do this only in a semi-concurrent game of order $k \geq m$. Since semi-concurrent games of order 1 are concurrent games, the following corollary follows.

Corollary 1   *For* $\triangle \in \{\Diamond, \square\}$, $\mathcal{M} \in \{sure, almost\}$, *and* $k > 1$, *there is a game structure* G *and a subset of states* $R \subseteq S$ *such that*

$$win_1(G, \triangle R, \mathcal{M} \mid \Pi_1^c, \Pi_2^c, PrbS) \subsetneq win_1(G, \triangle R, \mathcal{M} \mid \vec{\Pi}_1^{k_1}, \Pi_2^c, PrbS).$$

3.3   Player Two Reachability Games

We now consider the case when player 2 has to reach a region R, while player 1 is able to gather information about her choice of moves using a semi-concurrent strategy. Again, for winning mode *sure*, the solution of the game coincides with that of concurrent games. Informally, if player 2 must ensure that *all* outcome plays reach R (as opposed to a set of outcome plays with measure 1), player 1 does not need to get information about player 2's choice of moves: he can just guess it. For mode *almost*, semi-concurrent reachability games of order k is solved using a predecessor operator $VAPre_2^k : 2^S \times 2^S \mapsto 2^S$, that plays the same role as the operator $APre_2$ for concurrent games: for $X \subseteq Y \subseteq S$, the set $VAPre_2^k(Y, X) \subseteq S$ consists of the states from which player 2 can ensure a positive probability of going to X in one round, while staying in Y.

For $s \in S$ *and* $X, Y \subseteq S$, *we have* $s \in VAPre_2^k(Y, X)$ *iff for all* k-*partitions* $B_1, \ldots, B_k$ *of* $Safe_2(s, Y, \Gamma_2(s))$, *there is* $j \in \{1, \ldots, k\}$ *such that:*

$$\forall a \in \Gamma_1(s). \exists b \in B_j. \delta(s, a, b) \cap X \neq \emptyset.$$

The idea is as follows: the best strategy for player 2 at a state $s \in S$ consists in playing uniformly at random all moves in $Safe_2(s, Y, \Gamma_2(s))$; the above definition ensures that, if $s \in VAPre_2^k(Y, X)$, then a transition to X happens with positive probability, regardless of the partition of $\Gamma_2(s)$ chosen by player 1's spy strategy. We are now ready to state the following:

Theorem 8 *For all game structures G, $R \subseteq S$, and $k > 1$, the following holds:*

1. *We have $win_2(G, \Diamond R, sure \mid \vec{\Pi}_1^k, \Pi_2^c, PrbS) = win_1(G, \Diamond R, sure \mid \Pi_1^c, \Pi_2^c, Prb)$; as in the case of concurrent games, the above sets can be computed in time $\mathcal{O}(|G|)$ by (2). There are are always memoryless and deterministic winning strategies.*
2. *We have $win_2(G, \Diamond R, almost \mid \vec{\Pi}_1^k, \Pi_2^c, PrbS) = \nu Y.\mu X.(R \cup VAPre_2^k(Y, X))$; the fixpoint can be computed in time $\mathcal{O}(|G|^k)$. There are always winning strategies that are memoryless, but there may not be deterministic winning strategies.*

Theorem 9 *For $\mathcal{M} \in \{sure, almost\}$, the game type $(\Diamond, 1, \mathcal{M} \mid \vec{\Pi}_1^k, \Pi_2^c, PrbS)$ is determined.*

If we consider the order $k > 0$ to be part of the input, we obtain the following result (compare with Theorem 4). The result is proved by reducing Vertex Cover to non-membership in $VAPre_2^k(\cdot, \cdot)$.

Theorem 10 *Given input (k, G, R), the membership problem in $win_2(G, \Box R, almost \mid \vec{\Pi}_1^k, \Pi_2^c, PrbS)$ is co-NP-complete.*

4   Team Games

In *team games*, one of the players does not consist of a single player, but rather of a team, composed by *members*. At each state, each team member can choose a move; the resulting team move is a tuple consisting of the choices of the members. We assume that each team member has complete information about the past history of the game, and we explicitly model the coordination used by the team members in choosing their moves. In particular, we consider both *non-communicating* and *communicating* strategies for team members. When using non-communicating strategies, the team members must select their moves simultaneously and independently not only from the opposing player, but also from each other. When using communicating strategies, the team members are allowed to communicate some information before choosing the moves.

Formally, a (m_1, m_2)-*team game structure* is a concurrent game structure $G = (S, M, \Gamma_1, \Gamma_2, \tau)$, where $M = \prod_{j=1}^{m_1} M_1^j \cup \prod_{j=1}^{m_2} M_2^j$, and for all $s \in S$, $\Gamma_1(s) = \prod_{j=1}^{m_1} \Gamma_1^j(s)$ and $\Gamma_2(s) = \prod_{j=1}^{m_2} \Gamma_2^j(s)$. Intuitively, the game is played by teams 1 and 2, and the team $i \in \{1, 2\}$ is composed by members $1, \dots, m_i$. At a state $s \in S$, the set $\Gamma_i^j(s)$ contains the moves that can be chosen by member $j \in \{1, \dots, m_i\}$. A move of team i is thus a tuple $\langle a_1, \dots, a_{m_i} \rangle$, consisting of the choices of the members. We let Π_1^c and Π_2^c be the sets of concurrent strategies for G, as defined in Section 2.

4.1   Team Games with Non-communicating Strategies

A *non-communicating team strategy* for team $i \in \{1, 2\}$ is a function $\bar{\pi}_i : Hist \mapsto \mathcal{D}(M)$ that prescribes for each game history a move distribution to be played by the team. Since we forbid communication between the members of the team, the distributions chosen by the team members must be mutually independent. Hence, we require that there are m_i functions $\pi_i^j : Hist \mapsto \mathcal{D}(M_i^j)$, for $1 \leq j \leq m_i$, such that $\bar{\pi}_i(\sigma)(\langle a_1, \dots, a_{m_i} \rangle) = \prod_{j=1}^{m_i} \pi_i^j(\sigma)(a_j)$ for all $\sigma \in Hist$ and all $\langle a_1, \dots, a_{m_i} \rangle \in \Gamma_i(s)$. We denote by $\bar{\Pi}_i$ the set of all team strategies for team $i \in \{1, 2\}$. In a (m_1, m_2)-team game we have $\bar{\Pi}_i \subseteq \Pi_i^c$, and the inclusion is strict whenever $m_i > 1$ and there is a state $s \in S$ with $|\Gamma_i(s)| > 1$. The probability measure $\mathrm{PrbT}_s^{\pi_1, \pi_2}$ for $s \in S$ and $\pi_1 \in \bar{\Pi}_1$, $\pi_2 \in \bar{\Pi}_2$ can then be defined in a straightforward way, yielding the family of (m_1, m_2)-team strategies $\langle \bar{\Pi}_1, \bar{\Pi}_2, \mathrm{PrbT} \rangle$.

A team game with non-communicating strategies (also called *non-communicating team game*) differs from a concurrent game because, at each state, each team must choose a probability distribution over moves that can be written as the product of the distributions chosen by the team members. Since deterministic distributions can always be written in product form (if team i wants to play tuple $\langle a_1, \dots, a_{m_i} \rangle$, each member $1 \leq j \leq m_i$ simply plays a_j), for the games where the existence of deterministic winning strategies is assured, the winning states of non-communicating team games coincide with the winning states of concurrent games.

Theorem 11   *For all $m_1, m_2 > 0$, all (m_1, m_2)-team game structures G, sets $R \subseteq S$, and teams $i \in \{1, 2\}$, the following assertions hold:*

1. *For $\mathcal{M} \in \{sure, almost\}$, we have $win_i(G, \Box R, \mathcal{M} \mid \bar{\Pi}_1, \bar{\Pi}_2, \mathrm{PrbT}) = win_i(G, \Box R, \mathcal{M} \mid \Pi_1^c, \Pi_2^c, \mathrm{Prb})$. There are always winning strategies that are memoryless and deterministic.*
2. *We have $win_i(G, \Diamond R, sure \mid \bar{\Pi}_1, \bar{\Pi}_2, \mathrm{PrbT}) = win_i(G, \Diamond R, sure \mid \Pi_1^c, \Pi_2^c, \mathrm{Prb})$. There are always winning strategies that are memoryless and deterministic.*

Corollary 2   *For $\mathcal{M} \in \{sure, almost\}$ and $i \in \{1, 2\}$, the game type $(\Box, i, \mathcal{M} \mid \bar{\Pi}_1, \bar{\Pi}_2, \mathrm{PrbT})$ is determined. The game type $(\Diamond, i, sure \mid \bar{\Pi}_1, \bar{\Pi}_2, \mathrm{PrbT})$ is also determined.*

Hence, the interesting problem in team games consists in solving reachability games with probability 1, where the winning strategies need randomization in the general case [dAHK98]. Such games can be solved using the predecessor operator $TAPre_1$, defined as follows. In the following, we call *cube* any set $C \in \prod_{j=1}^{m_1}(2^{M_1^j} \setminus \{\emptyset\})$.

> Given two sets $X, Y \subseteq S$ and a state $s \in S$, we say that $s \in TAPre_1(Y, X)$ if and only if there exists a cube $C \in \prod_{j=1}^{m_1}(2^{M_1^j} \setminus \{\emptyset\})$. such that:

$$\forall b \in \Gamma_2(s). (\forall a \in C. \delta(s, a, b) \subseteq Y \text{ and } \exists a \in C. \delta(s, a, b) \subseteq X).$$

Theorem 12   *For all $(2,1)$-team game structures G and sets $R \subseteq S$, we have $win_1(G, \Diamond R, almost \mid \bar{\Pi}_1, \bar{\Pi}_2, PrbT) = \nu Y.\mu X.(R \cup TAPre_1(Y, X))$, and membership in this fixpoint is an NP-complete problem. Moreover, there are always winning strategies that are memoryless, but there may not be deterministic winning strategies.*

In order to prove the NP-hardness, a non-trivial reduction is developed, transforming the classical 3-CNF-SAT problem to membership in $TAPre_1(\cdot, \cdot)$.

By means of a counterexample, the following can be shown.

Theorem 13   *For $i \in \{1, 2\}$, the game type $(\Diamond, i, almost \mid \bar{\Pi}_1, \bar{\Pi}_2, PrbT)$ is not determined.*

4.2   Team Games with Communication

In this section we consider the case in which members of the same team are allowed to communicate some information in each turn of the game. To simplify the notation, rather than considering arbitrary flows of information between team members, we consider a team composed of only two members. This special case suffices to capture the interesting features of the general case.

Formally, given a $(2,1)$-team game structure, a *communicating team strategy of order* $k > 0$ for team 1 is a function $\hat{\pi}_1^k : Hist \mapsto \mathcal{D}(M)$ subject to the following requirements. There are three functions $\pi_1^t : Hist(G) \times \Sigma_k^* \mapsto \mathcal{D}(\Sigma_k)$, $\pi_1^1 : Hist(G) \times \Sigma_k^* \mapsto \mathcal{D}(M_1^1)$, and $\pi_1^2 : Hist(G) \times \Sigma_k^* \mapsto \mathcal{D}(M_1^2)$. The function π_1^t represents the generation of random symbols; these symbols are then communicated to the team members, which then use the functions π_1^1 and π_1^2 to choose their moves, on the basis of the game history and of teh received symbols. Note that if we consider both the generation of symbols π_1^t and the choice π_1^i to be done by team member i, for $i \in \{1, 2\}$, then communication effectively takes place from team member i to team member $3 - i$. For $n \geq 0$, $\langle r_0, \dots, r_n \rangle \in \Sigma_k^*$ and $\langle s_0, \dots, s_{n+1} \rangle \in Hist$, we set

$$Pr_{s_0}^{\pi_1^t}(\langle r_0, \dots, r_n \rangle \mid \langle s_0, \dots, s_n \rangle) = \prod_{j=0}^n \pi_1^t(\langle s_0, \dots, s_j \rangle, \langle r_0, \dots, r_{j-1} \rangle)(r_j),$$

with the convention that $r_0, \dots, r_{-1} = \varepsilon$ (where ε denotes the empty string). Then, for all $n \geq 0$, all $\sigma = \langle s_0, \dots, s_n \rangle \in Hist_{s_0}$, all $a_1 \in M_1^1$, and all $a_2 \in M_1^2$ we define the overall team strategy $\hat{\pi}_1^k$ by

$$\hat{\pi}_1^k(\sigma)(\langle a_1, a_2 \rangle) = \sum_{\rho \in \Sigma_k^n} Pr_{s_0}^{\pi_1^t}(\rho \mid \sigma) \cdot \pi_1^1(\sigma, \rho)(a_1) \cdot \pi_1^2(\sigma, \rho)(a_2).$$

We denote by $\widehat{\Pi}_i^k$ the set of communicating team strategies of order k for team i, and by PrbTC the probability measure on Ω induced by $\widehat{\Pi}_1^k$ and Π_2^c, defined as for concurrent games. We prove that, for all $k > 1$, team 1 has a winning strategy if and only if player 1 has a winning strategy in the corresponding concurrent game.

Theorem 14   *For all $(2,1)$-team game structures G, sets $R \subseteq S$, and $k > 1$, we have $win_1(G, \Diamond R, almost \mid \widehat{\Pi}_1^k, \Pi_2^c, PrbTC) = win_1(G, \Diamond R, almost \mid \Pi_1^c, \Pi_2^c, Prb)$.*

This theorem implies that, from the point of view of winning reachability games with probability 1, being able to communicate 1 bit per round, or even just sharing 1 bit of random information, is as good as the ability to communicate an arbitrary amount of information.

We outline the idea of the proof for $k = 2$, the case for a one-bit channel. Consider the solution $Y^* = win_1(G, \Diamond R, almost \mid \Pi_1^c, \Pi_2^c, Prb) = \mu X.(R \cup APre_i(Y^*, X))$ of the concurrent reachability game. As remarked in Section 2.2, if the team members could coordinate perfectly, they could play a sequence of $|Y^*|$ moves that leads to R with positive probability, and that does not leave Y^* otherwise. The problem, here, is that the two team members cannot communicate perfectly. To communicate a sequence of $|Y^*|$ moves, they need $|Y^*| \cdot \lceil \log |M| \rceil$ bits. However, the two team members can play a *deterministic* strategy to stay in Y^*.

The winning strategy of the team thus consists in the alternation of two phases, a *planning* phase, and an *execution* phase. In the planning phase, which lasts $|Y^*| \cdot \lceil \log |M| \rceil$ rounds, the two team members play a deterministic strategy to stay in Y^*. In the meantime, the symbol generator generates $|Y^*| \cdot \lceil \log |M| \rceil$ random bits (notice that the bits are not visible to the adversary). In the subsequent execution phase, which lasts $|Y^*|$ rounds, the team members use the sequence of bits to coordinate their actions, and play at each $s \in Y^*$ all the moves in $Safe_1(s, Y^*, \Gamma_2(s))$ with strictly positive probability. It is easy to see that each cycle consisting in a planning and in an execution phase results in (i) either reaching R, or (ii) in remaining in Y^*, and outcome (i) occurs with positive probability. This leads to the result. The result can be easily extended to the case of more than two team members.

References

[AHK97]   R. Alur, T.A. Henzinger, and O. Kupferman. Alternating-time temporal logic. In *Proc. 38th IEEE Symp. Found. of Comp. Sci.*, pages 100–109. IEEE Computer Society Press, 1997.

[BL69]   J.R. Büchi and L.H. Landweber. Solving sequential conditions by finite-state strategies. *Trans. Amer. Math. Soc.*, 138:295–311, 1969.

[dAH00]   L. de Alfaro and T.A. Henzinger. Concurrent omega-regular games. In *Proc. 15th IEEE Symp. Logic in Comp. Sci.*, pages 141–154, 2000.

[dAHK98]   L. de Alfaro, T.A. Henzinger, and O. Kupferman. Concurrent reachability games. In *Proc. 39th IEEE Symp. Found. of Comp. Sci.*, pages 564–575. IEEE Computer Society Press, 1998.

[dAHM00]   L. de Alfaro, T.A. Henzinger, and F.Y.C. Mang. The control of synchronous systems. In *CONCUR 00: Concurrency Theory. 11th Int. Conf.*, volume 1877 of *Lect. Notes in Comp. Sci.*, pages 458–473. Springer-Verlag, 2000.

[dAHM01] L. de Alfaro, T.A. Henzinger, and F.Y.C. Mang. The control of synchronous systems part II. In *CONCUR 01: Concurrency Theory. 12th Int. Conf.*, volume 2154 of *Lect. Notes in Comp. Sci.*, pages 566–581. Springer-Verlag, 2001.

[EJ91] E.A. Emerson and C.S. Jutla. Tree automata, mu-calculus and determinacy (extended abstract). In *Proc. 32nd IEEE Symp. Found. of Comp. Sci.*, pages 368–377. IEEE Computer Society Press, 1991.

[FV97] J. Filar and K. Vrieze. *Competitive Markov Decision Processes*. Springer-Verlag, 1997.

[GH82] Y. Gurevich and L. Harrington. Trees, automata, and games. In *Proc. 14th ACM Symp. Theory of Comp.*, pages 60–65. ACM Press, 1982.

[GJ79] M.R. Garey and D.S. Johnson. *Computers and Intractability: A Guide to the Theory of NP-Completeness*. Freeman and Co., 1979.

[KV97] O. Kupferman and M.Y. Vardi. Synthesis with incomplete informatio. In *2nd International Conference on Temporal Logic*, pages 91–106, Manchester, July 1997.

[KV01] O. Kupferman and M.Y. Vardi. Synthesizing distributed systems. In *Proc. 16th IEEE Symp. on Logic in Computer Science*, July 2001.

[PR79] G.L. Peterson and J.H. Reif. Multiple-person alternation. In *Proc. 20th IEEE Symp. Found. of Comp. Sci.*, pages 348–363, 1979.

[PR90] A. Pnueli and R. Rosner. Distributed-reactive systems are hard to synthesize. In *Proc. 31th IEEE Symp. Found. of Comp. Sci.*, pages 746–757, 1990.

[Rei84] J.H. Reif. The compexity of two-player games of incomplete information. *Journal of Computer and System Sciences*, 29:274–301, 1984.

[Sha53] L.S. Shapley. Stochastic games. *Proc. Nat. Acad. Sci. USA*, 39:1095–1100, 1953.

[Tho95] W. Thomas. On the synthesis of strategies in infinite games. In *Proc. of 12th Annual Symp. on Theor. Asp. of Comp. Sci.*, volume 900 of *Lect. Notes in Comp. Sci.*, pages 1–13. Springer-Verlag, 1995.

[Wil91] D. Williams. *Probability With Martingales*. Cambridge University Press, 1991.

[Zie98] Wiesaw Zielonka. Infinite games on finitely coloured graphs with applications to automata on infinite trees. *Theoretical Computer Science*, 200:135–183, June 1998.

Impact of Local Topological Information on Random Walks on Finite Graphs

Satoshi Ikeda[1], Izumi Kubo[2], Norihiro Okumoto[3], and Masafumi Yamashita[4]

[1] Department of Computer Science,
Tokyo University of Agriculture and Technology,
Naka-cho 2-24-16, Koganei, Tokyo, 184-8588, Japan.
bisu@cc.tuat.ac.jp
[2] Department of Environmental Design, Faculty of Environmental Studies,
Hiroshima Institute of Technology,
2-1-1 Miyake, Saeki-ku Hiroshima 731-5193, Japan.
kubo@cc.it-hiroshima.ac.jp
[3] Financial Information Systems Division, Hitachi,Ltd.
890 Kashimada, Saiwai-ku, Kawasaki, Kanagawa, 212-8567, Japan.
okumoto@itg.hitachi.co.jp
[4] Department of Computer Science and Communication Engineering,
Kyushu University,
Hakozaki, Higashi-ku, Fukuoka 812-8581, Japan.
mak@csce.kyushu-u.ac.jp

Abstract. It is just amazing that both of the mean hitting time and the cover time of a random walk on a finite graph, in which the vertex visited next is selected from the adjacent vertices at random with the same probability, are bounded by $O(n^3)$ for any undirected graph with order n, despite of the lack of global topological information. Thus a natural guess is that a better transition matrix is designable if more topological information is available. For any undirected connected graph $G = (V, E)$, let $P^{(\beta)} = (p_{uv}^{(\beta)})_{u,v \in V}$ be a transition matrix defined by

$$p_{uv}^{(\beta)} = \frac{\exp\left[-\beta U(u, v)\right]}{\sum_{w \in N(u)} \exp\left[-\beta U(u, w)\right]} \text{ for } u \in V, \ v \in N(u),$$

where β is a real number, $N(u)$ is the set of vertices adjacent to a vertex u, $\deg(u) = |N(u)|$, and $U(\cdot, \cdot)$ is a potential function defined as

$$U(u, v) = \log\left(\max\left\{\deg(u), \deg(v)\right\}\right) \text{ for } u \in V, \ v \in N(u).$$

In this paper, we show that for any undirected graph with order n, the cover time and the mean hitting time with respect to $P^{(1)}$ are bounded by $O(n^2 \log n)$ and $O(n^2)$, respectively. We further show that $P^{(1)}$ is best possible with respect to the mean hitting time, in the sense that the mean hitting time of a path graph of order n, with respect to *any* transition matrix, is $\Omega(n^2)$.

J.C.M. Baeten et al. (Eds.): ICALP 2003, LNCS 2719, pp. 1054–1067, 2003.
© Springer-Verlag Berlin Heidelberg 2003

1   Introduction

Random walks on finite graphs are rich source of attractive researches both in applied mathematics and in computer science. Blom et al. [3, Chap. 12] surveyed works devoted to the cover time, which is the expected number of moves necessary for a random walk on a finite undirected connected graph $G = (V, E)$ to visit all vertices, where the vertex visited next is selected from the adjacent vertices at random with the same probability (see also [1,4,5,6,7,11,12,14]). The transition matrix $P = (p_{uv})_{u,v \in V} \in [0,1]^{V \times V}$ on $V \times V$ is hence given by

$$p_{uv} = \begin{cases} 1/\deg(u) & \text{if } v \in N(u), \\ 0 & \text{otherwise}, \end{cases}$$

where $N(u)$ is the set of vertices adjacent to a vertex u, and $\deg(u) = |N(u)|$. For this transition rule, Aldous [1] showed, for any graph with order n and size m, an upper bound $2m(n-1)(= O(n^3))$ on the cover time, and on the other hand a lower bound $\Omega(n^3)$ is obtained for a lollipop graph L_n shown in Fig. 1. A lollipop graph L_n is a complete graph of order $\lceil n/2 \rceil$ with a tail (i.e., a path graph) of length $\lfloor n/2 \rfloor$. Let s and t be the two endpoints of the tail as shown in Fig. 1. Then the mean hitting time from s to t, i.e., the expected number of moves necessary for a random walk starting at s to reach t, is $\Omega(n^3)$ [12]. Thus both of the mean hitting time and the cover time are $\Theta(n^3)$ in this sense.[1]

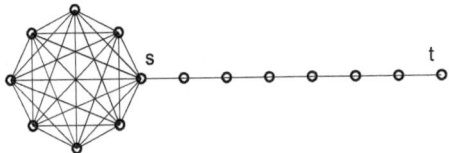

Fig. 1. A lollipop graph L_{15}.

Observing that P depends only on topological information $\deg(u)$ local to each vertex u, let us examine the following plausible claim:

> The cover time and the mean hitting time are (properly) reducible by using more topological information to construct a transition matrix.

The claim is indeed correct in many cases. For instance, consider a complete graph K_n, assuming that the whole G is available to construct an ideal transition matrix for G. Then there is a transition matrix Q that achieves the cover time $n-1$, since it has a Hamiltonian circuit,[2] while the cover time for K_n achieved

---

[1] For more notes, an upper bound $2m(n-1)(= O(n^3))$ on the cover time of any regular graph is first shown by Aleliunas et al. [2]. Both of the cover time and the mean hitting time are shown to be at most $4n^3/27$ [4,7].

[2] Let $\{0, 1, \ldots, n-1\}$ be the vertex set of K_n and define $Q = (q_{ij})$ by $q_{ij} = 1$ if $j \equiv i + 1 \pmod{n}$, and $q_{ij} = 0$ otherwise. Then the cover time for K_n with respect to Q is obviously $n-1$.

by P is $O(n \log n)$. However, there are of course some other cases in which even the complete information G does not help in reducing the cover time.

This paper shows that the bounds $\Theta(n^3)$ on the mean hitting time and the cover time are reducible to $\Theta(n^2)$ and $\Theta(n^2 \log n)$ respectively, in the sense mentioned in the first paragraph, if the topological information on the adjacent vertices is available. For any $\beta \in \mathbf{R}$, let $P^{(\beta)} = (p_{uv}^{(\beta)})_{u,v \in V}$ be a transition matrix defined by

$$
p_{uv}^{(\beta)} = \begin{cases} \dfrac{\exp\left[-\beta U(u,v)\right]}{\sum_{w \in N(u)} \exp\left[-\beta U(u,w)\right]} & \text{if } v \in N(u), \\ 1 - \displaystyle\sum_{w \in N(u)} p_{uw}^{(\beta)} & \text{if } u = v, \\ 0 & \text{otherwise,} \end{cases}
$$

which is known as the Gibbs distribution with respect to a local potential U in statistical mechanics. In this paper, we adopt

$$
U(u,v) = \log\left(\max\left\{\deg(u), \deg(v)\right\}\right).
$$

Observe that $P^{(\beta)}$ depends only on the topological information on $N(u)$, and that $P^{(0)} = P$.

We summarize our results.

1. For any transition matrix, the maximum mean hitting time (and hence the cover time) of a path graph with order n is $\Omega(n^2)$.
2. For $P^{(\beta)}$, the maximum mean hitting time is $O(n^{1+\beta})$ if $\beta \geq 1$, and $O(n^{3-\beta})$ if $\beta < 1$. The maximum mean hitting time of any graph with order n is hence always $O(n^2)$ for $P^{(1)}$, which means that $P^{(1)}$ is best possible with respect to the mean hitting time.
3. For $P^{(\beta)}$, the cover time is $O(n^{\beta+1} \log n)$ if $\beta \geq 1$, $O(n^{3-\beta} \log n)$ if $0 < \beta \leq 1$, and $O(n^{3-\beta})$ if $\beta \leq 0$. The cover time of any graph with order n is hence always $O(n^2 \log n)$ for $P^{(1)}$. There is still a possible gap from the lower bound $\Omega(n^2)$ in Item 1.
4. For $P^{(\beta)}$ with $\beta \geq 1$, the cover time of a "glitter star" S_n given in Fig. 3 is $\Omega(n^{1+\beta} \log n)$, i.e., for $P^{(\beta)}$ with $\beta \geq 1$, the cover time is $\Theta(n^{1+\beta} \log n)$.

For the sake of generality, we analyze random walks in a more general setting.

2   Preliminaries

Suppose that $G = (V, E)$ is a finite, undirected, simple connected graph with the order $n = |V|$ and the size $m = |E|$. For $u \in V$, by $N(u) = \left\{v : \{u, v\} \in E\right\}$ we denote the set of vertices adjacent to u. Note that $v \in N(u)$ iff $u \in N(v)$. The number of adjacent vertices, denoted by $\deg(u) = |N(u)|$, is called the degree of $u \in V$.

Let $\Omega = V^{\mathbf{N} \cup \{0\}}$ be the set of all infinite sequences of vertices, where $\mathbf{N}$ is the set of natural numbers. For $\omega = (\omega_0, \omega_1, \cdots) \in \Omega$, the $(i+1)$-st element w_i

is denoted by $X_i(\omega)$ for $i \geq 0$. By $\mathcal{M}(\Omega)$ we denote the space of the Markov measures on Ω. Put $\mu \in \mathcal{M}(\Omega)$ with an initial distribution (vector) $\mathbf{q} = (q_v) \in [0,1]^{V \times V}$ and a transition matrix $Q = (q_{uv}) \in (0,1]^{V \times V}$. That is, for $\omega = (\omega_0, \omega_1, \cdots) \in \Omega$, $t \in \mathbf{N} \cup \{0\}$,

$$X_t : \Omega \to V, \quad X_t(\omega) = \omega_t,$$

$$\sum_{v \in V} q_v = 1, \quad \mu(X_0(\omega) = v) = q_v \quad \text{for any } v \in V,$$

$$\sum_v q_{uv} = 1 \quad \text{for any } u \in V,$$

and for $u, v, x_0, x_1, \cdots, x_i \in V$ and $i \in \mathbf{N} \cup \{0\}$

$$\mu(X_{i+1}(\omega) = v | X_0(\omega) = x_0, X_1(\omega) = x_1, \cdots, X_i(\omega) = x_i = u)$$
$$= \mu(X_{i+1}(\omega) = v | X_i(\omega) = u) = q_{uv}.$$

As we are analyzing random walks on graph $G = (V, E)$, we assume without loss of generality that $q_{uv} > 0$ if $v \in N(u)$, and $q_{uv} = 0$ if $v \notin N(u) \cup \{u\}$. The space of Markov measures that meet the above requirement is denoted by $\mathcal{M}^+(\Omega)$, i.e.,

$$\mathcal{M}^+(\Omega) = \left\{ \mu \in \mathcal{M}(\Omega) : q_{uv} > 0 \text{ if } v \in N(u) \text{ and } q_{uv} = 0 \text{ if } v \notin N(u) \cup \{u\} \right\}.$$

To define the edge cost, let us now introduce a cost matrix $K = (k_{uv}) \in [0, \infty)^{V \times V}$. For $\omega \in \Omega$, put

$$n(\omega) = \inf \left\{ i \in \mathbf{N} : \{X_0(\omega), X_1(\omega), \cdots, X_i(\omega)\} = V \right\}.$$

If ω is an infinite legal token circulation on G, $n(\omega)$ denotes the minimum number of token moves necessary to visit all the vertices in V. We are interested in the circulation cost $k(\omega)$ incurred before the token visits all the vertices, i.e.,

$$k(\omega) = \sum_{i=0}^{n(\omega)-1} k_{X_i(\omega)X_{i+1}(\omega)}.$$

For any connected graph $G = (V, E)$, $\mu \in \mathcal{M}^+(\Omega)$, cost matrix K and $u, v \in V$, we define *the weighted mean hitting time* $H_\mu^{G,K}(u, v)$ from u to v with respect to μ and K by

$$H_\mu^{G,K}(u, v) = E_\mu \left[\sum_{i=0}^{t(\omega,v)-1} k_{X_i(\omega)X_{i+1}(\omega)} \middle| X_0(\omega) = u \right],$$

where

$$t(\omega, v) = \inf \left\{ i \geq 1 : X_i(\omega) = v \right\}.$$

In particular, $\max\limits_{u,v\in V} H_\mu^{G,K}(u,v)$ is called the *maximum weighted mean hitting time* of G with respect to μ and K. By the reason of asymmetry either of the graph G or of the cost matrix K, $H_\mu^{G,K}(u,v) \neq H_\mu^{G,K}(v,u)$ may hold. Finally, we define the *weighted cover time* $C_\mu(G,K)$ of G with respect to μ and K by

$$C_\mu(G,K) = \max_{u\in V} C_\mu(G,K,u), \quad C_\mu(G,K,u) = \mathbf{E}_\mu\Big[k(\omega)\big|X_0(\omega) = u\Big].$$

Let $Q = (q_{uv})_{u,v\in V}$ be a transition matrix for a Markov measure $\mu \in \mathcal{M}^+(\Omega)$ and $\pi = (\pi_v)_{v\in V}$ be its stationary distribution vector. Since

$$H_\mu^{G,K}(u,v) = \sum_{w\in V} q_{uw}(k_{uw} + H_\mu^{G,K}(w,v)) - q_{uv}H_\mu^{G,K}(v,v), \qquad (2.1)$$

we get

$$\sum_{u\in V} \pi_u H_\mu^{G,K}(u,v) = \sum_{w\in V} \pi_w H_\mu^{G,K}(w,v) + \sum_{u\in V}\sum_{w\in V} \pi_u q_{uw} k_{uw} - \pi_v H_\mu^{G,K}(v,v)$$

by the equality $\sum\limits_{u\in V} \pi_u q_{uv} = \pi_v$, which implies that

$$\pi_v H_\mu^{G,K}(v,v) = \overline{k} \quad \text{for all}\ \ v \in V, \qquad (2.2)$$

where

$$\overline{k} = \overline{k}(\mu, K, G) = \sum_{u\in V}\sum_{v\in V} \pi_u q_{uv} k_{uv}. \qquad (2.3)$$

We call $\overline{k} = \overline{k}(\mu, K, G)$ the *weighted average cost* with respect to μ and K, which is the mean value of $k_{X_0(\omega),X_1(\omega)}$ with respect to the stationary measure of $\{X_t\}$. By (2.1) and (2.2), we get

$$H_\mu^{G,K}(u,v) \leq \overline{k}(q_{vu}\pi_v)^{-1} \quad \text{for any}\ u \in V,\ v \in N(u). \qquad (2.4)$$

Let K^0 be a cost matrix such that $k_{uv} = 1$ if $\{u,v\} \in E$ or $u = v$. Then $\overline{k} = 1$ for K^0. Let $\pi^{(\beta)} = (\pi_v^\beta) \in [0,1]^V$ be the stationary distribution of transition matrix $P^{(\beta)} = (p_{uv}^{(\beta)})_{u,v\in V}$, that is, $P^{(\beta)}\pi^{(\beta)} = \pi^{(\beta)}$, $\sum_{v\in V}\pi_v^\beta = 1$, and $\pi_v^\beta \geq 0$ for all $v \in V$. Through out this paper, $\nu^{(\beta)}$ denotes the Markov measure on Ω with $\pi^{(\beta)}$ as the initial distribution and $P^{(\beta)}$ as the transition matrix.

3   Lower Bounds for Path Graph

This section proves a lower bound $\Omega(n^2)$ on the maximum mean hitting time and the cover time of a path graph of order n for any transition matrix.

Theorem 1. *Let $P_n = (V,E)$ be a path graph with order n. Then for any $\mu \in \mathcal{M}^+(\Omega)$ and cost matrix $K = (k_{uv})_{u,v\in V}$,*

$$\frac{1}{2}(\overline{k}n^2 - \sum_{w\in V} q_{vw}k_{vw}) \leq C_\mu(P_n, K) \leq 2 \max_{u,v\in V} H_\mu^{P_n,K}(u,v),$$

where $\overline{k} = \overline{k}(\mu, K, P_n)$ is defined by (2.3).

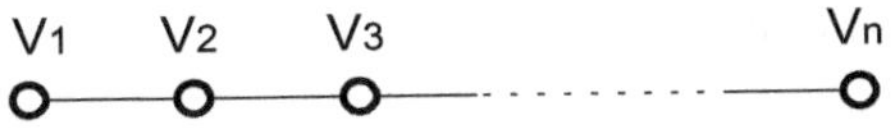

Fig. 2. A path graph with order n.

Proof. Suppose that $P_n = (V, E)$ is a path graph given in Fig. 2.

Let $Q = (q_{uv})_{u,v \in V}$ be any transition matrix for $\mu \in \mathcal{M}^+(\Omega)$ and $\pi = (\pi_v)_{v \in V}$ be its stationary distribution. Then by (2.2) we have

$$\pi_v H_\mu^{P_n, K}(v, v) = \overline{k} \quad \text{for any } v \in V. \tag{3.1}$$

By the definitions of $H_\mu^{P_n, K}$ and $C_\mu(P_n, K)$,

$$\max_{s,t \in V} H_\mu^{P_n, K}(s, t) = \max \left\{ H_\mu^{P_n, K}(v_1, v_n), H_\mu^{P_n, K}(v_n, v_1) \right\},$$

$$\max_{s,t \in V} H_\mu^{P_n, K}(s, t) \leq C_\mu(P_n, K),$$

and

$$C_\mu(P_n, K) \leq H_\mu^{P_n, K}(v_1, v_n) + H_\mu^{P_n, K}(v_n, v_1)$$

for any $\mu \in \mathcal{M}^+(\Omega)$ and cost matrix K. Hence

$$\frac{1}{2} \left\{ H_\mu^{P_n, K}(v_1, v_n) + H_\mu^{P_n, K}(v_n, v_1) \right\} \leq C_\mu(P_n, K) \leq 2 \max_{u,v \in V} H_\mu^{P_n, K}(u, v). \tag{3.2}$$

By putting $u = v$ in (2.1),

$$H_\mu^{P_n, K}(v, v) = \sum_{w \in N(v)} q_{vw} H_\mu^{P_n, K}(w, v) + \sum_{w \in V} q_{vw} k_{vw} \quad \text{for any } v \in V.$$

Thus we have

$$\overline{k} \sum_{v \in V} \pi_v^{-1} = \sum_{v \in V} H_\mu^{P_n, K}(v, v) = \sum_{v \in V} \left(\sum_{w \in N(v)} q_{vw} H_\mu^{P_n, K}(w, v) + \sum_{w \in V} q_{vw} k_{vw} \right)$$

$$\leq \sum_{v \in V} \sum_{w \in N(v)} H_\mu^{P_n, K}(w, v) + \sum_{v \in V} \sum_{w \in V} q_{vw} k_{vw},$$

by (3.1). By Markov property,

$$\sum_{v \in V} \sum_{w \in N(v)} H_\mu^{P_n, K}(w, v) = H_\mu^{P_n, K}(v_1, v_n) + H_\mu^{P_n, K}(v_n, v_1),$$

which implies that

$$H_\mu^{P_n, K}(v_1, v_n) + H_\mu^{P_n, K}(v_n, v_1) \geq \overline{k} \sum_{v \in V} \pi_v^{-1} - \sum_{w \in V} q_{vw} k_{vw} \geq \overline{k} n^2 - \sum_{w \in V} q_{vw} k_{vw}.$$

Together with (3.2), we have

$$\frac{1}{2}(\bar{k}n^2 - \sum_{w \in V} q_{vw}k_{vw}) \leq C_\mu(P_n, K) \leq 2 \max_{u,v \in V} H_\mu^{P_n, K}(u, v).$$

$$\square$$

By Theorem 1, we have the following lower bounds.

Corollary 1. *For any* $\mu \in \mathcal{M}^+(\Omega)$,

1. $\max\limits_{u,v \in V} H_\mu^{P_n, K^0}(u, v) = \Omega(n^2)$, *and*
2. $C_\mu(P_n, K^0) = \Omega(n^2)$.

4   Upper Bounds

In Section 3, we showed that both of the mean hitting time and the cover time are bounded from below by $\Omega(n^2)$ for any $\mu \in \mathcal{M}^+(\Omega)$. As mentioned, Aldous [1] showed that for $\nu^{(0)} \in \mathcal{M}^+(\Omega)$, both of them are bounded from above by $O(n^3)$. In this section, we show that for $\nu^{(1)} \in \mathcal{M}^+(\Omega)$, they are bounded by $O(n^2)$ and $O(n^2 \log n)$, respectively. This implies that $\nu^{(1)}$ is best possible with respect to the hitting time. Let us start this section with associating the weighted cover time with the weighted mean hitting time. The following theorem generalizes [11] and [12].

Theorem 2. *Let* $G = (V, E)$ *be a graph. Then for any* $\mu \in \mathcal{M}^+(\Omega)$ *and cost matrix* K,

$$H_{n-1} \min_{u \neq v \in V} H_\mu^{G,K}(u, v) \leq C_\mu(G, K) \leq H_{n-1} \max_{u \neq v \in V} H_\mu^{G,K}(u, v), \qquad (4.1)$$

where H_n *denotes the n-th harmonic number, i.e.,* $H_n = \sum\limits_{i=1}^{n} i^{-1}$.

Proof. Let S_V be the set of all permutations of V and ν be the uniform measure on S_V. For $\pi = (v_1, v_2, \ldots, v_n) \in S_V$, we put $\sigma_j(\pi) = v_j$.

For a fixed $u \in V$, let ν_u be the conditional measure of ν conditioned by the set $\{\pi : \sigma_1(\pi) = u\}$ and μ_u be the conditional measure of μ conditioned by the set $\{\omega : X_0(\omega) = u\}$. Let P_u be the product measure of μ_u and ν_u. Define $\tau(\omega, v)$, $T_j(\omega, \pi)$ and $\ell_j(\omega, \pi)$ by

$$\tau(\omega, v) = \inf\{t \geq 0 : X_t(\omega) = v\}, \quad T_j(\omega, \pi) = \max_{i \leq j} \tau(\omega, \sigma_i(\pi)) \quad \text{and} \quad \ell_j(\omega, \pi) = X_{T_j(\omega,\pi)}(\omega),$$

respectively. Then obviously, $T_{j-1}(\omega, \pi) < T_j(\omega, \pi)$ holds, iff $\ell_{j-1}(\omega, \pi) \neq \ell_j(\omega, \pi)$. Therefore we have that

$$P_u(\ell_{j-1}(\omega,\pi) \neq \ell_j(\omega,\pi)) = P_u(T_{j-1}(\omega,\pi) < T_j(\omega,\pi))$$
$$= P_u(\tau(\omega,\sigma_i(\pi)) < \tau(\omega,\sigma_j(\pi)),\ 2 \leq i < j)$$
$$= \int_\Omega \nu_u(\{\pi : \tau(\omega,\sigma_i(\pi)) < \tau(\omega,\sigma_j(\pi)),\ 2 \leq i < j\}) d\mu_u(\omega)$$
$$= \int_\Omega \frac{(n-1)!}{(j-1)!(n-j)!} \times \frac{(j-2)!(n-j)!}{(n-1)!} d\mu_u(\omega) = \frac{1}{j-1}$$

by Fubini's theorem.

Since $T_n(\omega,\pi) = n(\omega)$ holds for any $\pi \in S_n$, we see that

$$C_\mu(G,K,v)$$
$$= E_{P_u}\left[\sum_{s=0}^{T_n(\omega,\pi)-1} k_{X_s(\omega)X_{s+1}(\omega)}\right]$$
$$= \sum_{j=2}^n E_{P_u}\left[\sum_{s=0}^{T_j(\omega,\pi)-1} k_{X_s(\omega)X_{s+1}(\omega)} - \sum_{s=0}^{T_{j-1}(\omega,\pi)-1} k_{X_s(\omega)X_{s+1}(\omega)}\right]$$
$$= \sum_{j=2}^n E_{P_u}\left[\sum_{s=T_{j-1}(\omega,\pi)}^{T_j(\omega,\pi)-1} k_{X_s(\omega)X_{s+1}(\omega)} : \ell_{j-1}(\omega,\pi) \neq \ell_j(\omega,\pi)\right]$$
$$= \sum_{j=2}^n \sum_{\xi \neq \eta \in V} E_{P_u}\left[\sum_{s=T_{j-1}(\omega,\pi)}^{T_j(\omega,\pi)-1} k_{X_s(\omega)X_{s+1}(\omega)} : \ell_{j-1}(\omega,\pi) = \xi, \ell_j(\omega,\pi) = \eta\right].$$
$$= \sum_{j=2}^n \sum_{\xi \neq \eta \in V} H_\mu^{G,K}(\xi,\eta) P_u(\ell_{j-1}(\omega,\pi) = \xi, \ell_j(\omega,\pi) = \eta).$$
$$\leq \max\{H_\mu^{G,K}(\xi,\eta) : \xi \neq \eta \in V\} \sum_{j=2}^n \sum_{\xi \neq \eta} P_u(\ell_{j-1}(\omega,\pi) = \xi, \ell_j(\omega,\pi) = \eta)$$
$$= \max\{H_\mu^{G,K}(\xi,\eta) : \xi \neq \eta \in V\} \sum_{j=2}^n P_u(\ell_{j-1}(\omega,\pi) \neq \ell_j(\omega,\pi))$$

Thus we showed the right-hand side inequality of (4.1). The left-hand side inequality can be shown similarly. $\square$

Theorem 2 can be generalized further to obtain the following theorem. For a given $G = (V,E)$, cost matrix K, $\mu \in M^+(\Omega)$ and $V' \subseteq V$, we define the weighted cover time $C_\mu(G,V',K)$ *with respect to* V' by

$$C_\mu(G,V',K) = \max_{u \in V'} C_\mu(G,V',K,u),$$

where for $u \in V'$,

$$C_\mu(G,V',K,u) = E_\mu[k_{V'}(\omega)|X_0(\omega) = u].$$

Here $k_{V'}(\omega)$ is defined by

$$k_{V'}(\omega) = \sum_{i=0}^{n_{V'}(\omega)-1} k_{x_i(\omega)x_{i+1}(\omega)},$$

where

$$n_{V'}(\omega) = \inf\Big\{i \in \mathbf{N} \mid \{X_0(\omega), X_1(\omega), \cdots, X_{i-1}(\omega)\} = V'\Big\}.$$

Theorem 3. *Let $G = (V, E)$ be a graph and $V' \subseteq V$. Then for any $\mu \in \mathcal{M}^+(\Omega)$ and cost matrix K,*

$$H_{n'-1} \min_{u \neq v \in V'} H_\mu^{G,K}(u, v) \leq C_\mu(G, V', K) \leq H_{n'-1} \max_{u \neq v \in V'} H_\mu^{G,K}(u, v),$$

where $n' = |V'|$.

Let $\nu \in \mathcal{M}^+(\Omega)$ be a Markov measure with respect to a transition matrix $Q = (q_{uv})_{u,v \in V}$. By the definition of $\mathcal{M}^+(\Omega)$, $q_{vv} > 0$ may hold for some $v \in V$. We define $\hat{Q} = (\hat{q}_{uv})_{u,v \in V}$ by

$$\hat{q}_{uv} = \begin{cases} 0 & \text{if } u = v, \\ q_{uv}(1 + q_{vv})(1 - q_{vv})^{-1} & \text{otherwise.} \end{cases}$$

Let $\hat{\nu}$ be a Markov measure with respect to $\hat{Q}$. Then

$$H_{\hat{\nu}}^{G,K}(u, v) \leq H_\nu^{G,K}(u, v) \quad \text{for any } u \neq v \in V$$

holds[13]. We thus have the following lemma.

Lemma 1. *For a given undirected graph $G = (V, E)$ and cost matrix K, let $\nu, \mu \in \mathcal{M}^+(\Omega)$ be two Markov measures with respect to transition matrices $A = (a_{uv})_{u,v \in V}$ and $B = (b_{uv})_{u,v \in V}$, respectively. If there is a set of real numbers $\{c(u) \in (0, 1] : u \in V\}$ such that $a_{uv}c(u) = b_{uv}$ for any $u \in V, v \in N(u)$, then*

$$H_\nu^{G,K}(u, v) \leq H_\mu^{G,K}(u, v) \quad \text{for any } u \neq v.$$

Hence

$$\Big\{\min_{w \in V} c(w)\Big\} H_\mu^{G,K^0}(u, v) \leq H_\nu^{G,K^0}(u, v) \leq \Big\{\max_{w \in V} c(w)\Big\} H_\mu^{G,K^0}(u, v)$$

hold for any $u \neq v \in V$.

We are now ready to introduce our main results.

Theorem 4. *Let $G = (V, E)$ be a graph. Then the following two statements hold for any cost matrix K:*

(a) $\displaystyle\max_{u,v\in V} H^{G,K}_{\nu^{(\beta)}}(u,v) \le \begin{cases} 2\overline{k}_\beta n^\beta (3n-4) & \text{for } \beta \ge 1, \\ 2\overline{k}_\beta n^{2-\beta}(3n-4) & \text{for } 0 < \beta \le 1, \\ \overline{k}_\beta n^{2-\beta}(n-1) & \text{for } \beta \le 0. \end{cases}$

(b) $C_{\nu^{(\beta)}}(G,K) \le \begin{cases} 2\overline{k}_\beta n^\beta (3n-4)H_{n-1} & \text{for } \beta \ge 1, \\ 2\overline{k}_\beta n^{2-\beta}(3n-4)H_{n-1} & \text{for } 0 < \beta \le 1, \\ \overline{k}_\beta n^{2-\beta}(2n-3) & \text{for } \beta \le 0. \end{cases}$

Here $\overline{k}_\beta \equiv k(\hat{\nu}^{(\beta)}, K, G) = \dfrac{1}{n}\displaystyle\sum_{u\in V}\sum_{v\in V}\hat{p}^{(\beta)}_{uv}k_{uv}$ *and* $\hat{\nu}^{(\beta)}$ *is a Markov measure with*

respect to a symmetrical transition matrix $\hat{P}^{(\beta)} = (\hat{p}^{(\beta)}_{uv})_{u,v\in V}$ *defined by*

$$\hat{p}^{(\beta)}_{uv} = \begin{cases} \exp[-\beta U(u,v)] & \text{if } v \in N(u), \\ 1 - \sum_{w\in N(u)}\hat{p}^{(\beta)}_{uw} & \text{if } u = v, \\ 0 & \text{otherwise}, \end{cases}$$

for $\beta \ge 1$ *and*

$$\hat{p}^{(\beta)}_{uv} = \begin{cases} n^{\beta-1}\exp[-\beta U(u,v)] & \text{if } v \in N(u), \\ 1 - \sum_{w\in N(u)}\hat{p}^{(\beta)}_{uw} & \text{if } u = v, \\ 0 & \text{otherwise}, \end{cases}$$

for $\beta \le 1$.

Proof. By definition, $\hat{\nu}^{(\beta)}$'s stationary distribution $\hat{\pi}^{(\beta)} = (\hat{\pi}^{(\beta)}_v)_{v\in V}$ is uniform, that is, $\hat{\pi}^{(\beta)}_v = 1/n$ for all $v \in V$ and $\beta \in \mathbf{R}$.

Assume first $\beta \ge 1$. By (2.4) and Lemma 1,

$$H^{G,K}_{\nu^{(\beta)}}(u,v) \le H^{G,K}_{\hat{\nu}^{(\beta)}}(u,v) \le \overline{k}_\beta n \max\{\deg(u), \deg(v)\}^\beta,$$

which implies that

$$H^{G,K}_{\nu^{(\beta)}}(u,v) \le \overline{k}n^\beta \max\{\deg(u), \deg(v)\}. \tag{4.2}$$

We next assume $\beta \le 1$. Again by (2.4) and Lemma 1,

$$H^{G,K}_{\nu^{(\beta)}}(u,v) \le H^{G,K}_{\hat{\nu}^{(\beta)}}(u,v) \le \overline{k}n^{2-\beta}\max\{\deg(u), \deg(v)\}^\beta \quad \text{for } v \in N(u).$$

Since

$$n^{2-\beta}\max\{\deg(u), \deg(v)\}^\beta \le \begin{cases} n^{2-\beta}\max\{\deg(u), \deg(v)\} & \text{for } 0 < \beta \le 1, \\ n^{2-\beta} & \text{for } \beta \le 0, \end{cases}$$

together with (4.2), we get for $v \in N(u)$

$$H^{G,K}_{\nu^{(\beta)}}(u,v) \le \begin{cases} \overline{k}_\beta n^\beta \max\{\deg(u), \deg(v)\} & \text{for } \beta \ge 1, \\ \overline{k}_\beta n^{2-\beta}\max\{\deg(u), \deg(v)\} & \text{for } 0 < \beta \le 1, \\ \overline{k}_\beta n^{2-\beta} & \text{for } \beta \le 0. \end{cases} \tag{4.3}$$

Now, we evaluate the weighted mean hitting time. For given $u, v \in V$ with $u \neq v$, we choose a shortest path $u = v_0, v_1, \cdots, v_l = v$ satisfying

$$\left(N(v_i) \cup \{v_i\} \right) \bigcap \left(N(v_j) \cup \{v_j\} \right) = \emptyset \tag{4.4}$$

for $1 \leq i < i+2 < j \leq l$. The existence of such a path can be shown as follows. Suppose that for i and j with $j > i+2$, there exists a w in $N(v_i) \cap N(v_j)$, then we can take a shortcut

$$u = v_0, v_1, \cdots, v_i, w, v_j, \cdots, v_l = v,$$

whose length is less than l. Applying this procedure finitely many times, we get a path satisfying (4.4). For the path satisfying condition (4.4), we have

$$\sum_{i=0}^{l} \deg(v_i) \leq 3n - 4. \tag{4.5}$$

Since $H_\mu^{G,K}(x, y) \leq H_\mu^{G,K}(x, z) + H_\mu^{G,K}(z, y)$ for any $x, y, z \in V$ and $\mu \in \mathcal{M}^+(\Omega)$,

$$H_{\nu(\beta)}^{G,K}(u, v) \leq \sum_{i=0}^{l-1} H_{\nu(\beta)}^{G,K}(v_i, v_{i+1}) \tag{4.6}$$

holds. Together with (4.3),(4.5) and (4.6), we get

$$H_{\nu(\beta)}^{G,K}(u, v) \leq \begin{cases} 2\overline{k}_\beta n^\beta (3n - 4) & \text{for } \beta \geq 1, \\ 2\overline{k}_\beta n^{2-\beta}(3n - 4) & \text{for } 0 < \beta \leq 1, \\ \overline{k}_\beta n^{2-\beta}(n - 1) & \text{for } \beta \leq 0, \end{cases}$$

which imply (a).

As for (b), the inequality for $\beta > 0$ holds by (a) and Theorem 2. For $\beta \leq 0$, since there is a path of length $2n - 3$ that visits all vertices,

$$H_{\nu(\beta)}^{G,K}(u, v) \leq \overline{k}_\beta n^{2-\beta}(2n - 3)$$

by (4.3). Again by Theorem 2, we have inequality (b).   □

Recall that with respect to $P^{(0)}$, both of the mean hitting time and the cover time are $O(n^3)$. Theorem 4 generalizes this fact: With respect to $P^{(\beta)}$, both of them are $O(n^{3-\beta})$ if $\beta \leq 0$. A more important conclusion is that both of the mean hitting time and the cover time achieve the minimum values when $\beta = 1$. Since $\overline{k} = 1$ for K^0, we have the following corollary.

Corollary 2. *Let $G = (V, E)$ be a graph. Then the following two statements hold for any $\beta \in \mathbb{R}$:*

(a) $\displaystyle \max_{u,v \in V} H_{\nu(\beta)}^{G,K^0}(u, v) = \begin{cases} O(n^{1+\beta}) \text{ for } \beta \geq 1, \\ O(n^{3-\beta}) \text{ for } \beta \leq 1. \end{cases}$

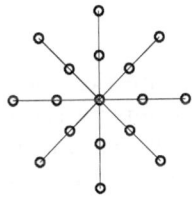

Fig. 3. A glitter star S_{17}.

(b) $C_{\nu(\beta)}(G, K^0) = \begin{cases} O(n^{\beta+1} \log n) \text{ for } \beta \geq 1, \\ O(n^{3-\beta} \log n) \text{ for } 0 < \beta \leq 1, \\ O(n^{3-\beta}) \quad\quad \text{ for } \beta \leq 0. \end{cases}$

We finally show that the cover time of a glitter star S_n introduced in Section 1 is $\Omega(n^{\beta+1} \log n)$, when $\beta \geq 1$.

Theorem 5. *For any $n = 2m + 1$, $m \in \mathbf{N}$ with $m \geq 3$, and $\beta \geq 1$,*

$$C_{\nu(\beta)}(S_n, K^0) = \Theta(n^{1+\beta} \log n).$$

Proof. By Corollary 2, it is sufficient to show

$$C_{\nu(\beta)}(S_n, K^0) = \Omega(n^{\beta+1} \log n).$$

Let V and V_O be the set of vertices of S_n and the set of pendant vertices of S_n. Hence $|V_O| = m$. Since $V_O \subseteq V$,

$$C_{\nu(\beta)}(S_n, K^0) \geq C_{\nu(\beta)}(S_n, V_O, K^0). \tag{4.7}$$

On the other hand, for any $u, v \in V_O$ with $u \neq v$, we can easily calculate that

$$H_{\nu(\beta)}^{S_n, K^0}(u, v) = 2(n+1)(n^\beta + 1) + \frac{2n}{n-1}.$$

Hence

$$C_{\nu(\beta)}(S_n, V_O, K^0) \geq \left\{ 2(n+1)(n^\beta + 1) + \frac{2n}{n-1} \right\} H_{m-1}$$

by Theorem 3. Together with (4.7), we have

$$C_{\nu(\beta)}(S_n, K^0) = \Omega(n^{1+\beta} \log n).$$

$\square$

By Theorem 1 and Corollary 2, $\nu^{(1)}$ is best possible with respect to the mean hitting time. As for the cover time, by Theorem 5, there is still a gap from the lower bound given by Theorem 1, as long as we adopt $\beta \geq 1$.

5   Conclusion

Random walks on finite graphs are rich source of attractive researches both in applied mathematics and in computer science. Despite of the lack of global topological information, both of the maximum mean hitting time and the cover time of a (conventional) random walk with respect to transition matrix $P = P^{(0)}$ can be bounded by $O(n^3)$. Hence a natural guess is that a better transition matrix is designable if more topological information is available. This paper showed that the guess is correct by investigating the maximum mean hitting time and the cover time with respect to $P^{(1)}$; the maximum mean hitting time of any graph with respect to $P^{(1)}$ is bounded by $O(n^2)$. Since the maximum mean hitting time of a path graph is shown to be $\Omega(n^2)$ for any transition matrix, $P^{(1)}$ is the best transition matrix as order, with respect to the mean hitting time. We also showed that the cover time of any graph with respect to $P^{(1)}$ is bounded by $\Theta(n^2 \log n)$.

There are many problems left unsolved. There is still a possible gap from a known lower bound $\Omega(n^2)$ on the cover time for a path graph. Looking for a matching bound seems to be challenging. We only investigated "universal" bounds on the mean hitting time and the cover time. Perhaps, there are some β values good for some classes of graphs.

The authors would like to thank an anonymous referee who contributes to Theorem 5 by pointing out a glitter star that achieves the matching lower bound, which graph makes the proof simpler than our original one.

References

1. D.J. Aldous, "On the time taken by random walks on finite groups to visit every state", *Z. Wahrsch. verw. Gebiete* 62 361–393, 1983.
2. R. Aleliunas, R.M Karp, R.J. Lipton, L. Lovász, and C. Rackoff, "Random walks, universal traversal sequences, and the complexity of maze problems", *Proc. 20th Ann. Symposium on Foundations of Computer Science*, 218–223, 1979.
3. G. Blom, L. Holst, and D. Sandell, "Problems and Snapshots from the World of Probability", Springer-Verlag, New York, NY, 1994.
4. G. Brightwell and P. Winkler, "Maximum hitting time for random walks on graphs", *J. Random Structures and Algorithms*, 3, 263–276, 1990.
5. A.Z. Broder and Karlin, "Bounds on covering times", *In 29th Annual Symposium on Foundations of Computer science*, 479–487, 1988.
6. D. Coppersmith, P. Tetali, and P. Winkler, "Collisions among random walks on a graph", *SIAM Journal on Discrete Mathematics*, 6, 3, 363–374, 1993.
7. U. Feige, "A tight upper bound on the cover time for random walks on graphs," *J. Random Structures and Algorithms*, 6, 4, 433–438, 1995.
8. S. Ikeda, I. Kubo, N. Okumoto and M. Yamashita, "Fair circulation of a token," *IEEE Trans. Parallel and Distributed Systems*, Vol.13, No.4, 367–372, 2002.
9. L. Isaacson and W. Madsen, "Markov chains: Theory and Application", Wiley series in probability and mathematical statistics, New York, 1976.
10. A. Israeli and M. Jalfon, "Token management schemes and random walks yield self stabilizing mutual exclusion", *Proc. of the 9th ACM Symposium on Principles of Distributed Computing*, 119–131, 1990.

11. P. Matthews, "Covering Problems for Markov Chain", *The Annals of Probability* Vol.16, No.3, 1215–1228, 1988.
12. R. Motowani and P. Raghavan, "Randomized Algorithms", Cambridge University Press, New York, 1995.
13. N. Okumoto, " A study on random walks of tokens on graphs ", M.E.Thesis, Hiroshima Univ., Higashi-Hiroshima, Japan, 1996.
14. J.L. Palacios, "On a result of Aleiliunas et al. concerning random walk on graphs," *Probability in the Engineering and Informational Sciences,* 4, 489–492, 1990.

Analysis of a Simple Evolutionary Algorithm for Minimization in Euclidean Spaces

Jens Jägersküpper*

FB Informatik, LS 2, Univ. Dortmund, 44221 Dortmund, Germany
jj@Ls2.cs.uni-dortmund.de

Abstract. Although evolutionary algorithms (EAs) are widely used in practical optimization, their theoretical analysis is still in its infancy. Up to now results on the (expected) runtime are limited to discrete search spaces, yet EAs are mostly applied to continuous optimization problems. So far results on the runtime of EAs for continuous search spaces rely on validation by experiments/simulations since merely a simplifying model of the respective stochastic process is investigated.

Here a first algorithmic analysis of the expected runtime of a simple, but fundamental EA for the search space $\mathbb{R}^n$ is presented. Namely, the so-called (1+1) Evolution Strategy ((1+1) ES) is investigated on unimodal functions that are monotone with respect to the distance between search point and optimum. A lower bound on the expected runtime is proven under the only assumption that isotropic distributions are used to generate the random mutation vectors. Consequently, this bound holds for any mutation adaptation mechanism. Finally, we prove that the commonly used "Gauss mutations" in combination with the so-called 1/5-rule for the mutation adaptation do achieve asymptotically optimal expected runtime.

Keywords: Evolutionary Algorithms, Black-Box Optimization, Continuous Search Space, Expected Runtime, Mutation Adaptation

1 Introduction

The optimization, here the minimization, of functions $f\colon S \to \mathbb{R}$ for some given search space S is one of the fundamental algorithmic problems. Discrete search spaces, e. g. $\{0, 1\}^n$, lead to combinatorial optimization problems like TSP, knapsack, or maximum matching. Mathematical optimization deals with continuous search spaces, usually $\mathbb{R}^n$. Here, problems are commonly defined by classes of functions, like polynomials of degree d, k-times differentiable functions, etc. Many problem-specific algorithms have been designed for each of these two scenarios. Since such algorithms are analyzed (in general), they can be compared and there is a theory on algorithms.

---

* supported by the Deutsche Forschungsgemeinschaft (DFG) as part of the collaborative research center "Computational Intelligence" (SFB 531)

J.C.M. Baeten et al. (Eds.): ICALP 2003, LNCS 2719, pp. 1068–1079, 2003.

If not enough resources are on hand to design a problem-specific algorithm, however, robust algorithms like randomized search heuristics are often a good alternative. Especially, if the knowledge about the function f to be optimized is not sufficient, classical mathematical optimization algorithms like the steepest descent method or the conjugate gradient method cannot be applied. In the extreme, for instance if f is only given implicitly, knowledge about f can solely be gathered by consecutively evaluating f at selected points. This situation is commonly named "black-box optimization." In this scenario, runtime is measured by the number of f-evaluations. Obviously, if we know nothing about f, a (reasonable) theoretical analysis of the runtime of some search heuristic like an evolutionary algorithm is impossible. Thus, to get insight into why such algorithms do often work quite well in practice, assumptions about the properties of f must be made, with respect to which the analysis is carried out.

This approach has been taken since the early 1990s for the discrete search space $\{0, 1\}^n$. Probably the first function that was analyzed is $\text{ONEMAX}(b) := b_1 + \cdots + b_n$, $b = (b_1, \ldots, b_n) \in \{0, 1\}^n$ (the name reflects that maximization was considered rather than minimization). The algorithm investigated was the so-called $(1+1)$ Evolutionary Algorithm $((1+1)\,\text{EA})$, which is in fact the discrete counterpart of the $(1+1)\,\text{ES}$ investigated here. Both algorithms use a population consisting of only one search point, called an individual in the field of Evolutionary Computation. Thus, recombination is precluded, and mutation is the only "evolutionary force." Within each beat of the evolution loop, the mutation of the current individual temporarily generates a second individual, and selection determines which one of both founds the next generation. An $O(n \log n)$ bound on the expected runtime of the $(1+1)\,\text{EA}$ for ONEMAX (if mutation consists in flipping each bit of the individual independently with probability $1/n$) is proved in [9]. Retrospectively, this bound is easy to obtain; yet more sophisticated papers on the $(1+1)\,\text{EA}$ have been published: In [2] linear functions are analyzed, in [14] quadratic polynomials, and in [13] monotone polynomials. Furthermore, [4] investigates the $(1+1)\,\text{EA}$ for the maximum matching problem. Even the effect of recombination has been analyzed for the search space $\{0, 1\}^n$ [7,8], and the number of papers on algorithmic analyses is increasing.

The situation for continuous search spaces is different: The vast majority of results on EAs are empirical, i. e., based on experiments and simulations. In the few papers that focus on theoretical analyses, however, either (global) convergence is investigated or local changes from one generation to the next. In the former case, one must recall that EAs for continuous search spaces merely approximate an optimum rather than optimize the respective function. Convergence deals with the question of whether the algorithm reaches the ε-neighborhood of some (global) optimum in a finite number of steps or not (e. g. [11]). However, the order of the number of steps necessary remains open—in particular with respect to the dimension of the search space. On the other hand, results dealing with local changes in one step, for instance convergence rates, (generally) do not enable statements on the long-time behavior of EAs. Normally, the effect of mutation/recombination depends on the location of the respective individual(s) in the search space. Consequently, the changes from one generation to the next one generally do not resemble the changes from the next generation to the second next. This is the reason why EAs for continuous search spaces apply so-called adaptation mechanisms, particularly mutation adaptation. The idea behind such adaptation mechanisms is to enable EAs to optimize as many types of functions as possible. Another idea behind mutation adaptation is that the mutative changes must in some way scale with the approximation quality. The rule of thumb reads: the closer the search approaches an optimum, the smaller the mutative changes. Unfortunately, (mutation) adaptation complicates the stochastic process an EA induces—and the analysis of the expected runtime.

The Scenario

As mentioned above, we will concentrate on the (1+1) ES which uses solely mutation because of a single-individual population. Let $c \in \mathbb{R}^n$ denote this current individual. For a given initialization of c, i. e., for a given starting point, the rough structure of the (1+1) ES is given by the following evolution loop:

1. Randomly choose the mutation vector $m \in \mathbb{R}^n$.
2. Generate the mutant $x \in \mathbb{R}^n$ by $x := c + m$.
3. Using $f(c)$ and $f(x)$, the selection rule determines whether this mutant becomes the current individual ($c := x$) or is discarded (c unchanged).
4. If the stopping criterion is met then output c else goto 1.

A single execution of the loop is called a *step* of the (1+1) ES, and if "$c := x$" is executed in a step, the mutation/mutant is said to be *accepted*, otherwise *rejected*. For a concrete instantiation of the (1+1) ES, the distribution of m, the selection rule, and the stopping criterion must be specified. Although the stopping criterion is important in practice, we investigate the (1+1) ES as an infinite process. Let $T_f \in \mathbb{N}$ denote the number of steps the (1+1) ES needs to reach some fixed approximation quality when optimizing f. Then we are interested in $\mathsf{E}[T_f]$ and in $\mathsf{P}\{T_f \leq \tau\}$ for a given number of steps τ. By defining an appropriate randomized selection rule, simulated annealing can be realized for instance. However, we will investigate the commonly and originally used elitist selection where the mutant x becomes/replaces the current individual c if and only if $f(x) \leq f(c)$. As this selection rule precludes worsenings, the (1+1) ES becomes a randomized hill-climber. If mutation adaptation is applied, obviously, the distribution of the mutation vector m is not fixed, but varies during the optimization process. Here we concentrate on mutation vectors that are isotropically distributed.

Definition 1. *For $m \in \mathbb{R}^n$, let $|m|$ denote its length, i. e., its L_2-norm, and $\widehat{m} := m/|m|$ the normalized vector. The distribution of the random vector m is isotropic if $|m|$ is independent of $\widehat{m}$ and $\widehat{m}$ is uniformly distributed upon the unit hyper-sphere $\{u \in \mathbb{R}^n \mid |u| = 1\}$.*

Under these two assumptions (elitist selection and isotropically distributed mutation vectors) the lower bound on the runtime will be proved. That is, in each step the mutation adaptation is free to choose an arbitrary isotropic distribution for m. Consequently, the lower bound particularly holds for so-called "Gauss mutations" which are very common in practice (cf. Lemma 5 for the isotropy).

Definition 2. *Let $\widetilde{m} \in \mathbb{R}^n$ be $(N_1(0,1), \ldots, N_n(0,1))$-distributed (each component is independently standard normal distributed). A mutation is called* Gauss mutation *if the mutation vector's distribution equals the one of $s \cdot \widetilde{m}$, $0 < s \in \mathbb{R}$.*

In particular, the upper bound on the runtime of the $(1+1)$ ES will be proved with respect to Gauss mutations.

This scenario, $(1+1)$ ES using elitist selection and Gauss mutations, has been introduced by Rechenberg, whose 1973 book *Evolutionsstragie* [10] is one starting point of evolutionary optimization. Rechenberg applied the $(1+1)$ ES to optimize the shape of some workpiece. Furthermore, he presents some rough calculations on what length of the mutation vector maximizes the expected spacial gain in one step. These calculations are carried out with respect to two different kinds of functions. On the one hand, the so-called corridor model is considered, and on the other hand, the SPHERE function, where $\text{SPHERE}(x) := x_1^2 + \cdots + x_n^2$ for $x = (x_1, \ldots, x_n) \in \mathbb{R}^n$. The calculations for the one-step behavior of the $(1+1)$ ES on SPHERE have been improved by Beyer and can be found in his 2001 book *The Theory of Evolution Strategies* [1]. As a conclusion, Rechenberg states that the length of a Gauss mutation vector should be adapted such that the *success probability of a step*, the probability that the mutation in this step is accepted, is about $1/5$. This led to the notion of the *1/5-rule* for mutation adaptation: The (expected) length of the Gauss mutation vectors are scaled by adapting the factor s in Definition 2 as follows. For a certain number of steps (originally $\Theta(n)$ many), the relative frequency of successful steps is observed without changing s. Subsequent to each observation phase, the relative share of successful steps in the respective phase is evaluated; if it is smaller than $1/5$, s is divided by some fixed constant greater than 1, and otherwise, s is multiplied by some fixed (possibly different) constant greater than 1. The upper bound on the runtime will be proved with respect to this 1/5-rule.

Finally, the class of functions we consider contains all unimodal $f : \mathbb{R}^n \to \mathbb{R}$, $n \in \mathbb{N}$, such that for $x, y \in \mathbb{R}^n$ and the respective optimum/minimum $o_f \in \mathbb{R}^n$: $|x - o_f| < |y - o_f| \Rightarrow f(x) < f(y)$. In other words, if an individual is closer to the optimum than some other, also its function value is better/smaller. We assume w.l.o.g. that the optimum o_f coincides with the origin, and thus, w.l.o.g. $|x| < |y| \Rightarrow f(x) < f(y)$ for $x, y \in \mathbb{R}^n$. Obviously, the L_2-norm itself and for instance SPHERE (as well as all their translations) bear this property.

Results

As mentioned above, the one-step behavior of $(1+1)$ ES on SPHERE has been investigated by Rechenberg and in great detail by Beyer. Unfortunately, at certain points within these calculations the limit $n \to \infty$ is taken without controlling the error terms; this is problematic in an algorithmic analysis, which exactly focuses on how the runtime depends on n. Thus, in Section 2 the n-dependence of the one-step behavior of the $(1+1)$ ES is investigated. The impact of the 1/5-rule on the convergence of the $(1+1)$ ES is investigated in [12] and [5] for instance; yet the order of the number of steps is not tackled. Applying methods and concepts known from the field of randomized algorithms, the main results mentioned in the abstract are shown in Section 3. Finally, we close with some concluding remarks. Note that more detailed proofs can be found in [6].

Notions and Notations

As mentioned in Definition 1, $|x|$ denotes the L_2-norm of the vector $x \in \mathbb{R}^n$, i.e., its length in Euclidean space, and $x_i \in \mathbb{R}$ its ith component. Furthermore, for instance, "n-sphere" abbreviates "n-dimensional sphere."

Definition 3. *A probability $p(n)$ is* exponentially small *in n if for a positive constant ε, $p(n) = \exp(-\Omega(n^\varepsilon))$. An event $A(n)$ happens with* overwhelming probability *(w.o.p.) with respect to n if $P\{\neg A(n)\}$ is exponentially small in n.*

2   One-Step Behavior

As we are interested in how fast the "evolving" individual of the $(1+1)$ ES approaches the optimum in the search space, the spatial gain towards the optimum in one step is the intermediate objective. Since the 1/5-rule for mutation adaptation is investigated, it is particularly interesting what length of the mutation vector results in the mutant being accepted with probability 1/5. Due to the independence of the random length of an isotropic mutation vector and its random direction (cf. Definition 1), we may assume that the length $\ell > 0$ of the mutation vector m is chosen according to $|m|$'s distribution first; then the mutant is uniformly distributed upon the n-sphere with radius ℓ centered at the current search point c.

The situation is depicted by the figure on the right. The left sphere $F := \{c' \in \mathbb{R}^n \mid |c'| = |c|\}$ will be called the *fitness sphere* since the properties of f imply that all points inside (resp. outside) the fitness sphere are better (resp. worse) than the current search point c. The potential mutants define the *mutation sphere* $M := \{x \in \mathbb{R}^n \mid |x - c| = \ell\}$.

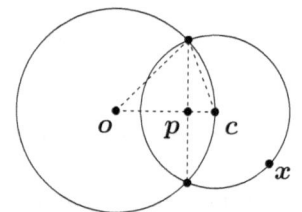

Let $I := F \cap M \subset \mathbb{R}^n$ denote the intersection of the two spheres. Obviously, if $\ell > 2|c|$, I is empty, and if $\ell = 2|c|$, I is a singelton, such that we concentrate on $\ell < 2|c|$. It is easy to see that I forms an $(n-1)$-sphere, and that the hyperplane $P \supset I$ is orthogonal to the line passing through c and o. (Let $p \in P$ denote the point where this line passes through P.) Hence, the mutation sphere's part lying inside the fitness sphere forms a hyper-spherical cap $C \subset M - I$, the missing boundary of which is I. Basic geometry shows that the distance between c and P equals $g := |c| - |p| = \ell^2/(2|c|)$ if $\ell \le \sqrt{2}|c|$.

Since the mutant x is uniformly distributed upon the mutation sphere M, for any (Lebesgue measurable) $S \subseteq M$, $P\{x \in S \mid |m| = \ell\}$ equals the ratio of the $(n-1)$-volume of S to the one of M, inducing a probability measure. Consequently, I is of zero measure, and since x is better than c if $x \in C$, and worse if $x \in M - (C \cup I)$, the probability that the mutant is accepted equals the ratio of the hypersurface area of C to the one of M. Now, the interesting question is how this ratio depends on $|c|$, ℓ, and, of course, n, the number of dimensions.

As the height of the mutation sphere's cap that is cut off by the fitness sphere equals $h := \ell - g = \ell - \ell^2/(2\,|c|)$, the *relative height* of C, the ratio h/ℓ, equals $1 - \ell/(2\,|c|)$. It can be shown (cf. [3, Appendix B] for instance) that

$$\frac{\text{hypersurface area of } C}{\text{hypersurface area of } M} = \frac{\Psi_{n-2}\big(\arccos(1 - h/\ell)\big)}{\Psi_{n-2}(\pi)}$$

in n-space, $n \geq 3$, where $\Psi_k(\gamma) := \int_0^\gamma (\sin \beta)^k d\beta$. Note that $1 - h/\ell = (\ell - h)/\ell = g/\ell$. This formula may be directly used to estimate a step's success probability, yet it can also be utilized more generally: The ratio $\Psi_{n-2}(\arccos(g/\ell))/\Psi_{n-2}(\pi)$ not only equals the probability that the mutation hits C, but also the one of "the spatial gain of an isotropic mutation vector m parallel to some fixed direction (for instance $\overline{co}$) is greater than g," under the condition $|m| = \ell$. Therefore, let G denote the random variable given by the spatial gain of an isotropic mutation m parallel to a fixed direction under the condition $|m| = \ell$. Then

$$\mathrm{P}\{G \leq g\} = 1 - \mathrm{P}\{G > g\} = 1 - \frac{\Psi_{n-2}(\arccos(g/\ell))}{\Psi_{n-2}(\pi)},$$

and hence, $F_n(x) := 1 - \Psi_{n-2}(\arccos(x/\ell))/\Psi_{n-2}(\pi)$ for $x \in [-\ell, \ell]$ is G's probability distribution over $[-\ell, \ell]$ in n-space. Since Ψ_k is continuous, the probability density of G at $g \in [-\ell, \ell]$ equals $\frac{dF_n(x)}{dx}(g) = F_n'(g)$,

$$\begin{aligned}
F_n'(x) &= \Psi_{n-2}(\pi)^{-1} \cdot (-1) \cdot \frac{d}{dx}\Psi_{n-2}(\arccos(x/\ell)) \\
&= \Psi_{n-2}(\pi)^{-1} \cdot (-1) \cdot \frac{d}{dx} \int_0^{\arccos(x/\ell)} (\sin \beta)^{n-2}\, d\beta \\
&= \Psi_{n-2}(\pi)^{-1} \cdot \ell^{-1} \cdot \left(1 - (g/\ell)^2\right)^{(n-3)/2}
\end{aligned}$$

for $n \geq 4$. To make things clear, this is the density of the spatial gain of an isotropically distributed mutation vector m parallel to an arbitrarily fixed direction—independently of the function optimized— if $|m|$ takes the value ℓ, not the one towards the optimum after selection.

With the help of this density function, we obtain an alternative formula for the success probability of a step, in which c is mutated using an isotropically distributed mutation vector m with $|m| = \ell$ (y substitutes g/ℓ):

$$\mathrm{P}\{x \text{ is accepted} \mid |m| = \ell\} = \mathrm{P}\{x \in C \mid |m| = \ell\} = \mathrm{P}\left\{G \geq \tfrac{\ell^2}{2|c|}\right\}$$

$$= \int_{\ell^2/(2|c|)}^{\ell} F_n'(g)\, dg = \frac{\ell}{\Psi_{n-2}(\pi) \cdot \ell} \int_{\ell/(2|c|)}^{1} \left(1 - y^2\right)^{(n-3)/2} dy$$

With respect to the 1/5-rule, which will be investigated for the upper bound on the expected runtime, we can now answer what length of the mutation vector results in a step of the $(1+1)$ ES having success probability 1/5. Note that, obviously, this probability approaches 1/2 as $\ell/|c| \to 0$.

Lemma 1. *In the scenario considered, the mutant $c + m \in \mathbb{R}^n$ is accepted with a constant probability greater than 0 and smaller than 1/2 if and only if $|m|$ takes a value $\ell = \Theta(|c|/\sqrt{n})$ in the respective step.*

Proof. The distance between c and P, the hyperplane containing the intersection of mutation sphere and fitness sphere, equals $\ell^2/(2|c|) = \ell \cdot (\lambda/\sqrt{n})$ with $\lambda = \Theta(1)$, i.e., the relative height of the cap C equals $1 - \lambda/\sqrt{n}$. Using the formula derived above, we must show that $\int_{\lambda/\sqrt{n}}^{1} (1 - y^2)^{(n-3)/2} dy$ as well as $\int_0^{\lambda/\sqrt{n}} (1 - y^2)^{(n-3)/2} dy$ are in $\Omega(\Psi_{n-2}(\pi))$, respectively. See [6]. $\square$

In other words, if the 1/5-rule was able to ensure a success probability of exactly 1/5 in a step, the length of the mutation vector would be $\Theta(|c|/\sqrt{n})$ in this step. Thus, the expected spatial gain towards the optimum in this situation is of particular interest and is estimated in the following.

Lemma 2. *If (in the scenario considered) $|m| = \Theta(|c|/\sqrt{n})$ in a step then the spatial gain towards the optimum is $\Omega(|m|/\sqrt{n}) = \Omega(|c|/n)$ with probability $\Omega(1)$ in this step, and thus, also the expected decrease in distance to the optimum in this step is $\Omega(|c|/n)$.*

Proof. As in Lemma 1, the assumptions imply that C has height $\ell \cdot (1 - \lambda/\sqrt{n})$ for $\lambda = \Theta(1)$. One result of that Lemma is that the mutation hits C with probability $\Omega(1)$. Let $A \subset C$ denote the cap with height $\ell \cdot (1 - 2\lambda/\sqrt{n})$ such that its pole coincides with the one of C. Then each point in A is at least $\ell \cdot (1 - \lambda/\sqrt{n}) - \ell \cdot (1 - 2\lambda/\sqrt{n}) = \ell \cdot \lambda/\sqrt{n}$ distance units closer to the optimum than a point belonging to the boundary of C. Since the boundary of C equals the intersection of mutation sphere and fitness sphere, the distance to the optimum is decreased by at least $\ell \cdot \lambda/\sqrt{n} = \Theta(|c|/n)$ distance units if the mutation hits A. This still happens with probability $\Omega(1)$ because the relative height of A equals $1 - \Theta(1/\sqrt{n})$ like the one of C. Since the properties of f in combination with the selection rule preclude a negative spatial gain, the expected decrease in distance to the optimum is $\Omega(|c|/n)$. $\square$

Consequently, if the 1/5-rule is capable of adjusting the mutation vector's length such that the success probability is close to 1/5, the distance to the optimum is expected to decrease by an $\Omega(1/n)$-fraction. Note that, e.g., an 1/8-rule or an 1/3-rule would lead to the same asymptotic expected gain. Naturally, one might ask if an expected spatial gain $\omega(|c|/n)$ is possible. We prove that in our scenario the expected spatial gain towards the optimum is $O(|c|/n)$ for any adaptation of the length of an isotropic mutation vector. Hence, the 1/5-rule indeed tries to adjust the mutation vector's length to have optimal order $\Theta(|c|/\sqrt{n})$ such that the expected spatial gain towards the optimum has maximum order $\Theta(|c|/n)$.

Obviously, the spatial gain of a step equals 0 if the mutation is rejected, and is upper bounded by the mutation's spatial gain parallel to $\overline{co}$, otherwise. A mutation is accepted (resp. rejected) if the spatial gain parallel to $\overline{co}$ is greater (resp. smaller) than $\ell^2/(2|c|)$. Using the probability density function obtained above, the expected spatial gain of a step, call it E[gain], is bounded above by

$$\int_{\ell^2/(2|c|)}^{\ell} g F'_{n-2}(g)\, dg \;=\; \frac{\ell}{\Psi_{n-2}(\pi)} \int_{\ell/(2|c|)}^{1} y \cdot (1-y^2)^{(n-3)/2}\, dy$$

$$=\; \frac{\ell}{\Psi_{n-2}(\pi)\cdot(n-1)} \cdot \left(1 - \left(\tfrac{\ell}{2|c|}\right)^2\right)^{(n-1)/2}$$

$$<\; \frac{\ell}{\sqrt{2\pi}\sqrt{n-1}} \cdot \left(1 - \left(\tfrac{\ell}{2|c|}\right)^2\right)^{(n-1)/2}$$

because for $n \geq 4$, $\int y \cdot (1-y^2)^{(n-3)/2}\, dy = (1-y^2)^{(n-1)/2}/(-(n-1))$ and $\Psi_{n-2}(\pi) > \sqrt{2\pi}/\sqrt{n-1}$ (cf. [6]). Consequently, E[gain] $= O(|m|/\sqrt{n})$ independently of the *scaled distance from the optimum* $|c|/|m|$ (remember that $|m| > 2|c|$ results in the mutant being rejected since it lies outside the fitness sphere). Furthermore, the inequality enables the proof that E[gain] $= O(|c|/n)$ for any adaptation of the mutation vector's length.

Lemma 3. *In the scenario considered, the expected spatial gain towards the optimum in a step is $O(|c|/n)$ —for any isotropic mutation.*

Proof. To prove this claim, we must show that E[gain]$/|c| = O(1/n)$ even if the mutation vector's length ℓ is chosen such that the expected spatial gain is maximized. Let $d := |c|/\ell$ denote the scaled distance from the optimum. Applying the upper bound on the expected spatial gain from above yields

$$\text{E[gain]}/|c| \;<\; \underbrace{\left(2\pi(n-1)\right)^{-1/2} \cdot (1/d) \cdot \left(1-(2d)^{-2}\right)^{(n-1)/2}}_{=:\; w_n(d)}.$$

Hence, an upper bound on E[gain]$/|c|$ can be derived by maximizing the function w_n. In fact, $w_n(d) = O(1/\sqrt{n})$ for $d > 0$ (cf. [6]), and thus,

$$\text{E[gain]}/|c| \;<\; w_n(d)/\sqrt{2\pi(n-1)} \;=\; O(1/\sqrt{n})/\sqrt{2\pi(n-1)} \;=\; O(1/n) \quad \square$$

3   Multi-step Behavior and Expected Runtime

Obviously, the multi-step behavior of the (1+1) ES crucially depends on the mutation adaptation used. For a lower bound on the expected runtime, however, optimal mutation adaptation may be assumed. Surprisingly, we need not prove explicitly what mutation adaptation is optimal. Furthermore, it is not evident what "runtime" means since f is merely approximated rather than optimized.

Due to the symmetry and scalability properties of f, linearity of expectation enables further statements if one knows (for an arbitrary starting point) the expected number of steps to halve the distance from the optimum using optimal mutation adaptation. Namely, the expected runtime to reduce the distance from the optimum to a $1/k$-fraction is lower bounded by $\lfloor \log_2 k \rfloor$ times the lower bound on the expected runtime to halve it. We apply the following modification of Wald's equation to prove the lower bound on the expected number of steps the $(1+1)$ ES needs to halve the distance from the optimum (cf. [6] for the proof of this lemma).

Lemma 4. *Let* $X_1, X_2, \ldots$ *denote random variables with bounded range and* T *the random variable defined by* $T = \min\{t \mid X_1 + \cdots + X_t \geq g\}$ *for a given* $g > 0$. *If* $\mathsf{E}[T]$ *exists and* $\mathsf{E}[X_i \mid T \geq i] \leq u$ *for* $i \in \mathbb{N}$ *then* $\mathsf{E}[T] \geq g/u$.

Theorem 1. *In the scenario considered, for any adaptation of isotropic mutations the expected number of steps to halve the distance to the optimum is* $\Omega(n)$.

Proof. For $i \geq 1$, let X_i denote the random variable that corresponds to the spatial gain towards the optimum in the ith step. Furthermore, let $a \in \mathbb{R}^n - \{o\}$ denote the starting point and T the (random) number of steps until $|c| \leq |a|/2$ for the first time. As mentioned previously, worsenings are precluded such that $X_i \geq 0$ and in particular $|c| \leq |a|$ in each step. Consequently, $X_i \leq |c| \leq |a|$, and according to Lemma 3, $\mathsf{E}[X_i \mid T \geq i] = O(|c|/n) = O(|a|/n)$. Choosing $g := |a|/2$ in Lemma 4, $\mathsf{E}[T] \geq (|a|/2)/O(|a|/n) = \Omega(n)$ if $\mathsf{E}[T]$ exists. If $\mathsf{E}[T]$ is not defined (due to improper adaptation), one may informally argue that "$\mathsf{E}[T] = \infty = \Omega(n)$" since T is positive.   □

This lower bound on the expected runtime holds independently of the mutation adaptation applied since theoretically optimal adaptation is (implicitly) assumed. For the upper bound, we concretize the lower-bound scenario by choosing Gauss mutations and the 1/5-rule for mutation adaptation. The following properties of Gauss-mutations are useful (and proved in [6]).

Lemma 5. *A Gauss-mutation* $m \in \mathbb{R}^n$ *is isotropically distributed, and moreover,* $\ell_\mathsf{E} := \mathsf{E}[\|m\|]$ *exists and* $\mathsf{P}\{|\|m\| - \ell_\mathsf{E}| \geq \delta \cdot \ell_\mathsf{E}\} \leq \delta^{-2}/(2n-1)$.
   Let $m_1, \ldots, m_n$ *denote independent copies of* m. *For any constant* $\lambda < 1$ *two positive constants* a_λ, b_λ *exist such that* $\#\{i \mid a_\lambda \ell_\mathsf{E} \leq |m_i| \leq b_\lambda \ell_\mathsf{E}\} \geq \lambda n$ *w. o. p.*

Furthermore, we investigate this instantiation of the 1/5-rule: The scaling factor s (cf. Definition 2) is adapted after every nth step: if less than $n/5$ of the respective last n steps were successful, s is halved, otherwise doubled. The asymptotic calculations we present, however, are valid for any 1/5-rule keeping s unchanged for $\Theta(n)$ steps, respectively, and using any two constants, each greater than 1, for the scaling of s.
   The run of the $(1+1)$ ES is partitioned into phases each of which lasts n steps such that $\mathsf{E}[\|m\|]$ is constant in each phase. Let s_i denote the scaling factor used throughout the ith phase and ℓ_i the corresponding $\mathsf{E}[\|m\|]$. A phase after which s is doubled is symbolized by "$\times$", and one after which s is halved by "$\div$". Furthermore, let d_i denote the distance from the optimum at the beginning of the ith phase; hence, $d_i - d_{i+1}$ equals the spatial gain in/of the ith phase.

Lemma 6. *In the scenario considered for the 1/5-rule for Gauss mutations:*

1. *if $\ell_i = \Theta(d_i/\sqrt{n})$ then $d_{i+1} = d_i - \Omega(d_i)$ w. o. p.,*
2. *if s is doubled after the ith phase then $\ell_i = O(d_i/\sqrt{n})$ w. o. p.,*
3. *if s is halved after the ith phase then $\ell_{i+1} = \Omega(d_{i+1}/\sqrt{n})$ w. o. p.*

Proof. Assume that the total spatial gain of the ith phase is not $\Omega(d_i)$. Then the distance from the optimum is $\Theta(d_i)$ in each step of the phase (remember that the distance is non-increasing). Lemma 5 yields that w. o. p. $|\boldsymbol{m}| = \Theta(d_i/\sqrt{n})$ in $0.9n$ steps. According to Lemma 2, in each such step the spatial gain is $\Omega(d_i/n)$ with probability $\Omega(1)$. Hence, we expect $\Omega(n)$ steps each of which reduces the distance by $\Omega(d_i/n)$. By Chernoff bounds, the number of such steps is $\Omega(n)$ w. o. p. Consequently, our initial assumption contradictorily implies that the total spatial gain of the ith phase is $\Omega(d_i)$ w. o. p.

For the second claim, assume ℓ_i is not $O(d_i/\sqrt{n})$. Since the distance from the optimum is non-increasing, ℓ_i is not $O(|\boldsymbol{c}|/\sqrt{n})$ in each step of the ith phase. Lemma 5 yields that $|\boldsymbol{m}|$ is not $O(|\boldsymbol{c}|/\sqrt{n})$ in $0.9n$ steps w. o. p. According to Lemma 1, the success probability of each such step is $o(1)$. Hence, the expected number of unsuccessful steps is lower bounded by $0.9n - o(n)$. By Chernoff bounds, w. o. p. more than $0.8n$ steps are not successful. Thus, the assumption "ℓ_i is not $O(d_i/\sqrt{n})$" contradictorily implies that s is halved w. o. p.

Assume ℓ_{i+1} is not $\Omega(d_{i+1}/\sqrt{n})$ for the third claim. Since $s_i = 2s_{i+1}$ also $\ell_i = 2\ell_{i+1}$. As the distance is non increasing, the assumption implies that ℓ_i is not $\Omega(|\boldsymbol{c}|/\sqrt{n})$ for each step of the $\div$-phase. Following the proof of the second claim with symmetric arguments, w. o. p. more than $0.8n$ steps are successful—contradictorily implying that the ith phase is a $\times$-phase w. o. p. $\qquad\square$

Now we can deal with sequences of phases in a run of the $(1+1)$ ES.

Lemma 7. *If (in the scenario considered) the 1/5-rule for Gauss mutations causes a sequence $\div\times^k$ of phases, $k = \text{poly}(n)$, then w. o. p. the distance from the optimum is k times reduced by a constant fraction in the respective phases.*

Proof. Let the $\div$-phase be the ith one. By Lemma 6, $\ell_{i+1} = \Omega(d_{i+1}/\sqrt{n})$ w. o. p. Since the adaptation yields $\ell_{i+w} \geq \ell_{i+1}$, $1 \leq w \leq k$, and the distance is non-increasing, w. o. p. $\ell_{i+w} = \Omega(d_{i+w}/\sqrt{n})$ for $1 \leq w \leq k$. Lemma 6 also yields that w. o. p. $\ell_{i+w} = O(d_{i+w}/\sqrt{n})$ for $1 \leq w \leq k$. Consequently, w. o. p. $\ell_{i+w} = \Theta(d_{i+w}/\sqrt{n})$ for $1 \leq w \leq k$, and finally, again according to Lemma 6, in each of the k $\times$-phases the distance is reduced by a constant fraction w. o. p. $\qquad\square$

Lemma 8. *If (in the scenario considered) the 1/5-rule for Gauss mutations causes a sequence $\times\div^k$ of phases, $k = \text{poly}(n)$, then w. o. p. the distance from the optimum is k times reduced by a constant fraction in the respective phases.*

Proof. Let the $\times$-phase be the ith one. For $k = 1$, assume that the total spatial gain of the ith and the $(i+1)$th phase is not $\Omega(d_i)$. According to Lemma 6, w. o. p. $\ell_i = O(d_i/\sqrt{n})$ and w. o. p. $\ell_{i+2} = \Omega(d_{i+2}/\sqrt{n})$. Hence, $\ell_i = \Theta(d_i/\sqrt{n})$

as well as $\ell_{i+1} = \Theta(d_{i+1}/\sqrt{n})$, and Lemma 6 contradictorily implies that in each of the two phases the distance is reduced by a constant fraction w. o. p. Consequently, w. o. p. these two phases yield $d_{i+2} = d_i - \Omega(d_i)$.

For $k \geq 2$, the adaptation yields $s_{i+w} = s_i\, 2^{2-w} = 4\, s_i/2^w$ for $1 \leq w \leq k$, and according to Lemma 6, for $2 \leq w \leq k$ w. o. p. $\ell_{i+w} = \Omega(d_{i+w}/\sqrt{n})$. If $d_{i+w} \leq d_i/2^w$ then by a simple accounting argument after the $(i+w)$th phase $d_{i+w+1} \leq d_{i+w} \leq d_i/2^w \leq d_i/\lambda^{w+1}$ for a constant $\lambda \geq \sqrt{2}$ and we are done. Thus, assume $d_{i+w} > d_i/2^w$. As $\ell_{i+w} = 4\ell_i/2^w$, in this case "w. o. p. $\ell_i = O(d_i/\sqrt{n})$" implies that w. o. p. $\ell_{i+w} = O(d_{i+w}/\sqrt{n})$. Since also $\ell_{i+w} = \Omega(d_{i+w}/\sqrt{n})$, Lemma 6 yields that the $(i+w)$th phase reduces the distance by a constant fraction w. o. p.

Altogether, the first two phases yield w. o. p. $d_{i+2} = d_i - \Omega(d_i)$, and for $2 \leq w \leq k$, either the distance from the optimum is reduced by a constant fraction in the $(i+w)$th phase w. o. p., or after this phase $d_{i+w+1} \leq d_i/\lambda^{w+1}$ for a constant $\lambda \geq \sqrt{2}$ even if there was no spatial gain in the $(j+w)$th phase.   □

Finally, the three preceding lemmas together with Theorem 1 yield the bound on the expected runtime, the expected number of steps the $(1+1)$ ES needs for a predefined reduction of the distance from the optimum $\boldsymbol{o}$ in the search space.

Theorem 2. *If (in the scenario considered) for the suboptimal initial search point $\boldsymbol{a} \in \mathbb{R}^n - \{\boldsymbol{o}\}$ and the initial scaling factor s_1, $|\boldsymbol{u} - \boldsymbol{v}|\,/s_1 = \Theta(n)$ then the expected number of steps to obtain a search point $\boldsymbol{c}$ such that $|\boldsymbol{c} - \boldsymbol{o}| \leq 2^{-t}\,|\boldsymbol{a} - \boldsymbol{o}|$ for $t \in poly(n)$ is $\Theta(t \cdot n)$.*

Proof. Assume w. l. o. g. that the optimum $\boldsymbol{o}$ coincides with the origin. The lower bound $\Omega(t \cdot n)$ follows immediately from Theorem 1.

If the sequence of phases starts with $\times\div$ or with $\div\times$, the two preceding lemmas yield that the number phases until $\mathsf{E}[|\boldsymbol{c}|] < 2^{-(t+1)}\,|\boldsymbol{a}|$ is $O(t)$. If the sequence starts with $\times^k$ or with $\div^k$ for $k \geq 2$, we must show that in these phases the distance is w. o. p. reduced k times by a constant fraction. The assumptions on the starting values ensure that in the first phase $\ell_1 = \mathsf{E}[|\widetilde{\boldsymbol{m}}|] \cdot s_1 = \Theta(\sqrt{n}) \cdot s_1 = \Theta(d_1/\sqrt{n})$ (cf. Definition 2 for $\widetilde{\boldsymbol{m}}$ and [6] for $\mathsf{E}[|\widetilde{\boldsymbol{m}}|] = \Theta(\sqrt{n})$). Therefore, the same argumentation as for $\div\times^k$ resp. $\times\div^k$ can be applied (without the preceding $\div$-phase resp. $\times$-phase).

Hence, the number of phases such that $\mathsf{E}[|\boldsymbol{c}|] < |\boldsymbol{a}| \cdot 2^{-t}/2$ is bounded by $O(t)$. By Markov's inequality, $\mathsf{P}\{|\boldsymbol{c}| \leq |\boldsymbol{a}| \cdot 2^{-t}\} \geq 1/2$ after these $O(t)$ phases. If this is not the case, after all $|\boldsymbol{c}| \leq |\boldsymbol{a}|$ such that again with probability at least $1/2$, $|\boldsymbol{c}| \leq |\boldsymbol{a}| \cdot 2^{-t}$ after another $O(t)$ phases. Repeating this argument, the expected number of phases is upper bounded by $\sum_{i \geq 1} 2^{-i} \cdot i \cdot O(t) = 2 \cdot O(t)$, and the expected number of steps is $O(t \cdot n)$.   □

For other starting conditions, the (expected) number of steps necessary to ensure the theorem's assumptions must be estimated before the theorem can be applied—for instance by estimating the number of steps until the scaling factor is halved and doubled at least once, respectively. This is a rather simple task when using the results presented.

4   Conclusion

For the first time, the (expected) runtime of a simple, but fundamental evolutionary algorithm for optimization in $\mathbb{R}^n$ is rigorously analyzed—not a simplifying model of it. In particular, this analysis shows that, in the scenario considered, the well-known 1/5-rule for mutation adaptation indeed results in asymptotically optimal expected runtime. As the analysis covers a wide range of realizations of the 1/5-rule, it additionally yields an interesting byproduct: Fine tuning the parameters of the 1/5-rule actually does not affect the order of the expected runtime; we could even replace 1/5 by 1/8 or by 1/3, for instance. This may be interpreted as an indicator for the robustness often ascribed to evolutionary algorithms; yet it is proved for the scenario considered only.

Acknowledgments. Thanks for productive discussions and for pointing out flaws especially go to Ingo Wegener, Carsten Witt, and Stefan Droste.

References

1. Beyer, H.-G. (2001). *The Theory of Evolution Strategies*. Springer, Berlin.
2. Droste, S., Jansen, T., Wegener, I. (2002). On the analysis of the (1+1) evolutionary algorithm. Theoretical Computer Science, 276, pp. 51–82.
3. Ericson, T., Zinoviev, V. (2001). *Codes on Euclidian Spheres*. Elsevier, Amsterdam.
4. Giel, O., Wegener, I. (2003). Evolutionary algorithms and the maximum matching problem. *Proceedings of the 20th International Symposium on Theoretical Computer Science (STACS 2003)*, LNCS 2607, pp. 415–426.
5. Greenwood, G. W., Zhu, Q. J. (2001). Convergence in evolutionary programs with self-adaptation. Evolutionary Computation, 9(2), pp. 147–157.
6. Jägersküpper, J. (2002). Analysis of a simple evolutionary algorithm for the minimization in euclidian spaces. Tech. Rep. CI-140/02, Univ. Dortmund, SFB 531, `http://sfbci.uni-dortmund.de/home/English/Publications/Reference/`.
7. Jansen, T., Wegener, I. (2001). Real royal road functions—where crossover provably is essential. *Proceedings of the 3rd Genetic and Evolutionary Computation Conference (GECCO 2001)*, Morgan Kaufmann, San Francisco, pp. 375–382.
8. Jansen, T., Wegener, I. (2002). The analysis of evolutionary algorithms—A proof that crossover really can help. Algorithmica, 34, pp. 47–66.
9. Mühlenbein, H. (1992). How genetic algorithmis really work: Mutation and hill-climbing. *Proceedings of the 2nd Parallel Problem Solving from Nature (PPSN II)*, North-Holland, Amsterdam, pp. 15–25.
10. Rechenberg, I. (1973). *Evolutionsstrategie*. Frommann-Holzboog, Stuttgart, Germany.
11. Rudolph, G. (1997). *Convergence Properties of Evolutionary Algorithms*. Verlag Dr. Kovač, Hamburg.
12. Rudolph, G. (2001). Self-adaptive mutations may lead to premature convergence. IEEE Transactions on Evolutionary Computation, 5(4), pp. 410–414.
13. Wegener, I. (2001). Theoretical aspects of evolutionary algorithms. *Proceedings of the 28th International Colloquium on Automata, Languages and Programming (ICALP 2001)*, LNCS 2076, pp. 64–78.
14. Wegener, I., Witt, C. (2003). On the analysis of a simple evolutionary algorithm on quadratic pseudo-boolean functions. Journal of Discrete Algorithms, to appear.

Optimal Coding and Sampling of Triangulations

Dominique Poulalhon and Gilles Schaeffer

LIX – CNRS, École polytechnique, 91128 Palaiseau Cedex, France,
{Dominique.Poulalhon,Gilles.Schaeffer}@lix.polytechnique.fr,
http://lix.polytechnique.fr/Labo/{Dominique.Poulalhon,Gilles.Schaeffer}

Abstract. We present a bijection between the set of plane triangulations (aka. maximal planar graphs) and a simply defined subset of plane trees with two leaves per inner node. The construction takes advantage of the minimal realizer (or Schnyder tree decomposition) of a plane triangulation.

This yields a simple interpretation of the formula for the number of plane triangulations with n vertices. Moreover the construction is simple enough to induce a linear random sampling algorithm, and an explicit information theory optimal encoding.

1 Introduction

This paper addresses three problems on finite *triangulations*, or *maximal planar graphs*: coding, counting, and sampling. The results are obtained as consequences of a new bijection, between triangulations endowed with their minimal realizer and trees in the simple class of plane trees with two leaves per inner node.

Coding. The coding problem was first raised in algorithmic geometry: find an encoding of triangulated geometries which is as compact as possible. As demonstrated by previous work, a very effective "structure driven" approach consists in distinguishing the encoding of the combinatorial structure, – that is, the triangulation – from the geometry – that is, vertex coordinates (see [26] for a survey and [16] for an opposite "coordinate driven" approach). Three main properties of the combinatorial code are then desirable: *compacity*, that is minimization of the bit length of code words, *linear complexity* of the complete coding and decoding procedure, and *locality*, that is the possibility to navigate efficiently (and to code the coordinates by small increments).

For the fundamental class $\mathcal{T}_n$ of triangulations of a sphere with $2n$ triangles, several codes of linear complexity were proposed, with various bit length $\alpha n(1 + o(1))$: from $\alpha = 4$ in [6,11,18], to $\alpha = 3.67$ in [21,29], and recently $\alpha = 3.37$ bits in [7]. The information theory bound on α is $\alpha_0 = \frac{1}{n}\log|T_n| \sim \frac{256}{27} \approx 3.245$ (see below). In some sense the compacity problem was given an optimal solution for general recursive classes of planar maps by Lu *et al.* [19,22]. For a fixed class, say triangulations, this algorithm does not use the knowledge of α_0, as expected for a generic algorithm, and instead relies on a cycle separator algorithm

J.C.M. Baeten et al. (Eds.): ICALP 2003, LNCS 2719, pp. 1080–1094, 2003.

and, at bottom levels of recursion, on an exponential optimal coding algorithm. This leads to an algorithm difficult to implement with low complexity constants. Moreover the implicit nature of the representation makes it unlikely that locality constraints can be dealt with in this framework: known methods to achieve locality require the code to be based on a spanning tree of the graph.

Counting. The exact enumeration problem for triangulations was solved by Tutte in the sixties [31]. The number of rooted triangulations with $2n$ triangles, $3n$ edges and $n + 2$ vertices is

$$T_n = \frac{2(4n - 3)!}{n!(3n - 1)!}. \tag{1}$$

(This formula gives the previous constant $\alpha_0 = \frac{256}{27}$.) More generally Tutte was interested in *planar maps*: embedded planar multigraphs considered up to homeomorphisms of the sphere. He obtained several elegant formulas akin to (1) for the number of planar maps with n edges and for several subclasses (bipartite maps, 2-connected maps, 4-regular maps). It later turned out that constraints of this kind lead systematically to explicit enumeration results for subclasses of maps (in the form of algebraic generating functions, see [5] and references therein). A natural question in this context is to find simple combinatorial proofs explaining these results, as opposed to the technical computational proofs *à la Tutte*. This was done in a very general setting for maps without restrictions on multiple edges and loops [9,27]. Two main ingredients are at the heart of this approach: dual breath-first search to go from maps to trees, and a closure operation for the inverse mapping. When loops are forbidden, the first ingredient is no longer suited, but it was shown that it can be replaced by bipolar orientations [24,28]. When multiple edges are forbidden as well, the situation appears completely different and neither of the previous methods directly apply.

It should be stressed that planar graphs have in general non-unique embeddings: a given planar graph may underlie many planar maps. This explains that, as opposed to the situation for maps, no exact formula is known for the number of planar graphs with n vertices (even the asymptotic growth factor is not known, see [7,23]). However according to Whitney's theorem, 3-connected planar graphs have an essentially unique embedding. In particular the class of triangulations is equivalent to the class of maximal planar graphs (a graph is maximal planar if no edge can be added without losing planarity).

Sampling. A perfect (resp. approximate) random sampling algorithm outputs a random triangulation chosen in $\mathcal{T}_n$ under the uniform distribution (resp. under an approximation thereof): the probability to output a specific rooted triangulation T with $2n$ vertices is (resp. is close to) $1/T_n$. Safe for an exponentially small fraction of them, triangulations have a trivial automorphism group [25], so that as far as polynomial parameters are concerned, the uniform distribution on rooted or unrooted triangulations are indistinguishable.

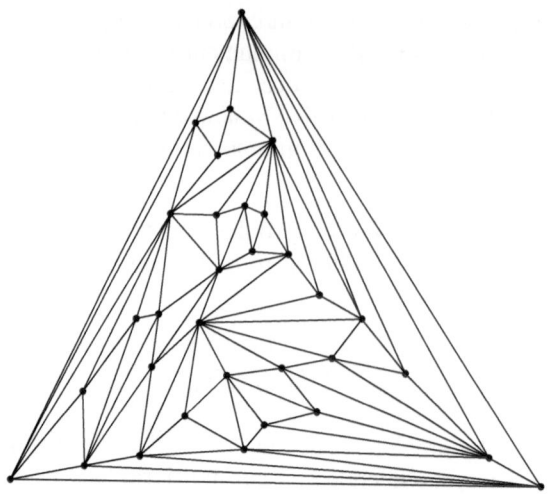

Fig. 1. A random triangulation with 60 triangles.

This question was first considered by physicists willing to test experimentally properties of two dimensional quantum gravity: it turns out that the proper discretization of a typical quantum universe is precisely obtained by sampling from the uniform distribution on rooted triangulations [4]. Several approximate sampling algorithms were thus developed by physicists for planar maps, including for triangulations [3]. Most of them are based on Markov chains, the mixing times of which are not known (see however [17] for a related study). A recursive perfect sampler was also developed for cubic maps, but has at least quadratic complexity [1]. More efficient and perfect samplers were recently developed for a dozen of classes of planar maps [5,29]. These algorithms are linear for triangular maps (with multiple edges allowed) but have average complexity $O(n^{5/3})$ for the class of triangulations.

Most random sampling algorithms are usually either based on Markov chains, or on enumerative properties. On the one hand, an algorithm of the first type perform a random walk on the set of configurations until it has (approximately) forgotten its start point. This is a very versatile method that requires little knowledge of the structures. It can even allow for perfect sampling in some restricted cases [32]. However in most cases it yields only approximate samplers of at least quadratic complexities. On the other hand, algorithms of the second type take advantage of exact counting results to construct directly a configuration from the uniform distribution [15]. As a result these perfect samplers often operate in linear time with little more than the amount of random bits required by information theory bounds to generate a configuration [2,13]. It is very desirable to obtain such an algorithm when the combinatorial class to be sampled displays simple enumerative properties, like Formula (1) for triangulations.

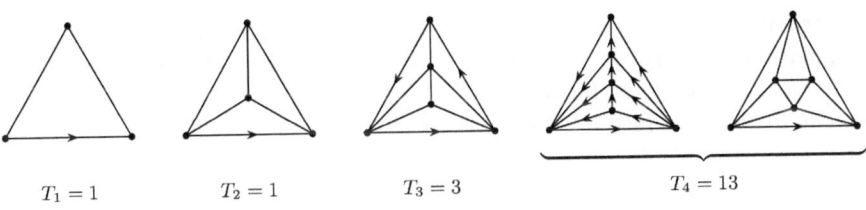

$T_1 = 1$ $T_2 = 1$ $T_3 = 3$ $T_4 = 13$

Fig. 2. The smallest triangulations with their inequivalent rootings.

New results. The central result of this paper is a one-to-one correspondence between the triangulations of $\mathcal{T}_n$ and the *balanced trees* of a new simple family $\mathcal{B}_n$ of plane trees. We give a linear *closure algorithm* that constructs a triangulation out of a balanced tree, and conversely, a linear *opening algorithm* that recovers a balanced tree as a special depth-first search spanning tree of a triangulation endowed with its minimal realizer. Realizers, or Schnyder tree decompositions, where introduced by Schnyder [30] to compute graph embeddings and have proved a fundamental tool in the study of planar graphs [8,10,14,20]. The role played in this paper by minimal realizers of triangulations is akin to the role of breadth-first search spanning trees in planar maps [9,27,29], and of minimal bipolar orientations in 2-connected maps [24,28], however the closure algorithm is very different from the closure used in the latter works. Our bijection allows us to address the three previously discussed problems.

From the coding point of view, our encoding in terms of trees preserves the entropy and satisfies linearity: each triangulation is encoded by one of the $\binom{4n}{n}$ bit strings of length $4n$ with sum of bits equal to n. The techniques of [18] to ensure locality apply to this $4n$ bit encoding. Optimal compacity can then be reached still within linear time, using for instance [7, Lemma 7].

From the exact enumerative point of view, the outcome of this work is a bijective derivation of Formula (1), giving it a simple interpretation in terms of trees. As far as we know, this is the first such bijective construction for a natural family of 3-connected planar graphs.

As far as random sampling is concerned, we obtain a linear time algorithm to sample random triangulations according to the (perfect) uniform distribution. In practice the speed we reach is about 100,000 vertices per second on a standard PC and triangulations with millions of vertices can be generated.

2   A One-to-One Correspondence

Let us first recall some definitions, illustrated by Figure 2.

Definition 1. *A planar map is an embedding of a connected planar graph in the oriented sphere. It is* rooted *if one of its edges it distinguished and oriented; this determines a root edge, a root vertex (its origin) and a root face (to its right), which is usually chosen as infinite face for drawing in the plane.*

A triangular map *is a rooted planar map with all faces of degree 3. It is a* triangulation *if moreover it has no loop or multiple edge. A triangular map of size n has 2n triangular faces, 3n edges and n + 2 vertices; the three vertices incident to the root face are called* external, *as opposed to the n − 1 internal other ones. The set of triangulations of size n is denoted by* $\mathcal{T}_n$.

2.1   From Trees to Triangulations

In view of Formula (1), it seems natural to ask for a bijection between triangulations and some kind of quaternary trees: indeed the number of such trees with n nodes is well known to be $\frac{(4n)!}{n!(3n+1)!}$. It proves however more interesting to consider the following less classical family of plane trees, illustrated by Fig. 3:

Definition 2. *Let* $\mathcal{B}_n$ *be the set of plane trees with n nodes each carrying two leaves and rooted on one of these leaves.*

It will prove useful to make a distinction between *nodes* (vertices of degree at least 2) and *leaves* (vertices of degree 1), and between *inner edges* (connecting two nodes) and *external edges* (connecting a node to a leaf).

The partial closure.   We introduce here a partial closure operation that merges leaves to nodes in order to create triangular faces.

Let B be a tree of $\mathcal{B}_n$. The border of the infinite face consists of inner and external edges. An *admissible triple* is a sequence (e_1, e_2, e_3) of two successive inner edges followed by an external one in counterclockwise direction around the infinite face. An admissible triple is thus formed of three edges $e_1 = (v, v')$, $e_2 = (v', v'')$ and $e_3 = (v'', \ell)$ such that v, v' and v'' are nodes and ℓ is a leaf. The *local closure* of such an admissible triple (e_1, e_2, e_3) consists in merging the leaf ℓ with the node v so as to create a bounded face of degree 3. The external edge $e_3 = (v'', \ell)$ then becomes an internal edge (v'', v). For instance the first three edges after the root around the infinite face of the tree of Figure 4(a) form an admissible triple, and the local closure of this triple produces the planar map of Figure 4(b). In turn, the first three edges of this map form a new admissible triple, and its local closure yields the map of Figure 4(c).

The *partial closure* $\tilde{B}$ of a tree B is the result of the greedy recursive application of local closure to all admissible triples available. The partial closure of the tree of Figure 4(a) is shown on Figure 4(d). At a given step of the construction,

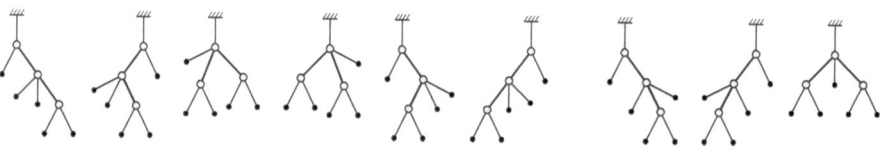

Fig. 3. The 9 elements of the set $\mathcal{B}_3$.

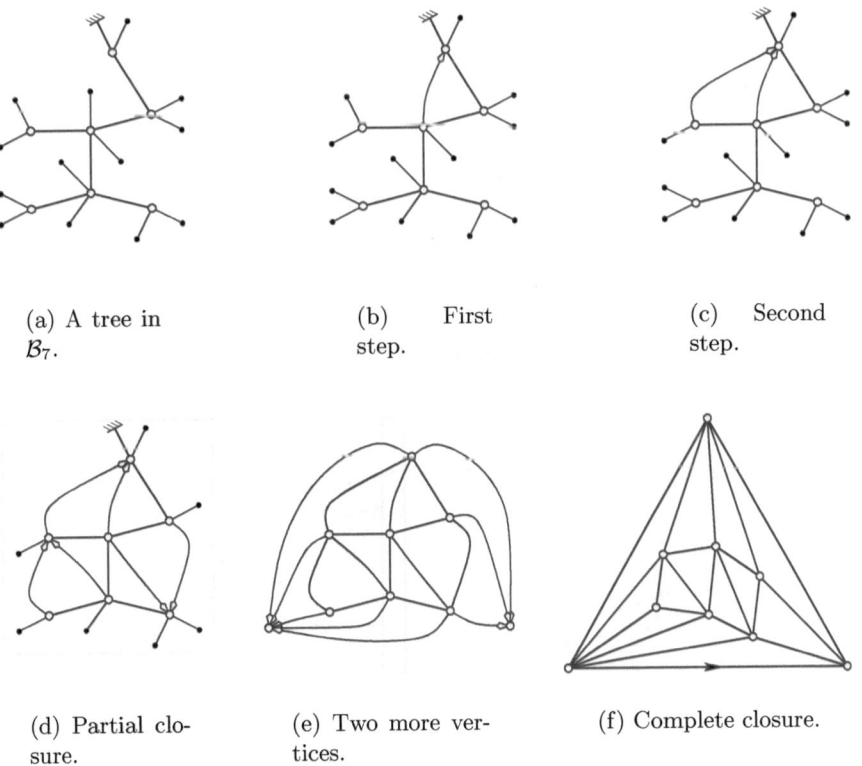

(a) A tree in $\mathcal{B}_7$.

(b)   First step.

(c)   Second step.

(d) Partial closure.

(e) Two more vertices.

(f) Complete closure.

Fig. 4. Complete closure construction on an element of $\mathcal{B}_7$.

there are usually several admissible triples, but their local closures are independent so that the order in which they are performed is irrelevant and the final map $\tilde{B}$ is uniquely defined.

In the tree B there are two more external edges than sides of inner edges in the infinite face, and this property is preserved by local closures. When the partial closure ends, there is no more admissible triple but some leaves remain unmatched. Hence in the infinite face of $\tilde{B}$ no two inner edges can be consecutive: each inner edge lies between two external edges, as illustrated by Figures 4(d) and 5(a) (ignore orientations and colors for the time being). More precisely the external edges and sides of inner edges alternate except at two special nodes: these two nodes v_0 and v_0' each carry two external edges with leaves ℓ_1, ℓ_2 and ℓ_1', ℓ_2' such that ℓ_1 (resp. ℓ_1') follows ℓ_2 (resp. ℓ_2') in the infinite face.

Observe that the partial closure of a tree is defined regardless of which of its leaves is the root. A tree B of $\mathcal{B}_n$ is said to be *balanced* if its root leaf is one of the two leaves ℓ_1 or ℓ_1' of its partial closure $\tilde{B}$. Let $\mathcal{B}_n^*$ be the subset of balanced trees of $\mathcal{B}_n$. The fourth, sixth and eighth trees in Figure 3 are balanced. The following immediate property shall be useful later on.

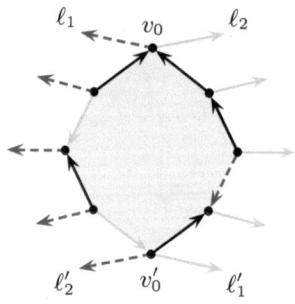

(a) After partial closure.

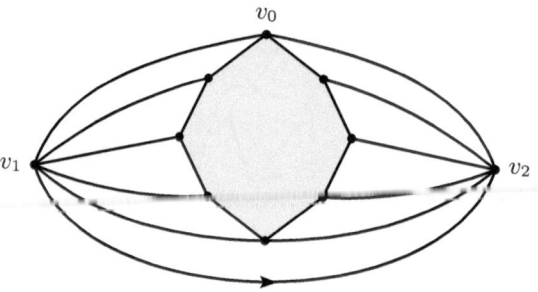

(b) After complete closure.

Fig. 5. Generic situation.

Property 1. Let B be a balanced tree; then each local closure is performed between a leaf ℓ and a vertex v that is before ℓ in the left-to-right preorder on B.

The complete closure. Let B be a balanced tree of $\mathcal{B}_n^*$, and call v_0 and v_0' the two special nodes of $\tilde{B}$ that carry the leaves ℓ_1, ℓ_2, ℓ_1' and ℓ_2'. The *complete closure* of B is obtained from its partial closure as follows (see Figures 4 and 5(b)):

1. merge ℓ_1, ℓ_2' and all leaves in between at a new vertex v_1;
2. merge ℓ_1', ℓ_2 and all leaves in between at a new vertex v_2;
3. add a root edge (v_1, v_2).

The result of this complete closure is clearly a triangular map, which we denote $\bar{B}$. Apart from the orientation of the root, the complete closure is more generally well defined for any tree of $\mathcal{B}_n$ and does not depend on which of the $2n$ leaves is the root of the tree. Since $2n$ rooted trees correspond to a given unrooted one B (or n in the exceptional case where B has a global symmetry),

Fig. 6. Local property of a realizer.

in general $2n$ trees have the same image. This image is a triangular map with a marked (non oriented) edge (v_1, v_2). However the use of the subset of balanced trees allows us to state a plain one-to-one correspondence rather than a "$2n$-to-2" one. We shall prove the following theorem in Section 3:

Theorem 1. *The complete closure is a one-to-one correspondence between the set $\mathcal{B}_n^*$ of balanced plane trees with n nodes with two leaves per node, and the set $\mathcal{T}_n$ of rooted triangulations of size n.*

Although the constructions are formally unrelated, the terminology we used here is reminiscent from [9,24,27], where bijections were proposed between some trees and planar maps with multiple edges.

2.2   From Triangulations to Trees

Minimal realizer.   We shall use the following notion, due to Schnyder [30].

Definition 3. *Let T be a triangulation, with root edge (v_1, v_2), and with v_0 its third external vertex. A realizer of T is a coloration of its internal edges in three colors c_0, c_1 and c_2 satisfying the following conditions:*

- *for each $i \in \{0, 1, 2\}$, edges of color c_i form a spanning oriented tree of $T \setminus \{v_{i+1}, v_{i+2}\}$ rooted on v_i; this induces an orientation of edges of color c_i toward v_i, such that each vertex has exactly one outgoing edge of color c_i;*
- *around each internal vertex, outgoing edges of each color always appear in the cyclic order shown on Figure 6, and entering edges of color c_i appear between outgoing edges of the two other colors.*

From now on, this second condition is referred to as Schnyder condition.

Realizers of triangulations satisfy a number of nice properties [12,14,30], among which we shall use the following ones:

Proposition 1.   – *Every triangulation has a realizer.*
- *The set of realizers of a triangulation can be endowed with an order for which there is a unique minimal (resp. maximal) element.*
- *The minimal realizer of a triangulation T is the unique realizer of T that contains no direct cycle.*
- *The minimal realizer of a triangulation can be computed in linear time.*

Depth-first search opening. Let T be a triangulation, endowed with its minimal realizer. Let (v_1, v_2) be its root edge, v_0 be the other external vertex. We construct a spanning tree of $T \setminus \{v_1, v_2\}$ using a right-to-left depth-first search traversal of T, modified to accept edges only if they are oriented toward the root:

1. delete (v_1, v_2), and detach (v_0, v_1) and (v_0, v_2) from v_1 and v_2 to form two new leaves ℓ_1, ℓ_2 attached to v_0;
2. set $v \leftarrow v_0$ and $e \leftarrow (v_0, \ell_2)$, and mark v and e;
3. as long as $e \neq (v_0, \ell_1)$, repeat:
   a) $e' \leftarrow (v, u)$, the edge after e around v in *clockwise direction*;
   b) *special orientation test:* if e' is oriented $v \to u$ and is not marked, mark e' and detach it from u to produce a leaf attached to v;
   c) otherwise, if u is marked and e' is not, set $e \leftarrow e'$;
   d) otherwise, mark both u and e' if necessary and set $e \leftarrow e'$ and $v \leftarrow u$.

Step 3c prevents the opening algorithm from forming a cycle of marked edges and ensures that it eventually terminates. Let $S(T)$ be the visited tree, containing all marked edges. Without Step 3b, the opening algorithm would be a standard right-to-left depth-first search, and $S(T)$ would be the corresponding spanning tree. We shall prove the following proposition:

Proposition 2. *For any triangulation T, the tree $S(T)$ is a spanning tree of $T \setminus \{v_1, v_2\}$. Moreover it is the unique balanced tree with complete closure T.*

Because of the minimal orientation of T (without counterclockwise circuit), we shall see that the condition of Step 3c is in fact never satisfied. This line of the algorithm could thus as well be ignored: it was included only to make clear the fact that the algorithm terminates.

3   Proofs

3.1   The Closure Produces a Triangulation

The closure construction adds edges to a planar map and only creates triangular faces. It is thus clear that the resulting map is a triangular map with external vertices v_0, v_1 and v_2, and with exactly two more vertices than B has nodes. Let us show that $\bar{B}$ is indeed a triangulation, *i.e.* has no multiple edge.

Let B be a balanced tree of $\mathcal{B}_n^*$. By definition the root leaf ℓ_1 of B is immediately followed around v_0 in clockwise direction by a second leaf ℓ_2. Set ℓ_1 in color c_1, ℓ_2 in color c_2, and other edges incident to v_0 in color c_0. Upon orienting all inner edges of B toward v_0 and all external edges toward their leaf, all vertices but v_0 have three outgoing edges. Since the tree B is acyclic, its orientation induces a unique coloration of edges satisfying the Schnyder condition (Figure 6) at all vertices but v_0.

Lemma 1. *The orientation and coloration of edges still satisfy the Schnyder condition on each node but v_0 after the partial closure of B.*

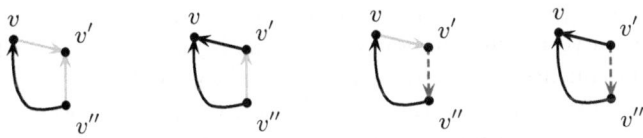

Fig. 7. The different cases of closure of a leaf.

Proof. This lemma is checked iteratively, by observing that each face created during the partial closure falls into one of the four types indicated on Figure 7 (up to cyclic permutation of colors). Indeed, consider an admissible triple (e_1, e_2, e_3). Assuming that the external edge e_3 to be closed is of color c_0, only two colors are possible for e_2 in view of the Schnyder condition at v''. In each case again, only two colors are possible for e_1. Finally in all four cases, the merging of ℓ into v does not contradict the Schnyder condition at v.

Property 2. If a (triangular) face of $\tilde{B}$ is oriented so that its sides form a circuit, then this circuit is necessarily oriented in the clockwise direction. More generally, each circuit in $\tilde{B}$ is created by the closure of a (last) leaf, the orientation of which imposes on the circuit to be clockwise.

Lemma 2. *After the complete closure, the Schnyder condition is satisfied at each internal vertex, and, apart from the three edges of the root face, each external vertex v_i is incident only to entering edges of color c_i.*

Proof. As illustrated by Figure 5(a), the Schnyder condition on nodes along the border of the partial closure implies that all external edges between ℓ_1 and ℓ'_2 (resp. ℓ_2 and ℓ'_1) are of color c_1 (resp. c_2). This is readily checked iteratively by a case analysis akin to the previous one.

The following proposition can be seen as an independant result on realizers.

Proposition 3. *A triangular map endowed with a colored 3-orientation satisfying the Schnyder condition on its inner vertices and the special condition on its three external vertices is in fact a triangulation endowed with a realizer.*

Proof. Let us first consider the color c_0 of the external vertex v_0. By Schnyder condition each inner vertex has exactly one outgoing edge of color c_0. In particular any cycle of edges of color c_0 is in fact a circuit. Moreover from each inner vertex originates a unique oriented path of color c_0, ending either in v_0 or on a circuit of color c_0.

Now consider two paths with distinct colors, say c_0 and c_1. In view of the Schnyder coloring, a crossing between these two paths is necessarily of the type:

Hence two such paths can only cross once. Here crossing is taken in the (weak) sense having a vertex in common, even if this is just the origin of the path.

As a consequence monochrome circuits are vertex disjoint, and thus ordered by inclusion with respect to the external face. Consider a vertex v on an innermost circuit C. The Schnyder condition at v provides an edge e going out of v into the inner region delimited by C. Since this region contains no cycle the oriented path extending e has to cross C a second time, in contradiction with the previous discussion. This excludes monochrome circuits and proves that, for each $i = 0, 1, 2$, edges of color c_i form an oriented tree rooted at v_i. In particular multiple edges are excluded, and the coloring satisfies the definition of a realizer.

Combining Proposition 3 and Property 2, we obtain the following corollary that concludes the first part of the proof.

Corollary 1. *Upon keeping colors, the closure maps a balanced trees B of $\mathcal{B}_n^*$ on a triangulation with $n + 2$ vertices endowed with its minimal realizer.*

3.2   The Depth-First Search Opening Is Inverse to Closure

The following Lemmas 3-6 imply Proposition 2, and, together with Corollary 1, conclude the proof of Theorem 1.

Lemma 3. *The depth-first search opening visits all vertices of $T \setminus \{v_1, v_2\}$.*

Proof. Assume that the inner vertex v is not visited by the opening algorithm, that is to say, v does not belong to $S(T)$. By definition of realizers, there is a unique oriented path P of color c_0 starting in v and ending in v_0. Let t be the last vertex on P that does not belong to $S(T)$, and $u \in S(T)$ the next vertex on P. Then the edge between t and u is oriented toward u and should have been included in the tree when u was visited.

Lemma 4. *The conditions of Step 3c are never satisfied.*

Proof. Consider the first time the conditions of Step 3c are satisfied. Up to that point an oriented tree S was constructed that contains v and u but not the edge (v, u). Since the unmarked edge (v, u) was not considered by Step 3b, it is oriented from u toward v.

Let E be the set of edges that were already cut by Step 3b. Then S is the initial part of the right-to-left depth-first search tree of $T \setminus E$. In particular since the edge (u, v) is probed from v, the vertex u is an ancestor of v in the tree. But then the tree contains an oriented path from v to u, that forms a couterclockwise circuit with (u, v). This contradicts the minimality of the orientation.

Lemma 5. *Edges that are cut by the opening algorithm lie on the left-hand side of the tree, as in Property 1. Hence the complete closure of $S(T)$ is T.*

Proof. As already observed, as the algorithm proceeds the tree which is constructed can be thought of as the right-to-left depth-first search tree of a submap of T. In particular when the algorithm probes an edge $e' = (v, u)$, this edge lies on the left-hand side of the tree, as in Property 1.

To check that the complete closure of $S(T)$ is T, it is sufficient to check that a cut edge would be properly replaced by local application of the closure algorithm. Since cut edges are bordered on one side by the infinite face and the final tree is a spanning tree, then the other face is bounded, that is triangular. Hence when $e' = (v, u)$ is cut, the vertex u lie two corners away from v along the infinite face in clockwise direction, as specified for admissible triples.

Lemma 6. *At most one spanning tree of $T \setminus \{v_1, v_2\}$ satisfies Property 1.*

Proof. Assume there are two such trees S and S'. Consider a left-to-right depth-first search traversal of both trees in parallel. Let $e = (v, u)$ be the first met edge that belongs to one of them – say S – and not to the other one. As the tree S' is also a spanning tree, there exists in S' a path from u to v_0, the first edge of which, (u, t), is oriented from u towards t. This orientation forbids to this edge to belong to the tree S; it corresponds thus in that tree to the closure of a leaf of u. But since the edge (v, u) has been visited before (u, t) in the depth-first search traversal, this contradicts Property 1; there is therefore only one covering tree of T that satisfies this property.

4   Applications

4.1   An Explicit Optimal Code for Triangulations

As a first byproduct of Theorem 1, we obtain a code of triangulations in $\mathcal{T}_n$ by balanced trees in $\mathcal{B}_n$. Since a triangulation can be endowed with its minimal realizer in linear time (Proposition 1), the tree code can be obtained in linear time. Another fundamental feature of our code is that the tree code is a spanning tree of the original triangulation, making locality amenable to the techniques of [18]. Elements of $\mathcal{B}_n$ can themselves be coded by bit strings of length $4n - 2$ and weight $n - 1$ using a trivial variant of the usual prefix code for trees.

Theorem 2. *A tree B of $\mathcal{B}_n$ can be linearly represented by the word $s(B)$ that is obtained by writing 1 for "down" steps along inner edges, and 0 for leaves and for "up" steps along inner edges, during left-to-right depth-first search traversal.*

Hence a code for triangulations which is a subset of the set S of bit strings with length $4n - 2$ and weight $n - 1$. According to [7, Lem. 7] it can be given in linear time a representation as a bit string of length $\log |S| + o(n) \sim \log \binom{4n}{n} \sim \frac{256}{27} n$.

4.2   A Bijective Proof of Formula (1)

Proposition 4. *The set* $\mathcal{B}_n$ *has cardinality* $\frac{2}{4n-2} \cdot \binom{4n-2}{n-1}$.

Proof. As for classical prefix code of trees, the code words corresponding to trees of $\mathcal{B}_n$ can be easily characterized: they are the bit strings of length $4n - 2$ with weight $n - 1$ such that any proper prefix satisfies $3|u|_1 - |u|_0 > -2$. Now the number of such bit strings is readily obtained by the cycle lemma: in each cyclic class of words with length $4n - 2$ and weight $n - 1$, exactly 2 elements among $4n - 2$ are codes words (or 1 among $2n - 1$ for symmetric classes).

Now as seen in Section 2.1, any tree in $\mathcal{B}_n$ has two particular leaves among its $2n$ ones, and it is balanced if and only if one of these two is its root. Hence the ratio of balanced trees in $\mathcal{B}_n$ is $\frac{2}{2n}$. From Theorem 1 we obtain:

Corollary 2. *The number of triangulations with $2n$ triangles, $3n$ edges and $n+2$ vertices is* $\frac{2}{2n} \cdot \frac{2}{4n-2} \cdot \binom{4n-2}{n-1}$, *which is exactly Formula (1).*

4.3   Linear Time Perfect Random Sampling of Triangulations

The closure construction provides a sampling algorithm with linear complexity:

1. generate a random bit string of length $4n - 2$ and weight $n - 1$;
2. choose randomly one of its two cyclic shift that code an element of $\mathcal{B}_n$;
3. decode this word to construct the corresponding tree;
4. construct its partial closure by turning around the tree; using a stack, this can be done in at most two complete turns, hence in linear time;
5. complete the closure and choose a random orientation for the edge (v_1, v_2).

Theorem 3. *This algorithm produces in linear time a random triangulation uniformly chosen in* $\mathcal{T}_n$.

Observe that Steps 1–3 correspond to a special case of the classical algorithm described *e.g.* in [2] for sampling trees.

Acknowledgments. We thank the authors of [7] for providing a draft of this work and for interesting discussions. In particular special thanks are due to Nicolas Bonichon for his invaluable knowledge of minimal realizers, and to Cyril Gavoille for pointing out Lemma 7 in [7].

References

1. M.E. Agishtein and A.A. Migdal. Geometry of a two-dimensional quantum gravity: numerical study. *Nucl. Phys. B*, 350:690–728, 1991.
2. L. Alonso, J.-L. Remy, and R. Schott. A linear-time algorithm for the generation of trees. *Algorithmica*, pages 162–183, 1997.

3. J. Ambjørn, P Białas, Z. Burda, J. Jurkiewicz, and B. Petersson. Effective sampling of random surfaces by baby universe surgery. *Phys. Lett. B*, 325:337–346, 1994.

4. J. Ambjørn, B. Durhuus, and T. Jonsson. *Quantum geometry*. Cambridge Monographs on Mathematical Physics. Cambridge University Press, Cambridge, 1997.

5. C. Banderier, P. Flajolet, G. Schaeffer, and M. Soria. Planar maps and Airy phenomena. In *ICALP*, pages 388–402, 2000.

6. N. Bonichon. A bijection between realizers of maximal plane graphs and pairs of non-crossing Dyck paths. In *FPSAC*, 2002.

7. N. Bonichon, C. Gavoille, and N. Hanusse. An information-theoretic upper bound of planar graphs using triangulations. In *STACS*, 2003.

8. N. Bonichon, B. Le Saëc, and M. Mosbah. Wagner's theorem on realizers. In *ICALP*, pages 1043–1053, 2002.

9. M. Bousquet-Melou and G. Schaeffer. The degree distribution in bipartite planar maps: applications to the Ising model. 2002, arXiv:math.CO/0211070.

10. E. Brehm. 3-orientations and Schnyder 3-tree-decompositions. Master's thesis, FB Mathematik und Informatik, Freie Universität Berlin, 2000.

11. R. C.-N. Chuang, A. Garg, X. He, M.-Y. Kao, and H.-I Lu. Compact encodings of planar graphs via canonical orderings. In *ICALP*, pages 118–129, 1998.

12. H. de Fraysseix and P. Ossona de Mendez. Regular orientations, arboricity and augmentation. In *Graph Drawing*, 1995.

13. P. Duchon, P. Flajolet, G. Louchard, and G. Schaeffer. Random sampling from Boltzmann principles. In *ICALP*, pages 501–513, 2002.

14. S. Felsner. Convex drawings of planar graphs and the order dimension of 3-polytopes. *Order*, 18:19–37, 2001.

15. P. Flajolet, P. Zimmermann, and B. Van Cutsem. A calculus for random generation of labelled combinatorial structures. *Theoret. Comput. Sci.*, 132(2):1–35, 1994.

16. P.-M. Gandoin and O. Devillers. Progressive lossless compression of arbitrary simplicial complexes. *ACM Transactions on Graphics*, 21(3):372–379, 2002.

17. Z. Gao and J. Wang. Enumeration of rooted planar triangulations with respect to diagonal flips. *J. Combin. Theory Ser. A*, 88(2):276–296, 1999.

18. X. He, M.-Y. Kao, and H.-I Lu. Linear-time succinct encodings of planar graphs via canonical orderings. *SIAM J. on Discrete Mathematics*, 12(3):317–325, 1999.

19. X. He, M.-Y. Kao, and H.-I Lu. A fast general methodology for information-theoretically optimal encodings of graphs. *SIAM J. Comput*, 30(3):838–846, 2000.

20. G. Kant. Drawing planar graphs using the canonical ordering. *Algorithmica*, 16:4–32, 1996. (also *FOCS'92*).

21. D. King and J Rossignac. Guaranteed 3.67v bit encoding of planar triangle graphs. In *CCCG*, 1999.

22. H.-I Lu. Linear-time compression of bounded-genus graphs into information-theoretically optimal number of bits. In *SODA*, pages 223–224, 2002.

23. D. Osthus, H.J. Prömel, and A. Taraz. On random planar graphs, the number of planar graphs and their triangulations. *J. Comb. Theory, Ser. B*, 2003. to appear.

24. D. Poulalhon and G. Schaeffer. A bijection for triangulations of a polygon with interior points and multiple edges. *Theoret. Comput. Sci.*, 2003. to appear.

25. L. B. Richmond and N. C. Wormald. Almost all maps are asymmetric. *J. Combin. Theory Ser. B*, 63(1):1–7, 1995.

26. J. Rossignac. Edgebreaker: Connectivity compression for triangle meshes. *IEEE Transactions on Visualization and Computer Graphics*, 5(1):47–61, 1999.

27. G. Schaeffer. Bijective census and random generation of Eulerian planar maps with prescribed vertex degrees. *Electron. J. Combin.*, 4(1):# 20, 14 pp., 1997.

28. G. Schaeffer. *Conjugaison d'arbres et cartes combinatoires aléatoires.* PhD thesis, Université Bordeaux I, 1998.
29. G. Schaeffer. Random sampling of large planar maps and convex polyhedra. In *STOC*, pages 760–769, 1999.
30. W. Schnyder. Embedding planar graphs on the grid. In *SODA*, pages 138–148, 1990.
31. W. T. Tutte. A census of planar triangulations. *Canad. J. Math.*, 14:21–38, 1962.
32. D. B. Wilson. Annotated bibliography of perfectly random sampling with Markov chains. `http://dimacs.rutgers.edu/~dbwilson/exact`.

Generating Labeled Planar Graphs Uniformly at Random

Manuel Bodirsky[1], Clemens Gröpl[2], and Mihyun Kang[1]

[1] Humboldt-Universität zu Berlin, Germany
{bodirsky,kang}@informatik.hu-berlin.de

[2] Freie Universität Berlin, Germany
groepl@inf.fu-berlin.de

Abstract. We present an expected polynomial time algorithm to generate a labeled planar graph uniformly at random. To generate the planar graphs, we derive recurrence formulas that count all such graphs with n vertices and m edges, based on a decomposition into 1-, 2-, and 3-connected components. For 3-connected graphs we apply a recent random generation algorithm by Schaeffer and a counting formula by Mullin and Schellenberg.

1 Introduction

A *planar graph* is a graph wich can be embedded in the plane, as opposed to a *map*, which is an embedded graph. There is a rich literature on the enumerative combinatorics of maps, starting with Tutte's census papers, e.g. [20], and an efficient random generation algorithm was recently obtained by Schaeffer [16]. Much less is known about random planar graphs, although they recently attracted much attention [3,12,14,6,9,5]. Even the expected number of edges for random planar graphs is not known (both in the labeled and in the unlabeled case), and the gap between known upper and lower bounds is still large [14,9,5]. There are also some results on the asymptotic number of labeled planar graphs [3,14]. If we had an efficient algorithm to generate a planar graph uniformly at random, we could experimentally verify conjectures about properties of the random planar graph. We could also use it to evaluate the average-case running times of algorithms on planar graphs. Denise, Vasconcellos and Welsh [6] introduced a Markov chain having the uniform distribution on all labeled planar graphs as its stationary distribution. However, the mixing time is unknown and seems hard to analyze, and is perhaps not even polynomial. Moreover, their algorithm only approximates the uniform distribution.

We obtain the first expected polynomial time algorithm to generate a labeled planar graph uniformly at random.

[1] This research was supported by the Deutsche Forschungsgemeinschaft within the European graduate program 'Combinatorics, Geometry, and Computation' (No. GRK 588/2).

[2] Most of this work was done while the author was supported by DFG grant Pr 296/3.

J.C.M. Baeten et al. (Eds.): ICALP 2003, LNCS 2719, pp. 1095–1107, 2003.

Theorem 1. *A random planar graph with n vertices and m edges can be generated uniformly at random in expected time $O(n^{13/2})$ after a deterministic preprocessing of running time $O(n^7 (\log n)^2 (\log \log n))$. The memory requirement is $O(n^5 \log n)$ bits.*

We believe that the actual generation is much faster in practice, see Section 6. Our result uses known graph decomposition and counting techniques [21, 24] to reduce the counting and generation of labeled planar graphs to the counting and generation of 3-connected *rooted planar maps*, also called *c-nets*.

Usually a planar graph has many embeddings which are non-isomorphic as maps, but some graphs have a unique embedding. A classical theorem of Whitney (see e.g. [7]) asserts that 3-connected planar graphs are *rigid* in the sense that all embeddings in the sphere are combinatorially equivalent. As *rooting* destroys any further symmetries, c-nets are closely related to 3-connected *labeled* planar graphs. Moreover, the 'degrees of freedom' of the embedding of a planar graph are governed by its connectivity structure. We exploit this fact by composing a planar graph out of 1-, 2-, and 3-connected components.

The generation procedure first determines the number of components, and how many vertices and edges they shall contain. Each connected component is generated independently from the others, but having the chosen numbers of vertices and edges. To generate a connected component with given numbers of vertices and edges, we decide for a decomposition into 2-connected subgraphs and how the vertices and edges shall be distributed among its parts. So far this approach is similar to the one used in [4], where the goal was to generate random outerplanar graphs. In the planar case we need to go one step further.

Trakhtenbrot [19] showed that every 2-connected graph is uniquely composed of special graphs (called *networks*) of three kinds. Such networks can be combined in series, in parallel, or using a 3-connected graph as a template (see Theorem 2 below). Using this composition we can then employ known results about the counting and generation of 3-connected planar maps.

The concept of rooting plays an important role for the enumeration of planar maps. A *face-rooted* map is one with a distinguished edge which lies on the outer face and to which a direction is assigned. The rooting forces isomorphisms to map the outer face to the outer face, keep the root edge incident to the outer face, and preserve its direction. The enumeration of 3-connected face-rooted unlabeled maps with given numbers of vertices and faces was achieved by Mullin and Schellenberg [13]. We invoke their closed formulas in order to count 3-connected labeled planar graphs with given numbers of vertices and edges. For the generation of 3-connected labeled planar graphs with given numbers of vertices and edges we employ a recent algorithm by Schaeffer [17] running in expected polynomial time.

When we apply the various counting and generation subroutines along the stages of the connectivity decomposition, we must branch with the right probabilities. Instead of explicit (closed-form) counting formulas, which seem difficult to obtain, we derive recurrence formulas that can be evaluated in polynomial time using dynamic programming. These recurrence formulas can be translated immediately into a generation procedure.

The paper is organized as follows: In the next section we give the graph theoretic background for the decomposition of planar graphs along their connectivity structure. This decomposition guides us when we derive the counting formulas for planar graphs in the following three sections. We analyze the running time and memory requirements of the corresponding generation procedure in Section 7. Some results from an implementation of the counting part are shown in Section 8. We conclude with a discussion of variations of the approach and how to derive a generation procedure for unlabeled planar graphs.

2   Decomposition by Connectivity

Let us recall and fix some terminology [7,22,23,21]. A *graph* will be assumed unoriented and *simple*, i.e., having no loops or multiple (also called *parallel*) edges; if multiple edges are allowed, the term *multigraph* will be used. We consider labeled graphs whose vertex sets are initial segments of $\mathbb{N}_0$.

Every connected graph can be decomposed into *blocks* by being split at cutvertices. Here a block is a maximal subgraph that is either 2-connected, or a pair of adjacent vertices, or an isolated vertex. The *block structure* of a graph G is a tree whose vertices are the cutvertices of G and the blocks (considered as vertices) of G, where adjacency is defined by containment. Conversely, we will *compose* connected graphs by identifying the vertex 0 of one part with an arbitrary vertex of the other. A formal definition of compose operations is given at the end of this section.

A *network* N is a multigraph with two distinguished vertices 0 and 1, called its *poles*, such that the multigraph N^* obtained from N by adding an edge between its poles is 2-connected. (The new edge is not considered a part of the network.) We can replace an edge uv of a network M with another network X_{uv} by identifying u and v with the poles 0 and 1 of X_{uv}, and iterate the process for all edges of M. Then the resulting graph G is said to have a *decomposition* with *core* M and *components* X_e, $e \in E(M)$.

Every network can be decomposed into (or composed out of) networks of three special types. A *chain* is a network consisting of 2 or more edges connected in *series* with the poles as its terminal vertices. A *bond* is a network consisting of 2 or more edges connected in *parallel*. A *pseudo-brick* is a network N with no edge between its poles such that N^* is 3-connected. (3-connected subgraphs are sometimes called bricks.) A network N is called an *h-network* (respectively, a *p-network*, or an *s-network*) if it has a decomposition whose core is a pseudo-brick (respectively, a bond, or a chain). Trakhtenbrot [19] formulated a canonical decomposition theorem for networks:

Theorem 2 (Trakhtenbrot).   *Any network with at least 2 edges belongs to exactly one of the 3 classes: h-networks, p-networks, s-networks. An h-network has a unique decomposition and a p-network (respectively, an s-network) can be* *uniquely decomposed into components which are not themselves p-networks (s-networks), where uniqueness is up to orientation of the edges of the core, and also up to their order if the core is a bond.*

A network is *simple* if it is a simple graph. Let $N(n, m)$ be the number of simple planar networks on n vertices and m edges. In view of Theorem 2 we introduce the functions $H(n, m)$, $P(n, m)$, and $S(n, m)$ that count the number of simple planar h-, p-, and s-networks on n vertices and m edges.

Let us define *compose operations* for the three stages $c = 0, 1, 2$ of the connectivity decomposition formally as follows. Assume that M and X are graphs on the vertex sets $[0 .. k - 1]$ and $[0 .. i - 1]$ and we want to compose them by identifying the vertices j of X with the vertices v_j of M, for $j = 0, \ldots, c - 1$, such that the resulting graph will have $n := k + i - c$ vertices. (No vertices are identified for $c = 0$.) Moreover, let S be a set of $i - c$ vertices from $[c .. n - 1]$ which are designated for the remaining part of X. Let M' be the graph obtained by mapping the vertices of M to the set $[0 .. n - 1] \setminus S$, retaining their relative order. Let X' be the graph obtained by mapping the vertices $[c .. i - 1]$ of X to the set S, retaining their relative order, and mapping j to the image of v_j in M' for $j = 0, \ldots, c - 1$. Then the result of the compose operation for the arguments M, $(v_0, \ldots, v_{c-1})$, X, and S is the graph with vertex set $[0 .. n - 1]$ and edge set $E(M') \cup E(X')$.

We use $G^{(c)}(n, m)$ to denote the number of c-connected planar graphs with n vertices and m edges.

3   Planar Graphs

We show how to count and generate labeled planar graphs with a given number of vertices and edges in three steps. A first easy recurrence formula reduces the problem to the case of connected graphs. In the next section, we will use the block structure to reduce the problem to the 2-connected case. This may serve as an introduction to the method before we go into the more involved arguments of Section 5.

Let $F_k(n, m)$ denote the number of planar graphs with n vertices and m edges having k connected components. Clearly, $F_1(n, m) = G^{(1)}(n, m)$ and $G^{(0)}(n, m) = \sum_{k=1}^{n} F_k(n, m)$. Moreover,

$$F_k(n, m) = 0 \quad \text{for } m + k < n.$$

We count $F_k(n, m)$ by induction on k. Every graph with $k \geq 2$ connected components can be decomposed into the connected component containing the vertex 0 and a remaining part, using the inverse of the compose operation for $c = 0$ as defined in Section 2. If the split off part has i vertices, then there are $\binom{n-1}{i-1}$ ways to choose its vertex set, as the vertex 0 is always contained in it. The remaining part has $k - 1$ connected components. We obtain the recurrence formula

$$F_k(n, m) = \sum_{i=1}^{n-1} \sum_{j=0}^{m} \binom{n-1}{i-1} G^{(1)}(i, j) F_{k-1}(n - i, m - j).$$

Thus it suffices to count connected graphs. But the counting recurrence also has an analogue for generation: Assume that we want to generate a planar graph G with n vertices and m edges uniformly at random. First, we choose $k \in [1 .. n]$ with probability proportional to $F_k(n, m)$. Then we choose the number of vertices i of the component containing the vertex 0 and its number of edges j with a joint probability proportional to $\binom{n-1}{i-1} G^{(1)}(i, j) F_{k-1}(n - i, m - j)$. We also pick an $(i-1)$-element subset $S' \subseteq [1 .. n - 1]$ uniformly at random and set $S := S' \cup \{0\}$. Then we compose G (as explained in Section 2) out of a random connected planar graph with parameters i and j, which is being mapped to the vertex set S, and a random planar graph with parameters $n-i$ and $m-j$ having $k - 1$ connected components, which is generated in the same manner.

4   Connected Planar Graphs

In this section we reduce the counting and generation of connected labeled planar graphs to the 2-connected case. Let $M_d(n, m)$ denote the number of connected labeled planar graphs in which the vertex 0 is contained in d blocks. Here we will call them m_d-*planars*. An m_1-planar is a planar graph in which 0 is not a cutvertex. Clearly, $G^{(1)}(n, m) = \sum_{d=1}^{n-1} M_d(n, m)$ and

$$M_d(n, m) = 0 \quad \text{for} \quad n < d \text{ or } m < d.$$

In order to count m_d-planars by induction on d (for $d \geq 2$), we split off the largest connected subgraph containing the vertex 1 in which 0 is not a cutvertex. This is done by performing the inverse of the compose operation for $c = 1$ as defined in Section 2. If the split off m_1-planar has i vertices, then there are $\binom{n-2}{i-2}$ possible choices for its vertex set, as the vertices 0 and 1 are always contained in it. The remaining part is an m_{d-1}-planar. Thus

$$M_d(n, m) = \sum_{i=2}^{n-d+1} \sum_{j=1}^{m-1} \binom{n-2}{i-2} M_1(i, j) M_{d-1}(n - i + 1, m - j),$$

and this immediately translates into a generation procedure.

Next we consider m_1-planars. The *root block* is the block containing the vertex 0. A recurrence formula for m_1-planars arises from splitting off the subgraphs attached to the root block at its cutvertices one at a time. Thus we consider m_1-planars such that the root block has b vertices and the c least labeled vertices in the root block are no cutvertices. Let us call them $m_{b,c}$-*planars* and denote the number of $m_{b,c}$-planars with n vertices and m edges by $M_{b,c}(n, m)$. Then $M_1(n, m) = \sum_{b=1}^{n} M_{b,1}(n, m)$. The initial cases are graphs without cutvertices. We have

$$M_{b,b}(n, m) = \begin{cases} G^{(2)}(n, m) & \text{for } b = n > 2 \\ 1 & \text{for } b = n \in \{1, 2\} \text{ and } m = n - 1 \\ 0 & \text{for } b \neq n. \end{cases}$$

To count $M_{b,c}$ using $M_{b,c+1}$, we split off the subgraph attached to the c-th least labeled vertex in the root block, if it is a cutvertex. This can be any connected planar graph. The remaining part is an $m_{b,c+1}$-planar. If the split off subgraph has i vertices, then there are $\binom{n-1}{i-1}$ ways to choose them, as the vertex 0 of the subgraph will be replaced with the cutvertex. We obtain the recurrence formula

$$M_{b,c}(n,m) = \sum_{i=1}^{n-1}\sum_{j=0}^{m-1}\binom{n-1}{i-1}G^{(1)}(i,j)M_{b,c+1}(n-i+1,m-j).$$

Again, the generation procedure is straightforward.

5   2-Connected Planar Graphs

In this section we show how to count and generate 2-connected planar graphs. Note that every labeled 2-connected planar graph with n vertices and m edges is obtained from some simple planar network with n vertices and $m-1$ edges by adding an edge between the poles, then choosing $0 \le x, y \le n-1$, $x \ne y$, and exchanging the vertices 0 with x and 1 with y. Thus

$$G^{(2)}(n,m) = \begin{cases} \dfrac{\binom{n}{2}}{m}N(n,m-1) & \text{for } n \ge 3, m \ge 3 \\ 0 & \text{otherwise.} \end{cases}$$

Now we derive recurrence formulas for the number N of simple planar networks. Trakhtenbrot's decomposition theorem implies

$$N(n,m) = \begin{cases} P(n,m) + S(n,m) + H(n,m) & \text{for } n \ge 3, m \ge 2 \\ 0 & \text{otherwise.} \end{cases}$$

p-Networks. Let us call a p-network with a core consisting of k parallel edges a p_k-*network*, and let $P_k(n,m)$ be the number of p_k-networks having n vertices and m edges. Clearly, $P(n,m) = \sum_{k=2}^m P_k(n,m)$. In order to count p_k-networks by induction on k, we split off the component containing the vertex labeled 2 by performing the inverse of the compose operation for $c = 2$ as defined in Section 2. Technically, it is convenient to consider the split off component as a p_1-*network*. But note that according to the canonical decomposition, a p_1-network is either an h- or an s-network. Thus

$$P_1(n,m) = \begin{cases} H(n,m) + S(n,m) & \text{for } n \ge 3, m \ge 2 \\ 0 & \text{otherwise.} \end{cases}$$

The remaining part is a p_{k-1}-network (even if $k = 2$). For $k \ge 2$ we have

$$P_k(n,m) = 0 \quad \text{if } n \le 2 \text{ or } m < k.$$

If a p-network with n vertices is split into a p_1-network with i vertices and a p_{k-1}-network, there are $\binom{n-3}{i-3}$ ways how the vertex labels $[0 .. n-1]$ can be distributed among both sides, as the labels 0, 1, and 2 are fixed. We obtain the recurrence formula

$$P_k(n, m) = \sum_{i=3}^{n} \sum_{j=2}^{m-1} \binom{n-3}{i-3} P_1(i, j) P_{k-1}(n - i + 2, m - j).$$

s-Networks. Let us call an s-network whose core is a path of k edges an s_k-*network*, and denote the number of s_k-networks which have n vertices and m edges by $S_k(n, m)$. Then $S(n, m) = \sum_{k=2}^{m} S_k(n, m)$. We use induction on k again, but for s_k-networks we split off the component containing the vertex labeled 0. Again it can be considered as an s_1-network, and it is either an h- or a p-network, according to the canonical decomposition. Thus

$$S_1(n, m) = \begin{cases} H(n, m) + P(n, m) & \text{for } n \geq 3, m \geq 2 \\ 1 & \text{for } n = 2, m = 1 \\ 0 & \text{otherwise}. \end{cases}$$

The remaining part is an s_{k-1}-network (even if $k = 2$). For $k \geq 2$ we have

$$S_k(n, m) = 0 \quad \text{if } n < k + 1 \text{ or } m < k.$$

Concerning the number of ways how the labels can be distributed among both parts, note that the labels 0 and 1 are fixed, hence the new 0-root for the remaining part can be one out of $n - 2$ vertices, and then the number of choices for the internal vertices of the split off s_1-network is $\binom{n-3}{i-2}$. We obtain the recurrence formula

$$S_k(n, m) = (n - 2) \sum_{i=2}^{n-1} \sum_{j=1}^{m-1} \binom{n-3}{i-2} S_1(i, j) S_{k-1}(n - i + 1, m - j).$$

h-Networks. Let us call an h-network whose core is a pseudo-brick on k edges an h_k-*network*, and denote the number of h_k-networks with n vertices and m edges by $H_k(n, m)$. Then $H(n, m) = \sum_{k=5}^{m} H_k(n, m)$, as the smallest pseudo-brick has 5 edges. We can order the edges of the core lexicographically by the vertex numbers. A recurrence formula similar to the p- and s-network case arises from replacing the edges of the core with components one at a time and in lexicographic order. To give names to the intermediate stages, let $H_{k,\ell}(n, m)$ be the number of $h_{k,\ell}$-networks with n vertices and m edges, where an $h_{k,\ell}$-network is an h_k-network in which the components corresponding to the first ℓ edges of the core are simple edges. Thus $H_{m,m}(n, m)$ is the number of pseudo-bricks with n vertices and m edges, and $H_{k,k}(n, m) = 0$ for $k \neq m$. Applying the recurrence formula derived below for $\ell = k - 1$ down to 0, we can calculate $H_k(n, m) = H_{k,0}(n, m)$, and hence, $H(n, m)$. For the initial case, we have

$$H_{m,m}(n, m) = \frac{(n-2)!}{2} Q(n, m + 1),$$

where $Q(n, m)$ denotes the number of c-nets, i.e., rooted 3-connected simple maps, with n vertices and m edges (see the next section): for we assign 0 to the root vertex, 1 to the other vertex of the root edge and the remaining labels to the remaining vertices, and neglect the orientation. To count $H_{k,\ell}$ using $H_{k,\ell+1}$, we split off the ℓ-th component of an $h_{k,\ell}$-network, i.e., the component replacing the ℓ-th edge of the core. This can be a network of any of the three kinds. Thus

$$
H_1(n, m) = \begin{cases} N(n, m) + N(n, m-1) & \text{for } n \geq 3, \, m \geq 2 \\ 1 & \text{for } n = 2, \, m = 1 \\ 0 & \text{otherwise}. \end{cases}
$$

The remaining part is an $h_{k,\ell+1}$-network. If the ℓ-th component has i vertices, then there are $\binom{n-2}{i-2}$ ways to choose them, as the vertices 0 and 1 are merged with the endpoints of the ℓ-th edge of the core, respecting their relative order. We obtain the recurrence formula

$$
H_{k,\ell}(n, m) = \sum_{i=2}^{n-2} \sum_{j=1}^{m-k+1} \binom{n-2}{i-2} H_1(i, j) H_{k,\ell+1}(n-i+2, m-j+1).
$$

6   c-Nets

In the preceding sections, we have shown how to count and generate random planar graphs assuming that we can do so for c-nets, i.e., 3-connected simple rooted maps.

A *counting* formula for $Q(n, m)$ was derived by Mullin and Schellenberg in [13] in terms of given numbers of vertices and faces. Using Euler's formula, it asserts that

$$
Q(n, m) = 0 \quad \text{for } n < 4 \text{ or } m < n + 2
$$

and otherwise

$$
Q(n, m) = -\sum_{i=2}^{n} \sum_{j=n}^{m} (-1)^{i+j-n} \binom{i+j-n}{i} \binom{i}{2}
$$
$$
\times \left[\binom{2m-2n+2}{n-i} \binom{2n-2}{m-j} - 4 \binom{2m-2n+1}{n-i-1} \binom{2n-3}{m-j-1} \right].
$$

This concludes the counting task.

A *generation* algorithm for c-nets with given numbers of vertices and edges running in expected polynomial time algorithm is due to Schaeffer et al. [1, 2, 17, 15, 16]. Here we only outline the method. The c-net is obtained by extracting the 3-connected core from a 2-connected map. There is a linear time algorithm to generate 2-connected maps [15], and the extraction is linear as well [16]. If the parameters of the 2-connected map are tuned appropriately, chances are good that the resulting c-net will have the desired parameters. Otherwise the sample is rejected and the procedure restarts. A map with n vertices and m edges is said to have an *imbalance* x which is defined by $n+1 = m(\frac{1}{2}+x)$. To obtain a core with m edges and imbalance x, one should select a 2-connected map with imbalance $3x$ and $m/\alpha_0(3x)$ edges, where the *tuning ratio* is $\alpha_0(x) = \frac{(1-2x)(1+2x)}{3(1-2x/3)(1+2x/3)}$ [16,2]. We have $\alpha_0(x) = \Omega(1/m)$ in the worst case. The expected number of iterations is $O(m^{2/3} + 1/p_\nu)$ for any given number of edges, where the probability p_ν that the core (whose size obeys a bimodal distribution) has around m edges is $p_\nu = \frac{16}{9}\alpha_0(3x)^2 = \Omega(1/m^2)$, and the $O(m^{2/3})$ term accounts for prescribing the exact number of edges. Prescribing also the number of vertices exactly (and not just up to a constant factor as in [16]) increases the running time by another factor $O(n^{1/2})$ (see [15, p. 140] and [17]). Thus a random c-net with m edges and imbalance x can be generated in expected time $O(m^{1+2+1/2+2}) = O(n^{11/2})$.

We conjecture that in fact a much faster generation should be possible based on two grounds: Most c-nets have an imbalance with $|x| \leq 1/2 - \varepsilon$, where $\varepsilon > 0$ is any constant. In this case the tuning ratio α_0 and hitting probability p_ν are bounded by constants and the expected running time reduces to $O(m^{1+2/3+1/2}) = O(n^{13/6})$. Moreover, if we are about to generate many planar graphs, we might store the rejected samples for future use, possibly resulting in a near-linear amortized running time at the expense of a larger (but still polynomial) memory requirement.

7 Running Time and Memory Requirements

In this section we establish a polynomial upper bound on the expected running time and the memory requirement of our algorithm.

A number of dynamic programming arrays has to be precalculated before the actual random generation starts. As an example, consider the recurrence formula for $H_{k,\ell}(n, m)$. The number of entries is $O(n^4)$ for all tables. All entries are bounded by the number of all planar graphs. Therefore the encoding length of each entry is $O(\log(n!\,38^n)) = O(n \log n)$ [14,6] and the total space requirement is $O(n^5 \log n)$ bit. The calculation of each entry involves a summation over $O(n^2)$ terms. Using a fast multiplication algorithm, the precomputation time is $O(n^7 (\log n)^2 (\log \log n))$.

We assume that we can obtain random bits at unit cost. In order to prepare for branching with the right probabilities, we can easily calculate the necessary partial sums in a second pass over the dynamic programming arrays. We can then perform random decisions with the right probabilities in time linear in the encoding length, i.e., in $O(n \log n)$.

The total expected time spent in all calls to Schaeffer's c-net generation algorithm is bounded by $O(n^{13/2})$ (but we believe it is much faster in practice, see Section 6). Similarly, the random decisions for the connectivity decomposition require $O(n^2 \log n)$ time in total. An h-element subset of a k-element ground set can be chosen in $O(h \log k)$ time, hence the total time spent for random decisions for the label assignments during the composition is $O(n^2 \log n)$ as well. The compose operation itself is linear and requires at most $O(n^2)$ total time.

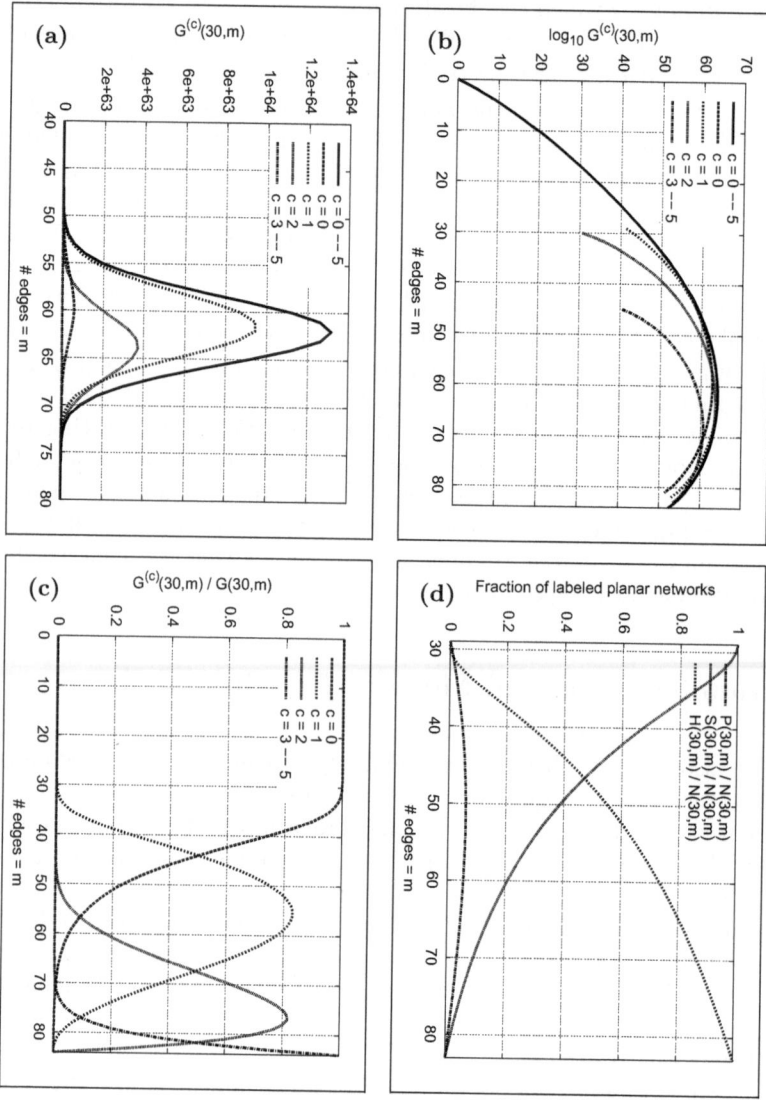

Fig. 1. Some counting results for labeled planar graphs on 30 vertices. The figures show the dependency on the number of edges m and the connectivity c. (a) Number of c-connected labeled planar graphs. (b) Similar in logarithmic scale. (c) Expected connectivity. (d) Expected type of a network (i.e., P, S, or H).

We see that the running time is dominated by $O(n^7(\log n)^2(\log \log n))$ for the preprocessing and $O(n^{13/2})$ (in expectation) for the random generation of c-nets. The space requirement is $O(n^5 \log n)$ bits due to the dynamic programming arrays.

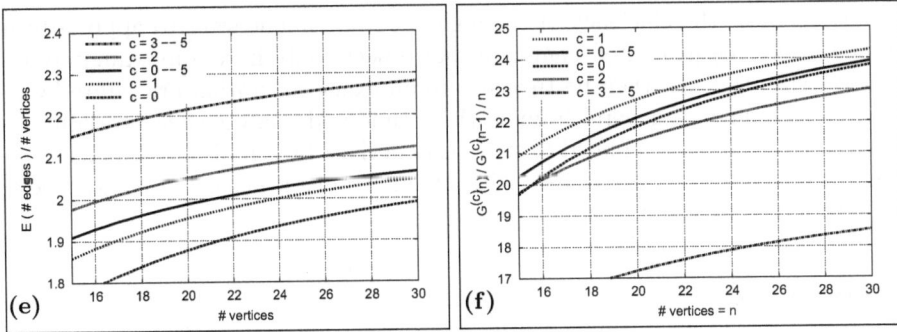

Fig. 2. (e) Edge density of a random labeled planar graph. The limit for general labeled planar graphs is known to be $\geq 13/6$ [9] and ≤ 2.54 [5]. (f) Growth rate of the number of labeled planar graphs.

8   Experimental Results

In this section we report on first computational results from an implementation of the counting formulas. The program was written in C++ using the GMP library for exact arithmetic [10]. A run for 30 vertices completed within one hour on a 1.3 GHz PC using 570 MB RAM. We also checked the recurrences and initial cases in Section 3-6 using an independent counting method. A list of all unlabeled planar graphs with up to 12 vertices was generated by a program of Köthnig [11]. From these the labeled planar graphs were enumerated by 'brute force'. The unlabeled numbers, in turn, were confirmed by entries in Sloane's encyclopedia of integer sequences [18] and [13].

Figures 1 and 2 are explained in the legend.

9   Conclusion

We have seen how to count and generate random planar graphs on a given number of vertices and edges using a recursive decomposition along the connectivity structure. Therefore a by-product of our result is that we can also generate *connected* and *2-connected* labeled planar graphs uniformly at random. Moreover it is easy to see that we can count and generate random planar *multigraphs* by only changing the initial values for planar networks as follows:

$$
\begin{aligned}
N(n,m) &= P(n,m) && \text{for } n=2\,, m\geq 2 \\
P_k(n,m) &= 1 && \text{for } n=2\,, m=k\,, k\geq 1\,.
\end{aligned}
$$

It seems difficult to simplify our counting recurrences to closed formulas. In this way one could eliminate the need for a preprocessing stage. Using generating functions Bender, Gao and Wormald obtained an *asymptotic* formula for the number of labeled 2-connected graphs [3].

To increase the efficiency of the algorithm one might want to apply a technique where the generated combinatorial objects only have approximately the correct size; this can then be turned into an exact generation procedure by rejection sampling. A general framework to tune and analyze such procedures is developed in [8,2] and applied to structures derived by e.g. disjoint unions, products, sequences and sets. To deal with planar graphs it needs to be extended to the compose operation used in this paper.

References

1. C. Banderier, P. Flajolet, G. Schaeffer, and M. Soria. Planar maps and Airy phenomena. In *ICALP'00*, number 1853 in LNCS, pages 388–402, 2000.
2. C. Banderier, P. Flajolet, G. Schaeffer, and M. Soria. Random maps, coalescing saddles, singularity analysis, and Airy phenomena. *Random Structures and Algorithms*, 19:194–246, 2001.
3. A. Bender, Z. Gao, and N. Wormald. The number of labeled 2-connected planar graphs. Preprint, 2000.
4. M. Bodirsky and M. Kang. Generating random outerplanar graphs. Presented at ALICE 03, 2003. Journal version submitted.
5. N. Bonichon, C. Gavoille, and N. Hanusse. An information-theoretic upper bound of planar graphs using triangulation. In *In 20th Annual Symposium on Theoretical Aspects of Computer Science (STACS)*, 2003.
6. A. Denise, M. Vasconcellos, and D. Welsh. The random planar graph. *Congressus Numerantium*, 113:61–79, 1996.
7. R. Diestel. *Graph Theory*. Springer–Verlag, New York, 1997.
8. P. Duchon, P. Flajolet, G. Louchard, and G. Schaeffer. Random sampling from Boltzmann principles. In *ICALP '02*, LNCS, pages 501–513, 2002.
9. S. Gerke and C. McDiarmid. On the number of edges in random planar graphs. Submitted.
10. The GNU multiple precision arithmetic library, version 4.1.2.
    http://swox.com/gmp/.
11. I. Köthnig. Personal communication. Humboldt-Universität zu Berlin, 2002.
12. C. McDiarmid, A. Steger, and D. J. Welsh. Random planar graphs. *Preprint*, 2001.
13. R. Mullin and P. Schellenberg. The enumeration of c-nets via quadrangulations. *Journal of Combinatorial Theory*, 4:259–276, 1968.
14. D. Osthus, H. J. Prömel, and A. Taraz. On random planar graphs, the number of planar graphs and their triangulations. *Jombinatorial Theory, Series B*, to appear.
15. G. Schaeffer. *Conjugaison d'arbres et cartes combinatoires aléatoires*. PhD thesis, Université Bordeaux I, 1998.
16. G. Schaeffer. Random sampling of large planar maps and convex polyhedra. In *Proc. of the thirty-first annual ACM symposium on theory of computing (STOC'99)*, pages 760–769, Atlanta, Georgia, May 1999. ACM press.
17. G. Schaeffer. Personal communication, 2002.
18. N. J. A. Sloane. The on-line encyclopedia of integer sequences.
    http://www.research.att.com/~njas/sequences/index.html, 2002.
19. B. A. Trakhtenbrot. Towards a theory of non-repeating contact schemes. *Trudi Mat. Inst. Akad. Nauk SSSR*, 51:226–269, 1958. [In Russian].
20. W. Tutte. A census of planar maps. *Canad. J. Math.*, 15:249–271, 1963.

21. T. Walsh. Counting labelled three-connected and homeomorphically irreducible two-connected graphs. *J. Combin. Theory*, 32:1–11, 1982.

22. T. Walsh. Counting nonisomorphic three-connected planar maps. *J. Combin. Theory*, 32:33–44, 1982.

23. T. Walsh. Counting unlabelled three-connected and homeomorphically irreducible two-connected graphs. *J. Combin. Theory*, 32:12–32, 1982.

24. T. Walsh and V. A. Liskovets. Ten steps to counting planar graphs. In *Eighteenth Southeastern International Conference on Combinatorics, Graph Theory, and Computing, Congr. Numer.*, volume 60, pages 269–277, 1987.

Online Load Balancing Made Simple: Greedy Strikes Back[*]

Pilu Crescenzi[1], Giorgio Gambosi[2], Gaia Nicosia[3], Paolo Penna[4][**], and
Walter Unger[5]

[1] Dipartimento di Sistemi ed Informatica, Università di Firenze, via C. Lombroso
6/17, I-50134 Firenze, Italy (piluc@dsi.unifi.it)
[2] Dipartimento di Matematica, Università di Roma "Tor Vergata", via della Ricerca
Scientifica, I-00133 Roma, Italy (gambosi@mat.uniroma2.it)
[3] Dipartimento di Informatica e Automazione, Università degli studi "Roma Tre",
via della Vasca Navale 79, I-00146 Roma, Italy (nicosia@dia.uniroma3.it)
[4] Dipartimento di Informatica ed Applicazioni "R.M. Capocelli", Università di
Salerno, via S. Allende 2, I-84081 Baronissi (SA), Italy (penna@dia.unisa.it),
[5] RWTH Aachen, Ahornstrasse 55, 52056 Aachen, Germany
(quax@cs.rwth-aachen.de)

Abstract. We provide a new simpler approach to the on-line load
balancing problem in the case of restricted assignment of temporary
weighted tasks. The approach is very general and allows to derive on
line distributed algorithms whose competitive ratio is characterized by
some combinatorial properties of the underlying graph representing the
problem.

The effectiveness of our approach is shown by the hierarchical server
model introduced by Bar-Noy *et al* '99. In this case, our method yields
simpler and distributed algorithms whose competitive ratio is at least as
good as the existing ones. Moreover, the resulting algorithms and their
analysis turn out to be simpler. Finally, in all cases the algorithms are
optimal up to a constant factor.

Some of our results are obtained via a combinatorial characterization
of those graphs for which our technique yields $O(\sqrt{n})$-competitive algorithms.

1 Introduction

Load balancing is a fundamental problem which has been extensively studied in
the literature because of its many applications in resource allocation, processor
scheduling, routing, network communication, and many others. The problem
is to assign tasks to a set of n processors, where each task has an associated

---

[*] A similar title is used in [13] for a facility location problem.
[**] supported by the European Project IST-2001-33135, Critical Resource Sharing for
Cooperation in Complex Systems (CRESCCO). Work partially done while at the
Dipartimento di Matematica, Università di Roma "Tor Vergata" and while at the
Institut für Theoretische Informatik, ETH Zentrum.

J.C.M. Baeten et al. (Eds.): ICALP 2003, LNCS 2719, pp. 1108–1122, 2003.
© Springer-Verlag Berlin Heidelberg 2003

load vector and duration. Tasks must be assigned immediately to exactly one processor, thereby increasing the load of that processor by the amount specified by the corresponding coordinate of the load vector for the duration of the task. Usually, the goal is to minimize the maximum load over all processors.

The *on-line* load balancing problem has several natural applications. For instance, consider the case in which processors represent channels and tasks are communication requests which arrive one by one. When a request is assigned to a channel, a certain amount of its bandwidth is reserved for the duration of the communication. Since channels have limited bandwidth, the maximum load is an important measure here.

Several variants have been proposed depending on the structure of the load vectors, whether we allow *preemption* (i.e., to reassign tasks), whether tasks remain in the system "forever" or not (i.e., permanent vs. temporary tasks), whether the (maximum) duration of the tasks is known, and so on [1,3,4,5,6,14, 16] (see also [2] for a survey).

In this paper, we study the on line load balancing problem in the case of *temporary tasks* with *restricted assignment* and *no preemption*, that is:

- Tasks arrive one by one and their duration is unknown.
- Each task can be assigned to one processor among a subset depending on the type of the task.
- Once a task has been assigned to a processor, it cannot be reassigned to another one.
- Assigning a task to a processor increases the corresponding load by an amount equal to the weight of the task.

The problem asks to find an assignment of the tasks to the processors which minimizes the maximum load over all processors and over time.

Among others, this variant has a very important application in the context of wireless networks. In particular, consider the case in which we are given a set of base stations and each mobile user can only connect to a subset of them: those that are "close enough". A user may unexpectedly appear in some spot and ask for a connection at a certain transmission rate (i.e., bandwidth). Also, the duration of this transmission is not specified (this is the typical case of a telephone call). Because of the application, it is desirable not to reassign users to other base stations (i.e., to avoid the handover), unless this becomes unavoidable because a user moves away from the transmission range of its current base (in the latter case, we can model this as a termination of the current request and a new request appearing in the new position).

As usual, we compare the cost of a solution computed by an on-line algorithm with the best off-line algorithm which minimizes the maximum load *knowing the entire sequence* of task arrivals and departures. Informally, an on-line algorithm is r-competitive if, at any instant, its maximum processor load is at most r times the optimal maximum processor load.

It is convenient to formulate our problem by means of a bipartite graph with vertices corresponding to processors and possible "task types". More formally, let $P = \{p_1, \ldots, p_n\}$ be a set of processors and let $\mathcal{T} \subseteq 2^P$ be a set of *task types*.

We represent the set of task types by means of an *associated bipartite graph* $G_{P,\mathcal{T}}(X_{\mathcal{T}} \cup P, E_{\mathcal{T}})$, where $X_{\mathcal{T}} = \{x_1, \ldots, x_{|\mathcal{T}|}\}$ and

$$E_{\mathcal{T}} = \{(x_i, p_j) \mid p_j \text{ belongs to the } i\text{-th element of } \mathcal{T}\}.$$

A *task t* is a pair (x, w), where $x \in X_{\mathcal{T}}$ and w is the positive integer weight of t. The set of processors to which t can be *assigned* is $P_t = \{p \mid (x, p) \in E_{\mathcal{T}}\}$, that is, the set of nodes of $G_{P,\mathcal{T}}(X_{\mathcal{T}} \cup P, E_{\mathcal{T}})$ that are adjacent to x. In our example of mobile users above, the type of a task corresponds to the user's position[1]. In general, we consider two tasks which can be assigned to the same set of processors as belonging to the same type.

We follow the intuition that "nice" graphs may yield better competitive ratios, as one can see with the following examples:

General case. We do not assume anything regarding the possible task types. So, the graph must contain all possible task types corresponding to any subset of processors, that is, $\mathcal{T} = 2^P$. Under this assumption, the best ratio achievable is $\Theta(\sqrt{n})$ [3,5] (see also [17]), while the greedy algorithm is exactly $\frac{3n^{2/3}}{2}(1 + o(1))$-competitive [3], thus not optimal.

Identical machines. There is only one task type since a task can be assigned to any of the machines. Therefore, the graph is the complete bipartite graph $K_{1,n}$ and the competitive ratio of the problem is $\Theta(2 - 1/n)$. This ratio is achieved by the greedy algorithm [11,12], which is optimal [4].

Hierarchical servers. Processors are totally ordered and the type of a task corresponds to the "rightmost" processor that can execute that task. The set $\mathcal{T}$ contains one node per processor and the i-th node of $\mathcal{T}$ is adjacent to all processors j, with $1 \leq j \leq i$. There exists a 5-competitive algorithm and the greedy algorithm is at least $\Omega(\log n)$-competitive.

Noticeably, one can consider the first two cases as the two extremes of our problem, because of both the (non-) optimality of the greedy algorithm and the (non-) constant competitive ratio of the optimal on-line algorithm. From this point of view, the latter problem is somewhat in between.

A related question is whether the greedy approach performs badly because of the fact that it must decide where to assign a task only based on *local* information (i.e., the current load of those processors that can execute that task). Indeed, the optimal algorithm in [5, Robin-Hood Algorithm] requires the computation of (an estimation of) the off-line optimum, which seems hard to compute in this local fashion. The algorithms in [7], too, require the computation of a quantity related to the optimum which depends on the current assignment of tasks of several types (see [7, Algorithm Continuous and Optimal Lemma]).

The idea of exploiting combinatorial properties of the graph $G_{P,\mathcal{T}}(X_{\mathcal{T}} \cup P, E_{\mathcal{T}})$ has been first used in [9]. In particular, the approach in [9] is based on the construction of a suitable subgraph which is used by the greedy algorithm (in

---

[1] Clearly, this is a simplification of the reality where other constraints must also be taken into account.

place of the original one). This subgraph is the union of a set of complete bipartite subgraphs (called clusters). So, this method can be seen as a modification of the greedy algorithm where the topology of the network is taken into account in order to limit the possible choices of the greedy algorithm[2] . Therefore, the resulting algorithms only use "local information" as the greedy one does.

Several topologies have been considered in [9] for which the method improves over the greedy algorithm and matches the lower bound of the problem(s). In all such cases, however, the improvement is only by a constant factor, since the greedy algorithm was already $O(1)$-competitive.

The main contribution of this paper (see Sect. 2) is a new approach to the problem based on the construction of a suitable subgraph to be used by the greedy algorithm. In this sense, our work is similar in spirit to [9]. However, the results here greatly improve over the method in that paper. Indeed, we show that:

- Some problems cannot be optimally solved with the solution in [9], while our approach does yield optimal competitive ratios.
- Our approach subsumes the one in [9] since the latter can be seen as a special case of the one presented here.

Also, our method yields the first example in which there is a significant improvement w.r.t. the greedy algorithm. This arises from the relevant case of hierarchical topologies, for which we attain a competitive ratio of 5 (4 for unweighted tasks[3]), while the greedy algorithm is at least $\Omega(\log n)$ in both cases. Table 1 summarizes the results obtained for these topologies. Even though, when $n \to \infty$, we achieve the same competitive ratio of [7], our algorithms and their analysis turn out to be much simpler. (Actually, for fixed n, our analysis yields strictly better ratios.)

We then turn our attention to the general case. In general, it might be desirable to automatically compute the best subgraph, as this would also give a simple way to test the goodness of our method w.r.t. a given graph. Unfortunately, one of our results is the NP-hardness of the problem of computing an optimal or even a c-approximate solution, for some constant $c > 1$.

In spite of this negative result, we demonstrate that a "sufficiently good" subgraph can be obtained very easily in many cases. We first provide a *sufficient condition* for obtaining $O(\sqrt{n})$-competitive algorithms with our technique: the existence of a b-matching in (a suitable subgraph of) the graph $G_{\mathcal{P},\mathcal{T}}$, for some constant b independent of n. Notice that, the lower bound for the general case is $\Omega(\sqrt{n})$, which applies to randomized algorithms [3] and to sequences of tasks of length polynomial in n [17]. By using this result, we obtain a $(2\sqrt{n} + 2)$-

---

[2] This approach is somewhat counterintuitive since the algorithm improves when adding further restrictions to it (not to the adversary). This is reminiscent of the well-known Braess' paradox [8,15], where the removal of some edges from a graph unexpectedly improves the latency of the flow at Nash equilibrium.

[3] We denote the version of these problems in which all tasks have weight one by *unweighted*.

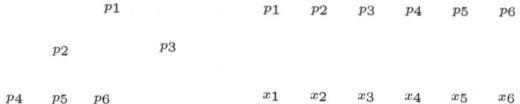

Fig. 1. An example of tree hierarchy (left) and the corresponding bipartite graph (right).

competitive distributed algorithm for the hierarchical server version in which processors are ordered as in a rooted tree (see the example in Fig. 1). A $\Omega(\sqrt{n})$ lower bound for centralized on-line algorithms also applies to this restriction [7], thus implying the optimality of our result. Additionally, we can achieve the same upper bound also when the ordering of the processors is given by *any* directed graph. This bound is only slightly worse than the $2\sqrt{n}+1$ given by the Robin-Hood algorithm in [5]. All algorithms obtained with our technique can be

Table 1. Performance of our method in the case of hierarchical servers.

	Our Method	Previous Best	Greedy
Weighted	$5n/(n+4)$ [Th. 5]	5 [7]	$\Omega(\log n)$ [folklore]
Unweighted	$4n/(n+4)$ [Th. 5]	4 [7]	$\Omega(\log n)$ [folklore]

considered distributed in that they compute a task assignment only based on the current load of those processors that can be used for that task. This is a rather appealing feature since, in applications like the mobile networks above, introducing a global communication among bases for every new request to be processed may turn out to be unfeasible. Additionally, in several cases considered here (linear and tree hierarchical topologies), the construction of the subgraph is used solely for the analysis, while the actual on-line algorithm does not require any pre-computation (although the algorithm is different from the greedy one and it implements the subgraph used in the analysis).

Finally, we believe that our analysis is per se interesting since it translates the notion of adversary into a combinatorial property of the subgraph we are able to construct during an off-line preprocessing phase. As a by-product, the analysis of our algorithms is simpler and more intuitive.

Roadmap. We introduce some notation and definitions in Sect. 1.1. The technique and its analysis are presented in Sect. 2. The hardness results are given in Sect. 2.2. We give a first application to the hierarchical topologies in Sect. 3. The application to the general case is described in Sect. 4 where we provide sufficient conditions for $O(\sqrt{n})$-competitiveness. These results are used in Sect. 4.1 where we obtain the results on generalized server hierarchies. Finally, in Sect. 5 we discuss some further features of our algorithms and present some open problems.

Due to lack of space, some of the proofs are only sketched or omitted in this version of the paper. The omitted proofs are contained in [10].

1.1   Preliminaries and Notation

An *instance* σ of the on-line load balancing problem with processors P and task types $\mathcal{T}$ is defined as a sequence of $\texttt{new}(\cdot,\cdot)$ and $\texttt{del}(\cdot)$ commands. In particular: (i) $\texttt{new}(x,w)$ means that a new task of weight w and type $x \in \mathcal{T}$ is created; (ii) $\texttt{del}(i)$ means that the task created by the i-th $\texttt{new}(\cdot,\cdot)$ command of the instance is deleted.

As already mentioned, we model the problem by means of a bipartite graph $G_{P,\mathcal{T}}(X_{\mathcal{T}} \cup P, E_{\mathcal{T}})$, where $\mathcal{T}$ depends on the problem version we are considering. For the sake of brevity, in the following we will always omit the subscripts '$P,\mathcal{T}$' and '$\mathcal{T}$' since the set of processors and the set of task types will be clear from the context. Given a graph $G(V,E)$, $\Gamma_G(v)$ denotes the *open* neighborhood of the node $v \in V$. So, a task of type x can be assigned to $\Gamma_G(x)$.

We will distinguish between the *unweighted* case, in which all tasks have weight 1, and the *weighted* case, in which the weights may vary from task to task. We also refer to (un-)weighted tasks to denote these variants.

Given an instance σ, a *configuration* is an assignment of the tasks of σ to the processors in P, such that each task t is assigned to a processor in P_t. Given a configuration C, we denote with $l_C(i)$ the *load* of processor p_i, that is, the sum of the weights of all tasks assigned to it. In the sequel, we will usually omit the configuration when it will be clear from the context. The load of C is defined as the maximum of all the processor loads and is denoted with $l(C)$. Given an instance $\sigma = \sigma_1 \cdots \sigma_n$ and given an on-line algorithm $\mathcal{A}$, let $C_h^{\mathcal{A}}$ be the configuration reached by $\mathcal{A}$ after processing the first h commands. Moreover, let C_h^{off} be the configuration reached by the optimal off-line algorithm after processing the first h commands. Let also $\text{opt}(\sigma) = \max_{1 \le h \le n} l(C_h^{\text{off}})$ and $l_{\mathcal{A}}(\sigma) = \max_{1 \le h \le n} l(C_h^{\mathcal{A}})$.

An on-line algorithm $\mathcal{A}$ is said to be *r-competitive* if there exists a constant b such that, for any instance σ, it holds that $l_{\mathcal{A}}(\sigma) \le r \cdot \text{opt}(\sigma) + b$. An on-line algorithm $\mathcal{A}$ is said to be *strictly r-competitive* if, for any instance σ, it holds that $l_{\mathcal{A}}(\sigma) \le r \cdot \text{opt}(\sigma)$.

A simple on-line algorithm for the load-balancing problem described above is the *greedy algorithm* that assigns a new task to the least loaded processor among those processors that can serve the task. That is, whenever a $\texttt{new}(x,w)$ command is encountered and the current configuration is C, the greedy algorithm looks for the processor p_i in $P_{t=(x,w)}$ such that $l_C(i)$ is minimal and assigns the new task $t = (x,w)$ to p_i. (Ties are broken arbitrarily.)

2   (Sub-)graphs and (Sub-)greedy Algorithms

In the sequel we will describe an on-line load balancing algorithm whose competitive ratio depends on some combinatorial properties of $G(X \cup P, E)$. The two main ideas used in our approach are the following:

1. We remove some edges on G and then we apply the greedy algorithm *to the resulting bipartite graph*;
2. While removing edges we try to balance the number of processors used by tasks of type $x \in X$ and the number of processors that the adversary can use to assign the same set of tasks in the original graph G.

First of all, let us observe that our method aims to obtain a good competitive ratio by *adding further constraints* to the original problem: this indeed corresponds to remove a suitable set of edges from G. Choosing which edges to remove depends on some combinatorial properties we want the resulting bipartite graph to satisfy. Before giving a formal description of such properties, we will describe the basic idea behind our approach.

The main idea. Let us consider a generic iteration of an algorithm that has to assign a task of type $x \in X$. Assume that our algorithm takes into account a set $U(x)$ of processors and assigns the task to the least loaded one. In order to evaluate the competitive ratio of this approach we need to know which set of processors $A(x)$ an adversary can use to assign the overall load currently in $U(x)$ (as we will see in the sequel the competitive ratio of our algorithm is roughly $|A(x)|/|U(x)|$). In the following, we will show how the set $A(x)$ is determined by the choices of our algorithm in the previous steps (see Fig. 2).

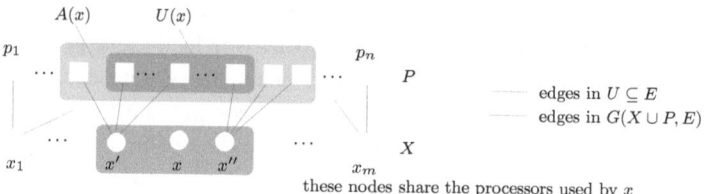

Fig. 2. The main idea of the sub-greedy algorithm is to balance the ratio between the number of used processors $|U(x)|$ and the number of processors available $|A(x)|$ to the adversary.

2.1   Analysis

In this section we formalize the idea above and we provide the performance analysis of the resulting algorithm.

Definition 1 (Used Processors). *For any $x \in X$, we define a non-empty set $U(x) \subseteq \Gamma_G(x)$ of used processors. Moreover, given a processor $p \in P$, we denote by $U^{-1}(p)$ those vertices in X that have p as an available processor, i.e. $U^{-1}(p) = \{x \mid p \in U(x)\}$.*

Definition 2 (Adversary Processors). *For any $x \in X$ we denote by $A(x)$ those processors that an off-line adversary can use to balance the load assigned to $U(x)$. In particular, $A(x) = \bigcup_{p \in U(x)} \bigcup_{x' \in U^{-1}(p)} \Gamma_G(x')$.*

Notice that the set $U(x)$ specifies a subset of the edges in $G(X \cup P, E)$ incident to x. By considering the union over all $x \in X$ of these edges and the resulting bipartite subgraph, we have the following:

Definition 3 (Sub-Greedy Algorithm). *For any bipartite graph $G(X \cup P, E)$ and for any subset of edges $U \subseteq E$, the sub-greedy algorithm is defined as the greedy on-line algorithm applied to $G_U = G(X \cup P, U)$.*

Remark 1. It is easy to see that the sub-greedy algorithm is a special case of the cluster algorithm in [9], since the latter imposes each connected component of G_U to be a complete bipartite graph [9, Definition of Cluster].

It is clear that the performance of the sub-greedy algorithm will depend on the choice of the set $U \subseteq E$. In particular, we can characterize its competitive ratio in terms of the ratio between the set of adversary processors $A(x)$ and the set of used processors $U(x)$. Let us consider the following quantities:

$$\rho_w(U) = \max_{x \in X}\left\{\frac{|A(x)| - 1}{|U(x)|}\right\}, \qquad \rho_u(U) = \max_{x \in X}\left\{\frac{|A(x)|}{|U(x)|}\right\}.$$

Then, the following two results hold.

Theorem 1. *The sub-greedy algorithm is strictly $(1 + \rho_w(U))$-competitive in the case of weighted tasks.*

Proof. Let p_i be the processor with the highest load and let $t = (w, x)$ be the last task assigned to p_i by the sub-greedy algorithm. Since t has been assigned to p_i whose load, before the arrival of t, was $l(i) - w$, we have that each processor in $U(x)$ had load at least $l(i) - w$. So, the overall load of $U(x)$ is at least $|U(x)|(l(i) - w) + w$. We now consider the number of processors that any off-line strategy can use to spread such load. Such a number is equal to $|A(x)|$, which implies that the optimal off-line solution has measure at least

$$l^* \geq \max\left\{\frac{|U(x)|l(i) - w(|U(x)| - 1)}{|A(x)|}, w\right\}.$$

The worst case is when we equate the two quantities, that is $l^* \geq w = \frac{|U(x)|l(i)}{|A(x)| + |U(x)| - 1}$, which implies the following bound on the competitive ratio

$$\frac{l(i)}{l^*} \leq \frac{|A(x)| + |U(x)| - 1}{|U(x)|} \leq 1 + \rho_w(U).$$

Hence, the theorem follows.

Theorem 2. *The sub-greedy algorithm is $\rho_u(U)$-competitive (resp., strictly $\lceil \rho_u(U) \rceil$-competitive) in the case of unweighted tasks.*

Proof. Let us consider a generic iteration of the sub-greedy algorithm in which a task t arising in $x \in X$ has been assigned to $p_i \in U(x)$. Since t has been assigned to p_i whose load, before the arrival of t, was $l(i) - 1$, we have that each processor in $U(x)$ had load at least $l(i) - 1$. This implies that the overall number of tasks in $U(x)$, after the arrival of t, was at least $|U(x)|(l(i)-1)+1$. Let us also observe that the number of processors to which the off-line optimal solution can assign these tasks is at most $|A(x)|$. Thus, the optimal off-line solution has measure at least

$$l^* \geq \frac{|U(x)|(l(i)-1)+1}{|A(x)|} \geq \frac{l(i)-1}{\rho_u(U)} + \frac{1}{|A(x)|}. \tag{1}$$

By contradiction, let us suppose that $l(i) > l^* \lceil \rho_u(U) \rceil$. Then, since both $l(i)$ and l^* have integer values, we have that $l(i) - 1 \geq l^* \lceil \rho_u(U) \rceil \geq l^* \rho_u(U)$. This leads to the following contradiction:

$$l^* \geq \frac{l(i)-1}{\rho_u(U)} + \frac{1}{|A(x)|} \geq \frac{l^* \rho_u(U)}{\rho_u(U)} + \frac{1}{|A(x)|} > l^*.$$

We have thus proved that the sub-greedy algorithm is strictly $\lceil \rho_u(U) \rceil$-competitive. Finally, Eq. 1 implies

$$l(i) < l^* \rho_u(U) + \frac{1}{\rho_u(U)} \leq l^* \rho_u(U) + 1,$$

where the last inequality follows from the fact that $\rho_u(U) \geq 1$. So, the sub-greedy algorithm is also $\rho_u(U)$-competitive. Hence, the theorem follows.

We next show the limits of our approach as it is and we generalize it in order to handle more cases. First, consider the bipartite graph $G(X \cup P, E)$ with $X = \{x_1, x_2, \ldots, x_n\} \cup \{x_0\}$ and $E = \{(x_i, p_i) \mid 1 \leq i \leq n\} \cup \{(x_0, p_i) \mid 1 \leq i \leq n\}$. It is easy to see that any subset of edges U yields $\rho_w(U) = n - 1$. However, a rather simple idea might be to separate the high-degree vertex x_0 from the low-degree vertices $x_1, x_2, \ldots, x_n$. So, tasks of type x_0 are processed *independently* from tasks of type $x_1, x_2, \ldots, x_n$. It is possible to prove that this algorithm has a constant competitive ratio. This idea leads to the following:

Definition 4 (sub-greedy*). *Let $X_1, X_2, \ldots, X_k$ be any partition of the set X of the task types vertices of $G(X \cup P, E)$. Also let $G_i = G(X_i \cup P, E_i)$ be the corresponding induced subgraph, and let $U_i \subseteq E_i$ for $1 \leq i \leq n$. We denote by* sub-greedy* *the algorithm assigning tasks of type in X_i as the sub-greedy algorithm on the subgraph of G_i corresponding to U_i, and with only these tasks as input (i.e., independently of tasks of other types).*

In the sequel we denote by $\rho_w(U_i, G_i)$ the quantity $\rho_w(U)$ computed w.r.t. the graph G_i and subset of edges $U_i \subseteq E_i$.

Theorem 3. *The sub-greedy* algorithm is strictly $(k + \rho_w^*(U))$-competitive in the case of weighted tasks, where k is the number of subgraphs and $\rho_w^*(U) = \sum_{i=1}^{k} \rho_w(U_i, G_i)$.*

Proof. Given a sequence of tasks σ, let $\sigma(i)$ denote the subsequence containing tasks whose type is in X_i. Also let $l(j)$ denote the load of processor p_j at some time step and $l^i(j)$ the load at the same time step w.r.t. tasks corresponding to X_i only. Then, the definition of sub-greedy* and Theorem 1 imply $\max_{1 \le j \le n} l^i(j) \le \mathrm{opt}(\sigma(i))(1 + \rho_w(U_i, G_i))$, for $1 \le i \le k$. It then holds that

$$\max_{1 \le j \le n} l(j) \le \sum_{i=1}^{k} \max_{1 \le j \le n} l^i(j) \le \sum_{i=1}^{k} \mathrm{opt}(\sigma(i))(1 + \rho_w(U_i, G_i)) \qquad (2)$$

$$\le k \cdot \mathrm{opt}(\sigma) + \mathrm{opt}(\sigma) \sum_{i=1}^{k} \rho_w(U_i, G_i), \qquad (3)$$

where the last inequality follows from the fact that $\mathrm{opt}(\sigma(i)) \le \mathrm{opt}(\sigma)$.

The above theorem will be a key-ingredient in deriving algorithms for the general case (see Sect. 4).

2.2   Computing Good Subgraphs

From Theorems 1-2 it is clear that, in order to attain a good competitive ratio, it is necessary to select a subset of edges $U \subseteq E$ such that $\rho(U)$ is as small as possible. Similarly, Theorem 3 implies that U should minimize $k + \rho_w^*(U)$, when considering the sub-greedy* algorithm.

We now rewrite the sets $A(x)$ and $U(x)$ by looking at the *open* neighborhood operator $\Gamma_G(\cdot)$. In particular, we have that $U(x) = \Gamma_{G_U}(x)$ and $A(x) = \Gamma_G(\Gamma_{G_U}(\Gamma_{G_U}(x)))$. When considering the weighted case, this leads to the following optimization problem:

Problem 5 *Min Weighted Adversary Subgraph (MWAS).*
Instance: *A bipartite graph $G(X \cup P, E)$.*
Solution: *A subgraph $G_U = G(X \cup P, U)$, such that $U \subseteq E$ and, for every $x \in X$, $|\Gamma_{G_U}(x)| \ge 1$.*
Measure: $\rho_w(U, G) = \max_{x \in X} \frac{|\Gamma_G(\Gamma_{G_U}(\Gamma_{G_U}(x)))| - 1}{|\Gamma_{G_U}(x)|}.$

Problem 6 *Min Weighted Adversary Multi-Subgraph (MWAMS).*
Instance: *A bipartite graph $G(X \cup P, E)$.*
Solution: *A partition $X_1, X_2, \ldots, X_k$ of X and a collection $U = \{U_1, \ldots, U_k\}$ of subsets of edges $U_i \subseteq E_i$, where E_i denotes the set of edges of the subgraph $G_i = G(X_i \cup P, E_i)$ induced by X_i, such that, for every $1 \le i \le k$ and $x \in X_i$, $|\Gamma_{G_{U_i}}(x)| \ge 1$.*
Measure: $k + \rho_w^*(U, G) = k + \sum_{i=1}^{k} \rho_w(U_i, G_i).$

Similarly, the Min Unweighted Adversary Subgraph (MUAS) problem and the Min Unweighted Adversary Multi-Subgraph (MUAMS) problem are defined by replacing ρ_w with ρ_u in the two definitions above, respectively.

It is possible to construct a reduction showing the NP-hardness of all these problems (see [10]). Also, the same reduction is a gap-creating reduction, thus implying the non existence of a PTAS for any such problem. In particular, we obtain the following result:

Theorem 4. *The MUAS and MWAS problems cannot be approximated within a factor smaller than 7/6 and 3/2, respectively, unless* P = NP. *Moreover MUAMS and MWAMS cannot be approximated within a factor smaller than 11/10 and 5/4, respectively, unless* P = NP.

3   Application to Hierarchical Server Topologies

In this section we apply our method to the hierarchical server topologies introduced in [7]. In particular, we consider the *linear* hierarchical topology: Processors are ordered from p_1 (the most capable processor) to p_n (the least capable processor) in decreasing order with respect to their capabilities. So, if a task can be assigned to processor p_i, for some i, then it can also be assigned to any p_j with $1 < j < i$. We can therefore consider task types corresponding to the intervals $\{p_1, p_2, \ldots, p_i\}$, for each $1 \leq i \leq n$.

The resulting bipartite graph $G(X \cup P, E)$ is given by $X = \{x_1, \ldots, x_n\}$, $P = \{p_1, \ldots, p_n\}$ and $E = \{(x_i, p_j) \mid x_i \in X, p_j \in P, j \leq i\}$. We denote this graph as K_n^{hst}.

We next provide an efficient construction of subgraphs of K_n^{hst}.

Lemma 1. *For any positive integer n, there exists a U such that $\rho_w(U, K_n^{hst}) \leq (4n-2)/(n+2)$ and $\rho_u(U, K_n^{hst}) \leq (4n)/(n+2)$. Moreover, the set U can be computed in linear time.*

Proof. For each $1 \leq i \leq n$, we define the set $U(x_i)$ as

$$U(x_i) = \{p_{\lceil i/2 \rceil}, p_{\lceil i/2 \rceil + 1}, \ldots, p_i\} = \begin{cases} \{p_{i/2}, p_{i/2+1}, \ldots, p_i\} & \text{if } i \text{ is even,} \\ \{p_{(i+1)/2}, p_{(i+1)/2+1}, \ldots, p_i\} & \text{otherwise.} \end{cases}$$

Clearly, $|U(x_i)| = i/2 + 1$ if i is even, and $|U(x_i)| = (i+1)/2$ otherwise. Moreover, $|A(x_i)| = \max_{i \leq j \leq n}\{j \mid U(x_i) \cap U(x_j) \neq \emptyset\}$. It is easy to see that $|A(x_i)| \leq 2i$, thus implying

$$\rho_w(U) = \max_{1 \leq i \leq n} \frac{|A(x_i)| - 1}{|U(x_i)|} \leq \max_{1 \leq i \leq n} \frac{4i-2}{i+1} \leq \frac{4n-2}{n+1}.$$

A better bound can be obtained by distinguishing two cases: 1) $i \leq n/2$ and 2) $i \geq n/2 + 1$. In case 1) we apply the bound above, while for 2) we simply use $|A(i)| \leq n$; in both cases we obtain $\rho_w(U) \leq (4n-2)/(n+2)$.

With a similar proof we can show that the same construction yields $\rho_u(U) \leq (4n)/(n+2)$.

An immediate application of Lemma 1 combined with Theorems 1-2 is the following result:

Theorem 5. *For linear hierarchy topologies the sub-greedy algorithm is strictly $(5n)/(n+2)$-competitive in the case of weighted tasks, and $(4n)/(n+2)$-competitive and strictly 4-competitive for unweighted tasks.*

Notice that our approach improves over the 5-competitive (respectively, 4-competitive) algorithm for weighted (respectively, unweighted) tasks given in [7].

Remark 2. Observe that, if we impose the subgraph to be a set of complete bipartite graphs as in [9], then K_n^{hst} does not admit a construction yielding $O(1)$-competitive algorithms. So, for these topologies, the sub-greedy algorithm constitutes a significant improvement w.r.t. the result in [9].

4 The General Case

In this section we provide a sufficient condition for obtaining $O(\sqrt{n})$-competitive algorithms. This applies to the hierarchical server model when the order of the servers is a tree. Thus, in this case our result is optimal because of the $\Omega(\sqrt{n})$ lower bound [7].

We first define an overall strategy to select the set $U(x)$ depending on the degree $\delta(x)$ of x:

High degree (easy case): $\delta(x) \geq \sqrt{n}$. In this case we use *all of its adjacent vertices in P*. Since $|U(x)| = \delta(x) \geq \sqrt{n}$, we have $|A(x)|/|U(x)| \leq \sqrt{n}$.

Low degree (hard case): $\delta(x) < \sqrt{n}$. For low degree vertices our strategy will be to choose a *single* processor p_x^* in $\Gamma_G(x) \subseteq P$. The choice of this element must be carried out carefully so to guarantee $|A(x)| \leq \sqrt{n}$. For instance, it would suffice that p_x^* does not appear in any other set $U(x')$.

Then, our next idea will be to partition the graph $G(X \cup P, E)$ into two subgraphs $G_l(X_l \cup P, E_l)$ and $G_h(X_h \cup P, E_h)$ containing low and high degree vertices, respectively. Notice that, if we are able to have a $f(n)$-competitive algorithm for the low degree graph, then we have a $O(\sqrt{n} + f(n))$-competitive algorithm for our problem (see Theorem 3).

We next focus on low degree graphs and we provide sufficient conditions for $O(\sqrt{n})$-competitive algorithms.

Theorem 6. *If $G_l(X_l \cup P, E_l)$ admits a b-matching, then the sub-greedy* algorithm is at most $((b+1)\sqrt{n}+2)$-competitive.*

Proof. Let U be a b-matching for G_l. It is easy to see that in G_l $|A(x)| \leq b\sqrt{n}$, for all $x \in X_l$. Thus, $\rho_w(U, G_l) \leq b\sqrt{n}$. By definition of G_h, $\rho_w(E_h, G_h) \leq \sqrt{n}$. We can thus apply Theorem 3 with $k = 2$, $U_1 = U$ and $U_2 = E_h$. Hence the theorem follows.

Theorem 7. *If $G(X \cup P, E)$ admits a b-matching, then the sub-greedy* algorithm is at most $(2\sqrt{bn} + 2)$-competitive.*

Proof Sketch. Define low-degree vertices as those $x \in X$ such that $\Gamma_G(x) \leq \sqrt{n/b}$. Subgraphs G_l and G_h are defined accordingly. The existence of a b-matching U yields $\rho_w(U, G_l) \leq b\sqrt{n/b}$. By definition of G_h, $\rho_w(E_h, G_h) \leq \sqrt{bn}$. We can thus apply Theorem 3 with $k = 2$, $U_1 = U$ and $U_2 = E_h$. Hence the theorem follows.

Theorem 8. *If $G(X \cup P, E)$ admits a matching, then the sub-greedy algorithm is at most δ_{max}-competitive, where $\delta_{max} = \max_{x \in X} |\Gamma_G(x)|$.*

Proof. Let U be a matching for G. Then, $|A(x)| \leq |\Gamma_G(x)|$ and $|U(x)| = 1$, for any $x \in X$.

4.1   Generalized Hierarchical Server Topologies

We now apply these results to the hierarchical model in the case in which the ordering of the servers forms a tree. Figure 1 shows an example of this problem version: processors are arranged on a rooted tree and there is a task type x_i for each node p_i of the tree; a task of type x_i can be assigned to processor p_i or to any of its ancestors.

We first generalize this problem version to a more general setting:

Definition 7. *Let $H(P, F)$ be a directed graph. The associated bipartite graph $G_H(X \cup P, E)$ is defined as $X = \{x_1, x_2, \ldots, x_n\}$, and $(x_i, p_j) \in E$ if and only if $i = j$ or there exists a directed path in H from p_i to p_j.*

We can model a tree hierarchy by considering a rooted tree T whose edges are directed *upward*. We then obtain the following:

Theorem 9. *For any rooted tree $T(P, E)$ the corresponding graph $G_T(X \cup P, E)$ admits a matching. In this case, the sub-greedy* algorithm is always at most $(2\sqrt{n} + 2)$-competitive. Moreover, the sub-greedy algorithm is at most h-competitive, where h is the height of T.*

Proof. It is easy to see that $M = \{(x_i, p_i) \mid 1 \leq i \leq n\}$ is a matching for $G_T(X \cup P, E)$. We can thus apply Theorem 6 with $b = 1$.

Theorem 10. *Let $H(P, F)$ be any directed graph representing an ordering among processors, and let $G_H(X \cup P, E)$ be the corresponding bipartite subgraph. Then, the sub-greedy* algorithm is at most $(2\sqrt{n} + 2)$-competitive.*

Proof Sketch. We first reduce every strongly connected component of H to a single vertex, since processors of this component are equally powerful: if a task can be assigned to a processor of this component, then it can also be assigned to any other processor of the same component. (Equivalently, this transformation does not affect G_H.) So, we can assume H being not acyclic. We then greedily construct a matching U by repeating the following three steps: 1) for a p_i with no outgoing edges, include the edge (x_i, p_i) in U; 2) remove p_i and x_i from both H and G_H.

Since H is acyclic, such a vertex p_i must exist. Moreover, in G_H, p_i is adjacent to x_i only (otherwise, p_i must have one out-going edge in H). Removing p_i and x_i from G_H yields the graph corresponding to $H \setminus \{p_i\}$. After step 3), we are left with a new H' and $G_{H'}$ which enjoys the same property as H. So, we can iterate this procedure until all vertices in H are removed. Since the number of task types equals the number of vertices of H, this method yields a matching for G_H.

5   Conclusions and Open Problems

We have presented a novel technique which allows to derive on-line algorithms with a simple modification of the greedy one. This modification preserves the good feature of deciding where to assign a task solely based on the current load of processors to which that task can be potentially assigned to. Indeed, the pre-computation of the subgraph required by our approach is performed off-line given the graph representing the problem constraints. Additionally, for several cases we have considered here, this subgraph is only used in the analysis, while the resulting algorithms are simple modifications of the greedy *implementing* the subgraph: the construction of Lemma 1 yields an algorithm performing a greedy choice on the rightmost half of the available processors $\Gamma_G(x_i) = \{p_1, \ldots, p_i\}$. So, this algorithm can be implemented even without knowing n. A similar argument applies to the sub-greedy* algorithm with the subgraph of Theorem 9: in this case knowing n is enough to decide whether a vertex as "low-degree" or not; in the latter case the matching (x_i, p_i) yields a fixed assignment for tasks corresponding to type x_i.

In general, the adopted strategy of the sub-greedy* algorithm depends on the type of the task and on the current load of the adjacent processors in the appropriate subgraph. Since the algorithm assigns tasks corresponding to different subgraphs *independently*, it must be able to compute the load of a processor w.r.t. a subset X_i; this can be easily done whenever tasks are specified as pairs (x, w).

So, our algorithms are distributed and, for the generalized hierarchical topologies, their competitive ratio is only slightly worse than the $2\sqrt{n}+1$ upper bound provided by the Robin-Hood algorithm [5]. Also, for tree hierarchical topologies, our analysis yields a much better ratio whenever the height h of the tree is $o(\sqrt{n})$ (e.g., for balanced trees).

An interesting direction for future research might be that of characterizing the competitive ratio of distributed algorithms under several assumptions on the graph $G(X \cup P, E)$: 1) G is *unknown*, 2) G is uniquely determined by n, but n is unknown, 3) G is known.

A related question is: under which hypothesis does our technique yield optimal competitive ratios?

Acknowledgements. The fourth author wishes to thank Amotz Bar-Noy for a useful discussion and for bringing the work [7] to his attention.

References

1. S. Albers. Better bounds for on-line scheduling. *Proc. of the 29th ACM Symp. on Theory of Computing (STOC)*, pages 130–139, 1997.
2. Y. Azar. *On-line load balancing*, chapter in "On-line Algorithms - The state of the Art", A. Fiat and G. Woeginger (eds.). Springer Verlag, 1998.
3. Y. Azar, A. Broder, and A. Karlin. Online load balancing. *Theoretical Computer Science*, 130:73–84, 1994.
4. Y. Azar and L. Epstein. On-line load balancing of temporary tasks on identical machines. *Proc. of the 5th Israeli Symposium on Theory of Computing and Systems (ISTCS)*, pages 119–125, 1997.
5. Y. Azar, B. Kalyanasundaram, S. Plotkin, K. Pruhs, and O. Waarts. Online load balancing of temporary tasks. *Journal of Algorithms*, 22:93–110, 1997.
6. Y. Azar, J. Naor, and R. Rom. The competitiveness of online assignments. *Journal of Algorithms*, 18:221–237, 1995.
7. A. Bar-Noy, A. Freund, and J. Naor. On-line load balancing in a hierarchical server topology. *SIAM Journal on Computing*, 31(2):527–549, 2001. Preliminary version in Proc. of the 7th Annual European Symposium on Algorithms, ESA'99.
8. D. Braess. Ueber ein Paradoxon aus der Verkehrsplanung. *Unternehmensforschung*, 12:258–268, 1968.
9. P. Crescenzi, G. Gambosi, and P. Penna. On-line algorithms for the channel assignment problem in cellular networks. In *Proc. of the 4th ACM International Workshop on Discrete Algorithms and Methods for Mobile Computing (DIALM)*, pages 1–7, 2000. Full version to appear in Discrete Applied Mathematics.
10. P. Crescenzi, G. Gambosi, and P. Penna. On-line load balancing made simple: Greedy strikes back. Technical report, Università di Salerno, 2003. Electronic version available at http://www.dia.unisa.it/~penna.
11. R. Graham. Bounds for certain multiprocessor anomalies. *Bell System Technical Journal*, 45:1563–1581, 1966.
12. R. Graham. Bounds on multiprocessor timing anomalies. *SIAM J. Appl. Math.*, 17:263–269, 1969.
13. S. Guha and S. Khuller. Greedy strikes back: Improved facility location algorithms. In *ACM-SIAM Symposium on Discrete Algorithms (SODA)*, 1998.
14. E. Tardòs J.K. Lenstra, D.B. Shmoys. Approximation algorithms for scheduling unrelated parallel machines. *Math. Programming*, 46:259–271, 1990.
15. J. D. Murchland. Braess's paradox of traffic flow. *Transportation Research*, 4:391–394, 1070.
16. S. Phillips and J. Westbrook. Online load balancing and network flow. *Algorithmica*, 21(3):245–261, 1998.
17. Serge Y. Ma and A. Plotkin. An improved lower bound for load balancing of tasks with unknown duration. *Information Processing Letters*, 62(6):301–303, 1997.

Real-Time Scheduling with a Budget

Joseph (Seffi) Naor[1], Hadas Shachnai[2*], and Tami Tamir[3]

[1] Computer Science Dept., Technion, Haifa 32000, Israel
naor@cs.technion.ac.il,
[2] Bell Labs, Lucent Technologies, 600 Mountain Ave., Murray Hill, NJ 07974.
hadas@research.bell-labs.com,
[3] Dept. of Computer Science and Eng., Box 352350, Univ. of Washington, Seattle,
WA 98195.
tami@cs.washington.edu

Abstract. Suppose that we are given a set of jobs, where each job has a processing time, a non-negative weight, and a set of possible time intervals in which it can be processed. In addition, each job has a processing cost. Our goal is to schedule a feasible subset of the jobs on a single machine, such that the total weight is maximized, and the cost of the schedule is within a given budget. We refer to this problem as *budgeted real-time scheduling (BRS)*. Indeed, the special case where the budget is *unbounded* is the well-known real-time scheduling problem. The second problem that we consider is *budgeted real-time scheduling with overlaps (BRSO)*, in which several jobs may be processed simultaneously, and the goal is to maximize the time in which the machine is utilized. Our two variants of the real-time scheduling problem have important applications, in vehicle scheduling, linear combinatorial auctions and QoS management for Internet connections. These problems are the focus of this paper.
Both BRS and BRSO are strongly NP-hard, even with unbounded budget. Our main results are $(2 + \varepsilon)$-approximation algorithms for these problems. This ratio coincides with the best known approximation factor for the (unbudgeted) real-time scheduling problem, and is slightly weaker than the best known approximation factor of $e/(e - 1)$ for the (unbudgeted) real-time scheduling with overlaps, presented in this paper. We show that better ratios (or simpler approximation algorithms) can be derived for some special cases, including instances with unit-costs and the budgeted *job interval selection problem (JISP)*. Budgeted JISP is shown to be APX-hard even when overlaps are allowed and with unbounded budget. Finally, our results can be extended to instances with multiple machines.

1 Introduction

In the well-known *real-time scheduling* problem (also known as the *throughput maximization* problem), we are given a set of n jobs; each job J_j has a processing

---

* On leave from the Computer Science Dept., Technion, Haifa 32000, Israel.

J.C.M. Baeten et al. (Eds.): ICALP 2003, LNCS 2719, pp. 1123–1137, 2003.
© Springer-Verlag Berlin Heidelberg 2003

time p_j, a non-negative weight w_j, and a set of time intervals in which it can be processed (given as either a window with release and due-dates or as a discrete set of possible processing intervals). The goal is to schedule a feasible subset of the jobs on a single machine, such that the overall weight of the scheduled jobs is maximized. In this paper we consider two variants of this problem.

In the *budgeted real-time scheduling (BRS)* problem, each job J_j has a processing cost c_j. A budget B is given, and the goal is to find a maximum weight schedule, among the feasible schedules whose total processing cost is at most B. In *real-time scheduling with overlaps (RSO)*, the jobs are scheduled on a single *non-bottleneck* machine, which can process simultaneously several jobs. The goal is to maximize the overall time in which the machine is utilized (i.e., processes at least one job).[1] In the budgeted case *(BRSO)*, each job J_j has a processing cost c_j. The goal is to maximize the time in which the machine is utilized, among the schedules with total processing cost at most B.

In our study of BRS, RSO and BRSO, we distinguish between *discrete* and *continuous* instances. In the discrete case, each job J_j can be scheduled to run in one of a given set of n_j intervals $I_{j,\ell}$ ($\ell = 1, \ldots, n_j$). The special case where each job has at most k intervals, i.e, $\forall j, n_j \le k$, is called $JISP_k$. In the continuous case, job J_j has release date r_j, due date d_j, and a processing time p_j. It is possible to schedule J_j in any interval $[s_j, e_j]$ such that $s_j \ge r_j$, $e_j \le d_j$, and $e_j = s_j + p_j$. We consider also $JISP_1$, where each job can be processed in the single interval $I_{j,1} = [r_j, d_j]$, and $p_j = d_j - r_j$.

We consider general (discrete and continuous) instances, where each job has processing time p_j, a weight w_j, and a processing cost c_j. For some variants we study also classes of instances in which (*i*) jobs have unit-costs (that is, $c_j = 1 \ \forall j$), or (*ii*) for all the jobs $w_j = p_j$.

The BRS and BRSO problems extend the classic real-time scheduling problem to model the natural goal of gaining the maximum available service for a given budget. In particular, the following practical scenarios yield instances of our problems.[2]

Multi-Vehicle Scheduling on a Path: The vehicle scheduling problem arises in many applications, including robot handling in manufacturing systems and secondary storage management in computer systems (see e.g. [KN-01]). Suppose that a fleet of vehicles needs to service requests on a path. There is an operation cost to each vehicle, and a segment on the line in which the vehicle can provide service. Our objective is to assign the vehicles to service requests on line segments such that the total length of the union of the line segments, i.e., the part of the line which is covered, is maximized, yet the overall cost is within some budget constraints.

Combinatorial Auctions: In auctions used in e-commerce, a buyer needs to complete an order for a given set of goods. There is a collection of *sellers*, each

---

[1] Note that job weights have no effect on the objective function.
[2] Other applications, including transmission of continuous-media data and crew scheduling, are given in [NST-03].

offers a subset (or *bundle*) of the goods at some cost. Each of the goods g_i is associated with a weight w_i, which indicates its priority in the order. The buyer needs to satisfy a fraction of the order of maximum weight, by selecting a subset of the offers, such that the total cost is bounded by the buyer's budget, B. In auctions for *linear* goods (see, e.g., in [T-00]), we have an ordered list of m goods $g_1, \ldots, g_m$, and the offers should refer to bundles of the form $g_i, g_{i+1}, \ldots, g_{j-1}, g_j$. Note that while selecting a subset of the offers we allow overlaps, i.e., the buyer may acquire more than the needed amount from some good; however, this does not decrease the cost of any of the offers. Thus, we get an instance of the *BRSO* problem, where any job J_j can be processed in one possible time interval.

QoS Upgrade in a Network: Consider an end-to-end connection between s and t that uses several Internet service providers (ISP). Each ISP provides a basic service (for free) and to upgrade the service one needs to pay; that is, an ISP can decrease the delay in its part of the path for a certain cost. (See, e.g., [LORS-00,LO-02].) The end-to-end delay is additive (over all ISP-s). We have a budget and we need to decide on how to distribute it between the ISP-s. In certain scenarios, an ISP may need to choose to upgrade only a portion of the part of the $s - t$ path that it controls, however, it has the freedom to choose which portion. In this problem instance, "jobs" (upgraded segments) are allowed to overlap.

1.1   Our Results

We give hardness results and approximation algorithms for *BRS*, *RSO*, and *BRSO*. Specifically, we show that continuous *RSO* is strongly NP-hard.[3] In the discrete case, both *BRS* and *BRSO* are shown to be APX-hard, already for instances where $\forall j, n_j \leq k$ ($JISP_k$), and where all the intervals corresponding to a job have the same length, for any $k \geq 3$. In Section 3, we present a $(2 + \varepsilon)$-approximation algorithm for *BRS* (both discrete and continuous). We build on the framework of Jain and Vazirani [Va-01] for using Lagrangian relaxation in developing approximation algorithms. Our algorithm is based on a novel combination of Lagrangian relaxation with efficient search on the set of feasible solutions. We show that a simple Greedy algorithm yields a 4-approximation for *BRS* with unit costs, where $w_j = p_j$ $\forall j$.

In Section 4, we give a $(2 + \varepsilon)$-approximation algorithm for continuous inputs of *BRSO*, and a $(3 + \varepsilon)$-approximation for discrete inputs, using the Lagrangian relaxation technique. For *RSO* we present a Greedy algorithm that achieves the ratio of 2. An improved ratio of $e/(e-1)$ is obtained by a randomized algorithm (where e denotes the base of the natural logarithm). For $JISP_1$, we obtain an optimal solution for instances of *BRSO* with unit costs, and a *fully polynomial time approximation scheme (FPTAS)* for arbitrary costs. (Note that $JISP_1$ is weakly NP-hard. This can be shown by reduction from Knapsack [GJ-79].) Finally, in Section 5 our results are shown to extend to instances of *BRS* and *BRSO* in which the jobs can be scheduled on multiple machines.

---

[3] The continuous real-time scheduling problem (with no overlaps) is known to be strongly NP-hard [GJ-79].

The approximation technique that we use for deriving our $(2 + \varepsilon)$-approximation results (see in Section 2) is shown to apply to a fundamental class of budgeted maximization problems, including *throughput maximization in a system of dependent jobs*, which generalizes the *BRS* problem (see in [NST-03]). We show that, using the technique, any problem in the class which has an LP based ρ-approximation with unbounded budget, can be approximated within factor $\rho + \varepsilon$ in the budgeted case, for any $B \geq 1$.

Due to space constraints we state some of the results without proofs.[4]

1.2   Related Work

To the best of our knowledge, the *budgeted* real-time scheduling problem is studied here for the first time. There has been extensive work on real-time scheduling, both in the discrete and the continuous models. Garey and Johnson (cf. [GJ-79]) showed that the continuous case is strongly NP-hard, while the discrete case, JISP, was shown by Spieksma [S-99] to be APX-hard, already for instances of $JISP_k$, where $k \geq 2$. Bar-Noy et al. [BG+99,BB+00] and independently Berman and DasGupta [BD-00] presented 2-approximation algorithms for the discrete case,[5] and $(2 + \varepsilon)$ ratio in the continuous case. As shown in [BB+00], this ratio holds for arbitrary number of machines. While none of the existing techniques has been able to improve upon the 2 and $(2+\varepsilon)$ ratios for general instances of the real-time scheduling problem, improved bounds were obtained for some special cases. In particular, Chuzhoy et al. [COR-01] considered the unweighted version, for which they gave an $(e/(e-1) + \varepsilon)$-approximation algorithm, where ε is any constant. For other special cases, they developed polynomial time approximation schemes. Finally, some special cases of JISP were shown to be polynomially solvable (see, e.g., in [AS-87,B-99]).

We are not aware of previous work on the *RSO* and *BRSO* problems. Since overlaps are allowed and the goal is to maximize the overall time in which the machine is utilized, these problems can be viewed as maximum coverage problems. In previous work on budgeted covering (see, e.g., [KMN-99]), the covering items are *sets*; once a set is selected, the covered elements are uniquely defined. In contrast, in *RSO* (and *BRSO*) the covering items are *jobs*, and we can choose the time segments (= elements) that will be covered by a job, by determining the time interval in which this job is processed.

2   Approximation via Lagrangian Relaxation

We describe below the general approximation technique that we use for deriving our results for *BRS* and *BRSO*. Our approach builds on the framework

---

[4] The detailed proofs are given in [NST-03].

[5] A 2-approximation for *unweighted* JISP is obtained by a Greedy algorithm, as shown in [S-99].

developed by Jain and Vazirani [Va-01][pp. 250-251] (see also [Ga-96]), for using Lagrangian relaxations in approximation algorithms. Our approach applies to the following class of *subset selection* problems. The input for any problem in the class consists of a set of elements $A = \{a_1, \ldots, a_n\}$; each element $a_j \in A$ is associated with a weight w_j, and the cost of adding a_j to the solution set is $c_j \geq 1$. We have a budget $B \geq 1$. The goal is to find a subset of the elements $A' \subseteq A$ satisfying a given set of constraints (including the budget constraint), such that the total weight is maximized. We assume that any problem Π in the class satisfies the following property. **(P1)** Let A' be a feasible solution for Π; then, any subset $A'' \subseteq A'$ is also a feasible solution.

Denote by $x_j \in \{0, 1\}$ the indicator variable for the selection of a_j. The integer program for Π has the following form.

$$(\Pi) \qquad \text{maximize} \qquad \sum_{a_j \in A} w_j x_j$$

$$\text{subject to:} \qquad \text{Constraints}: C_1, \ldots, C_r$$

$$\sum_j c_j x_j \leq B.$$

In the linear relaxation we have $x_j \in [0, 1]$. The Lagrangian relaxation of this program is

$$(L - \Pi(\lambda)) \quad \text{maximize} \qquad \lambda \cdot B + \sum_{a_j \in A} (w_j - c_j \lambda) x_j$$

$$\text{subject to:} \qquad \text{Constraints}: C_1, \ldots, C_r$$

Assume that $\mathcal{A}_\pi$ is a ρ-approximation algorithm to the optimal integral solution for $L - \Pi(\lambda)$, for any value of $\lambda > 0$. Thus, there exist values $\lambda_1 < \lambda_2$ such that $\mathcal{A}_\pi$ finds integral ρ-approximate solutions $\mathbf{x}_1, \mathbf{x}_2$ for $L - \Pi(\lambda_1), L - \Pi(\lambda_2)$, respectively, and the budgets used in these solutions are B_1, B_2, where

$$B_2 < B < B_1. \tag{1}$$

Let W_1, W_2 denote the weights of the solutions $\mathbf{x}_1, \mathbf{x}_2$, then $W_i = \lambda_i B + \sum_{a_j \in A} (w_j - c_j \lambda_i) \mathbf{x}_{ij}$, $i \in \{1, 2\}$, $1 \leq j \leq n$. W.l.o.g, we assume that $W_1, W_2 \geq 1$.

Following the framework of [Va-01], we require that $\mathcal{A}_\pi$ satisfies the following property. Let $\alpha = (B - B_2)/(B_1 - B_2)$, then the convex combination of the solutions $\mathbf{x}_1, \mathbf{x}_2$, namely, $\mathbf{x} = \alpha \mathbf{x}_1 + (1 - \alpha)\mathbf{x}_2$ is a (fractional) ρ-approximate solution that uses the budget B. This is indeed the case if, for example, the solutions $\mathbf{x}_1, \mathbf{x}_2$ are obtained from a primal-dual algorithm. In this case, a convex combination of the dual solutions corresponding to $\mathbf{x}_1$ and $\mathbf{x}_2$ can be used to prove this property. This will be heavily used in our algorithms for the *BRS* and *BRSO* problems. Our goal is to find a feasible integral solution whose weight is close to the weight of $\mathbf{x}$. We show that for the class of subset selection problem that we consider here, by finding 'good' values of λ_1, λ_2, we obtain an integral solution that is within factor $\rho + \varepsilon$ from the optimal. The running time of our algorithm is dominated by the complexity of the search for λ_1, λ_2 and the running time of $\mathcal{A}_\pi$.

We now summarize the steps of the algorithm, $\mathcal{A}_L$, which gets as input the set of elements $a_1, \ldots, a_n$, an accuracy parameter $\varepsilon > 0$ and the budget $B \geq 1$. Let $c = \sum_j c_j$ denote the total cost of the instance.

1. Let $\varepsilon' = \varepsilon/c$.
2. Define the *modified weight* of an element a_j to be $w'_j = w_j/c_j$.
   Let $\omega_1 \leq \cdots \leq \omega_R$ be the set of R distinct values of modified weights.
3. Find in $(0, \omega_R)$ the values of $\lambda_1 < \lambda_2$
   satisfying (1), such that $\lambda_2 - \lambda_1 \leq \varepsilon'$.
4. Output the (feasible) integral solution found by $\mathcal{A}_\pi$ for $L - \Pi(\lambda_2)$.

Analysis: The heart of our approximation technique is the following theorem.

Theorem 1. *For any $0 < \varepsilon'$ and λ_1, λ_2 satisfying (1), if $0 < \lambda_2 - \lambda_1 < \varepsilon'$, then $W_2 \geq W_1 - \varepsilon'c$, where c is the total cost of the instance.*

Proof. We note that for a fixed value of λ we can omit from the input elements a_j for which $w'_j = w_j/c_j \leq \lambda$. We denote by S_i the *feasible* set of modified weights for λ_i, i.e., the set of values ω_ℓ satisfying $\omega_\ell \geq \lambda_i$; then $S_2 \subseteq S_1$. Let $A_i \subseteq S_i$ be the set of elements selected by $\mathcal{A}_\pi$ for the solution, for the given value λ_i. Then, $W_1 = \lambda_1 B + \sum_{A_1}(w_j - c_j\lambda_1)$ and $W_2 = \lambda_2 B + \sum_{A_2}(w_j - c_j\lambda_2)$. We handle two cases separately.

(i) The feasible sets for λ_1, λ_2 are identical, that is, $S_1 = S_2$. Then

$$W_2 \geq \lambda_2 B + \sum_{A_1}(w_j - c_j\lambda_2) \geq \lambda_1 B + \sum_{A_1}(w_j - c_j(\lambda_1 + \varepsilon')) = W_1 - \varepsilon' \sum_{A_1} c_j \geq W_1 - \varepsilon'c$$

The leftmost inequality follows from the fact that all the elements in A_1 were feasible also with λ_2. (Note that we can guarantee that the inequality is satisfied, by comparing W_2 with the weight of A_1 at λ_2, and by taking the subset which gains the maximum of these two weights.)

(ii) The feasible set for λ_1 contains some elements whose modified weights that are not contained in S_1, that is, $S_2 \subset S_1$. For simplicity, we assume that $S_1 = \{\omega_{\ell+1}, \omega_{\ell+2}, \ldots, \omega_R\}$, while $S_2 = \{\omega_{\ell+2}, \ldots, \omega_R\}$, that is, for some $1 \leq \ell < R$, $\omega_{\ell+1} \in S_1$ and $\omega_{\ell+1} \notin S_2$. In general, several modified weight values may be contained in S_1 but not in S_2. A similar argument can be applied for this case. Denote by $\hat{A}_1$ the subset of elements in A_1 whose modified weights are equal to $\omega_{\ell+1}$. Then,

$$W_2 \geq \lambda_2 B + \sum_{A_1 \setminus \hat{A}_1}(w_j - c_j\lambda_2) \geq \lambda_1 B + \sum_{A_1 \setminus \hat{A}_1}(w_j - c_j(\lambda_1 + \varepsilon'))$$

$$= \lambda_1 B + \sum_{A_1}(w_j - \lambda_1 c_j) - \sum_{\hat{A}_1}(w_j - \lambda_1 c_j) - \varepsilon' \sum_{A_1 \setminus \hat{A}_1} c_j$$

$$= W_1 - \sum_{\hat{A}_1} c_j(\omega_{\ell+1} - \lambda_1) - \varepsilon' \sum_{A_1 \setminus \hat{A}_1} c_j \geq W_1 - \varepsilon' \sum_{A_1} c_j \geq W_1 - \varepsilon'c$$

The first inequality is due to the fact that the set of elements $A_1 \setminus \hat{A}_1$ was available with λ_2, and that Π satisfies property **(P1)**; the second inequality follows from the difference $(\lambda_2 - \lambda_1)$ being bounded by ε'; the last inequality follows from $(\omega_{\ell+1} - \lambda_1) < \varepsilon'$. This completes the proof.   □

Let $0 < \varepsilon < 1$ be an input parameter, then taking $\varepsilon' = \varepsilon/c$, we get from Theorem 1 that

$$W_2 \geq (W_1 - \varepsilon'c)\alpha + W_2(1 - \alpha) \geq (W_1\alpha + W_2(1 - \alpha)) - \varepsilon'c \geq (W_1\alpha + W_2(1 - \alpha))(1 - \varepsilon).$$

Finally, since $\mathbf{x}$ gives a ρ-approximation to the optimal, we get

Theorem 2. *Algorithm $\mathcal{A}_L$ achieves an approximation factor of $(\rho + \varepsilon)$ for Π.*

Implementation: Note that to obtain a $(\rho + \varepsilon)$-approximation, we need to find the values of $\lambda_1, \lambda_2 \in (0, \omega_R)$ that satisfy (1), such that $(\lambda_2 - \lambda_1) < \varepsilon/c$. As $\omega_R = \max_j w_j/c_j$ may be arbitrarily large, a naive search may require exponential number of steps. We show that by allowing a small increase (of ε) in the approximation ratio, we can implement this search in polynomial time. (i) Initially, we guess the weight of an optimal integral solution, W^*, to within factor $(1 - \varepsilon)$. This can be done in $O(\lg(n/\varepsilon))$ steps, since $\max_j w_j \leq W^* \leq n \cdot \max_j w_j$. We then omit from the input any element a_j whose weight is smaller than $\varepsilon W^*/n$. We scale the weights of the remaining elements, so that all the weights are in the range $[1, n/\varepsilon]$. (ii) For any element a_j with $c_j < \varepsilon B/n$, we round up c_j to $\varepsilon B/n$. We scale the other costs, such that all costs are in $[1, n/\varepsilon]$. (iii) We scale accordingly the size of the interval $(0, \omega_R)$.

Now, we argue that in the above scaling and rounding we only slightly decrease the weight of the solution. Indeed, by omitting elements with 'small' weights, we decrease the total weight of the elements selected by $\mathcal{A}_\pi$ at most by factor of ε. Also, by rounding up the 'small' costs to $\varepsilon B/n$, we get that the total weight obtained by $\mathcal{A}_\pi$ at λ_2 is at least $\lambda_2 B + \sum_{a_j \in A_2} \left(w_j - (c_j + \frac{\varepsilon B}{n})\lambda_2\right) \geq W(1 - \varepsilon)$. Thus, overall we lose a factor of 2ε in the approximation ratio. The overall running time of our search procedure is $O(\lg(n/\varepsilon) \cdot \lg(n^3/\varepsilon^3)) = O(\lg^2(n/\varepsilon^3))$. It follows that the running time of $\mathcal{A}_L$ is $O(\lg^2(n/\varepsilon^3))$ times the running time of algorithm $\mathcal{A}_\pi$.

3   Approximating Bounded Real-Time Scheduling

3.1   A Greedy Algorithm

Consider first the special case of unit cost jobs, where $w_j = p_j \; \forall \; j$. Suppose that the budget is $B = k$; thus, we need to select a subset of k non-overlapping jobs, such that machine utilization is maximized. For such instances, we can obtain a constant approximation using an $O(n \log n)$ greedy algorithm. The algorithm $\mathcal{A}_G$ (formulated for continuous inputs) first sorts the jobs in non-increasing order by their processing times. Then, $\mathcal{A}_G$ schedules at most k jobs by scanning the sorted list; that is, while there is available budget, the next job J_j is scheduled in the earliest available time interval in $[r_j, d_j]$.

Theorem 3. $\mathcal{A}_G$ *is a* $(4+\varepsilon)$*-approximation for BRS with unit costs where* $w_j = p_j \;\forall\; j$. *In the discrete case,* $\mathcal{A}_G$ *achieves the ratio of* 4.

As we show below, better ratio can be achieved, for *general* instances, by using the Lagrangian relaxation technique.

3.2   A $(2 + \varepsilon)$-Approximation Algorithm

In the following we derive a $(2+\varepsilon)$-approximation for discrete instances of BRS. A similar result can be obtained for the continuous case, by discretizing the instance. Recall that in the discrete case, any job J_j can be scheduled in the intervals $I_{j,1}, \ldots, I_{j,n_j}$. We define a variable $x(j, \ell)$ for each interval $I_{j,\ell}$, $1 \le j \le n, 1 \le \ell \le n_j$. Then the integer program for the problem is:

$$(BRS) \qquad \text{maximize} \qquad \sum_{j=1}^{n}\sum_{\ell=1}^{n_j} w_j x(j, \ell)$$

$$\text{subject to}: \qquad \forall j: \quad \sum_{\ell=1}^{n_j} x(j, \ell) \le 1$$

$$\forall t: \quad \sum_{t \in I_{j,\ell}} x(j, \ell) \le 1$$

$$\sum_{j=1}^{n}\sum_{\ell=1}^{n_j} c_j x(j, \ell) \le B.$$

In the linear program $x_j \in [0, 1]$. Taking the Lagrangian relaxation, we get an instance of the *throughput maximization* problem. As shown in [BB+00], an algorithm based on the local ratio technique yields a 2-approximation for this problem, in $O(n^2)$ steps. This algorithm has a primal-dual interpretation; thus, we can apply the technique in Section 2 to obtain an algorithm, $\mathcal{A}$, which uses the algorithm for throughput maximization as a procedure.[6]

Theorem 4. *Algorithm* $\mathcal{A}$ *yields a* $(2 + \varepsilon)$*-approximation for BRS, in* $O(n^2 \lg^2(n/\varepsilon^2))$ *steps.*

4   Approximation Algorithms for *RSO* and *BRSO*

In this section, we present approximation algorithms for the *RSO* and *BRSO* problems. In Section 4.1 we consider *RSO*. We first give a randomized $e/(e-1)$-approximation algorithm for discrete inputs; then, we describe a greedy algorithm that achieves a ratio of $(2 - \epsilon)$ for continuous inputs, and $(3 - \epsilon)$ for discrete inputs. In Section 4.2, we show that the greedy algorithm can be interpreted equivalently as a primal-dual algorithm. This allows us to apply the Lagrangian relaxation framework (Section 2) and to achieve a $(3 + \varepsilon)$-approximation for *BRSO* in the discrete case, where all the intervals corresponding to a job have the same length. For continuous inputs we obtain a $(2 + \varepsilon)$-approximation.

---

[6] Note that since $W_1, W_2 \ge \max_j w_j$, in our search for λ_1, λ_2 we can take $\varepsilon' = \varepsilon/n$.

4.1   The *RSO* Problem

In the RSO problem, we may select *all* the jobs, and the problem reduces to scheduling the jobs optimally so as to maximize the coverage of the line. Clearly, when $\forall j, p_j = d_j - r_j$, i.e., each job has only one possible interval, the schedule in which all the jobs are selected is optimal. When $\forall j, p_j \leq d_j - r_j$, the problem becomes hard to solve (see in [NST-03]).

Theorem 5. *The RSO problem is strongly NP-hard.*

A Randomized $e/(e-1)$-approximation algorithm. We start with a linear programming formulation of *RSO*. Assume that the input is given in a discrete fashion, and let $b_0, \ldots, b_m$ denote the set of start and end points (in sorted order), called *breakpoints*, of the time intervals $I_{j,\ell}$, $j = 1, \ldots, n$, $\ell = 1, \ldots, n_j$. We have a variable $x(j, \ell)$ for each interval $I_{j,\ell}$. For any pair of consecutive breakpoints b_{i-1} and b_i, the objective function gains $(b_i - b_{i-1})$ times the "coverage" of the interval $[b_{i-1}, b_i]$. Note that we take the minimum between 1 and the total cover, since we gain nothing if some interval is covered by more than one job.

$$(L - RSO) \quad \text{maximize} \quad \sum_{i=1}^{m} \min \left(\sum_{j=1}^{n} \sum_{I_{j,\ell} \ni [b_{i-1}, b_i]} x(j, \ell), 1 \right) \cdot (b_i - b_{i-1})$$

$$\text{subject to:} \qquad \text{For all jobs } J_j: \quad \sum_{\ell=1}^{n_j} x(j, \ell) \leq 1$$

$$\text{For all } (j, \ell), \ell = 1, \ldots, n_j: \quad x(j, \ell) \geq 0.$$

We compute an optimal (fractional) solution to $(L-RSO)$. Clearly, the value of this solution is an upper bound on the value of an optimal integral solution. To obtain an integral solution, we apply randomized rounding to the optimal fractional solution. That is, for every job J_j, the probability that J_j is assigned to interval $I_{j,\ell}$ is equal to $x(j, \ell)$. If $\sum_{\ell=1}^{n_j} x(j, \ell) < 1$, then with probability $1 - \sum_{\ell=1}^{n_j} x(j, \ell)$ job J_j is not assigned to any interval.

We now analyze the randomized rounding procedure. Consider two consecutive breakpoints b and b'. Define for each job J_j, $y_j = \sum_{I_{j,\ell} \ni [b,b']} x(j, \ell)$. Clearly, $\sum_{j=1}^{n} y_j = \sum_{j=1}^{n} \sum_{I_{j,\ell} \ni [b,b']} x(j, \ell)$. W.l.o.g., since each job J_j is assigned to a single interval, we can think of all the intervals of J_j that cover $[b, b']$ as a single (virtual) interval that is chosen with probability y_j. The probability that none of the virtual intervals is chosen is $P_0 = \prod_{j=1}^{n}(1 - y_j)$. Let $r = \min(\sum_{j=1}^{n} y_j, 1)$. Then,

$$P_0 \leq \prod_{j=1}^{n} \left(1 - \frac{\sum_{i=1}^{n} y_i}{n} \right) = \left(1 - \frac{\sum_{i=1}^{n} y_i}{n} \right)^n < e^{-\sum_{i=1}^{n} y_i} \leq e^{-r}.$$

Hence, the probability that $[b, b']$ is covered is $1 - P_0 \geq 1 - e^{-r} \geq \left(1 - \frac{1}{e}\right) \cdot r \geq \left(1 - \frac{1}{e}\right) \cdot \min\left(\sum_{j=1}^{n} y_j, 1\right)$. Therefore, the expected contribution to the objective function of any interval $[b_{i-1}, b_i]$ is $(1 - 1/e) \cdot \min(\sum_{j=1}^{n} y_j, 1) \cdot (b_i - b_{i-1})$. By linearity of expectation, the expected value of the objective function after applying randomized rounding is

$$\left(1 - \frac{1}{e}\right) \cdot \sum_{i=1}^{m} \min\left(\sum_{j=1}^{n} \sum_{I_{j,\ell} \ni [b_{i-1}, b_i]} x(j, \ell), 1\right) \cdot (b_i - b_{i-1}),$$

yielding an approximation factor of $1 - 1/e \approx 0.63212$.

A Greedy Approximation Algorithm. We now describe a greedy algorithm, which yields a $(2 - \varepsilon)$-approximation for continuous instances of RSO, and $(3 - \varepsilon)$-approximation for discrete instances. Assume that $\min_j r_j = 0$, and let $T = \max_j d_j$. Let I be the set of all the jobs in the instance; U is the set of unscheduled jobs. Denote by s_j, e_j the start-time and completion time of the job J_j in the greedy schedule, respectively. Given a partial schedule, we say that J_ℓ is *redundant* if we can remove J_ℓ from the schedule without decreasing the machine utilization. Algorithm Greedy proceeds in the interval $[0, T]$. At time t, we select an arbitrary job, among the jobs J_i with $r_i < t$ and $d_i > t$. We schedule this job such that its contribution to the utilization, starting at time t, is maximized. The following is a pseudocode of the algorithm.

1. $U = I$, $t = 0$;
2. Let $J_j \in U$ be a job having $d_j > t$ and $r_j \leq t$.
    Schedule J_j such that its completion time, e_j, is $\min(t + p_j, d_j)$.
    Remove J_j from U.
    For any redundant job J_ℓ, omit J_ℓ from the schedule and return it to U.
3. Let $F \subseteq U$ be the set of unscheduled jobs, i, having $d_i > e_j$.
    Let $t_F = \min_{J_i \in F} r_i$, and let $t = \max(e_j, t_F)$.
    If $F \neq \emptyset$ and $t < T$ go to step 2.

We use in the analysis the following properties of the greedy schedule.

Property 1. Once an interval $[x_1, x_2] \in [0, T]$ is covered by Greedy, it remains covered until the end of the schedule.

Property 2. When the algorithm considers time t, some job will be selected and scheduled such that for some $\varepsilon > 0$, the machine is utilized in the interval $[t, t+\varepsilon]$.

Property 3. Consider the set U of non-scheduled jobs at the end of the schedule. For any $J_j \in U$ the machine is utilized in the time interval $[r_j, d_j)$.

Theorem 6. *The Greedy algorithm yields a $(2-\varepsilon)$-approximation for the RSO problem.*

Proof. Let $S = I \setminus U$ denote the set of jobs scheduled by Greedy, and let $O \subseteq S$ denote the set of scheduled job such that $J_j \in O$ iff J_j overlaps with another scheduled job, J_k, and $e_j > e_k$. (i) By Property 3, for any $J_j \in U$ the machine is utilized in the time interval $[r_j, d_j]$. (ii) For any $J_j \in S$ the machine is utilized in the time interval $[r_j, s_j]$, otherwise Greedy would have scheduled it earlier. (iii) For any $J_j \in O$ the machine is utilized in the time interval $[r_j, d_j]$. This follows from (ii) and from the fact that $e_j = d_j$, otherwise J_j would not overlap with a job with an earlier completion time.

Given the schedule of Greedy, we allow OPT to add jobs of U and to shift the scheduled jobs of S in any way that increases the utilization. Consider the three *disjoint* sets of jobs, $U, O, S \setminus O$. By the above discussion, the utilization can be increased only by shifting to the left (i.e., scheduling earlier) the jobs of $S \setminus O$. Note that in any time $t \in [0, T]$ there is at most one job, J_j, of $S \setminus O$ (if two or more jobs overlap, then only the one with the earliest completion time is in $S \setminus O$). Let $0 < \varepsilon_j \le 1$ be such that J_j overlaps in the Greedy schedule in $(1 - \varepsilon_j)$ of its length. Then OPT can shift J_j into an interval in which it does not overlap at all. Hence, OPT can increase the amount of time the machine is utilized in the greedy schedule at most by factor of $(2 - \varepsilon)$. ☐

The analysis for discrete inputs is similar, only that in this case, by selecting for two jobs overlapping at time t different intervals, and by adding a non-scheduled job to run at t, OPT may triple the utilization obtained by Greedy, thus we get the bound of $(3 - \varepsilon)$.

4.2   The *BRSO* Problem

As *BRSO* generalizes the *RSO* problem, Theorem 5 implies that it is strongly NP-hard. For discrete inputs we show the following (see in [NST-03]).

Theorem 7. *The discrete BRSO is APX-hard, already for instances where $\forall j, n_j \le k$ (JISP$_k$), for any $k \ge 3$.*

A Primal-Dual Algorithm. We first present a primal-dual algorithm for RSO, and show that an execution of the Greedy algorithm given in Section 4.1 can be equivalently interpreted as an execution of the primal-dual algorithm. Thus, the primal-dual algorithm finds a 3-approximate solution to RSO. The primal LP is equivalent to $L - RSO$, given in Section 4.1. We have a variable $x(j, \ell)$ for each interval $I_{j,\ell}$, and a variable z_i, $i = 1, \ldots, m$ for each interval $[b_{i-1}, b_i]$ defined by consecutive breakpoints. In the dual LP we have a variable y_j for each job J_j, and two variables, p_i and q_i for each interval $[b_{i-1}, b_i]$ defined by consecutive breakpoints.

$(L - RSO - Primal)$ maximize $\sum_{i=1}^{m} z_i \cdot (b_i - b_{i-1})$

subject to :   For all jobs J_j: $\sum_{\ell=1}^{n_j} x(j, \ell) \leq 1$

For all $i = 1, \ldots, m$: $z_i \leq 1$

For all $i = 1, \ldots, m$: $z_i - \sum_{I_{j,\ell} \ni [b_{i-1}, b_i]} x(j, \ell) \leq 0$

For all j, ℓ, i: $x(j, \ell), z_i \geq 0.$

$(L - RSO - Dual)$ minimize $\sum_{j=1}^{n} y_j + \sum_{i=1}^{m} p_i$

subject to :   For all $(j, \ell), \ell = 1, \ldots, n_j$: $y_j - \sum_{I_{j,\ell} \ni [b_{i-1}, b_i]} q_i \geq 0$

For all $i = 1, \ldots, m$: $p_i + q_i \geq (b_i - b_{i-1})$

For all j, i: $y_j, p_i, q_i \geq 0.$

Given an integral solution for $L - RSO - Primal$, we say that an interval I belongs to it if there is a job that is assigned to I. An integral solution for $L - RSO - Primal$ is *maximal*, if it cannot be extended and if no interval belonging to it is contained in the union of other intervals belonging to it.

Lemma 1. *Any maximal integral solution (x, z) to $L - RSO - Primal$ is a 3-approximate solution.*

Proof. If $[b_{i-1}, b_i]$ is covered by (x, z), then set $p_i = b_i - b_{i-1}$, otherwise set $q_i = b_i - b_{i-1}$. Clearly, this defines a feasible dual solution in which $\sum_{i=1}^{m} p_i = \sum_{i=1}^{m} z_i \cdot (b_i - b_{i-1})$. Thus, it remains to bound $\sum_{j=1}^{n} y_j$ in this solution.

For each job J_j that is not assigned to any interval in (x, z), i.e., its intervals are contained in intervals of other jobs, we can set $y_j = 0$. Suppose that for job J_j, $x(j, \ell) = 1$. Consider, for example, an interval $I_{j,\ell'}$, $\ell \neq \ell'$, that contains two consecutive breakpoints b_{i-1} and b_i such that $[b_{i-1}, b_i]$ is not covered by any job. In this case $q_i = b_i - b_{i-1}$ and $y_j \geq q_i$. Thus, in order to bound $\sum_{j=1}^{n} y_j$, we say that the values of q_i-s that determine the y_j-s "charge" the p_i values corresponding to the breakpoints covered by $I_{j,\ell}$. This can be done since all the intervals in which J_j can be scheduled have the same length. Since our primal solution is maximal, any point is covered by at most two intervals to which jobs are assigned, and therefore any variable p_i can be "charged" by intervals belonging to at most two different jobs. Thus, $\sum_j y_j \leq 2 \sum_i p_i$, proving that $\sum_{i=1}^{m} p_i = \sum_{i=1}^{m} z_i \cdot (b_i - b_{i-1}) \leq \sum_{j=1}^{n} y_j + \sum_{i=1}^{m} p_i \leq 3 \cdot \sum_{i=1}^{m} p_i$, meaning that (x, z) is a 3-approximate solution. □

For continuous instances, we can discretize time such that each time slot is of size ϵ. This will incur a $(1 + \epsilon')$ degradation in the objective function where $\epsilon' = \text{poly}(\epsilon, n)$. We can show that for a discrete input obtained from a discretization of a continuous input instance, the primal-dual algorithm yields a 2-approximate solution. Applying the Lagrange relaxation technique (presented in Section 2), we get the following.

Theorem 8. *BRSO can be approximated within factor $(2+\varepsilon)$ in the continuous case, and $(3+\varepsilon)$ in the discrete case, in $O(n^2 \cdot \lg^2(n/\varepsilon^2))$ steps.*

4.3   An FPTAS for $JISP_1$

For instances of $BRSO$ where $p_j = d_j - r_j$ ($JISP_1$), we use a reduction to the *budgeted longest path problem* in acyclic graph, to obtain an optimal polynomial time algorithm for unit costs, and an FPTAS for general instances. In the budgeted longest path problem, we are given an acyclic graph, $G(V, E)$; each edge $e \in E$ has a length $\ell(e)$, and a cost $c(e)$. Our goal is to find the longest path in G connecting two given vertices s, t, whose price is bounded by a given budget B. The problem is polynomially solvable for unit edge costs, and has an FPTAS for arbitrary costs [Ha-92].

Given an instance of $BRSO$ where $\forall j$, $p_j = d_j - r_j$, we construct the following graph, G. Each job j is represented by a vertex; there is an edge $e = (i, j)$ iff $d_i < d_j$ and $r_i \le r_j$. The length of the edge is $\ell(e) = \min(d_j - d_i, p_j)$, and its cost is c_j. Note that $\ell(e)$ reflects the machine utilization gained if the deadlines of J_i, J_j are adjacent to each other in the schedule. In addition, each vertex j is connected to the source, s, where $\ell(s, j) = p_j$, $c(s, j) = c_j$, and to a sink t, where $\ell(j, t) = 0$ and $c(s, j) = 0$.

Theorem 9. *There is a schedule achieving utilization of u time units and having cost $b \le B$ if and only if G contains a path of length u and price b.*

Proof. For a given schedule, sort the jobs in the schedule such that $d_{j_1} \le d_{j_2} \le \ldots \le d_{j_w}$. W.l.o.g, the schedule does not include two jobs J_i, J_j such that $r_j < r_i$ and $d_i < d_j$, since in such a schedule we gain nothing from processing J_i. Thus, we can assume that $r_{j_i} \le r_{j_{i+1}}$, $\forall 1 \le i \le w$. This implies that the graph G contains the path $s, j_1, j_2, \ldots, j_w, t$ (the first and last edges in this path exist by the structure of G). Suppose that the utilization of the schedule is u and its cost is b. We show that the length of the corresponding path in G is u and its cost is b. Recall that the edge (j_w, t) has length 0 and costs nothing, thus, we consider only the first w edges in this path. The utilization we gain from scheduling j_i is p_{j_i} if $i = 1$ and $\min(p_{j_i}, d_{j_i} - d_{j_{i-1}})$ if $1 < i \le w$. This is exactly $\ell(j_{i-1}, j_i)$ (or $\ell(s, j_1)$ for the first vertex in the path). Also, the cost of the schedule is the total processing cost of the scheduled jobs, which is identical to the total cost of edges in the path.

For a given directed path in G, we schedule all the jobs whose corresponding vertices appear on the path. Note that the price of the path consists of the price of the participating vertices, thus, b is also the price of the schedule. Also, as discussed above $\ell(i,j)$ reflects the contribution of the corresponding job to the utilization, thus the path induces a schedule with the correct utilization and cost.   □

5   Multiple Machines

Suppose that we have m machines, and a budget B, which can be distributed in any way among the machines. It can be shown that this model is equivalent to the single machine case, by concatenating the schedules on the m machines to a single schedule in the interval $[0, mT]$, on a single machine. Thus, all of our results carry over to this model. When we have a budget specified for each machine, we show that any approximation algorithm $\mathcal{A}$ for a single machine can be run iteratively on the machines and the remaining jobs, to obtain a similar approximation ratio. Denote this algorithm $\mathcal{A}^*$.

Theorem 10. *If $\mathcal{A}$ is an r-approximation then the iterative algorithm $\mathcal{A}^*$ is an $(r+1)$-approximation.*

Note that in most cases $\mathcal{A}^*$ performs better. For example, when we iterate Greedy (Section 4.1) for the RSO problem, it can be seen that the proof for a single machine is valid also for multiple machines, thus Greedy is $(2-\varepsilon)$-approximation. In the full version of the paper, we show that our results can be extended to apply also for the case where the processing costs of the jobs are machine dependent, that is, the cost of processing J_j on the k-th machine is c_{jk}, $1 \leq j \leq n$, $1 \leq k \leq m$.

Acknowledgments. We thank Shmuel Zaks for encouraging us to work on RSO and its variants. We also thank Magnús Halldórsson and Baruch Schieber for valuable discussions.

References

[AS-87]   E.M. Arkin and E.B. Sliverberg. "Scheduling jobs with fixed start and end times". *Discrete Applied Math.*, 18:1–8, 1987.

[B-99]   P. Baptiste. Polynomial time algorithms for minimizing the weighted number of late jobs on a single machine with equal processing times. *J. of Scheduling*, 2:245–252, 1999.

[BB+00]   A. Bar-Noy, R. Bar-Yehuda, A. Freund, J. Naor, and B. Schieber. A unified approach to approximating resource allocation and scheduling. *J. of the ACM*, 48:1069–1090, 2001.

[BG+99]   A. Bar-Noy, S. Guha, J. Naor and B. Schieber Approximating the throughput of real-time multiple machine scheduling. *SIAM J. on Computing*, 31:331–352, 2001.

[BD-00]   P. Berman and B. DasGupta. "Multi-phase algorithms for throughput maximization for real-time scheduling." *J. of Combinatorial Optimization*, 4:307–323, 2000.

[COR-01]   J. Chuzhoy, R. Ostrovsky, and Y. Rabani. "Approximation algorithms for the job interval selection problem and related scheduling problems." *FOCS*, 2001.

[GJ-79]   M.R. Garey and D.S. Johnson. *Computers and intractability: a guide to the theory of NP-completeness.* W.H. Freeman, 1979.

[Ga-96]   N. Garg, A 3-approximation for the minimum tree spanning k vertices, Proceedings of FOCS, 1996.

[Ha-92]   R. Hassin, Approximation schemes for the restricted shortest path problem. In *Mathematics of Operations Research*, 17(1): 36–42, 1992.

[K-91]   V. Kann.   Maximum bounded 3-dimensional matching is max SNP-complete. *Information Processing Letters*, 37:27–35, 1991.

[KN-01]   Y. Karuno and H. Nagamochi, A 2-approximation algorithm for the multi-vehicle scheduling problem on a path with release and handling times. *ESA*, 2001.

[KMN-99]   S. Khuller, A. Moss, J. Naor, The budgeted maximum coverage problem. *Information Processing Letters* 70(1): 39–45, 1999.

[LO-02]   D. H. Lorenz and A. Orda, Optimal partition of QoS requirements on unicast paths and multicast trees, IEEE/ACM Trans. on Networking, 10(1), 102–114, 2002.

[LORS-00]   D. H. Lorenz, A. Orda, D. Raz, Y. Shavitt, Efficient QoS partition and routing of unicast and multicast, *8th Int. Workshop on Quality of Service*, Pittsburgh, 2000.

[NST-03]   J. Naor, H. Shachnai and T. Tamir Real-time Scheduling with a Budget, http://www.cs.technion.ac.il/~hadas/PUB/rtbudget.ps.

[S-99]   F. C. R. Spieksma. "On the approximability of an interval scheduling problem." *J. of Scheduling*, 2:215–227, 1999.

[T-00]   M. Tennenholtz. Some tractable combinatorial auctions. *AAAI/IAAI*, 98–103, 2000.

[Va-01]   V. Vazirani, Approximation algorithms, Springer Verlag, 2001.

Improved Approximation Algorithms for Minimum-Space Advertisement Scheduling*

Brian C. Dean and Michel X. Goemans

M.I.T., Cambridge, MA 02139, USA,
`bdean@theory.lcs.mit.edu`, `goemans@math.mit.edu`

Abstract. We study a scheduling problem involving the optimal placement of advertisement images in a shared space over time. The problem is a generalization of the classical scheduling problem $P||C_{max}$, and involves scheduling each job on a specified number of parallel machines (not necessarily simultaneously) with a goal of minimizing the makespan. In 1969 Graham showed that processing jobs in decreasing order of size, assigning each to the currently-least-loaded machine, yields a 4/3-approximation for $P||C_{max}$. Our main result is a proof that the natural generalization of Graham's algorithm also yields a 4/3-approximation to the minimum-space advertisement scheduling problem. Previously, this algorithm was only known to give an approximation ratio of 2, and the best known approximation ratio for any algorithm for the minimum-space ad scheduling problem was 3/2. Our proof requires a number of new structural insights, which leads to a new lower bound for the problem and a non-trivial linear programming relaxation. We also provide a pseudo-polynomial approximation scheme for the problem (polynomial in the size of the problem and the number of machines).

1 Introduction

We study a scheduling problem whose application is the optimal placement of advertisement images within a shared space over time, typically on a web page. Roughly \$3 billion was spent on web advertising in the first half of 2002 alone [7], so improvements in algorithms for ad placement are of both economic and theoretical interest. In the model we consider here, we have a set $\mathcal{A} = [n] := \{1, \cdots, n\}$ of n ads (images of fixed width and varying heights) to schedule within a shared vertical space, typically in the margin of a web page. We must determine the subset of ads to display in each of T occurences of the page, or T time slots. Ad i has a height h_i and a display count $c_i \leq T$ which represents the number of time slots out of T in which the ad must appear. The goal is to assign each ad i to a set of c_i distinct time slots so as to minimize the maximum height of any occurrence of the page, as illustrated graphically in Figure 1. Mathematically, this means that we need to find $A_i \subseteq [n]$ for $i \in [n]$ such that (i) $|A_i| = c_i$ for every i and (ii) $\max_{t \in [T]} \sum_{i : t \in A_i} h_i$ is minimized. This problem was first posed

---

* This work was supported by NSF contracts CCR-0098018 and ITR-0121495.

J.C.M. Baeten et al. (Eds.): ICALP 2003, LNCS 2719, pp. 1138–1152, 2003.

in 1998 by Adler, Gibbons, and Matias [1]. Notice that we do not care about the vertical ordering of the ads within a single time slot.

When $c_i = 1$ for every i, the problem reduces to the classical NP-Hard scheduling problem $P||C_{max}$ with the time slots corresponding to machines, the ads corresponding to jobs and the height h_i corresponding to the processing time p_i. In fact, our ad scheduling problem is very similar to $P||C_{max}$ with *high-multiplicity encoding* of jobs with the same processing time, except that we require that these additional copies be scheduled on different machines. The high-multiplicity encoding of $P||C_{max}$, denoted by $P|M|C_{max}$ (see Clifford and Posner [2]) has been the focus of some study in the 90's, but there are still many open questions surrounding this problem, in particular whether or not the complexity of an optimal solution is of size polynomial in the input. See McCormick et al. [9] and Clifford and Posner [2] for further discussion.

For $P||C_{max}$, Graham [4] shows that processing the jobs in any order and greedily placing each job in sequence on the currently-least-loaded machine is a 2-approximation algorithm and that this is tight. He further shows that greedy processing of jobs with largest processing time first (LPT) is a 4/3-approximation algorithm and this is also tight. For the ad scheduling problem, Adler, Gibbons, and Matias [1] introduce the analogue of LPT, the *Largest Size Least Full (LSLF)* algorithm which processes ads in non-increasing order of height and greedily schedules ad i on the c_i currently-least-loaded time slots, and show that it is a 2-approximation algorithm, leaving open the question whether a better approximation factor could be proved. Subsequently, Dawande, Kumar, and Sriskandarajah [3] show that any list processing of ads in a greedy fashion is a 2-approximation algorithm, and that this is tight. Dawande et al. also show that rounding the optimum solution of a trivial linear programming relaxation of the problem leads to a 2-approximation algorithm and that a more sophisticated relaxation can be used to obtain a 3/2-approximation algorithm. Previous to this paper, these were the best known approximation algorithms for the ad scheduling problem. One of the main results of this paper is to show that LSLF is a 4/3-approximation algorithm, thereby matching Graham's bound for $P||C_{max}$ (this is tight, due Graham's analysis). Our proof is, however, much more elaborate than for the case of $P||C_{max}$.

Theorem 1. *LSLF is a 4/3-approximation algorithm for the ad scheduling problem.*

Regarding approximation schemes for $P||C_{max}$, for case when the number of machines is a constant Horowitz and Sahni [6] gave a $(1 + \epsilon)$-approximation algorithm for any $\epsilon > 0$ (PTAS). This was improved by Hochbaum and Shmoys [5] to all instances of $P||C_{max}$. Since the running time of this algorithm depends polynomially on the number of machines, it is only a pseudopolynomial approximation scheme for $P|M|C_{max}$. We introduce a similar pseudopolynomial approximation scheme for the ad scheduling problem, whose running time is polynomial in n and T.

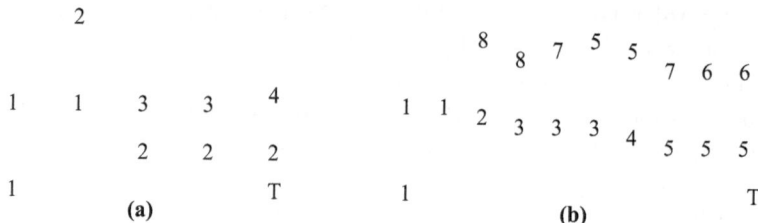

Fig. 1. Sample schedules. The vertical axis represents vertical space and each time slot corresponds to a column along the horizontal axis. (a) LSLF schedule for which Graham's proof fails, (b) Illustration of wrap-around behavior of LSLF when scheduling large ads.

Discussion of Our Results. Graham's proof for the performance guarantee of 4/3 for LPT differentiates *large* jobs with processing times greater than $OPT/3$ (OPT denotes the makespan of an optimal schedule), from the remaining *small* jobs. He observes the following: (i) if the makespan of the LPT schedule is defined by a machine containing only large jobs, then the LPT schedule will in fact be optimal; (ii) on the other hand, if the makespan of the LPT schedule is defined by a machine whose final job j is small, then the makespan can be written as $L + p_j$, where L denotes the load of that machine prior to j's scheduling. The 4/3 bound then follows since $p_j \leq OPT/3$ and since $L \leq OPT$ because greedy scheduling ensures that every other machine must have a load of at least L.

We will show that (i) also holds for the ad scheduling problem: if a maximum-height slot in the LSLF schedule contains only large ads (having heights greater than $OPT/3$), then the LSLF schedule is optimal. However, in (ii) the assumption that $L \leq OPT$ no longer holds for the ad scheduling case, as shown in Figure 1(a).

Regarding (i), we will prove in Theorem 2 the following stronger result: if all ads are large, then LSLF not only minimizes the height of the largest time slot, but it also minimizes the sum of the heights of the largest k time slots for any k over all schedules that place at most 2 ads per time slot (and this is the case indeed for the optimum schedule). To the best of our knowledge, this structural result was not even known for $P||C_{max}$. We use this to devise a stronger lower bound based on the following intuition: consider a set P_k comprising k time slots into which LSLF places the greatest amount of large ad volume. By our structural result, the total large ad volume placed in P_k by LSLF is a lower bound on the large ad volume found in P_k in an optimal schedule. Further, we can lower bound the amount of small ad volume that must appear in P_k by noting that for each small ad i, we must schedule at least $\max(c_i - (T - k), 0)$ copies of i in P_k. Therefore, if we are able to locate such a set P_k for which (i) every small ad i is scheduled by LSLF in exactly $\max(c_i - (T - k), 0)$ slots in P_k, and (ii) the maximum and minimum slot heights in P_k differ by at most $OPT/3$, then the 4/3 approximation bound would follow. Unfortunately, there are problem instances in which no such set P_k satisfies these conditions. In these cases, we manage to prove that there exists a subset $\mathcal{A}'$ of our set $\mathcal{A}$ of ads and a value k such that (i) the values of LSLF on $\mathcal{A}$ and $\mathcal{A}'$ are equal, $LSLF(\mathcal{A}) = LSLF(\mathcal{A}')$, and (ii) the above property holds for $\mathcal{A}'$ and k.

We will also present a different 4/3-approximation algorithm based on rounding a new linear programming relaxation built on our structural results above. We will show an efficient combinatorial technique for solving this LP, and that its solution is exactly the closed-form expression of our aforementioned lower bound.

Finally, we describe a PTAS based on a combination of dynamic programming and LP rounding, which is similar to other approximation schemes for $P||C_{max}$. Its running time is polynomial in n and T, and therefore only pseudopolynomial. Our two 4/3-approximation algorithms, in contrast, can be implemented to run in time polynomial in n with no dependence on T; details are left for the full version. For that purpose, the output is condensed by listing every distinct assignment of ads to a time slot along with the number of time slots with this assignment.

The extended abstract is structured as follows. In the next section, we study the LSLF schedule and shows that it schedules the large ads extremely well. We also deduce our new lower bound. In section 3, we prove that LSLF produces a schedule within a factor of 4/3 of the optimum. In Section 4, we introduce our linear programming relaxation, show that its value is equal to our lower bound and derive rounding-based 4/3-approximation algorithms. Finally, in Section 5, we sketch a pseudo-polynomial approximation scheme for the problem.

2   Large Ads in the LSLF Schedule

We declare ad i to be *large* if $h_i > OPT/3$, and *small* if $h_i \leq OPT/3$. Observe that the total number of large ad copies in our problem instance cannot exceed $2T$ since any optimal solution must contain no more than two large ads per slot.

Lemma 1. *The LSLF algorithm places at most two large ads in every slot.*

Proof. Consider a division of the large ads into two sets: the set of *huge* ads containing ads i for which $h_i > 2OPT/3$, and the remaining set which we call the *medium* ads. LSLF first schedules the huge ads, one copy per slot. Let P denote the set of slots occupied by these ads, and Q denote the remaining slots. LSLF then schedules the medium ads one by one. We claim by induction that the following two invariants hold during this process: (i) all medium ads are placed solely within slots in Q, and (ii) at most two medium ads are placed in each slot in Q.

Suppose our two invariants hold up to the point where LSLF schedules a particular medium ad i. We must have $c_i \leq |Q|$, otherwise this ad would share a slot with some huge ad in every optimal solution, which clearly cannot happen. Therefore, in principle we have sufficient room to fit all copies of ad i within only the slots in Q. We now claim that there must be c_i slots in Q which currently contain at most one medium ad. If this were not the case, then the total number of medium ads would exceed $2|Q|$, and this would imply that in every optimal solution either (i) a huge and medium ad must share a slot, or (ii) there must be a slot containing more than two medium ads; both of these situations are clearly impossible. All slots in Q having at most one medium ad are currently shorter than slots in Q having two medium ads and also shorter than slots in P, so LSLF selects these slots for the placement of ad i. This maintains both of our invariants.

Not only does LSLF place at most two large ads in any slot, but it arranges these ads according to a very specific structure. LSLF schedules one row of large ads and then wraps around to schedule another row of large ads in reverse. The wrap-around point is a bit delicate though, as shown in Figure 1(b), because we must avoid placing two copies of the same ad in a common slot. One can argue that the LSLF schedule for the large ads can be constructed in the following simple recursive way. (i) If an ad exists with multiplicity T, then we schedule this ad in every slot and then any remaining large ad one copy per slot; (ii) if there are $2T$ large ad copies, we pair up a copy of one of the maximum-height large ads with a copy of one of the minimum-height large ads, decrease T, update the multiplicities and then recurse; (iii) if there are fewer than $2T$ large ad copies, we place a copy of a maximum-height large ad by itself and recurse.

We refer to the part of a schedule consisting of only large ads as the *base* of the schedule. We have just described a simple recursive way to construct the base of the LSLF schedule. For any schedule S, let $b_1(S) \geq b_2(S) \geq \cdots \geq b_T(S)$ denote the height occupied by the base, namely the large ads, in every time slot, ordered in a non-increasing fashion. Also, for $k = 1, \cdots, T$, define $B_k(S) = \sum_{i=1}^{k} b_i(S)$. We claim that LSLF schedules the large ads in a very balanced way, at least as well as in the optimal schedule. This is formalized in the following theorem.

Theorem 2. *Let LSLF denote the schedule produced by the LSLF algorithm, and let Σ be an optimum schedule. Then for all values of k, $B_k(LSLF) \leq B_k(\Sigma)$.*

Proof. The proof is by contradiction. Suppose we have a schedule S generated by $LSLF$, an optimal schedule Σ, and a value k for which the lemma fails. We will show how to transform the base of Σ into the base of S without increasing $B_k(\Sigma)$. Based on our recursive characterization of how LSLF schedules the large ads, we will make the bases of S and Σ agree in one slot — for example by making an exchange in Σ so a maximum-height large ad is paired with a minimum-height large ad. We will then recursively transform the remaining $T - 1$ slots of the base of Σ into the remaining $T - 1$ slots of the base of S. Our transformation must consider these three cases:

1. If a large ad i exists for which $c_i = T$, then LSLF will schedule this ad in every slot, and the remaining large ads one copy per slot. In Σ, we will also find ad i scheduled once in every slot, and every other large ad scheduled one copy per slot, since the optimal schedule cannot have more than two large ads per slot. Therefore, if a large ad of multiplicity T exists, then the

base of Σ will be the same as the base of S. We can terminate our recursive transformation once we reach this case.

2. Suppose that $c_i < T$ for every large ad i and that there are exactly $2T$ large ad copies. If there exists a slot in Σ in which a maximum-height large ad is paired with a minimum-height large ad, then we can hold this slot fixed, decrement T and the display counts of the two ads in this slot, and continue our recursive transformation on the remaining slots. Otherwise, suppose we can locate distinct large ads i and j such that a maximum-height large ad is paired with a copy of i and a minimum-height large ad is paired with a copy of j. Swapping ad i and the minimum-height large ad with each-other will change the pairing from (Max, i), (Min, j) to (Max, Min), (i, j), and will not decrease $B_k(\Sigma)$ for any k. We may then continue our recursive transformation. If none of the above cases apply, then every instance of a maximum-height or minimum-height large ad must be paired with the same large ad l. However, since $c_l < T$ there must be some slot in which two large ads i and j different from l are paired. Assume $h_i \le h_j$. If $h_i < h_l$, then we make a swap to change the pairing (Max, l), (i, j) to (Max, i), (l, j). If $h_i \ge h_l$, we make a swap to change (Min, l), (i, j) into (Min, j), (i, l). This exchange also will not decrease $B_k(\Sigma)$ for any k, and will place us in the prior situation where one more swap suffices to pair up the mininimum-height and maximum-height large ads.

3. Suppose that $c_i < T$ for every large ad i and that there are fewer than $2T$ large ad copies. If there exists a slot in the base of Σ in which the only large ad is a maximum-height large ad, then we hold this slot fixed, decrement T and the display count of the ad in this slot, and continue our recursive transformation on the remaining slots. Otherwise, we find a slot containing a maximum-height large ad and swap away the remaining large ad in its slot. The details are essentially the same as in case 2.

One can easily prove the following corollaries to Theorem 2.

Corollary 1. *If all ads are large, LSLF produces an optimal schedule.*

Corollary 2. *Consider the schedule produced by LSLF. If a maximum-height slot in this schedule contains only large ads, then this schedule is optimal.*

2.1   A New Lower Bound on OPT

In order to argue approximation bounds we need to find suitable lower bounds on OPT. The maximum ad height, h_{max}, and the average slot height in any schedule, $\sum_i c_i h_i / T$, are both trivial lower bounds on OPT. These lower bounds are sufficient to show that LSLF is a 2-approximation algorithm [3]; however, in order to show that LSLF is a 4/3-approximation algorithm, we introduce a new, more powerful lower bound based on Theorem 2.

Theorem 3. *Let $\mathcal{A}_S$ denote the set of small ads in our problem instance. Then for every k, we have $L_k \leq OPT$, where*

$$L_k = \frac{1}{k}\left(B_k(LSLF) + \sum_{i \in \mathcal{A}_S} h_i \max(c_i - (T - k), 0)\right).$$

Proof. Consider k slots of maximum base height in the optimum schedule Σ. We know from Theorem 2 that the total base height $B_k(\Sigma)$ in these k slots is greater or equal to $B_k(LSLF)$. Now consider any small ad i. The smallest number of copies of i that Σ could possibly assign to these k slots is $\max(c_i - (T - k), 0)$ (since $T - k$ is an upper bound on the number of copies outside of these k slots). Thus the total height in Σ in these k slots is at least $B_k(LSLF) + \sum_{i \in \mathcal{A}_S} h_i \max(c_i - (T - k), 0)$ and the maximum height of any of these slots must be at least the average value L_k.

Observe that the two trivial lower bounds mentioned earlier are dominated by this bound: $h_{max} \leq L_1$ and $\sum_i c_i h_i / T = L_T$.

Although we do not know which ads are large and which are small, we can nevertheless compute a lower bound based on Theorem 3. If we know the index j of the smallest large ad, we can compute the bound $LB(j) = \max_k L_k$ given by the above Theorem. Since we do not know j, compute j^* to be the largest index such that $LB(j^*) \leq 3h_{j^*}$ or $j^* = 0$ if no such index exists, in which case there are no large ads. We claim that $LB(j^*)$ is a lower bound on OPT, and it is at least as good as the one if we knew which were the large ads. Indeed, it is a lower bound since either $h_{j^*} > OPT/3$ in which case j^* is a large ad and $LB(j^*)$ is a lower bound on OPT by Theorem 3, or $h(j^*) \leq OPT/3$ in which case $3h(j^*)$ is a lower bound on OPT and therefore so is $LB(j^*)$. Furthermore, $j^* + 1$ is not the index of a large ad since otherwise $LB(j^* + 1) \leq OPT < 3h_{j^* + 1}$, contradicting the choice of j^*. Thus the unknown index j of the smallest large ad satisfies $j \leq j^*$. Finally, $LB(\cdot)$ can be shown to be non-decreasing (either directly from the formula for L_k given above, or using Theorem 5 in Section 4 and arguing that the way LSLF places job l in $LB(l)$ is feasible for the linear program corresponding to $LB(l - 1)$). The fact that $j \leq j^*$ then implies that $LB(j^*)$ is at least as good a lower bound as $LB(j)$.

3   Analysis of the LSLF Algorithm

Throughout this section, we assume that the time slots are indexed in non-increasing order of base heights: $b_1(LSLF) \geq b_2(LSLF) \geq \cdots \geq b_T(LSLF)$. We first need several definitions. If P is a set of slots, we say that a small ad i is P-minimal if exactly $\max(c_i - (T - |P|), 0)$ copies of i are scheduled in P by LSLF. Thus, in the above argument we would like to have that all small ads are P-minimal, where $P = [k]$ is a prefix of the slots. For a time slot t, let $H_i(t)$ denote the height of t immediately after LSLF schedules ad i. For a set of slots P we then define $Min_i(P) = \min\{H_i(t) : t \in P\}$ and

$Max_i(P) = \max\{H_i(t) : t \in P\}$. Finally, we define the *range* of a set of slots as $Range_i(P) = Max_i(P) - Min_i(P)$. For notational simplicity, we will omit the subscript on these quantities when we speak of the final schedule produced by LSLF, so for example $Min(P) = Min_n(P)$.

As alluded to in the introduction, if we could find a value k such that (i) $Range(\lfloor k \rfloor) \le OPT/3$ and (ii) every small ad is $[k]$-minimal then it would be easy to show that LSLF produces a schedule of value at most $L_k + OPT/3 \le \frac{4}{3}OPT$; this is formalized in Lemma 4 below. Unfortunately, such a value k does not always exist, but we will show that we can find a subset of the ads for which the value of LSLF does not change and the above property is satisfied.

In order to guarantee that $Range(P) \le OPT/3$ for certain sets P, we use the following Lemma and Corollary.

Lemma 2. *Let t and u be slots satisfying $t < u$. If i is a small ad or the last large ad processed, then $H_i(u) \le H_i(t) + OPT/3$.*

Proof. Since $t < u$, we know that $b_t(LSLF) \ge b_u(LSLF)$, so u will be no taller than t just before scheduling all the small ads. As LSLF schedules small ads, as long as u remains no taller than t the lemma is certainly satisfied. So let us assume that at some point in time we schedule a small ad i in u but not in t, resulting in u becoming taller than t. However, at this point, u will be taller by at most $h_i \le OPT/3$, so we have $Range_i(\{t, u\}) \le OPT/3$. We can now observe that $H_i(u) - H_i(t)$ cannot increase as we increment i as long as $H_i(t)$ does not overcome $H_i(u)$. This shows that $Range(\{t, u\}) \le OPT/3$ at termination. ∎

Corollary 3. *Let k be a maximum-height slot in the final LSLF schedule. If a small ad was scheduled in k, then $Range(\lfloor k \rfloor) \le OPT/3$.*

We need one more concept. As the algorithm progresses, we will differentiate between *heavy* and *light* time slots with respect to a prefix $P = [k]$ of time slots. We assume that LSLF has already scheduled the large ads; so from now on, we focus on small ads only. At the beginning (so just before scheduling any small ad), we designate all time slots to be heavy; more formally, we say that all slots are $(P, 0)$-heavy. If we now consider the point in time right after LSLF schedules some (small) ad i, we say that a time slot is t is (P, i)-light if:

- Rule I: Slot t is $(P, i - 1)$-light.
- Rule II: Ad i is small, not P-minimal, and is scheduled in some slot u for which $H_i(u) \ge H_i(t)$.
- Rule III: Ad i is small, P-minimal, and is scheduled in some $(P, i - 1)$-light slot u for which $H_i(u) \ge H_i(t)$.

If none of these three conditions apply, then we say slot t is (P, i)-heavy. Let us briefly build some intuition about this definition. As soon as a slot becomes light, it will remain light forever. If a small ad i is not P-minimal, this immediately forces all slots receiving a copy of i to become light. This means that, at the end, all slots are P-heavy if and only if all small ads are P-minimal. Additionally, a slot becomes light any time we notice that another light slot matches or exceeds it in height, so the heavy slots always dominate the light slots in height. This is formalized by the following lemma.

Lemma 3. *If slot t is (P, i)-heavy and slot u is (P, i)-light, then $H_i(t) > H_i(u)$.*

Proof. Suppose the lemma fails for two slots t and u. Consider the smallest i for which t is (P, i)-heavy, u is (P, i)-light, and $H_i(t) \leq H_i(u)$. We know i is a small ad and that t is $(P, i-1)$-heavy by rule I. Consider the following cases:

1. **Slot u is $(P, i-1)$-light, and i is not P-minimal.** Since we picked i to be minimal, we know $H_{i-1}(t) > H_{i-1}(u)$, so i must be scheduled in u but not in t. Rule II thus implies that t must be (P, i)-light.
2. **Slot u is $(P, i-1)$-light, and i is P-minimal.** Again, since we picked i to be minimal, $H_{i-1}(t) > H_{i-1}(u)$ and i must be scheduled in u but not in t. Rule III thus implies that t must be (P, i)-light.
3. **Slot u is $(P, i-1)$-heavy, and i is not P-minimal.** By rule II, we know i is scheduled in some slot v for which $H_i(v) \geq H_i(u)$. However, since $H_i(v) \geq H_i(u) \geq H_i(t)$, rule II implies that t must be (P, i)-light.
4. **Slot u is $(P, i-1)$-heavy, and i is P-minimal.** By rule III, we know i is scheduled in some $(P, i-1)$-light slot v for which $H_i(v) \geq H_i(u)$. However, since $H_i(v) \geq H_i(u) \geq H_i(t)$, rule III again implies that t must be (P, i)-light.

In all cases, we conclude that t is (P, i)-light, a contradiction.

The following lemma formalizes our informal discussion earlier in this section of a case for which the 4/3 bound follows easily. As we said earlier, such a prefix of slots does not always exist.

Lemma 4. *If there exists a prefix $P = [k]$ of slots for which $Range(P) \leq OPT/3$ and for which all slots are (P, n)-heavy, then $Max([T]) \leq 4/3OPT$.*

Proof. Consider the lower bound L_k from Theorem 3. We argue that L_k will be the average slot height within P, since every small ad i must be P-minimal. If any small ad weren't P-minimal, all slots receiving it would have become light. Since the minimum slot height in P is a lower bound on the average slot height in P, we have $Min(P) \leq L_k$, and since $L_k \leq OPT$ by Theorem 3, $Min(P) \leq OPT$. Finally, since $Range(P) \leq OPT/3$, we have $Max(P) \leq Min(P) + OPT/3 \leq 4/3OPT$.

We are now equipped to give the main result of this paper, a proof of Theorem 1.

Proof. We first describe the core argument at a high level, postponing discussion of a few key technical lemmas. Consider the schedule produced by LSLF on some instance I. Let P be the largest possible prefix of slots one can form such that $Range(P) \leq OPT/3$. By Corollary 3, we know that P can at least be made large enough to capture every maximum-height slot. Consider now the following three cases.

1. **All slots are (P,n)-heavy.** In this case, Lemma 4 says that the makespan of the LSLF schedule is at most $4/3OPT$.
2. **All slots are (P,n)-light.** This case is impossible. If all slots are P-light, then Lemma 5 below applied to the last small ad and the slot achieving the maximum height implies that we can extend P to a larger prefix P' for which $Range(P') \leq OPT/3$, thereby contradicting the maximality of P.
3. **Some slots are (P,n)-heavy and some are (P,n)-light.** In this case, we reduce our problem instance I to a strictly smaller instance I' by deleting a carefully chosen subset of the ads. Since I' contains fewer ads, we have $OPT(I') \leq OPT(I)$, and Lemma 6 below shows how to construct I' such that $LSLF(I') = LSLF(I)$. We now claim by induction on the size of our instance that $LSLF(I') \leq 4/3OPT(I')$, so $LSLF(I) = LSLF(I') \leq 4/3OPT(I') \leq 4/3OPT(I)$.

This completes the proof. All that remains is to argue lemmas 5 and 6.

Lemma 5. *If slot t is (P,i)-light where $P = [k]$ then $H_i(t) - H_i(k+1) \leq OPT/3$.*

Proof. We proceed by induction on i and consider 4 different cases.

1. **Ad i is scheduled on t but not on $k+1$.** In this case, we have

$$H_i(t) = H_{i-1}(t) + h_i \leq H_{i-1}(k+1) + h_i \leq H_i(k+1) + h_i \leq H_i(k+1) + OPT/3.$$

2. **We are not in case 1, and t is $(P,i-1)$-light.** We know that $H_{i-1}(t) - H_{i-1}(k+1) \leq OPT/3$ by induction. Furthermore, since we are not in case 1, we have that $(H_i(t) - H_{i-1}(t)) - (H_i(k+1) - H_{i-1}(k+1)) \leq 0$, which summed up with the previous inequality gives the statement of the lemma.
3. **Slot t is $(P,i-1)$-heavy and i is P-minimal.** Since t is (P,i)-light, rule III must have applied. This means that i is scheduled on a $(P,i-1)$-light slot u with $H_i(u) \geq H_i(t)$. Since u must fall into either case 1 or 2 above, we have $H_i(u) - H_i(k+1) \leq OPT/3$. Therefore, $H_i(t) \leq H_i(u) \leq H_i(k+1) + OPT/3$.
4. **Slot t is $(P,i-1)$-heavy and i is not P-minimal.** So, rule II must have applied and i is scheduled on a slot u with $H_i(u) \geq H_i(t)$. Since i is not P-minimal, there exists $v \geq k+1$ such that i is not scheduled on v. We further consider two cases.
   a) If v can be chosen to be $k+1$ then

   $$H_i(t) \leq H_i(u) = H_{i-1}(u) + h_i \leq H_{i-1}(k+1) + h_i \leq H_i(k+1) + OPT/3.$$

   b) If not, then we can assume that i is scheduled on $k+1$ and then we have

   $$H_i(t) \leq H_i(u) = H_{i-1}(u) + h_i \leq H_{i-1}(v) + h_i$$
   $$\leq H_{i-1}(k+1) + OPT/3 + h_i = H_i(k+1) + OPT/3,$$

   where the last *inequality* follows from Lemma 2 since $v > k+1$.

Lemma 6. *Fix an instance I and a prefix P of slots. Suppose that both (P, n)-heavy and (P, n)-light slots exist. Create a new problem instance I' by deleting all small ads except those ads i having at least one copy scheduled in some (P, i)-heavy slot. Then (i) I' will be a strictly smaller instance than I, (ii) $OPT(I') \leq OPT(I)$, and (iii) $LSLF(I') = LSLF(I)$.*

Proof. We argue that (i) follows from the existence of (P, n)-light slots. In order for (P, n)-light slots to exist, there must be some non P-minimal small ad i, and every slot receiving a copy of i will be (P, i)-light by rule II. Thus i will be deleted when forming I'. Point (ii) is also straightforward, as deletion of ads can never cause OPT to increase. We therefore focus our attention on (iii).

Consider applying LSLF in parallel to simultaneously construct schedules for I and I'; we will compare corresponding slots in the two schedules as we do so. At any point in time after scheduling all copies of ad i, let $\mathcal{H}(i)$ denote the set of (P, i)-heavy slots (with respect to the schedule for I) and let $\mathcal{L}(i)$ denote the set of (P, i)-light slots (also with respect to the schedule for I). We inductively argue the following: after scheduling any ad i,

1. $H_i^I(t) = H_i^{I'}(t)$ for all $t \in \mathcal{H}(i)$, and
2. $H_i^I(t) \geq H_i^{I'}(t)$ for all $t \in \mathcal{L}(i)$.

The superscripts I and I' above refer to the schedule in which we are measuring the height of a slot. Otherwise stated, the heavy slots in the schedules I and I' will always agree in their heights, while the light slots in I's schedule will always upper bound their corresponding slots in the schedule for I'. Since every slot t maximizing $H_n(t)$ (in either schedule) will belong to $\mathcal{H}(n)$, this will imply $LSLF(I) = LSLF(I')$.

The large ads all belong to both I and I', and will be identically-scheduled for both instances. Consider therefore the insertion of an arbitrary small ad i, assuming that our inductive hypothesis holds for $i - 1$.

- $i \in I'$. Consider the schedule for instance I. Since i was not deleted, it is scheduled in some slot in $\mathcal{H}(i)$, and therefore also in some slot in $\mathcal{H}(i - 1)$. By lemma 3, i is scheduled in every slot in $\mathcal{L}(i-1)$. The inductive hypothesis and lemma 3 together imply that $H_{i-1}^{I'}(t) < H_{i-1}^{I'}(u)$ for every $t \in \mathcal{L}(i - 1)$ and $u \in \mathcal{H}(i-1)$; hence, i will also be scheduled in every slot in $\mathcal{L}(i-1)$ in the schedule for I'. This ensures the invariant is maintained for all $t \in \mathcal{L}(i-1)$. Since the heights of slots in $\mathcal{H}(i - 1)$ agree between the two schedules at time $i - 1$, ad i will be scheduled in analogous slots in $\mathcal{H}(i-1)$ among both schedules. Some of these slots will be in $\mathcal{H}(i)$; the rest will move to $\mathcal{L}(i)$. In either case, since the heights of corresponding slots agree, the invariant will also be maintained for these slots.

- $i \notin I'$. In this case, i is not scheduled in any slot in $\mathcal{H}(i)$, so it can only appear in I's schedule in slots in $\mathcal{L}(i - 1)$ and in $\mathcal{H}(i - 1) \backslash \mathcal{H}(i)$. By the invariant, heights of slots in these two sets in I's schedule are already upper bounds on their corresponding slots in the schedule for I', and we can only be strengthening this upper bound. Therefore the invariant is maintained.

4 A New Linear Programming Relaxation

Theorem 2 allows us to give a new linear programming relaxation for the ad scheduling problem. By rounding the solution to this linear program we obtain another 4/3-approximation algorithm. The linear program optimally (and fractionally) assigns the *small* ads on top of the base obtained by LSLF so as to minimize the tallest slot. Since we do not know OPT, we do not know which ads are large and which are small, but if the ads are sorted in decreasing order of height, the large ads must comprise some prefix of this list, so we must run our rounding algorithm on every prefix and take the best result. Henceforth, we can therefore assume the large ads are known.

The linear program is the following, where $\mathcal{A}_S$ denotes the set of small ads.

$LP = $ Minimize z

subject to:

$$\sum_{t \in [T]} x_{it} = c_i \qquad\qquad i \in \mathcal{A}_S$$

$$\sum_{i \in [n]} x_{it} h_i \leq z - b_t(LSLF) \qquad t \in [T] \qquad\qquad (1)$$

$$0 \leq x_{it} \leq 1 \qquad\qquad i \in [n], t \in [T].$$

It is not straightforward that this linear program is a lower bound on the optimum, since the optimum schedule might schedule the base differently than LSLF. We know of two ways of arguing that LP is a lower bound on OPT. The first proof is based on (contra)-polymatroids and is given below for completeness; the second follows from Theorems 5 and 6 where a simple combinatorial algorithm is shown to solve the LP optimally and with optimal value $\max_k L_k$, the lower bound given in Theorem 3.

Theorem 4. $LP \leq OPT$.

Proof. Let C denote the set of all vectors $l \in R^T$ such that the linear program with the right-hand-side of equation (1) replaced by l_t (from $z - b_t(LSLF)$) is feasible. By definition of Σ, we have that $s \in C$ where $s_t = OPT - b_t(\Sigma)$. By symmetry of the time slots and the fact that C is a convex set, this implies that

$$C \supseteq P := conv \left\{ x : x_i = s_{\sigma(i)} \text{ for some permutation } \sigma \text{ of } [T] \right\} + R_+^T.$$

This latter polyhedral set P is a contra-polymatroid (see [10]) and can be completely described by inequalities in the following way:

$$P = \{ x \in R^T : x(S) \geq g(|S|) \text{ for all } S \subseteq [T] \},$$

where $x(S) = \sum_{t \in S} x_t$ and $g(k)$ is the sum of the k smallest s_t, i.e. $g(k) = kOPT - B_k(\Sigma)$ using the notation introduced before Theorem 2. We claim that

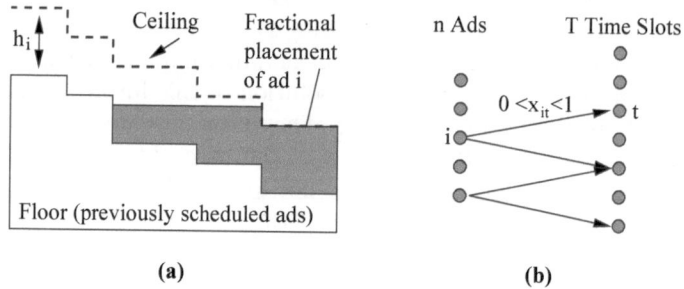

Fig. 2. Combinatorially solving the LP. (a) "Fluid" interpretation of fractional ad placement, (b) Bipartite graph of fractional assignments used for rounding.

the right-hand-side of (1) for $z = OPT$, i.e. $OPT - b_t(LSLF)$, is in P, hence in C, and therefore $LP \leq OPT$. Indeed, for any set S,

$$\sum_{t \in S}(OPT - b_t(LSLF)) \geq |S|OPT - B_{|S|}(LSLF) \geq |S|OPT - B_{|S|}(\Sigma) = g(|S|)$$

by Theorem 2.

In the next theorem, we show that LP is equal to the lower bound given in Theorem 3. The proof of that theorem actually shows that, for any k, $LP \geq L_k$. This implies that $LP \geq \max_{k \in [T]} L_k$. The converse also holds:

Theorem 5. $LP = \max_{k \in [T]} L_k$.

In order to prove this theorem, we give a simple combinatorial algorithm that solves the LP and show that it has the right value. The algorithm first initializes the height H_t of each slot t to be $b_t(LSLF)$ and processes the small ads in any order. The process of scheduling each small ad i can be thought of as filling the top of the schedule with a fluid, as shown in Figure 2(a). Ad i is fractionally "poured" onto the slot with minimum height until this slot catches up to the second-shortest slot and both are filled together uniformly, and so on. However, we must prevent the height of this fluid in each slot from exceeding h_i; this is done by imposing a "ceiling" at height h_i over the top of each slot, at which the fluid stops. When a total of c_i units of ad i have been fractionally filled in, we update $H_1 \dots H_T$ and the ceilings, and continue filling in the next ad.

Theorem 6. *The "fluid" algorithm generates an optimal solution to LP.*

Proof. In order to show that this algorithm optimally solves LP, consider the fractional solution it obtains at the end of its execution and the heights H_t of the time slots. By construction, $z = H_1 \geq H_2 \geq \cdots \geq H_T$. Let k be the maximum slot such that $H_k = H_1 = z$. During the execution of the algorithm, if an ad is ever fractionally assigned to any two adjacent slots t and $t + 1$ (that is, if strictly less than its full height is assigned to both slots) then the heights of t

and $t + 1$ will remain equal for the remainder of the algorithm. In other words, since $H_k > H_{k+1}$, no ad was fractionally assigned to k and $k+1$ simultaneously. Furthermore since the algorithm assigns at least as much of an ad to time slot $t+1$ compared to t, no ad is assigned to any time slot in $[k]$ unless it is assigned fully to each of the time slots $k + 1, \cdots, T$. This means that every small ad is $[k]$-minimal and we thus have that $H_1 + H_2 + \cdots + H_k = kL_k$. The fact that $H_1 = H_2 = \cdots = H_k$ by our choice of k implies that $z = L_k$. This simultaneously proves the correctness of the algorithm for solving the LP and of Theorem 5.

Approximation algorithms with performance guarantee of $4/3$ can be obtained by solving this linear program and rounding the solution using classical rounding schemes such as the ones by Lenstra, Shmoys and Tardos [8] or by Shmoys and Tardos [11]. For example, consider in Figure 2(b) the bipartite graph corresponding to the fractional assignments produced by LP. That is, we have an edge from ad i to slot t if $0 < x_{it} < 1$. We assume without loss of generality that the edges in this subgraph form a forest, for if there were an alternating cycle among these edges we could "augment the flow" appropriately (while preserving the amount assigned to any time slot) around such a cycle, maintaining feasibility and optimality of our solution, until the x_{it} value on one of its edges reached either 0 or 1, thereby breaking the cycle. In this graph, the outdegree of every ad i must be at least 2, since we have $\sum_t x_{it} = c_i$ and c_i is an integer. Therefore, we must be able to find an alternating path with endpoints at two different slots t and t', such that both t and t' have indegree of 1. We round our solution by augmenting flow on such alternating paths until all x_{it} values eventually become integral. During the process we only increase the flow entering a slot t if there is a single fractional edge (i, t) directed into t, so the total increase in height of any such slot will be at most $h_i \leq OPT/3$. Therefore, rounding increases the makespan of our solution by at most the maximum height of a fractionally-scheduled ad, which in this case is at most $OPT/3$. As described, our rounding approach takes time polynomial in n and T, but it is straightforward to eliminate the dependence on T by appropriate grouping of time slots. Further details are left for the full version.

5   A Pseudopolynomial $(1 + \epsilon)$-Approximation Scheme

We describe briefly a $(1 + \epsilon)$-approximation scheme whose running time is polynomial in n and T. The algorithm is similar to that of Hochbaum and Shmoys [5] for $P||C_{max}$. Let $\alpha = \epsilon/2$. For this section, we designate ads with heights larger than αOPT large, and the remaining ads small. As in the LP rounding case, we do not know which ads are large, so we must try our algorithm on every prefix of the sorted ads and take the best result. Let us therefore assume the large ads are known.

Our approximation scheme schedules the large ads via dynamic programming. We first run the greedy $LSLF$ algorithm to obtain approximate bounds on OPT, so $3LSLF/4 \leq OPT \leq LSLF$. The heights of the large ads are first rounded up to the nearest multiple of $3\alpha^2 LSLF/4$. Since large ads have heights at least αOPT, this inflates the height of each large ad (and hence also the optimal makespan) by a factor of at most $1 + \alpha$. After rounding, we have at most $\lceil \frac{LSLF}{3\alpha^2 LSLF/4} \rceil = O(1)$ distinct large ad sizes. We can encode each base shape as a vector $(n_0, n_1, \cdots, n_{\lceil 4/3\alpha^2 \rceil})$ where n_i gives the number of slots having base height $3i\alpha^2 LSLF/4$, and $\sum n_i = T$. The number of distinct base shapes will therefore be at most $T^{\lceil 4/3\alpha^2 \rceil} = T^{O(1)}$, a polynomial in T. All achievable base shapes can be enumerated via dynamic programming in time polynomial in n and T. We omit further details. We henceforth assume that the base of a $(1 + \alpha)$-approximate solution is known.

Fixing such a base, we solve LP to fractionally schedule the small ads. The optimal value of LP gives a lower bound on $(1 + \alpha)OPT$, and rounding of the small ads results in an increase of at most αOPT. Thus, our final schedule has makespan at most $(1 + \epsilon)OPT$.

Acknowledgements. We wish to thank the reviewers of this paper for their insightful comments.

References

1. M. Adler, P.B. Gibbons, and Y. Matias (1998). "Scheduling Space-Sharing for Internet Advertising". *Journal of Scheduling*. To appear.
2. J. Clifford, M. Posner (2001). "Parallel Machine Scheduling with High Multiplicity". *Mathematical Programming Ser. A* 89, 359–383.
3. M. Dawande, S. Kumar, and C. Sriskandarajah (2001). "Algorithms for Scheduling Advertisements on a Web Page: New and Improved Performance Bounds" *Journal of Scheduling*. To appear.
4. R. Graham (1969). "Bounds on Multiprocessing Timing Anomalies". *SIAM Journal on Applied Mathematics* 17, 416–429.
5. D. Hochbaum and D. Shmoys (1987). "Using Dual Approximation Algorithms for Scheduling Problems: Theoretical and Practical Results". *Journal of the ACM* 34, 144–162.
6. E. Horowitz and S. Sahni (1976). "Exact and Approximate Algorithms for Scheduling Nonidentical Processors". *Journal of the ACM* 23, 317–327.
7. Interactive Advertising Bureau (www.iab.net). IAB Internet Advertising Revenue Report.
8. J.K. Lenstra, D.B. Shmoys, and É. Tardos (1990). "Approximation algorithms for scheduling unrelated parallel machines". *Mathematical Programming* 46, 259–271.
9. S. T. McCormick, S. Smallwood, and F. Spieksma (2001). "A Polynomial Algorithm for Multiprocessor Scheduling with Two Job Lengths". *Mathematics of Operations Research* 26(1), 31–49.
10. A. Schrijver (2003). "Combinatorial Optimization - Polyhedra and Efficiency". Springer-Verlag.
11. D.B. Shmoys and É. Tardos (1993). "An Approximation Algorithm for the Generalized Assignment Problem". *Mathematical Programming* 62, 461–474.

Anycasting in Adversarial Systems: Routing and Admission Control

Baruch Awerbuch[*,1], André Brinkmann[**,2], and Christian Scheideler[1]

[1] Department of Computer Science, Johns Hopkins University, 3400 N. Charles Street, Baltimore, MD 21218, USA, {baruch,scheideler}@cs.jhu.edu
[2] Heinz Nixdorf Institute and Department of Electrical Engineering, University of Paderborn, 33102 Paderborn, Germany, brinkman@hni.upb.de

Abstract. In this paper we consider the problem of routing packets in dynamically changing networks, using the anycast mode. In anycasting, a packet may have a set of destinations but only has to reach any one of them. This set of destinations may just be given implicitly by some anycast address. For example, each service (such as DNS) may be given a specific anycast address identifying it, and computers offering this service will associate themselves with this address. This allows communication to be made transparent from node addresses, which makes anycasting particularly interesting for dynamic networks, in which redundancy and transparency are vital to cope with a dynamically changing set of nodes. However, so far not much is known from a theoretical point of view about how to efficiently support anycasting in dynamic networks. This paper formalizes the anycast routing and admission control problem for arbitrary traffic in arbitrary dynamic networks, and provides first competitive solutions. In particular, we show that a simple local load balancing approach allows to achieve a near-optimal throughput if the available buffer space is sufficiently large compared to an optimal algorithm. Furthermore, we show via lower bounds and instability results that allowing admission control (i.e. dropping some of the injected packets) tremendously helps in keeping the buffer resources necessary to compete with optimal algorithms low.

Keywords: Adversarial routing, anycasting, online algorithms, load balancing, dynamic networks

[*] Supported by DARPA grant F306020020550 "A Cost Benefit Approach to Fault Tolerant Communication" and DARPA grant F30602000-2-0526 "High Performance, Robust and Secure Group Communication for Dynamic Coalitions".
[**] Supported in part by the DFG-Sonderforschungsbereich 376 "Massive Parallelität: Algorithmen, Entwurfsmethoden, Anwendungen". Part of the research was done while visiting the Johns Hopkins University, supported by a scholarship from the German Academic Exchange Service (DAAD Doktorandenstipendium im Rahmen des gemeinsamen Hochschulsonderprogramms III von Bund und Ländern).

J.C.M. Baeten et al. (Eds.): ICALP 2003, LNCS 2719, pp. 1153–1168, 2003.

1   Introduction

This paper studies the problem of supporting anycasting in adversarial networks. The notion of anycasting was first standardized in RFC 1546 [16]. In this RFC, IP anycast is defined as a network service that allows a sender to access the nearest of a group of receivers that share the same anycast address, where "nearest" is defined according to the routing system's measure of distance. Usually, the receivers in the anycast group are replicas, able to support the same service (e.g. mirrored web servers). RFC 1546 proposes anycast as a means to discover a service location and provide host auto-configuration. For example, by assigning the same anycast address to a set of replicated FTP servers, a user downloading a file need not choose the best server manually from the list of mirrors. The user can simply use the anycast address to directly download the file from the nearest server. In order to aid host auto-configuration, all DNS servers may be given the same anycast address. In this case, a host that is moved to a new network need not be reconfigured with the local DNS address. The host can simply use the global anycast address to access a local DNS server. Service discovery and auto-configuration are seen as vital components of protocols for dynamic networks, and therefore anycasting is seen as a crucial mechanism to ensure robust support for networking services in mobile networks. Since its introduction, anycasting has received considerable attention in the systems community and it has been adopted by all proposed successors of IPv4 (e.g. Pip, SIPP, and IPv6). However, to our surprise, it seems that anycasting has not been investigated by the theory community so far.

Since in highly dynamic networks it may be very hard to predict which may be the nearest server belonging to some anycast address, it seems to be a formidable problem to efficiently support anycasting in dynamic networks, especially for those that are under adversarial control. However, we demonstrate in this paper that even if both the network and the packet injections are under adversarial control, distributed routing strategies can be found for anycasting with a close to optimal throughput. Thus, in principle, anycasting can even be supported in such networks as mobile ad-hoc networks, where connections between users may change quickly and unpredictably.

1.1   Our Approach and Related Results

We measure the performance of our protocols by comparing them with a best possible strategy that knows all actions of the adversary in advance. The performance is measured in terms of communication throughput and space overhead. In order to ensure a high throughput efficiency in dynamic networks, several challenging tasks have to be solved:

- *Routing*: What is the next edge to be traversed by a packet?
- *Queueing*: What is the next packet to be transmitted on an edge? In particular, which destination should be preferred?
- *Admission control*: What is the packet to be dropped if a buffer is full?

The study of adversarial models was initiated, in the context of *queueing alone*, by Borodin et al. [12]. Other work on queueing includes [6,13,14,15,17, 18]. In these papers it is assumed that the adversary has to provide a path for every injected packet and reveals these paths to the system. The paths have to be selected so that they do not overload the system. Hence, it remains to find the right queueing discipline (such as furthest-to-go) to ensure that the number of packets in the system (resp. the time needed by packets to reach their destination) is bounded. However, the bounds on the buffer size given in these papers to avoid dropping any packet usually depend on the network size and are sometimes unrealistically high. This motivated Aiello et al. [5] to study the throughput performance of queueing disciplines under the assumption that the routing buffers have a fixed size (i.e. that is independent of network parameters), using an adversary that can inject an unbounded number of packets. In this case, of course, a queueing discipline cannot guarantee the delivery of every injected packet. So the goal is rather to find a queueing strategy whose throughput is as close as possible to a best possible throughput. Aiello et al. show among other results that there are queueing disciplines that are guaranteed to achieve an $\Omega(1/(d \cdot m))$ fraction of the best possible throughput achievable with the same buffer size, where m is the number of edges and d is the longest path injected by the adversary. This upper bound and their lower bound of $O(\sqrt{m})$ for the line that holds for arbitrary greedy protocols seem to indicate that online protocols cannot compete well with best possible protocols when using the same buffer size.

The study of adversarial models was initiated, in the context of *routing*, by Awerbuch, Mansour and Shavit [11] and further refined by [4,7,9,10,14]. In these papers the model is used that the adversary *does not* reveal the paths to the system, and therefore the routing protocol has to figure out paths for the packets by itself. Based on work by Awerbuch and Leighton [10], Aiello et al. [4] show that there is a simple distributed routing protocol that keeps the number of packets in transit bounded in a dynamic network if, roughly speaking, in each window of time the paths selected for the injected packets require a capacity that is below what the available network capacities can handle in the same window of time. Awerbuch et al. [7] generalize this to an adversarial model in which the adversary is allowed to control the network topology and packet injections as it likes, as long as for every injected packet it can provide a schedule to reach its destination. They show that even for the case that the network capacity is fully exploited, if all packets have the same destination, the number of packets in transit is bounded at any time.

With the exception of [5], the weakness of the adversarial models above is that they assume that the adversary never overloads the system with packets. In static networks this may be a reasonable restriction, since one can imagine that in principle it is possible to perform some kind of admission control *before* injecting a packet into the system. However, in highly dynamic networks such as mobile ad-hoc networks, this may not be possible without being too conservative and therefore wasting too much of the already scarce bandwidth. Hence, for dynamic networks it would be highly desirable to have protocols that can handle not only the routing and queueing part but also packet-level admission control, i.e. dropping packets from either input or intermediate buffers.

Also, we note that all of the above work on adversarial queueing and routing only considered the *unicasting* mode (every packet has a single destination). We consider the more general anycasting mode, using a very general adversarial model that gets rid of somewhat artificial restrictions of previously suggested models for dynamic networks. In fact, the only limiting assumptions left in our model are that packets are of atomic nature (i.e. they cannot be split or compressed) and that packets cannot be killed by the adversary. Thus, our upper bounds also apply to other adversarial routing and queueing models suggested so far.

Finally, we note that all approaches in the adversarial routing area, including this current paper, are based on simple load balancing schemes first pioneered by Awerbuch, Mansour and Shavit [11], and refined in [1,2,3,4,7,9,10] for various routing purposes. Our achievement is to demonstrate that balancing even works for anycasting. Also, we use a much more general adversarial network model then was used in previous papers, and we consider the admission control problem.

In order to state our analytical results, we need some notation.

1.2   The Anycast Routing and Admission Control Model

First, we describe the basics of our network model and injection model. We assume that $V = \{1, \ldots, n\}$ represents the set of nodes in the system. The selection of the edges is under adversarial control and can change from one time step to the next. We assume that all edges are directed. This does not exclude the undirected edge case, since an undirected edge can be viewed as consisting of two directed edges, one in each direction. Each edge can forward at most one packet in a time step. Each node can have at most Δ incoming and at most Δ outgoing edges at any time. Δ can be seen as the maximum number of active (logical or physical) connections a node can handle at the same time (due to, for example, its hardware restrictions). Apart from this restriction, the adversary can interconnect the nodes in an arbitrary way in each time step. This includes the possibility of connecting the same pair of nodes via several edges.

The adversary does not only control the topology of the network but also the injection of packets. Each anycast packet is given a fixed anycast group at the time of its injection. We allow this group just to be specified implicitly (for example, by an anycast address). Note that for implicitly specified groups, the nodes in the network may have no knowledge about their size. It may even be possible that the group is empty. Thus, our anycast algorithm has to cope with this situation.

The adversary can inject an arbitrary number of packets and can activate an arbitrary number of edges in each time step as long as the number of incoming or outgoing edges at a node does not exceed Δ. In this case, only some of the injected packets may be able to reach their destination, even when using a best possible strategy. Each time an anycast packet reaches one of its destinations, we count it as one *delivery*. The number of deliveries that is achieved by an algorithm is called its *throughput*. We are interested in maximizing the throughput. Since the adversary is allowed to inject an unbounded number of packets, we will allow routing algorithms to drop packets so that a high throughput can be achieved with a buffer size that is as small as possible.

In order to compare the performance of a best possible strategy with our online strategies, we will use competitive analysis. We assume that both the optimal and the online algorithm are allowed to allocate one buffer in each node for each type of packet. Thus, if there are b different anycast addresses, then a node can allocate up to b different buffers. This will simplify the comparison. However, our competitive results also work if every node only has a single buffer (or a fixed number of buffers). In this case, the buffer overhead for our online algorithm has to be multiplied by b.

Given any sequence of edge activations and packet injections σ, let $\mathrm{OPT}_B(\sigma)$ be the maximum possible throughput (i.e. the maximum number of deliveries) when using a buffer size of B (i.e. each buffer can store up to B packets), and let $A_{B'}(\sigma)$ be the throughput achieved by some given online algorithm A with buffer size B'. We call an online algorithm A (c, s)-competitive if for all σ and all B, A can guarantee that

$$A_{s' \cdot B}(\sigma) \geq c \cdot \mathrm{OPT}_B(\sigma) - r$$

for any $s' \geq s$, where $r \geq 0$ is some value that is independent of σ (but may depend on s, B and n). $c \in [0, 1]$ denotes here the fraction of the best possible throughput that can be achieved by A and s denotes the space overhead necessary to achieve this. If c can be brought arbitrarily close to 1, A is also called $s(\epsilon)$-competitive (or simply competitive), where $s(\epsilon)$ reflects the relationship between s and ϵ with $c = 1 - \epsilon$. Obviously, it always holds that $s(\epsilon) \geq 1$, and the smaller $s(\epsilon)$, the better is the algorithm A.

In the following, B will always mean the buffer size of an optimal routing algorithm.

1.3   New Results

Our new results are arranged in two sections. In Section 2, we demonstrate that if it is allowed to drop packets, a near-optimal throughput can be achieved with a low space overhead. In particular, we present a simple algorithm for anycasting, called *T-balancing algorithm*, that achieves the following result:

For every $T \geq B + 2(\Delta - 1)$, the T-balancing algorithm is $1 + (1 + (T + \Delta)/B)L/\epsilon$-competitive, where L is the *average* path length used by successful packets in an optimal solution. For $B \geq \Delta$ and $T = O(B)$, this boils down to a competitive ratio of $O(L/\epsilon)$. The result is sharp up to a constant factor.

In Section 3, we demonstrate with the help of lower bounds and instability results that even if the adversary is friendly (i.e. it only injects packets that can be delivered when using a buffer size of B), routing without the ability to drop packets may have a poor performance both with respect to throughput and space overhead.

Some of the proofs are only sketched due to space limitations. Please see [8] for details.

2   Adversarial Anycasting

Let $h_{v,a,t}$ denote the amount of packets in the buffer for anycast address a in node v at the beginning of time step t. $h_{v,a,t}$ will also be called the *height* of the corresponding buffer. The maximum height a buffer can have is denoted by H.

We now present a simple balancing strategy that extends the balancing strategies used by Aiello et al. [4] and Awerbuch et al. [7] by a rule for deleting packets. In every time step $t \geq 1$ the T-balancing algorithm performs the following operations.

1. For every edge (v, w), determine the anycast address a with maximum $h_{v,a,t} - h_{w,a,t}$ and check whether $h_{v,a,t} - h_{w,a,t} > T$. If so, send a packet for a from v to w (otherwise do nothing).
2. Receive all incoming packets and absorb all packets that reached the destination. Afterwards, receive all newly injected packets. If a packet cannot be stored in a buffer because its height is already H, delete it.

Note that if T is large enough compared to Δ, then packets are guaranteed never to be deleted at intermediate buffers but only at the source. This provides the sources with a very easy rule to perform admission control: if a packet cannot be stored because its buffer is already full, delete it.

Let L denote an upper bound on the (best possible) average path length used by the successful packets in an optimal algorithm with buffer size B, and let Δ denote the maximum number of edges leaving or leading to a node that can be active at any time. We do not demand that these edges have to connect different pairs of nodes. Hence, the result below also extends to dynamic networks with non-uniform edge capacities.

Theorem 1. *For any $\epsilon > 0$ and any $T \geq B + 2(\Delta - 1)$, the T-balancing algorithm is $1 + (1 + (T + \Delta)/B)L/\epsilon$-competitive.*

Proof. To simplify the analysis, we prove the competitive ratio for a more general model than our anycast model, called *option set model*.

In the option set model, we have a set of nodes V' with a single buffer each, and all injected packets want to go to the same destination $d \in V'$. The adversary can inject an arbitrary number of packets in each time step. Also, it can activate an arbitrary collection of edge sets $E_1, \ldots, E_k \subseteq V' \times V'$, called *option sets*, in each step as long as every node $v \in V' \setminus \{d\}$ has an incoming or outgoing edge in at most Δ many sets. For each option set E_i, the algorithm is allowed to use only one edge in E_i for the transmission of a packet.

This model is indeed more general than our anycast model.

Lemma 1. *Any algorithm that is c-competitive in the option set model is also c-competitive in the anycast model.*

Proof. For this it suffices to show how to transform the anycast model into the option set model. Suppose that A is the set of all anycast addresses. Then we define $V' = V \times A$, i.e. each buffer in the original model represents a node in the option set model. Each edge $e = (v, w)$ that is activated in the anycast model can then be represented as option set $E_e = \{((v, a), (w, a)) \mid a \in A\}$. Since all packets reaching their destination buffers in the anycast model are absorbed, we can view all of these buffers as a single node in the option set model without affecting the throughput.   □

Hence, in the following we only work with the option set model.

Let N be the number of non-destination nodes in the option set model, and let node 0 represent the destination node. The height of node 0 is always 0, since any packet reaching 0 will be absorbed. For each of the remaining nodes we assume that it has H slots to store packets. The slots are numbered in a consecutive way starting from below with 1. Every slot can store at most one packet. After every step of the balancing algorithm we assume that if a node holds h packets, then its first h slots are occupied. The *height* of a packet is defined as the number of the slot in which it is currently stored. If a new packet is injected, it will obtain the lowest slot that is available after all packets that are moved to that node from another node have been placed.

For each successful packet in an optimal algorithm, a schedule can be identified. A schedule $S = (t_0, (e_1, t_1), \ldots, (e_\ell, t_\ell))$ consists of a sequence of movements by which the injected packet P is sent from its source node to the destination. It has the property that P is injected at time t_0, the edges $e_1, \ldots, e_\ell$ form a connected path, with the starting point of e_1 being the source of P and the endpoint of e_ℓ being the destination of P, the time steps have the ordering $t_0 < t_1 < \ldots < t_\ell$, and edge e_i was available in some option set at time t_i for all $1 \le i \le \ell$. Certainly, no two schedules are allowed to use the same option set at the same time. A schedule $S = (t_0, (e_1, t_1), \ldots, (e_\ell, t_\ell))$ is called *active* at time t if $t_0 \le t \le t_\ell$. The *position* of a schedule at time t is the node at which its corresponding packet would be if it is moved according to S. An edge in an option set is called a *schedule edge* if it belongs to a schedule of a packet. Suppose that we want to compare the performance of the balancing algorithm with an optimal algorithm that uses a buffer size of B. Then the following fact obviously holds.

Fact 1 *At every time step, at most B schedules can have their current position at some node v.*

Next we introduce some further notation. We will distinguish between three kinds of packets: *representatives*, *zombies*, and *losers*. During their lifetime, the packets have to fulfill certain rules. (These rules will be crucial for our analysis. The balancing algorithm, of course, cannot and does not distinguish between these types of packets.) Every injected packet that has a schedule (i.e. that will be delivered by the optimal algorithm) will initially be a representative. Every other injected packet will initially be a zombie. The goal of a representative is to stay with its schedule, and the goal of a zombie is to stay at a slot of height more than $H - B$. Whenever this cannot be fulfilled, the packet is transformed into a loser. Together with Fact 1, this implies the following fact.

Fact 2 *At any time, the number of zombies and representatives stored in a node is at most B.*

If a packet is injected into a full node, then the highest available loser will be selected to take over its role (Fact 2 implies that this is always possible if $H > B$).

Our goal for the analysis is to ensure that a representative always stays with its schedule as long as this is possible. That is, each time the schedule moves, the representative tries to move with it, and otherwise it tries to stay at the current position of the schedule. This implies the following rules for a representative R when the adversary offers an option set containing one of its schedule edges $e = (v, w)$:

1. A packet is sent along e: Then we always select R to be moved along e.
2. No packet is sent along edge e: If w has a loser, then the representative exchanges its role with the highest available loser in w. In this case we will also talk about a *virtual* movement. Otherwise, the representative is simply transformed into a loser. In this case, we will disregard the rest of the schedule (i.e. we will not select a representative for it afterwards and the rest of the schedule edges will simply be treated as non-schedule edges).

Furthermore, if a packet is sent along a non-schedule edge $e = (v, w)$, then we always make sure that none of the representatives is moved out of v but only a loser (which always exists if T is large enough).

The three types of packets are stored in the slots in a particular order. The lowest slots are always occupied by the losers, followed by the zombies and finally the representatives.

Let $h_{v,t}$ be the *height* of node v (i.e. the number of packets stored in the buffer represented by v) at the beginning of time step t, and let $h'_{v,t}$ be its height when considering only the losers. The *potential* of node v at step t is defined as $\phi_{v,t} = \sum_{j=1}^{h'_{v,t}} j = \binom{h'_{v,t}+1}{2}$ and the potential of the system at step t is defined as $\Phi_t = \sum_v \phi_{v,t}$.

First, we study how the potential can change in a single step. Since schedules are not allowed to overlap, every option set contains either one or no schedule edge. To simplify the consideration of these two cases, we consider the option sets given in a time step one by one, starting with option sets without a schedule edge and always assuming the worst case concerning previously considered option sets. Also, when processing these option sets, we always use the (worst case) rule that if a loser is moved to some node w, it will for the moment be put on top of all old packets in w. This will simplify the consideration of option sets with a schedule edge. At the end, we then move all losers down to fulfill the ordering condition for the representatives, zombies, and losers. This will certainly only decrease the potential. Using this strategy, we can show the following result.

Lemma 2. *If $T \geq B + 2(\Delta - 1)$, then any option set that does not contain a schedule edge does not increase the potential of the system.*

Proof. Consider any fixed option set without a schedule edge. If no edge in the given option set is used by a packet, the lemma is certainly true. Otherwise, let $e = (v, w)$ be the edge along which a packet is sent. Note that in this case, $h_{v,t} - h_{w,t} > T$. If $T \geq B + 2(\Delta - 1)$, then even after $\Delta - 1$ removals of packets from v and the arrival of $\Delta - 1$ packets at w, there are still losers left in v, and the height of the highest of these is higher than the height of w. Hence, we can avoid moving any representative away from the position of its schedule and instead move a loser from v to w without increasing the potential.   □

For option sets with a schedule edge (i.e. an edge that still has a representative associated with it), only a slight increase in the potential is caused.

Lemma 3. *If $T \geq B + 2(\Delta - 1)$, then every option set that contains a schedule edge increases the potential of the system by at most $T + B + \Delta$.*

Proof. Consider some fixed option set with a schedule edge $e = (v, w)$. If e is selected for the transmission of a packet, then we can send the corresponding representative along e, which has no effect on the potential.

Otherwise, it must be that either $\delta_e = h_{v,t} - h_{w,t} \leq T$ or $\delta_e > T$ and another edge was preferred. In both cases, the representative R for e has to be moved virtually or transformed into a loser.

First of all, note that our rule of placing new losers on top of the old packets makes sure that the height of the representative in v does not increase. Furthermore, there are two ways for w to lose losers before considering e: either an unused schedule edge to w forced a virtual movement of a representative to w, or a used non-schedule edge from w forced to move a loser out of w. Let s be the number of edges with the former property and ℓ be the number of edges with the latter property. If w had r representatives (and zombies) at the beginning of t, then it must hold that $r + s - (\Delta - \ell) + 1 \leq B$ to ensure that at the end of step t, w has at most B representatives (the $+1$ is due to e). Thus, $r + s + \ell \leq B + \Delta - 1$. Hence, if there is still a loser left in w when considering e, the highest of these must have a height of at least $h_{w,t} - (B + \Delta - 1)$. Therefore, if $h_{w,t} \geq B + \Delta$, then it is possible to exchange places between R and a loser in w so that the potential increases by at most

$$h_{v,t} - (h_{w,t} - (B + \Delta - 1)) = \delta_e + B + \Delta - 1 . \tag{1}$$

If $\delta_e \leq T$, this is at most $T + B + \Delta$. If $h_{w,t} < B + \Delta$, then it may be necessary to convert R into a loser. However, since $h_{v,t} - h_{w,t} \leq T$, this increases the potential also by at most $T + B + \Delta$.

Otherwise, δ_e might be quite big, but in this case there must be some other edge $e' = (v', w')$ that won against e because $\delta_{e'} \geq \delta_e$. Since $\delta_{e'} > T$, v' must have a loser even if $\Delta - 1$ losers already left v' and the maximum possible number of losers in v' was converted into representatives. In fact, similar to w above, the height of the highest of the remaining losers in v' must be at least $h_{v',t} - (B + \Delta - 1)$. On the other hand, w' can receive at most $\Delta - 1$ other packets before receiving the packet sent by e'. So the potential drop due to moving the highest available loser in v' to w' is at least

$$(h_{v',t} - B - \Delta + 1) - (h_{w',t} + \Delta) = (h_{v',t} - h_{w',t}) - B - 2\Delta + 1 \geq \delta_e - B - 2\Delta + 1 . \tag{2}$$

Subtracting (2) from (1), the increase in potential due to the given option set is at most $(\delta_e + B + \Delta - 1) - (\delta_e - B - 2\Delta + 1) = 2B + 3\Delta - 2$. If $T \geq B + 2(\Delta - 1)$, this is at most $T + B + \Delta$.   □

In addition to option sets, also injection events and the transformation of a zombie into a loser can influence the potential. This will be considered in the next two lemmata.

Lemma 4. *Every deletion of a newly injected packet decreases the potential by at least $H - B$.*

Proof. According to Fact 2, the highest available loser in a full node must have a height of at least $H - B$. Since the deletion of a newly injected packet causes this loser to be transformed into a representative or zombie, this decreases the potential by at least $H - B$. (Note that in case of a zombie, it might be directly afterwards converted back into a loser, but this will be considered in the next lemma.)   □

If an injected packet is not deleted, this will initially not affect the potential, since it will either become a representative or a zombie. However, a zombie may be converted into a loser.

Lemma 5. *Every zombie can increase the potential by at most $H - B$.*

Proof. Note that zombies do not count for the potential. Hence, the only time when a zombie influences the potential is the time when it is transformed into a loser. Since we allow this only to happen if the height of a zombie is at most $H - B$, the lemma follows.

Now we are ready to prove an upper bound on the number of packets that are deleted by the balancing algorithm.

Lemma 6. *Let σ be an arbitrary sequence of edge activations and packet injections. Suppose that in an optimal strategy, s of the injected packets have schedules and the other z packets do not. Let L be the average length of the schedules. If $H \geq B + 2(\Delta - 1)$, then the number of packets that are deleted by the balancing algorithm is at most*

$$s \cdot \frac{L(T + B + \Delta)}{H - B} + z \ .$$

Proof. First of all, note that only newly injected packets get deleted. Let p denote the number of option sets with a schedule edge and d denote the number of packets that are deleted by the balancing algorithm. Since

- due to Lemma 2 option sets without a schedule edge do not increase the potential,
- due to Lemma 3 every option set with a schedule edge increases the potential by at most $T + B + \Delta$,
- due to Lemma 4 every deletion of a newly injected packet decreases the potential by at least $H - B$, and
- due to Lemma 5 every zombie increases the potential by at most $H - B$,

it holds for the potential Φ after executing σ that $\Phi \le p \cdot (T + B + \Delta) + z \cdot (H - B) - d \cdot (H - B)$. Since on the other hand $\Phi \ge 0$, it follows that

$$d \le \frac{p \cdot (T + B + \Delta)}{H - B} + z .$$

Using in this inequality the fact that the average number of edges used by successful packets is at most L, and therefore the number of injected packets with a schedule, s, satisfies $s \ge p/L$, concludes the proof of the lemma.   □

From Lemma 6 it follows that the number of packets that are successfully delivered to their destination by the balancing algorithm must be at least

$$s + z - \left(s \cdot \frac{L(T + B + \Delta)}{H - B} + z \right) - H \cdot N = s \cdot \left(1 - \frac{L(T + B + \Delta)}{H - B} \right) - H \cdot N ,$$

where N is the number of (virtual) non-destination nodes. For $H \ge L(T + B + \Delta)/\epsilon + B$ this is at least $(1 - \epsilon)s - r$ for some value r independent of the number of packets successful in an optimal schedule.   □

Next we demonstrate that the analysis of the T-balancing algorithm is essentially tight, even when using just a single destination.

Theorem 2. *For any $\epsilon > 0$, $T > 0$, and $L \ge 1$, the T-balancing algorithm requires a buffer size of at least $T \cdot (L - 1)/\epsilon$ to achieve a more than $1 - \epsilon$ fraction of the best possible throughput.*

Proof. Consider a source node s that is connected to a destination d via two paths: one of length 1 and one of length $(L - 1)/\epsilon$. Further suppose packets are injected at s so that $1 - \epsilon$ of the injected packets have a schedule along the short path and ϵ of the packets have a schedule along the long path. Then the average path length is $1(1 - \epsilon) + ((L - 1)/\epsilon) \cdot \epsilon \le L$. Since each time a packet is moved forward along a node its height (i.e. slot number) must decrease by at least T, a packet can only reach the destination along the long path if s has a buffer of size $H \ge T \cdot (L - 1)/\epsilon$. Hence, such a buffer size is necessary to achieve a throughput of more than $1 - \epsilon$.   □

3   Unicasting without Admission Control

In this section we demonstrate that routing without admission control mechanisms seems to be very difficult if not impossible, even in the adversarial unicast setting, and even if an unbounded (or extremely high) amount of resources for the buffering is available.

We will start by defining some properties of online routing algorithms which intuitively seem to be necessary for the successful online delivery of packets. A *priority function* $f : \mathbb{N}_0 \times \mathbb{N}_0 \to \mathbb{N}_0$ gets as arguments two buffer heights and outputs a number determining the priority with which a packet should be sent from one to the other buffer. In a balancing algorithm that uses a priority function f, the pair with the highest priority wins. We call f *monotonic* if for all $h_1, h_2 \in \mathbb{N}_0$, $f(h_1 + 1, h_2) > f(h_1, h_2)$ and $f(h_1 + 1, h_2 + 1) \geq f(h_1, h_2)$.

Consider a routing algorithm that uses a monotonic priority function to determine a winning buffer pair (h_1, h_2) for each activated edge in the unicast model. If $h_1 \leq h_2$, no packet is allowed to be sent. Otherwise, a packet for that pair (or none) may be sent, but if the buffer corresponding to h_2 is a destination buffer, a packet has to be sent for that pair. Intuitively, these rules seem to be reasonable to ensure a high throughput, and we will therefore call this class of routing algorithms *natural* algorithms.

We start with an observation demonstrating that for adversaries that are unbounded in their injections it is necessary to drop packets in a natural routing algorithm in order to make sure that *any* of the injected packets can be delivered, even if only two different destinations are used. Note that when we speak about algorithms that do not drop packets, this implies that they must have sufficient space to accommodate all injected packets.

Claim. For every natural algorithm that does not drop packets, there is an adversary for unicast injections using just two different destinations that can force the algorithm never to deliver a packet, no matter how high the throughput of an optimal strategy can be.

Proof. The adversary will simply pick one destination as the so-called *dead destination* and will inject so many packets into the system that whenever an edge is offered, a packet will be sent for the dead destination. Hence, the adversary can prevent packets from reaching the good destination, although there may be plenty of opportunities, had the good packets been chosen. On the other side, the adversary will never offer an edge directly to the dead destination. Hence, no packet will ever get delivered.   □

Thus, unbounded adversaries seem to be difficult to handle without allowing packets to get dropped. However, what about "friendly" adversaries, i.e. adversaries that only inject packets so that when using an optimal algorithm, only a bounded number of packets are in transit at any time without deleting any? We show that also in this case some natural algorithms have severe problems if packets cannot be dropped.

Theorem 3. *If the adversary is allowed to inject packets for more than one destination, then the adversary can force the T-balancing algorithm to store by a factor of $\Theta(2^{n/4})$ more packets in a buffer than for an optimal algorithm.*

Proof. (Sketch) For the proof it is sufficient to use two destinations, a and b, and to set $B = 1$. Given a node v, the height of its a-buffer is denoted by $h_a(v)$ and the height of its b-buffer is denoted by $h_b(v)$.

We show the theorem by complete induction. Suppose that we can construct a scheme using $2(5 + 2i)$ nodes with two nodes $v_i^{(a)}$ and $v_i^{(b)}$ so that $h_a(v_i^{(a)}) \geq H_i$ and $h_b(v_i^{(a)}) = 0$, and $h_a(v_i^{(b)}) = 0$ and $h_b(v_i^{(b)}) \geq H_i$, where $H_i = 2^i \max\{4T, 3\} - (2^i - 1)(2T + 1)$. Then we can show that 4 more nodes suffice to identify nodes so that the hypothesis above also holds for $i + 1$. The basic

idea is to create "copies" u_a and u_b of $v_i^{(a)}$ and $v_i^{(b)}$ and then to inject schedules for a-packets (resp. b-packets) with path $(v_{i+1}^{(a)}, u_b, u_a, a)$ (resp. $(v_{i+1}^{(b)}, u_a, u_b, b)$).
□

The theorem implies together with the results in [7] that only for the case that we have a single destination, the T-balancing algorithm without a dropping rule can be space-efficient under friendly adversaries.

What about other natural algorithms studied in the literature such as algorithms based on exponential priority functions (e.g. [10])? A routing algorithm is called *stable* if the number of packets in transit does not grow unboundedly with time. In order to investigate the stability of natural algorithms, we start with an important property of natural algorithms that allows us to study instability in the option set model (suggested in the proof of Theorem 1) instead of the original unicast model, which is much more difficult to handle.

Theorem 4. *For any natural deterministic algorithm it holds: If it is not stable in the option set model, it is also not stable in the adversarial unicast model.*

Proof. (Sketch) We only show how to get from the anycast to the unicast model. See [8] for details.

Consider any natural deterministic algorithm A that is instable in the anycast setting. Let V be the set of nodes and let $\mathcal{D} = (D_1, \ldots, D_N)$ be the set of anycast sets. To prove instability for A in the unicast model, we extend V to $V \cup \{d_1, \ldots, d_N\}$, where d_i is the new and only destination node for packets originally having destination set D_i. Let S be the strategy that caused instability for A in the anycast model. We simulate S until a packet of type i is supposed to reach one of its destination nodes D_i. Instead, we will offer now an edge to d_i. If this edge is taken by a packet of type i, we continue with the simulation.

Otherwise, it follows from the definition of a natural algorithm that another packet must have been sent to d_i. This causes the total number of packets stored in the buffers in $\{d_1, \ldots, d_N\}$ to increase by one. Then, we remove all packets from V by offering again and again edges to destinations d_i and start from the beginning with the simulation of S.

Thus, either we obtain a perfect simulation of S for the unicast case, in which case A will be instable, or we increase the number of packets in the buffers in $\{d_1, \ldots, d_N\}$ in every failed simulation attempt, which will also cause A to become instable. This completes the proof.
□

The theorem allows us to show the following result.

Theorem 5. *Natural routing algorithms which are based on exponential priority functions are not stable.*

Proof. By algorithms with exponential priority function we mean algorithms using the potential drop $f(h_1, h_2) = (\phi(h_1) + \phi(h_2)) - (\phi(h_1 - 1)) + \phi(h_2 + 1))$ with $\phi(h) = \sum_{i=1}^{h} e^{\alpha \cdot i}$ for some $\alpha > 0$ to determine the priority of a packet movement.

We will show that this rule can cause packets not to be delivered under certain circumstances, and that these situations can be generated arbitrarily often. We assume that the nodes are sorted according to their heights with $h_{N-1} \geq h_{N-2} \geq \ldots \geq h_1 \geq h_0$ and that node 0 is the destination node.

Lemma 7. *If the height difference between node $(N-1)$ and node 1 is $h_{N-1} - h_1 \geq \frac{\ln(e+1)}{\alpha}$, a new packet can be injected without any packet leaving the system.*

Proof. We assume that the adversary injects a new packet into node 1, which stores the lowest number of packets and has height h_1 before the injection of the packet. Then the adversary offers an option set with the two links $\{(N-1,1),(1,0)\}$. This option set is a valid schedule for the newly injected packet. The algorithm, however, will choose link $(N-1,1)$ if $e^{\alpha \cdot h_{N-1}} - e^{\alpha \cdot (h_1+1)} > e^{\alpha \cdot h_1} - e^{\alpha}$, which is true if $h_{N-1} - h_1 > \frac{\ln(e+1)}{\alpha}$. $\qquad \square$

The important observation from the previous lemma is that the necessary height difference between the two nodes does not depend on the actual height of these nodes. If it is always possible to create this fixed height difference for a given algorithm and a given number of packets in the system, then the algorithm is not stable. In the next lemma we will show that this is the case.

Lemma 8. *Given a network with at least $\Delta + 1$ nodes it is possible to achieve a difference in height of at least Δ packets between the node with the highest number of packets and the non-destination node with the lowest number of packets without reducing the number of packets, or the algorithm is instable.*

Proof. Pick a set S of Δ non-destination nodes, and consider the following strategy: Suppose that there are two nodes in S of equal height, say v and w. Then we inject a packet in v and offer the option set $\{(v,w)\}$. If the algorithm sends a packet, we offer the option set $\{(v,0)\}$ and otherwise the option set $\{(w,0)\}$. This ensures that the injected packet will have a schedule in any case and that the number of packets in S does not change. Furthermore, using the potential function $\phi_u = \sum_{i=1}^{h_u} i$, one can show that this operation increases the potential in S.

Hence, either the number of packets in S goes to infinity or there cannot be two nodes in S of the same height any more. In the latter case, this means that the highest and lowest node in S must have a difference of at least Δ. $\qquad \square$

We now assume that we have a network with at least $\frac{\ln(e+1)}{\alpha} + 1$ nodes. From Lemma 8 we know that in this case a height difference of at least $\frac{\ln(e+1)}{\alpha}$ can be created.

After this, the adversary repeats the strategies in Lemma 7 and Lemma 8 again and again. With every iteration, the number of packets in the system will increase by one, which proves the theorem. $\qquad \square$

The proof of the theorem immediately implies the following result.

Corollary 1. *Natural routing algorithms which always prefer the buffer with the largest number of packets are not stable.*

We conjecture that any (natural) online algorithm is either instable or requires an exponential buffer size to be stable under friendly adversaries, which would imply together with Theorem 1 that the ability to drop packets can tremendously improve the performance of routing algorithms.

4 Conclusions and Open Problems

In this paper we presented a simple balancing algorithm for anycasting in adversarial systems. Many open questions remain. Although our space overhead is already reasonably low (essentially, $O(L/\epsilon)$), the question is whether it can still be reduced. For example, could knowledge about the location of a destination or structural properties of the network (for instance, it has to form a planar graph) help to get better bounds? Or are there other protocols that can achieve a lower space overhead in general?

References

1. Y. Afek, B. Awerbuch, E. Gafni, Y. Mansour, A. Rosen, and N. Shavit. Slide – the key to polynomial end-to-end communication. *Journal of Algorithms*, 22(1):158–186, 1997.
2. Y. Afek and E. Gafni. End-to-end communication in unreliable networks. In *PODC '88*, pages 131–148, 1988.
3. W. Aiello, B. Awerbuch, B. Maggs, and S. Rao. Approximate load balancing on dynamic and synchronous networks. In *STOC '93*, pages 632–641, 1993.
4. W. Aiello, E. Kushilevitz, R. Ostrovsky, and A. Rosén. Adaptive packet routing for bursty adversarial traffic. In *STOC '98*, pages 359–368, 1998.
5. W. Aiello, R. Ostrovsky, E. Kushilevitz, and A. Rosén. Dynamic routing on networks with fixed-size buffers. In *SODA '03*, 2003.
6. M. Andrews, B. Awerbuch, A. Fernández, J. Kleinberg, T. Leighton, and Z. Liu. Universal stability results for greedy contention-resolution protocols. In *FOCS '96*, pages 380–389, 1996.
7. B. Awerbuch, P. Berenbrink, A. Brinkmann, and C. Scheideler. Simple routing strategies for adversarial systems. In *FOCS '01*, pages 158–167, 2001.
8. B. Awerbuch, A. Brinkmann, and C. Scheideler. Anycasting and multicasting in adversarial systems. Technical report, Dept. of Computer Science, Johns Hopkins University, March 2002. See http://www.cs.jhu.edu/~scheideler.
9. B. Awerbuch and F. Leighton. A simple local-control approximation algorithm for multicommodity flow. In *FOCS '93*, pages 459–468, 1993.
10. B. Awerbuch and F. Leighton. Improved approximation algorithms for the multi-commodity flow problem and local competitive routing in dynamic networks. In *STOC '94*, pages 487–496, 1994.
11. B. Awerbuch, Y. Mansour, and N. Shavit. End-to-end communication with polynomial overhead. In *FOCS '89*, pages 358–363, 1989.
12. A. Borodin, J. Kleinberg, P. Raghavan, M. Sudan, and D. P. Williamson. Adversarial queueing theory. In *STOC '96*, pages 376–385, 1996.
13. D. Gamarnik. Stability of adversarial queues via fluid models. In *FOCS '98*, pages 60–70, 1998.
14. D. Gamarnik. Stability of adaptive and non-adaptive packet routing policies in adversarial queueing networks. In *STOC '99*, pages 206–214, 1999.
15. A. Goel. Stability of networks and protocols in the adversarial queueing model for packet routing. In *SODA '99*, pages 911–912, 1999.
16. C. Partridge, T. Mendez, and W. Milliken. Rfc 1546: Host anycasting service, November 1993.

17. C. Scheideler and B. Vöcking. From static to dynamic routing: Efficient transformations of store-and-forward protocols. In *STOC '99*, pages 215–224, 1999.
18. P. Tsaparas. Stability in adversarial queueing theory. Master's thesis, Dept. of Computer Science, University of Toronto, 1997.

Dynamic Algorithms for Approximating Interdistances

Sergei Bespamyatnikh[1] and Michael Segal[2]

[1] Department of Computer Science, University of Texas at Dallas,
Richardson, TX 75083, USA
besp@utdallas.edu, http://www.utdallas.edu/~besp
[2] Department of Communication Systems Engineering,
Ben-Gurion University of the Negev, Beer-Sheva 84105, Israel
segal@cs.bgu.ac.il, http://www.cs.bgu.ac.il/~segal

Abstract. In this paper we present efficient dynamic algorithms for approximation of kth, $1 \leq k \leq \binom{n}{2}$ distance defined by some pair of points from a given set S of n points in d-dimensional space, for every fixed d. Our technique is based on dynamization of well-separated pair decomposition proposed in [11], computing approximate nearest and farthest neighbors [23,26] and use of persistent search trees [18].

1 Introduction

Let S be a set of n points in $\mathbb{R}^d, d \geq 1$ and let $1 \leq k \leq \frac{n(n-1)}{2}$. Let $d_1 \leq d_2 \leq \ldots \leq d_{\binom{n}{2}}$ be the L_p-distances determined by the pairs of points in S. In this paper we consider the dynamic version of the following optimization problem:

- **Distance selection.** Compute the k-th smallest Euclidean distance between a pair of points of S.

In the dynamic version of the distance selection problem points are allowed to be inserted or deleted and given a number k, $1 \leq k \leq \binom{|S|}{2}$ one wants to answer efficiently what is the k-th smallest distance between a pair of points of S (by $|S|$ we denote the cardinality of the current set of points).

The distance selection problem above received a lot of attention during the past decade. The solution to the distance selection problem can be obtained using a parametric searching. The decision problem is to compute, for a given real r, the sum $\Sigma_{p \in S} |D_r(p) \cap (S - \{p\})|$, where $D_r(p)$ is the closed disk of radius r centered at p. Agarwal et al. [1] gave an $O(n^{\frac{4}{3}} \log^{\frac{4}{3}} n)$ expected-time randomized algorithm for the decision problem, which yields an $O(n^{\frac{4}{3}} \log^{\frac{8}{3}} n)$ expected-time algorithm for the distance selection problem. Goodrich [22] derandomized this algorithm, at a cost of an additional polylogarithmic factor in the runtime. Katz and Sharir [27] obtained an expander-based $O(n^{4/3} \log^{2+\varepsilon} n)$-time deterministic algorithm for this problem. By applying a randomized approach Chan [13] was able to obtain an $O(n \log n + n^{2/3} k^{1/3} \log^{5/3} n)$ expected time algorithm for this problem. Bespamyatnikh and Segal [9] considered an approximation version of the distance

J.C.M. Baeten et al. (Eds.): ICALP 2003, LNCS 2719, pp. 1169–1180, 2003.

selection problem. For a distance d determined by some pair of points in S and for any fixed $0 < \delta_1 \leq 1$, $\delta_2 \geq 1$, the value d' is the (δ_1, δ_2)-*approximation* of d, if $\delta_1 d \leq d' \leq \delta_2 d$. They [9] present an $O(n \log^3 n/\varepsilon^2)$ runtime solution for the distance selection problem that computes a pair of points realizing distance d' that is either $(1, 1+\varepsilon)$ or $(1-\varepsilon, 1)$-approximation of the actual k-th distance, for any fixed $\varepsilon > 0$. They also present an $O(n \log n/\varepsilon^2)$ time algorithm for computing the $(1 - \varepsilon, 1 + \varepsilon)$-approximation of k-th distance and show how to extend their solution in order to answer efficiently the queries approximating k-th distance for a static set of points. Agarwal et al. [1] considers a similar problem, where one want to identify approximate "median" distance, that is, a pair of points $p, q \in S$ with the property that there exist absolute constants c_1 and c_2 such that $0 < c_1 < \frac{1}{2} < c_2 < 1$ and the rank of the distance determined by p and q is between $c_1 \binom{n}{2}$ and $c_2 \binom{n}{2}$. They [1] showed how to solve this problem in $O(n \log n)$ time. Arya and Mount[4] introduced a *balanced box-decomposition tree* (BBD tree) in order to answer efficiently approximate range searching queries. They obtained $O(\log n + \frac{1}{\varepsilon^d})$ query time for d-dimensional point sets using linear space after $O(n \log n)$ preprocessing time. Their results also can be used to solve the decision version of the distance selection problem with $(1-\varepsilon, 1+\varepsilon)$-approximation in $O(n \log n + \frac{n}{\varepsilon^2})$ runtime.

We call an algorithm a *almost-linear-time approximation scheme with almost logarithmic update time (ALTAS-LOG) of order* (c_1, c_2) if it has a preprocessing time of the form $O(n \log^{l_1} n/\varepsilon^{c_1})$, for some constant $l_1 > 0$ and update time of the form $O(\log^{l_2} n/\varepsilon^{c_2})$, for some constant $l_2 > 0$.

In this paper we show an ALTAS-LOG algorithm of order $(2, 2)$ such that given number k, $1 \leq k \leq \binom{|S|}{2}$ it outputs in $O(\log n)$ time a pair of points realizing distance which is the $(1 - \varepsilon^2, 1 + \varepsilon)$-approximation (or $(1 - \varepsilon, 1 + \varepsilon^2)$-approximation) of kth distance. More precisely, we show how to construct a data structure in $O(n \log n/\varepsilon^2 + n \log^4 n/\gamma)$ time that dynamically maintains a set of n points in the plane in $O(\log^4 n/\gamma)$ time under insertions and deletions, for any fixed $\gamma > 0$, such that given number k, $1 \leq k \leq \binom{|S|}{2}$ one can compute in $O(\log n)$ time a pair of points realizing distance which is the $(1 - \gamma, 1 + \varepsilon)$ $((1 - \varepsilon, 1 + \gamma))$-approximation of kth distance. We also show how to obtain dynamic $(1 - \varepsilon, 1 + \varepsilon)$-approximation of kth distance by simpler ALTAS-LOG algorithm of order $(2, 0)$ with slightly faster preprocessing time. It should be noted here that approximating the actual k-th distance within the factor $1 + \varepsilon^2$ (or $1 - \varepsilon^2$) is considerable harder than getting $1 + \varepsilon$ (resp. $1 - \varepsilon$) approximation with the same ε dependency in the running time of algorithm. We also generalize our algorithms to work in higher dimensions.

For our best knowledge, the dynamic problem of maintaining exact and approximate k^{th} distance is not studied in literature, except the famous closest pair problem (1st distance selection) with optimal $O(\log n)$ worst-case update time [7] and diameter problem (farthest pair selection) with $O(n^\varepsilon)$ worst-case update time [19], expected $O(\log n)$ update time [20] and $O(b \log n)$ update time [25] that maintains the approximation diameter (the approximation factor depends on the integer constant $b > 0$). One may find our algorithms useful in parametric searching applications, where a set of candidate solutions is defined by the distances between pairs of points of dynamic set S. For example, Agarwal and Procopiuc [3] (see also [2,14,29]) studied various k-center problems in R^d under L_∞ and L_2 metric: combinations of exact and approximate, continuous and discrete, uncapacitated and capacitated versions. Typically an algorithm performs a search (for example, binary search) on the sorted list of interdistances between data points. Our algorithms provide fast implementation of the search if an approximate solution suffices.

The main contribution of this paper is by developing efficient approximating dynamic algorithm for the well known distance selection problem using an approach that based on well separated pairs decomposition introduced by Callahan and Kosaraju [11] (see also [17]), computing approximate nearest and farthest neighbors [23,26] and persistent binary search trees introduced by Driscoll et al. [18].

This paper is organized as follows. In the next section we briefly describe well-separated pair decomposition. Section 3 is dedicated to the approximating dynamic distance selection problem. Finally we conclude in Section 4.

2   Well-Separated Pair Decomposition

In this section we shortly describe the well-separated pair decomposition proposed by Callahan and Kosaraju [11]. Let A and B be two sets of points in d-dimensional space ($d \geq 1$) of size n and m, respectively. Let s be some constant strictly greater than 0 and let $R(A)$ (resp. $R(B)$) be the smallest axisparallel bounding box that encloses all the points of A (resp. B). We say that point sets A and B are *well-separated* with respect to s, if $R(A)$ and $R(B)$ can be each contained in d-dimensional ball of some radius r, such that the distance between these two balls is at least sr. One can easily show that for a given two well-separated sets A and B, if $p_1, p_4 \in A$, $p_2, p_3 \in B$ then $dist(p_1, p_2) \leq (1 + \frac{2}{s})dist(p_1, p_3)$ and $dist(p_1, p_2) \leq (1 + \frac{4}{s})dist(p_4, p_3)$. (For general L_p metric inequality may differ by some multiplicative constant.) Let S be a set of d-dimensional points, and let $s > 0$. A *well-separated pair decomposition* (WSPD) for S with respect to S is a set of pairs $\{(A_1, B_1), (A_2, B_2), \ldots, (A_p, B_p)\}$ such that:

(i) $A_i \subseteq S$ and $B_i \subseteq S$, for all $i = 1, \ldots, p$.
(ii) $A_i \cap B_i = \emptyset$, for all $i = 1, \ldots, p$.
(iii) A_i and B_i are well-separated with respect to s.
(iv) for any two distinct points p and q in S, there is exactly one pair (A_i, B_i) such that either $p \in A_i$ and $q \in B_i$ or $p \in B_i$ and $q \in A_i$.

The main idea of the algorithm for construction WSPD is to build a binary fair split tree T whose leaves are points of S, with internal nodes corresponding to subsets of S. More precisely, split tree of S is a binary tree, constructed recursively as follows. If $|S| = 1$, its unique split tree consists of the node of S. Otherwise a split tree is any tree with root S and two subtrees that are split trees of the subsets formed by a split of S by an axis-parallel hyperplane into two non-empty subsets. For any node A in the tree, denote its parent (if exists) by $p(A)$. The outer rectangle of A, denoted by $R(A)$ is either (if A is a root) an open d-cube centered at the center of the bounding box of S with the side size that equals to the largest side $l_{max}(S)$ of the bounding box of S, or we have a situation when the splitting hyperplane used for the split of $p(A)$ divides $R(p(A))$ into two open rectangles. Let $R(A)$ be the one that contains A. A fair split of A is a split in which the splitting hyperplane is at distance of at least $l_{max}(A)/3$ from each of the two boundaries of $R(A)$ parallel to it. A split tree formed using only fair splits is called a fair split tree.

Each pair (A_i, B_i) in WSPD is represented by two nodes $v, u \in T$, such that all the leaves in the subtree rooted at by v correspond to the points of A_i and all the leaves in the subtree rooted at by u correspond to the points of B_i. The paper of Callahan and Kosaraju [11] presents an algorithm that implicitly constructs WSPD for a given set S and separation value $s > 0$ in $O(n \log n + s^d n)$ time such that the number of pairs (A_i, B_i) is $O(s^d n)$. Moreover, Callahan [10] showed to compute $WSPD$ in which at least one of the sets A_i, B_i of each pair (A_i, B_i) contains exactly one point of S. The running time remains the same; however, the number of pairs increases to $O(s^d n \log n)$.

3   Approximating k-th Distance

Our algorithm consists of several stages. At the first stage we compute a WSPD for S with separation constant $s = \frac{12}{\varepsilon}$. From each (A_i, B_i) we take any pair $(a_i, b_i) \in (A_i, B_i), 1 \le i \le p, p = O(n)$. Our task now is to find the smallest index j in the sorted list of (a_i, b_i) pairs , such that the sum of cardinalities of all pairs (A_i, B_i) that correspond to this prefix is at least k. Therefore, we sort the distances d_i' between a_i and b_i, $1 \le i \le p$. We assume that the pairs (A_i, B_i) are in order of increasing d_i'. Next, for each pair (A_i, B_i), $1 \le i \le p, p = O(n)$ we compute the $\alpha_i = |A_i||B_i|$ value, i.e. α_i is the total number of distinct pairs (a, b), $a \in A_i$, $b \in B_i$. Let

- $m_i = \min_{a \in A_i, b \in B_i} dist(a, b)$.
- $M_i = \max_{a \in A_i, b \in B_i} dist(a, b)$ and

Let also $l_i, 1 \le i \le p$ be a number such that $(1 - \gamma)M_i \le l_i \le M_i$, for arbitrary fixed $\gamma > 0$.

As we said above, for a particular k we compute the smallest j such that $\sum_{i=1}^{j} \alpha_i \ge k$. Let $M' = \max_{i=1}^{j} M_i$ and let $l' = \max_{i=1}^{j} l_i$. We claim that l' is the $(1 - \gamma, 1 + \varepsilon)$-approximation of k-th distance.

In what follows we prove the correctness of our algorithm and show how to implement it efficiently.

Lemma 1. $(1 - \gamma)d_k \le l' \le (1 + \varepsilon)d_k$.

Proof. We observe that the total number of distances defined by pairs (A_i, B_i), $1 \le i \le j$ is at least k because $\sum_{i=1}^{j} \alpha_i \ge k$. Since M' is the maximum of these distances $M' \ge d_k$ follows. Thus, from $l' \ge (1 - \gamma)M'$ it follows that $l' \ge (1 - \gamma)d_k$. Our goal now it to prove that $M' \le (1 + \varepsilon)d_k$. We recall that all possible pairs of points of S are uniquely represented by pairs (A_i, B_i) in

WSPD. Consider the set of pairs $D = \{(a,b)|a \in A_i, b \in B_i, i \geq j\}$. There is an index r, $j \leq r \leq p$ such that m_r is the smallest distance defined by pairs of D. The total number of pairs in D is larger than $\binom{n}{2} - k$. Therefore, $d_k \geq m_r$. Let t, $1 \leq t \leq j$ be the index such that $M' = M_t$. From the observation in previous section it follows that $M_t \leq (1 + \frac{2}{s})d'_t = (1 + \varepsilon/3)d'_t$. Thus, $M' \leq (1 + \varepsilon/3)d'_j \leq (1 + \varepsilon/3)d'_r$, since the sequence d'_i, $j \leq i \leq p$ is non-decreasing. It follows that $(1 + \varepsilon/3)d'_r \leq (1 + \varepsilon/3)(1 + \varepsilon/3)m_r \leq (1 + \varepsilon)d_k$. So, $l' \leq M' \leq (1 + \varepsilon)d_k$.

Remark 1. Using the similar approximation scheme with decreasing list of d'_i distances and by taking $M_i = \min_{a \in A_i, b \in B_i} dist(a,b)$ and l_i such $(1 + \gamma)M_i \geq l_i \geq M_i$ we can obtain $(1 - \varepsilon, 1 + \gamma)$-approximation of the kth distance.

Remark 2. If, instead of computing l_i, we choose d'_j as the value returned by the algorithm, we obtain $(1 - \varepsilon, 1 + \varepsilon)$-approximation of the kth distance. This is based on fact that $(1 + \varepsilon)d'_j = \max_{1 \leq i \leq j}(1 + \varepsilon)d'_i \geq \max_{1 \leq i \leq j} M_i - M' \geq d_k$.

It remains to show how to implement this algorithm efficiently, i.e. how to compute the values l_i, α_i, $1 \leq i \leq p$. First we show how to compute α_i. In other words we need to compute the cardinalities of A_i and B_i, $1 \leq i \leq p$. Recall that each pair (A_i, B_i) in WSPD is represented by two nodes v_i, u_i of the split tree T. The cardinality of $A_i(B_i)$ equals to the number of leaves in the subtree of T rooted at $v_i(u_i)$. Thus, by postorder traversal of T we are able to compute all the required cardinalities. Bespamyatnikh and Segal [9] showed how to compute the values m_i, M_i, $1 \leq i \leq p$ exactly using Voronoi diagrams [6] and Bentley's [5] logarithmic method. By assuming that the singleton set of each pair (A_i, B_i) in WSPD is $A_i = \{a_i\}$ they reduce the original problem of computing m_i and M_i values to the problem of computing for each a_i, $1 \leq i \leq p$ the nearest and the farthest neighbor in corresponding B_i. Since the computing of all Voronoi Diagrams may lead to undesired $O(n^2)$ runtime factor, they maintain dynamically Voronoi Diagrams while traversing a split tree T in a bottom-up fashion. Let S_v be a subset of S associated with a node v in T. By traversing a split tree T in a postorder fashion starting from leaves they use a partition R_v of S_v into disjoint sets $S_v^1, \ldots, S_v^q$ and maintain the Voronoi Diagram VD with corresponding point location data structure PL for each set S_v^j, $1 \leq j \leq q$ in R_v. The sizes of the sets in R_v are different and restricted to be the powers of two. As the consequence the number of such sets is at most $\log n$, i.e. $q \leq \log n$. It can be shown that the total time needed to spend for all described operations is $O(n \log^3 n)$.

Dynamic Updates

The main drawback of the above scheme is the fact that during processing of T the Voronoi Diagram data structures are destroyed, so that at the end of the process we know only the Voronoi Diagram for the entire set S. Suppose that now we insert or delete some leaf from T. It may have influence on a number of other internal nodes. How we can determine now the new values of m_i and

M_i? Basically, we have two major problems. The first one is how to store the Voronoi Diagram in each one of the internal nodes of T and the second one is how to update it quickly when T changes it's structure by insertion of a new point or deletion of an existing point from T. In order to solve the first problem we will use a fully persistent binary search trees described by Driscoll et al. [18]. A fully persistent structure supports any sequence of operations in which each operation applied to any previously existing version. The result of the update is an entirely new version, distinct from all others. Unfortunately we cannot represent a Voronoi Diagram as a collection of a sublinear number of binary search trees and therefore, we need to find a way of computing the values m_i and M_i using another strategy. In fact we are interested in computing values l_i. Let us first consider the L_∞ metric. The points defining M_i should lye on the boundary of the smallest axis-parallel bounding box of set $A_i \cup B_i$. Recall that A_i and B_i are well separated and, thus, the L_∞ diameter of $A_i \cup B_i$ is defined by a pair (p, q) such that $p \in A_i$ and $q \in B_i$.

The computation of $m_i, 1 \leq i \leq p$ can be done similarly to the approach described in [8]. Suppose we use a WSPD with $p = O(n \log n)$ and assume $A_i = \{a_i\}$, $1 \leq i \leq p$. For each point a_i we need to find the closest neighbor in corresponding B_i. Consider, for example, the planar case. Let l_1 be a line whose slope is 45° passing through the a_i and l_2 a a line whose slope is 135° passing through the a_i. These lines define four wedges: $Q_{top}, Q_{bottom}, Q_{left}, Q_{right}$. For any point p lying in $Q_{left} \cup Q_{right}(Q_{bottom} \cup Q_{top})$ the L_∞-distance to a_i is defined by the x-distance (y-distance, resp.) to a_i. We perform four range queries, using orthogonal range tree [6] data structure (in coordinate system defined by lines l_1, l_2), each of them corresponding to the appropriate wedge. For each node in a secondary data structure we keep four values $x_{min}, x_{max}, y_{min}, y_{max}$ (computed in the initial coordinate system) of points in corresponding range. Consider for a example wedge Q_{right}. Our query corresponding to Q_{right} marks $O(\log^2 n)$ nodes. The minimum of x_{min} values stored in these nodes define the closest neighbor point to a_i lying in Q_{right}. We proceed similarly with the other wedges. We maintain orthogonal range tree data structures dynamically in a bottom-up fashion while traversing split tree T. In order to merge two data structures we simply insert all the points stored in the smaller range tree into the larger one. However, we are interested in the values of m_i computed for the Euclidean metric. We will use the following two results in order to accomplish our task. The first result has been proposed by Kapoor and Smid [26] that finds, for a given query point $p \in \mathbb{R}^d$ a $(1+\gamma)$-approximate L_2-neighbor of p in a given set of n points in $O(\log^{d-1} n / \gamma^{d-1})$ time using a data structure of $O(n \log^{d-2} n)$ space. They [26] store a set S in a constant number of a range trees, where each range tree stores the points according to its own coordinate system using the construction of Yao [32]. Then, for a given p, they use all the range trees to compute L_∞ nearest neighbors of p in all coordinate systems. One of these L_∞ neighbors is $(1 + \gamma)$-approximate L_2 nearest neighbor of p. But we still need to compute the values of M_i. The second result is due to Indyk [23] that shows how to compute $(1 - \gamma)$-approximate farthest neighbor of a given point p by performing a constant number of $(1+\gamma)$-approximate nearest neighbors queries. The idea is to construct a set of a constant number of concentric disks (balls) around the origin. Each point is rounded to the nearest circle (sphere). For each disk (ball) we build an $(1 + \gamma)$-approximate nearest neighbor data structure for the set of points on corresponding circle (sphere). Next, for each point $p \in S$ and each disk (ball) B_i, the "antipode" p^i of p with respect to B_i is defined as follows. Let p_1 and p_2 be the two points of the intersection of the circle (sphere) of B_i with the line passing through p and origin. Let h_p denote the hyperplane through the origin that is perpendicular to the line through p and origin. The point p^i is the one of the points p_1, p_2 which lies on the side of h_p different from the side containing p. In order to find the furthest neighbor of q, we issue $(1+\gamma)$-approximate nearest neighbor query with the point q^i in the data structure for the points on each one of the circles (spheres). Among the points found, we return the one furthest from q. Preprocessing time is $O(d^{O(1)}n)$ plus the cost of initiating a constant number of data structures for $(1 + \gamma)$-approximate nearest neighbor queries. The query time is bounded by the the query time for the $(1 + \gamma)$-approximate nearest neighbor query.

The good thing in the described algorithms is the fact that all of them can be implemented using orthogonal range search trees, or in other words, binary search trees. This will allow us to make all of them fully persistent using Driscoll et al. [18] algorithm, thus solving our task of storing the appropriate data structure for each of the nodes of T without being destroyed. Generally speaking, ordinary data structures are ephemeral in sense that making a change to the structure destroys the old version, leaving only the new one. In a fully persistent data structure, past versions of the data structure are remembered and can be queried and updated. In [18] a method termed *node copying with displaced storage of changes* was developed that could make red-black tree data structure to become fully persistent, in worst-case time per operation of $O(\log n)$ and worst-case space cost of $O(1)$ per insertion or deletion. Instead of indicating a change to an ephemeral node x by storing the change in the corresponding persistent node x', Driscoll et al. [18] stores information about the change in some possibly different node that lies on the access path to x' in the new version. Thus the record of the change is in general displaced from the node to which the change applies. The path from the node containing the change information to the affected node is called the *displacement path*. By copying nodes judiciously, Driscoll et al. [18] were able to keep the displacement paths sufficiently disjoint to guarantee an $O(1)$ worst-case space bound per insertion or deletion while having $O(\log n)$ worst-case time bound per access, insertion or deletion.

While traversing a tree T, we maintain all the described data structures for computing $(1+\gamma)$-approximate nearest neighbor and $(1-\gamma)$-approximate farthest neighbor. We use again Bentley's [5] logarithmic method as described before. Notice, that each point in S can be inserted at most $O(\log n)$ times into the data structures while traversing T in a bottom-up fashion. Each insertion takes $O(\log^3 n)$ time. To give access to the persistent structure, the access pointers to the roots of the various versions must be stored in a balanced search tree, ordered by index. The total time for maintaining the range trees and computing $l_i, 1 \le i \le p$ is $O(n \log^4 n)$, since $p = O(n \log n)$, each query takes $O(\log^2 n)$ time and each node contains a logarithmic number of the related data structures. The above computation can also be generalized to d-dimensional space, $d > 2$. Thus, we have

Theorem 1. *Given a set S of n points in $\mathbb{R}^d$, a number k, $1 \leq k \leq \binom{n}{2}$, $\varepsilon > 0$, $\gamma > 0$ a pair of points realizing $(1 - \gamma, 1 + \varepsilon)$ $((1 - \varepsilon, 1 + \gamma))$-approximation of d_k can be determined in $O(n \log n / \varepsilon^d + n \log^{d+2} n / \gamma^{d-1})$ time.*

Remark 3. Notice that we can obtain better running time (by logarithmic factor) using orthogonal range trees with the fractional cascading technique [16]. However, in order to allow persistence for the future dynamic updates we use orthogonal range trees avoiding this technique.

Remark 4. We can use a simpler strategy in order to compute the M_i values. We maintain the bounding boxes for sets of points corresponding to the nodes of T. The new bounding box can be computed in $O(1)$ time using the information from the previous steps. It results in a very fast algorithm with $(1 - \frac{1}{\sqrt{2}}, 1 + \varepsilon)$-approximation of kth distance which can be made dynamic fairly easy.

Remark 5. The runtime of the algorithm presented in [9] and the approximation factor achieved by that algorithm is better than in Theorem 1 for $d = 2$. Moreover, we should note that there is a more efficient algorithm even for $d > 2$. Instead of using Kapoor and Smid data structure [26] for querying approximate nearest neighbor, we can use either Kleinberg [28] or Indyk and Motwani [24] or Kushilevitz et al. [30] or Chan's [15] data structures for the same purpose. For example, using the result by Chan [15] that gave an ALTAS-LOG algorithm of order $(\frac{d-1}{2}, \frac{d-1}{2})$ that achieves $(1 + \varepsilon)$-approximation for nearest neighbor query instead of Kapoor and Smid [26] ALTAS-LOG algorithm of order $(d - 1, d - 1)$ we obtain a better runtime of the entire algorithm. Unfortunately, the algorithm in [9] and also [15,24,28,30] data structures cannot be made dynamic with a polylogarithmic update time. As we will see later, the result in Theorem 1 can be extended to deal with the dynamic point sets.

Following Remark 2 we also can conclude

Theorem 2. *Given a set S of n points in $\mathbb{R}^d$, a number k, $1 \leq k \leq \binom{n}{2}$, $\varepsilon > 0$, a pair of points realizing $(1 - \varepsilon, 1 + \varepsilon)$-approximation of d_k can be determined in $O(n \log n / \varepsilon^d)$ time.*

It remains to check what happens with the tree T when a new point is inserted or some existing point is deleted. By $\sigma(v)$, $v \in T$ we denote the subset of points associated with v at some instance in the sequence of updates. If v has two children w_1 and w_2 then $\sigma(v) = \sigma(w_1) \cup \sigma(w_2)$. If v is a leaf, then $|\sigma(v)| = 1$. Since the fair split property depends on the value of $l_{max}(\sigma(v))$. Each time we insert a new point, this may increase the value of $l_{max}(\sigma(v))$ for all its ancestors in T, and the fair split property may be violated. Deletion of a point will not increase the value of $l_{max}(\sigma(v))$ for any of its ancestors, and hence can be performed on any fair split tree without restructuring. Callahan [10] shows that we can deal with the updates by maintaining a labeled binary tree T in which each node satisfies the following invariants:

1. For all internal nodes v with children w_1 and w_2, there is a fair cut that partitions $R(v)$ into two rectangles R_1 and R_2, such that $\sigma(w_1) = \sigma(v) \cap R_1$, $\sigma(w_2) = \sigma(v) \cap R_2$, $R(w_1)$ can be constructed from R_1 by applying a sequence of fair splits and $R(w_2)$ can be constructed from R_2 by applying a sequence of fair splits.

2. For all leaves v, $\sigma(v) = \{p\}$, and $R(v) = p$.

To insert a point p into this structure, we first retrieve the deepest internal node v in T such that $p \in R(v)$, ignoring the case in which p lies outside the rectangle at the root node. Let R_1 and w_1 have the same meaning as in the first invariant. Assume w.l.o.g. that $p \in R_1$. The way we chose v guarantees that $p \notin R(w_1)$. Now we introduce a new internal node u, which replaces w_1 as a child of v. We insert w_1 along with its subtree as a child of u, and insert a new leaf u' as the other child of u, where $\sigma(u') = \{p\}$. Finally we construct a rectangle $R(u)$ satisfying the first invariant. To delete the point p, we simply find a leaf v such that $\sigma(v) = \{p\}$, delete v, and compress out the internal node $p(v)$. Callahan [10] proves that once we have determined where to insert a point p, we may perform such an insertion in constant time, while preserving the invariants of the tree. Using the directed topology tree of Frederickson [21], Callahan has been able to maintain T in $O(\log n)$ time, where n is the current size of the point set. Generally speaking only $O(\log n)$ nodes of T can be affected during insertion or deletion of a point and therefore we can maintain the persistent structures associated with these nodes at sublinear cost.

Another problem that we have to deal with is the fact that introduction of a single new point can require the creation of many new pairs. Callahan [10] proposed an idea to predict all but a constant number of the new pairs ahead of time. The way to do it is to introduce *dummy points* where appropriate. Let $\bar{S}$ be a set of dummy points. Such points will not be counted in $\sigma(v)$ for any $v \in T$, but the tree T will have the same structure and rectangle labels as a fair split tree of $S \cup \bar{S}$. For efficiency we introduce only a constant number of dummy points for each well-separated pair $\{v, w\}$, such that $\sigma(v)$ and $\sigma(w)$ are not-empty. Since the number of new pairs is constant we can compute and maintain the relevant persistent structures efficiently.

The only missing thing is how to perform a query, i.e. how, for a given value of k, we can find the approximate k^{th} distance? We maintain a balanced binary search tree T' for distances d'_i as defined before. Suppose that we build a binary tree T' with the leaves corresponding to $d'_1, \ldots, d'_p$. Each internal node $v \in T_r$ will keep three values: $\Sigma_{i=q_1}^{q_2} \alpha_i$, $\Sigma_{i=q_2+1}^{q_3} \alpha_i$, where $\alpha_{q_1}, \ldots, \alpha_{q_2}$ ($\alpha_{q_2+1}, \ldots, \alpha_{q_3}$) are the values that correspond to the leaves of the left subtree (resp. right subtree) of a tree rooted by v, and the third value $L_v = \max_{i=q_1}^{q_3} l_i$ (or $R_v = \min_{i=q_1}^{q_3} r_i$, $(1+\gamma)m_i \geq r_i \geq m_i$). Clearly, the construction of this tree T' with the augmented values can be computed in $O(p)$ time. We associate with each node $v \in T'$ an index j_v, such that d'_{j_v} corresponds to the rightmost leaf in the subtree rooted at v. Given a value k, we traverse T' starting from the root towards its children. We need to find a node u, with the smallest j_u such that $\Sigma_{i=1}^{j_u} \alpha_i \geq k$. It can be done in $O(\log n)$ time, by simple keeping the total number of nodes to the left of the current searching path. At each node where the path goes right, we collect the value $L_v(R_v)$ stored in the left subtree. At the end, we report the maximal (minimal) of the collected $L_v(R_v)$ values. If T' is implemented as a balanced binary search tree then the update of the values r_i and l_i can be done in logarithmic time. Moreover, while updating T the new pairs may appear (and the previous pairs may disappear). Thus, we need to update the corresponding d'_i values in T' together with L_i, R_i, α_i values. The whole process in the plane can be accomplished in $O(\log^4 n)$ time since we have a logarithmic number of affected nodes in T, each query/update takes $O(\log^2 n)$ time and each node contains at most logarithmic number of associated data structures.

Therefore we can conclude by the following.

Theorem 3. *Given a set S of n points in $\mathbb{R}^d$, $\varepsilon > 0$, $\gamma > 0$ we can construct a data structure in time $O(n \log n/\varepsilon^d + n \log^{d+2} n/\gamma^{d-1})$ and $O(n \log n/\varepsilon^d)$ space with $O(\log^{d+2} n/\gamma^{d-1})$ update time for insertions/deletions of points such that given a number k, $1 \leq k \leq \binom{n}{2}$, a pair of points realizing $(1 - \gamma, 1 + \varepsilon)$ $((1 - \varepsilon, 1 + \gamma))$-approximation of d_k can be determined in $O(\log n)$ time.*

Theorem 4. *Given a set S of n points in $\mathbb{R}^d$, $\varepsilon > 0$, we can construct a data structure in time $O(n \log n/\varepsilon^d)$ and $O(n/\varepsilon^d)$ space such that given a number k, $1 \leq k \leq \binom{n}{2}$, a pair of points realizing $(1 - \varepsilon, 1 + \varepsilon)$-approximation of d_k can be determined in $O(\log n)$ time under insertions and deletions of points.*

4   Conclusions

We studied the dynamic problem for computing k-th Euclidean interdistance between n points in $\mathbb{R}^d$. The dynamization makes the problem more complicated. We are not aware of any other algorithms for exact or approximate solutions. We designed two efficient algorithms for maintaining a set of points and answering distance queries. The algorithms are based on the well-separated pair decomposition by Callahan and Kosaraju [11] and persistent data structures for approximate nearest/farthest neighbor. Both algorithms answer the queries in $O(\log n)$ time. The first algorithm provides $(1 - \varepsilon, 1 + \varepsilon)$ approximation and the second one provides a two parameters approximation $(1 - \varepsilon, 1 + \gamma)$ (or $(1 - \gamma, 1 + \varepsilon)$). It would be interesting to reduce the dependence of our algorithms on ε and γ.

References

1. P. Agarwal, B. Aronov, M. Sharir, S. Suri, "Selecting distances in the plane", *Algorithmica*, 9, pp. 495–514, 1993.
2. P. Agarwal, M. Sharir, E. Welzl "The discrete 2-center problem", *Proc. 13th ACM Symp. on Computational Geometry*, pp. 147–155, 1997.
3. P.K. Agarwal and C.M. Procopiuc, "Exact and Approximation Algorithms for Clustering", in *Proc. SODA '98*, pp. 658–667, 1998.
4. S. Arya and D. Mount, "Approximate range searching", in *Proc. 11th ACM Symp. on Comp. Geom.*, pp. 172–181, 1995.
5. J. Bentley, "Decomposable searching problems", *Inform. Process. Lett.*, 8, pp. 244–251, 1979.

6. M. de Berg, M. van Kreveld, M. Overmars, O. Schwarzkopf, "Computational Geometry: Algorithms and Applications", Springer-Verlag, 1997.
7. S. Bespamyatnikh, "An Optimal Algorithm for Closest-Pair Maintenance", *Discrete Comput. Geom.*, 19, pp. 175–195, 1998.
8. S. Bespamyatnikh, K. Kedem, M. Segal and A. Tamir "Optimal Facility Location under Various Distance Function", in *Workshop on Algorithms and Data Structures'99*, pp. 318–329, 1999.
9. S. Bespamyatnikh and M. Segal "Fast algorithm for approximating distances", *Algorithmica*, 33(2), pp. 263–269, 2002.
10. P. Callahan "Dealing with higher dimensions: the well-separated pair decomposition and its applications", Ph.D thesis, Johns Hopkins University, USA, 1995.
11. P. Callahan and R. Kosaraju "A decomposition of multidimensional point sets with applications to k-nearest neighbors and n-body potential fields", *Journal of ACM*, 42(1), pp. 67–90, 1995.
12. P. Callahan and R. Kosaraju "Faster Algorithms for Some Geometric Graph Problems in Higher Dimensions", in *Proc. SODA'93*, pp. 291–300, 1993.
13. T. Chan "On enumerating and selecting distances", *International Journal of Computational Geometry and Applications*, 11, pp. 291–304, 2001.
14. T. Chan "Semi-online maintenance of geometric optima and measures", in *Proc. 13th ACM-SIAM Symp. on Discrete Algorithms*, pp. 474–483, 2002.
15. T. Chan "Approximate nearest neighbor queries revised", *Discrete and Computational Geometry*, 20, pp. 359–373, 1998.
16. B. Chazelle and L. Guibas, "Fractional Cascading: I. A data structuring technique, II. Applications", *Algorithmica*, 1, pp. 133–162, 163–192, 1986.
17. S. Govindarajan, T. Lukovzski, A. Maheshwari and N. Zeh, "I/O Efficient Well-Separated Pair Decomposition and its Applications", In *Proc. of the 8th Annual European Symposium on Algorithms*, pp. 220–231 , 2000.
18. J. Driscoll, N. Sarnak, D. Sleator and R. Tarjan "Making data structures persistent", *Journal of Computer and System Sciences*, 38, pp. 86–124, 1989.
19. D. Eppstein, "Dynamic Euclidean minimum spanning trees and extrema of binary functions", *Discrete and Computational Geometry*, 13, pp. 111–122, 1995.
20. D. Eppstein, "Average case analysis of dynamic geometric optimization", in *Proc. 5th ACM-SIAM Symp. on Discrete Algorithms*, pp. 77–86, 1994.
21. G. Frederickson "A data structure for dynamically maintaining rooted trees", in *Proc. 4th Annu. Symp. on Disc. Alg.*, pp. 175–184, 1993.
22. M. Goodrich, "Geometric partitioning made easier, even in parallel", *Proc. 9th Annu. ACM Sympos. Comput. Geom.*, pp. 73–82, 1993.
23. P. Indyk, "High-dimensional computational geometry", Ph.D. thesis, Stanford University, pp. 68–70, 2000.
24. P. Indyk and R. Motwani "Approximate nearest neighbors: towards removing the curse of dimensionality", in *Proc. 30th ACM Symp. Theory of Comput.*, 1998.
25. R. Janardan, "On maintaining the width and diameter of a planar point-set online", *Int. J. Comput. Geom. Appls.*, 3, pp. 331–344, 1993.
26. S. Kapoor and M. Smid, "New techniques for exact and approximate dynamic closest-point problems", *SIAM J. Comput.*, 25, pp. 775–796, 1996.
27. M. Katz and M. Sharir, "An expander-based approach to geometric optimization", *SIAM J. Comput.*, 26(5), pp. 1384–1408, 1997.
28. J. Kleinberg "Two algorithms for nearest-neighbor search in high dimensions", in *Proc. 29th ACM Symp. Theory of Comput.*, pp. 599–608, 1997.
29. D. Krznaric "Progress in hierarchical clustering and minimum weight triangulation", *Ph. D. thesis*, Lund University, 1997.

30. E. Kushelevitz, R. Ostrovsky and Y. Rabani "Efficient search for approximate nearest neighbor in high dimensional spaces", in *Proc. 30th ACM Symp. Theory of Comput.*, 1998.

31. J. Salowe, "L_∞ interdistance selection by parametric searching", *Inf. Process. Lett.*, 30, pp. 9–14, 1989.

32. A. C. Yao "On constructing minimum spanning trees in k-dimensional spaces and related problems", in *SIAM Journal on Computing*, 11, pp. 721–736, 1982.

Solving the Robots Gathering Problem

Mark Cieliebak[1], Paola Flocchini[2], Giuseppe Prencipe[3], and Nicola Santoro[4]

[1] ETH Zurich, cieliebak@inf.ethz.ch
[2] University of Ottawa, flocchini@site.uottawa.ca
[3] University of Pisa, prencipe@di.unipi.it
[4] Carleton University, santoro@scs.carleton.ca

Abstract. Consider a set of $n > 2$ simple autonomous mobile robots (decentralized, asynchronous, no common coordinate system, no identities, no central coordination, no direct communication, no memory of the past, deterministic) moving freely in the plane and able to sense the positions of the other robots. We study the primitive task of gathering them at a point not fixed in advance (GATHERING PROBLEM). In the literature, most contributions are simulation-validated heuristics. The existing algorithmic contributions for such robots are limited to solutions for $n \leq 4$ or for restricted sets of initial configurations of the robots. In this paper, we present the first algorithm that solves the GATHERING PROBLEM for *any* initial configuration of the robots.

1 Introduction

We consider a distributed system of autonomous mobile robots that are able to freely move in the two-dimensional plane. Due to their autonomy, the coordination mechanisms used by the robots to perform a task (i.e., solve a problem) must be totally *decentralized*, i.e., no central control is used. The problem we consider is *gathering* (or rendez-vous, or point-formation): all robots must gather at one point; the choice of the point is not fixed in advance.

Gathering is one of the basic interaction primitives in systems of autonomous mobile robots, and has been studied in robotics and in artificial intelligence [4,9, 11]. Mostly, the problem is approached from an experimental point of view: algorithms are designed using mainly heuristics, and then tested either by means of computer simulations or with real robots. Neither proofs of correctness of the algorithms, nor any analysis of the relationship between the problem to be solved, the capabilities of the robots employed, and the robots' knowledge of the environment are given. Recently, concerns on computability and complexity of coordination problems have motivated *algorithmic* investigations, and the problems have also been approached from a *computational* point of view [2,7,8,12, 14].

The solution to the GATHERING PROBLEM obviously depends on the capabilities of the robots. The research interest is on a very weak model of autonomous robots: the robots are anonymous (i.e., identical), have no common coordinate system, are oblivious (i.e., they do not remember previous observations and calculations), and have no means of direct communication. Initially, they are in

J.C.M. Baeten et al. (Eds.): ICALP 2003, LNCS 2719, pp. 1181–1196, 2003.

a waiting state. They wake up independently and asynchronously, observe the other robots' positions, compute a point in the plane, move towards this points (but may not reach it[1]), and become waiting again. Details of the model are given in Section 2. For these robots, the GATHERING PROBLEM is defined as follows:

Definition 1. *Given n robots $r_1, \ldots, r_n$, arbitrarily placed in the plane, with no two robots at the same position, make them gather at one point in a finite number of cycles.*

This GATHERING PROBLEM is *unsolvable* for such weak robots [13]; this is rather surprising considering the fact that a variety of other tasks (e.g. forming a circle) are solvable. Also, if the robots are asked only to move "very close" to each other, this task is easily solved: each robot computes the center of gravity[2] of all robots, and moves towards it.

The reason the same solution (i.e., moving towards the center of gravity) does not work for the GATHERING PROBLEM is because the center of gravity is not invariant with respect to robots' movements towards it. Recall that the robots act independently and asynchronously from each other, and that they have no memory of the past; once a robot makes a move towards the center of gravity, the position of the center of gravity changes; hence a robot (even the same one) observing the new configuration will compute and move towards a different point.

An obvious solution strategy would then be to choose as destination a point that, unlike the center of gravity, is *invariant* with respect to the robots' movements towards it. The only known point with such a property is the *Weber* (or *Fermat* or *Torricelli*) *point*: the unique point in the plane that minimizes the sum of the distances between itself and all positions of the robots [10,15]. This point does not change when moving any of the robots straight towards it. Unfortunately, it has been proven in [3] that the Weber point is not expressible as an algebraic expression involving radicals since its computation requires finding zeroes of high-order polynomials even for the case $n = 5$ (see also [6]). In other words, the Weber point is *not computable* by radicals for $n \geq 5$ [3], and thus it cannot be used to solve the GATHERING PROBLEM.

The problem becomes solvable if we change the nature of the robots: if we assume a common coordinate system, gathering is possible even with limited visibility [8]; if the robots are synchronous and movements are instantaneous, gathering has a simple solution [14] and can be achieved even with limited visibility [2]. On the other hand, without changing the robots' nature, they clearly must have some additional ability to solve the GATHERING PROBLEM. One such ability is *multiplicity detection*: a robot can detect whether at a point there is none, one, or more than one robot; if there is more than one robot, we say that

---

[1] That is, a robot can stop before reaching its destination point, e.g. because of limits to the robot's motion energy.

[2] For n points $p_1, \ldots, p_n$ in the plane, the center of gravity is $c = \frac{1}{n} \sum_{i=1}^{n} p_i$.

there is *strict multiplicity* at that point. In the following, we will assume that the robots can detect multiplicities.

Even with multiplicity detection, the problem is surprisingly difficult and was, up to now, unsolved. It is actually unsolvable for $n = 2$ robots [13,14]. Simple solution algorithms exist for $n = 3$ and $n = 4$ robots. For $n \geq 5$ there are two *partial* solutions [5], i.e., algorithms that work for restricted sets of initial configurations. In particular, the first one works if the robots are initially in a biangular configuration (i.e., there exists a point c, and ordering of the robots, and two angles α, β such that the angles between adjacent robots w.r.t. c are either α or β, and the angles alternate; refer to Section 2 and Figure 2); the second algorithm works if in the initial scenario the positions of the robots do not form a regular n-gon (i.e., all robots are on a circle and the distances between each two adjacent robots are equal). Although the two sets of configurations together cover all possible input configurations, the two algorithms can not be integrated nor combined to solve the GATHERING PROBLEM in general.

In this paper, we present the first algorithm that solves the GATHERING PROBLEM for *any* initial configuration of the robots; all calculations performed by the robots can be computed by radicals. Due to space limitations, we only sketch the algorithm and the main ideas for its correctness. The complete algorithm and detailed proofs can be found in the full version of this paper.

2   Terminology, Notation, and Basic Tools

In this section, we introduce terminology and notation, and define the basic concepts used in our algorithm.

Autonomous Mobile Robots

A robot is a mobile computational unit provided with sensors, and it is viewed as a point in the plane. Once activated, the sensors return the set of all points in the plane occupied by at least one robot. In particular, for each such point the sensor outputs whether one or more than one robot is located there (*multiplicity detection*). This forms the current *local view* of the robot. The local view of each robot also includes a unit of length, an origin (which we assume w.l.o.g. to be the position of the robot in its current observation), and a coordinate system (e.g. Cartesian). There is no a priori agreement among the robots on the unit of length, the origin, or the coordinate systems.

A robot is initially in a *waiting* state (*Wait*). Asynchronously and independently from the other robots, it *observes* the environment (*Look*) by activating its sensors. The sensors return a snapshot of the world, i.e., the set of all points that are occupied by at least one other robot, with respect to the local coordinate system. The robot then *calculates* its destination point (*Compute*) according to its deterministic algorithm (the same for all robots), based only on its local view of the world. It then *moves* towards the destination point (*Move*); if the destination point is the current location, the robot stays still. A move may stop before

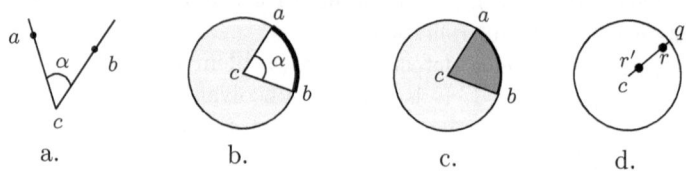

Fig. 1. (a) Convex angle $\alpha = \sphericalangle(a, c, b)$. (b) Arc (thick line) and (c) sector (grey part) defined by $\sphericalangle(a, c, b)$. (d) Two robots, r and r', on the same radius.

the robot reaches its destination, e.g. because of limits to the robot's motion energy. The robot then returns to the waiting state. The sequence *Wait - Look - Compute - Move* forms a *cycle* of a robot.

The robots are *fully asynchronous*, i.e., the amount of time spent in each state of a cycle is finite but otherwise unpredictable. In particular, the robots do not have a common notion of time. As a result, robots can be seen by other robots while moving, and thus computations can be made based on obsolete observations. The robots are *oblivious*, meaning that they do not remember any observation nor computations performed in previous cycles. The robots are *anonymous*, meaning that they are a priori indistinguishable by their appearance, and they do not have any kind of identifiers that can be used during the computation. Finally, the robots have *no means of direct communication*: any communication occurs in a totally implicit manner, by observing the other robots' positions.

There are two limiting assumptions concerning *infinity*: **(A1)** The amount of time required by a robot to complete a cycle is not infinite, nor infinitesimally small. **(A2)** The distance traveled by a robot in a cycle is not infinite, nor infinitesimally small (unless it brings the robot to the destination point). As no other assumptions on space exist, the distance traveled by a robot in a cycle is unpredictable.

Notation and Definitions

Basic Notation. In general, r indicates any robot in the system (when no ambiguity arises, r is used also to represent the point in the plane occupied by that robot). A *configuration* of the robots at a given time instant t is the set of positions in the plane occupied by the robots at time t.

For the following definitions, refer also to Figure 1. Given two distinct points a and b in the plane, $[a, b)$ denotes the half-line that starts in a and passes through b, and $[a, b]$ denotes the line segment between a and b. Given two half-lines $[c, a)$ and $[c, b)$, we denote by $\sphericalangle(a, c, b)$ the convex angle (i.e., the angle that is at most $180°$) centered in c and with sides $[c, a)$ and $[c, b)$. The intersection between the circumference of a circle C and an angle α at the center of C is denoted by $arc(\alpha)$, and the intersection between α and C is denoted by $sector(\alpha)$.

Given a circle C with center c and radius Rad, and a robot r, we say that r *is on C* if $dist(rc) = Rad$, where $dist(ab)$ denotes the Euclidean distance between

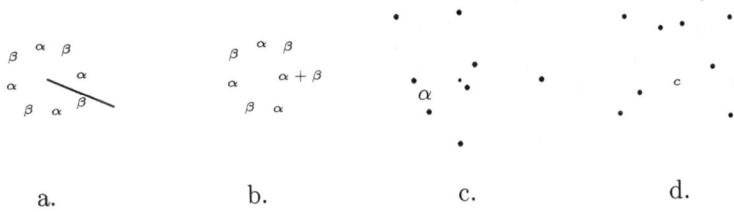

Fig. 2. (a) General biangular and (b) degenerated biangular configuration of 8 points. (c) General equiangular configuration. (d) The smallest enclosing circle of 10 points on the plane.

point a and b (i.e., r is on the circumference of $\mathcal{C}$); if $dist(rc) < Rad$, we will say that r *is inside* $\mathcal{C}$. Given two distinct robots r and r', with r inside $\mathcal{C}$, let q be the intersection between the circumference of $\mathcal{C}$ and $[c, r)$. We say that r and r' are on the same radius if $r' \in [c, q]$.

Biangular Configurations. A set of n robots is in *general biangular configuration* if there exists a point b, the *center*, an ordering of the robots, and two angles $\alpha, \beta > 0$, such that each two adjacent robots form an angle α or β w.r.t. b, and the angles alternate (see Figure 2.a). The robots are in *degenerated biangular configuration* if there is a robot r, an ordering of the other robots, and two angles $\alpha, \beta > 0$, such that each two adjacent robots (without r) form an angle α or β w.r.t. r, and the angles alternate, except for one "gap" where the angle is $\alpha + \beta$ (see Figure 2.b). A general biangular configuration becomes degenerated if one of the robots, namely r, moves to the center b.

Similarly, we say that the robots are in a *general equiangular configuration* if there exists a point e, the *center*, an ordering of the robots, and an angle α such that each two adjacent robots form an angle α w.r.t. e (see Figure 2.c). Note that equiangular configurations can be "almost" considered a special case of biangular configurations: the only difference is that in a biangular configuration there is always an even number of robots, while in an equiangular configuration there can be an odd number of robots. Hence, from now on we will only refer to biangular configurations.

If a set of $n \geq 3$ points P is in general or degenerated biangular configuration, then the center of biangularity b is unique, can be computed in polynomial time, and is invariant under straight movement of any of the points in its direction; that is, it does not change if any of the points move towards b [1].

Smallest Enclosing Circles. Given a set of n distinct points P in the plane, the *smallest enclosing circle* of the points is the circle with minimum radius such that all points from P are inside or on the circle (see Figure 2.d). We denote it by $SEC(P)$, or SEC if set P is unambiguous from the context. The smallest enclosing circle of a set of n points is unique and can be computed in polynomial time [16].

Obviously, the smallest enclosing circle of P remains invariant if we remove all or some of the points from P that are inside $SEC(P)$. In fact, the following lemma shows that we can even remove all but at most three points from P without changing $SEC(P)$.

Lemma 1. *Given a set P of n points, there exists a subset $S \subseteq P$ such that $|S| \leq 3$ and $SEC(S) = SEC(P)$.*

String of Angles. Given n distinct points $p_1, \ldots, p_n$ in the plane, let SEC be the smallest enclosing circle of the points, and c be its center. For an arbitrary point p_k, $1 \leq k \leq n$, we define the *string of angles* $SA(p_k)$ by the following algorithm (refer to Figure 3.a):

Compute_SA(p_k)

$\quad p := p_k, i := 1;$
$\quad$**While** $i \neq n+1$ **Do**
$\quad\quad p' := \text{Succ}(p);$
$\quad\quad SA[i] := \sphericalangle(p, c, p');$
$\quad\quad p := p'; i := i+1;$
$\quad$**Return** SA.

Here, all angles are oriented clockwise (note that the robots do not have a common coordinate system; however, each robot can locally distinguish between a clockwise and counterclockwise orientation). The successor of p, computed by $\text{Succ}(p)$, is (refer to Figure 3.b)

- either the point $p_i \neq p$ on $[c, p)$, such that $dist(c, p_i)$ is minimal among all points $p_j \neq p$ on $[c, p)$ with $dist(c, p_j) > dist(c, p)$, if such a point exists; or
- the point $p_i \neq p$ such that there is no other point inside $sector(\sphericalangle(p, c, p_i))$, and there is no other point on the line segment $[c, p_i]$.

Instead of $SA(p_k)$, we write SA if we do not consider a specific point p_k. Given p_k, procedure Succ() defines unique successors, and thus Compute_SA(p_k) defines a unique string of angles. Given two starting points p_k and p_ℓ, then $SA(p_k)$ is a cyclic shift of $SA(p_\ell)$. Given an angle α in SA, then we can associate it with its defining point; i.e., if $\alpha = \sphericalangle(p, c, p')$, then we say that α is *associated* to p, and we write $p = \mathfrak{r}(\alpha)$. Alternatively, since α is stored in SA, say at position i (i.e., $SA[i] = \alpha$), we denote the point associated to α by $\mathfrak{r}(i)$, saying that $\mathfrak{r}(i)$ is the point associated to position i in SA. We define the *reverse string of angles* $revSA$ in an analogous way: it is the string of angles with all angles counterclockwise oriented (i.e., $revSA$ is the reverse of SA). We say that SA (resp. $revSA$) is *general* if it does not contain any zeros; otherwise, at least two points are on a line starting in c (a radius), and we call the string of angles *degenerated*.

Given two strings $s = s_1, \ldots, s_n$ and $t = t_1, \ldots, t_n$, we say that s is *lexicographically smaller* than t if there exists an index $k \in \{1, \ldots, n\}$ such that $s_i = t_i$ for all $1 \leq i < k$, and $s_k < t_k$. We write $s <_{lex} t$. Let *LexMinString* be the lexicographically minimal string among all strings of angles (in both orientations),

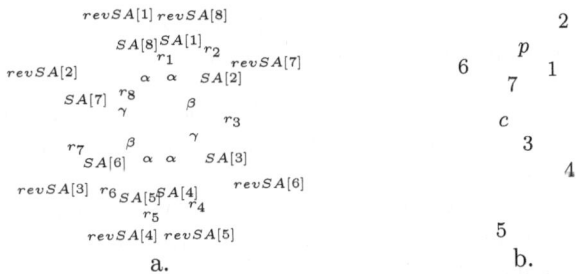

Fig. 3. (a) String of angles computed by `Compute_SA`(r_1). With $\alpha = 25°$, $\beta = 60°$ and $\gamma = 70°$, we have $SA(r_1) = \langle \alpha, \beta, \gamma, \alpha, \alpha, \beta, \gamma, \alpha \rangle = \langle 25°, 60°, 70°, 25°, 25°, 60°, 70°, 25° \rangle$, $LexMinString = \langle \alpha, \alpha, \beta, \gamma, \alpha, \alpha, \beta, \gamma \rangle$, $\mathfrak{r}(SA[3]) = \mathfrak{r}(\gamma) = \mathfrak{r}(3) = r_3$, $StartSet = \{4, 8\}$, and $revStartSet = \emptyset$. (b) Routine `Succ`(p) with clockwise orientation. The points are numbered according to routine `Succ`(); that is `Succ`(1)=2, `Succ`(2)=3, and so on. Note that `Succ`(7)=1

i.e., $LexMinString := min(\{SA(p_i) \mid 1 \leq i \leq n\} \cup \{revSA(p_i) \mid 1 \leq i \leq n\})$. Let $StartSet$ be the set of all indices in SA where $LexMinString$ starts, i.e., $StartSet = \{i \mid 1 \leq i \leq n, SA(p_i) = LexMinString\}$, and let $revStartSet$ be the set of all indices in $revSA$ where $LexMinString$ starts.

Robot Motion and Critical Points

In our algorithm, we use four different types of "move" operations; in each, when a robot moves, it moves in a straight line.

The basic operation is `moveTo`(p), where a robot r moves towards point p (recall that, although restricted by assumption **A2**, the robot may enter the waiting state before reaching p).

In the operation `moveToIfFreeWay`(p), the robot r moves towards p only if no other robot is between r and p; otherwise, r does not move at all. This operation is used to avoid that the moving robot creates an (unintended) point with strict multiplicity. Note that, if all robots in the system are moving towards p and only this type of moves are executed, then strict multiplicity can only occur at p.

The remaining two types of movement are crucial to control the swap of a non-biangular configuration into a biangular one, due to robots' movements. To introduce them, we need the notion of *critical points*, defined as follows:

Definition 2. *Given n robots and a point p in the plane, a point x is a critical point for the movement of robot r towards p if $x = p$, or if x is on the half-line from p to r and the configuration of the robots becomes biangular when r is at position x.*
A pair of points (y, z) is a critical pair for the movements of robots r' and r'' towards destinations p' and p'' respectively, if $(y, z) = (p', p'')$, or if y is on the

half-line from p' to r', z is on the half-line from p'' to r'', and the configuration of the robots is biangular when r' is at position y and r'' is at position z.

The operation moveStepwiseTo(p) requires the robot r to first compute all critical points for its movement towards p, and then to move towards the first critical point on its way towards and stop there.

With operation moveStepwiseTo($(r', p'), (r'', p'')$) we coordinate the movement of two robots r' and r'' which move in the direction of points p' and p'', respectively. We compare the number of critical points between the robots and their destinations. The robot with most critical points ahead is allowed to move; if they have the same number, they both move. Once allowed, if the robot is between two critical points it moves to the next one; if it is already at a critical point it moves towards half the distance to the next critical point.

Finally, given a circle C with center c, we extend our four types of move operations and allow robots to move onto or away from C. In particular, we say that a robot moves to circle C (moveTo(C), moveToIfFreeWay(C), moveStepwiseTo(C)) if the destination point of the robot is the intersection of C and the half-line starting from the center of C and going through the position of the robot (note that the robot does not move at all if it is already at this intersection point). Moreover, we define what a movement into the inside of circle C is (moveTo(into C), moveToIfFreeWay(into C), moveStepwiseTo(into C)): if the robot is already inside C, it does not move at all. Otherwise, it moves to the point p that is half on its way towards c, the center of the circle. For two robots r' and r'', we define moveStepwiseTo(r', r'', C) and moveStepwiseTo(r', r'', into C) accordingly.

3   The Solution Algorithm

In this section, we describe the algorithm that solves the GATHERING PROBLEM for arbitrary initial configurations of $n \geq 5$ robots, and discuss its correctness.

3.1   Description

At a high level, the strategy of the algorithm is as follows. Initially all robots are in distinct locations; that is, in the initial configuration, there is no point with strict multiplicity. Our algorithm ensures that at any time during the execution there is at most one point with strict multiplicity; moreover, such a point will eventually be generated. Once this occurs, the robots that are already at that point remain there, while all other robots move towards this unique point.

If the (initial) configuration is biangular, then all robots move towards the center of biangularity. The future configuration remains biangular until two (or more) robots reach the center. When this occurs, a unique point with strict multiplicity has been created.

In all other configurations, we select a strict subset of the robots; the selection is done using the string of angles of the robots w.r.t. the center of their smallest enclosing circle. If we can elect a unique robot, it will go to some other robot creating a unique point with strict multiplicity. Otherwise, the selected robots move towards the center of the smallest enclosing circle, ensuring that the circle does not change because of these movements. If no biangular configuration is created during these movements, two (or more) robots reach the center of the circle, and we have a unique point with strict multiplicity.

One of the difficult and crucial components of the algorithm is the use of appropriate move operations to ensure the following: if a biangular configuration is created during the movements of some robots, then *all* robots have to become aware of it in their next *Look* state, ensuring that they will gather at the center of biangularity. The difficulty arises from the asynchrony, obliviousness and autonomy of the robots; the component is crucial to avoid that some robots move towards the center of biangularity while others still move towards the center of the circle (possibly destroying biangularity).

The main algorithm is shown in Algorithm 1. In the algorithm we use four different subroutines; their behavior differs depending on the value of s, the cardinality of the set $StartSet \cup revStartSet$ (therefore, s denotes the number of starting positions of $LexMinString$ in SA and $revSA$).

Algorithm 1 Algorithm GATHER

$\mathcal{Z} :=$ Observed Configuration;

$SEC :=$ Smallest Enclosing Circle of all robots;

$c :=$ Center of SEC;

$InnerC :=$ Circle with center c and radius $\dfrac{\text{radius of } SEC}{2}$;

5: **Case $\mathcal{Z}$ Is Such That:**

   • *There is One point m with strict multiplicity:*

     moveToIfFreeWay(m).

   • *The robots are in general (resp. degenerated) biangular configuration:*

     $b :=$ Center of general (resp. degenerated) biangularity;

10:     moveToIfFreeWay(b).

   • *default:*

     If No robot is at c **Then**

       $SA :=$ (Compute_SA); %String of angles of all robots%

       $StartSet, revStartSet :=$ Indices Where Lex. Minimal String Starts;

15:       $s := |StartSet \cup revStartSet|$;

       If SA is general **Then Routine1. Else Routine2.**

     Else %One robot r is at c%

       $r :=$ robot at c;

       $SA^- :=$ String of angles of all robots except r;

20:       $StartSet^-, revStartSet^- :=$ Indices Where Lex. Minimal String Starts;

       If SA^- is general **Then Routine3. Else Routine4.**

In the following, we first discuss the main properties of $LexMinString$, and then we sketch the correctness proof of the algorithm.

3.2   Properties of *LexMinString*

1. One Starting Position of LexMinString ($s = 1$): Let $StartSet \cup revStartSet = \{x\}$ and $SA(x) = \alpha_1, \ldots, \alpha_n$; then $revSA(x) = \alpha_n, \ldots, \alpha_1$, and the following holds:

Lemma 2. *If $StartSet \cup revStartSet = \{x\}$, then either $SA(x) = LexMinString$ or $revSA(x) = LexMinString$.*

This implies that there is a unique starting position and a unique direction for *LexMinString*, yielding a unique ordering of the robots. If all robots are on *SEC*, then we can use this ordering and Lemma 1 to define operation ElectOne() to elect the first robot r such that *SEC* remains invariant if r is moved to the inside of *SEC*. If more than one robot is inside *SEC*, then ElectOneInside() is used to elect a unique robot that is already inside *SEC* (again, using the uniqueness of *LexMinString*).

2. Two Starting Positions of LexMinString ($s = 2$): Let $StartSet \cup revStartSet = \{x, y\}$. The following lemma shows that *LexMinString* can start in each position in only one direction.

Lemma 3. *If $StartSet \cup revStartSet = \{x, y\}$, then it is not possible that $SA(x) = revSA(x) = LexMinString$ or $SA(y) = revSA(y) = LexMinString$.*

If *LexMinString* starts in x and y in the same direction, then the angle between these two positions w.r.t. c is $180°$. Moreover, for every robot there is a partner such that their angle is $180°$. Recall that $\mathfrak{r}(x)$ is the robot associated with index x. Using the starting positions and the direction of *LexMinString*, we define ElectTwo() as follows: if $\mathfrak{r}(x)$ and $\mathfrak{r}(y)$ are on *SEC*, then we elect the "next" pair of robots with an angle of $180°$; otherwise we elect $\mathfrak{r}(x)$ and $\mathfrak{r}(y)$ themselves.

If *LexMinString* starts in x and y in opposite direction, say $x \in StartSet$ and $y \in revStartSet$, then let γ be the angle between $\mathfrak{r}(x)$ and $\mathfrak{r}(y)$ w.r.t. c. If $\gamma = 180°$, then ElectTwo() elects the first two robots, according to the starting positions and directions of *LexMinString*, that are not both on *SEC*. If $\gamma < 180°$, we define the *opposite robots* of $\mathfrak{r}(x)$ and $\mathfrak{r}(y)$ to be one or two robots in the half of *SEC* where $\mathfrak{r}(x)$ and $\mathfrak{r}(y)$ are not (see Figure 4): let ℓ be the line that bisects γ. Then ℓ is a symmetry line for the angles of the robots w.r.t. c. We choose either the robot r that is on line ℓ, if such a robot exists, or we choose the two robots u and v that are closest to ℓ (in terms of their angles w.r.t. c). Observe that the construction of the opposite robots guarantees that c is inside the convex hull of $\mathfrak{r}(x)$, $\mathfrak{r}(y)$ and their opposite robot(s). Thus, ElectTwo() can elect two appropriate robots such that *SEC* remains invariant if they move (using Lemma 1).

Finally, we define routine ElectPairInside() that elects the "first" pair of robots that is inside *SEC*. Again, the ordering of the robots is given by the starting positions and orientations of *LexMinString*.

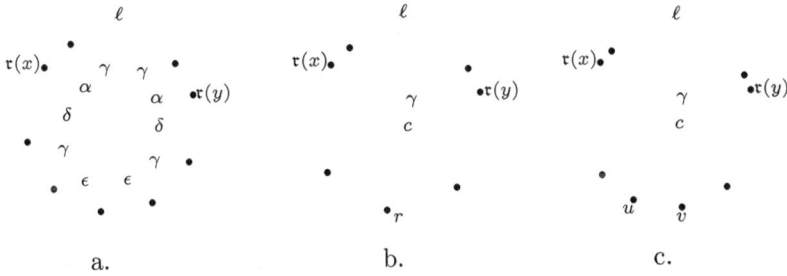

Fig. 4. (a) The line ℓ that runs through c and bisects $\gamma = \sphericalangle(\mathfrak{r}(x), c, \mathfrak{r}(y)) = 2\epsilon + 2\gamma$ is a symmetry axis for the angles that the robots form w.r.t. c. In the depicted example, $x \in StartSet$, $y \in revStartSet$, and $SA(x) = revSA(y) = LexMinString = \langle \alpha, \gamma, \gamma, \alpha, \delta, \gamma, \epsilon, \epsilon, \gamma, \delta \rangle$. (b) One robot r opposite to $\mathfrak{r}(x)$ and $\mathfrak{r}(y)$. (c) Two robots u and v opposite to $\mathfrak{r}(x)$ and $\mathfrak{r}(y)$.

3. Many Starting Positions of LexMinString (s > 2): Let $StartSet = \{x_1, \ldots, x_l\}$. Then SA and $revSA$ are periodic. Moreover, if k is the minimum length of a period of SA, then we can divide SA into $\frac{n}{k}$ equal periods, and the angles in each period sum up to $\gamma = \frac{360°}{\frac{n}{k}}$. If the period length is one or two, then the configuration is biangular; hence we can exclude this case in the following, since it is covered in Lines 8–10 of the main algorithm.

We say that two robots r and r' are *equivalent (modulo periodic shift)* if they have the same position in different periods, i.e., if $\sphericalangle(r, c, r')$ is a multiple of γ (see the example depicted in Figure 5). If all robots are on SEC, then for any robot r, there are $\frac{n}{k} - 1$ equivalent robots, and they form a regular $\frac{n}{k}$-gon with c inside. Thus, if at least one robot and all its equivalent robots remain on SEC, then SEC remains invariant (using Lemma 1).

Lemma 4. *If $StartSet \neq \emptyset$, all robots are on SEC, and the minimum period length of SA is $k \geq 3$, then SEC remains invariant when all robots $\mathfrak{r}(x)$, with $x \in \{StartSet \cup revStartSet\}$, move inside SEC.*

Observe that if all robots are on SEC, then equivalent robots cannot be distinguished, hence they act in the same way. In the case of one or two starting positions of $LexMinString$, we were able to elect one or two robots to move, and we used stepwise movements to ensure that these robots stop when the configuration becomes biangular. If there are many starting positions of $LexMinString$, we do not need to apply stepwise movements, as shown by the following lemma.

Lemma 5. *Given a non biangular configuration of the robots such that SA is periodic, then moving any subset of the robots towards c cannot make the configuration become biangular.*

To see this, observe that c is the Weber point of the robots, and that the center of biangularity, if it exists, is the Weber point as well. Thus, since the

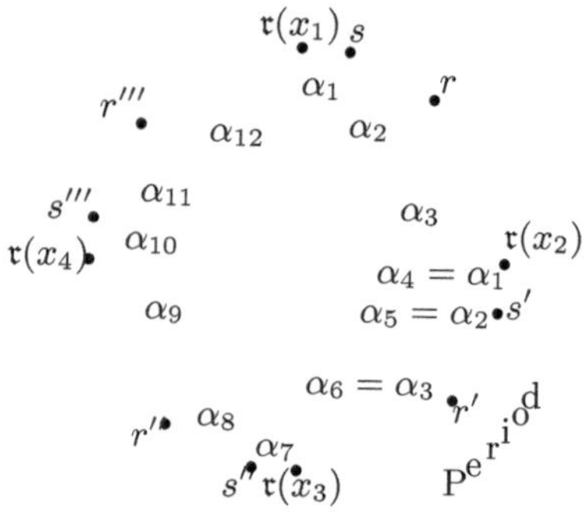

Fig. 5. Example with $|StartSet|=4$, $SA(x_1) = SA(x_i) = LexMinString = \alpha_1, \ldots, \alpha_{12}$, and the period of $SA(x_1)$ is $\langle \alpha_1, \alpha_2, \alpha_3 \rangle$. There are $\frac{n}{k} = \frac{12}{3}$ periods, and $\gamma = \alpha_1 + \alpha_2 + \alpha_3 = \frac{360°}{4}$. Thick lines represent the starting points of each of the four periods. Robots r, r', r'' and r''' are *equivalent*, as well as $\mathfrak{r}(x_1)$, $\mathfrak{r}(x_2)$, $\mathfrak{r}(x_3)$ and $\mathfrak{r}(x_4)$, and s, s', s'' and s'''.

Weber point is unique, the robots cannot swap into a biangular configuration if there was none before. This lemma implies that the robots cannot create a biangular configuration while they move towards or away from c, hence we do not need to introduce a stepwise movement in this case.

Correctness Proof (Sketch)

The first thing a robot does when it starts its computation is to check whether there is a point p in the plane with strict multiplicity. If this is the case, the robot simply moves there. Point p will be the final gathering point (Lines 6–7).

Otherwise, the robots check whether the observed configuration of the robots is biangular. In this case, the center of biangularity b is computed, and the robots move there using `moveToIfFreeWay(b)`. As long as none of the robots reaches b, the configuration remains general biangular; hence the algorithm continues to move all robots towards b. By Assumptions A1 and A2, in a finite number of cycles, at least one robot reaches b. If only one robot reaches b, then the configuration becomes degenerated biangular. In this case, the center of degenerated biangularity[3] is again b, and all robots continue moving towards b. As soon as two robots reach b, there is a unique point with strict multiplicity, and all robots will gather there.

If the observed configuration is not biangular, then the SEC and its center c are computed. The algorithm distinguishes four cases.

---

[3] If a general biangular configuration with center b turns into a degenerated biangular configuration because one of the robots reaches b, then the center of the degenerated biangular configuration is again b.

(A) **There is no robot at c, and SA is general.** Routine1 is called, which behaves differently depending on the value of s, the cardinality of $StartSet \cup revStartSet$.

If $s = 1$, then a unique robot r is elected, and it moves stepwise[4] towards c. Robot r is chosen such that SEC does not change during its movement. When the movement stops, either the configuration is biangular, and Line 8 of the main algorithm applies; or Routine1 is called again (with r – the only robot inside SEC – again elected), until r reaches c, and Routine3 applies.

If $s = 2$, at first all robots that are inside SEC move to the circumference of SEC (by repeatedly calling ElectPairInside()). Afterwards, only the two robots elected by ElectTwo() are allowed to move, and they move towards c. All movements are stepwise, and there are always at most two moving robots, either the robots run into a biangular configuration and stop (Line 8 of the main algorithm then applies), or one of them reaches c and Routine3 is called, or the two elected robots reach c simultaneously, and c becomes the unique point with strict multiplicity. In the last two cases, c will be the final gathering point.

If $s > 2$, first all robots associated to indices in $StartSet \cup revStartSet$ are elected. Then, all robots that are not elected and that are inside SEC are moved towards the circumference of SEC. Afterwards, all elected robots (and only these) move towards[5] c (without changing SEC, by Lemma 4), with the only restriction that an elected robot can reach c only if all other elected robots are already inside SEC (note that two robots inside SEC would be sufficient). This is achieved by first calling routine moveTo(into C). In a finite number of cycles at least two robots reach c, creating strict multiplicity there.

(B) **There is no robot at c, and SA is degenerated.** Routine2 is called. Recall that, if SA is degenerated, then there is at least one radius of SEC with more than one robot on it. Therefore, due to our definition of SA, the lexicographically minimal string of angles always start with zeros. Moreover, on each radius with at least two robots, one robot is already inside SEC. Similarly to previous cases, different actions are taken depending on the value of s.

If $s = 1$, then the subroutine elects a unique radius rad that has at least two robots lying on it. Let $StartSet = \{x\}$ (the case $revStartSet = \{x\}$ is handled similarly), and rad_x be the radius where $\mathfrak{r}(x)$ lies (i.e., $[c, \mathfrak{r}(x)]$). Then rad can be chosen as the first radius with at least two robots on it, starting from rad_x and according to the ordering of the robots established by SA. Let r and r' be the first two innermost robots on rad. Then r moves stepwise towards r', while all other robots do not move. In a finite number of cycles, either a biangular configuration is reached (r stops at the first critical

---

[4] Recall that stepwise movement implies that r stops at its first critical point.
[5] Some of them can already be inside SEC, and others are still on the circumference of SEC.

point on its path towards r') and Line 8 of the main algorithm applies, or r reaches r' and a unique point with strict multiplicity is created.

If $s = 2$, the algorithm works similar to Case (A), except that all operations are done with respect to $InnerC$ instead of SEC. In particular, first the robots that are inside $InnerC$ move out of $InnerC$. If we would move these robots simply to the circumference of $InnerC$, we could obtain unintended points with strict multiplicity, since all robots on the same radius would end up at the same point on $InnerC$. Therefore, we define a sufficient number of distinct positions "just outside" $InnerC$ (using the radius of SEC) where we move the robots that are on the same radius. Thereby, we ensure that the innermost robots will end up on $InnerC$. Afterwards, the two robots elected by ElectTwo() are allowed to move, and they move stepwise towards c, and in a finite number of cycles at least one of them reaches c.

If $s > 2$, then SA is periodic (see paragraph on $s > 2$). Again, the algorithm works similar to Case (A), except that all operations are done with respect to $InnerC$ instead of SEC.

(C) **There is exactly one robot r at c, and SA^- (the string of angles of all robots except r) is general.** Routine3 is called. If r is the only robot inside SEC, then r chooses an arbitrary robot q on SEC and moves stepwise towards it. By this movement, the string of angle becomes degenerated, since r and q are on the same radius. Hence, by (D), r continues to move towards q. If no critical points are on its path towards q, in a finite number of cycles r reaches q and a unique point with strict multiplicity is obtained. Otherwise, r stops at the first critical point it meets. Then a biangular configuration is obtained, and Line 8 of the main algorithm applies.

If there are only two robots r and r' inside SEC (with r at c), then r' moves stepwise towards c. The argument follows similarly to the previous paragraph.

If more than two robots are inside SEC and SA^- is periodic except for one gap[6], then all robots inside SEC move towards c. By Lemma 5, no biangularity can occur.

If more than two robots are inside SEC and SA^- is not periodic except for one gap, then the routine behaves similarly to Routine1. The only difference is that in this case all the operations are done using SA^- instead of SA; that is, robot r is ignored.

(D) **There is exactly one robot r at c, and SA^- (the string of angles of all robots except r) is degenerated.** Routine4 is called. If r is not the only robot inside $InnerC$, then this routine is similar to Routine 3, except that all operations refer to $InnerC$ instead of SEC. Otherwise, if r is the only robot inside $InnerC$, then it chooses some an arbitrary index q in $StartSet^- \cup revStartSet^-$. Note that q is always associated to a position in SA where $LexMinString$ starts. Robot r moves stepwise towards $\mathfrak{r}(q)$, while all other robots do not move. As soon as r leaves c, a unique starting

---

[6] That is, the string would be periodic if r — the robot at c — would not be at c, but somewhere inside SEC.

position of *LexMinString* is obtained in the positions associated to r, since an angle with $0°$ has been added. Thus, SA is degenerated with no robot at c, and r will be chosen again to move on to q due to Case (B) above.

4 Conclusion

We have presented a deterministic algorithm for the GATHERING PROBLEM for $n \geq 5$ robots that works for all initial configurations of the robots. Some interesting questions are still open. For example, it is not known which other abilities, other than multiplicity detection, would allow the weak robots to solve the GATHERING PROBLEM.

It is known that changing the nature of the robots (e.g. by synchronizing them, or by adding common knowledge on the coordinate system) enables solvability. It is still not known if (and how) removing obliviousness would have the same effect. It would be interesting to explore the relationship between memory and solvability or, for that matter, to study the impact of (weak) explicit communication among the robots.

Acknowledgments. We would like to thank all people that have offered their ideas, comments, suggestions, and (conflicting) conjectures on this problem over the years. Especially, we would like to thank Elmo Welzl and Peter Widmayer.

References

1. L. Anderegg, M. Cieliebak, and G. Prencipe. A Linear Time Algorithm to Identify Equiangular and Biangular Point Configurations. Technical Report TR-03-01, Università di Pisa, 2003.
2. H. Ando, Y. Oasa, I. Suzuki, and M. Yamashita. A Distributed Memoryless Point Convergence Algorithm for Mobile Robots with Limited Visibility. *IEEE Transaction on Robotics and Automation*, 15(5):818–828, 1999.
3. C. Bajaj. The Algebraic Degree of Geometric Optimization Problems. *Discrete and Computational Geometry*, 3:177–191, 1988.
4. T. Balch and R. C. Arkin. Behavior-based Formation Control for Multi-robot Teams. *IEEE Transaction on Robotics and Automation*, 14(6), 1998.
5. M. Cieliebak and G. Prencipe. Gathering Autonomous Mobile Robots. In *SIROCCO 9*, pages 57–72, 2002.
6. E. J. Cockayne and Z. A. Melzak. Euclidean Constructibility in Graph-minimization Problems. *Mathematical Magazine*, 42:206–208, 1969.
7. P. Flocchini, G. Prencipe, N. Santoro, and P. Widmayer. Hard Tasks for Weak Robots: The Role of Common Knowledge in Pattern Formation by Autonomous Mobile Robots. In *ISAAC '99*, volume 1741 of *LNCS*, pages 93–102, 1999.
8. P. Flocchini, G. Prencipe, N. Santoro, and P. Widmayer. Gathering of Autonomous Mobile Robots With Limited Visibility. In *STACS 2001*, volume 2010 of *LNCS*, pages 247–258, 2001.
9. D. Jung, G. Cheng, and A. Zelinsky. Experiments in Realising Cooperation between Autonomous Mobile Robots. In *ISER*, 1997.

10. Y. Kupitz and H. Martini. Geometric Aspects of the Generalized Fermat-Torricelli Problem. *Intuitive Geometry*, 6:55–127, 1997.
11. M. J. Matarić. Designing Emergent Behaviors: From Local Interactions to Collective Intelligence. In *From Animals to Animats 2: Int. Conf. on Simulation of Adaptive Behavior*, pages 423–441, 1993.
12. G. Prencipe. CORDA: Distributed Coordination of a Set of Autonomous Mobile Robots. In *ERSADS 2001*, pages 185–190, 2001.
13. G. Prencipe. Instantaneous Actions vs. Full Asynchronicity: Controlling and Coordinating a Set of Autonomous Mobile Robots. In *ICTCS 2001*, pages 185–190, 2001.
14. I. Suzuki and M. Yamashita. Distributed Anonymous Mobile Robots: Formation of Geometric Patterns. *Siam Journal of Computing*, 28(4):1347–1363, 1999.
15. E. Weiszfeld. Sur le Point Pour Lequel la Somme Des Distances de n Points Donnés Est Minimum. *Tohoku Mathematical*, 43:355–386, 1936.
16. E. Welzl. Smallest Enclosing Disks (Balls and Ellipsoids). In H. Maurer, editor, *New Results and New Trends in Computer Science*, LNCS, pages 359–370. Springer, 1991.

Author Index